Studies in Classification, Data Analysis, and Knowledge Organization

Titles in the Series:

E. Diday, Y. Lechevallier, and O. Opitz (Eds.) Ordinal and Symbolic Data Analysis. 1996

R. Klar and O. Opitz (Eds.) Classification and Knowledge Organization. 1997

C. Hayashi, N. Ohsumi, K. Yajima, Y. Tanaka, H.-H. Bock, and Y. Baba (Eds.) Data Science, Classifaction, and Related Methods. 1998

I. Balderjahn, R. Mather, and M. Schader (Eds.) Classification, Data Analysis, and Data Highways. 1998

A. Rizzi, M. Vichi, and H.-H. Bock (Eds.) Advances in Data Science and Classification. 1998

M. Vichi and O. Optiz (Eds.) Classification and Data Analysis. 1999

W. Gaul and H. Locarek-Junge (Eds.) Classification in the Information Age. 1999

H.-H. Bock and E. Diday (Eds.) Analysis of Symbolic Data. 2000

H. A. L. Kiers, J.-P. Rasson, P.J.F. Groenen, and M. Schader (Eds.) Data Analysis, Classification, and Related Methods. 2000

W. Gaul, O. Opitz, M. Schader (Eds.) Data Analysis. 2000

R. Decker and W. Gaul (Eds.) Classification and Information Processing at the Turn of the Millenium. 2000

S. Borra, R. Rocci, M. Vichi, and M. Schader (Eds.) Advances in Classification and Data Analysis. 2000

W. Gaul and G. Ritter (Eds.) Classification, Automation, and New Media. 2002

K. Jajuga, A. Sokolowski, and H.-H. Bock (Eds.) Classification, Clustering and Data Analysis. 2002

M. Schwaiger and O. Opitz (Eds.) Exploratory Data Analysis in Empirical Research. 2003

M. Schader, W. Gaul, and M. Vichi (Eds.) Between Data Science and Applied Data Analysis. 2003

H.-H. Bock, M. Chiodi, and A. Mineo (Eds.) Advances in Multivariate Data Analysis. 2004

D. Banks, L. House, F.R. McMorris, P. Arabie, and W. Gaul (Eds.) Classification, Clustering, and Data Minig Applications. 2004

D. Baier and K.-D. Wernecke (Eds.) Innovations in Classification, Data Science, and Information Systems. 2005

M. Vichi, P. Monari, S. Mignani, and A. Montanari (Eds.) New Developments in Classification and Data Analysis. 2005

D. Baier, R. Decker, and L. Schmidt-Thieme (Eds.) Data Analysis and Decision Support. 2005

C. Weihs and W. Gaul (Eds.) Classification - the Ubiquitous Challenge. 2005

M. Spiliopoulou, R. Kruse, C. Borgelt, A. Nürnberger, and W. Gaul (Eds.) From Data and Information Analysis to Knowledge Engineering. 2006

V. Batagelj, H.-H. Bock, A. Ferligoj, and A. Žiberna (Eds.) Data Science and Classification. 2006

S. Zani, A. Cerioli, M. Riani, M. Vichi (Eds.) Data Analysis, Classification and the Forward Search. 2006

P. Brito, P. Bertrand, G. Cucumel, F. de Carvalho (Eds.) Selected Contributions in Data Analysis and Classification. 2007

R. Decker, H.-J. Lenz (Eds.) Advances in Data Analysis. 2007

C. Preisach, H. Burkhardt, L. Schmidt-Thieme, R. Decker (Eds.) Data Analysis, Machine Learning and Applications. 2008

Christine Preisach · Hans Burkhardt
Lars Schmidt-Thieme · Reinhold Decker
(Editors)

Data Analysis, Machine Learning and Applications

Proceedings of the 31st Annual Conference
of the Gesellschaft für Klassifikation e.V.,
Albert-Ludwigs-Universität Freiburg,
March 7–9, 2007

With 226 figures and 96 tables

Springer

Editors

Christine Preisach
Institute of Computer Science and
Institute of Business Economics and
Information Systems
University of Hildesheim
Marienburgerplatz 22
31141 Hildesheim
Germany

Professor Dr. Hans Burkhardt
Lehrstuhl für Mustererkennung und
Bildverarbeitung
Universität Freiburg
Gebäude 052
79110 Freiburg i. Br.
Germany

Professor Dr. Dr. Lars Schmidt-Thieme
Institute of Computer Science and
Institute of Business Economics and
Information Systems
Marienburgerplatz 22
31141 Hildesheim
Germany

Professor Dr. Reinhold Decker
Fakultät für Wirtschaftswissenschaften
Lehrstuhl für Betriebswirtschaftslehre,
insbes. Marketing
Universitätsstraße 25
33615 Bielefeld
Germany

ISBN: 978-3-540-78239-1 e-ISBN: 978-3-540-78246-9

Library of Congress Control Number: 2008925870

Cover Design: WMX Design GmbH, Heidelberg, Germany

Printed on acid-free paper

5 4 3 2 1 0

springer.com

Preface

This volume contains the revised versions of selected papers presented during the 31st Annual Conference of the German Classification Society (Gesellschaft für Klassifikation – GfKl). The conference was held at the Albert-Ludwigs-University in Freiburg, Germany, in March 2007. The focus of the conference was on Data Analysis, Machine Learning, and Applications, it comprised 200 talks in 36 sessions. Additionally 11 plenary and semi-plenary talks were held by outstanding researchers. With 292 participants from 19 countries in Europe and overseas this GfKl Conference, once again, provided an international forum for discussions and mutual exchange of knowledge with colleagues from different fields of interest. From altogether 120 full papers that had been submitted for this volume 82 were finally accepted.

With the occasion of the 30st anniversary of the German Classification Society the associated societies Sekcja Klasyfikacji i Analizy Danych PTS (SKAD), Vereniging voor Ordinatie en Classificatie (VOC), Japanese Classification Society (JCS) and Classification and Data Analysis Group (CLADAG) have sponsored the following invited talks: Paul Eilers - Statistical Classification for Reliable High-volume Genetic Measurements (VOC); Eugeniusz Gatnar - Fusion of Multiple Statistical Classifiers (SKAD); Akinori Okada - Two-Dimensional Centrality of a Social Network (JCS); Donatella Vicari - Unsupervised Multivariate Prediction Including Dimensionality Reduction (CLADAG).

The scientific program included a broad range of topics, besides the main theme of the conference, especially methods and applications of data analysis and machine learning were considered. The following sessions were established:

I. Theory and Methods

Supervised Classification, Discrimination, and Pattern Recognition (G. Ritter); Cluster Analysis and Similarity Structures (H.-H. Bock and J. Buhmann); Classification and Regression (C. Bailer-Jones and C. Hennig); Frequent Pattern Mining (C. Borgelt); Data Visualization and Scaling Methods (P. Groenen, T. Imaizumi, and A. Okada); Exploratory Data Analysis and Data Mining (M. Meyer and M. Schwaiger); Mixture Analysis in Clustering (S. Ingrassia, D. Karlis, P. Schlattmann and W. Seidel); Knowledge Representation and Knowledge Discovery (A. Ultsch); Statistical Relational Learning (H. Blockeel and K. Kersting); Online Algorithms and Data Streams (C. Sohler); Analysis of Time Series, Longitudinal and Panel Data (S. Lang); Tools for Intelligent Data Analysis (M. Hahsler and K. Hornik); Data Preprocessing and Information Extraction (H.-J. Lenz); Typing for Modeling (W. Esswein).

II. Applications

Marketing and Management Science (D. Baier, Y. Boztug, and W. Steiner); Banking and Finance (K. Jajuga and H. Locarek-Junge); Business Intelligence and Personalization (A. Geyer-Schulz and L. Schmidt-Thieme); Data Analysis in Retailing (T. Reutterer); Econometrics and Operations Research (W. Polasek); Image and Signal Analysis (H. Burkhardt); Biostatistics and Bioinformatics (R. Backofen, H.-P. Klenk and B. Lausen); Medical and Health Sciences (K.-D. Wernecke); Text Mining, Web Mining, and the Semantic Web (A. Nürnberger and M. Spiliopoulou); Statistical Natural Language Processing (P. Cimiano); Linguistics (H. Goebl and P. Grzybek); Subject Indexing and Library Science (H.-J. Hermes and B. Lorenz); Statistical Musicology (C. Weihs); Archaeology and Archaeometry (M. Helfert and I. Herzog); Psychology (S. Krolak-Schwerdt); Data Analysis in Higher Education (A. Hilbert).

Contributed Sessions (by CLADAG and SKAD)

Latent class models for classification (A. Montanari and A. Cerioli); Classification and models for interval-valued data (F. Palumbo); Selected Problems in Classification (E. Gatnar); Recent Developments in Multidimensional Data Analysis between research and practice I (L. D'Ambra); Recent Developments in Multidimensional Data Analysis between research and practice II (B. Simonetti).

The editors would like to emphatically thank all the section chairs for doing such a great job regarding the organization of their sections and the associated paper reviews.

Cordial thanks also go to the members of the scientific program committee for their conceptual and practical support as well as for the paper reviews: D. Baier (Cottbus), H.-H. Bock (Aachen), H. Bozdogan (Tennessee), J. Buhmann (Zürich), H. Burkhardt (Freiburg), A. Cerioli (Parma); R. Decker (Bielefeld), W. Gaul (Karlsruhe), A. Geyer-Schulz (Karlsruhe), P. Groenen (Rotterdam), T. Imaizumi (Tokyo), K. Jajuga (Wroclaw), R. Kruse (Magdeburg), S. Lang (Innsbruck), B. Lausen (Erlangen-Nürnberg), H.-J. Lenz (Berlin), F. Murtagh (London), H. Ney (Aachen), A. Okada (Tokyo), L. Schmidt-Thieme (Hildesheim), C. Schnoerr (Mannheim), M. Spiliopoulou (Magdeburg), C. Weihs (Dortmund), D. A. Zighed (Lyon).

Furthermore we would like to thank the additional reviewers: A. Hotho, L. Marinho, C. Preisach, S. Rendle, S. Scholz, K. Tso.

The great success of this conference would not have been possible without the support of many people mainly working in the backstage. We would like to particularly thank M. Temerinac (Freiburg), J. Fehr (Freiburg), C. Findlay (Freiburg), E. Patschke (Freiburg), A. Busche (Hildesheim), K. Tso (Hildesheim), L. Marinho (Hildesheim) and the student support team for their hard work in the preparation of this conference, for the support during the event and the post-processing of the conference.

The GfKl Conference 2007 would not have been possible in the way it took place without the financial and/or material support of the following institutions and companies (in alphabetical order): Albert-Ludwigs-University Freiburg – Faculty of Applied Sciences, Gesellschaft für Klassifikation e.V., Microsoft München and Springer Verlag. We express our gratitude to all of them. Finally, we would like to thank Dr. Martina Bihn from Springer Verlag, Heidelberg, for her support and dedication to the production of this volume.

Hildesheim, Freiburg and Bielefeld, February 2008

Christine Preisach
Hans Burkhardt
Lars Schmidt-Thieme
Reinhold Decker

Contents

Part I Classification

Part I

Classification

Distance-based Kernels for Real-valued Data

Lluís Belanche[1], Jean Luis Vázquez[2] and Miguel Vázquez[3]

[1] Dept. de Llenguatges i Sistemes Informàtics
Universitat Politècnica de Catalunya
08034 Barcelona, Spain
belanche@lsi.upc.edu

[2] Departamento de Matemáticas
Universidad Autónoma de Madrid.
28049 Madrid, Spain
juanluis.vazquez@uam.es

[3] Dept. Sistemas Informáticos y Programación
Universidad Complutense de Madrid
28040 Madrid, Spain
mivazque@fdi.ucm.es

Abstract. We consider distance-based similarity measures for real-valued vectors of interest in kernel-based machine learning algorithms. In particular, a *truncated Euclidean* similarity measure and a *self-normalized* similarity measure related to the Canberra distance. It is proved that they are positive semi-definite (p.s.d.), thus facilitating their use in kernel-based methods, like the Support Vector Machine, a very popular machine learning tool. These kernels may be better suited than standard kernels (like the RBF) in certain situations, that are described in the paper. Some rather general results concerning positivity properties are presented in detail as well as some interesting ways of proving the p.s.d. property.

1 Introduction

One of the latest machine learning methods to be introduced is the Support Vector Machine (SVM). It has become very widespread due to its firm grounds in statistical learning theory (Vapnik (1998)) and its generally good practical results. Central to SVMs is the notion of *kernel function*, a mapping of variables from its original space to a higher-dimensional Hilbert space in which the problem is expected to be easier. Intuitively, the kernel represents the *similarity* between two data observations. In the SVM literature there are basically two common-place kernels for real vectors, one of which (popularly known as the RBF kernel) is based on the Euclidean distance between the two collections of values for the variables (seen as vectors).

Obviously not all two-place functions can act as kernel functions. The conditions for being a kernel function are very precise and related to the so-called *kernel matrix* being positive semi-definite (p.s.d.). The question remains, how should the similarity between two vectors of (positive) real numbers be computed? Which of these similarity measures are valid kernels? There are many interesting possibilities that come from well-established distances that may share the property of being p.s.d. There has been little work on this subject, probably due to the widespread use of the initially proposed kernel and the difficulty of proving the p.s.d. property to obtain additional kernels.

In this paper we tackle this matter by examining two alternative distance-based similarity measures on vectors of real numbers and show the corresponding kernel matrices to be p.s.d. These two distance-based kernels could better fit some applications than the normal Euclidean distance and derived kernels (like the RBF kernel). The first one is a truncated version of the standard Euclidean metric in $\boldsymbol{R}$, which additionally extends some of Gower's work in Gower (1971). This similarity yields more sparse matrices than the standard metric. The second one is inversely related to the Canberra distance, well-known in data analysis (Chandon and Pinson (1971)). The motivation for using this similarity instead of the traditional Euclidean-based distance is twofold: (a) it is self-normalised, and (b) it scales in a log fashion, so that similarity is smaller if the numbers are small than if the numbers are big.

The paper is organized as follows. In Section 2 we review work in kernels and similarities defined on real numbers. The intuitive semantics of the two new kernels is discussed in Section 3. As main results, we intend to show some interesting ways of proving the p.s.d. property. We present them in full in Sections 4 and 5 in the hope that they may be found useful by anyone dealing with the difficult task of proving this property. In Section 6 we establish results for positive vectors which lead to kernels created as a combination of different one-dimensional distance-based kernels, thereby extending the RBF kernel.

2 Kernels and similarities defined on real numbers

We consider kernels that are similarities in the classical sense: strongly reflexive, symmetric, non-negative and bounded (Chandon and Pinson (1971)). More specifically, kernels k for positive vectors of the general form:

$$k(\mathbf{x},\mathbf{y}) = f\left(\sum_{j=1}^{n} g_j(d_j(x_j,y_j))\right), \tag{1}$$

where x_j, y_j belong to some subset of $\boldsymbol{R}$, $\{d_j\}_{j=1}^{n}$ are metric distances and $\{f, g_j\}_{j=1}^{n}$ are appropriate continuous and monotonic functions in $\boldsymbol{R}^+ \cup \{0\}$ making the resulting k a valid p.s.d. kernel. In order to behave as a similarity, a natural choice for the kernels k is to be distance-based. Almost invariably, the choice for distance-based real number comparison is based on the standard metric in $\boldsymbol{R}$. The aggregation of a number n of such distance comparisons with the usual 2-norm leads to Euclidean distance in $\boldsymbol{R}^n$. It is known that there exist inverse transformations of this quantity (that can thus be seen as similarity measures) that are valid kernels. An example of this is the kernel:

$$k(\mathbf{x},\mathbf{y}) = \exp\{-\frac{||\mathbf{x}-\mathbf{y}||^2}{2\sigma^2}\},\ \mathbf{x},\mathbf{y} \in \boldsymbol{R}^n, \sigma \neq 0 \in \boldsymbol{R}, \quad (2)$$

popularly known as the RBF (or Gaussian) kernel. This particular kernel is obtained by taking $d(x_j,y_j) = |x_j - y_j|, g_j(z) = z^2/(2\sigma_j^2)$ for non-zero σ_j^2 and $f(z) = e^{-z}$. Note that nothing prevents the use of different scaling parameters σ_j for every component. The decomposition need not be unique and is not necessarily the most useful for proving the p.s.d. property of the kernel.

In this work we concentrate on upper-bounded metric distances, in which case the partial kernels $g_j(d_j(x_j,y_j))$ are lower-bounded, though this is not a necessary condition in general. We list some choices for partial distances:

$$d_{TrE}(x_i,y_i) = \min\{1, |x_i - y_i|\} \qquad \text{(Truncated Euclidean)} \quad (3)$$

$$d_{Can}(x_i,y_i) = \frac{|x_i - y_i|}{x_i + y_i} \qquad \text{(Canberra)} \quad (4)$$

$$d(x_i,y_i) = \frac{|x_i - y_i|}{\max(x_i,y_i)} \qquad \text{(Maximum)} \quad (5)$$

$$d(x_i,y_i) = \frac{(x_i - y_i)^2}{x_i + y_i} \qquad \text{(squared } \chi^2\text{)} \quad (6)$$

Note the first choice is valid in $\boldsymbol{R}$, while the others are valid in $\boldsymbol{R}^+$. There is some related work worth mentioning, since other choices have been considered elsewhere: with the choice $g_j(z) = 1 - z$, a kernel formed as in (1) for the distance (5) appears as p.s.d. in Shawe-Taylor and Cristianini (2004). Also with this choice for g_j, and taking $f(z) = e^{z/\sigma}, \sigma > 0$ the distance (6), leads to a kernel that has been proved p.s.d. in Fowlkes et al. (2004).

3 Semantics and applicability

The distance in (3) is a truncated version of the standard metric in $\boldsymbol{R}$, which can be useful when differences greater than a specified threshold have to be ignored. In similarity terms, it models situations wherein data examples can become more and more similar until they are suddenly indistinguishable. Otherwise, it behaves like the standard metric in $\boldsymbol{R}$. Notice that this similarity may lead to more sparse matrices than those obtainable with the standard metric. The distance in (4) is called the Canberra distance (for one component). It is self-normalised to the real interval $[0,1)$, and is multiplicative rather than additive, being specially sensitive to small changes near zero. Its behaviour can be best seen by a simple example: let a variable stand for the number of children, then the distance between 7 and 9 is not the same "psychological" distance than that between 1 and 3 (which is triple); however, $|7-9| = |1-3|$. If we would like the distance between 1 and 3 be much greater than that between 7 and 9, then this effect is captured. More specifically, letting $z = x/y$, then $d_{Can}(x,y) = g(z)$, where $g(z) = |z-1|/(z+1)$ and thus $g(z) = g(1/z)$. The Canberra distance has been used with great success in content-based image retrieval tasks in Kokare et al. (2003).

4 Truncated Euclidean similarity

Let x_i be an arbitrary finite collection of n different real points $x_i \in \boldsymbol{R}$, $i = 1, \ldots, n$. We are interested in the $n \times n$ similarity matrix $\mathbf{A} = (a_{ij})$ with

$$a_{ij} = 1 - d_{ij}, \quad d_{ij} = \min\{1, |x_i - x_j|\}, \tag{7}$$

where the usual Euclidean distances have been replaced by *truncated Euclidean distances*. We can also write $a_{ij} = (1-d_{ij})_+ = \max\{0, 1-|x_i - x_j|\}$.

Theorem 1. *The matrix* $\mathbf{A}$ *is positive definite (p.s.d.).*

PROOF. We define the bounded functions $X_i(x)$ for $x \in \boldsymbol{R}$ with value 1 if $|x - x_i| \leq 1/2$, zero otherwise. We calculate the interaction integrals

$$l_{ij} = \int_{\boldsymbol{R}} X_i(x) X_j(x) dx.$$

The value is the length of the interval $[x_i - 1/2, x_i + 1/2] \cap [x_j - 1/2, x_j + 1/2]$. It is easy to see that $l_{ij} = 1 - d_{ij}$ if $d_{ij} < 1$, and zero if $|x_i - x_j| \geq 1$ (i.e., when there is no overlapping of supports). Therefore, $l_{ij} = a_{ij}$ if $i \neq j$. Moreover, for $i = j$ we have

$$\int_{\boldsymbol{R}} X_i(x) X_j(x)\, dx = \int X_i^2(x)\, dx = 1.$$

We conclude that the matrix $\mathbf{A}$ is obtained as the interaction matrix for the system of functions $\{X_i\}_{i=1}^N$. These interactions are actually the dot products of the functions in the functional space $L^2(\boldsymbol{R})$. Since a_{ij} is the dot product of the inputs cast into some Hilbert space it forms, by definition, a p.s.d. matrix.

Notice that rescaling of the inputs would allow us to substitute the two "1" (one) in equation (7) by any arbitrary positive number. In other words, the kernel with matrix

$$a_{ij} = (s - d_{ij})_+ = \max\{0, s - |x_i - x_j|\} \tag{8}$$

with $s > 0$ is p.s.d. The classical result for general Euclidean similarity in Gower (1971) is a consequence of this Theorem when $|x_i - x_j| \leq 1$ for all i, j.

5 Canberra distance-based similarity

We define the Canberra similarity between two points as follows

$$s_{Can}(x_i,x_j) = 1 - d_{Can}(x_i,x_j), \quad d_{Can}(x_i,x_j) = \frac{|\, x_i - x_j \,|}{x_i + x_j}, \tag{9}$$

where $d_{Can}(x_i,x_j)$ is called the *Canberra distance*, as in (4). We establish next the p.s.d. property for Canberra distance matrices, for $x_i, x_j \in \boldsymbol{R}^+$.

Theorem 2. *The matrix* $A = (a_{ij})$ *with* $a_{ij} = s_{Can}(x_i,x_j)$ *is p.s.d.*

PROOF. *First step.* Examination of equation (9) easily shows that for any $x_i, x_j \in \boldsymbol{R}^+$ (not including 0) the value of $s_{Can}(x_i,x_j)$ is the same for every pair of points x_i, x_j that have the same quotient x_i/x_j. This gives us the idea of taking logarithms on the input and finding an equivalent kernel for the translated inputs. From now on, define $x \equiv x_i, z \equiv x_j$, for clarity. We use the following straightforward result:

Lemma 1. *Let* K' *be a p.s.d. kernel defined in the region* $B \times B$*, let* Φ *be map from a region* A *into* B*, and let* K *be defined on* $A \times A$ *as* $K(x,z) = K'(\Phi(x), \Phi(z))$*. Then the kernel* K *is p.s.d.*

PROOF. Clearly Φ is a restriction of B, and K' is p.s.d in all $B \times B$.

Here, we take $K = s_{Can}$, $A = \boldsymbol{R}^+$, $\Phi(x) = \log(x)$, so that B is $\boldsymbol{R}$. We now find what K' would be by defining $x' = \log(x)$, $\quad z' = \log(z)$, so that distance d_{Can} can be rewritten as

$$d_{Can}(x,z) = \frac{|\, x - z \,|}{x + z} = \frac{|\, e^{x'} - e^{z'} \,|}{e^{x'} + e^{z'}}.$$

As we noted above, $d_{Can}(x,z)$ is equivalent for any pair of points $x, z \in \boldsymbol{R}^+$ with the same quotients x/z or z/x. Assuming that $x > z$ without loss of generality, we write this as a *translation invariant* kernel by introducing the increment in logarithmic coordinates $h = |\, x' - z' \,| = x' - z' = \log(x/z)$:

$$d_{Can}(x,z) = \frac{e^{z'} e^h - e^{z'}}{e^{z'} e^h + e^{z'}} = \frac{e^h - 1}{e^h + 1}.$$

Substitution on $K = s_{Can}$ gives

$$s_{Can}(x,z) = 1 - \frac{e^h - 1}{e^h + 1} = \frac{2}{e^h + 1}$$

Therefore, for $x', z' \in \boldsymbol{R}$, $x' = z' + h$, we have

$$K'(x',z') = K'(x' - z') = \frac{2}{e^h + 1} = F(h). \tag{10}$$

Note that F is a convex function of $h \in [0, \infty)$ with $F(0) = 1$, $F(\infty) = 0$.

Second step. To prove our theorem we now only have to prove the p.s.d. property for kernel K' satisfying equation (10).

A direct proof uses an integral representation of convex functions that proceeds as follows. Given a twice continuously differentiable function F of the real variable $s \geq 0$, integrating by parts we find the formula

$$F(x) = -\int_x^\infty F'(s)\,ds = \int_x^\infty F''(s)(s-x)\,ds,$$

valid for all $x > 0$ on the condition that $F(s)$ and $sF'(s) \to 0$ as $s \to \infty$. The formula can be written as

$$F(x) = \int_0^\infty F''(s)(s-x)_+\,ds,$$

which implies that whenever $F'' > 0$, we have expressed $F(x)$ as an integral combination with positive coefficients of functions of the form $(s-x)_+$. This is a non-trivial, but commonly used, result in convex theory.

Third step. The functions of the form $(s-x)_+$ are the building blocks of the Truncated Euclidean Similarity kernels (7). Our kernel K' is represented as an integral combination of these functions with positive coefficients. In the previous Section we have proved that functions of the form (8) are p.s.d. We know that the sum of p.s.d. terms is also p.s.d., and the limit of p.s.d. kernels is also p.s.d. Since our expression for K' is, like all integrals, a limit of positive combinations of functions of the form $(s-x)_+$, the previous argument proves that equation (10) is p.s.d., and by Lemma 1 our theorem is proved. More precisely, what we say is that, as a convex function, F can be arbitrarily approximated by sums of functions of the type

$$f_n(x) = \max\{0, a_n - \frac{x}{r_n}\} \tag{11}$$

for $n \in [0,\ldots,N]$, and the r_n equally spaced in the range of the input (so that the bigger the N the closer we get to (10)). Therefore, we can write

$$\frac{2}{e^h+1} = \lim_{n\to\infty} \sum_{i=0}^{n} (a_i - \frac{h}{r_i})_+, \tag{12}$$

where each term in the succession (12) is of the form (11), equivalent to (8).

6 Kernels defined on real vectors

We establish now a result for positive vectors that leads to kernels analogous to the Gaussian RBF kernel. The reader can find useful additional material on positive and negative definite functions in Berg et al. 1984 (esp. Ch. 3).

Definition 1 (Hadamard function). *If $A = [a_{ij}]$ is a $n \times n$ matrix, the function $f : A \to f(A) = [f(a_{ij})]$ is called a Hadamard function (actually, this is the simplest type of Hadamard function).*

Theorem 3. *Let a p.s.d. matrix $A = [a_{ij}]$ and a Hadamard function f be given. If f is an analytic function with positive radius of convergence $R > |a_{ij}|$ and all the coefficients in its power series expansion are non-negative, then the matrix $f(A)$ is p.s.d. as proved in Horn and Johnson (1991).*

Definition 2 (p.s.d. function). *A real symmetric function $f(x,y)$ of real variables will be called p.s.d. if for any finite collection of n real numbers $x_1,\dots,x_n$, the $n\times n$ matrix A with entries $a_{ij} = f(x_i,x_j)$ is p.s.d.*

Lemma 2. *Let $b > 1 \in \mathbf{R}, c \in \mathbf{R}$ and let $c - f(x,y)$ be a p.s.d. function. Then $b^{-f(x,y)}$ is a p.s.d. function.*

PROOF. The function $x \longrightarrow b^x$ is analytic with infinite radius of convergence and all the coefficients in its power series expansion are non-negative in case $b > 1$. By theorem (3) the function $b^{c-f(x,y)}$ is p.s.d.; then so is $b^c b^{-f(x,y)}$ and consequently $b^{-f(x,y)}$ is p.s.d. (since b^c is a positive constant).

Theorem 4. *The following function*

$$k(\mathbf{x},\mathbf{y}) = exp\left(-\sum\nolimits_{i=1}^{n}\frac{d(x_i,y_i)}{\sigma_i}\right), \qquad x_i,y_i,\sigma_i \in \mathbf{R}^+$$

where d is any of (3), (4), (5), is a valid p.s.d. kernel.

PROOF. For simplicity, make $d_i \equiv d(x_i,y_i)$. We know $1-d_i$ is a p.s.d. function, for the choices of d_i defined in (3), (4), (5). Therefore, $(1-d_i)/\sigma_i$ for $\sigma_i > 0 \in R$ is also p.s.d. Making $c \equiv \sum_{i=1}^{n} 1/\sigma_i$ and $f \equiv d_i/\sigma_i$, by lemma (2), the function $exp(-d_i/\sigma_i)$ is p.s.d. The product of p.s.d. functions is p.s.d., and thus $\prod\limits_{i=1}^{n} exp(-d_i/\sigma_i) = exp\left(-\sum\limits_{i=1}^{n}\frac{d_i}{\sigma_i}\right)$ is p.s.d.

This result is useful since it establishes new kernels analogous to the Gaussian RBF kernel but based on alternative metrics. Computational considerations should not be overlooked: the use of the exponential function considerably increases the cost of evaluating the kernel. Hence, kernels not involving this function are specially welcome.

Proposition 1. *Let $d(x_i,x_j) = \frac{|x_i-x_j|}{x_i+x_j}$ be the Canberra distance. Then $k(x_i,x_j) = 1 - d(x_i,x_j)/\sigma$ is a valid p.s.d. kernel if and only if $\sigma \geq 1$.*

PROOF. Let $d_{ij} \equiv d(x_i,x_j)$. We know $\sum_{i=1}^{n}\sum_{j=1}^{n} c_i c_j (1-d_{ij}) \geq 0$ for all $c_i,c_j \in \mathbf{R}$. We have to show that $\sum_{i=1}^{n}\sum_{j=1}^{n} c_i c_j (1-\frac{d_{ij}}{\sigma}) \geq 0$. This can be expressed as $\sigma(\sum_{i=1}^{n}\sum_{j=1}^{n} c_i c_j) \geq \sum_{i=1}^{n}\sum_{j=1}^{n} c_i c_j d_{ij}$.

This result is a generalization of Theorem (2), valid for $\sigma = 1$. It is immediate that the following function (the *Canberra kernel*) is a valid kernel:

$$k(\mathbf{x},\mathbf{y}) = 1 - \frac{1}{n}\sum_{i=1}^{n}\frac{d_i(x_i,y_i)}{\sigma_i},\ \sigma_i \geq 1$$

The inclusion of the σ_i (acting as *learning parameters*) has the purpose of adding flexibility to the models. Concerning the truncated Euclidean distance, a corresponding kernel can be obtained in a similar way. Let $d(x_i,x_j) = min\{1,|x_i - x_j|\}$ and denote for a real number a, $a_+ \equiv 1 - min(1,a) = max(0,1-a)$. Then $\sigma - min\{\sigma, |x_i - x_j|\}$ is p.s.d. by Theorem (1) and so is $max\{0, 1 - \frac{|x_i - x_j|}{\sigma}\}$. In consequence, it is immediate to affirm that the following function (the *Truncated Euclidean kernel*) is again a valid kernel:

$$k(\mathbf{x},\mathbf{y}) = \frac{1}{n}\sum_{i=1}^{n}\left(\frac{d_i(x_i,y_i)}{\sigma_i}\right)_+,\ \sigma_i > 0$$

7 Conclusions

We have considered distance-based similarity measures for real-valued vectors of interest in kernel-based methods, like the Support Vector Machine. The first is a truncated Euclidean similarity and the second a self-normalized similarity. Derived real kernels analogous to the RBF kernel have been proposed, so the kernel toolbox is widened. These can be considered as suitable alternatives for a proper modeling of data affected by multiplicative noise, skewed data and/or containing outliers. In addition, some rather general results concerning positivity properties have been presented in detail.

Acknowledgments

Supported by the Spanish project CICyT CGL2004-04702-C02-02.

References

BERG, C. CHRISTENSEN, J.P.R. and RESSEL, P. (1984): *Harmonic Analysis on Semigroups: Theory of Positive Definite and Related Functions,* Springer.

CHANDON, J.L. and PINSON, S. (1981): *Analyse Typologique. Théorie et Applications,* Masson, Paris.

FOWLKES, C., BELONGIE, S., CHUNG, F., and MALIK. J. (2004): Spectral Grouping Using the Nyström Method. *IEEE Trans. on PAMI*, 26(2), 214–225.

GOWER. J.C. (1971): A general coefficient of similarity and some of its properties, *Biometrics* 27, 857–871.

HORN, R.A. and JOHNSON, C.R. (1991): *Topics in Matrix Analysis*, Cambridge University Press.

KOKARE, M., CHATTERJI, B.N. and BISWAS, P.K. (2003): Comparison of similarity metrics for texture image retrieval. In: *IEEE Conf. on Convergent Technologies for Asia-Pacific Region*, 571–575.

SHAWE-TAYLOR, J. and CRISTIANINI, N. (2004): *Kernel Methods for Pattern Analysis*, Cambridge University Press.

VAPNIK. V. (1998): *The Nature of Statistical Learning Theory*. Springer-Verlag.

Fast Support Vector Machine Classification of Very Large Datasets

Janis Fehr[1], Karina Zapién Arreola[2] and Hans Burkhardt[1]

[1] University of Freiburg, Chair of Pattern Recognition and Image Processing
79110 Freiburg, Germany
fehr@informatik.uni-freiburg.de

[2] INSA de Rouen, LITIS
76801 St Etienne du Rouvray, France

Abstract. In many classification applications, Support Vector Machines (SVMs) have proven to be highly performing and easy to handle classifiers with very good generalization abilities. However, one drawback of the SVM is its rather high classification complexity which scales linearly with the number of Support Vectors (SVs). This is due to the fact that for the classification of one sample, the kernel function has to be evaluated for all SVs. To speed up classification, different approaches have been published, most which of try to reduce the number of SVs. In our work, which is especially suitable for very large datasets, we follow a different approach: as we showed in (Zapien et al. 2006), it is effectively possible to approximate large SVM problems by decomposing the original problem into linear subproblems, where each subproblem can be evaluated in $\Omega(1)$. This approach is especially successful, when the assumption holds that a large classification problem can be split into mainly easy and only a few hard subproblems. On standard benchmark datasets, this approach achieved great speedups while suffering only sightly in terms of classification accuracy and generalization ability. In this contribution, we extend the methods introduced in (Zapien et al. 2006) using not only linear, but also non-linear subproblems for the decomposition of the original problem which further increases the classification performance with only a little loss in terms of speed. An implementation of our method is available in (Ronneberger and et al.) Due to page limitations, we had to move some of theoretic details (e.g. proofs) and extensive experimental results to a technical report (Zapien et al. 2007).

1 Introduction

In terms of classification-speed, SVMs (Vapnik 1995) are still outperformed by many standard classifiers when it comes to the classification of large problems. For a non-linear kernel function k, the classification function can be written as in Eq. (1). Thus, the classification complexity lies in $\Omega(n)$ for a problem with n SVs. However, for linear problems, the classification function has the form of Eq. (2), allowing classification in $\Omega(1)$ by calculating the dot product with the normal vector $\mathbf{w}$ of the hyperplane. In addition, the SVM has the problem that the complexity of a SVM model always scales with the most difficult samples, forcing an increase in Support Vectors. However, we observed that many large scale problems can easily be divided in a large set of rather simple subproblems and only a few difficult ones. Following this assumption, we propose a classification method based on a tree whose nodes consist mostly of linear SVMs (Fig.(1)).

$$f(\mathbf{x}) = \text{sign}\left(\sum_{i=1}^{m} y_i \alpha_i k(\mathbf{x_i}, \mathbf{x}) + b\right) \tag{1}$$

$$f(\mathbf{x}) = \text{sign}\left(\langle \mathbf{w}, \mathbf{x} \rangle + b\right) \tag{2}$$

This paper is structured as follows: first we give a brief overview of related work. Section 2 describes our initial linear algorithm in detail including a discussion of the zero solution problem. In section 3, we introduce a non-linear extension to our initial algorithm, followed by Experiments in section 4.

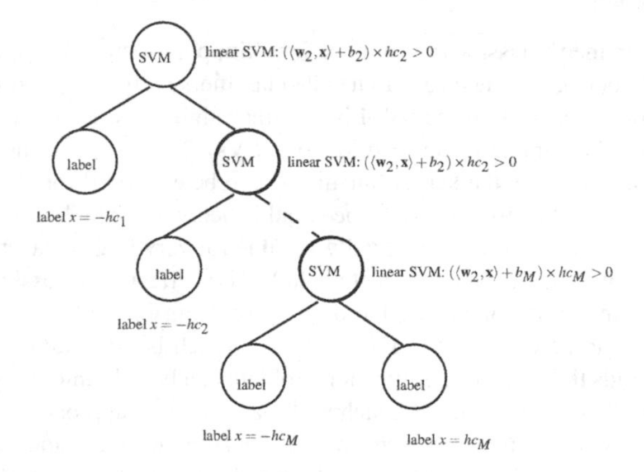

Fig. 1. Decision tree with linear SVM

1.1 Related work

Recent work on SVM classification speedup mainly focused on the reduction of the decision problem: A method called *RSVM* (Reduced Support Vector Machines) was proposed by Lee and Mangasarian (2001), it preselects a subset of training samples as SVs and solves a smaller Quadratic Programming problem. Lei and Govindaraju (2005) introduced a reduction of the feature space using principal component analysis and Recursive Feature Elimination. Burges and Schoelkopf (1997) proposed a method to approximate $\mathbf{w}$ by a list of vectors associated with coefficients α_i. All these methods yield good speedup, but are fairly complex and computationally expensive. Our approach, on the other hand, was endorsed by the work of Bennett and Bredensteiner (2000) who experimentally proved that inducing a large margin in decision trees with linear decision functions improved the generalization ability.

2 Linear SVM trees

The algorithm is described for binary problems, an extension to multiple-class problems can be realized with different techniques like one vs. one or one vs. rest (Hsu and Lin 2001) (Zapien et al. 2007).
At each node i of the tree, a hyperplane is found that correctly classifies all samples in one class (this class will be called the "hard"' class, denoted hc_i). Then, all correctly classified samples of the other class (the "soft" class) are removed from the problem, Fig. (2). The decision of which class is to be assigned "hard" is taken

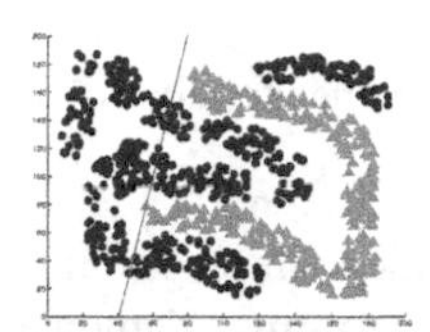
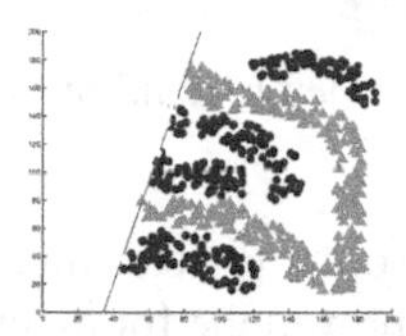

Fig. 2. Problem fourclass (Schoelkopf and Smola 2002). Left: hyperplane for the first node. Right: Problem after first node ("hard" class = triangles).

in a greedy manner for every node (Zapien et al. 2007). The algorithm terminates when the remaining samples all belong to the same class. Fig.(3) shows a training sequence. We will further extend this algorithm, but first we give a formalization for the basic approach.
Problem Statement. Given a two class problem with $m = m_1 + m_{-1}$ samples $\mathbf{x}_i \in \mathbb{R}^n$ with labels y_i, $i \in CC$ and $CC = \{1,...,m\}$. Without loss of generality we define a ***Class 1 (Positive Class)*** $CC_1 = \{1,...,m_1\}$, $y_i = 1$ for all $i \in CC_1$, with a global penalization value D_1 and individual penalization values $C_i = D_1$ for all $i \in CC_1$ as well as an analog ***Class -1 (Negative Class)*** $CC_{-1} = \{m_1+1,...,m_1+m_{-1}\}$, $y_i = -1$ for all $i \in CC_{-1}$, with a global penalization value D_{-1} and individual penalization values $C_i = D_{-1}$ for all $i \in CC_{-1}$.

2.1 Zero vector as solution

In order to train a SVM using the previous definitions, taking one class to be "hard" in a training step, e.g. CC_{-1} is the "hard" class, one could simply set $D_{-1} \rightarrow \infty$ and $D_1 << D_{-1}$ in the primal SVM optimization problem:

$$\underset{\mathbf{w}\in\mathcal{H},b\in\mathbb{R},\xi\in\mathbb{R}^m}{\text{minimize}} \quad \tau(\mathbf{w},\xi) = \frac{1}{2}\|\mathbf{w}\|^2 + \sum_{i=1}^{m} C_i\xi_i, \tag{3}$$

$$\text{subject to} \quad y_i(\langle\mathbf{x}_i,\mathbf{w}\rangle + b) \geq 1 - \xi_i, \; i = 1,..,m, \tag{4}$$

$$\xi_i \geq 0, \; i = 1,..,m. \tag{5}$$

Unfortunately, in some cases the optimization process converges to a trivial solution: the zero vector. We used the convex hull interpretation of SVMs (Bennett and

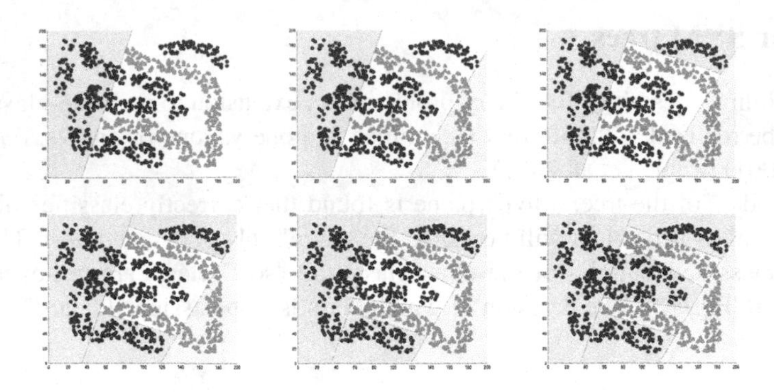

Fig. 3. Sequence (left to right) of hyperplanes for nodes 1-6 of the tree.

Bredensteiner 2000), in order to determine under which circumstances the trivial solution is occurring and proved the following theorems (Zapien et al. 2007):

Theorem 1: If the convex hull of the "hard" class CC_1 intersects the convex hull of the "soft" class CC_{-1}, then $\mathbf{w} = \mathbf{0}$ is a feasible point for the primal Problem (4) if $D_{-1} \geq \max_{i \in CC_1}\{\lambda_i\} \cdot D_1$, where λ_i are such that

$$\mathbf{p} = \sum_{i \in CC_1} \lambda_i \mathbf{x}_i,$$

is a convex combination for a point $\mathbf{p}$ that belongs to both convex hulls.

Theorem 2: If the center of gravity $\mathbf{s}_{-1}$ of class CC_{-1} is inside the convex hull of class CC_1, then it can be written as

$$\mathbf{s}_{-1} = \sum_{i \in CC_1} \lambda_i \mathbf{x}_i \quad \text{and} \quad \mathbf{s}_{-1} = \sum_{j \in CC_{-1}} \frac{1}{m_{-1}} \mathbf{x}_j$$

with $\lambda_i \geq 0$ for all $i \in CC_1$ and $\sum_{i \in CC_1} \lambda_i = 1$. If additionally, $D_1 \geq \lambda_{\max} D_{-1} m_{-1}$, where $\lambda_{\max} = \max_{i \in CC_1}\{\lambda_i\}$, then $\mathbf{w} = \mathbf{0}$ is a feasible point for the primal Problem.

Please refer to (Zapien et al. 2007) for detailed proofs of both theorems.

2.2 H1-SVM problem formulation

To avoid the zero vector, we proposed a modification of the original SVM optimization problem, which is taking advantage of the previous theorems: the H1-SVM (H1 for one hard class).

H1-SVM Primal Problem

$$\min_{\mathbf{w}\in\mathbb{R}^n, b\in\mathbb{R}} \quad \tfrac{1}{2}\|\mathbf{w}\|^2 - \textstyle\sum_{i\in CC_{\bar{k}}} y_i(\langle \mathbf{x}_i, \mathbf{w}\rangle + b) \tag{6}$$

$$\text{subject to} \quad y_i(\langle \mathbf{x}_i, \mathbf{w}\rangle + b) \geq 1 \text{ for all } i \in CC_k, \tag{7}$$

where $k = 1$ and $\bar{k} = -1$, or $k = -1$ and $\bar{k} = 1$.
This new formulation constraints Eq. (7) to classify all samples in the class CC_k perfectly, forcing a "hard" convex hull (H1) for CC_k. The number of misclassification on the other class $CC_{\bar{k}}$ is added to the objective function, hence the solution is a trade-off between a maximal margin and a minimum number of misclassifications in the "soft" class $CC_{\bar{k}}$.

H1-SVM Dual Formulation

$$\max_{\alpha\in\mathbb{R}^m} \textstyle\sum_{i=1}^m \alpha_i - \tfrac{1}{2}\sum_{i,j=1}^m \alpha_i\alpha_j y_i y_j \langle \mathbf{x}_i, \mathbf{x}_j\rangle \tag{8}$$

$$\text{subject to} \qquad 0 \leq \alpha_i \leq C_i,\ i \in CC_k, \tag{9}$$

$$\alpha_j = 1,\ j \in CC_{\bar{k}}, \tag{10}$$

$$\textstyle\sum_{i=1}^m \alpha_i y_i = 0, \tag{11}$$

where $k = 1$ and $\bar{k} = -1$, or $k = -1$ and $\bar{k} = 1$.
This problem can be solved in a similar way as the original SVM Problem using the SMO algorithm (Schoelkopf and Smola 2002)(Zapien et al. 2007), and adding some modifications to force $\alpha_i = 1\ \forall i \in CC_{\bar{k}}$.

Theorem 3: For the H1-SVM the zero solution can only occur if $|CC_k| \geq (n-1)$ and there exists a linear combination of the sample vectors in the "hard" class $\mathbf{x}_i \in CC_k$ and the sum of the sample vectors in the "soft" class, $\sum_{i\in CC_{\bar{k}}} \mathbf{x}_i$.
Proof: Without loss of generality, let the "hard" class be class CC_1. Then,

$$\begin{aligned}\mathbf{w} = \sum_{i=1}^m \alpha_i y_i \mathbf{x}_i &= \sum_{i\in CC_1} \alpha_i \mathbf{x}_i - \sum_{i\in CC_{-1}} \alpha_i \mathbf{x}_i \\ &= \sum_{i\in CC_1} \alpha_i \mathbf{x}_i - \sum_{i\in CC_{-1}} \mathbf{x}_i.\end{aligned} \tag{12}$$

If we define $\mathbf{z}_i = \sum_{i\in CC_{-1}} \mathbf{x}_i$ and $|CC_1| \geq (n-1) = dim(\mathbf{z}_i) - 1$, there exist $\{\alpha_i\}, i \in CC_1, \alpha_i \not\equiv 0$ such that

$$\mathbf{w} = \sum_{i\in CC_1} \alpha_i \mathbf{x}_i - \mathbf{z}_i \quad = 0.$$

The usual threshold calculation ((Keerthi et al. 1999) and (Schoelkopf and Smola 2002)) can no longer be used to define the hyperplane, please refer to (Zapien et al. 2007) for details on the threshold computation.
The basic algorithm can be improved with some heuristics for greedy "hard"-class determination and tree pruning, shown in (Zapien et al. 2007).

3 Non-linear extension

In order to classify a sample, one simply runs it down the SVM-tree. When using only linear nodes, we already obtained good results (Zapien et al. 2006), but we also observed that first of all, most errors occur in the last node, and second, that over all only a few samples will reach the last node during the classification procedure. This motivated us to add a non-linear node (e.g. using RBF kernels) to the end of the tree. Training of this extended SVM-tree is analogous to the original case. First a pure

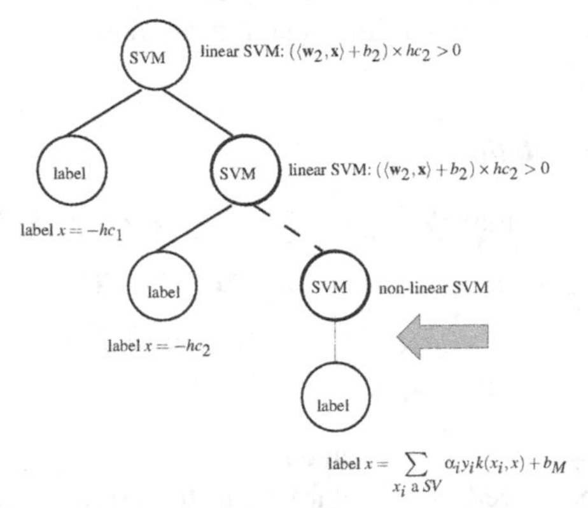

Fig. 4. SVM tree with non-linear extesion

linear tree is build. Then we use a heuristic (trade-off between average classification depth and accuracy) to move the final, non-linear node from the last node up the tree. It is very important to notice, that to avoid overfitting, the final non-linear SVM has to be trained on the entire initial training set, and not only on the samples remaining after the last linear node. Otherwise the final node is very likely to suffer from strong overfitting. Of cause, then the final model will have many SVs, but since only a few samples will reach the final node, our experiments indicate that the average classification depth will be hardly affected.

4 Experiments

In order to show the validity and classification accuracy of our algorithm we performed a series of experiments on standard benchmark data sets. These experiments were conducted[1] e.g. on *Faces* (Carbonetto) (9172 training samples, 4262 test samples, 576 features) and *USPS* (Hull 1994) (18063 training samples, 7291 test samples, 256 features) as well as on several other data sets. More and detailed experiments can be found in (Zapien et al. 2007). The data was split into training and test sets and normalized to minimum and maximum feature values (Min-Max) or standard deviation (Std-Dev).

[1] These experiments were run on a computer with a P4, 2.8 GHz and 1G in Ram.

Faces (Min-Max)	RBF Kernel	H1-SVM	H1-SVM *Gr-Heu*	RBF/H1	RBF/H1 *Gr-Heu*
Nr. SVs or Hyperplanes	2206	4	4	551.5	551.5
Training Time	14:55.23	10:55.70	14:21.99	1.37	1.04
Classification Time	03:13.60	00:14.73	00:14.63	***13.14***	***13.23***
Classif. Accuracy %	95.78 %	91.01 %	91.01 %	1.05	1.05

USPS (Min-Max)	RBF Kernel	H1-SVM	H1-SVM *Gr-Heu*	RBF/H1	RBF/H1 *Gr-Heu*
Nr. SVs or Hyperplanes	3597	49	49	73.41	73.41
Training Time	00:44.74	00:22.70	02:09.58	1.97	0.35
Classification Time	01:58.59	00:19.99	00:20.07	***5.93***	***5.91***
Classif. Accuracy %	95.82 %	93.76 %	93.76 %	1.02	1.02

Comparisons to related work are difficult, since most publications (Bennett and Bredensteiner 2000), (Lee and Mangasarian 2001) used datasets with less than 1000 samples, where the training and testing time are negligible. In order to test the performance and speedup on very large datasets, we used our own *Cell Nuclei Database* (Zapien et al. 2007) with 3372 training samples, 32 features each, and about **16 million** test samples:

	RBF-Kernel	linear tree H1-SVM	non-linear tree H1-SVM
training time	≈1s	≈3s	≈5s
Nr. SVs or Hyperplanes	980	86	86
average classification depth	-	7.3	8.6
classifiaction time accuracy	≈1.5h 97.69%	≈2 min 95.43%	≈2 min 97.5%

5 Conclusion

We have presented a new method for fast SVM classification. Compared to non-linear SVM and speedup methods our experiments showed a very competitive speedup while achieving reasonable classification results (loosing only marginal when we apply the non-linear extension compared to non-linear methods). Especially if our initial assumption holds , that large problems can be split in mainly easy and only a few hard problems, our algorithm achieves very good results. The advantage of this approach clearly lies in its simplicity since no parameter has to be tuned.

References

V. VAPNIK (1995): The Nature of Statistical Learning Theory, New York: Springer Verlag.

Y. LEE and O. MANGASARIAN (2001): RSVM: Reduced Support Vector Machines, Proceedings of the first SIAM International Conference onData Mining, 2001 SIAM International Conference, Chicago, Philadelphia.

H. LEI and V. GOVINDARAJU (2005): Speeding Up Multi-class SVM Evaluation by PCA andFeature Selection, Proceedings of the Workshop on Feature Selection for DataMining: Interfacing Machine Learning and Statistics, 2005 SIAM Workshop.

C. BURGES and B. SCHOELKOPF (1997): Improving Speed and Accuracy of Support Vector Learning Machines, Advances in Neural Information Processing Systems9, MIT Press, MA, pp 375-381.

K. P. BENNETT and E. J. BREDENSTEINER (2000): Duality and Geometry in SVM Classifiers, Proc. 17th International Conf. on Machine Learning, pp 57-64.

C. HSU and C. LIN (2001): A Comparison of Methods for Multi-Class Support Vector Machines, Technical report, Department of Computer Science and Information Engineering, National Taiwan University, Taipei, Taiwan.

T. K. HO AND E. M. KLEINBERG (1996): Building projectable classifiers of arbitrary complexity, Proceedings of the 13th International Conference onPattern Recognition, pp 880-885, Vienna, Austria.

B. SCHOELKOPF and A. SMOLA (2002): Learning with Kernels, The MIT Press, Cambridge,MA, USA.

S. KEERTHI and S. SHEVADE and C. Bhattacharyya and K. Murthy (1999): Improvements to Platt's SMO Algorithm for SVM Classifier Design, Technical report, Dept. ofCSA, Banglore, India.

P. CARBONETTO: Face data base, http://www.cs.ubc.ca/ pcarbo/, University of British Columbia Computer Science Deptartment.

J. J. HULL (1994): A database for handwritten text recognition research, IEEE Transactions on Pattern Analysis and Machine Intelligence, Vol 16, No 5, pp 550-554.

K. ZAPIEN, J. FEHR and H. BURKHARDT (2006): Fast Support Vector Machine Classification using linear SVMs, in Proceedings: ICPR, pp. 366- 369 ,Hong Kong 2006.

O. RONNEBERGER and et al.: SVM template library, University of Freiburg, Department of Computer Science, Chair of Pattern Recognition and Image Processing, http://lmb.informatik.uni-freiburg.de/lmbsoft/libsvmtl/index.en.html

K. ZAPIEN, J. FEHR and H. BURKHARDT (2007): Fast Support Vector Machine Classification of very large Datasets, Technical Report 2/2007, University of Freiburg, Department of Computer Science, Chair of Pattern Recognition and Image Processing. http://lmb.informatik.uni-freiburg.de/people/fehr/svm_tree.pdf

Fusion of Multiple Statistical Classifiers

Eugeniusz Gatnar

Institute of Statistics, Katowice University of Economics,
Bogucicka 14, 40-226 Katowice, Poland
egatnar@ae.katowice.pl

Abstract. In the last decade the classifier ensembles have enjoyed a growing attention and popularity due to their properties and successful applications.

A number of combination techniques, including majority vote, average vote, behavior-knowledge space, etc. are used to amplify correct decisions of the ensemble members. But the key of the success of classifier fusion is diversity of the combined classifiers.

In this paper we compare the most commonly used combination rules and discuss their relationship with diversity of individual classifiers.

1 Introduction

Fusion of multiple classifiers is one of the recent major advances in statistics and machine learning. In this framework, multiple models are built on the basis of training set and combined into an ensemble or a committee of classifiers. Then the component models determine the predicted class.

Classifier ensembles proved to be high performance classification systems in numerous applications, e.g. pattern recognition, document analysis, personal identification, data mining etc.

The high accuracy of the ensemble is achieved if its members are "weak" and diverse. The term "weak" refers to unstable classifiers, such as classification trees, and neural nets. Diversity means that the classifiers are different from each other (independent, uncorrelated). This is usually obtained by using different training subsets, assigning different weights to instances or selecting different subsets of features.

Tumer and Ghosh (1996) have shown that the ensemble error decreases with the reduction in correlation between component classifiers. Therefore, we need to assess the level of indpendence of the members of the ensemble, and different measures of diversity have been proposed so far.

The paper is organised as follows. In Section 2 we give some basics on classifier fusion. Section 3 contains a short description of selected diversity measures. In Section 4 we discuss the fusion methods (combination rules). The problems related to assessment of performance of combination rules and their relationship with diversity measures are presented in Section 5. Section 6 gives a brief description of our experiments and the obtained results. The last section contains some conclusions.

2 Classifier fusion

A classifier C is any mapping $C : \mathbf{X} \to Y$ from the feature space $\mathbf{X}$ into a set of class labels $Y = \{l_1, l_2, \ldots, l_J\}$.

The classifier fusion consists of two steps. In the first step the set of M individual classifiers $\{C_1, C_2, \ldots, C_M\}$ is designed on the basis of the training set $T = \{(\mathbf{x}_1, y_1), (\mathbf{x}_2, y_2), \ldots, (\mathbf{x}_N, y_N)\}$.

Then, in the second step, their predictions are combined into an ensemble $\hat{C}^*$ using a combination function F:

$$\hat{C}^* = F(\hat{C}_1, \hat{C}_2, \ldots, \hat{C}_M). \tag{1}$$

Various combinatorial rules have been proposed in the literature to approximate the function F, and some of them will be discussed in Section 4.

3 Diversity of ensemble members

In order to assess the mutual independence of individual classifiers, different measures have been proposed. The simplest ones are pairwise measures defined between two classifiers, and the overall diversity of the ensemble is the average of the diversities (ρ) between all pairs of the ensemble members:

$$Diversity(C^*) = \frac{2}{M(M-1)} \sum_{m=1}^{M-1} \sum_{k=m+1}^{M} \rho(m,k). \tag{2}$$

The relationship between a pair of classifiers C_i and C_j can be shown in the form of the 2×2 contingency table (Table 1).

Table 1. A 2×2 contingency table for the two classifier outputs.

Classifiers	C_j is correct	C_j is wrong
C_i is correct	a	b
C_i is wrong	c	d

The well known measure of classifier dependence is the binary version of the Pearson's correlation coefficient:

$$r(i,j) = \frac{ad - bc}{\sqrt{(a+b)(c+d)(a+c)(b+d)}}. \tag{3}$$

Partridge and Yates (1996) have used a measure named *within-set generalization diversity*. This measure is simply the *kappa* statistics:

$$\kappa(i,j) = \frac{2(ac-bd)}{(a+b)(c+d)+(a+c)(b+d)}. \tag{4}$$

Skalak (1996) reported the use of the *disagreement measure*:

$$DM(i,j) = \frac{b+c}{a+b+c+d}. \tag{5}$$

Giacinto and Roli (2000) have introduced a measure based on the compound error probability for the two classifiers, and named *compound diversity*:

$$CD(i,j) = \frac{d}{a+b+c+d}. \tag{6}$$

This measure is also named "double-fault measure" because it is the proportion of the examples that have been misclassified by both classifiers.

Kuncheva et al. (2000) strongly recommended the Yule's Q statistics to evaluate the diversity:

$$Q(i,j) = \frac{ad-bc}{ad+bc}. \tag{7}$$

Unfortunately, this measure has two disadvantages. In some cases its value may be undefined. e.g. when $a=0$ and $b=0$, and it cannot distinguish between different distributions of classifier outputs.

In order to overcome the drawbacks of the Yule's Q statistics, Gatnar (2005) proposed the diversity measure based on the Hamann's coefficient:

$$H(i,j) = \frac{(a+d)-(b+c)}{a+b+c+d}. \tag{8}$$

Several non-pairwise measures have been also developed to evaluate the level of diversity between all members of the ensemble.

Cunningham and Carney (2000) suggested using the *entropy function*:

$$EC = -\frac{1}{N}\sum_{i=1}^{N} L(\mathbf{x}_i)log(L(\mathbf{x}_i)) - \frac{1}{N}\sum_{i=1}^{N}(M-L(\mathbf{x}_i))log(M-L(\mathbf{x}_i)), \tag{9}$$

where $L(\mathbf{x})$ is the number of classifiers that correctly classified the observation $\mathbf{x}$. Its simplified version was introduced by Kuncheva and Whitaker (2003):

$$E = \frac{1}{N}\sum_{i=1}^{N}\frac{1}{M-\lceil M/2\rceil} min\{L(\mathbf{x}_i), M-L(\mathbf{x}_i)\}. \tag{10}$$

Kohavi and Wolpert (1996) used their variance to evaluate the diversity:

$$KW = \frac{1}{NM^2} \sum_{i=1}^{N} L(\mathbf{x}_i)(M - L(\mathbf{x}_i)). \tag{11}$$

Also Dietterich (2000) proposed the measure to assess the level of agreement between classifiers. It is the *kappa* statistics:

$$\kappa = 1 - \frac{\frac{1}{M}\sum_{i=1}^{N} L(\mathbf{x}_i)(M - L(\mathbf{x}_i))}{N(M-1)\bar{p}(1-\bar{p})}. \tag{12}$$

Hansen and Salamon (1990) introduced the measure of difficulty θ. It is simply the variance of the random variable $Z = L(\mathbf{x})/M$:

$$\theta = Var(Z). \tag{13}$$

Two measures of diversity have been proposed by Partridge and Krzanowski (1997) for evaluation of the software diversity. The first one is the *generalized diversity measure*:

$$GD = 1 - \frac{p(2)}{p(1)}, \tag{14}$$

where $p(k)$ is the probability that k randomly chosen classifiers will fail on the observation $\mathbf{x}$. The second measure is named *coincident failure diversity*:

$$CFD = \begin{cases} 0 & \text{where } p_0 = 1 \\ \frac{1}{1-p_0}\sum_{m=1}^{M} \frac{M-m}{M-1} p_m & \text{where } p_0 < 1 \end{cases}, \tag{15}$$

where p_m is the probability that exactly m out of M classifiers will fail on an observation $\mathbf{x}$.

4 Combination rules

Once we have produced the set of individual classifiers of desired level of diversity, we combine their predictions to amplify their correct decisions and cancel out the wrong ones. The combination function F in (1) depends on the type of the classifier outputs.

There are three different forms of classifier output. The classifier can produce a single class label (abstract level), rank the class labels according to their posterior probabilities (rank level), or produce a vector of posterior probabilities for classes (measurement level).

Majority voting is the most popular combination rule for class labels[1]:

$$\hat{C}^*(\mathbf{x}) = \arg\max_j \left\{ \sum_{m=1}^{M} I(\hat{C}_m(\mathbf{x}) = l_j) \right\}. \tag{16}$$

[1] In the **R** statistical environment we obtain class labels using the command `predict(...,type="class")`.

It can be proved that it is optimal if the number of classifiers is odd, they have the same accuracy, and the classifier's outputs are independent. If we have evidence that certain models are more accurate than others, weighing the individual predictions may improve the overall performance of the ensemble.

Behavior Knowledge Space developed by Huang and Suen (1995) uses a look-up table that keeps track of how often each class combination is produced by the classifiers during training. Then, during testing, the winner class is the most frequently observed class in the BKS table for the combination of class labels produced by the set of classifiers.

Wernecke (1992) proposed a method similar to BKS, that uses the look-up table with 95% confidence intervals of the class frequencies. If the intervals overlap, the least wrong classifier gives the class label.

Naive Bayes combination introduced by Domingos and Pazzani (1997) also needs training to estimate the prior and posterior probabilities:

$$s_j(\mathbf{x}) = P(l_j) \prod_{m=1}^{M} P(\hat{C}_m(\mathbf{x})|l_j). \tag{17}$$

Finally, the class with the highest value of $s_j(\mathbf{x})$ is chosen as the ensemble prediction.

On the measurement level, each classifier produces a vector of posterior probabilities[2] $\hat{C}_m(\mathbf{x}) = [c_{m1}(\mathbf{x}), c_{m2}(\mathbf{x}), \ldots, c_{mJ}(\mathbf{x})]$. And combining predictions of all models, we have a matrix called *decision profile* for an instance $\mathbf{x}$:

$$DP(\mathbf{x}) = \begin{bmatrix} c_{11}(\mathbf{x}) & c_{12}(\mathbf{x}) & \ldots & c_{1J}(\mathbf{x}) \\ \ldots & \ldots & \ldots & \ldots \\ c_{M1}(\mathbf{x}) & c_{M2}(\mathbf{x}) & \ldots & c_{MJ}(\mathbf{x}) \end{bmatrix} \tag{18}$$

Based on the decision profile we calculate the support for each class ($s_j(\mathbf{x})$), and the final prediction of the ensemble is the class with the highest support:

$$\hat{C}^*(\mathbf{x}) = \arg\max_j \{s_j(\mathbf{x})\}. \tag{19}$$

The most commonly used is the average (mean) rule:

$$s_j(\mathbf{x}) = \frac{1}{M} \sum_{m=1}^{M} c_{mj}(\mathbf{x}). \tag{20}$$

There are also other algebraic rules that calculate median, maximum, minimum and product of posterior probabilities for the j-th class. For example, the product rule is:

$$s_j(\mathbf{x}) = \frac{1}{M} \prod_{m=1}^{M} c_{mj}(\mathbf{x}). \tag{21}$$

Kuncheva et al. (2001) proposed a combination method based on Decision Templates, that are averaged decision profiles for each class (DT_j). Given an instance $\mathbf{x}$,

[2] We use the command `predict(...,type="prob")`.

its decision profile is compared to the decision templates of each class, and the class whose decision template is closest (in terms of the Euclidean distance) is chosen as the ensemble prediction:

$$s_j(\mathbf{x}) = 1 - \frac{1}{MJ} \sum_{m=1}^{M} \sum_{k=1}^{J} (DT_j(m,k) - c_{mk}(\mathbf{x}))^2. \tag{22}$$

There are other combination functions using more sophisticated methods, such as fuzzy integrals (Grabisch, 1995), Dempster-Shafer theory of evidence (Rogova, 1994) etc.

The rules presented above can be divided into two groups: trainable and non-trainable. In trainable rules we determine the values of their parameters using the training set, e.g. cell frequencies in the BKS method, or Decision Templates for classes.

5 Open problems

There are several problems that remain open in classifier fusion. In this paper we only focus on two of them. We have shown above ten combination rules, so the first problem is the search for the best one, i.e. the one that gives the more accurate ensembles.

And the second problem is concerned with the relationship between diversity measures and combination functions. If there is any, we would be able to predict the ensemble accuracy knowing the level of diversity of its members.

6 Results of experiments

In order to find the best combination rule and determine relationship between combination rules and diversity measures we have used 10 benchmark datasets, divided into learning and test parts, as shown in Table 2.

For each dataset we have generated 100 ensembles of different sizes: $M = 10, 20, 30, 40, 50$, and we used classification trees[3] as the base models.

We have computed the average ranks for the combination functions, where rank 1 was for the best rule, i.e. the one that produced the most accurate ensemble, and rank 10 - for the worst one. The ranks are presented in Table 3.

We found that the mean rule is simple and has consistent performance for the measurement level, and majority voting is a good combination rule for class labels. Maximum rule is too optimistic, while minimum rule is too pessimistic.

If the classifier correctly estimates the posterior probabilities, the product rule should be considered. But it is sensitive to the most pessimistic classifier.

[3] In order to grow trees, we have used the `Rpart` procedure written by Therneau and Atkinson (1997) for the **R** environment.

Table 2. Benchmark datasets.

Dataset	Number of cases in training set	Number of cases in test set	Number of predictors	Number of classes
DNA	2124	1062	180	3
Letter	16000	4000	16	26
Satellite	4290	2145	36	6
Iris	100	50	4	3
Spam	3000	1601	57	2
Diabetes	512	256	8	2
Sonar	138	70	60	2
Vehicle	564	282	18	4
Soybean	455	228	34	19
Zip	7291	2007	256	10

Table 3. Average ranks for combination methods.

Method	Rank
mean	2.98
vote	3.50
prod	4.73
med	4.91
min	6.37
bayes	6.42
max	7.28
DT	7.45
Wer	7.94
BKS	8.21

Figure 1 illustrates the comparison of performance of the combination functions for the Spam dataset, which is typical of the datasets used in our experiments. We can observe that the fixed rules perform better than the trained rules.

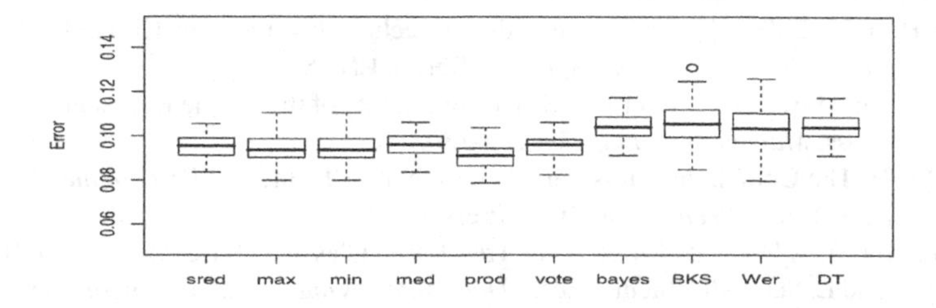

Fig. 1. Boxplots of combination rules for the Spam dataset.

We have also noticed that the mean, median and vote rules give similar results. Moreover, cluster analysis has shown that there are three more groups of rules of similar performance: minimum and maximum, Bayes and Decision Templates, BKS and Wernecke's combination method.

In order to find the relationship between the combination functions and the diversity measures, we have calculated Pearson correlations. Correlations are moderate (greater than 0.4) between mean, mode, product, and vote rules and Compound Diversity (6) as the only pairwise measure of diversity.

For non-pairwise measures correlations are strong (greater than 0.6) only between average, median, and vote rules, and Theta (13).

7 Conclusions

In this paper we have compared ten functions that combine outputs of the individual classifiers into the ensemble. We have also studied the relationships between the combination rules and diversity measures.

In general, we have observed that trained rules, such as BKS, Wernecke, Naive Bayes and Decision Templates, perform poorly, especially for large number of component classifiers (M). This result is contrary to Duin (2002), who argued that trained rules are better than fixed rules.

We have also found that the mean rule and the voting rule are good for the measurement level and abstract level, respectively.

But there are not strong correlations between the combination functions and the diversity measures. This means that we can not predict the ensemble accuracy for the particular combination method.

References

CUNNINGHAM, P. and CARNEY J. (2000): Diversity versus quality in classification ensembles based on feature selection, In: *Proc. of the European Conference on Machine Learning*, Springer, Berlin, LNCS 1810, 109-116.

DIETTERICH, T.G. (2000): Ensemble methods in machine learning, In: Kittler J., Roli F. (Eds.), *Multiple Classifier Systems*, Springer, Berlin, LNCS 1857, 1-15,

DOMINGOS, P. and PAZZANI, M. (1997): On the optimality of the simple Bayesian classifier under zero- loss, *Machine Learning, 29, 103-130.*

DUIN, R. (2002): The Combining Classifier: Train or Not to Train?, In: *Proc. of the 16th Int. Conference on Pattern Recognition*, IEEE Press.

GATNAR, E. (2005): A Diversity Measure for Tree-Based Classifier Ensembles. In: D. Baier, R. Decker, and L. Schmidt-Thieme (Eds.): *Data Analysis and Decision Support*. Springer, Heidelberg New York.

GIACINTO, G. and ROLI, F. (2001): Design of effective neural network ensembles for image classification processes. *Image Vision and Computing Journal, 19, 699–707.*

GRABISCH M. (1995): On equivalence classes of fuzzy connectives -the case of fuzzy integrals, *IEEE Transactions on Fuzzy Systems, 3(1), 96-109.*

HANSEN, L.K. and SALAMON, P. (1990): Neural network ensembles. *IEEE Transactions on Pattern Analysis and Machine Intelligence 12, 993–1001.*

HUANG, Y.S. and SUEN, C.Y. (1995): A method of combining multiple experts for the recognition of unconstrained handwritten numerals, *IEEE Transactions on Pattern Analysis and Machine Intelligence, 17, 90-93.*

KOHAVI, R. and WOLPERT, D.H. (1996): Bias plus variance decomposition for zero-one loss functions, In: Saitta L. (Ed.), *Machine Learning: Proceedings of the Thirteenth International Conference*, Morgan Kaufmann, 275- 283.

KUNCHEVA, L. and WHITAKER, C. (2003): Measures of diversity in classifier ensembles, *Machine Learning, 51, 181-207.*

KUNCHEVA, L., WHITAKER, C., SHIPP, D. and DUIN, R. (2000): Is independence good for combining classifiers? In: J. Kittler and F. Roli (Eds.): *Proceedings of the First International Workshop on Multiple Classifier Systems.* LNCS 1857, Springer, Berlin.

KUNCHEVA, L., BEZDEK, J.C., and DUIN, R. (2001): Decision Templates for Multiple Classifier Fusion: An Experimental Comparison. *Pattern Recognition 34, 299-314.*

PARTRIDGE, D. and YATES, W.B. (1996): Engineering multiversion neural-net systems. *Neural Computation 8, 869–893.*

PARTRIDGE, D. and KRZANOWSKI, W.J. (1997): Software diversity: practical statistics for its measurement and exploitation. *Information and software Technology, 39, 707-717.*

ROGOVA, (1994): Combining the results of several neural network classifiers, *Neural Networks, 7, 777-781.*

SKALAK, D.B. (1996): The sources of increased accuracy for two proposed boosting algorithms. *Proceedeings of the American Association for Artificial Intelligence AAAI-96*, Morgan Kaufmann, San Mateo.

THERNEAU, T.M. and ATKINSON, E.J. (1997): *An introduction to recursive partitioning using the RPART routines*, Mayo Foundation, Rochester.

TUMER, K. and GHOSH, J. (1996): Analysis of decision boundaries in linearly combined neural classifiers. *Pattern Recognition 29, 341–348.*

WERNECKE K.-D. (1992): A coupling procedure for discrimination of mixed data, *Biometrics, 48, 497-506.*

Calibrating Margin–based Classifier Scores into Polychotomous Probabilities

Martin Gebel[1] and Claus Weihs[2]

[1] Graduiertenkolleg Statistische Modellbildung,
Lehrstuhl für Computergestützte Statistik,
Universität Dortmund, D-44221 Dortmund, Germany
magebel@statistik.uni-dortmund.de

[2] Lehrstuhl für Computergestützte Statistik,
Universität Dortmund, D-44221 Dortmund, Germany
weihs@statistik.uni-dortmund.de

Abstract. Margin–based classifiers like the SVM and ANN have two drawbacks. They are only directly applicable for two–class problems and they only output scores which do not reflect the assessment uncertainty. K–class assessment probabilities are usually generated by using a reduction to binary tasks, univariate calibration and further application of the pairwise coupling algorithm. This paper presents an alternative to coupling with usage of the Dirichlet distribution.

1 Introduction

Although many classification problems cover more than two classes, the *margin–based* classifiers such as the *Support Vector Machine* (*SVM*) and *Artificial Neural Networks* (*ANN*), are only directly applicable to binary classification tasks. Thus, tasks with number of classes K greater than 2 require a reduction to several binary problems and a following combination of the produced binary assessment values to just one assessment value per class.
Before this combination it is beneficial to generate comparable outcomes by calibrating them to probabilities which reflect the assessment uncertainty in the binary decisions, see Section 2. Analyzes for calibration of dichotomous classifier scores show that the calibrators using Mapping with Logistic Regression or the Assignment Value idea are performing best and most robust, see Gebel and Weihs (2007). Up to date, pairwise coupling by Hastie and Tibshirani (1998) is the standard approach for the subsequent combination of binary assessment values, see Section 3. Section 4 presents a new multi–class calibration method for margin–based classifiers which combines the binary outcomes to assessment probabilities for the K classes. This method based on the Dirichlet distribution will be compared in Section 5 to the coupling algorithm.

2 Reduction to binary problems

Regard a classification task based on training set $T := \{(\mathbf{x}_i, c_i), i = 1, \ldots, N\}$ with $\mathbf{x}_i$ being the ith observation of random vector $\mathbf{X}$ of p feature variables and respective class $c_i \in C = \{1, \ldots, K\}$ which is the realisation of random variable C determined by a supervisor. A classifier produces an assessment value or score $S_{\text{METHOD}}(C = k|\mathbf{x}_i)$ for every class $k \in C$ and assigns to the class with highest assessment value. Some classification methods generate assessment values $P_{\text{METHOD}}(C = k|\mathbf{x}_i)$ which are regarded as probabilties that represent the assessment uncertainty. It is desirable to compute these kind of probabilities, because they are useful in cost–sensitive decisions and for the comparison of results from different classifiers.
To generate assessment values of any kind, margin–based classifiers need to reduce multi–class tasks to seveal binary classfication problems. Allwein et al. (2000) generalize the common methods for reducing multi–class into B binary problems such as the *one–against–rest* and the *all–pairs* approach with using so–called *error–correcting output coding* (*ECOC*) matrices. The way classes are considered in a particular binary task $b \in \{1, \ldots, B\}$ is incorporated into a code matrix Ψ with K rows and B columns. Each column vector ψ_b determines with its elements $\psi_{k,b} \in \{-1, 0, +1\}$ the classes for the bth classification task. A value of $\psi_{k,b} = 0$ implies that observations of the respective class k are ignored in the current task b while -1 and $+1$ determine whether a class k is regarded as the negative and the positive class, respectively.

One–against–rest approach

In the one–against–rest approach the number of binary classification tasks B is equal to the number of classes K. Each class is considered once as positive while all the remaining classes are labeled as negative. Hence, the resulting code matrix Ψ is of size $K \times K$, displaying $+1$ on the diagonal while all other elements are -1.

All–pairs approach

In the all–pairs approach one learns for every single pair of classes a binary task b in which one class is considered as positive and the other one as negative. Observations which do not belong to either of these classes are omitted in the learning of this binary task. Thus, Ψ is a $K \times \binom{K}{2}$–matrix with each column b consisting of elements $\psi_{k_1,b} = +1$ and $\psi_{k_2,b} = -1$ corresponding to a distinct class pair (k_1, k_2) while all the remaining elements are 0.

3 Coupling probability estimates

As described before, the reduction approaches apply to each column ψ_b of the code matrix Ψ, i. e. binary task b, a classification procedure. Thus, the output of the reduction approach consists of B score vectors $\mathbf{s}_{+,b}(\mathbf{x}_i)$ for the associated positive class.

To each set of scores separately one of the univariate calibration methods described in Gebel and Weihs (2007) can be applied. The outcome is a calibrated assessment probability $\mathbf{p}_{+,b}(\mathbf{x}_i)$ which reflects the probabilistic confidence in assessing observation $\mathbf{x}_i$ for task b to the set of positive classes $\mathfrak{K}_{b,+} := \{k; \psi_{k,b} = +1\}$ as opposed to the set of negative classes $\mathfrak{K}_{b,-} := \{k; \psi_{k,b} = -1\}$. Hence, this calibrated assessment probability can be regarded as function of the assessment probabilities involved in the current task:

$$p_{+,b}(\mathbf{x}_i) = \frac{\sum_{k \in \mathfrak{K}_{b,+}} P(C = k|\mathbf{x}_i)}{\sum_{k \in \mathfrak{K}_{b,+} \cup \mathfrak{K}_{b,-}} P(C = k|\mathbf{x}_i)} \,. \tag{1}$$

The values $P(C = k|\mathbf{x}_i)$ solving equation (1) would be the assessment probabilities that reflect the assessment uncertainty. However, considering the additional constraint to assessment probabilities

$$\sum_{k=1}^{K} P(C = k|\mathbf{x}_i) = 1 \tag{2}$$

there exist only $K - 1$ free parameters $P(C = k|\mathbf{x}_i)$ but at least K equations for the one–against–rest approach and even more for all–pairs $(K(K-1)/2)$. Since the number of free parameters is always smaller than the number of constraints, no unique solution for the calculation of assessment probabilities is possible and an approximative solution has to be found instead. Therefore, Hastie and Tibshirani (1998) supply the coupling algorithm which finds the estimated conditional probabilities $\hat{p}_{+,b}(\mathbf{x}_i)$ as realizations of a Binomial distributed random variable with an expected value $\mu_{b,i}$ in a way that

- $\hat{p}_{+1,b}(\mathbf{x}_i)$ generate unique assessment probabilities $\hat{P}(C = k|\mathbf{x}_i)$,
- $\hat{P}(C = k|\mathbf{x}_i)$ meet the probability constraint (2) and
- $\hat{p}_{+1,b}(\mathbf{x}_i)$ have minimal Kullback–Leibler divergence to observed $p_{+1,b}(\mathbf{x}_i)$.

4 Dirichlet calibration

The idea underlying the following multivariate calibration method is to transform the combined binary classification task outputs into realizations of a Dirichlet distributed random vector $\mathbf{P} \sim \mathcal{D}(h_1, \ldots, h_K)$ and regard the elements as assessment probabilities $P_k := P(C = k|\mathbf{x})$.

Due to the concept of *well–calibration* by DeGroot and Fienberg (1983), we want to achieve that the confidence in the assignment to a particular class converges to the probability for this class. This requirement can be easily attained with a Dirichlet distributed random vector by choosing parameters h_k proportional to the a–priori probabilities $\pi_1, \ldots, \pi_K$ of classes, since elements P_k have expected values $\mathbf{E}(P_k) = h_k / \sum_{j=1}^{K} h_j$.

Dirichlet distribution

A random vector $\mathbf{P} = (P_1, \ldots, P_K)'$ generated by

$$P_k = \frac{S_k}{\sum_{j=1}^{K} S_j} \quad (k = 1, 2, \ldots, K)$$

with K independently χ^2–distributed random variables $S_k \sim \chi^2(2 \cdot h_k)$ is Dirichlet distributed with parameters $h_1, \ldots, h_K$, see Johnson et al. (2002).

Dirichlet calibration

Initially, instead of applying a univariate calibration method we normalize the output vectors $s_{i,+1,b}$ by dividing them by their range and add half the range so that boundary values $(s = 0)$ lead to boundary probabilities $(p = 0.5)$:

$$p_{i,+1,b} := \frac{s_{i,+1,b} + \rho \cdot max_i |s_{i,+1,b}|}{2 \cdot \rho \cdot max_i |s_{i,+1,b}|}, \tag{3}$$

since the doubled maximum of absolute values of scores is the range of scores. It is required to use a smoothing factor $\rho = 1.05$ in (3) so that $p_{i,+1,b} \in]0,1[$, since we calculate in the following the geometric mean of associated binary proportions for each class $k \in \{1, \ldots, K\}$

$$r_{i,k} := \left[\prod_{b:\psi_{k,b}=+1} p_{i,+1,b} \cdot \prod_{b:\psi_{k,b}=-1} (1 - p_{i,+1,b}) \right]^{\frac{1}{\#\{\psi_{k,b} \neq 0\}}}.$$

This mean confidence is regarded as a realization of a Beta distributed random variable $R_k \sim \mathcal{B}(\alpha_k, \beta_k)$ and parameters α_k and β_k are estimated from the training set by the method of moments. We prefer the geometric to the arithmetic mean of proportions, since the product is well applicable for proportions, especially when they are skewed. Skewed proportions are likely to occur when using the one–against–rest approach in situations with high class numbers, since here the number of negative strongly outnumber the positive class observations.

To derive a multivariate Dirichlet distributed random vector, the $r_{i,k}$ can be transformed to realizations of a uniformly distributed random variable

$$u_{i,k} := F_{\mathcal{B},\hat{\alpha}_k,\hat{\beta}_k}(r_{i,k}).$$

By using the inverse of the χ^2–distribution function these uniformly distributed random variables are further transformed into χ^2–distributed random variables. The realizations of a Dirichlet distributed random vector $\mathbf{P} \sim \mathcal{D}(h_1, \ldots, h_K)$ with elements

$$\hat{p}_{i,k} := \frac{F^{-1}_{\chi^2,h_k}(u_{i,k})}{\sum_{j=1}^{K} F^{-1}_{\chi^2,h_j}(u_{i,j})}$$

are achieved by normalizing. New parameters $h_1, \ldots, h_K$ should be chosen proportional to frequencies $\pi_1, \ldots, \pi_K$ of the particular classes. In the optimization procedure we choose the factor $m = 1, 2, \ldots, 2 \cdot N$ with respective parameters $h_k = m \cdot \pi_k$ which score highest on the training set in terms of performance, determined by the geometric mean of measures (4), (5) and (6).

5 Comparison

This section supplies a comparison of the presented calibration methods based on their performance. Naturally, the precision of a classification method is the major characteristic of its performance. However, a comparison of classification and calibration methods just on the basis of the precision alone, results in a loss of information and would not include all requirements a probabilistic classifier score has to fulfill. To overcome this problem, calibrated probabilities should satisfy the two additional axioms:

- Effectiveness in the assignment and
- Well–calibration in the sense of DeGroot and Fienberg (1983).

Precision

The correctness rate

$$\mathbf{Cr} = \frac{1}{N} \sum_{i=1}^{N} \mathbf{I}_{[\hat{c}(\mathbf{x}_i) = c_i]}(\mathbf{x}_i) \tag{4}$$

where $\mathbf{I}$ is the indicator function, is the key performance measure in classification, since it mirrors the quality of the assignment to classes.

Effective assignment

Assessment probabilities should be effective in their assignment, i. e. moderately high for true classes and small for false classes. An indicator for such an effectiveness is the complement of the *Root Mean Squared Error*:

$$1 - \mathbf{RMSE} := 1 - \frac{1}{N} \sum_{i=1}^{N} \sqrt{\frac{1}{K} \sum_{k=1}^{K} \left[\mathbf{I}_{[c_i = k]}(\mathbf{x}_i) - P(c_i = k | \mathbf{x}) \right]^2} \,. \tag{5}$$

Well–calibrated probabilities

DeGroot and Fienberg (1983) give the following definition of a well–calibrated forecast: “If we forecast an event with probability p, it should occur with a relative frequency of about p.” To transfer this requirement from forecasting to classification we partition the training/test set according to the assignment to classes into K groups $T_k := \{(c_i, \mathbf{x}_i) \in T : \hat{c}(\mathbf{x}_i) = k\}$ with $N_{T_k} := |T_k|$ observations. Thus, in a partition T_k

the forecast is class k.
Predicted classes can differ from true classes and the remaining classes $j \not\equiv k$ can actually occur in a partition T_k. Therefore, we estimate the average confidence $\mathbf{Cf}_{k,j} := \frac{1}{N_{T_k}} \sum_{x_i \in T_k} P(k|\hat{c}(\mathbf{x}_i) = j)$ for every class j in a partition T_k. According to DeGroot and Fienberg (1983) this confidence should converge to the average correctness $\mathbf{Cr}_{k,j} := \frac{1}{N_{T_k}} \sum_{x_i \in T_k} \mathbf{I}_{[c(\mathbf{x}_i)=j]}$. The average closeness of these two measures

$$\mathbf{WCR} := 1 - \frac{1}{K^2} \sum_{k=1}^{K} \sum_{j=1}^{K} \left| \mathbf{Cf}_{k,j} - \mathbf{Cr}_{k,j} \right| \tag{6}$$

indicates how well–calibrated the assessment probabilities are.
On the one hand, the minimizing "probabilities" for the **RMSE** (5) can be just the class indicators especially if overfitting occurs in the training set. On the other hand, vectors of the individual correctness values maximize the **WCR** (6). To overcome these drawbacks, it is convenient to combine the two calibration measures by their geometric mean to the calibration measure

$$\mathbf{Cal} := \sqrt{(1 - \mathbf{RMSE}) \cdot \mathbf{WCR}}\,. \tag{7}$$

Experiments

The following experiments are based on the two three–class data sets *Iris* and *balance–scale* from the UCI ML–Repository as well as the four–class data set *B3*, see Newman et al. (1998) and Heilemann and Münch (1996), respectively.
Recent analyzes on risk minimization show that the minimization of a risk based on the hinge loss which is usually used in SVM leads to scores without any probability information, see Zhang (2004). Hence, the *L*2–SVM, see Suykens and Vandewalle (1999), with using the quadratic hinge loss function and thus squared slack variables is preferred to standard SVM. For classification we used the *L*2–SVM with radial–basis Kernel function and a Neural Network with one hidden layer, both with the one–against–rest and the all–pairs approach. In every binary decision a separate 3–fold cross–validation grid search was used to find optimal parameters.
The results of the analyzes with 10–fold cross–validation for calibrating *L*2–SVM and ANN are presented in Tables 1–2, respectively.
Table 1 shows that for *L*2–SVM no overall best calibration method is available. For the Iris data set all–pairs with mapping outperforms the other methods, while for B3 the Dirichlet calibration and the all–pairs method without any calibration are performing best. Considering the balance–scale data set, no big differences according to correctness occur for the calibrators.
However, comparing these results to the ones for ANN in Table 2 shows that the ANN, except the all–pairs method with no calibration, yields better results for all data sets.

Here, the one–against–rest method with usage of the Dirichlet calibrator outperforms all other methods for Iris and B3. Considering **Cr** and **Cal** for balance–scale,

Table 1. Results for calibrating *L*2–SVM–scores

	Iris		B3		balance	
	Cr	**Cal**	**Cr**	**Cal**	**Cr**	**Cal**
$P_{\text{all–pairs,no}}$	0.853	0.497	0.720	0.536	0.877	0.486
$P_{\text{all–pairs,map}}$	0.940	0.765	0.688	0.656	0.886	0.859
$P_{\text{all–pairs,assign}}$	0.927	0.761	0.694	0.677	0.886	0.832
$P_{\text{all–pairs,Dirichlet}}$	0.893	0.755	0.720	0.688	0.888	0.771
$P_{\text{1–v–rest,no}}$	0.833	0.539	0.688	0.570	0.885	0.464
$P_{\text{1–v–rest,map}}$	0.873	0.647	0.682	0.563	0.878	0.784
$P_{\text{1–v–rest,assign}}$	0.867	0.690	0.701	0.605	0.885	0.830
$P_{\text{1–v–rest,Dirichlet}}$	0.880	0.767	0.726	0.714	0.880	0.773

Table 2. Results for calibrating ANN–scores

	Iris		B3		balance	
	Cr	**Cal**	**Cr**	**Cal**	**Cr**	**Cal**
$P_{\text{all–pairs,no}}$	0.667	0.614	0.490	0.573	0.302	0.414
$P_{\text{all–pairs,map}}$	0.973	0.909	0.752	0.756	0.970	0.946
$P_{\text{all–pairs,assign}}$	0.960	0.840	0.771	0.756	0.954	0.886
$P_{\text{all–pairs,Dirichlet}}$	0.953	0.892	0.777	0.739	0.851	0.619
$P_{\text{1–v–rest,no}}$	0.973	0.618	0.803	0.646	0.981	0.588
$P_{\text{1–v–rest,map}}$	0.973	0.942	0.803	0.785	0.978	0.921
$P_{\text{1–v–rest,assign}}$	0.973	0.896	0.796	0.752	0.976	0.829
$P_{\text{1–v–rest,Dirichlet}}$	0.973	0.963	0.815	0.809	0.971	0.952

Table 3. Comparing to direct classification methods

	Iris		B3		balance	
	Cr	**Cal**	**Cr**	**Cal**	**Cr**	**Cal**
$P_{\text{ANN,1–v–rest,Dirichlet}}$	0.973	0.963	0.815	0.809	0.971	0.952
P_{LDA}	0.980	0.972	0.713	0.737	0.862	0.835
P_{QDA}	0.980	0.969	0.771	0.761	0.914	0.866
$P_{\text{Logistic Regression}}$	0.973	0.964	0.561	0.633	0.843	0.572
P_{tree}	0.927	0.821	0.427	0.556	0.746	0.664
$P_{\text{Naive Bayes}}$	0.947	0.936	0.650	0.668	0.904	0.710

one–against–rest with mapping performs best, but with correctness just slightly better than the Dirichlet calibrator.
Finally, the comparison of the one–against–rest ANN with Dirichlet calibration to other direct classification methods in Table 3 shows that for Iris LDA and QDA are the best classifiers, since the Iris variables are more or less multivariate normally distributed. Considering the two further data sets the ANN yields highest performance.

6 Conclusion

In conclusion it is to say that calibration of binary classification outputs is beneficial in most cases, especially for an ANN with the all–pairs algorithm.
Comparing classification methods to each other, one can see that the ANN with one–against–rest and Dirichlet calibration performs better than other classifiers, except LDA and QDA on Iris. Thus, the Dirichlet calibration is a nicely performing alternative, especially for ANN. The Dirichlet calibration yields better results with usage of one–against–all, since combination of outputs with their geometric mean is better applicable in this case where outputs are all based on the same binary decisions. Furthermore, the Dirichlet calibration has got the advantage that here only one optimization procedure has to be computed instead of the two steps for coupling with an incorporated univariate calibration of binary outputs.

References

ALLWEIN, E. L. and SHAPIRE, R. E. and SINGER, Y. (2000): Reducing Multiclasss to Binary: A Unifying Approach for Margin Classifiers. *Journal of Machine Learning Research 1, 113–141.*

DEGROOT, M. H. and FIENBERG, S. E. (1983): The Comparison and Evaluation of Forecasters. *The Statistician 32, 12–22.*

GEBEL, M. and WEIHS, C. (2007): Calibrating classifier scores into probabilities. In: R. Decker and H. Lenz (Eds.): *Advances in Data Analysis*. Springer, Heidelberg, 141–148.

HASTIE, T. and TIBSHIRANI, R. (1998): Classification by Pairwise Coupling. In: M. I. Jordan, M. J. Kearns and S. A. Solla (Eds.): *Advances in Neural Information Processing Systems 10*. MIT Press, Cambridge.

HEILEMANN, U. and MÜNCH, J. M. (1996): West german business cycles 1963–1994: A multivariate discriminant analysis. *CIRET–Conference in Singapore, CIRET–Studien 50.*

JOHNSON, N. L. and KOTZ, S. and BALAKRISHNAN, N. (2002): *Continuous Multivariate Distributions 1, Models and Applications, 2nd edition.* John Wiley & Sons, New York.

NEWMAN, D.J. and HETTICH, S. and BLAKE, C.L. and MERZ, C.J. (1998): *UCI Repository of machine learning databases* [http://www.ics.uci.edu/~learn/MLRepository.html]. University of California, Department of Information and Computer Science, Irvine.

SUYKENS, J. A. K. and VANDEWALLE, J. P. L. (1999): Least Squares Support Vector Machine classifiers. *Neural Processing Letters 9:3,93–300.*

ZHANG, T. (2004): Statistical behavior and consitency of classification methods based on convex risk minimization. *Annals of Statistics 32:1, 56–85.*

Classification with Invariant Distance Substitution Kernels

Bernard Haasdonk[1] and Hans Burkhardt[2]

[1] Institute of Mathematics, University of Freiburg
Hermann-Herder-Str. 10, 79104 Freiburg, Germany
haasdonk@mathematik.uni-freiburg.de,
[2] Institute of Computer Science, University of Freiburg
Georges-Köhler-Allee 52, 79110 Freiburg, Germany
burkhardt@informatik.uni-freiburg.de

Abstract. Kernel methods offer a flexible toolbox for pattern analysis and machine learning. A general class of kernel functions which incorporates known pattern invariances are *invariant distance substitution (IDS)* kernels. Instances such as tangent distance or dynamic time-warping kernels have demonstrated the real world applicability. This motivates the demand for investigating the elementary properties of the general IDS-kernels. In this paper we formally state and demonstrate their invariance properties, in particular the adjustability of the invariance in two conceptionally different ways. We characterize the definiteness of the kernels. We apply the kernels in different classification methods, which demonstrates various benefits of invariance.

1 Introduction

Kernel methods have gained large popularity in the pattern recognition and machine learning communities due to the modularity of the algorithms and the data representations by kernel functions, cf. (Schölkopf and Smola (2002)) and (Shawe-Taylor and Cristianini (2004)). It is well known that prior knowledge of a problem at hand must be incorporated in the solution to improve the generalization results. We address a general class of kernel functions called IDS-kernels (Haasdonk and Burkhardt (2007)) which incorporates prior knowledge given by pattern invariances.

The contribution of the current study is a detailed formalization of their basic properties. We both formally characterize and illustratively demonstrate their adjustable invariance properties in Sec. 3. We formalize the definiteness properties in detail in Sec. 4. The wide applicability of the kernels is demonstrated in different classification methods in Sec. 5.

2 Background

Kernel methods are general nonlinear analysis methods such as the *kernel principal component analysis, support vector machine, kernel perceptron, kernel Fisher discriminant*, etc. (Schölkopf and Smola (2002)) and (Shawe-Taylor and Cristianini (2004)). The main ingredient in these methods is the kernel as a similarity measure between pairs of patterns from the set $\mathcal{X}$.

Definition 1 (Kernel, definiteness). *A function* $k : \mathcal{X} \times \mathcal{X} \to \mathbb{R}$ *which is symmetric is called a* kernel. *A kernel k is called* positive definite (pd), *if for all n and all sets of observations* $(x_i)_{i=1}^n \in \mathcal{X}^n$ *the kernel matrix* $\mathbf{K} := (k(x_i, x_j))_{i,j=1}^n$ *satisfies* $\mathbf{v}^T\mathbf{K}\mathbf{v} \geq 0$ *for all* $\mathbf{v} \in \mathbb{R}^n$*. If this only holds for all* $\mathbf{v}$ *satisfying* $\mathbf{v}^T\mathbf{1} = 0$*, the kernel is called* conditionally positive definite (cpd).

We denote some particular l^2-inner-product $\langle \cdot, \cdot \rangle$ and l^2-distance $\|\cdot - \cdot\|$ based kernels by $k^{\text{lin}}(\mathbf{x}, \mathbf{x}') := \langle \mathbf{x}, \mathbf{x}' \rangle, k^{\text{nd}}(\mathbf{x}, \mathbf{x}') := -\|\mathbf{x} - \mathbf{x}'\|^\beta$ for $\beta \in [0, 2]$, $k^{\text{pol}}(\mathbf{x}, \mathbf{x}') := (1 + \gamma \langle \mathbf{x}, \mathbf{x}' \rangle)^p, k^{\text{rbf}}(\mathbf{x}, \mathbf{x}') := e^{-\gamma\|\mathbf{x}-\mathbf{x}'\|^2}$ for $p \in \mathbb{N}, \gamma \in \mathbb{R}_+$. Here, the linear k^{lin}, polynomial k^{pol} and Gaussian radial basis function (rbf) k^{rbf} are pd for the given parameter ranges. The negative distance kernel k^{nd} is cpd (Schölkopf and Smola (2002)). We continue with formalizing the prior knowledge about pattern variations and corresponding notation:

Definition 2 (Transformation knowledge). *We assume to have transformation knowledge for a given task, i.e. the knowledge of a set* $T = \{t : \mathcal{X} \to \mathcal{X}\}$ *of transformations of the object space including the identity, i.e.* $\text{id} \in T$*. We denote the set of transformed patterns of* $x \in \mathcal{X}$ *as* $T_x := \{t(x) | t \in T\}$ *which are assumed to have identical or similar inherent meaning as x.*

The set of concatenations of transformations from two sets T, T' is denoted as $T \circ T'$. The n-fold concatenation of transformations t are denoted as $t^{n+1} := t \circ t^n$, the corresponding sets denoted as $T^{n+1} := T \circ T^n$. If all $t \in T$ are invertible, we denote the set of inverted functions as T^{-1}. We denote the semigroup of transformations generated by T as $\bar{T} := \bigcup_{n \in \mathbb{N}} T^n$. The set $\bar{T}$ induces an equivalence relation on $\mathcal{X}$ by $x \sim x' :\Leftrightarrow$ there exist $\bar{t}, \bar{t}' \in \bar{T}$ such that $\bar{t}(x) = \bar{t}'(x')$. The equivalence class of x is denoted with E_x and the set of all equivalence sets is $\mathcal{X}/_\sim$.

Learning targets can often be modeled as functions of several input objects, for instance depending on the training data and the data for which predictions are required. We define the desired notion of invariance:

Definition 3 (Total Invariance). *We call a function* $f : \mathcal{X}^n \to \mathcal{H}$ totally invariant *with respect to T, if for all patterns* $x_1, \ldots, x_n \in \mathcal{X}$ *and transformations* $t_1, \ldots, t_n \in T$ *holds* $f(x_1, \ldots, x_n) = f(t_1(x_1), \ldots, t_n(x_n))$.

As the IDS-kernels are based on distances, we define:

Definition 4 (Distance, Hilbertian Metric). *A function $d : \mathcal{X} \times \mathcal{X} \to \mathbb{R}$ is called a* distance, *if it is symmetric and nonnegative and has zero diagonal, i.e. $d(x,x) = 0$. A distance is a* Hilbertian metric *if there exists an embedding into a Hilbert space $\Phi : \mathcal{X} \to \mathcal{H}$ such that $d(x,x') = \|\Phi(x) - \Phi(x')\|$.*

So in particular the triangle inequality does not need to be valid for a distance function in this sense. Note also that a Hilbertian metric can still allow $d(x,x') = 0$ for $x \neq x'$.

Assuming some distance function d on the space of patterns $\mathcal{X}$ enables to incorporate the invariance knowledge given by the transformations T into a new dissimilarity measure.

Definition 5 (Two-Sided invariant distance). *For a given distance d on the set $\mathcal{X}$ and some cost function $\Omega : T \times T \to \mathbb{R}_+$ with $\Omega(t,t') = 0 \Leftrightarrow t = t' = \mathrm{id}$, we define the* two-sided invariant distance *as*

$$d_{2S}(x,x') := \inf_{t,t' \in T} d(t(x),t'(x')) + \lambda\Omega(t,t'). \tag{1}$$

For $\lambda = 0$ the distance is called *unregularized.* In the following we exclude artificial degenerate cases and reasonably assume that $\lim_{\lambda\to\infty} d_{2S}(x,x') = d(x,x')$ for all x,x'. The requirement of precise invariance is often too strict for practical problems. The points within T_x are sometimes not to be regarded as identical to x, but only as similar, where the similarity can even vary over T_x. An intuitive example is optical character recognition, where the similarity of a letter and its rotated version is decreasing with growing rotation angle. This approximate invariance can be realized with IDS-kernels by choosing $\lambda > 0$.

With the notion of invariant distance we define the *invariant distance substitution kernels* as follows:

Definition 6 (IDS-Kernels). *For a* distance-based kernel, *i.e.* $k(\mathbf{x},\mathbf{x}') = f(\|\mathbf{x} - \mathbf{x}'\|)$, *and the invariant distance measure d_{2S} we call $k_{IDS}(x,x') := f(d_{2S}(x,x'))$ its* invariant distance substitution kernel *(IDS-kernel). Similarly, for an* inner-product-based *kernel k, i.e. $k(\mathbf{x},\mathbf{x}') = f(\langle \mathbf{x},\mathbf{x}' \rangle)$, we call $k_{IDS}(x,x') := f(\langle x,x' \rangle^O)$ its IDS-kernel, where $O \in \mathcal{X}$ is an arbitrary* origin *and a generalization of the inner product is given by $\langle x,x' \rangle^O := -\frac{1}{2}(d_{2S}(x,x')^2 - d_{2S}(x,O)^2 - d_{2S}(x',O)^2)$.*

The IDS-kernels capture existing approaches such as tangent distance or dynamic time-warping kernels which indicates the real world applicability, cf. (Haasdonk (2005)) and (Haasdonk and Burkhardt (2007)) and the references therein.

Crucial for efficient computation of the kernels is to avoid explicit pattern transformations by using or assuming some additional structure on T. An important computational benefit of the IDS-kernels must be mentioned, which is the possibility to precompute the distance matrices. By this, the final kernel evaluation is very cheap and ordinary fast model selection by varying kernel or training parameters can be performed.

3 Adjustable invariance

As first elementary property, we address the invariance. The IDS-kernels offer two possibilities for controlling the transformation extent and thereby interpolating between the invariant and non-invariant case. Firstly, the size of T can be adjusted. Secondly, the regularization parameter λ can be increased to reduce the invariance. This is summarized in the following:

Proposition 1 (Invariance of IDS-Kernels).

i) *If* $T = \{\mathrm{id}\}$ *and* d *is an arbitrary distance, then* $k_{IDS} = k$.
ii) *If all* $t \in T$ *are invertible, then distance-based unregularized IDS-kernels* $k_{IDS}(\cdot, x)$ *are constant on* $(T^{-1} \circ T)_x$.
iii) *If* $T = \bar{T}$ *and* $\bar{T}^{-1} = \bar{T}$, *then unregularized IDS-kernels are totally invariant with respect to* $\bar{T}$.
iv) *If* d *is the ordinary Euclidean distance, then* $\lim_{\lambda \to \infty} k_{IDS} = k$.

Proof. Statement i) is obvious from the definition, as $d_{2S} = d$ in this case. Similarly, iv) follows as $\lim_{\lambda \to \infty} d_{2S} = d$. For statement ii), we note that if $x' \in (T^{-1} \circ T)_x$, then there exist transformations $t, t' \in T$ such that $t(x) = t'(x')$ and consequently $d_{2S}(x, x') = 0$. So any distance-based kernel k_{IDS} is constant on this set $(T^{-1} \circ T)_x$. For proving iii) we observe that for $\bar{t}, \bar{t}' \in \bar{T}$ holds $d_{2S}(\bar{t}(x), \bar{t}'(x')) = \inf_{t,t'} d(t(\bar{t}(x)), t'(\bar{t}'(x'))) \geq \inf_{t,t'} d(t(x), t'(x')) = d_{2S}(x, x')$. Using the same argumentation with $\bar{t}(x)$ for x, $\bar{t}^{-1}$ for t and similar replacements for $x', \bar{t}'$ yields $d_{2S}(x, x') \geq d_{2S}(\bar{t}(x), \bar{t}'(x'))$, which gives the total invariance of d_{2S} and thus for all unregularized IDS-kernels.

Points i) to iii) imply that the invariance can be adjusted by the size of T. Point ii) implies that the invariance occasionally exceeds the set T_x. If for instance T is closed with respect to inversions, i.e. $T = T^{-1}$, then the set of constant values is $(T^2)_x$. Point iii) and iv) indicate that λ can be used to interpolate between the full invariant and non-invariant case.

We give simple illustrations of the proposed kernels and these adjustability mechanisms in Fig. 1. For the illustrations, our objects are simply points in two dimensions and several transformations define sets of points to be regarded as similar. We fix one argument $\mathbf{x}'$ (denoted with a black dot) of the kernel, and the other argument $\mathbf{x}$ is varying over the square $[-1, 2]^2$ in the Euclidean plane. We plot the different resulting kernel values $k(\mathbf{x}, \mathbf{x}')$ in gray-shades. All plots generated in the sequel can be reproduced by the MATLAB library *KerMet-Tools* (Haasdonk (2005)).

In Fig. 1 a) we focus on a linear shift along a certain slant direction while increasing the transformation extent, i.e. the size of T. The figure demonstrates the behaviour of the linear unregularized IDS-kernel, which perfectly aligns to the transformation direction as claimed by Prop. 1 i) to iii). It is striking that the captured transformation range is indeed much larger than T and very accurate for the IDS-kernels as promised by Prop. 1 ii).

The second means for controlling the transformation extent, namely increasing the regularization parameter λ, is also applicable for discrete transformations such

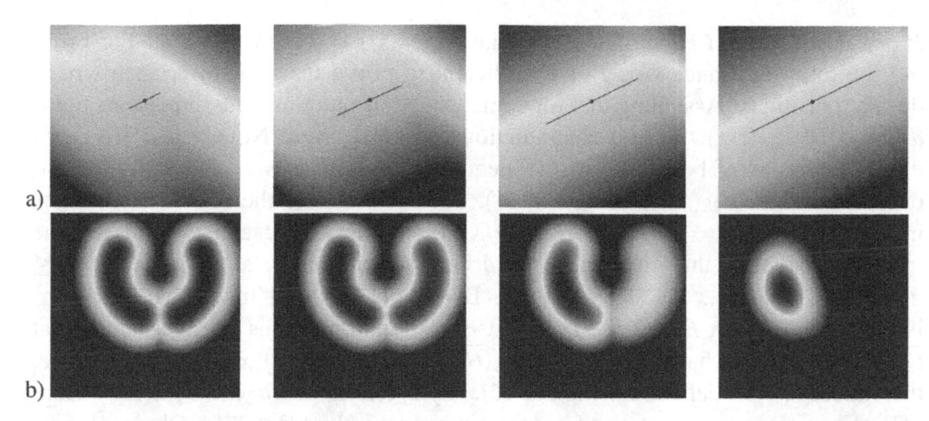

Fig. 1. Adjustable invariance of IDS-kernels. a) Linear kernel k_{IDS}^{lin} with invariance wrt. linear shifts, adjustability by increasing transformation extent by the set T, $\lambda = 0$, b) kernel k_{IDS}^{rbf} with combined nonlinear and discrete transformations, adjustability by increasing regularization parameter λ.

as reflections and even in combination with continuous transformations such as rotations, cf. Fig. 1 b). We see that the interpolation between the invariant and non-invariant case as claimed in Prop. 1 ii) and iv) is nicely realized. So the approach is indeed very general concerning types of transformations, comprising discrete, continuous, linear, nonlinear transformations and combinations thereof.

4 Positive definiteness

The second elementary property of interest, the positive definiteness of the kernels, can be characterized as follows by applying a finding from (Haasdonk and Bahlmann (2004)):

Proposition 2 (Definiteness of Simple IDS-Kernels). *The following statements are equivalent:* *i)* d_{2S} *is a Hilbertian metric*

ii) k_{IDS}^{nd} *is cpd for all* $\beta \in [0,2]$ *iii)* k_{IDS}^{lin} *is pd*

iv) k_{IDS}^{rbf} *is pd for all* $\gamma \in \mathbb{R}_+$ *v)* k_{IDS}^{pol} *is pd for all* $p \in N, \gamma \in \mathbb{R}_+$.

So the crucial property, which determines the (c)pd-ness of IDS-kernels is, whether the d_{2S} is a Hilbertian metric. A practical criterion for disproving this is a violation of the triangle inequality. A precise characterization for d_{2S} being a Hilbertian metric is obtained from the following.

Proposition 3 (Characterization of d_{2S} as Hilbertian Metric). *The unregularized* d_{2S} *is a Hilbertian metric if and only if* d_{2S} *is totally invariant with respect to* $\overline{T}$ *and* d_{2S} *induces a Hilbertian metric on* $\mathcal{X}/_\sim$.

Proof. Let d_{2S} be a Hilbertian metric, i.e. $d_{2S}(x,x') = \|\Phi(x) - \Phi(x')\|$. For proving the total invariance wrt. $\bar{T}$ it is sufficient to prove the total invariance wrt. T due to transitivity. Assuming that for some choice of patterns/transformations holds $d_{2S}(x,x') \neq d_{2S}(t(x),t'(x'))$ a contradiction can be derived: Note that $d_{2S}(t(x),x')$ differs from one of both sides of the inequality, without loss of generality the left one, and assume $d_{2S}(x,x') < d_{2S}(t(x),x')$. The definition of the two-sided distance implies $d_{2S}(x,t(x)) = \inf_{t',t''} d(t'(x),t''(t(x))) = 0$ via $t' := t$ and $t'' = \text{id}$. By the triangle inequality, this gives the desired contradiction $d_{2S}(x,x') < d_{2S}(t(x),x') \leq d_{2S}(t(x),x) + d_{2S}(x,x') = 0 + d_{2S}(x,x')$. Based on the total invariance, $d_{2S}(\cdot,x'')$ is constant on each $E \in \mathcal{X}/_{\sim}$: For all $x \sim x'$ transformations $\bar{t},\bar{t}'$ exist such that $\bar{t}(x) = \bar{t}'(x')$. So we have $d_{2S}(x,x'') = d_{2S}(\bar{t}(x),x'') = d_{2S}(\bar{t}'(x'),x'') = d_{2S}(x',x'')$, i.e. this induces a well defined function on $\mathcal{X}/_{\sim}$ by $\bar{d}_{2S}(E,E') := d_{2S}(x(E),x(E'))$. Here $x(E)$ denotes one representative from the equivalence class $E \in \mathcal{X}/_{\sim}$. Obviously, $\bar{d}_{2S}$ is a Hilbertian metric. via $\bar{\Phi}(E) := \Phi(x(E))$. The reverse direction of the proposition is clear by choosing $\Phi(x) := \bar{\Phi}(E_x)$.

Precise statements for or against pd-ness can be derived, which are solely based on properties of the underlying T and base distance d:

Proposition 4 (Characterization by d and T).

i) If T is too small compared to $\bar{T}$ in the sense that there exists $x' \in \bar{T}_x$, but $d(T_x, T_{x'}) > 0$, then the unregularized d_{2S} is not a Hilbertian metric.

ii) If d is the Euclidean distance in a Euclidean space $\mathcal{X}$ and T_x are parallel affine subspaces of $\mathcal{X}$ then the unregularized d_{2S} is a Hilbertian metric.

Proof. For i) we note that $d(T_x, T_{x'}) = \inf_{t,t' \in T} d(t(x),t'(x')) > 0$. So d_{2S} is not totally invariant with respect to $\bar{T}$ and not a Hilbertian metric due to Prop. 3. For statement ii) we can define the orthogonal projection $\Phi : \mathcal{X} \to \mathcal{H} := (T_O)^{\perp}$ on the orthogonal complement of the linear subspace through the origin O, which implies that $d_{2S}(x,x') = d(\Phi(x),\Phi(x'))$ and all sets T_x are projected to a single point $\Phi(x)$ in $(T_O)^{\perp}$. So d_{2S} is a Hilbertian metric.

In particular, these findings allow to state that the kernels on the left of Fig. 1 are not pd as they are not totally invariant wrt. $\bar{T}$. On the contrary, the extension of the upper right plot yields a pd kernel, as soon as T_x are complete affine subspaces. So these criteria can practically decide about the pd-ness of IDS-kernels.

If IDS-kernels are involved in learning algorithms, one should be aware of the possible indefiniteness, though it is frequently no relevant disadvantage in practice. Kernel principal component analysis can work with indefinite kernels, the SVM is known to tolerate indefinite kernels and further kernel methods are developed that accept such kernels. Even if an IDS-kernel can be proven by the preceding to be non-(c)pd in general, for various kernel parameter choices or a given dataset, the resulting kernel matrix can occasionally still be (c)pd.

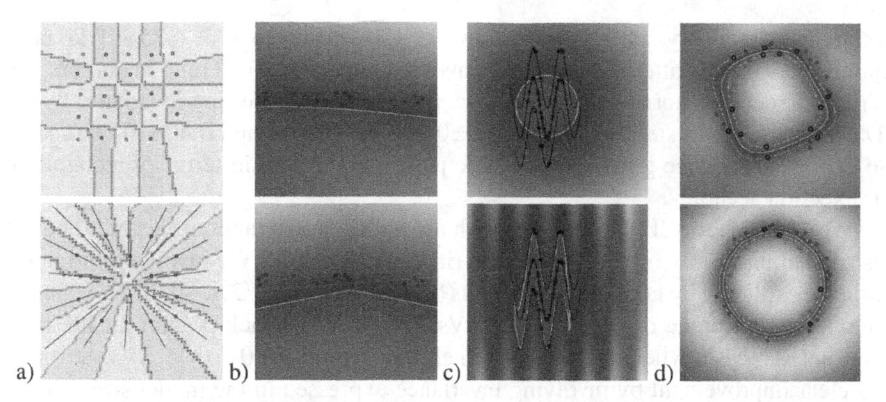

Fig. 2. Illustration of non-invariant (upper row) versus invariant (lower row) kernel methods. a) Kernel k-nn classification with k^{rbf} and scale-invariance, b) kernel perceptron with k^{pol} of degree 2 and y-axis reflection-invariance, c) one-class-classification with k^{lin} and sine-invariance, d) SVM with k^{rbf} and rotation invariance.

5 Classification experiments

For demonstration of the practical applicability in kernel methods, we condense the results on classification with IDS-kernels from (Haasdonk and Burkhardt (2007)) in Fig. 2. That study also gives summaries of real-world applications in the fields of optical character recognition and bacteria-recognition.

A simple kernel method is the kernel nearest-neighbour algorithm for classification. Fig. 2 a) is the result of the kernel 1-nearest-neighbour algorithm with the k^{rbf} and its scale-invariant k^{rbf}_{IDS} kernel, where the scaling sets T_x are indicated with black lines. The invariance properties of the kernel function obviously transfer to the analysis method by IDS-kernels.

Another aspect of interest is the convergence speed of online-learning algorithms exemplified by the kernel perceptron. We choose two random point sets of 20 points each lying uniformly distributed within two horizontal rectangular stripes indicated in Fig. 2 b). We incorporate the y-axis reflection invariance. By a random data drawing repeated 20 times, the non-invariant kernel k^{pol} of degree 2 results in 21.00 ± 6.59 update steps, while the invariant kernel k^{pol}_{IDS} converges much faster after 11.55 ± 4.54 updates. So the explicit invariance knowledge leads to improved convergence properties.

An unsupervised method for novelty detection is the optimal enclosing hypersphere algorithm (Shawe-Taylor and Cristianini (2004)). As illustrated in Fig. 2 c) we choose 30 points randomly lying on a sine-curve, which are interpreted as normal observations. We randomly add 10 points on slightly downward/upward shifted curves and want these points to be detected as novelties. The linear non-invariant k^{lin}

results in an ordinary sphere, which however gives an average of 4.75 ± 1.12 false alarms, i.e. normal patterns detected as novelties, and 4.35 ± 0.93 missed outliers, i.e. outliers detected as normal patterns. As soon as we involve the sine-invariance by the IDS-kernel we consistently obtain 0.00 ± 0.00 false alarms and 0.40 ± 0.50 misses. So explicit invariance gives a remarkable performance gain in terms of recognition or detection accuracy.

We conclude the 2D experiments with the SVM on two random sets of 20 points distributed uniformly on two concentric rings, cf. Fig. 2 d). We involve rotation invariance explicitly by taking T as rotations by angles $\phi \in [-\pi/2, \pi/2]$. In the example we obtain an average of 16.40 ± 1.67 SVs (indicated as black points) for the non-invariant k^{rbf} case, whereas the IDS-kernel only returns 3.40 ± 0.75 SVs. So there is a clear improvement by involving invariance expressed in the model size. This is a determining factor for the required storage, number of test-kernel evaluations and error estimates.

6 Conclusion

We investigated and formalized elementary properties of IDS-kernels. We have proven that IDS-kernels offer two intuitive ways of adjusting the total invariance to approximate invariance until recovering the non-invariant case for various discrete, continuous, infinite and even non-group transformations. By this they build a framework interpolating between invariant and non-invariant machine learning. The definiteness of the kernels can be characterized precisely, which gives practical criteria for checking positive definiteness in applications.

The experiments demonstrate various benefits. In addition to the model-inherent invariance, when applying such kernels, further advantages can be the convergence speed in online-learning methods, model size reduction in SV approaches, or improvement of prediction accuracy. We conclude that these kernels indeed can be valuable tools for general pattern recognition problems with known invariances.

References

HAASDONK, B. (2005): *Transformation Knowledge in Pattern Analysis with Kernel Methods - Distance and Integration Kernels.* PhD thesis, University of Freiburg.

HAASDONK, B. and BAHLMANN, B. (2004): Learning with distance substitution kernels. In: *Proc. of 26th DAGM-Symposium.* Springer, 220–227.

HAASDONK, B. and BURKHARDT, H. (2007): Invariant kernels for pattern analysis and machine learning. *Machine Learning*, 68, 35–61.

SCHÖLKOPF, B. and SMOLA, A. J. (2002): *Learning with Kernels: Support Vector Machines, Regularization, Optimization and Beyond.* MIT Press.

SHAWE-TAYLOR, J. and CRISTIANINI, N. (2004): *Kernel Methods for Pattern Analysis.* Cambridge University Press.

Applying the Kohonen Self-organizing Map Networks to Select Variables

Kamila Migdał Najman and Krzysztof Najman

University of Gdańsk, Poland
K.Najman@panda.bg.univ.gda.pl

Abstract. The problem of selection of variables seems to be the key issue in classification of multi-dimensional objects. An optimal set of features should be made of only those variables, which are essential for the differentiation of studied objects. This selection may be made easier if a graphic analysis of an U-matrix is carried out. It allows to easily identify variables, which do not differentiate the studied objects. A graphic analysis may, however, not suffice to analyse data when an object is described with hundreds of variables. The authors of the paper propose a procedure which allows to eliminate variables with the smallest discriminating potential based on the measurement of concentration of objects on the Kohonen self organising map networks.

1 Introduction

An intensive development of computer technologies in recent years lead i.a. to an enormous increase in the size of available databases. The question refers not only to an increase in the number of recorded cases. An essential, qualitative change is the increase of the number of variables describing a particular case. There are databases where one object is described by over 2000 attributes. Such a great number of variables meaningfully changes the scale of problems connected with the analysis of such databases. It results, inter alia, in problems of separation of the group structure of studied objects. According to i.a. Milligan (1994, 1996, p. 348) the approach frequently applied by the creators of databases who strive to describe the objects with the possibly large number of variables is not only unnecessary but essentially erroneous. Adding several irrelevant variables to the set of studied variables may limit or even eliminate the possibility of discovering the group structure of studied objects. In the set of variables only such variables should be included, which (cf: Gordon 1999, p. 3), contribute to:

- an increase in the homogeneity of separate clusters,
- an increase in the heterogeneity among clusters,
- easier interpretation of features of clusters which were set apart.

The reduction of the space of variables would also contribute to a considerable reduction of time of analyses and to apply much more refined, but at the same time more sophisticated and time consuming methods of data analysis.

The problem of reduction of the set of variables is extremely important while solving the classification problems. That is why a considerable attention was devoted to it in literature (cf.: Gnanadieskian, Kettenring, Tsao, 1995). It is possible to distinguish three approaches to the development of an optimal set of variables:

1. weighing the variables – where each variable is given a weight which is related to its relative importance in description of the studied problem,
2. selection of variables – consisting in the elimination of variables with the smallest discriminating potential from the set of variables; this approach may be considered as a special case of the first approach where some variables are assigned the weight of 0 – in the case of rejected variables and the weight of 1 in the case of selected variables,
3. replacement of the original variables with artificial variables – this is a classical statistical approach based on the analysis of principal components.

In the present paper a method of selecting variables based on the neural SOM network belonging to the second of the above types of methods will be presented.

2 A proposition to reduce the number of variables

The Kohonen SOM network is a very attractive method of classifying multidimensional data. As shown by Deboeck G. and Kohonen T. (1998) it is an efficient method of sorting out complex data. It is also an excellent method of visualisation of multidimensional data, examples supporting this supposition may be found in Vesanto J. (1997). One of important properties of the SOM network is the possibility of visualisation of shares of particular variables in a matrix of unified distances (an U-matrix). Joint activation of particular neurons of the network is the sum of activations resulting from activation of particular variables. Since those components may be recorded in a separate data vector, they may be analysed independently from one another.

Let us consider two simple examples. Figure 2 shows a set of 200 objects described with 2 variables. It is possible to identify a clear structure of 4 clusters, each made of 50 objects. The combination of both variables clearly differentiates the clusters.

A SOM network was built for the above dataset with a hexagonal structure, with a dimension of 17x17 neurons with a Gaussian neighbour function. The visualisation of the matrix of unified distances (the U-matrix) is shown in Fig. 2. The colour of particular segments indicates the distance, in which a given neuron is located in relation to its neighbours. Since some neurons identify the studied objects, this colour shows at the same time the distances between objects in the space of features. The "wall" of higher distances is clearly visible. Large distances separate objects which create clear clusters (concentrations). The share of both variables in the matrix of unified distances (U-matrix) is presented in Fig. 2. It can be clearly observed, that

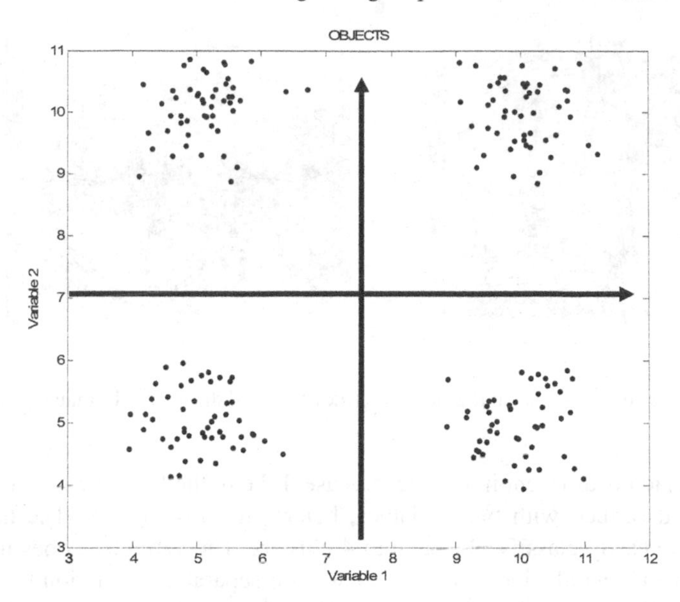

Fig. 1. An exemplary dataset - set 1

variables 1 and 2 separate the set of objects, each variable dividing the set into two parts. Both parts of the graph indicate extreme distances between objects located there. This observation allows to say, that both variables are characterised with a similar potential of discrimination of the studied objects. Since the boundary between both parts is so "acute" it may be considered, that both variables have a considerable potential to discriminate the studied objects.

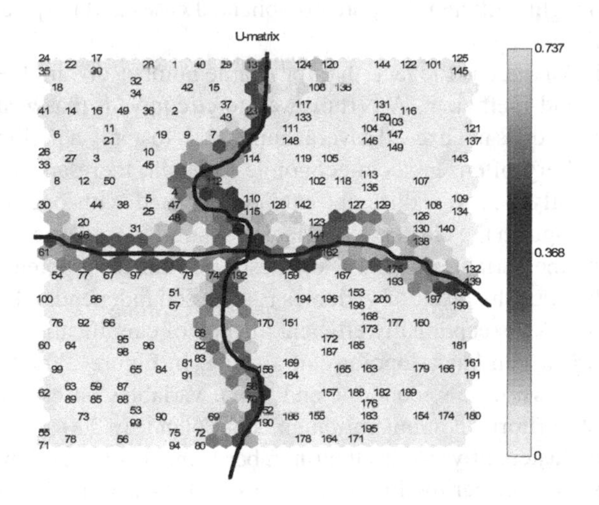

Fig. 2. The matrix of unified distances for the dataset 1

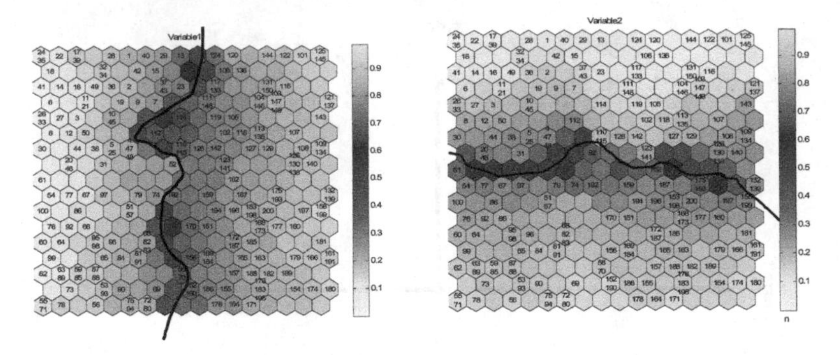

Fig. 3. The share of variable 1 and 2 in the matrix of unified distances (U-matrix) - dataset 1

The situation is different in the second case. Like in the former case we observe 200 objects described with two variables, belonging to 4 clusters. The first variable allows to easily classify objects into 4 clusters. The variable 2 does not have, however, such potential, since the clusters are non-separable in relation to it. Fig. 2 presents the objects, while Fig. 2 shows the share of particular variables in the matrix of unified distances (the U-matrix) based on the SOM network.

The analysis of distance between objects with the use of the two selected variables suggests, that variable 1 discriminates the objects very well. The borders between clusters are clear and easily discernible. It may be said that variable 1 has a great discriminating potential. Variable 2 has, however, much worse properties. It is not possible to identify clear clusters. Objects are rather uniformly distributed over the SOM network. We can say that variable 2 does not have the discriminating potential.

The application of the above procedure to assess the discriminating potential of variables is also highly efficient in more complicated cases and may be successfully applied in practice.

Its essential weakness is the fact, that for a large number of variables it becomes time consuming and inefficient. A certain way to circumvent that weakness, if the number of variables does not exceed several hundred, is to apply a preliminary grouping of variables. Very often, in socio-economic research, there are many variables which are differently and to a different extent correlated with one another. If we preliminarily distinguish the clusters of variables of similar properties, it will be possible to eliminate the variables with the smallest discriminating potential from each cluster of variables. Each cluster of variables is analysed independently, what makes the analysis easier. An exceptionally efficient method of classification of variables is the SOM network which has a topology of a chain. In Figure 2 the SOM network is shown, which classifies 58 economic and social variables describing 307 Polish poviats (smallest territorial administration units in Poland) in 2004.

In particular clusters of variables their number is much smaller than in the entire dataset and it is much easier to eliminate those variables with the smallest discriminating potential. At the same time this procedure does not allow to eliminate all

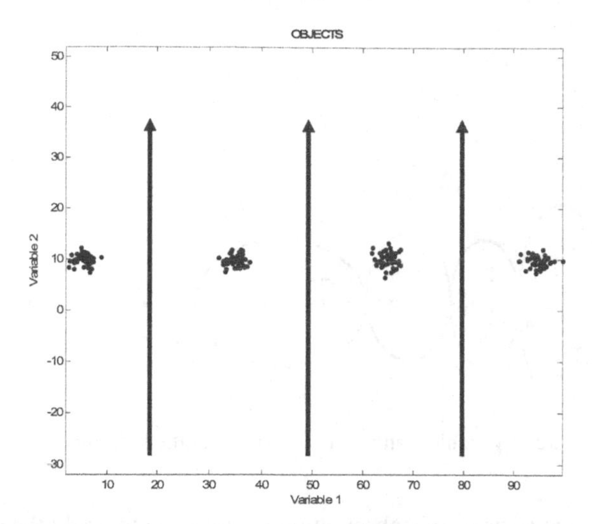

Fig. 4. An exemplary dataset - set no. 2

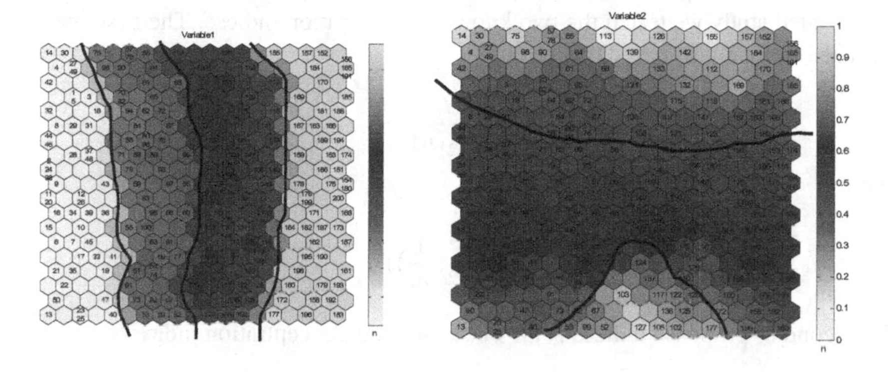

Fig. 5. The share of variable 1 and 2 in a matrix of unified distances - dataset 2

variables with similar properties, because they are located in one, not empty cluster. Quite frequently, because of certain factual reasons we would like to retain some variables, or prefer to retain at least one variable for each cluster.

For a great number of variables, above 100, a solely graphic analysis of discriminating potential of variables would be inefficient. Thus it seems justified to look for an analytical method of assessment of the discriminating potential of variables based on the SOM network and the above observations.

One of the possible solutions results from the observation of the location of objects on the map of unified distances for variables. It can be observed, that the variables with a great discriminating potential are characterised with a higher object concentration on the map than the variables with a small potential. The variables with a small discriminating potential are to an important extent rather uniformly located

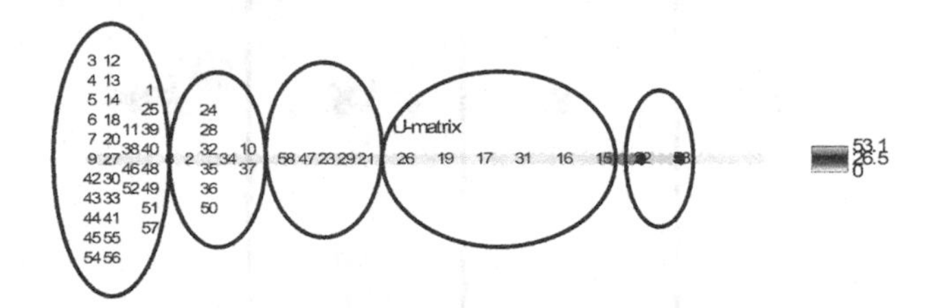

Fig. 6. The share of variable 1 and 2 in a matrix of unified distances (an U-matrix)

on the map. On the basis of this observation we propose to apply the concentration indices on the SOM map in the assessment of discriminating potential of variables. In the presented study we tested the two known concentration indices. The first one is the concentration index based on entropy:

$$K_e = 1 - \frac{H}{\log_2(n)} \tag{1}$$

where:

$$H = \sum_{i=1}^{n} (p_i \log_2(\frac{1}{p_i})) \tag{2}$$

The second of proposed indices is the classical Gini concentration index:

$$K = \frac{1}{100n} [\sum_{i=1}^{n} (i-1) p_i^{cum} - \sum_{i=2}^{n} i p_{i-1}^{cum}] \tag{3}$$

Both indices were written in the form appropriate for individual data. It seems that higher values of those coefficients should suggest variables with a greater discriminating potential.

3 Applications and results

As a result of application of the proposed indices in the first example, the values recorded in Table 1 were received (SOM network the same like in Fig 2).

The value of discriminating potential was initially assessed as high for both variables. The values of concentration coefficients for both variables were also similar[1].

[1] It is worth to note, that the value of coefficients is of no relevance here. The differences between values of particular variables are more important.

Table 1. Values of concentration coefficients for set 1.

Variable	K_e	*Gini*
1	0.0412	0.3612
2	0.0381	0.3438

The values of indices for variables from the second example are given in Table 2 (SOM network the same like in Fig 2). As it is possible to observe, the second variable is characterised with much smaller values of concentration coefficients than the first variable.

Table 2. Values of concentration coefficients for set 2.

Variable	K_e	*Gini*
1	0.0411	0.3568
2	0.0145	0.2264

It is compatible with observations based on graphic analysis, since the discriminating potential of the first variable was assessed as high, while the potential of the second variable was assessed as low. The procedure of elimination of variables of a low discriminating potential may be connected with a procedure of classification of variables. Thus a situation may be prevented, where all variables of a given type would be eliminated, if they were located in one cluster of variables only. Such property will be desirable in many cases. A full procedure of elimination of variables is presented in Fig. 3. It is a procedure consisting in several stages. In the first stage the SOM network is built on the basis of all variables. Then the values of concentration coefficients are determined. In the second stage variables are classified on the basis of the SOM network with a chain topology. Then, variables with the smallest value of concentration coefficient are eliminated from each cluster of variables. In the third stage a new SOM network is built for a reduced set of variables. In order to assess, whether the elimination of particular variables leads to an improvement in the resulting group structure, the value of one index of the quality of classification should be identified. Among the better known ones it is possible to mention the Calinski-Harabasz, Davies-Bouldin[2], and Silhouette[3] indices. In the quoted research the value of the Silhouette index was determined. Apart from its properties that allow for a good assessment of the group structure of objects, this index allows to visualise the belonging of objects to particular clusters, what is compatible with the idea of studies based on graphic analysis proposed here. This procedure is repeated until the number of variables in a cluster of variables is not smaller than a certain number

[2] Compare: Milligan G.W., Cooper M.C. (1985), *An examination of procedures for determining the number of clusters in data set*. Psychometrika, 50(2), p. 159-179.

[3] Rousseeuw P.J. (1987), *Silhouettes: a graphical aid to the interpretation and validation of cluster analysis*. J. Comput. Appl. Math. 20, p. 53-65.

determined in advance and the value of the Silhouette index increases. The application of the above procedure (compare Fig. 3) for the determination of an optimal set of variables in the description of Polish poviats is presented in Table 3. In the presented analysis the reduction of variables was carried out on the basis of the Ke concentration coefficient since it manifested several times higher differentiation of particular variables than the Gini coefficient. The value of the Silhouette index for the classification of poviats on the basis of all variables adopts the value of -0.07. It suggests, that the group structure is completely false. Elimination of the variable no. 24[4] clearly improves the group structure. In the subsequent iterations subsequent variables are systematically eliminated, increasing the value of the Silhouette index.

After six iterations the highest value of the Silhouette index is achieved and the elimination of further variables does not result in an improvement of the resulting cluster structure. The cluster structure obtained after the reduction of 14 variables is not very strong, but it is meaningfully better than the one resulting from the consideration of all variables. The resulting classification of poviats is factually justified, it is possible then to well interpret the clusters[5].

Table 3. Values of the Silhouette index after the reduction of variables

Step	*Removed Variables*	*Global Silhouette Index*
0	all var.	-0.07
1	24	0.10
2	36	0.11
3	18, 43	0.11
4	1, 2, 3, 6	0.13
5	3, 15, 26, 39	0.28
6	4, 17	0.39
7	5, 20, 23	0.38

4 Conclusions

The proposed method of selection of variables has numerous advantages. It is a fully automatic procedure, compatible with the Data Mining philosophy of analyses. Substantial empirical experience of the authors suggest, that it leads towards a considerable improvement in the obtained group structure in comparison with the analysis of the whole data set. It is more efficient the greater is the number of variables studied.

[4] After each iteration the variables are renumbered anew, that is why in subsequent iterations the same numbers of variables may appear.

[5] Compare: Migdał Najman K., Najman K. (2003), *Zastosowanie sieci neuronowej typu SOM w badaniu przestrzennego zróżnicowania powiatów (Application of the SOM neural network in studies of spatial differentiation of poviats)*, Wiadomości Statystyczne, 4/2003, p. 72-85

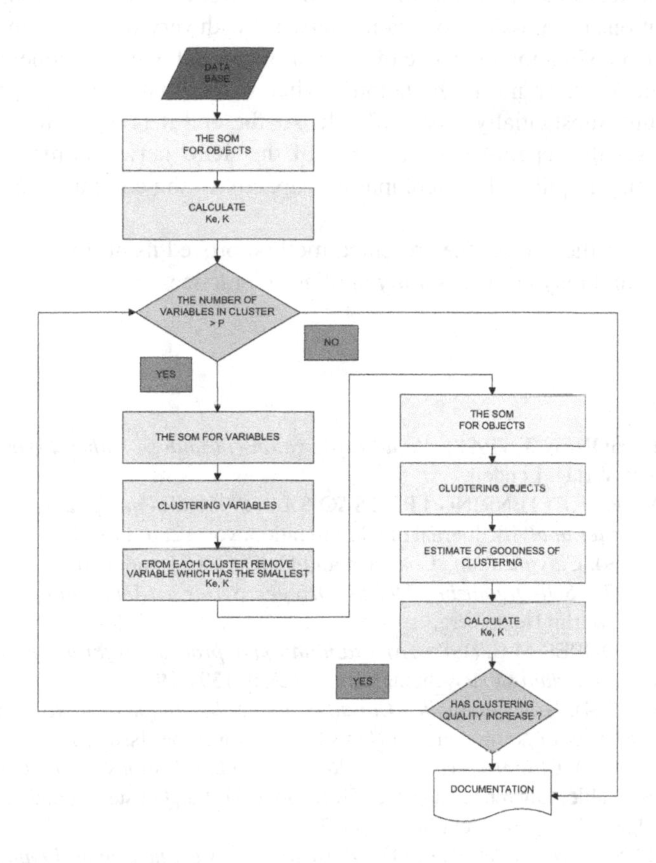

Fig. 7. Procedure of determination of an optimal set of variables

This procedure may be also applied together with other methods of data classification as a preprocessor. It is also possible to apply other measures of discriminating potential than the concentration coefficients. It is also possible to use the measures based on the distance between objects on the SOM map.

The proposed method is, however, not devoid of flaws. Its application should be preceded with a subjective determination of a minimum number of variables in a single cluster of variables. There are no factual indications, how great that number should be. This method is also very sensitive to the quality of the SOM network itself[6]. Since the learning algorithm of the SOM network is not deterministic, in subsequent iterations it is possible to obtain a network with very weak discriminating properties. In such a situation the value of the Silhouette index in subsequent stages of variable reduction may not be monotone, what would make the interpretation of obtained results substantially more difficult. At the end it is worth to note that for large databases the repetitive construction of the SOM networks may be time consuming and may require a large computing capacity of the computer equipment used.

In the opinion of the authors the presented method proved its utility in numerous empirical studies and may be successfully applied in practice.

References

DEBOECK G., KOHONEN T. (1998), *Visual explorations in finance with Self-Organizing Maps*, Springer-Verlag, London.

GNANADESIKAN R., KETTENRING J.R., TSAO S.L. (1995), *Weighting and selection of variables for cluster analysis*, Journal of Classification, vol. 12, p. 113-136.

GORDON A.D. (1999), *Classification* , Chapman and Hall / CRC, London, p.3

KOHONEN T. (1997), *Self-Organizing Maps, Springer Series in Information Sciences*, Springer-Verlag, Berlin Heidelberg.

MILLIGAN G.W., COOPER M.C. (1985), *An examination of procedures for determining the number of clusters in data set*. Psychometrika, 50(2), p. 159-179.

MILLIGAN G.W. (1994), *Issues in Applied Classification: Selection of Variables to Cluster*, Classification Society of North America News Letter, November Issue 37.

MILLIGAN G.W. (1996), *Clustering validation: Results and implications for applied analyses*. In Phipps Arabie, Lawrence Hubert & G. DeSoete (Eds.), Clustering and classification, River Edge, NJ: World Scientific, p. 341-375.

MIGDAŁ NAJMAN K., NAJMAN K. (2003), *Zastosowanie sieci neuronowej typu SOM w badaniu przestrzennego zróżnicowania powiatów*, Wiadomości Statystyczne, 4/2003, p. 72-85.

ROUSSEEUW P.J. (1987), *Silhouettes: a graphical aid to the interpretation and validation of cluster analysis*. J. Comput. Appl. Math. 20, p. 53-65.

VESANTO J. (1997), *Data Mining Techniques Based on the Self Organizing Map*, Thesis for the degree of Master of Science in Engineering, Helsinki University of Technology.

[6] The quality of the SOM network is assessed on the basis of the following coefficients: topographic, distortion and quantisation.

Computer Assisted Classification of Brain Tumors

Norbert Röhrl[1], José R. Iglesias-Rozas[2] and Galia Weidl[1]

[1] Institut für Analysis, Dynamik und Modellierung, Universität Stuttgart
Pfaffenwaldring 57, 70569 Stuttgart, Germany
roehrl@iadm.uni-stuttgart.de

[2] Katharinenhospital, Institut für Pathologie, Neuropathologie
Kriegsbergstr. 60, 70174 Stuttgart, Germany
jr.iglesias@katharinenhospital.de

Abstract. The histological grade of a brain tumor is an important indicator for choosing the treatment after resection. To facilitate objectivity and reproducibility, Iglesias et al. (1986) proposed to use a standardized protocol of 50 histological features in the grading process.

We tested the ability of Support Vector Machines (SVM), Learning Vector Quantization (LVQ) and Supervised Relevance Neural Gas (SRNG) to predict the correct grades of the 794 astrocytomas in our database. Furthermore, we discuss the stability of the procedure with respect to errors and propose a different parametrization of the metric in the SRNG algorithm to avoid the introduction of unnecessary boundaries in the parameter space.

1 Introduction

Although the histological grade has been recognized as one of the most powerful predictors of the biological behavior of tumors and significantly affects the management of patients, it suffers from low inter- and intraobserver reproducibility due to the subjectivity inherent to visual observation. The common procedure for grading is that a pathologist looks at the biopsy under a microscope and then classifies the tumor on a scale of 4 grades from I to IV (see Fig. 1). The grades roughly correspond to survival times: a patient with a grade I tumor can survive 10 or more years, while a patient with a grade IV tumor dies with high probability within 15 month. Iglesias et al. (1986) proposed to use a standardized protocol of 50 histological features in addition to make grading of tumors reproducible and to provide data for statistical analysis and classification.

The presence of these 50 histological features (Fig. 2) was rated in 4 categories from 0 (not present) to 3 (abundant) by visual inspection of the sections under a microscope. The type of astrocytoma was then determined by an expert and the corresponding histological grade between I and IV is assigned.

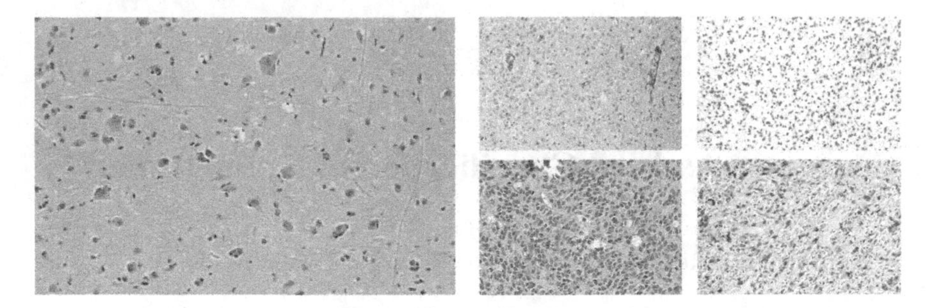

Fig. 1. Pictures of biopsies under a microscope. The larger picture is healthy brain tissue with visible neurons. The small pictures are tumors of increasing grade from left top to right bottom. Note the increasing number of cell nuclei and increasing disorder.

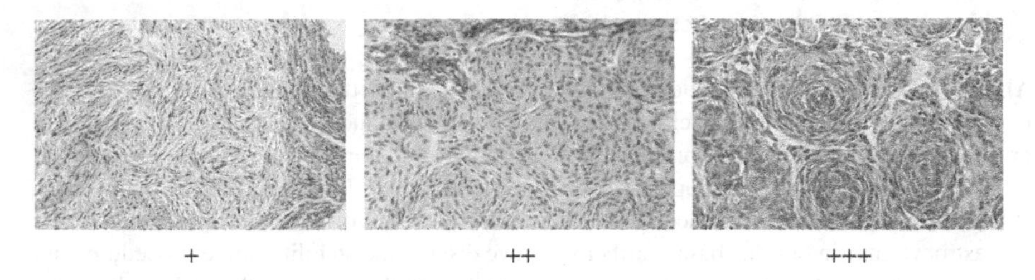

Fig. 2. One the 50 histological features: Concentric arrangement. The tumor cells build concentric formations with different diameters.

2 Algorithms

We chose LVQ (Kohonen (1995)), SRNG (Villmann et al. (2002)) and SVM (Vapnik (1995)) to classify this high dimensional data set, because the generalization error (expectation value of misclassification) of these algorithms does not depend on the dimension of the feature space (Barlett and Mendelson (2002), Crammer et al. (2003), Hammer et al. (2005)).

For the computations we used the original LVQ-PAK (Kohonen et al. (1992)), LIBSVM (Chan and Lin (2001)) and our own implementation of SRNG, since to our knowledge there exists no freely available package. Moreover for obtaining our best results, we had to deviate in some respects from the description given in the original article (Villmann et al. (2002)). In order to be able to discuss our modification we briefly formulate the original algorithm.

2.1 SRNG

Let the feature space be $\mathbb{R}^n$ and fix a discrete set of labels $\mathcal{Y}$, a training set $T \subseteq \mathbb{R}^n \times \mathcal{Y}$ and a prototype set $C \subseteq \mathbb{R}^n \times \mathcal{Y}$.

The distance in feature space is defined to be

$$d_\lambda(x,\tilde{x}) = \sum_{i=1}^{n} \lambda_i |x_i - \tilde{x}_i|^2 .$$

with parameters $\lambda = (\lambda_1,\dots,\lambda_n) \in \mathbb{R}^n$, $\lambda_i \geq 0$ and $\sum \lambda_i = 1$. Given a sample $(x,y) \in T$, we define denote its distance to the closest prototype with a different label by $d_\lambda^-(x,y)$,

$$d_\lambda^-(x,y) := \min\{d(x,\tilde{x}) | (\tilde{x},\tilde{y}) \in C, y \neq \tilde{y}\} .$$

We denote the set of all prototypes with label y by

$$W_y := \{(\tilde{x},y) \in C\}$$

and enumerate its elements $(\tilde{x},\tilde{y})$ according to their distance to (x,y)

$$\mathrm{rg}_{(x,y)}(\tilde{x},\tilde{y}) := \left|\{(\hat{x},\hat{y}) \in W_y | d(\hat{x},x) < d(\tilde{x},x)\}\right| .$$

Then the loss of a single sample $(x,y) \in T$ is given by

$$L_{C,\lambda}(x,y) := \frac{1}{c} \sum_{(\tilde{x},y) \in W_y} \exp\left(\gamma^{-1} \mathrm{rg}_{(x,y)}(\tilde{x},y)\right) \mathrm{sgd}\left(\frac{d_\lambda(x,\tilde{x}) - d_\lambda^-}{d_\lambda(x,\tilde{x}) + d_\lambda^-}\right) ,$$

where γ is the neighborhood range, $\mathrm{sgd} = (1+\exp(-x))^{-1}$ the sigmoid function and

$$c = \sum_{n=0}^{|W_y|-1} e^{\gamma^{-1} n}$$

a normalization constant. The actual SRNG algorithm now minimizes the total loss of the training set $T \subset X$

$$L_{C,\lambda}(T) = \sum_{(x,y) \in T} L_{C,\lambda}(x,y) \tag{1}$$

by stochastic gradient descent with respect to the prototypes C and the parameters of the metric λ, while letting the neighborhood range γ approach zero. This in particular reduces the dependence on the initial choice of the prototypes, which is a common problem with LVQ.

Stochastic gradient descent means here, that we compute the gradients $\nabla_C L$ and $\nabla_\lambda L$ of the loss function $L_{C,\lambda}(x,y)$ of a single randomly chosen element (x,y) of the training set and replace C by $C - \varepsilon_C \nabla_C L$ and λ by $\lambda - \varepsilon_\lambda \nabla_\lambda L$ with small learning rates $\varepsilon_C > 10\varepsilon_\lambda > 0$. The different magnitude of the learning rates is important, because classification is primarily done using the prototypes. If the metric is allowed to change too quickly, the algorithm will in most cases end in a suboptimal minimum.

2.2 Modified SRNG

In our early experiments and while tuning SRNG for our task, we found two problems with the distance used in feature space.

The straight forward parametrization of the metric comes at the price of introducing the boundaries $\lambda_i \geq 0$, which in practice are often hit too early and knock out the corresponding feature. Also, artificially setting negative λ_i to zero does slow down the convergence process.

The other point is, that by choosing different learning rates ε_C and ε_λ for prototypes and metric parameters, we are no longer using the gradient of the given loss function (1), which can also be problematic in the convergence process.

We propose using the following metric for measuring distance in feature space

$$d_\lambda(x,\tilde{x}) = \sum_{i=1}^{n} e^{r\lambda_i}|x_i - \tilde{x}_i|^2 ,$$

where the dependence on λ_i is exponential and we introduce a scaling factor $r > 0$. This definition avoids explicit boundaries for λ_i and r allows to adjust the rate of change of the distance function relative to the prototypes. Hence this parametrization enables us to minimize the loss function by stochastic gradient descent without treating prototypes and metric parameters separately.

3 Results

To test the prediction performance of the algorithms (Table 3), we divided the 794 cases (grade I: 156, grade II: 362, grade III: 238, grade 4: 38) into 10 subsets of equal size and grade distribution for cross validation.

For SVM we used a RBF kernel and let LIBSVM choose its two parameters. LVQ performed best with 700 prototypes (which is roughly equal to the size of the training set), a learning rate of 0.1 and 70000 iterations.

Choosing the right parameters for SRNG is a bit more complicated. After some experiments using cross validation, we got the best results using 357 prototypes, a learning rate of 0.01, a metric scaling factor $r = 0.1$ and a fixed neighborhood range $\gamma = 1$. We stopped the iteration process once the classification results for the training set got worse. An attempt to choose the parameters on a grid by cross validation over the training set yielded a recognition rate of 77.47%, which is almost 2% below our best result.

For practical applications, we also wanted to know how good the performance in the presence of noise would be. If we prepare the testing set such that in 5% of the features uniformly over all cases, a feature is ranked one class higher or lower with equal probability, we still get 76.6% correct predictions using SVM and 73.1% with SRNG. At 10% noise, the performance drops to 74.3% (SVM) resp. 70.2% (SRNG).

Table 1. The classification results. The columns show how many cases of grade i where classified as grade j. For example, in SRNG grade 1 tumors were classified as grade 3 in 2.26% of the cases.

LVQ	1	2	3	4
4	0.00	0.00	4.20	48.33
3	1.92	8.31	70.18	49.17
2	26.83	79.80	22.26	0.00
1	71.25	11.89	3.35	2.50

SRNG	1	2	3	4
4	0.00	0.28	2.10	50.83
3	2.62	3.87	77.30	46.67
2	28.83	88.41	18.06	2.50
1	68.54	7.44	2.54	0.00

SVM	1	2	3	4
4	0.00	0.56	2.08	53.33
3	0.67	3.60	81.12	44.17
2	28.21	85.35	15.54	2.50
1	71.12	10.50	1.25	0.00

Total	LVQ	SRNG	SVM
good	73.69	79.36	79.74

4 Conclusions

We showed that the histological grade of the astrocytomas in our database can be reliably predicted with Support Vector Machines and Supervised Relevance Neural Gas from 50 histological features rated on a scale from 0 to 3 by a pathologist. Since the attained accuracy is well above the concordance rates of independent experts (Coons et al. (1997)), this is a first step towards objective and reproducible grading of brain tumors.

Moreover we introduced a different distance function for SRNG, which in our case improved convergence and reliability.

References

BARLETT, PL. and MENDELSON, S. (2002): Rademacher and Gaussian Complexities: Risk Bounds and Structural Results. *Journal of Machine Learning, 3, 463–482.*

COONS, SW., JOHNSON, PC., SCHEITHAUER, BW., YATES, AJ., PEARL, DK. (1997): Improving diagnostic accuracy and interobserver concordance in the classification and grading of primary gliomas. *Cancer, 79, 1381–1393.*

CRAMMER, K., GILAD-BACHRACH, R., NAVOT, A. and TISHBY A. (2003): Margin Analysis of the LVQ algorithm. In: *Proceedings of the Fifteenth Annual Conference on Neural Information Processing Systems (NIPS).* MIT Press, Cambridge, MA 462–469.

HAMMER, B., STRICKERT, M., VILLMANN, T. (2005): On the generalization ability of GRLVQ networks. *Neural Processing Letters, 21(2), 109–120.*

IGLESIAS, JR., PFANNKUCH, F., ARUFFO, C., KAZNER, E. and CERVÓS-NAVARRO, J. (1986): Histopathological diagnosis of brain tumors with the help of a computer: mathematical fundaments and practical application. *Acta. Neuropathol. , 71, 130–135.*

KOHONEN, T., KANGAS, J., LAAKSONEN, J. and TORKKOLA, K. (1992): LVQ-PAK: A program package for the correct application of Learning Vector Quantization algorithms. In: *Proceedings of the International Joint Conference on Neural Networks.* IEEE, Baltimore, 725–730.

KOHONEN, T. (1995): *Self-Organizing Maps*. Springer Verlag, Heidelberg.

VAPNIK, V. (1995): *The Nature of Statistical Learning Theory*. Springer Verlag, New York, NY.

VILLMANN, T., HAMMER, B. and STRICKERT, M. (2002): Supervised neural gas for learning vector quantization. In: D. Polani, J. Kim, T. Martinetz (Eds.): *Fifth German Workshop on Artificial Life*. IOS Press, 9–18

VILLMANN, T., SCHLEIF, F-M. and HAMMER, B. (2006): Comparison of Relevance Learning Vector Quantization with other Metric Adaptive Classification Methods.*Neural Networks, 19(5), 610–622.*

Model Selection in Mixture Regression Analysis – A Monte Carlo Simulation Study

Marko Sarstedt and Manfred Schwaiger

Institut for Market-Based Management, Ludwig-Maximilians-Universität München, Germany
{sarstedt, schwaiger}@bwl.uni-muenchen.de

Abstract. Mixture regression models have increasingly received attention from both marketing theory and practice, but the question of selecting the correct number of segments is still without a satisfactory answer. Various authors have considered this problem, but as most of available studies appeared in statistics literature, they aim to exemplify the effectiveness of new proposed measures, instead of revealing the performance of measures commonly available in statistical packages. The study investigates how well commonly used information criteria perform in mixture regression of normal data, with alternating sample sizes. In order to account for different levels of heterogeneity, this factor was analyzed for different mixture proportions. As existing studies only evaluate the criteria's relative performance, the resulting success rates were compared with an outside criterion, so called chance models. The findings prove helpful for specific constellations.

1 Introduction

In the field of marketing, finite mixture models have recently received increasing attention from both a practical and theoretical point of view. In the last years, traditional mixture models have been extended by various multivariate statistical methods such as multidimensional scaling, exploratory factor analysis (DeSarbo et al. (2001)) or structural equation models (Jedidi et al. (1979); Hahn et al. (2002)), whereas regression models (Wedel and Kamakura, (1999), p. 99) for normally distributed data are the most common analysis procedure in marketing context, e.g. in terms of conjoint and market response models (Andrews et al. (2002); Andrews and Currim (2003b), p. 316). Correspondingly, mixture regression models are prevalent in marketing literature. Despite their widespread use and the importance of retaining the true number of segments in order to reach meaningful conclusions from any analysis, model selection is still an unresolved problem (Andrews and Currim (2003a), p. 235; Wedel and Kamakura (1999), p. 91). Choosing the wrong number of segments results in an under- or oversegmentation, thus leading to flawed management decisions on e.g. customer targeting, product positioning or the determination of the optimal marketing mix (Andrews and Currim (2003a), p. 235). Therefore the objective of this paper is to give recommendations on which criterion should be considered at what combination of sample/segment size in order to identify the true number of segments in a given data set.

Various authors have considered the problem of choosing the number of segments in mixture models in different context. But as most of the available studies appeared in statistics literature, they aim at exemplifying the effectiveness of new proposed measures, instead of revealing the performance of measures commonly available in statistical packages. Despite its practical importance, this topic has not been thoroughly considered for mixture regression models. An exception in this area are the studies by Hawkins et al. (2001), Andrews and Currim (2003b) and Oliveira-Brochado and Martins (2006), that examine the performance of various information criteria against several factors such as measurement level of predictors, number of predictors, separation of the segments or error variance. Regardless of the broad scope of questions covered in these studies, they do not profoundly investigate the criteria's performance against the one factor best influenceable by the marketing analyst, namely the sample size. From an application-oriented point of view, it is desirable to know which sample size is necessary in order to guarantee validity when choosing a model with a certain criterion. Furthermore, the sample size is a key differentiator between different criteria, having a large effect on the criteria's effectiveness. Therefore, the first objective of this study is to determine how well the information criteria perform in mixture regression of normal data with alternating sample sizes. Another factor that is closely related to this problem concerns segment size ratio, as past research suggests the mixture proportions to have a significant effect on the criteria's performance (Andrews and Currim (2003b)). Even though a specific sample size might prove beneficial in order to guarantee a satisfactory performance of the information criteria in general, the presence of niche segments might lead to a reduced heterogeneity and thus to a wrong decision in choosing the number of segments. That is why the second objective is to measure the information criteria's performance in order to be able to assess the validity of the criteria chosen when specific segment and sample sizes are present. These factors are evaluated for a three-segment solution by conducting a Monte Carlo simulation.

2 Model selection in mixture models

Assessing the number of segments in a mixture model is a difficult but important task. Whereas it is well known that conventional χ^2-based goodness of fit tests and likelihood ratio tests are unsuitable for making this determination (Aitkin and Rubin (1985)), the decision on what model selection statistic should be used still remains unsolved (McLachlan and Peel (2000)). Different test procedures, designed to circumnavigate implementation problems of classical χ^2-tests exist, but haven't yet found their way into widely used software applications for mixture model estimation (Sarstedt (2006), p. 8). Another main approach for deciding on the number of segments is based on a penalized form of the likelihood. These so called information criteria. Information criteria for model selection simultaneously take into account the goodness-of-fit (likelihood) of a model and the number of parameters used to achieve that fit.

The simulation study focuses on four of the most representative and widely applied model selection criteria. In a recent study by Oliveira-Brochado and Martins (2006), the authors report that in 37 published studies, the Akaike's Information Criterion (AIC) (Akaike, 1973) was used 15 times, the Consistent Akaike's Information criterion (CAIC) (Bozdogan (1987)) was used 13 times and the Bayes Information Criterion (BIC) (Schwarz (1978)) was used 11 times (multiple selections possible). In another meta-study of all major marketing journals, Sarstedt (2006) observes that BIC, AIC, CAIC and the Modified AIC with factor three (AIC_3) (Bozdogan (1994)) are the selection statistics most frequently used in mixture regression analysis. In none of the studies examined by Sarstedt did the author draw back on statistical tests to decide on the number of segments in the mixture. This report narrows its focus on presenting the simulation results for AIC, BIC, CAIC and AIC_3. Furthermore, the Adjusted BIC (Rissanen, 1978) is considered because the authors expect an increased usage due to its implementation into the increasingly popular software for estimating mixture models, Mplus. For a detailed discussion on the statistical properties of the criteria, the reader is referred to the references cited above.

3 Simulation design

The strategy for this simulation consists of initially drawing observations derived from an ordinary least squares regression and applying these to the FlexMix algorithm (Leisch, 2004; Grün and Leisch (2006)). FlexMix is a general framework for finite mixtures of regression models using the EM algorithm (Dempster et al., 1977) which is available as an extension package for the statistical computing software R. In this simulation study, models with alternating observations and three continuous predictors were considered for the OLS regression. First, $\mathbf{Y} = \beta'\mathbf{X}$ was computed for each observation, where $\mathbf{X}$ was drawn from a normal distribution. Subsequently an error term derived from a standard normal distribution was added to the true values. Each simulation set up was run with 1.000 iterations. The main parameters controlling the simulation were:

- The number of segments: $K = 3$
- The regression coefficients in each segment which were specified as follows:
  - Segment 1: $\beta_1 = (1, 1, 1.5, 2.5)'$
  - Segment 2: $\beta_2 = (1, 2.5, 1.5, 4)'$
  - Segment 3: $\beta_3 = (2, 4.5, 2.5, 4)'$
- Sample sizes which were varied in a hundred-step interval of [100;1.000]. For each of the sample sizes the simulation was run for three types of mixture proportions. To allow for a high level of heterogeneity, two small and one large segment were generated.
  - Minor proportions: $\pi_1^1 = \pi_2^1 = 0.1$ and $\pi_3^1 = 0.8$
  - Intermediate proportions: $\pi_1^1 = \pi_2^1 = 0.2$ and $\pi_3^1 = 0.6$

  - Near-uniform proportions: $\pi_1^1 = \pi_2^1 = 0.3$ and $\pi_3^1 = 0.4$
- Each simulation run was carried out five times for $k = 1, \ldots, 5$ segments.

The likelihood was maximized using the EM algorithm. As a limitation of the algorithm is its convergence to local maxima (Wedel and Kamakura (1999), p. 88), it was run repeatedly with 10 replications, totalling in 50 runs per iteration. For each number of segments, the best solution was picked.

4 Results summary

The performance of each criterion was measured by their success rate, or by the percentages of iterations in which the criterion succeeded in determining the true number of segments in the model. As indicated above, previous studies only observe the criteria's relative performance, ignoring the question whether the criteria perform any better than chance. To gain a deeper understanding of the criteria's absolute performance one has to compare the success rates with an ex-ante specified chance model. In order to verify whether the criteria are adequate, the predictive accuracy of each criterion with respect to chance is measured using the following chance models derived from discriminant analysis (Morrison (1969)): Random chance, proportional chance and maximum chance criterion. In order to be able to apply these criteria, the researcher has to have prior knowledge or make presumptions concerning the underlying model: For a given data set, let M_j be a model with K_j segments from a consideration set with C competing models $K = \{M_1, \ldots, M_C\}$ and ρ_j be the prior probability to observe M_j, $(j = 1, \ldots, C)$ and $\sum_{j=1}^{C} \rho_j = 1$. The random chance criterion is $\mathrm{CM_{ran}} = \frac{1}{\mathrm{C}} = \rho$, which indicates that each of the competing models has an equal prior probability. The proportional chance criterion is $\mathrm{CM_{prop}} = \sum_{\mathrm{j}=1}^{\mathrm{C}} \rho_{\mathrm{j}}^2$, which has been used mainly as a point of reference for subjective evaluation (Morrison (1969)), rather than the basis of a statistical test to determine if the expected proportion differs from the observed proportion of models that is correctly classified. The maximum chance criterion is $\mathrm{CM_{max}} = \max(\rho_1, \ldots, \rho_\mathrm{C})$, which defines the maximum prior probability to observe model j in a given consideration set as being the benchmark for a criterion's success rate. Since $\mathrm{CM_{ran}} < \mathrm{CM_{prop}} < \mathrm{CM_{max}}$, $\mathrm{CM_{max}}$ denotes the strictest of the three chance model criteria. If a criterion cannot do better than $\mathrm{CM_{max}}$, one might disregard the model selection statistics and choose M_j where $\max(\rho_j)$. But as model selection criteria may defy the odds by pointing at a model i where $\rho_i < \max(\rho_j)$, in most situations $\mathrm{CM_{prop}}$ should be used.

Relating to the focus of this article, an information criterion is adequate for a certain factor level combination when the success rate is greater than the value of a given chance model criterion. If this is not the case, a researcher shoud rather revert to practical considerations as for example segment identifiability when choosing the number of segments. To make use of the idea of chance models, one can define a consideration set $K = \{M_1, M_2, M_3\}$ where M_1 denotes a model with $K = 2$ segments (underfitting), M_2 a model with $K = 3$ segments (success) and M_3 a model with $K \geq 4$ segments (overfitting), thus leading to the random chance criterion $\mathrm{CM_{ran}} \approx 0.33$.

Suppose a researcher has the following prior probabilities to observe one of the models, $\rho_1 = 0.5$, $\rho_2 = 0.3$, and $\rho_3 = 0.2$ the proportional chance criterion for each factor level combination is $CM_{prop} = 0.38$ and the maximum chance criterion is $CM_{max} = 0.5$. The following figures illustrate the findings of the simulation run. Line charts are used to show the success rates for all sample/segment size combinations. Vertical dotted lines illustrate the boundaries of the previously mentioned chance models with $K = \{M_1, M_2, M_3\}$: $CM_{ran} \approx 0.33$ (lower dotted line), $CM_{prop} = 0.38$ (medial dotted line) and $CM_{max} = 0.5$ (upper dotted line). These boundaries are just exemplary and need to be specified by the researcher in dependence of the analysis at hand. Figure 1 illustrates the success rates of the five information criteria with re-

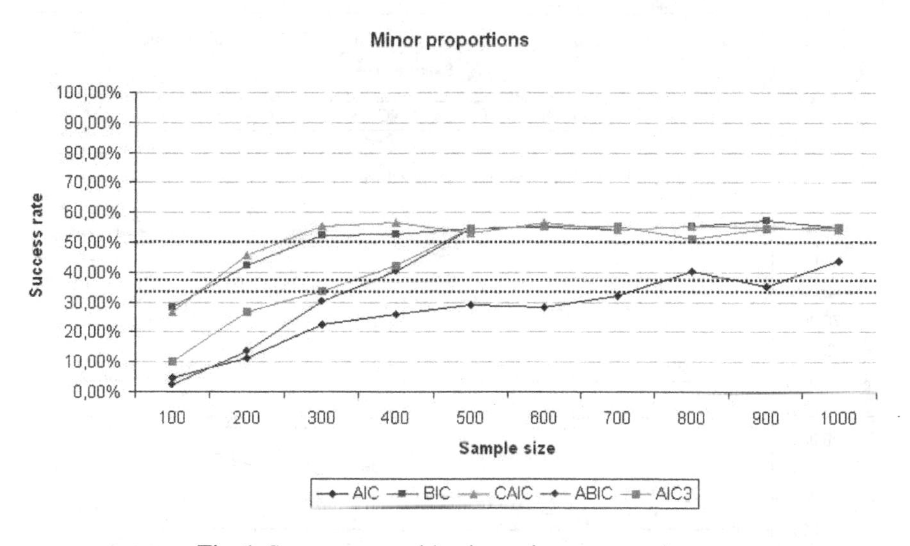

Fig. 1. Success rates with minor mixture proportions

spect to minor mixture proportions. Whereas AIC demonstrates a poor performance across all levels of sample size, CAIC outperforms the other criteria across almost all factor levels. The criterion performs favorably in recovering the true number of segments, meeting exemplary chance boundaries for sample sizes of approximately 150 (random chance, proportional chance) and 250 (maximum chance), respectively. The results in figure 2 from intermediate and near-uniform mixture proportions confirm the previous findings and underline the CAIC's strong performance in small sample size situations, quickly achieving success rates of over 90%. However as sample sizes increase to 400, both ABIC and AIC3 perform advantageously. Even with near-unifrom mixture proportions, AIC fails to any meet chance boundaries used in this set-up. In contrast to previous findings by Andrews and Currim (2003b), CAIC outperforms BIC across almost all sample/segment size combinations, whereupon the deviation is marginal in the minor mixture proportion case.

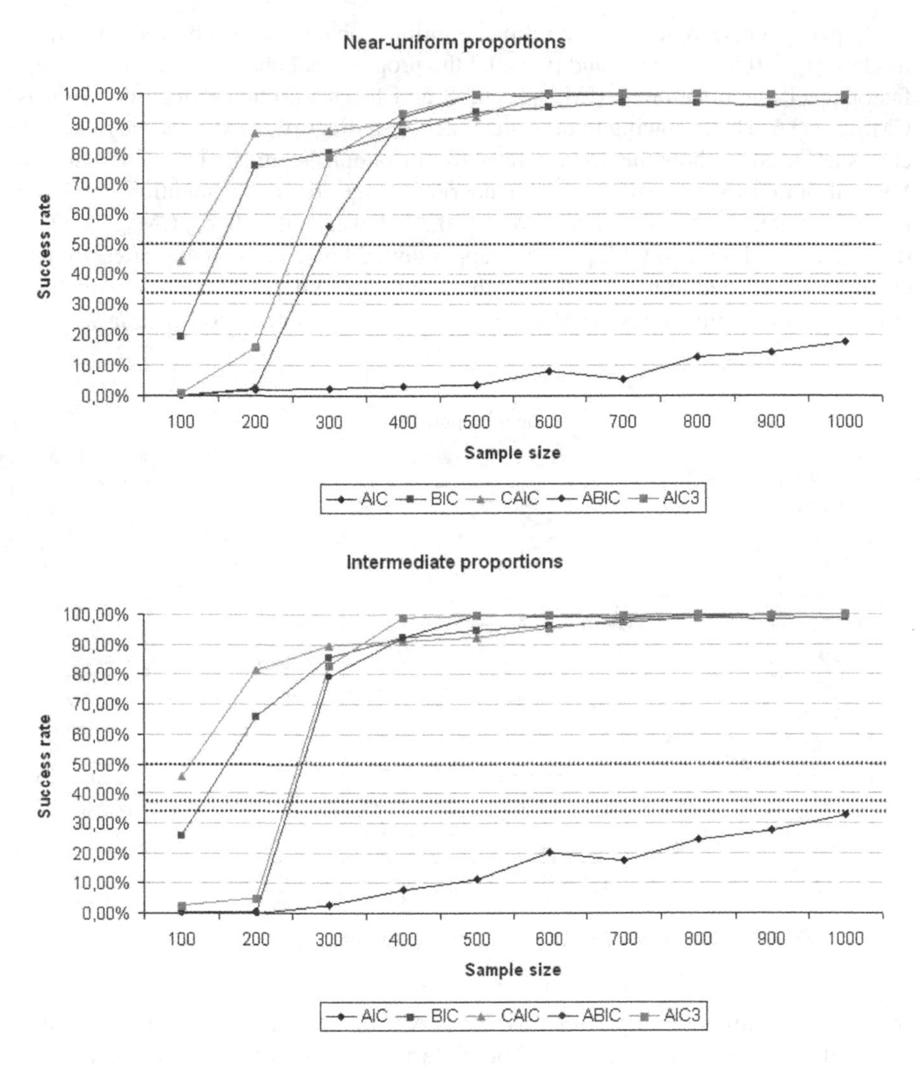

Fig. 2. Success rates with intermediate and near-uniform mixture proportions

5 Key contributions and future research directions

The findings presented in this paper are relevant to a large number of researchers building models using mixture regression analysis. This study extends previous studies by evaluating how the interaction of sample and segment size affects the performance of five of the most widely used information criteria for assessing the true number of segments in mixture regression models. For the first time the quality of these criteria was evaluated for a wide spectrum of possible sample/segment-size constellations. AIC demonstrates an extremely poor performance across all simulation situations. From an application-oriented point of view, this proves to be problematic, taking into account the high percentage of studies relying on this criterion to assess the number of segments in the model. CAIC performs favourably, showing slight weaknesses in determining the true number of segments for higher sample sizes, in comparison to ABIC and AIC_3. Especially in the context of intermediate and near-uniform mixture proportions AIC_3 performs well, quickly achieving high success rates.

A continued research on the performance of model selection criteria is needed in order to provide practical guidelines for disclosing the true number of segments in a mixture and to guarantee accurate conclusions for marketing practice. In the present study, only three combinations of mixture proportions were considered, but as the results show that market characteristics (i.e. different segment sizes) affect the performance of the criteria, future studies could allow for a greater variation of these proportions. However, considering the high number of research projects, one generally has to be critical with the idea of finding a unique measure that can be considered optimal in every simulation design or even practical applications, as indicated in other studies. Model selection decisions should rather be based on various evidences, not only derived from the data at hand but also from theoretical considerations.

References

AITKIN, M., RUBIN, D.B. (1985): Estimation and Hypothesis Testing in Finite Mixture Models. *Journal of the Royal Statistical Society, Series B (Methodological)*, 47 (1), 67-75.

AKAIKE, H. (1973): Information Theory and an Extension of the Maximum Likelihood Principle. In B. N. Petrov; F. Csaki (Eds.), *Second International Symposium on Information Theory* (267-281). Budapest: Springer.

ANDREWS, R., ANSARI, A., CURRIM, I. (2002): Hierarchical Bayes Versus Finite Mixture Conjoint Analysis Models: A Comparison of Fit, Prediction and Pathworth Recovery. *Journal of Marketing Research*, 39 (1), 87-98.

ANDREWS, R., CURRIM, I. (2003a): A Comparison of Segment Retention Criteria for Finite Mixture Logit Models. *Journal of Marketing Research*, 40 (3), 235-243.

ANDREWS, R., CURRIM, I. (2003b): Retention of Latent Segments in Regression-based Marketing Models. *International Journal of Research in Marketing*, 20 (4), 315-321.

BOZDOGAN, H. (1987): Model Selection and Akaike's Information Criterion (AIC): The General Theory and its Analytical Extensions. *Psychometrika*, 52 (3), 346-370.

BOZDOGAN, H. (1994): Mixture-model Cluster Analysis using Model Selection Criteria and a new Information Measure of Complexity. *Proceedings of the First US/Japan Conference on Frontiers of Statistical Modelling: An Informational Approach*, Vol. 2 (69-113). Boston: Kluwer Academic Publishing.

DEMPSTER, A. P., LAIRD, N. M., RUBIN, D. B. (1977): Maximum Likelihood from Incomplete Data via the EM-Algorithm. *Journal of the Royal Statistical Society, Series B (Methodological)*, 39 (1), 1-39.

DESARBO, W. S., DEGERATU, A., WEDEL, M., SAXTON, M. (2001): The Spatial Representation of Market Information. *Marketing Science*, 20 (4), 426-441.

GRÜN, B., LEISCH, F. (2006): Fitting Mixtures of Generalized Linear Regressions in R. *Computational Statistics and Data Analysis*, in press.

HAHN, C., JOHNSON, M. D., HERRMANN, A., HUBER, F. (2002): Capturing Customer Heterogeneity using a Finite Mixture PLS Approach. *Schmalenbach Business Review*, 54 (3), 243-269.

HAWKINS, D. S., ALLEN, D. M., STROMBERG, A. J. (2001): Determining the Number of Components in Mixtures of Linear Models. *Computational Statistics & Data Analysis*, 38 (1), 15-48.

JEDIDI, K., JAGPAL, H. S., DESARBO, W. S. (1979): Finite-Mixture Structural Equation Models for Response-Based Segmentation and Unobserved Heterogeneity. *Marketing Science*, 16 (1), 39-59.

LEISCH, F. (2004): FlexMix: A General Framework for Finite Mixture Models and Latent Class Regresion in R. *Journal of Statistical Software*, 11 (8), 1-18.

MANTRALA, M. K., SEETHARAMAN, P. B., KAUL, R., GOPALAKRISHNA, S., STAM, A. (2006): Optimal Pricing Strategies for an Automotive Aftermarket Retailer. *Journal of Marketing Research*, 43 (4), 588-604.

MCLACHLAN, G. J., PEEL, D. (2000): *Finite Mixture Models*, New York: Wiley.

MORRISON, D. G. (1969): On the Interpretation of Discriminant Analysis, *Journal of Marketing Research*, Vol. 6, 156-163.

OLIVEIRA-BROCHADO, A., MARTINS, F. V. (2006): *Examining the Segment Retention Problem for the "Group Satellite" Case*. FEP Working Papers, 220. www.fep.up.pt/investigacao/workingpapers/06.07.04_WP220_brochadomartins .pdf

RISSANEN, J. (1978): *Modelling by Shortest Data Description*. Automatica, 14, 465-471.

SARSTEDT, M. (2006): *Sample- and Segment-size specific Model Selection in Mixture Regression Analysis*. Münchener Wirtschaftswissenschaftliche Beiträge, 08-2006. Available electronically from http://epub.ub.uni-muenchen.de/archive/00001252/01/2006_08_LMU_sarstedt.pdf

SCHWARZ, G. (1978): Estimating the Dimensions of a Model. *The Annals of Statistics*, 6 (2), 461-464.

WEDEL, M., KAMAKURA, W. A. (1999): *Market Segmentation. Conceptual and Methodological Foundations* (2nd ed.), Boston, Dordrecht & London: Kluwer.

Comparison of Local Classification Methods

Julia Schiffner and Claus Weihs

Department of Statistics, University of Dortmund,
44221 Dortmund, Germany
schiffner@statistik.uni-dortmund.de

Abstract. In this paper four local classification methods are described and their statistical properties in the case of local data generating processes (LDGPs) are compared. In order to systematically compare the local methods and LDA as global standard technique, they are applied to a variety of situations which are simulated by experimental design. This way, it is possible to identify characteristics of the data that influence the classification performances of individual methods. For the simulated data sets the local methods on the average yield lower error rates than LDA. Additionally, based on the estimated effects of the influencing factors, groups of similar methods are found and the differences between these groups are revealed. Furthermore, it is possible to recommend certain methods for special data structures.

1 Introduction

We consider four *local classification methods* that all use the Bayes decision rule. The Common Components and the Hierarchical Mixture Classifiers, as well as Mixture Discriminant Analysis (MDA), are based on mixture models. In contrast, the Localized LDA (LLDA) relies on locally adaptive weighting of observations. Application of these methods can be beneficial in case of *local data generating processes* (LDGPs). That is, there is a finite number of sources where each one can produce data of several classes. The local data generation by individual processes can be described by *local models*. The LDGPs may cause, for example, a division of the data set at hand into several clusters containing data of one or more classes. For such data structures global standard methods may lead to poor results. One way to obtain more adequate methods is *localization*, which means to extend global methods for the purpose of local modeling. Both MDA and LLDA can be considered as localized versions of Linear Discriminant Analysis (LDA).
In this paper we want to examine and compare some of the statistical properties of the four methods. These are questions of interest: Are the local methods appropriate to classification in case of LDGPs and do they perform better than global methods? Which data characteristics have a large impact on the classification performances and which methods are favorable to special data structures? For this purpose, in a

simulation study the local methods and LDA as widely-used global technique are applied systematically to a large variety of situations generated and simulated by experimental design.
This paper is organized as follows: First the four local classification methods are described and compared. In section 3 the simulation study and its results are presented. Finally, in section 4 a summary is given.

2 Local classification methods

2.1 Common Components Classifier – CC Classifier

The CC Classifier (Titsias and Likas (2001)) constitutes an adaptation of a radial basis function (RBF) network for class conditional density estimation with full sharing of kernels among classes. Miller and Uyar (1998) showed that the decision function of this RBF Classifier is equivalent to the Bayes decision function of a classifier where class conditional densities are modeled by mixtures with common mixture components.
Assume that there are K given classes denoted by $c_1,\dots,c_K$. Then in the common components model the conditional density for class c_k is

$$f_\theta(x\,|\,c_k) = \sum_{j=1}^{G_{CC}} \pi_{jk} f_{\theta_j}(x\,|\,j) \quad \text{for } k = 1,\dots,K, \tag{1}$$

where θ denotes the set of all parameters and π_{jk} represents the probability $P(j\,|\,c_k)$. The densities $f_{\theta_j}(x\,|\,j)$, $j = 1,\dots,G_{CC}$, with θ_j denoting the corresponding parameters, do not depend on c_k. Therefore all class conditional densities are explained by the same G_{CC} mixture components.
This implicates that the data consist of G_{CC} groups that can contain observations of all K classes. Because all data points in group j are explained by the same density $f_{\theta_j}(x\,|\,j)$ classes in single groups are badly separable. The CC Classifier can only perform well if individual groups mainly contain data of a unique class. This is more likely if the parameter G_{CC} is large. Therefore the classification performance depends heavily on the choice of G_{CC}.
In order to calculate the class posterior probabilities the parameters θ_j and the priors π_{jk} and $P_k := P(c_k)$ are estimated based on maximum likelihood and the EM algorithm. Typically, $f_{\theta_j}(x\,|\,j)$ is a normal density with parameters $\theta_j = \{\mu_j,\Sigma_j\}$. A derivation of the EM steps for the gaussian case is given in Titsias and Likas (2001), p. 989.

2.2 Hierarchical Mixture Classifier – HM Classifier

The HM Classifier (Titsias and Likas (2002)) can be considered as extension of the CC Classifier. We assume again that the data consist of G_{HM} groups. But additionally, we suppose that within each group j, $j = 1,\dots,G_{HM}$, there are class-labeled subgroups that are modeled by the densities $f_{\theta_{kj}}(x \,|\, c_k, j)$ for $k = 1, \ldots, K$, where θ_{kj} are the corresponding parameters. Then the unconditional density of x is given by a three-level hierarchical mixture model

$$f_\theta(x) = \sum_{j=1}^{G_{\text{HM}}} \pi_j \sum_{k=1}^{K} P_{kj} f_{\theta_{kj}}(x \,|\, c_k, j) \tag{2}$$

with π_j representing the group prior probability $P(j)$ and P_{kj} denoting the probability $P(c_k \,|\, j)$. The class conditional densities take the form

$$f_{\theta_k}(x \,|\, c_k) = \sum_{j=1}^{G_{\text{HM}}} \pi_{jk} f_{\theta_{kj}}(x \,|\, c_k, j) \quad \text{for } k = 1, \ldots, K, \tag{3}$$

where θ_k denotes the set of all parameters corresponding to class c_k. Here, the mixture components $f_{\theta_{kj}}(x \,|\, c_k, j)$ depend on the class labels c_k and hence each class conditional density is described by a separate mixture. This resolves the data representation drawback of the common components model.
The hierarchical structure of the model is maintained when calculating the class posterior probabilities. In a first step, the group membership probabilities $P(j \,|\, x)$ are estimated and, in a second step, based on $\hat{P}(j \,|\, x)$ estimates for π_j, P_{kj} and θ_{kj} are computed. For calculating $\hat{P}(j \,|\, x)$ the EM algorithm is used. Typically, $f_{\theta_{kj}}(x \,|\, c_k, j)$ is the density of a normal distribution with parameters $\theta_{kj} = \{\mu_{kj}, \Sigma_{kj}\}$. Details on the EM steps in the gaussian case can be found in Titsias and Likas (2002), p. 2230. Note that the estimate $\hat{\theta}_{kj}$ is only provided if $\hat{P}_{kj} \gg 0$. Otherwise, it is assumed that group j does not contain data of class c_k and the associated subgroup is pruned.

2.3 Mixture Discriminant Analysis – MDA

MDA (Hastie and Tibshirani (1996)) is a localized form of Linear Discriminant Analysis (LDA). Applying LDA is equivalent to using the Bayes rule in case of normal populations with different means and a common covariance matrix. The approach taken by MDA is to model the class conditional densities by gaussian mixtures. Suppose that each class c_k is artificially divided into S_k subclasses denoted by c_{kj}, $j = 1, \ldots, S_k$, and define $S := \sum_{k=1}^{K} S_k$ as total number of subclasses. The subclasses are modeled by normal densities with different mean vectors μ_{kj} and, similar to LDA, a common covariance matrix Σ. Then the class conditional densities are

$$f_{\mu_k, \Sigma}(x \,|\, c_k) = \sum_{j=1}^{S_k} \pi_{jk} \phi_{\mu_{kj}, \Sigma}(x \,|\, c_k, c_{kj}) \quad \text{for } k = 1, \ldots, K, \tag{4}$$

where μ_k denotes the set of all subclass means in class c_k and π_{jk} represents the probability $P(c_{kj} \,|\, c_k)$. The densities $\phi_{\mu_{kj}, \Sigma}(x \,|\, c_k, c_{kj})$ of the mixture components depend on c_k. Hence, as in the case of the HM Classifier, the class conditional densities are described by separate mixtures.

Parameters and priors are estimated based on maximum likelihood. In contrast to the hierarchical approach taken by the HM Classifier, the MDA likelihood is maximized directly using the EM algorithm.
Let $x \in \mathbb{R}^p$. LDA can be used as a tool for dimension reduction by choosing a subspace of rank $p^* \leq \min\{p, K-1\}$ that maximally separates the class centers. Hastie and Tibshirani (1996), p. 160, show that for MDA a dimension reduction similar to LDA can be achieved by maximizing the log likelihood under the constraint $\text{rank}\{\mu_{kj}\} = p^*$ with $p^* \leq \min\{p, S-1\}$.

2.4 Localized LDA – LLDA

The Localized LDA (Czogiel et al. (2006)) relies on an idea of Tutz and Binder (2005). They suggest the introduction of locally adaptive weights to the training data in order to turn global methods into observation specific approaches that build individual classification rules for all observations to be classified. Tutz and Binder (2005) consider only two class problems and focus on logistic regression. Czogiel et al. (2006) extend their concept of localization to LDA by introducing weights to the n nearest neighbors $x_{(1)}, \ldots, x_{(n)}$ of the observation x to be classified in the training data set. These are given as

$$w\left(x, x_{(i)}\right) = W\left(\frac{\left\|x_{(i)} - x\right\|}{d_n(x)}\right) \tag{5}$$

for $i = 1, \ldots, n$, with W representing a kernel function. The Euclidean distance $d_n(x) = \left\|x_{(n)} - x\right\|$ to the farthest neighbor $x_{(n)}$ denotes the kernel width. The obtained weights are locally adaptive in the sense that they depend on the Euclidean distances of x and the training observations $x_{(i)}$.
Various kernel functions can be used. For the simulation study we choose the kernel $W_\gamma(y) = \exp(-\gamma y)$ that was found to be robust against varying data characteristics by Czogiel et al. (2006). The parameter $\gamma \in \mathbb{R}^+$ has to be optimized.
For each x to be classified we obtain the n nearest neighbors in the training data and the corresponding weights $w\left(x, x_{(i)}\right)$, $i = 1, \ldots, n$. These are used to compute weighted estimates of the class priors, the class centers and the common covariance matrix required to calculate the linear discriminant function. The relevant formulas are given in Czogiel et al. (2006), p. 135.

3 Simulation study

3.1 Data generation, influencing factors and experimental design

In this work we compare the local classification methods in the presence of local data generating processes (LDGPs). In order to simulate data for the case of K classes and M LDGPs we use the mixture model

Table 1. The chosen levels, coded by -1 and 1, of the influencing factors on the classification performances determine the data generating model (equation (6)). The factor *PUVAR* defines the proportion of useless variables that have equal class means and hence do not contribute to class separation.

influencing factor		model	factor level	
			-1	$+1$
LP	number of **LDGPs**	M	2	4
PLP	prior **p**robabilities of **LDGPs**	π_j	unequal	equal
DLP	**d**istance between **LDGP** centers	μ_{kj}	large	small
CL	number of **cl**asses	K	3	6
PCL	(conditional) prior **p**robabilities of **cl**asses	P_{kj}	unequal	equal
DCL	**d**istance between **cl**ass centers	μ_{kj}	large	small
VAR	number of **var**iables	μ_{kj}, Σ_{kj}	4	12
PUVAR	**p**roportion of **u**seless **var**iables	μ_{kj}	0%	25%
DEP	**dep**endency in the variables	Σ_{kj}	no	yes
DND	**d**eviation from the **n**ormal **d**istribution	T	no	yes

$$f_{\mu,\Sigma}(x) = \sum_{j=1}^{M} \pi_j \sum_{k=1}^{K} P_{kj} T\left(\phi_{\mu_{kj},\Sigma_{kj}}(x \mid c_k, j)\right) \tag{6}$$

with μ and Σ denoting the sets of all μ_{kj} and Σ_{kj} and priors π_j and P_{kj}. The jth LDGP is described by the local model $\sum_{k=1}^{K} P_{kj} T\left(\phi_{\mu_{kj},\Sigma_{kj}}(x \mid c_k, j)\right)$. The transformation of the gaussian mixture densities by the function T allows to produce data from non-normal mixtures. In this work we use the system of densities by Johnson (1949) to generate deviations from normality in skewness and kurtosis. If T is the identity the data generating model equals the hierarchical mixture model in equation (2) with gaussian subgroup densities and $G_{\text{HM}} = M$.
We consider ten influencing factors which are given in Table 1. These factors determine the data generating model. For example the factor *PLP*, defining the prior probabilities of the LDGPs, is related to π_j in equation (6) (cp. Table 1). We fix two levels for every factor, coded by -1 and $+1$, which are also given in Table 1. In general the low level is used for classification problems which should be of lower difficulty, whereas the high level leads to situations where the premises of some methods are not met (e.g. nonnormal mixture component densities) or the learning problem is more complicated (e.g. more variables). For more details concerning the choice of the factor levels see Schiffner (2006).
We use a fractional factorial 2^{10-3}-design with tenfold replication leading to 1280 runs. For every run we construct a training data set with 3000 and a test data set containing 1000 observations.

3.2 Results

We apply the local classification methods and global LDA to the simulated data sets and obtain 1280 test data error rates $r_i, i = 1, \ldots, 1280$, for every method. The chosen

Table 2. Bayes errors and error rates of all classification methods with the specified parameters and mixture component densities on the 1280 simulated test data sets. R^2 denotes the coefficients of determination for the linear regressions of the classification performances on the influencing factors in Table 1.

method	parameters	mixture component densities	error rate minimum	error rate mean	error rate maximum	R^2
Bayes error	-	-	0.000	0.026	0.193	-
LDA	-	-	0.000	0.148	0.713	0.901
CC M	$G_{CC} = M$	$f_{\theta_j} = \phi_{\mu_j,\Sigma_j}$	0.000	**0.441**	0.821	0.871
CC MK	$G_{CC} = M \cdot K$	$f_{\theta_j} = \phi_{\mu_j,\Sigma_j}$	0.000	0.054	0.217	0.801
LLDA	$\gamma = 5,\ n = 500$	-	0.000	**0.031**	0.207	0.869
MDA	$S_k = M$	-	0.000	0.042	0.205	0.904
HM	$G_{HM} = M$	$f_{\theta_{kj}} = \phi_{\mu_{kj},\Sigma_{kj}}$	0.000	0.036	0.202	0.892

parameters, the group and subgroup densities assumed for the HM and CC Classifiers and the resulting test data error rates are given in Table 2. The low Bayes errors (cp. also Table 2) indicate that there are many easy classification problems. For the data sets simulated in this study, in general, the local classification methods perform much better than global LDA. An exception is the CC Classifier with M groups, CC M, which probably suffers from the common components assumption in combination with the low number of groups. The HM Classifier is the most flexible of the mixture based methods. The underlying model is met in all simulated situations where deviations from normality do not occur. Probably for this reason the error rates for the HM Classifier are lower than for MDA and the CC Classifiers.
In order to measure the influence of the factors in Table 1 on the classification performances of all methods we estimate their main and interaction effects by linear regressions of $\ln(\text{odds}(1-r_i)) = \ln((1-r_i)/r_i) \in \mathbb{R}$, $i = 1,\ldots,1280$, on the coded factors. Then an estimated effect of 1, e.g. of factor *DND*, can be interpreted as an increase in proportion of hit rate to error rate by $e \approx 2.7$.
The coefficients of determination, R^2, indicate a good fit of the linear models for all classification methods (cp. Table 2), hence the estimated factor effects are meaningful. The estimated main effects are shown in Figure 1. For the most important factors *CL, DCL* and *VAR* they indicate that a small number of classes, a big distance between the class centers and a high number of variables improve the classification performances of all methods.
To assess which classification methods react similarly to changes in data characteristics they are clustered based on the Euclidean distances in their estimated main and interaction effects. The resulting dendrogram in Figure 2 shows that one group is formed by the HM Classifier, MDA and LLDA which also exhibit similarities in their theoretical backgrounds. In the second group there are global LDA and the local CC Classifier with MK groups, CC MK. The factors mainly revealing differences between CC M, which is isolated in the dendrogram, and the remaining methods are *CL, DCL, VAR* and *LP* (cp. Figure 1). For the first three factors the absolute effects for CC M are much smaller. Additionally, CC M is the only method with a positive

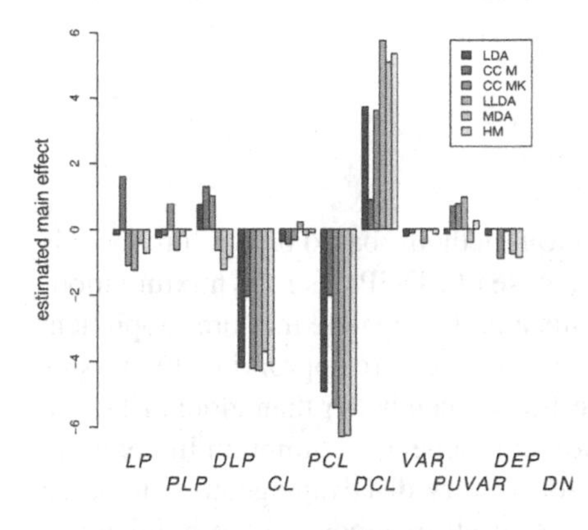

Fig. 1. Estimated main effects of the influencing factors in Table 1 on the classification performances of all methods

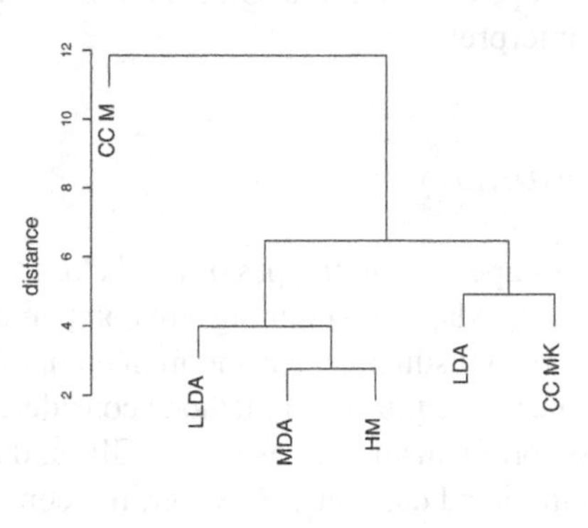

Fig. 2. Hierarchical clustering of the classification methods using average linkage based on the estimated factor effects

estimated effect of *LP*, the number of LDGPs, which probably indicates that a larger number of groups improves the classification performance (cp. the error rates of CC MK in Table 2). The factor *DLP* reveals differences between the two groups found in the dendrogram. In contrast to the remaining methods, for both CC Classifiers as well as LDA small distances between the LDGP centers are advantageous. Local modeling is less necessary, if the LDGP centers for individual classes are close together and hence, the global and common components based methods perform better than in other cases.

Based on theoretical considerations, the estimated factor effects and the test data error rates, we can assess which methods are favorable to some special situations. The estimated effects of factor *LP* and the error rates in Table 2 show that application of the CC Classifier can be disadvantageous and is only beneficial in conjunction with a big number of groups G_{CC} which, however, can make the interpretation of the results very difficult. However, for large M, problems in the E step of the classical EM algorithm can occur for the CC and the HM Classifiers in the gaussian case due to singular estimated covariance matrices. Hence, in situations with a large number of LDGPs MDA can be favorable because it yields low error rates and is insensible to changes of M (cp. Figure 1), probably thanks to the assumption of a common covariance matrix and dimension reduction.

A drawback of MDA is that the numbers of subclasses for all K classes have to be specified in advance. Because of subgroup-pruning for the HM Classifier only one parameter G_{HM} has to be fixed.

If deviations from normality occur in the mixture components LLDA can be recommended since, like CC M, the estimated effect of *DND* is nearly zero and the test data error rates are very small. In contrast to the mixture based methods it is applicable to data of every structure because it does not assume the presence of groups, subgroups or subclasses. On the other hand, for this reason, the results of LLDA are less interpretable.

4 Summary

In this paper different types of local classification methods, based on mixture models or locally adaptive weighting, are compared in case of LDGPs. For the mixture models we can distinguish the common components and the separate mixtures approach. In general the four local methods considered in this work are appropriate to classification problems in the case of LDGPs and perform much better than global LDA on the simulated data sets. However, the common components assumption in conjunction with a low number of groups has been found very disadvantageous. The most important factors influencing the performances of all methods are the numbers of classes and variables as well as the distances between the class centers. Based on all estimated factor effects we identified two groups of similar methods. The differences are mainly revealed by the factors *LP* and *DLP*, both related to the LDGPs. For a large number of LDGPs MDA can be recommended. If the mixture components are not gaussian LLDA appears to be a good choice. Future work can consist in considering robust versions of the compared methods that can better deal, for example, with deviations from normality.

References

CZOGIEL, I., LUEBKE, K., ZENTGRAF, M. and WEIHS, C. (2006): Localized Linear Discriminant Analysis. In: R. Decker, H.-J. Lenz (Eds.): *Advances in Data Analysis.* Springer, Berlin, 133–140.

HASTIE, T. J. and TIBSHIRANI, R. J. (1996): Discriminant Analysis by Gaussian Mixtures. *Journal of the Royal Statistical Society B, 58, 155–176.*

JOHNSON, N. L. (1949): Systems of Frequency Curves generated by Methods of Translation. *Biometrika, 36, 149–176.*

MILLER, D. J. and UYAR, H. S. (1998): Combined Learning and Use for a Mixture Model Equivalent to the RBF Classifier. *Neural Computation, 10, 281–293.*

SCHIFFNER, J. (2006): *Vergleich von Klassifikationsverfahren für lokale Modelle.* Diploma Thesis, Department of Statistics, University of Dortmund, Dortmund, Germany.

TITSIAS, M. K. and LIKAS, A. (2001): Shared Kernel Models for Class Conditional Density Estimation. *IEEE Transactions on Neural Networks, 12(5), 987–997.*

TITSIAS, M. K. and LIKAS, A. (2002): Mixtures of Experts Classification Using a Hierarchical Mixture Model. *Neural Computation, 14, 2221–2244.*

TUTZ, G. and BINDER H. (2005): Localized Classification. *Statistics and Computing, 15, 155–166.*

Incorporating Domain Specific Information into Gaia Source Classification

Kester W. Smith, Carola Tiede and Coryn A.L. Bailer-Jones

Max–Planck–Institut für Astronomie,
Königstuhl 17, 69117 Heidelberg, Germany
smith@mpia-hd.mpg.de

Abstract. Astronomy is in the age of large scale surveys in which the gathering of multidimensional data on thousands of millions of objects is now routine. Efficiently processing these data - classifying objects, searching for structure, fitting astrophysical models - is a significant conceptual (not to mention computational) challenge. While standard statistical methods, such as Bayesian clustering, k-nearest neighbours, neural networks and support vector machines, have been successfully applied to some areas of astronomy, it is often difficult to incorporate domain specific information into these. For example, in astronomy we often have good physical models for the objects (e.g. stars) we observe. That is, we can reasonably well predict the observables (typically, the stellar spectrum or colours) from the astrophysical parameters (APs) we want to infer (such as mass, age and chemical composition). This is the "forward model": The task of classification or parameter estimation is then an inverse problem. In this paper, we discuss the particular problem of combining astrometric information, effectively a measure of the distance of the source, with spectroscopic information.

1 Introduction

Gaia is an ESA astronomical satellite that will be launched in 2011. Its mission is to build a three dimensional map of the positions and velocities of a substantial part of our galaxy. In addition to the basic position and velocity data, the astrophysical nature of the detected objects will be determined. Since Gaia is expected to detect upwards of a billion individual objects of various types, and since the mission will not use an input catalogue, automated classification and parameterization based on the dataset is a crucial part of the mission.

1.1 Astronomical context

From galactic rotation curves and other evidence it is believed that most material in the universe is comprised of so-called dark matter. The nature of this material is a fundamental current question in astronomy. The distribution and properties of the dark matter at the time of the formation of our galaxy should leave traces in the distribution and dynamics of the stellar population that is observed today. Since heavy elements are formed by nucleosynthesis in the centres of massive stars, and are therefore scarce at early epochs, their relative abundances in stellar atmospheres can be used to discriminate between stellar populations on the basis of age. By building up a complete picture of a large portion of our galaxy, such tracers of galactic evolution can be studied in unprecedented detail.

1.2 Basic properties of the dataset

Gaia will detect all point sources down to a fixed limiting brightness. This limit corresponds to the brightness of the Sun if observed at a distance of approximately 11,000 parsecs (35,000 light years, compared the accepted distance to the Galactic centre of 26,000 light years). The vast majority of detected sources will be stars, but the sample will also include several million galaxies and quasars, which are extragalactic objects, and many objects from within our own solar system.

The positions of the various sources on the sky can of course be measured very easily. Radial velocities are determined from Doppler shifts of spectral lines observed with an onboard spectrometer. Transverse motions on the sky are of the order of a few milliarcseconds per year, scaling with distance, and these motions must be mapped over the timescale of the mission (5–6 years). Distances are a priori not known and are in fact one of the most difficult, and most crucially important, measurements in astronomy. Gaia is designed to measure the parallaxes of the stellar sources in order to determine distances to nearby stars. The parallax in question is the result of the changing viewpoint of the satellite as the Earth orbits the Sun. An object displaying a parallax of one arcsecond relative to distant, negligable-parallax stars, has by definition a distance of 1 parsec (3.26 light years). This distance happens to correspond roughly to the distance to the nearest stars. The parallax scales linearly with distance so that the Sun at a distance of 11,000 parsec (the approximate brightness limit of such an object for Gaia) would display a parallax of about 90 microarcseconds (μas)). Gaia is designed to measure parallaxes with a standard error of around 25 μas, so that the parallax-limit roughly corresponds to the brightness-limit for solar type stars.

As well as position, parallax and transverse motion (proper motion), and the high resolution spectra used to determine the radial velocities, the Gaia satellite will return low resolution spectra with approximately 96 resolution elements spanning the range 300-1000 nanometres (roughly from the ultraviolet to the near infrared range). These spectra can be used to classify objects according to basic type (galaxies, quasars, stars etc) and to then determine the basic parameters of the object (e.g. for stars, the effective temperature of the atmosphere). This information is important because the nature of the stellar population coupled with the kinematic information constrains models of galaxy formation and evolution.

2 Classification and parametrization

As the sky is continuously scanned by the satellite's detectors, sources are detected on board and the data (position, low resolution spectra and high resolution spectrum) are extracted from the raw detector output, processed into an efficient form and returned to the ground station. As the mission proceeds, repeated visits to the same area of sky allow the measurement of variations in the positions of sources, which are used to build up a model of the proper motions and parallaxes for the full set of sources. This leads to a distinction for the data processing between early mission data, consisting of the spectra and positions, and late mission data, which includes parallaxes and proper motions. The sources should be classified into broad astronomical classes on the basis of the spectra alone in the early mission, and on the basis of the spectra combined with astrometric information in the later part of the mission. This classification is important for the astrophysics, but also for the astrometric solution, since the distant quasars form a distant, essentially fixed (zero parallax and zero proper motion, plus or minus measurement errors) population. The early mission classifier should feed back the identified extragalactic objects to the astrometric processing, and the purer this sample, the better.

Once the classification is made, sources are fitted with astrophysical models to recover various parameters, such as effective surface temperature or atmospheric element abundances for stars. The algorithms for this classification and regression are in the early stages of development by the data processing consortium. For the classification, the algorithm mostly used at this stage is a Support Vector Machine (SVM) after Vapnik (1995), taken from the library libSVM (Chang and Lin (2001)), with a radial basis function (RBF) kernel. The decision to use SVM for classification is of course provisional and other methods may be considered. Synthetic data for training and testing the classifier is produced using standard models of various astronomical source classes. The multi-class SVM used returns a probability vector containing the probabilities that a particular source belongs to each class (Wu and Weng (2005)). Sources are classified according to the highest component of the probability vector. We are now incorporating into the simulated data values for the parallax and proper motion, indicating a distance. The current task is to incorporate this information into the classification and regression schemes.

3 Classification results

For current purposes, we consider only four classes of astrophysical object; single stars and binary stars, both of which belong to the set of objects within our own galaxy, and galaxies and quasars, both of which are extragalactic. Two datasets were generated, each with a total of 5000 sources split evenly between the four classes (i.e. 1250 of each). One set was used as a trianing set for the SVM, the other is a test set from which the statistics are compiled. The classification results for the basic SVM classifier running on the spectrum only are shown in Table 1. Here, and in subsequent experiments, the input data are scaled to have mean of zero and standard deviation of one for each bin. The classifier achieves an overall correct classification rate of approximately 93%. The main confusion is between single stars and binaries.

The parallaxes of the simulated data for stars and quasars are shown in Figure 1. The parallax could be included directly into the classifier as a 97th data point for each

Table 1. Confusion matrix for the SVM classifier, working on the spectral data without any astrometric information. Reading row by row, the matrix shows the percentage of test sources which are of a particular type, for example Stars, which are classified as each possible output type. The leading diagonal shows the sources that are correctly classified (true positives). The off-diagonal elements show the level of contamination (false positives) as a percentage of the input source sample. In this test case, the numbers of each class of source were roughly equal (just over 1000 each). In the real mission, the number of stars is expected to be three orders of magnitude greater than the number of galaxies or quasars.

	Stars	Binaries	Quasars	Galaxies
Stars	88.21	9.27	2.43	0.09
Binaries	8.67	91.13	0.00	0.20
Quasars	2.04	0.90	95.77	1.28
Galaxies	0.00	0.00	0.62	99.38

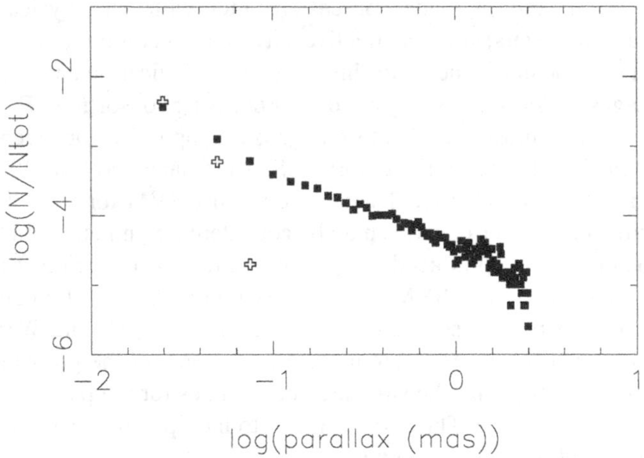

Fig. 1. The distribution of simulated parallaxes for stars (filled squares) and quasars (+ signs).

object, alongside the 96 spectral bins. Such a classifier would be expected to perform significantly better than spectrum-only version, and indeed it does (Table 2). It might, however, be possible to include the parallax in the classification in a way that utilises our knowledge of the astrophysical significance of the quantity. Significant values of parallax are expected for a subset of the galactic sources, i.e. the stars and binaries. Not all stars and binaries will have a detectable parallax, but none of the extragalactic sources will. This then suggests a split in the data, based on parallax, into objects that are certainly galactic and objects that may belong to any class.

To implement such a two-stage classifier, we trained two separate SVMs, one with all four classes, and the other with the galactic sources (stars and binaries) only. These SVMs were trained on the spectral data only, not including the parallax. We then classified the entire test set with each classifier. For each object, the output from each classifier is a four-component probability vector, in the case of the classifier

Table 2. Confusion matrix obtained by using the SVM with the parallax information included as an additional input.

	Stars	Binaries	Quasars	Galaxies
Stars	93.52	6.03	0.45	0.00
Binaries	6.38	93.62	0.00	0.00
Quasars	0.76	0.14	98.91	0.19
Galaxies	0.00	0.00	0.41	99.59

trained only on galactic sources (stars and binaries), the probabilities for the quasars and galaxies are necessarily always zero. Finally, we combined the output probability vectors of the two SVMs using a weighting function based on the parallax value.

If P_1 and P_2 are the probability vectors for the galactic and general SVM classifier respectively, they are combined to form the output probability as follows;

$$P = wP_1 + (1-w)P_2, \tag{1}$$

$$w = 0.5\,(1 + \tanh((\alpha \times SNR) + \delta)) \tag{2}$$

where SNR is the significance of the measured parallax, estimated by assuming that the standard error is 25μas. The parameter α is set to 1. and the value of δ to -5. With these values, the function does not produce significant weighting (w $\approx$ 0.1) toward exclusively galactic sources until the parallax rises to four times the standard error.

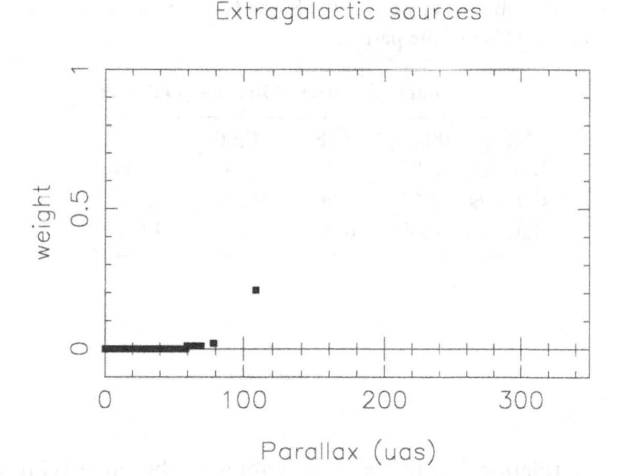

Fig. 2. The weighting function applied to the extragalactic sources.

The results of the two-stage classification are shown in Table 3. The leading diagonal shows that the completeness at each class is not as good as in the case of the single SVM classifier with parallax as discussed above (Table 2), however the contamination of the extragalactic sources with misidentified galactic sources has

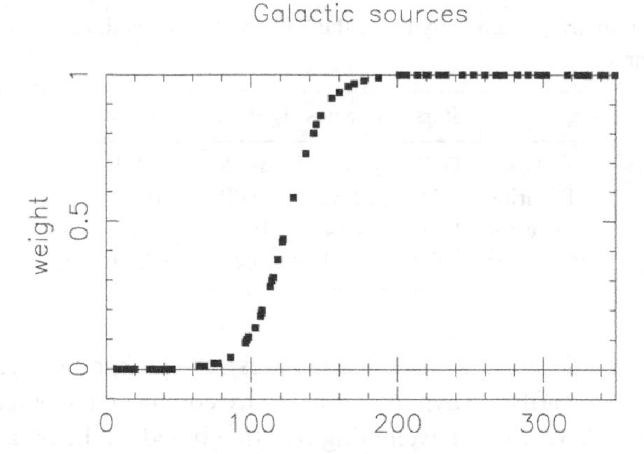

Fig. 3. The weighting function for the galactic sources. These sources are distributed through a range of parallaxes.

been strongly reduced - in fact falling to zero for the test sample of 5000 objects. As noted above, this is a significant advantage when the galaxies and quasars form important classes for determining the astrometric solution, and when there will be several hundred times more stars than extragalactic objects in the final sample.

Table 3. Confusion matrix obtained by using the SVM twice then combining the probabilities weighted according to the value of the parallax.

	Stars	Binaries	Quasars	Galaxies
Stars	90.82	9.18	0.00	0.00
Binaries	8.87	91.13	0.00	0.00
Quasars	2.04	0.90	95.77	1.28
Galaxies	0.00	0.00	0.62	99.38

4 Summary

Since we know the relationship of the observables to the underlying nature of the objects in the sample, we are in a position to incorporate this knowledge into the classification or regression problems in an informed way, making maximum use of this physical knowledge. The goal of this is twofold; Firstly, the addition of domain-specific information should improve the predictive accuracy. Second, but not unimportant, is that it allows an interpretation of how the model works: the sensitivities of the model observables to a given underlying parameter provide an explicit (and unique) weighting function of the observables. Apart from making the model more acceptable (and less like a "black box"), this allows us to identify where we gather higher quality data in order to improve performance further.

References

CHANG, C.-C. and LIN C.-J. (2001) : Libsvm: A library for support vector machines. *(Technical report)* URL `http://www.csie.ntu.edu.tw/~cjlin/libsvm`

VAPNIK, V. (1995) *The Nature of Statistical Learning Theory*. Springer Verlag, New York.

WU T.-F. and WENG R.C., (2004): Probability estimates for multi-class classification by pairwise coupling. *Journal of Machine Learning Research, 5:975–1005* [illegible] functions of the observables' input [illegible] like a "black box", this allows [illegible] [illegible] performance.

References

[illegible]

Identification of Noisy Variables for Nonmetric and Symbolic Data in Cluster Analysis

Marek Walesiak and Andrzej Dudek

Wroclaw University of Economics, Department of Econometrics and Computer Science, Nowowiejska 3, 58-500 Jelenia Gora, Poland
{marek.walesiak, andrzej.dudek}@ae.jgora.pl

Abstract. A proposal of an extended version of the HINoV method for the identification of the noisy variables (Carmone et al. (1999)) for nonmetric, mixed, and symbolic interval data is presented in this paper. Proposed modifications are evaluated on simulated data from a variety of models. The models contain the known structure of clusters. In addition, the models contain a different number of noisy (irrelevant) variables added to obscure the underlying structure to be recovered.

1 Introduction

Choosing variables is the one of the most important steps in a cluster analysis. Variables used in applied clustering should be selected and weighted carefully. In a cluster analysis we should include only those variables that are believed to help to discriminate the data (Milligan (1996), p. 348). Two classes of approaches, while choosing the variables for cluster analysis, can facilitate a cluster recovery in the data (e.g. Gnanadesikan et al. (1995); Milligan (1996), pp. 347–352):

– variable selection (selecting a subset of relevant variables),

– variable weighting (introducing relative importance of the variables according to their weights).

Carmone et al. (1999) discussed the literature on the variable selection and weighting (the characteristics of six methods and their limitations) and proposed the HINoV method for the identification of the noisy variables, in the area of the variable selection, to remedy problems with these methods. They demonstrated its robustness with metric data and k-means algorithm. The authors suggest further studies of the HINoV method with different types of data and other clustering algorithms on p. 508.

In this paper we propose extended version of the HINoV method for nonmetric, mixed, and symbolic interval data. The proposed modifications are evaluated for eight clustering algorithms on simulated data from a variety of models.

2 Characteristics of the HINoV method and its modifications

Algorithm of Heuristic Identification of Noisy Variables (HINoV) method for metric data (Carmone et al. (1999)) is following:

1. A data matrix $[x_{ij}]$ containing n objects and m normalized variables measured on a metric scale ($i = 1,\ldots,n;\ j = 1,\ldots,m$) is a starting point.

2. Cluster, via kmeans method, the observed data separately for each j-th variable for a given number of clusters u. It is possible to use clustering methods based on a distance matrix (pam or any hierarchical agglomerative method: single, complete, average, mcquitty, median, centroid, Ward).

3. Calculate adjusted Rand indices R_{jl} ($j,l = 1,\ldots,m$) for partitions formed from all distinct pairs of the m variables ($j \neq l$). Due to a fact that adjusted Rand index is symmetrical we need to calculate $m(m-1)/2$ values.

4. Construct $m \times m$ adjusted Rand matrix (parim). Sum rows or columns for each j-th variable $R_{j\bullet} = \sum_{l=1}^{m} R_{jl}$ (topri):

$$\begin{array}{ccc} \text{Variable} & \text{parim} & \text{topri} \\ \begin{bmatrix} M_1 \\ M_2 \\ \vdots \\ M_m \end{bmatrix} & \begin{bmatrix} & R_{12} & \ldots & R_{1m} \\ R_{21} & & \ldots & R_{2m} \\ \vdots & \vdots & \vdots & \vdots \\ R_{m1} & R_{m2} & \ldots & \end{bmatrix} & \begin{bmatrix} R_{1\bullet} \\ R_{2\bullet} \\ \vdots \\ R_{m\bullet} \end{bmatrix} \end{array}$$

5. Rank topri values $R_{1\bullet}, R_{2\bullet},\ldots, R_{m\bullet}$ in a decreasing order (stopri) and plot the scree diagram. The size of the topri values indicate a contribution of that variable to the cluster structure. A scree diagram identifies sharp changes in the topri values. Relatively low-valued topri variables (the noisy variables) are identified and eliminated from the further analysis (say h variables).

6. Run a cluster analysis (based on the same classification method) with the selected $m-h$ variables.

The modification of the HINoV method for nonmetric data (where number of objects is much more than a number of categories) differs in steps 1, 2, and 6 (Walesiak (2005)):

1. A data matrix $[x_{ij}]$ containing n objects and m ordinal and/or nominal variables is a starting point.

2. For each j-th variable we receive natural clusters, where the number of clusters equals the number of categories for that variable (for instance five for Likert scale or seven for semantic differential scale).

6. Run a cluster analysis with one of clustering methods based on a distance appropriate to nonmetric data (GDM2 for ordinal data – see Jajuga et al. (2003); Sokal and Michener distance for nominal data) with the selected $m-h$ variables.

The modification of the HINoV method for symbolic interval data differs in steps 1 and 2:

1. A symbolic data array containing n objects and m symbolic interval variables is a starting point.

2. Cluster the observed data with one of clustering methods (pam, single, complete, average, mcquitty, median, centroid, Ward) based on a distance appropriate to the symbolic interval data (e.g. Hausdorff distance – see Billard and Diday (2006), p. 246) separately for each j-th variable for a given number of clusters u.

Functions HINoV.Mod and HINoV.Symbolic of clusterSim computer program working in R allow adequately using mixed (metric, nonmetric), and the symbolic interval data. The proposed modifications of the HINoV method are evaluated on simulated data from a variety of models.

3 Simulation models

We generate data sets in eleven different scenarios. The models contain the known structure of clusters. In the models 2-11 the noisy variables are simulated independently from the uniform distribution.

Model 1. No cluster structure. 200 observations are simulated from the uniform distribution over the unit hypercube in 10 dimensions (see Tibshirani et al [2001], p. 418).

Model 2. Two elongated clusters in 5 dimensions (3 noisy variables). Each cluster contains 50 observations. The observations in each of the two clusters are independent bivariate normal random variables with means (0, 0), (1, 5), and covariance matrix $\sum$ ($\sigma_{jj} = 1$, $\sigma_{jl} = -0.9$).

Model 3. Three elongated clusters in 7 dimensions (5 noisy variables). Each cluster is randomly chosen to have 60, 30, 30 observations, and the observations are independently drawn from bivariate normal distribution with means (0, 0), (1.5, 7), (3, 14) and covariance matrix $\sum$ ($\sigma_{jj} = 1$, $\sigma_{jl} = -0.9$).

Model 4. Three elongated clusters in 10 dimensions (7 noisy variables). Each cluster is randomly chosen to have 70, 35, 35 observations, and the observations are independently drawn from multivariate normal distribution with means (1.5, 6, –3), (3, 12, –6), (4.5, 18, –9), and identity covariance matrix $\sum$, where $\sigma_{jj} = 1$ $(1 \leq j \leq 3)$, $\sigma_{12} = \sigma_{13} = -0.9$, and $\sigma_{23} = 0.9$.

Model 5. Five clusters in 3 dimensions that are not well separated (1 noisy variable). Each cluster contains 25 observations. The observations are independently drawn from bivariate normal distribution with means (5, 5), (–3, 3), (3, –3), (0, 0), (–5, –5), and identity covariance matrix $\sum$ ($\sigma_{jj} = 1$, $\sigma_{jl} = 0.9$).

Model 6. Five clusters in 5 dimensions that are not well separated (2 noisy variables). Each cluster contains 30 observations. The observations are independently drawn from multivariate normal distribution with means (5, 5, 5), (–3, 3, –3), (3, –3, 3), (0, 0, 0), (–5, –5, –5), and covariance matrix $\sum$, where $\sigma_{jj} = 1$ $(1 \leq j \leq 3)$, and $\sigma_{jl} = 0.9$ $(1 \leq j \neq l \leq 3)$.

Model 7. Five clusters in 10 dimensions (8 noisy variables). Each cluster is randomly chosen to have 50, 20, 20, 20, 20 observations, and the observations are independently drawn from bivariate normal distribution with means (0, 0), (0, 10), (5, 5), (10, 0), (10, 10), and identity covariance matrix $\sum$ ($\sigma_{jj} = 1$, $\sigma_{jl} = 0$).

Model 8. Five clusters in 9 dimensions (6 noisy variables). Each cluster contains 30 observations. The observations are independently drawn from multivariate normal distribution with means (0, 0, 0), (10, 10, 10), (–10, –10, –10), (10, –10, 10), (–10, 10, 10), and identity covariance matrix $\sum$, where $\sigma_{jj} = 3$ $(1 \leq j \leq 3)$, and $\sigma_{jl} = 2$ $(1 \leq j \neq l \leq 3)$.

Model 9. Four clusters in 6 dimensions (4 noisy variables). Each cluster is randomly chosen to have 50, 50, 25, 25 observations, and the observations are independently drawn from bivariate normal distribution with means (–4, 5), (5, 14), (14, 5), (5, –4), and identity covariance matrix $\sum$ $(\sigma_{jj} = 1, \sigma_{jl} = 0)$.

Model 10. Four clusters in 12 dimensions (9 noisy variables). Each cluster contains 30 observations. The observations are independently drawn from multivariate normal distribution with means (–4, 5, –4), (5, 14, 5), (14, 5, 14), (5, –4, 5), and identity covariance matrix $\sum$, where $\sigma_{jj} = 1$ $(1 \leq j \leq 3)$, and $\sigma_{jl} = 0$ $(1 \leq j \neq l \leq 3)$.

Model 11. Four clusters in 10 dimensions (9 noisy variables). Each cluster contains 35 observations. The observations on the first variable are independently drawn from univariate normal distribution with means –2, 4, 10, 16 respectively, and identity variance $\sigma_j^2 = 0.5$ $(1 \leq j \leq 4)$.

Ordinal data. The clusters in models 1-11 contain continuous data and a discretization process is performed on each variable to obtain ordinal data. The number of categories k determines the width of each class intervals: $\left[\max_i\{x_{ij}) - \min_i\{x_{ij}\}\right] / k$. Independently for each variable each class interval receive category $1, \ldots, k$ and the actual value of variable x_{ij} is replaced by these categories. In simulation study $k = 5$ (for $k = 7$ we have received similar results).

Symbolic interval data. To obtain symbolic interval data the data were generated for each model twice into sets A and B and minimal (maximal) value of $\{a_{ij}, b_{ij}\}$ is treated as the beginning (the end) of an interval.

Fifty realizations were generated from each setting.

4 Discussion on the simulation results

In testing the robustness of the HINoV modified algorithm using simulated ordinal or symbolic interval data, the major criterion was the identification of the noisy variables. The HINoV-selected variables contain variables with the highest topri values. In models 2-11 the number of nonnoisy variables is known. Due to this fact, in simulation study, the number of the HINoV-selected variables equals the number of nonnoisy variables in each model. When the noisy variables were identified, the next step was to run the one of clustering methods based on distance matrix (pam, single, complete, average, mcquitty, median, centroid, Ward) with the nonnoisy subset of variables (HINoV-selected variables) and with all variables. Then each clustering result was compared with the known cluster structure from models 2-11 using Hubert and Arabie's [1985] corrected Rand index (see Table 1 and 2).

Some conclusions can be drawn from the simulations results:

Table 1. Cluster recovery for all variables and HINoV-selected subsets of variables for ordinal data (five categories) by experimental model and clustering method

Model		Clustering method							
		pam	ward	single	complete	average	mcquitty	median	centroid
2	*a*	0.38047	0.53576	0.00022	0.11912	0.42288	0.25114	0.00527	0.00032
	b	0.84218	0.90705	0.72206	0.12010	0.99680	0.41796	0.30451	0.89835
3	*a*	0.27681	0.34071	0.00288	0.29392	0.40818	0.35435	0.04625	0.00192
	b	0.85946	0.60606	0.36121	0.61090	0.68223	0.51487	0.49199	0.61156
4	*a*	0.35609	0.44997	0.00127	0.43860	0.53509	0.47083	0.04677	0.00295
	b	0.83993	0.87224	0.56313	0.56541	0.80149	0.62102	0.54109	0.80156
5	*a*	0.54746	0.60139	0.27610	0.46735	0.58050	0.49842	0.33303	0.50178
	b	0.91071	0.84888	0.48550	0.73720	0.81317	0.79644	0.72899	0.74462
6	*a*	0.61074	0.60821	0.13400	0.53296	0.61037	0.56426	0.35113	0.47885
	b	0.83880	0.87183	0.56074	0.75584	0.86282	0.81395	0.71085	0.79018
7	*a*	0.10848	0.11946	0.00517	0.09267	0.10945	0.11883	0.00389	0.00659
	b	0.80072	0.87399	0.27965	0.87892	0.94882	0.77503	0.74141	0.91638
8	*a*	0.31419	0.43180	0.00026	0.29529	0.40203	0.36771	0.00974	0.00023
	b	0.95261	0.96372	0.58026	0.95596	0.96627	0.95507	0.93701	0.96582
9	*a*	0.37078	0.45915	0.01123	0.12128	0.50198	0.31134	0.04326	0.00709
	b	0.99966	0.98498	0.93077	0.96993	0.99626	0.98024	0.95461	0.99703
10	*a*	0.29727	0.41152	0.00020	0.22358	0.41107	0.34663	0.00030	0.00007
	b	1.00000	1.00000	0.99396	0.99911	1.00000	1.00000	0.99867	1.00000
	$\bar{b}$	0.89378	0.88097	0.60858	0.73259	0.89642	0.76384	0.71212	0.85838
	$\bar{r}$	0.53130	0.44119	0.56066	0.44540	0.45403	0.39900	0.61883	0.74730
ccr		98.22%	98.00%	94.44%	90.67%	97.11%	89.56%	98.89%	98.44%
11	*a*	0.04335	0.04394	0.00012	0.04388	0.03978	0,03106	0,00036	0.00009
	b	0.14320	0.08223	0.12471	0.08497	0.10373	0,12355	0,04626	0,06419

a (*b*) – values represent Hubert and Arabie's adjusted Rand indices averaged over fifty replications for each model with all variables (with HINoV-selected variables); $\bar{r} = \bar{b} - \bar{a}$; *ccr* – corrected cluster recovery.

1. The cluster recovery that used only the HINoV-selected variables for ordinal data (Table 1) and symbolic interval data (Table 2) was better than the one that used all variables for all models 2-10 and each clustering method.

2. Among 450 simulated data sets (nine models with 50 runs) the HINoV method was better (see *ccr* in Table 1 and 2):

– from 89.56% (mcquitty) to 98.89% (median) of runs for ordinal data,

– from 91.78% (ward) to 99,78% (centroid) of runs for symbolic interval data.

3. Figure 1 shows the relationship between the values of adjusted Rand indices averaged over fifty replications and models 2-10 with the HINoV-selected variables ($\bar{b}$) and values showing an improvement ($\bar{r}$) of average adjusted Rand indices (cluster recovery with the HINoV selected variables against all variables) separately for eight clustering methods and types of data (ordinal, symbolic interval). Based on adjusted

Table 2. Cluster recovery for all variables and HINoV-selected subsets of variables for symbolic interval data by experimental model and clustering method

Model		Clustering method							
		pam	ward	single	complete	average	mcquitty	median	centroid
2	a	0.86670	0.87920	0.08006	0.28578	0.32479	0.49424	0.02107	0.00004
	b	0.99920	0.97987	0.91681	0.99680	0.99524	0.98039	0.85840	0.95739
3	a	0.41934	0.39743	0.00368	0.37361	0.38831	0.36597	0.00088	0.00476
	b	1.00000	1.00000	1.00000	1.00000	1.00000	1.00000	0.99062	1.00000
4	a	0.04896	0.01641	0.00269	0.01653	–0.00075	0.01009	0.00177	0.00023
	b	1.00000	1.00000	1.00000	1.00000	1.00000	1.00000	1.00000	1.00000
5	a	0.71543	0.70144	0.73792	0.47491	0.60960	0.53842	0.34231	0.28338
	b	0.99556	0.99718	0.98270	0.91522	0.99478	0.99210	0.90252	0.97237
6	a	0.75308	0.67237	0.33392	0.47230	0.67817	0.55727	0.18194	0.10131
	b	0.99631	0.99764	0.99169	0.95100	0.98809	0.97881	0.84463	0.99866
7	a	0.36466	0.51262	0.00992	0.32856	0.33905	0.39823	0.00527	0.00681
	b	1.00000	0.99974	1.00000	0.98493	0.99954	1.00000	0.99974	0.99954
8	a	0.74711	0.85104	0.01675	0.50459	0.51029	0.61615	0.00056	0.00023
	b	1.00000	0.99966	0.99932	0.99966	0.99966	0.99843	0.99835	1.00000
9	a	0.86040	0.90306	0.30121	0.26791	0.54639	0.62620	0.00245	0.00419
	b	1.00000	1.00000	1.00000	1.00000	1.00000	1.00000	1.00000	1.00000
10	a	0.70324	0.91460	0.00941	0.48929	0.47886	0.54275	0.00007	0.00004
	b	1.00000	1.00000	1.00000	1.00000	1.00000	1.00000	1.00000	1.00000
	$\bar{b}$	0.99900	0.99712	0.98783	0.98306	0.99747	0.99441	0.95491	0.99199
	$\bar{r}$	0.39023	0.34732	0.82166	0.62601	0.56687	0.53337	0.89310	0.94744
ccr		94.67%	91.78%	97.33%	99.11%	96.22%	96.44%	99.56%	99.78%
11	a	0.05334	0.04188	0.00007	0.03389	0.02904	0.03313	0.00009	0.00004
	b	0.12282	0.04339	0.04590	0.08259	0.08427	0.14440	0.04380	0.08438

a (b); $\bar{r} = \bar{b} - \bar{a}$; *ccr* – see Table 1.

Rand indices averaged over fifty replications and models 2-10 the improvements in cluster recovery (HINoV selected variables against all variables) are varying:

– for ordinal data from 0.3990 (mcquitty) to 0.7473 (centroid),

– for symbolic interval data from 0.3473 (ward) to 0.9474 (centroid).

5 Conclusions

The HINoV algorithm has limitations for analyzing nonmetric and symbolic interval data almost the same as the ones mentioned in Carmone et al. (1999) article for metric data.

First, the HINoV is of a little use with a nonmetric data set or a symbolic data array in which all variables are noisy (no cluster structure – see model 1). In this situation topri values are similar and close to zero (see Table 3).

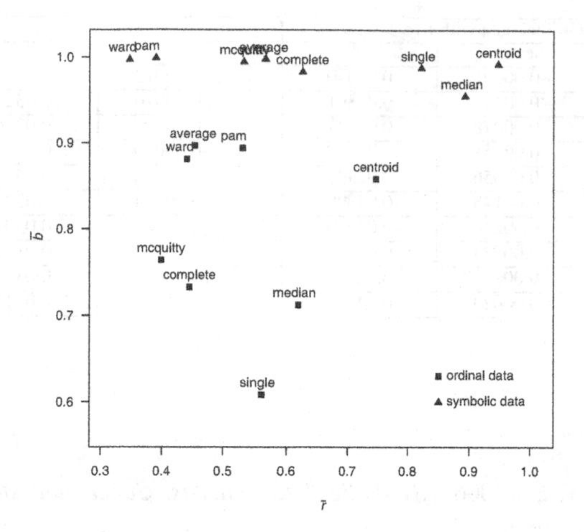

Fig. 1. The relationship between values of $\bar{b}$ and $\bar{r}$
Source: own research

Table 3. Mean and standard deviation of topri values for 10 variables in model 1

Variable	Ordinal data with five categories		Symbolic data array	
	mean	sd	mean	sd
1	−0.00393	0.01627	0.00080	0.02090
2	−0.00175	0.01736	0.00322	0.02154
3	0.00082	0.02009	0.00179	0.01740
4	−0.00115	0.01890	−0.00206	0.02243
5	0.00214	0.02297	−0.00025	0.02074
6	0.00690	0.02030	−0.00312	0.02108
7	−0.00002	0.02253	−0.00440	0.02044
8	0.00106	0.01754	0.00359	0.01994
9	0.00442	0.01998	0.00394	0.02617
10	−0.00363	0.01959	0.00023	0.02152

Second, the HINoV method depends on the relationship between pairs of variables. If we have only one variable with a cluster structure and the others are noisy, the HINoV will not be able to isolate this nonnoisy variable (see Table 4).

Third, if all variables have the same cluster structure (no noisy variables) the topri values will be large and similar for all variables. The suggested selection process using a scree diagram will be ineffective.

Fourth, an important problem is to decide on a proper number of clusters in stage two of the HINoV algorithm with symbolic interval data. To resolve this problem we should initiate the HINoV algorithm with a different number of clusters.

Table 4. Mean and standard deviation of topri values for 10 variables in model 11

Variable	Ordinal data with five categories		Symbolic data array	
	mean	sd	mean	sd
1	–0.00095	0.03050	0.00012	0.02961
2	–0.00198	0.02891	0.00070	0.03243
3	0.00078	0.02937	–0.00206	0.02969
4	–0.00155	0.02950	–0.00070	0.03185
5	0.00056	0.02997	–0.00152	0.03157
6	0.00148	0.03090	–0.00114	0.03064
7	–0.00246	0.02959	–0.00203	0.03019
8	–0.00274	0.03137	–0.00186	0.03021
9	–0.00099	0.02975	0.00088	0.03270
10	0.00023	0.02809	–0.00181	0.03126

References

BILLARD, L., DIDAY, E. (2006): *Symbolic data analysis. Conceptual statistics and data mining*, Wiley, Chichester.

CARMONE, F.J., KARA, A. and MAXWELL, S. (1999): HINoV: a new method to improve market segment definition by identifying noisy variables, *Journal of Marketing Research*, vol. 36, November, 501-509.

GNANADESIKAN, R., KETTENRING, J.R., and TSAO, S.L. (1995): Weighting and selection of variables for cluster analysis, *Journal of Classification*, vol. 12, no. 1, 113-136.

HUBERT, L.J., ARABIE, P. (1985): Comparing partitions, *Journal of Classification*, vol. 2, no. 1, 193-218.

JAJUGA, K., WALESIAK, M., BAK, A. (2003): On the General Distance Measure, In: M., Schwaiger, and O., Opitz (Eds.), *Exploratory data analysis in empirical research*, Springer-Verlag, Berlin, Heidelberg, 104-109.

MILLIGAN, G.W. (1996): Clustering validation: results and implications for applied analyses, In: P., Arabie, L.J., Hubert, G., de Soete (Eds.), *Clustering and classification*, World Scientific, Singapore, 341–375.

TIBSHIRANI, R., WALTHER, G., HASTIE, T. (2001): Estimating the number of clusters in a data set via the gap statistic, *Journal of the Royal Statistical Society*, ser. B, vol. 63, part 2, 411–423.

WALESIAK, M. (2005): *Variable selection for cluster analysis – approaches, problems, methods*, Plenary Session of the Committee on Statistics and Econometrics of the Polish Academy of Sciences, 15, March, Wroclaw.

Part II

Clustering

Families of Dendrograms

Patrick Erik Bradley

Institut für Industrielle Bauproduktion,
Englerstr. 7, 76128 Karlsruhe, Germany
bradley@ifib.uni-karlsruhe.de

Abstract. A conceptual framework for cluster analysis from the viewpoint of p-adic geometry is introduced by describing the space of all dendrograms for n datapoints and relating it to the moduli space of p-adic Riemannian spheres with punctures using a method recently applied by Murtagh (2004b). This method embeds a dendrogram as a subtree into the Bruhat-Tits tree associated to the p-adic numbers, and goes back to Cornelissen et al. (2001) in p-adic geometry. After explaining the definitions, the concept of classifiers is discussed in the context of moduli spaces, and upper bounds for the number of hidden vertices in dendrograms are given.

1 Introduction

Dendrograms are ultrametric spaces, and ultrametricity is a pervasive property of observational data, and by Murtagh (2004a) this offers computational advantages and a well understood basis for developping data processing tools originating in p-adic arithmetic. The aim of this article is to show that the foundations can be laid much deeper by taking into account a natural object in p-adic geometry, namely the *Bruhat-Tits tree*. This locally finite, regular tree naturally contains the dendrograms as subtrees which are uniquely determined by assigning p-adic numbers to data. Hence, the classification task is conceptionally reduced to finding a suitable p-adic data encoding. Dragovich and Dragovich (2006) find a 5-adic encoding of DNA-sequences, and Bradley (2007) shows that strings have natural p-adic encodings.

The geometric approach makes it possible to treat time-dependent data on an equal footing as data that relate only to one instant of time by providing the concept of *family of dendrograms*. Probability distributions on families are then seen as a convenient way of describing classifiers.

Our illustrative toy data set for this article is given as follows:

Example 1.1 *Consider the data set* $D = \{0, 1, 3, 4, 12, 20, 32, 64\}$ *given by* $n = 8$ *natural numbers. We want to hierarchically classify it with respect to the 2-adic norm* $|\cdot|_2$ *as our distance function, as defined in Section 2.*

2 A brief introduction to p-adic geometry

Euclidean geometry is modelled on the field $\mathbb{R}$ of real numbers which are often represented as decimals, i.e. expanded in powers of the number 10^{-1}:

$$x = \sum_{\nu=m}^{\infty} a_\nu 10^{-\nu}, \quad a_\nu \in \{0,\dots,9\}, \quad m \in \mathbb{Z}.$$

In this way, $\mathbb{R}$ completes the field $\mathbb{Q}$ of rational numbers with respect to the absolute norm $|x| = \begin{cases} x, & x \geq 0 \\ -x, & x < 0 \end{cases}$. On the other hand, the p-adic norm on $\mathbb{Q}$ with

$$|x|_p = \begin{cases} p^{-\nu_p(x)}, & x \neq 0 \\ 0, & x = 0 \end{cases}$$

is defined for $x = \frac{a_1}{a_2}$ by the difference $\nu_p(x) = \nu_p(a_1) - \nu_p(a_2) \in \mathbb{Z}$ in the multiplicities with which numerator and denominator of x are divisible by the prime number p: $a_i = p^{\nu_p(a_i)} u_i$, and u_i not divisible by p, $i = 1,2$.

The p-adic norm satisfies the *ultrametric triangle inequality*

$$|x+y|_p \leq \max\{|x|_p, |y|_p\}.$$

Completing $\mathbb{Q}$ with respect to the p-adic norm yields the field $\mathbb{Q}_p$ of *p-adic numbers* which is well known to consist of the power series

$$x = \sum_{\nu=m}^{\infty} a_\nu p^\nu, \quad a_\nu \in \{0,\dots,p-1\}, \quad m \in \mathbb{Z}. \tag{1}$$

Note, that the p-adic expansion is in increasing powers of p, whereas in the decimal expansion, it is the powers of 10^{-1} which increase arbitrarily. An introduction to p-adic numbers is e.g. Gouvêa (2003).

Example 2.1 *For our toy data set D, we have* $|0|_2 = 0$, $|1|_2 = |3|_2 = 1$, $|4|_2 = |12|_2 = |20|_2 = 2^{-2}$, $|32|_2 = 2^{-5}$, $|64|_2 = 2^{-6}$, *i.e.* $|\cdot|_2$ *is maximally* 1 *on D. Other examples:* $|3/2|_3 = |6/4|_3 = 3^{-1}$, $|20|_5 = 5^{-1}$, $|p^{-1}|_p = |p|_p^{-1} = p$.

Consider the unit disk $\mathbb{D} = \{x \in \mathbb{Q}_p \mid |x|_p \leq 1\} = B_1(0)$. It consists of the so-called *p-adic integers*, and is often denoted as $\mathbb{Z}_p$ when emphasizing its ring structure, i.e. closedness under addition, subtraction and multiplication. A p-adic number x lies in an arbitrary closed disk $B_{p^{-r}}(a) = \{x \in \mathbb{Q}_p \mid |x-a|_p \leq p^{-r}\}$, where $r \in \mathbb{Z}$, if and only if $x - a$ is divisible by p^r. This condition is equivalent to x and a having the first r terms in common in their p-adic expansions (1). The possible radii are all integer powers of p, so the disjoint disks $B_{p^{-1}}(0), B_{p^{-1}}(1), \dots, B_{p^{-1}}(p-1)$ are the maximal proper subdisks of $\mathbb{D}$, as they correspond to truncating the power series (1) after the constant term. There is a unique minimal disk in which $\mathbb{D}$ is contained properly, namely $B_p(0) = \{x \in \mathbb{Q}_p \mid |x|_p \leq p\}$. These observations hold true for arbitrary

p-adic disks, i.e. any disk $B_{p^{-r}}(x)$, $x \in \mathbb{Q}_p$, is partitioned into precisely p maximal subdisks and lies properly in a unique minimal disk. Therefore, if we define a graph $\mathscr{T}_{\mathbb{Q}_p}$ whose vertices are the p-adic disks, and edges are given by minimal inclusion, then every vertex of $\mathscr{T}_{\mathbb{Q}_p}$ has precisely $p+1$ outgoing edges. In other words, $\mathscr{T}_{\mathbb{Q}_p}$ is a $p+1$-*regular* tree, and p is the size of the residue field $\mathbb{F}_p = \mathbb{Z}_p/p\mathbb{Z}_p$.

Definition 2.2 The tree $\mathscr{T}_{\mathbb{Q}_p}$ is called the *Bruhat-Tits tree* for $\mathbb{Q}_p$.

Remark 2.3 *Definition 2.2 is not the usual way to define $\mathscr{T}_{\mathbb{Q}_p}$. The problem with this ad-hoc definition is that it does not allow for any action of the projective linear group* $\mathrm{PGL}_2(\mathbb{Q}_p)$. *A definition invariant under projective linear transformations can be found e.g. in Herrlich (1980) or Bradley (2006).*

An important observation is that any infinite descending chain

$$B_1 \supseteq B_2 \supseteq \dots \tag{2}$$

of strictly decreasing p-adic disks converges to a unique p-adic number $\{x\} = \bigcap_n B_n$. A chain (2) defines a halfline in the Bruhat-Tits tree $\mathscr{T}_{\mathbb{Q}_p}$. Halflines differing only by finitely many vertices are said to be *equivalent*, and the equivalence classes under this equivalence relation are called *ends*. Hence the observation means that the p-adic numbers correspond to ends of $\mathscr{T}_{\mathbb{Q}_p}$. There is a unique end $B_1 \subseteq B_2 \subseteq \dots$ coming from any strictly increasing sequence of disks. This end corresponds to the point at infinity in the p-adic projective line $\mathbb{P}^1(\mathbb{Q}_p) = \mathbb{Q}_p \cup \{\infty\}$, whence the well known fact:

Lemma 2.4 *The ends of $\mathscr{T}_{\mathbb{Q}_p}$ are in one-to-one correspondance with the $\mathbb{Q}_p$-rational points of the p-adic projective line $\mathbb{P}^1$, i.e. with the elements of $\mathbb{P}^1(\mathbb{Q}_p)$.*

From the viewpoint of geometry, it is important to distinguish between the p-adic projective line $\mathbb{P}^1$ as a p-adic manifold and its set $\mathbb{P}^1(\mathbb{Q}_p)$ of $\mathbb{Q}_p$-rational points, in the same way as one distinguishes between the affine real line $\mathbb{A}^1$ as a real manifold and its rational points $\mathbb{A}^1(\mathbb{Q}) = \mathbb{Q}$, for example. One reason for distinguishing between a space and its points is:

Lemma 2.5 *Endowed with the metric topology from $|\cdot|_p$, the topological space $\mathbb{Q}_p$ is totally disconnected.*

The usual approaches towards defining more useful topologies on p-adic spaces are by introducing more points. Such an approach is the *Berkovich topology*, which we will very briefly describe. More details can be found in Berkovich (1990).

The idea is to allow disks whose radii are arbitrary positive real numbers, not merely powers of p as before. Any strictly descending chain of such disks gives a point in the sense of Berkovich. For the p-adic line $\mathbb{P}^1$ this amounts to:

Theorem 2.6 (Berkovich) $\mathbb{P}^1$ *is non-empty, compact, hausdorff and arc-wise connected. Every point of $\mathbb{P}^1 \setminus \{\infty\}$ corresponds to a descending sequence $B_1 \supseteq B_2 \supseteq \dots$ of p-adic disks such that $B = \bigcap B_n$ is one of the following:*

1. *a point x in* $\mathbb{Q}_p$,
2. *a closed p-adic disk with radius* $r \in |\mathbb{Q}_p|_p$,
3. *a closed p-adic disk with radius* $r \notin |\mathbb{Q}_p|_p$,
4. *empty.*

Points of types 2. to 4. are called *generic*, points of type 1. *classical*. We remark that Berkovich's definition of points is technically somewhat different and allows to define more general p-adic spaces. Finally, the Bruhat-Tits tree $\mathscr{T}_{\mathbb{Q}_p}$ is recovered inside $\mathbb{P}^1$:

Theorem 2.7 (Berkovich) *$\mathscr{T}_{\mathbb{Q}_p}$ is a retract of $\mathbb{P}^1 \setminus \mathbb{P}^1(\mathbb{Q}_p)$, i.e. there is a map $\mathbb{P}^1 \setminus \mathbb{P}^1(\mathbb{Q}_p) \to \mathscr{T}_{\mathbb{Q}_p}$ whose restriction to $\mathscr{T}_{\mathbb{Q}_p}$ is the identity map on $\mathscr{T}_{\mathbb{Q}_p}$.*

3 p-adic dendrograms

ν_2	0	1	3	4	12	20	32	64
0	∞	0	0	2	2	2	5	6
1	0	∞	1	0	0	0	0	0
3	0	1	∞	0	0	0	0	0
4	2	0	0	∞	3	4	2	2
12	2	0	0	3	∞	3	2	2
20	2	0	0	4	3	∞	2	1
32	5	0	0	2	2	2	∞	5
64	6	0	0	2	2	1	5	∞

Fig. 1. 2-adic valuations for D.

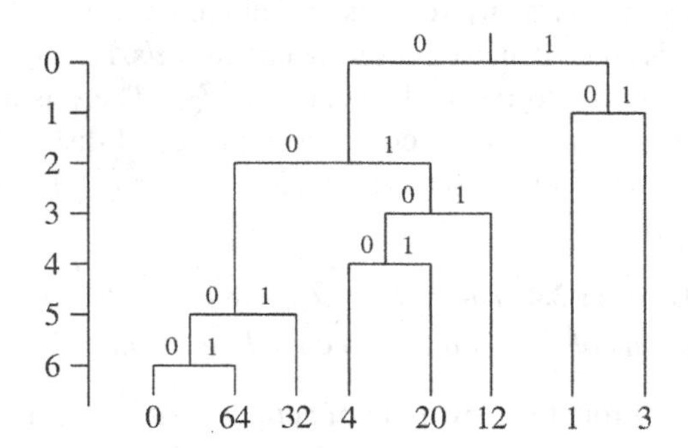

Fig. 2. 2-adic dendrogram for $D \cup \{\infty\}$.

Example 3.1 *The 2-adic distances within D are encoded in Figure 1, where* $\operatorname{dist}(i,j) = 2^{-\nu_2(i,j)}$, *if* $\nu_2(i,j)$ *is the corresponding entry in Figure 1, using* $2^{-\infty} = 0$. *Figure 2 is the dendrogram for D using* $|\cdot|_2$*: the distance between disjoint clusters equals the distances between any of their representatives.*

Let $X \subseteq \mathbb{P}^1(\mathbb{Q}_p)$ be a finite set. By Lemma 2.4, a point of X can be considered as an end in $\mathscr{T}_{\mathbb{Q}_p}$.

Definition 3.2 The smallest subtree $\mathscr{D}(X)$ of $\mathscr{T}_{\mathbb{Q}_p}$ whose ends are given by X is called the *p-adic dendrogram* for X.

Cornelissen et al. (2001) use p-adic dendrograms for studying p-adic symmetries, cf. also Cornelissen and Kato (2005). We will ignore vertices in $\mathscr{D}(X)$ from which precisely two edges emanate. Hence, for example, $\mathscr{D}(\{0,1,\infty\})$ consists of a unique vertex $v(0,1,\infty)$ and three ends. The dendrogram for a set $X \subseteq \mathbb{N} \cup \{\infty\}$ containing $\{0,1,\infty\}$ is a rooted tree with root $v(0,1,\infty)$.

Example 3.3 *The 2-adic dendrogram in Figure 2 is nothing but $\mathscr{D}(X)$ for $X = D \cup \{\infty\}$ and is in fact inspired by the first dendrogram of Murtagh (2004b). The path from the top cluster to x_i yields its binary representation $[\cdot]_2$ which easily translates into the 2-adic expansion:* $0 = [0000000]_2$, $64 = [1000000]_2 = 2^6$, $32 = [0100000]_2 = 2^5$, $4 = [0000100]_2 = 2^2$, $20 = [0010100]_2 = 2^2 + 2^4$, $12 = [0001100]_2 = 2^2 + 2^3$, $1 = [0000001]_2$, $3 = [0000011]_2 = 1 + 2^1$.

Any encoding of some data set M which assigns to each $x \in M$ a p-adic representation of an integer including 0 and 1, yields a p-adic dendrogram $\mathscr{D}(M \cup \{\infty\})$ whose root is $v(0,1,\infty)$, and any dendrogram for real data can be embedded in a non-unique way into $\mathscr{T}_{\mathbb{Q}_p}$ as a p-adic dendrogram in such a way that $v(0,1,\infty)$ represents the top cluster, if p is large enough. In particular, any binary dendrogram is a 2-adic dendrogram. However, a little algebra helps to find sufficiently large 2-adic Bruhat-Tits trees $\mathscr{T}_K$ which allow embeddings of arbitrary dendrograms into $\mathscr{T}_K$. In fact, by K we mean a finite extension field of $\mathbb{Q}_p$. The p-adic norm $|\cdot|_p$ extends uniquely to a norm $|\cdot|_K$ on K, for which it is a complete field, called a *p-adic number field*. The *integers of* K are again the unit disk $O_K = \{x \in K \mid |x|_K \leq 1\}$, and the role of the prime p is played by a so-called *uniformiser* $\pi \in O_K$. It has the property that $O_K/\pi O_K$ is a finite field with $q = p^f$ elements and contains $\mathbb{F}_p$. Hence, if some dendrogram has a vertex with maximally $n \geq 2$ children, then we need K large enough such that $2^f \geq n$. This is possible by the results of number theory. Restricting to the prime characteristic 2 has not only the advantage of avoiding the need to switch the prime number p in the case of more than p children vertices, but also the arithmetic in 2-adic number fields is known to be computationally simpler, especially as in our case the so-called *unramified* extensions, i.e. where $\dim_{\mathbb{Q}_2} K = f$, are sufficient.

Example 3.4 *According to Bradley (2007), strings over a finite alphabet can be encoded in an unramified extension of $\mathbb{Q}_p$, and hence be classified p-adically.*

4 The space of dendrograms

From now on, we will formulate everything for the case $K = \mathbb{Q}_p$, bearing in mind that all results hold true for general p-adic number fields K. Let $S = \{x_1, \ldots, x_n\} \subseteq \mathbb{P}^1(\mathbb{Q}_p)$ consist of n distinct classical points of $\mathbb{P}^1$ such that $x_1 = 0$, $x_2 = 1$, $x_3 = \infty$. Similarly as in Theorem 2.7, the p-adic dendrogram $\mathscr{D}(S)$ is a retract of the marked projective line $X = \mathbb{P}^1 \setminus S$. We call $\mathscr{D}(S)$ the *skeleton* of X. The space of all projective lines with n such markings is denoted by $\mathfrak{M}_n$, and the space of corresponding p-adic dendrograms by $\mathfrak{D}_{n-1}$. $\mathfrak{M}_n$ is a p-adic space of dimension $n-3$, its skeleton $\mathfrak{D}_{n-1}$ is a cw-complex of real polyhedra whose cells of maximal dimension $n-3$ consist of the binary dendrograms. Neighbouring cells are passed through by contracting bounded edges as the $n-3$ "free" markings "move" about $\mathbb{P}^1$ without colliding. For example, $\mathfrak{M}_3$ is just a point corresponding to $\mathbb{P}^1 \setminus \{0,1,\infty\}$. $\mathfrak{M}_4$ has one free marking λ which can be any $\mathbb{Q}_p$-rational point from $\mathbb{P}^1 \setminus \{0,1,\infty\}$. Hence, the skeleton $\mathfrak{D}_3$ is

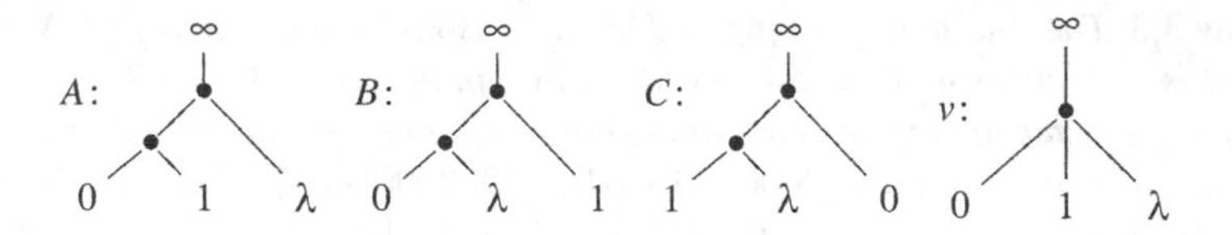

Fig. 3. Dendrograms representing the different regions of $\mathfrak{D}_3$.

itself a binary dendrogram with precisely one vertex v and three unbounded edges A,B,C (cf. Figure 3). For $n \geq 3$ there are maps

$$f_{n+1}\colon \mathfrak{M}_{n+1} \to \mathfrak{M}_n, \quad \phi_{n+1}\colon \mathfrak{D}_n \to \mathfrak{D}_{n-1},$$

which forget the $(n+1)$-st marking. Consider a $\mathbb{Q}_p$-rational point $x \in \mathfrak{M}_n$, corresponding to $\mathbb{P}^1 \setminus S$ with skeleton d. Its fibre $f_{n+1}^{-1}(x)$ corresponds to $\mathbb{P}^1 \setminus S'$ for all possible S' whose first n entries constitute S. Hence, the extra marking $\lambda \in S' \setminus S$ can be taken arbitrarily from $\mathbb{P}(\mathbb{Q}_p) \setminus S$. In this way, the space $f_{n+1}^{-1}(x)$ can be considered as $\mathbb{P}^1 \setminus S$, and $\phi_{n+1}^{-1}(d)$ as the p-adic dendrogram for S. What we have seen is that taking fibres recovers the dendrograms corresponding to points in the space $\mathfrak{D}_n$. Instead of fibres of points, one can take fibres of arbitrary subspaces:

Definition 4.1 A *family of dendrograms with n data points over a space Y* is a map $Y \to \mathfrak{D}_n$ from some p-adic space Y to $\mathfrak{D}_n$.

For example, take $Y = \{y_1,\ldots,y_T\}$. Then a family $Y \to \mathfrak{D}_n$ is a time series of n collision-free particles, if $t \in \{1,\ldots,T\}$ is interpreted as time variable. It is also possible to take into account colliding particles by using compactifications of $\mathfrak{M}_n$ as described in Bradley (2006).

5 Distributions on dendrograms

Given a dendrogram $\mathscr{D}$ for some data $S = \{x_1,\ldots,x_n\}$, the idea of a classifier is to incorporate a further datum $x \notin S$ into the classification scheme represented by $\mathscr{D}$. Often this is done by assigning probabilities to the vertices of $\mathscr{D}$, depending on x. The result is then a family of possible dendrograms for $S \cup \{x\}$ with a certain probability distribution. It is clear that, in the case of p-adic dendrograms, this family is nothing but $\phi_{n+1}^{-1}(d) \to \mathfrak{D}_n$, if $d \in \mathfrak{D}_{n-1}$ is the point representing $\mathscr{D}$. This motivates the following definition:

Definition 5.1 A *universal p-adic classifier C for n given points* is a probability distribution on $\mathfrak{M}_{n+1}$.

Here, we take on $\mathfrak{M}_{n+1}$ the Borel σ-algebra associated to the open sets of the Berkovich topology. If $x \in \mathfrak{M}_n$ corresponds to $\mathbb{P}^1 \setminus S$, then C induces a distribution on $f_{n+1}^{-1}(x)$, hence (after renormalisation) a probability distribution on $\phi_{n+1}^{-1}(d)$, where $d \in \mathfrak{D}_{n-1}$ is the point corresponding to the dendrogram $\mathscr{D}(S)$. The similar holds true for general families of dendrograms, e.g. time series of particles.

6 Hidden vertices

A vertex v in a p-adic dendrogram $\mathscr{D}$ is called *hidden*, if the class corresponding to v is not the top class and does not directly contain data points but is composed of non-trivial subclasses. The subforest of $\mathscr{D}$ spanned by its hidden vertices will be denoted by $\mathscr{D}^h$, and is called the *hidden part* of $\mathscr{D}$. The number b_0^h of connected components of $\mathscr{D}^h$ measures how the clusters corresponding to non-hidden vertices are spread within the dendrogram $\mathscr{D}$. We give bounds for b_0^h and the number v^h of hidden vertices, and refer to Bradley (2006) for the combinatorial proofs (Theorems 8.3 and 8.5).

Theorem 6.1 *Let $\mathscr{D} \in \mathfrak{D}_n$. Then*

$$v^h \leq \frac{n+2-b_0^h}{2} \quad \text{and} \quad b_0^h \leq \frac{n-4}{3},$$

where the latter bound is sharp.

7 Conclusions

Since ultrametricity is the natural property which allows classification and is pervasive in observational data, the techniques of ultrametric analysis and p-adic geometry are at ones disposal for identifying and exploiting ultrametricity. A p-adic encoding of data provides a way to investigate arithmetic properties of the p-adic numbers representing the data.

It is our aim to lay the geometric foundation towards p-adic data encoding. From the geometric point of view it is natural to perform the encoding by embedding its underlying dendrogram into the Bruhat-Tits tree. In fact, the dendrogram and its embedding are uniquely determined by the p-adic numbers representing the data. For this end, we give an account of p-adic geometry in order to define p-adic dendrograms as subtrees of the Bruhat-Tits tree.

In the next step we introduce the space of all dendrograms for a given number of data points which, by p-adic geometry, is contained in the space $\mathfrak{M}_n$ of all marked projective lines, an object appearing in the context of the classification of Riemann surfaces. The advantages of considering the space of dendrograms rely on the fact that a conceptual formulation of moving particles as families of dendrograms is made possible, and its simple geometry as a polyhedral complex. Also, assigning distributions on $\mathfrak{M}_n$ allows for probabilistic incorporation of further data to a given dendrogram. At the end, we give bounds for the numbers of hidden vertices and hidden components of dendrograms.

What remains to do is to computationally exploit the foundations laid in this article by developping a code along these lines and apply it to Fionn Murtagh's task of finding ultrametricity in data.

Acknowledgements

The author is supported by the Deutsche Forschungsgemeinschaft through the research project *Dynamische Gebäudebestandsklassifikation* BR 3513/1-1, and thanks Hans-Hermann Bock for suggesting to include a toy dataset, and an unknown referee for many valuable remarks.

References

BERKOVICH, V.G. (1990): Spectral Theory and Analytic Geometry over Non-Archimedean Fields. *Mathematical Surveys and Monographs, 33, AMS.*

BRADLEY, P.E. (2006): Degenerating families of dendrograms. *Preprint.*

BRADLEY, P.E. (2007): Mumford dendrograms. *Preprint.*

CORNELISSEN, G. and KATO, F. (2005): The p-adic icosahedron. *Notices of the AMS, 52, 720–727.*

CORNELISSEN, G., KATO, F. and KONTOGEORGIS, A. (2001): Discontinuous groups in positive characteristic and automorphisms of Mumford curves. *Mathematische Annalen, 320, 55–85.*

DRAGOVICH, B. AND DRAGOVICH, A. (2006): A p-Adic Model of DNA-Sequence and Genetic Code. *Preprint* arXiv:q-bio.GN/0607018.

GOUVÊA, F.Q. (2003): *p-adic numbers: an introduction.* Universitext, Springer.

HERRLICH, F (1980): Endlich erzeugbare p-adische diskontinuierliche Gruppen. *Archiv der Mathematik, 35, 505–515.*

MURTAGH, F. (2004): On ultrametricity, data coding, and computation. *Journal of Classification, 21, 167–184.*

MURTAGH, F. (2004): Thinking ultrametrically. In: D. Banks, L. House, F.R. McMorris, P. Arabie, and W. Gaul (Eds.): *Classification, Clustering and Data Mining*, Springer, 3–14.

Mixture Models in Forward Search Methods for Outlier Detection

Daniela G. Calò

Department of Statistics, University of Bologna,
Via Belle Arti 41, 40126 Bologna, Italy
danielagiovanna.calo@unibo.it

Abstract. Forward search (FS) methods have been shown to be usefully employed for detecting multiple outliers in continuous multivariate data (Hadi, (1994); Atkinson *et al.*, (2004)). Starting from an outlier-free subset of observations, they iteratively enlarge this good subset using Mahalanobis distances based only on the good observations. In this paper, an alternative formulation of the FS paradigm is presented, that takes a mixture of $K > 1$ normal components as a null model. The proposal is developed according to both the graphical and the inferential approach to FS-based outlier detection. The performance of the method is shown on an illustrative example and evaluated on a simulation experiment in the multiple cluster setting.

1 Introduction

Mixtures of multivariate normal densities are widely used in cluster analysis, density estimation and discriminant analysis, usually resorting to maximum likelihood (ML) estimation, via the EM algorithm (for an overview, see McLachlan and Peel, (2000)). When the number of components K is treated as fixed, ML estimation is not robust against outlying data: a single extreme point can make the parameter estimation of at least one of the mixture components break down. Among the solutions presented in the literature, the main computable approaches in the multivariate setting are: the addition of a noise component modelled as a uniform distribution on the convex hull of the data, implemented in the software `MCLUST` (Fraley and Raftery, (1998)); a mixture of t-distributions instead of normal distributions, implemented in the software `EMMIX` (McLachlan and Peel, (2000)). According to Hennig, both the alternatives " ... do not possess a substantially better breakdown behavior than estimation based on normal mixtures" (Hennig, (2004)).

An alternative approach to the problem is based on the idea that a good outlier detection method defines a robust estimation method, that works by omitting the observations nominated as outliers and computing a standard non-robust estimate on the remaining observations. Here, attention is focussed on the so-called *forward search* (FS) methods, which have been usefully employed for detecting multiple outliers in continuous multivariate data. These methods are based on the assumption that non-outlying data stem form a multivariate normal distribution or they are roughly elliptically symmetric.

In this paper, an alternative formulation of the FS algorithm is proposed, which is specifically designed for situations where non-outlying data stem from a mixture of a known number of normal components. It could not only enlarge the applicability of FS outlier detection methods, but could also provide a possible strategy for robust fitting in multivariate normal mixture models.

2 The Forward Search

The Forward search (FS) is a powerful general method for detecting multiple masked outliers in continuous multivariate data (Hadi, (1994); Atkinson, (1993)). The search starts by fitting the multivariate normal model to a small subset S_m, consisting of $m = m_0$ observations, that can be safely presumed to be free of outliers: it can be specified by the data analyst or obtained by an algorithm. All n observations are ordered by their Mahalanobis distance and S_m is updated as the set of the $m+1$ observations with the smallest Mahalanobis distances. Then, the number m is increased by 1 and the search goes on, by fitting the normal model to the current subset S_m and updating S_m as stated above – so that its size is increased by one unit at a time – until S_m includes all n observations (that is, $m = n$).

By ordering the data according to their closeness to the fitted model (by means of Mahalanobis distance), the various steps of the search provide subsets which are designed to be outlier-free, until there remain only outliers to be included. The inclusion of outlying observations can be signalled by following two main approaches. The former consists in graphically monitoring the values of suitable statistics during the search, such as the minimum squared Mahalanobis distance amongst units not included in subset S_m (for m ranging from m_0 to n): if it is large, it means that an outlier is going to join the subset (for a presentation of FS exploratory techniques, see Atkinson *et al.*, (2004)). The latter approach consists in testing the maximum squared Mahalanobis distance amongst the observations included in S_m: if it exceeds a given χ^2 cutoff, then the search stops (before its natural ending) and the tested observation is nominated as an outlier together with all observations not yet included in S_m (see Hadi, (1994)), for a presentation of the method).

When non-outlying data stem from a mixture distribution, the Mahalanobis distance cannot be generally used as a measure of discrepancy. A proper criterion for ordering the units by closeness to the assumed model is required, together with a consistent method for finding the starting subset of observations. In this paper a novel algorithm of sequential point addition is proposed, designed for situations where non-outlying data come from a mixture of $K > 1$ normal components, with K assumed to be known. Two possible formulations are presented, each related to one of the two aforementioned approaches to FS-based outlier detection, hereafter called "graphical" and "inferential", respectively.

3 Forward Search and Normal Mixture Models: the graphical approach

We assume that the d-dimensional random vector X is distributed according to a K component Normal mixture model:

$$p(x) = \sum_{k=1}^{K} w_k \phi(x|\mu_k, \Sigma_k), \tag{1}$$

where each Gaussian density $\phi(\cdot)$ is parameterized by its mean vector $\mu_k \in \mathbb{R}^d$ and covariance matrix Σ_k, belonging to the set of positive definite $d \times d$ matrices, and w_k $(k = 1,\dots,K)$ are mixing proportions; we suppose that some contamination is present in the sample. Because of the zero breakdown-point of ML estimators, the FS graphical approach can still be useful for outlier detection in normal mixtures, provided that the three aspects that make up the search are properly modified: the choice of an initial subset, the way we progress in the search and the statistic to be monitored during the search.

Subset S_{m_0} could be defined as the union of K subsets, each located well inside a single mixture component: each set could be determined by using robust bi-variate boxplots or robustly centered ellipses (both described in Atkinson *et al.*, (2004)) on a distinct element of the data partition provided by some robust clustering method. This requires that model (1) is a clustering model. As a more general solution, we propose to define S_{m_0} as a subset of high-density observations, since it is unlike that outliers lye in high-density regions of $\mathbb{R}^d$. For this purpose, a nonparametric density estimate is built on the whole data set and the observations x_i $(i = 1,\dots,n)$ are sorted in decreasing order of estimated density. Denoting by $x_{[i],0}$ the observation with the i–th ordered density (estimated at step 0), we take:

$$S_{m_0} = \{x_{[i],0} : i = 1,\dots,m_0\}. \tag{2}$$

It is worth noting that nonparametric density estimation is used here in order to dampen the effect of outliers. Its use limits the applicability of the proposed method to large medium-dimensional datasets; anyway, it is well known that nonparametric density estimation is less sensitive to the curse of dimensionality just in the region(s) around the mode(s).

In order to define how to progress in the search, the following criterion is proposed, for m ranging from m_0 to n. Given the current subset S_m, model (1) is fitted by the EM algorithm and the parameter estimates $\{\hat{w}_{k,m}, \hat{\mu}_{k,m}, \hat{\Sigma}_{k,m}; k = 1,\dots,K\}$ are obtained. For each observation x_i, the corresponding estimated value of the mixture density function

$$\hat{p}(x_i) = \sum_{k=1}^{K} \hat{w}_{k,m} \phi(x_i|\hat{\mu}_{k,m}, \hat{\Sigma}_{k,m}) \tag{3}$$

is taken as a measure of closeness of x_i to the fitted model. The density values $\hat{p}(x_i)$ are then ordered from largest to smallest and the $m+1$ observations with the highest values are taken to form the new subset S_{m+1}. This sorting criterion is coherent with (2); moreover, when $K = 1$ it is equivalent, but opposite, to that defined by the normalized squared Mahalanobis distance:

$$D^*(x_i; \hat{\mu}_m, \hat{\Sigma}_m) = \frac{1}{2}[d \ln(2\pi) + \ln(|\hat{\Sigma}_m|) + (x_i - \hat{\mu}_m)^T \hat{\Sigma}_m^{-1}(x_i - \hat{\mu}_m)]. \quad (4)$$

In elliptical K-means clustering, (4) is preferred to the squared Mahalanobis distance because of stability reasons.

In our experiments we found that the inclusion of outlying points can be well monitored by plotting the values of the following statistic:

$$s_m = -\ln(max\{\hat{p}(x_i); i \notin S_m\}). \quad (5)$$

It is the negative natural logarithm of the maximum density estimate amongst observations not included in the current subset: if an outlier is about to enter, the value of s_m will be large relative to the previous ones. When $K = 1$, monitoring (5) is equivalent to monitor the minimum value of (4) amongst observations not included in S_m.

The proposed procedure is illustrated on an artificial bi-variate dataset, reported by Cuesta-Albertos *et al.* (available at *http://personales.unican.es/cuestaj/RobustEstimationMixtures.pdf*) as an example where the t-mixture model can fail. The main stages of the procedure are shown in Figure 1: m_0 was set equal to 200 and density estimation has been carried out on the whole data set through a Gaussian kernel estimator with "rule of thumb" bandwidth. The forward plot of (5) is reported only for the last 100 steps of the search, so that its final part is more legible: it signals the introduction of the first outlying influential observation with a sharp peak, just after the inclusion of 600 units in S_m. Stopping the search before the peak provides a robust fitting of the mixture, since it is estimated on all observations but the outlying ones. Good results were obtained also in case of symmetrical contamination.

It could be objected that a 4-component mixture would work as well in the example above. However, in our experience we observed also situations where the cluster of outliers can be hardly identified by fitting a $K+1$-component mixture, since it tends to be "picked-up" by a flat component accounting for generic noise (see, for instance, Example 3.2 in Cuesta-Albertos *et al.*).

Anyway, the graphical exploration technique presented above is prone to errors, because not every data set will give rise to an obvious separation between extreme points which are outliers and extreme points which are not outliers. For this reason, a formulation of the FS in normal mixtures according to the "inferential approach" (mentioned in Section 2) should be devised. In the following section, a FS procedure involving a test about the outlyingness of a point with respect to a mixture is presented.

4 Forward Search and Normal Mixture Models: the inferential approach

The problem of outlier detection from a mixture is considered in McLachlan and Basford (1988). Attention is focused on the assessment of whether an observation is

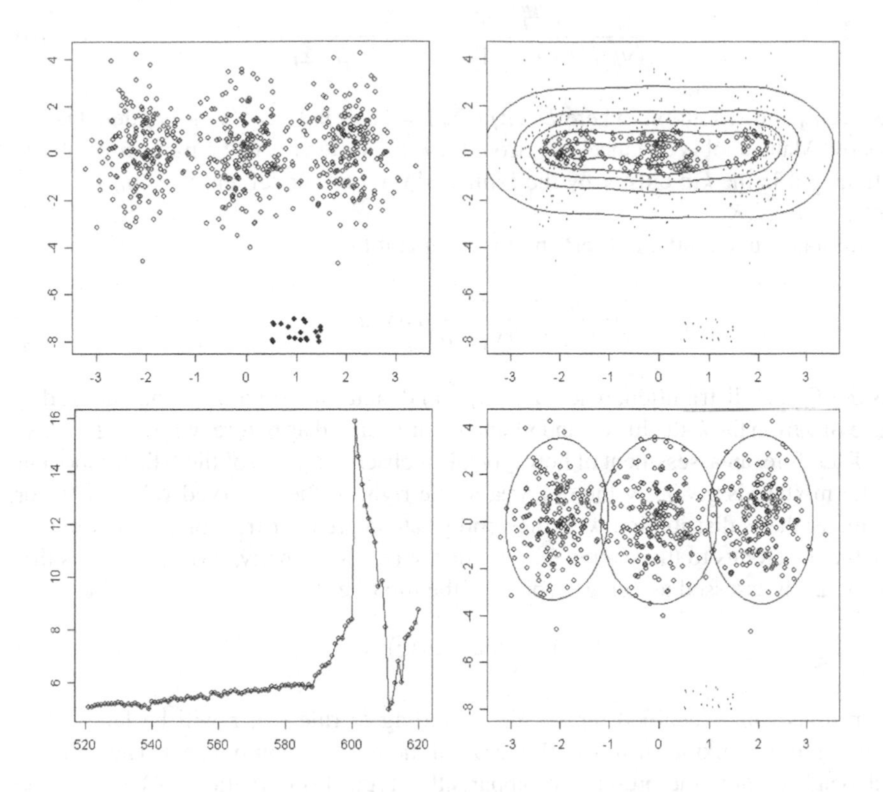

Fig. 1. The example from Cuesta-Albertos *et al.*: 20 outliers are added to a sample of 600 observations. Top right panel shows the contour plot of the density estimate and the $m_0 = 200$ (circled) observations belonging to the starting subset. Bottom left panel reports the monitoring plot of (5) for $m = 520, \ldots, 620$. The 95% ellipses of the mixture components fitted to S_{600} are plotted in the last panel.

atypical of a mixture of K normal populations, $P_1, \ldots, P_K$, on the basis of a set of m observations $\{x_{hk}; h = 1, \ldots, m_k, k = 1, \ldots, K\}$, where x_{hk} are known to come from P_k and $\sum_{k=1}^{K} m_k = m$. The problem is tackled by assessing how typical the observation is of each P_k in turn.

In case of unclassified data $\{x_j; j = 1, \ldots, m\}$ – like the one considered in the present paper – McLachlan and Basford suggest that the m observations should be first clustered by fitting a K-component heteroscedastic normal mixture model. Then, the aforementioned comparison of the tested observation to each of the mixture components in turn is applied to the resulting K clusters as if they represent a "true classification" of the data. The approach is based on the following distributional results, which are derived under the assumption that model (1) is valid:

for the generic sample observation x_j, the quantity

$$\frac{(\frac{\nu_k m_k}{d})D(x_j;\hat{\mu}_k,\hat{\Sigma}_k)}{(\nu_k+d)(m_k-1)-m_k D(x_j;\hat{\mu}_k,\hat{\Sigma}_k)} \tag{6}$$

has the F_{d,ν_k} distribution, where $D(x_j;\hat{\mu}_k,\hat{\Sigma}_k)=(x_j-\hat{\mu}_k)^T\hat{\Sigma}_k^{-1}(x_j-\hat{\mu}_k)$ denotes the squared Mahalanobis distance of x_j from the k-th cluster, m_k is the number of observations put in the kth cluster by the estimated mixture model and $\nu_k=m_k-d-1$, with $k=1,\ldots,K$;

for a new unclassified observation y, the quantity

$$\frac{m_k(\nu_k+1)}{(m_k+1)d(\nu_k+d)}D(y;\hat{\mu}_k,\hat{\Sigma}_k) \tag{7}$$

has the F_{d,ν_k+1} distribution, where $D(y;\hat{\mu}_k,\hat{\Sigma}_k)$ denotes the squared Mahalanobis distance of y from the k-th cluster, and ν_k and m_k are defined as before, with $k=1,\ldots,K$.

Therefore, an assessment of how typical an observation z is of the k-th component of the mixture is given by the tail area to the right of the observed value of (6) or (7) under the F distribution with the appropriate degrees of freedom, depending on whether z belongs to the sample ($z=x_j$) or not ($z=y$). Finally, if $a_k(z)$ denotes this tail area, z is assessed as being atypical of the mixture if

$$a(z)=\max_{k=1,\ldots,K} a_k(z)\leq\alpha, \tag{8}$$

where α is some specified threshold. According to rule (8), z will be labelled as outlying of the mixture if it is outlying of all the mixture components. The value of α depends on how the presence of apparently atypical observations is handled: the more protection is desired against the possible presence of outliers, the higher the value of α.

We present a FS algorithm using the typicality index $a(z)$ as a measure of "closeness" of a generic observation z to the fitted mixture model. For the sake of simplicity, the same criterion for selecting S_{m_0} described in Section 3 is employed. Then, at each step of the search, a K-component normal mixture model is fitted to the current subset S_m and the typicality index is computed for each observation $x_i (i=1,\ldots,n)$ by means of (6) or (7), depending on whether the observation is an element of S_m or an element of the remainder of the sample in step m. Then, observations are sorted in decreasing order of typicality: denoting by $x_{[i],m}$ the observation with the i-th ordered typicality value (computed on subset S_m), subset S_m is updated as the set of the $m+1$ most typical observations: $S_{m+1}=\{x_{[i],m}: i=1,\ldots,m+1\}$.

If the least typical observation in the newly created subset, that is $x_{[m+1],m}$, is assessed as being atypical according to rule (8), then the search stops: the tested observation is nominated as an outlier, together with all the observations not included in the subset. The performance of the FS-procedure based on the "inferential" approach has been compared with that of an outlier detection method for clustering in the presence of outliers (Hardin and Rocke, 2004). The method starts from a robust clustering of the data and involves a testing procedure about the outlyingness of the data, which exploits a distributional result for squared Mahalanobis distances based on minimum covariance determinant estimates of location and shape parameters. The comparison has been carried out on a simulation experiment reported in Hardin and Rocke's paper, with $N = 100$ independent replicates. In d=4 dimensions, two groups of 300 observations each are simulated from $N(\mathbf{0}, I)$ and $N(2c\mathbf{1}, I)$, respectively, where $c = \sqrt{\chi^2_{d;0.99}/d}$ and $\mathbf{1}$ is a vector of d ones. Sixty outliers stemming from $N(4c\mathbf{1}, I)$ are planted to each dataset, thus placing the cluster of outliers at the same distance the clean clusters are separated. By separating two clusters of standard normal data at a distance of $2c$, we have clusters that do not overlap with high probability. The following measures of performance have been used:

$$A = \frac{\sum_{j=1}^{N} Out_j}{N n_{out}}, \quad B = \frac{\sum_{j=1}^{N} TrueOut_j}{N n_{out}}, \tag{9}$$

where n_{out}=60 is the number of planted outliers and Out_j ($TrueOut_j$) is the number of observations (planted outliers) declared as outliers in the j-th replicate. Perfect performance occurs when $A = B = 1$.

Table 1. Results of the simulation experiment. In both the compared procedures $\alpha = 0.01$. The first row is taken from Hardin and Rocke's paper.

Technique	Measures of performance	
	$(A-1) \cdot 100$	$(B-1) \cdot 100$
Hardin and Rocke	4.03	-0.17
FS-based	0.01	-0.05

In Table 1 the measures of performance are given in terms of distance from 1. Both the methods identify all the planted outliers in nearly all replicates. However, Hardin and Rocke's technique seems to have some tendency in identifying a non-planted observation as an outlier. The FS-based method performs generally better, probably because it exploits the normality assumption on the components of the parental mixture density, by means of the typicality measure $a(\cdot)$. It is expected to be preferable also in case of highly overlapping mixture components, since Hardin and Rocke's algorithm may fail for clusters with significant overlap - as the Authors themselves point out.

5 Concluding remarks and open issues

One critical aspect of the proposed procedure (and of any FS method, indeed) is the choice of the size m_0 of the initial subset: it should be relatively small so as to avoid the initial inclusion of outliers, but also large enough to make stable estimates of the mixture parameters. Moreover, McLachlan and Basford's test for outlier detection is known to have poor control over the overall significance level; we dealt with the problem by using Bonferroni bounds. The test for outlier detection from a mixture proposed by Wang *et al.* (1997) does not suffer from this drawback but requires bootstrap techniques, thus its use in the FS algorithm would increase the computational burden of the whole procedure.

FS methods are naturally computer-intensive methods. In our FS algorithm, time savings could come from using the estimation results of step m as an initial value for the EM in step $m+1$. A possible drawback of this solution is that the results of one step irreversibly influence the following ones. The problem of improving computational efficiency while preserving effectiveness deserves further attention. Finally, we assume that the number of mixture components, K, is both fixed and known. In our experience, the first assumption seems to be not crucial: when subset S_0 does not contain data from one component, say g, the first observation from g may be signalled by the forward plot, but it can't appear like an outlier since its inclusion does not occur in the final steps of the search. On the contrary, generalizing the procedure for K unknown is a rather challenging task, which we are presently working on.

References

ATKINSON, A.C. (1993): Stalactite plots and robust estimation for the detection of multivariate outliers. In: E. Ronchetti, E. Morgenthaler, and W. Stahel (Eds.): *New Directions in Statistical Data Analysis and Robustenss.*, Birkhäuser, Basel.

ATKINSON, A.C., RIANI, C. and CERIOLI A. (2004): *Exploring Multivariate Data with the Forward Search.* Springer, New York.

FRALEY, C. and RAFTERY, A.E. (1998): How may clusters? Which clustering method? Answers via model-based cluster analysis. *The Computer Journal*, 41, 578-588.

HADI, A.S. (1994): A modification of a method for the detection of outliers in multivariate samples. *J R Stat Soc, Ser B*, 56, 393-396.

HARDIN, J. and ROCKE D.M. (2004): Outlier detection in the multiple cluster setting using the minimum covariance determinant estimator. *Computational Statistics and Data Analysis*, 44, 625-638.

HENNIG, C. (2004): Breakdown point for maximum likelihood estimators of location-scale mixtures. *The Annals of Statistics*, 32, 1313-1340.

MCLACHLAN, G.J. and BASFORD K.E. (1988): *Mixture Models: Inference and Applications to Clustering.* Marcel Dekker, New York.

MCLACHLAN, G.J. and PEEL, D. (2000): *Finite Mixture Models.* Wiley, New York.

WANG S. *et al.* (1997): A new test for outlier detection from a multivariate mixture distribution, *Journal of Computational and Graphical Statistics*, 6, 285-299.

On Multiple Imputation Through Finite Gaussian Mixture Models

Marco Di Zio and Ugo Guarnera

Istituto Nazionale di Statistica,
via Cesare Balbo 16, 00184 Roma, Italy
{dizio, guarnera}@istat.it

Abstract. Multiple Imputation is a frequently used method for dealing with partial nonresponse. In this paper the use of finite Gaussian mixture models for multiple imputation in a Bayesian setting is discussed. Simulation studies are illustrated in order to show performances of the proposed method.

1 Introduction

Imputation is a common approach to deal with nonresponse in surveys. It consists in substituting missing items with plausible values. This approach has been widely used because it allows to work with a complete data set so that standard analysis can be applied. Despite of this important advantage, the introduction of imputed values is not a neutral task. In fact, imputed values are not really observed and this should be explicitly taken into account in statistical inference based on the completed data set. If standard methods are applied as if the imputed values were really observed, there would be a general overestimate of the precision of the results, resulting, for instance, in too narrow confidence intervals. Multiple imputation (Rubin, (1987)) is a methodology for dealing with this problem. It essentially consists in imputing a certain number of times the incomplete data set following specific rules. The resulting completed data set is analysed by standard methods and results are combined in order to yield estimates and assessing their precision including the additional source of variability due to nonresponse. The multiplicity of completed data sets has the role of reflecting the variability due to the imputation mechanism. Although in multiple imputation data normality is frequently assumed, this assumption does not fit all situations (e.g., multimodal distributions). Moreover, the analyst who works on the completed data set not necessarily will or must be aware of the model used for imputation. Thus, problems may arise when the models used by the analyst and by the imputer are different. Meng (1994) suggests to use a model for imputation that is reasonably accurate and general to overcome this difficulty. To this aim, an interesting work is that of Paddock (2002) who proposes a nonparametric multiple imputation technique based on Polya trees. This technique is appealing since it allows to treat continuous and ordinal data, and in some circumstances also categorical variables. However, in Paddok's paper it is shown that, even with nonnormal data, in some case the technique based on normality is still quite better. Nonnormal data can be dealt with by using finite mixtures of Gaussian distributions (GMM) since they are flexible enough to approximate a wide class of density functions with a limited number of parameters. These models can be seen as generalizations of the general location model used by Little and Rubin (2002) to model partially observed data with mixed categorical and continuous variables. Unlike in the latter case, however, in the present approach categorical variables are latent variables ('class labels' that are never observed), and their role is merely to allow better approximation of the true data distribution. The performance of GMM in a likelihood based approach for single imputation is evaluated in Di Zio et al. (2007). In this paper we discuss the use of finite mixtures of Gaussian distributions for multiple imputation in a Bayesian framework. The paper is structured as follows. Section 2 describes multiple imputation through mixture models. In Section 3, the problem of label switching is discussed. Section 4 is devoted to the description and discussion of the experiments carried out in order to assess the performance of the proposed method.

2 Multiple imputation

Multiple imputation has been proposed for both frequentist and Bayesian analyses. Nevertheless, the theoretical justification is most easily understood from the Bayesian perspective. In this setting, the ultimate goal is to fill in missing values $\mathbf{Y}_{mis}$ with values $\mathbf{y}_{mis}$ drawn from the predictive distribution that, once an appropriate prior distribution for Φ is set, can be written as

$$P(\mathbf{Y}_{mis}|\mathbf{y}_{obs}) = \int P(\mathbf{Y}_{mis}|\mathbf{y}_{obs},\Phi)P(\Phi|\mathbf{y}_{obs})d\Phi \tag{1}$$

where $\mathbf{Y}_{mis}$ are the missing values and $\mathbf{Y}_{obs}$ the observed ones. The imputation process is repeated m times, so m completed data sets are obtained. These m different data sets incorporate the uncertainty about the missing imputed values. Let us suppose that $Q(\mathbf{Y})$ is the quantity of interest, e.g., a population mean, and that an estimate $\hat{Q}(\mathbf{Y})^{(i)}$ is computed on the ith completed data set, for $i = 1,\ldots,m$. The final estimate $\hat{Q}$ is defined by $\hat{Q} = \frac{1}{m}\sum_{i=1}^{m}\hat{Q}(\mathbf{Y})^{(i)}$. The estimate $\hat{T}$ of the variance of $\hat{Q}$ can be obtained by combining a within component term $\hat{U}$ and a between component term $\hat{B}$. The former is the average of the m standard variance estimates $\hat{U}^{(i)}$ for complete data computed on the ith completed data set, for $i = 1,\ldots,m$: $\hat{U} = \frac{1}{m}\sum_{i=1}^{m}\hat{U}^{(i)}$. The between variance is the variance of the m estimates, i.e. $\hat{B} = \frac{1}{m-1}\sum_{i=1}^{m}(\hat{Q}^{(i)} - \hat{Q})^2$. Finally, the total variance of $\hat{Q}$ is estimated by $\hat{T} = \hat{U} + (1+m^{-1})\hat{B}$, and a 95% confidence interval for Q is given by $\hat{Q} \pm t_{\nu,0.975}\hat{T}^{1/2}$, where the degrees of freedom are $\nu = (m-1)\{1+[(1+m^{-1})\hat{B}]^{-1}\hat{U}\}$, (see Rubin, 1987).

Since it is often difficult to obtain a closed form for the observed posterior distribution $P(\Phi|\mathbf{y}_{obs})$, the data augmentation algorithm may be used (Tanner and Wong, 1987). This algorithm consists of iterating the two following steps:

1. I-step - draw $\tilde{\mathbf{y}}_{mis}$ from $P(\mathbf{Y}_{mis}|\mathbf{y}_{obs},\tilde{\Phi})$
2. P-step - draw $\tilde{\Phi}$ from $P(\Phi|\tilde{\mathbf{y}}_{mis},\mathbf{y}_{obs})$.

This is a Gibbs sampling algorithm and, after convergence, the resulting sequence of values $\tilde{\mathbf{y}}_{mis}$ can be thought of as generated from $P(\mathbf{Y}_{mis}|\mathbf{y}_{obs})$. Data augmentation is explicitly described by Schafer (1997) when data follow a Gaussian distribution. We study the case when data are generated from a finite mixture of K Gaussian distributions, i.e., when each observation $\mathbf{y}_i$ for $i = 1,\ldots,n$ is supposed to be a realization of a p-dimensional r.v. $\mathbf{Y}_i$ with density:

$$f(\mathbf{y}_i|\Phi) = \sum_{k=1}^{K} \pi_k N_p(\mathbf{y}_i|\theta_k), \quad \mathbf{y} \in \mathbb{R}^p$$

where $\sum_k \pi_k = 1, \pi_k \geq 0$ for $k = 1,\ldots,K$, and $N_p(\mathbf{y}_i|\theta_k)$ is the Gaussian density with parameters $\theta_k = (\mu_k, \Sigma_k)$. Note that Φ denotes the full set of parameters: $\Phi = (\pi_1,\ldots\pi_K;\theta_1,\ldots,\theta_K)$.

Mixture models have a natural missing data formulation if we suppose that each observation $\mathbf{y}_i$ comes from a specific but unknown component k of the mixture, and introduce, for each unit i, an indicator or allocation variable Z_i, taking values in $\{1,\ldots,K\}$, with $z_i = k$ if individual i belongs to group k. The discrete variables Z_i are independently distributed according to $P(Z_i = k|\Phi) = \pi_k$, $(i = 1,\ldots,n; k = 1,\ldots,K)$. Furthermore, conditional on $Z_i = k$, the observations $\mathbf{y}_i$ are supposed to be *i.i.d.* from the density $N_p(\mathbf{y}_i|\theta_k)$. Thus, if some items are missing for the ith unit, the relevant distribution, conditional on $Z_i = k$, is $P(\mathbf{Y}_{mis}|\mathbf{y}_{obs},\theta_k)$, while the classification probabilities, expressed in terms of $\mathbf{y}_{i,obs}$, are:

$$\tau_{gi} = P(Z_i = g|\mathbf{y}_{i,obs},\Phi) = \frac{\pi_g N_p(\mathbf{y}_{i,obs}|\theta_g)}{\sum_{k=1}^{K} \pi_k N_p(\mathbf{y}_{i,obs}|\theta_k)}, \quad g = 1,\ldots,K \tag{2}$$

where $N_p(\mathbf{y}_{i,obs}|\theta_g)$ is the Gaussian marginal distribution of the gth mixture component of the variables observed in the ith unit.

The previous formulation leads to a data augmentation algorithm consisting, at the tth iteration, of the following two steps:

- I-step: for $i = 1,\ldots,n$
  - draw a random value of the allocation variable $z_i^{(t)}$ from the distribution $P(Z_i|\mathbf{y}_{i,obs},\Phi^{(t-1)})$, i.e., select a value in $\{1,\ldots,K\}$ using the probabilities $\tau_{1i},\ldots,\tau_{Ki}$ defined in formula (2) expressed in terms of the current value of vector $\Phi^{(t-1)}$;
  - draw $\mathbf{y}_{i,mis}^{(t)}$ (the missing part of the ith vector $\mathbf{y}_i^{(t)}$) from $P(\mathbf{y}_{i,mis}|z_i^{(t)},\mathbf{y}_{i,obs},\Phi^{(t)})$.
- P-step:
  draw $\Phi^{(t)}$ from the distribution $P(\Phi|\mathbf{y}_{obs},\mathbf{y}_{mis}^{(t)})$.

The above scheme produces a sequence $(z^{(t)}, \mathbf{y}_{mis}^{(t)}, \Phi^{(t)})$ which is a Markov chain with stationary distribution $P(Z, \mathbf{Y}_{mis}, \Phi|\mathbf{y}_{obs})$. The convergence properties of the algorithm have been studied by Diebolt and Robert (1994) in the case of completely observed data.

The choice of an appropriate prior is a critical issue in Gaussian mixture models. For instance, *reference priors* lead to improper priors for the specific component parameters that are independent across the mixture components. This situation is problematic insofar posterior distributions remain improper for configurations where no units are assigned to some components. In this paper we follow a hierarchical Bayesian approach, based on weakly informative priors, as introduced by Richardson and Green (1997) for univariate mixtures, and generalized to the multivariate case by Stephen (2000). In this approach it is assumed that the prior distribution for μ_k is rather flat over an interval of variation of the data. The hierarchical structure of the prior distributions for a K-component p-variate Gaussian mixture is given by:

$$\begin{aligned} \mu_k &\sim N(\xi, \Psi^{-1}) \\ \Sigma_k^{-1}|\beta &\sim W(2\alpha, (2\beta)^{-1}) \\ \beta &\sim W(2\delta, (2h)^{-1}) \\ \pi &\sim D(\gamma), \end{aligned}$$

where W and D denote the Wishart and Dirichlet distributions respectively, and the hyperparameters $\xi, \Psi, \alpha, \delta, h, \gamma$, are constants defined below. Let R_j be the length of the observed interval of variation (range) of the obtained valu s for the variable Y_j, and ξ_j the corresponding midpoint ($j = 1, \ldots, p$). Then, ξ is the p-vector: $(\xi_1, \ldots, \xi_p)$, while the matrix Ψ is the diagonal matrix whose element ψ_{jj} is R_j^{-2}. The other hyperparameters are specified as follows:

$$\alpha = p+1, \;\; \delta = \alpha/10, \;\; h = 10\Psi, \;\; \gamma = (1, \ldots, 1).$$

The P-step described in general above in this section, with $\Phi^{(t)} = (\beta^{(t)}, \pi_1^{(t)}, \ldots, \pi_K^{(t)}; \mu_1^{(t)}, \ldots, \mu_K^{(t)}; \Sigma_1^{(t)}, \ldots, \Sigma_K^{(t)})$ can be implemented by sampling from the appropriate posterior distributions as follows:

$$\begin{aligned} \beta^{(t+1)}|\cdots &\sim W\left(2\delta + 2g\alpha, (2h + 2\sum_{k=1}^{K} \Sigma_k^{(t)^{-1}})^{-1}\right), \\ \pi^{(t+1)}|\cdots &\sim D(\gamma + n_1, \ldots, \gamma + n_K), \\ \mu_k^{(t+1)}|\cdots &\sim N\left((n_k \Sigma_k^{(t)^{-1}} + \Psi)^{-1}(n_k \Sigma_k^{(t)^{-1}} \bar{\mathbf{y}}_k + \Psi\xi), (n_k \Sigma_k^{(t)^{-1}} + \Psi)^{-1}\right), \\ \Sigma_k^{-1(t+1)}|\cdots &\sim W\left(2\alpha + n_k, (2\beta^{(t+1)} + \sum_{i:z_i=k} (\mathbf{y}_i - \mu_k^{(t+1)})(\mathbf{y}_i - \mu_k^{(t+1)})')^{-1}\right), \end{aligned}$$

where $|\cdots$ denotes conditioning on all other variables. In the previous formulas n_k denotes the number of units assigned to the kth mixture component at the tth step, and $\bar{\mathbf{y}}_k$ is the mean: $\sum_{i:z_i=k} \mathbf{y}_i / n_k$.

3 Label switching

Label switching is a typical problem in Bayesian estimation of finite mixture models (Stephens, (2000)). When using symmetric priors (i.e., invariant with respect to permutations of the components), the posterior distributions are still symmetric and thus the marginal posterior distributions for the parameters will be identical for all the mixture components. Inference based on MCMC is meaningless, because it results in averaging over different mixture components. Nevertheless, this problem does not affect inference on parameters that are independent of label components. For instance, if the parameter to be estimated is the population mean, as often required in official statistics, the target quantity is independent of the component labels. Moreover, in multiple imputation, the estimate is computed on the observed and imputed values, and the imputed values are drawn from $P(Y_{mis}|\mathbf{y}_{obs})$ that is invariant with respect to permutation of component labels. As an illustrative example, we have drawn 200 random samples from the two-component mixture $f(y) = 0.5N(1.3, 0.1) + 0.5N(2, 0.15)$ in $\mathbb{R}^1$, and nonresponse is artificially introduced with a 20% missing rate. This dataset is multiply imputed according to the algorithm previously described. In Figure 1 the trace plot of the component means obtained via data augmentation, and of the sample mean that is used to produce multiple imputation estimates are shown (5000 iterations). In the figure, the component means of the generating mixture distribution (dashed lines) are also reported. Moreover vertical lines, corresponding to label switching, are depicted. It is worth to note that the label switching of the component means does not affect the target estimate that in fact is stable.

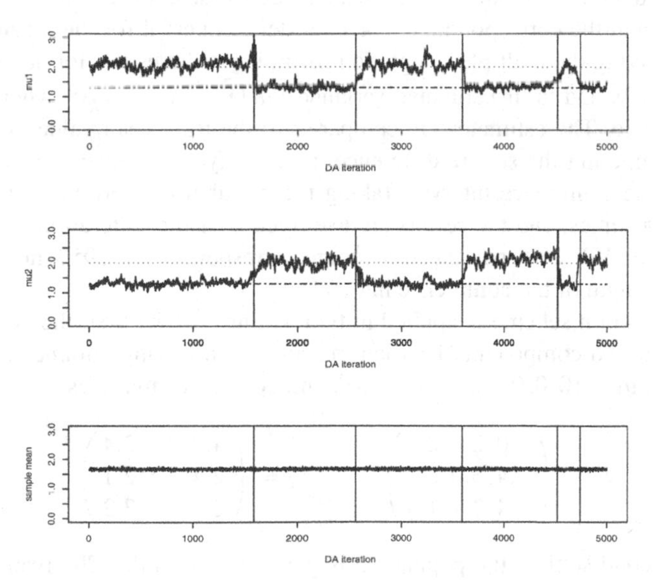

Fig. 1. Trace plot of the two-component means and the sample means computed through the data augmentation algorithm.

4 Simulation study and results

We present a simulation study to assess the performance of Bayesian GMM for multiple imputation. In order to mimic the situtation in official statistics, a sample of $N = 50000$ units (representing the finite population) with three variables (Y_1, Y_2, Y_3) is drawn from a probability model. The target parameter is the mean of the variables in the finite population. A random sample u without replacement of $n = 1000$ units is drawn from the reference population. This sample is corrupted by the introduction of missing values according to a Missing at Random mechanism (MAR). Missing items are introduced for the variables (Y_2, Y_3) depending on the observed values y_1 of the variable Y_1 under the assumption that the higher the value of Y_1 the higher is the nonresponse propensity. Denoting by q_i the ith quartile of the empirical distribution of Y_1, the nonresponse probabilities for (Y_2, Y_3) are 0.1 if $y_1 < q_1$, 0.2 if $y_1 \in [q_1, q_2)$, 0.4 if $y_1 \in [q_2, q_3)$ and 0.5 if $y_1 \geq q_3$.

The sample u is multiply imputed (m=5) via GMM. Data augmentation algorithm is initialized by using maximum likelihood estimates (MLE) obtained through the EM algorithm as described in Di Zio et al. (2007). After a burn-in period of 500 iterations, multiple imputation is performed by subsampling the chain every t iterations, that is, the $\mathbf{Y}_{mis}$ used for imputation are those referring to the iterations $(t, 2t, \ldots, 5t)$. Subsampling is used to avoid dependent samples, as suggested by Schafer (1997). Although the burn-in period may appear to be not very long, as again suggested by Schafer (1997), the initialization of the algorithm with a good starting point (e.g., through MLE) may speed up the convergence of the chain. This is also confirmed by analysing the trace plot of the parameters.

Once the data set is imputed, for each analysed variable, the estimate of the mean, its variance, and the corresponding 95% confidence interval for the mean are computed by applying the multiple imputation formulas to the usual Horvitz-Thompson estimator $\tilde{Y} = \bar{y}$, and to its estimated variance $\widetilde{Var}(\tilde{Y}) = (\frac{1}{n} - \frac{1}{N})s^2$, where s^2 is the sample variance. The estimates are compared to the true mean value of the population by computing the square difference, and verifying whether the true value is included in the confidence interval. Taking the population fixed, the experiment is repeated 1000 times, and the results are averaged over these iterations. The results give simulated MSE, bias, simulated coverage corresponding to a 95% nominal level, and average length of the confidence intervals.

This simulation scheme is applied in two settings. In the first, the population is drawn from a two-component Gaussian mixture, with mixing parameter $\pi = 0.75$, mean vectors $\mu_1 = (0,0,0)'$, $\mu_2 = (3,5,8)'$, and covariance matrices

$$\Sigma_1 = \begin{pmatrix} 3.0 & 2.4 & 2.4 \\ 2.4 & 3.0 & 2.1 \\ 2.4 & 2.1 & 1.3 \end{pmatrix}, \qquad \Sigma_2 = \begin{pmatrix} 4.0 & 2.4 & 2.4 \\ 2.4 & 3.5 & 2.1 \\ 2.4 & 2.1 & 3.2 \end{pmatrix}.$$

In the second setting, the population is generated from the Cheriyan and Rambhadran's multivariate Gamma distribution described in Kotz et al. (2000) pp. 454-456. In order to draw a sample of a 3-variate random vector (Y_1, Y_2, Y_3) from such a distribution the following procedure is adopted. First, we consider 4 independent random variables X_i in $\mathbb{R}^1$ for $i = 0, 1, 2, 3$ that are distributed according to Gamma distributions characterised by different parameters θ_i. Then, the 3-variate random vector is obtained combining the X_i so that $Y_i = X_0 + X_i$ for $i = 1, 2, 3$. The values of the parameters are $\theta = (1, 0.2, 0.2, 0.4)'$.

In the two-component Gaussian mixture population, multiple imputation is carried out according to a plain normal model (hereafter *NM*) and a mixture of two Gaussian components (M_2). The results for the variable Y_3 are illustrated in Table 1. For the Gamma population, multiple imputation is performed by using the plain normal model (*NM*) and a K-component mixture M_K for $K = 2, 3, 4$. Results for the variable Y_3 are provided in Table 2.

Table 1. Results of the experiment where population is based on a two-component Gaussian mixture

Mod	bias	MSE	S.Cov	Length
NM	-0.0144	0.1323	93.7%	0.5000
M_2	0.0014	0.1316	94.9%	0.5163

Table 2. Results of the experiment where population is based on Multivariate Gamma

Mod	bias	MSE	S.Cov	Length
NM	0.0015	0.0431	93.8%	0.1604
M_2	0.0052	0.0437	94.0%	0.1661
M_3	0.0043	0.0435	94.0%	0.1651
M_4	0.0059	0.0442	94.1%	0.1655

Results show that confidence intervals are close to the nominal coverage. In particular, in the first experiment, the confidence interval computed by the mixture models is better than that computed through a Gaussian distribution. The improvement is due to the fact that the model used for estimation is correctly specified. This suggests the need of improving estimation of unknown distribution by means of mixture models. To this aim it could be an important step to consider the number of mixture components as a random variable, thus incorporating the model uncertainty in the estimation phase.

References

DIEBOLT, J. and ROBERT, C.P. (1994): Estimation of finite mixture distributions through Bayesian sampling. *Journal of the Royal Statistical Society B, 56, 363–375.*

DI ZIO, M., GUARNERA, U. and LUZI, O. (2007): Imputation through finite Gaussian mixture models. *Computational Statistics and Data Analysis, 51, 5305–5316.*

KOTZ, S., BALAKRISHNAN, N. and JOHNSON, N.L. (2000): *Continuous multivariate distributions.* Vol.1, 2nd ed. Wiley, New York.

LITTLE, R.J.A. and RUBIN, D.B. (2002): *Statistical analysis with missing data.* Wiley, New York.

MENG, X.L. (1994): Multiple-imputation inferences with uncongenial sources of input (with discussion). *Statistical Science, 9, 538–558.*

PADDOCK, S.M. (2002): Bayesian nonparametric multiple imputation of partially observed data with ignorable nonresponse. *Biometrika, 89, 529–538.*

RICHARDSON, S. and GREEN, P.J. (1997): On Bayesian analysis of mixtures with an unknown number of components.*Journal of the Royal Statistical Society B, 59, 731–792.*

RUBIN, D.B. (1987): *Multiple imputation for nonresponse in surveys.* Wiley, New York.

SCHAFER, J.L. (1997): *Analysis of incomplete multivariate data.* Chapman & Hall, London.

STEPHENS, M. (2000): Bayesian analysis of mixture models with an unknown number of components-an alternative to reversible jump methods. *Annals of Statistics, 28, 40–74.*

TANNER, M.A. and WONG, W.H. (1987): The calculation of posterior distribution by data augmentation (with discussion). *Journal of the American Statistical Association, 82, 528–550.*

Mixture Model Based Group Inference in Fused Genotype and Phenotype Data

Benjamin Georgi[1], M.Anne Spence[2], Pamela Flodman[2], Alexander Schliep[1]

[1] Max-Planck-Institute for Molecular Genetics, Department of Computational Molecular Biology, Ihnestrasse 73, 14195 Berlin, Germany
[2] University of California, Irvine, Pediatrics Department, 307 Sprague Hall, Irvine, CA 92697, USA

Abstract. The analysis of genetic diseases has classically been directed towards establishing direct links between cause, a genetic variation, and effect, the observable deviation of phenotype. For complex diseases which are caused by multiple factors and which show a wide spread of variations in the phenotypes this is unlikely to succeed. One example is the Attention Deficit Hyperactivity Disorder (ADHD), where it is expected that phenotypic variations will be caused by the overlapping effects of several distinct genetic mechanisms. The classical statistical models to cope with overlapping subgroups are mixture models, essentially convex combinations of density functions, which allow inference of descriptive models from data as well as the deduction of groups. An extension of conventional mixtures with attractive properties for clustering is the context-specific independence (CSI) framework. CSI allows for an automatic adaption of model complexity to avoid overfitting and yields a highly descriptive model.

1 Introduction

The attention deficit hyperactivity disorder (ADHD) is diagnosed in 3% – 5% of all children in the US and is considered to be the most common neurobehavioral disorder in children. Today ADHD is known to be influenced by a multitude of factors such as genetic disposition, neurological properties and environmental conditions (Swanson et al. (2000a), Woodruff et al. (2004)). The phenotypes usually associated with ADHD fall into the general categories inattentiveness, hyperactivity and impulsivity. This is only a partial list of symptoms associated with ADHD and it is noteworthy that most patients will only show some of these behaviors, with differing degrees. This wide spread of observable symptoms associated with ADHD supports the notion that possible ADHD subtypes will have complex characteristics and may contain overlaps. Since ADHD has a complex non-mendelian mode of inheritance a partition of phenotypes into clearly separated groups cannot be expected. Rather some phenotypic variations will be caused by several distinct genetic mechanisms. The neurotransmitter dopamine and the genes involved in dopamine function are known to be relevant to ADHD (Gill et al. (1997)). According to the prevalent theory (Cook et al. (1995)), the contribution of the dopamine metabolism to ADHD is based on over-activity of dopamine transporters in the pre-synaptic membrane which leads to reduced dopamine concentrations in the synaptic gap. There have been studies linking the disposition towards ADHD with the genotypes of a *variable number of tandem repeats* (VNTR) region on the third exon of the dopamine receptor gene DRD4 (Swanson et al. (2000b)). Considering all this, it seems promising to explore the influences of different dopamine receptor haplotypes on ADHD related phenotypes and the sub group decompositions implicit in these relationships. For complex genetic diseases such as ADHD for which the degree of diagnostic uncertainty with respect to presence of the disease and determination of the disease subtype is large, the search for simple, direct causalities between different factors is likely to fail (Luft (2000)). Rather one would expect to find correlations in the form of changes in *disposition* for a specific disease feature. When attempting to cluster data from such a complex disease, it is important that the clustering method can accommodate this kind of uncertainty. The classical statistical approach in this situation is mixture modelling. An extension of the conventional mixture framework are the context-specific independence (CSI) mixture models (Barash and Friedman (2002), Georgi and Schliep (2006)). In a CSI model the number of parameters used, i.e. the model complexity, is automatically adapted to match the level of variability present in the data.

In this paper we present a CSI mixture model-based clustering of a data set of ADHD patients that consists of both genotypic and phenotypic features. The data set includes 134 samples with 91 genotypic variables and 27 phenotypic variables each. The genotype variables contain *variable number of tandem repeats* (VNTR) information on the DRD4 gene as well as *Single Nucleotide Polymorphism* (SNP) data on four dopamine receptor (DRD1-DRD3,DRD5) and one dopamine transporter (DAT1) genes. The DRD family proteins are G-protein coupled dopamine receptors located in the plasma membrane. DAT1 encodes for a dopamine transporter located in the presynaptic membrane. The phenotypes are represented by two IQ and three achievement test scores, as well as 21 diagnoses for various comorbid behavioral disorders.

2 Methods

Let $X_1,...,X_p$ be discrete random variables. Given a data set D of N realizations, $D = x_1,...,x_N$ with $x_i = (x_{i1},...,x_{ip})$ a conventional mixture density (see McLachlan and D. Peel (2000) for details) is given by:

$$P(x_i) = \sum_{k=1}^{K} \pi_k \ f_k(x_i;\theta_k), \tag{1}$$

where the π_k are non-negative the mixture coefficients, $\sum_{k=1}^{K} \pi_k = 1$ and each component distribution

$$f_k(x_i;\theta_k) = \prod_{j=1}^{p} f_{kj}(x_{ij};\theta_{kj}) \tag{2}$$

is a product distribution over $X_1,...,X_p$ parameterized by parameters $\theta_k = (\theta_{k1},...,\theta_{kp})$. In other words, we assume conditional independence between features within the mixture components and adopt the *Naive Bayes* model as component distributions. All component distribution parameters are denoted by $\theta_M = (\theta_1,...,\theta_K)$ Finally, the complete parameterizations of the mixture M is then given by $M = (\pi,\theta_M)$. The likelihood of data set D under the mixture M is given by

$$P(D|M) = \prod_{i=1}^{N} P(x_i). \tag{3}$$

that is we have the usual assumption of independence between samples.

The standard technique for learning the parameters Θ is the *Expectation Maximization* (EM) algorithm (Dempster et al. (1977)). The central quantity for the EM based parameter estimation is the posterior of component membership given by

$$\tau_{ik} = \frac{\pi_k\ f_k(x_i;\theta_k)}{\sum_{k=1}^{K} \pi_k\ f_k(x_i;\theta_k)}, \tag{4}$$

i.e. τ_{ik} is the probability that a sample x_i was generated by component k. Moreover, this posterior is used for assigning samples to clusters (i.e. components). This is done by assigning a sample to the component with maximal posterior.

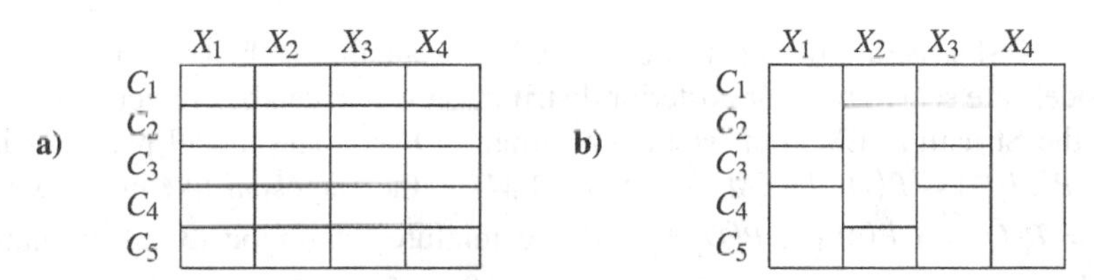

Fig. 1. Model structure matrices for a) conventional mixture model and b) CSI mixture model

The conventional mixture model defined above requires the estimation of one set of distribution parameters θ_{kj} per feature and distribution. This is visualized in the matrix in Fig. 1 a). This example shows a model with five components and four features. Each cell in the matrix represents one θ_{kj}. The central idea of the *context-specific independence* extension of the mixture framework is that for many data sets it will not be necessary to estimate separate parameters in each feature for all components. Rather one should learn only as many parameters as is justified by the variability found in the data. This leads to the kind of matrix shown in Fig. 1 b). Here each cell spanning multiple rows represents a single set of parameters for multiple components. For instance, for feature X_1, C_1 and C_2 share the same parameters, for feature X_2, $C_2 - C_4$ have the same parameters and for X_4 all components share a single set of parameters. This modification of the conventional mixture framework has a number of attractive properties: The model complexity is reduced as there are less free parameters to estimate. Also, if a feature has only a single set of parameters assigned for all components, (such as X_4 in Fig. 1 b), its contribution in (4) will cancel out and it will not affect the clustering. This amounts to a feature selection in which the impact of noisy features is negated as an integral part of model training. Hence, we can expect a more robust clustering in which the risk of overfitting is greatly reduced. Finally, the model structure matrix yields a highly descriptive model which facilitates the analysis of a clustering. For instance, the matrix 1 b) shows that clusters C_4 and C_5 are only distinguished by feature X_2.

Formally the CSI mixture model is defined as follows: Given the set of component indexes $C = \{1,..,K\}$ and features $X_1,...,X_p$ let $G = \{g_j\}_{(j=1,...,p)}$ be the CSI structure of the model M. Then $g_j = (g_{j1},...g_{jZ_j})$ such that Z_j is the number of subgroups for X_j and each $g_{jr}, r = 1,...,Z_j$ is a subset of component indexes from C. That means, each g_j is a partition of C into disjunct subsets where each g_{jr} represents a subgroup of components with the same distribution for X_j. The CSI mixture distribution is then obtained by replacing $f_{kj}(x_{ij};\theta_{kj})$ with $f_{kj}(x_{ij};\theta_{g_j(k)j})$ in (2) where $g_j(k) = r$ such that $k \in g_{jr}$. Accordingly $\theta_M = (\pi, \theta_{X_1|g_{1r}}, ..., \theta_{X_p|g_{pr}})$ is the model parametrization. Where $\theta_{X_j|g_{jr}}$ denotes the different parameter sets in the structure for feature j. The complete CSI model M is then given by $M = (G, \theta_M)$. Note that we have covered the CSI mixture model and the structure learning algorithm in some more detail in a previous publication (Georgi and Schliep (2006)).

2.1 Structure Learning

To learn the CSI structure from data we took a Bayesian approach. That means different models are scored by their posterior distribution which can be efficiently computed in the Structural EM framework (Friedman (1998)). The model posterior is given by $P(M|D) \propto P(D|M)P(M)$ where $P(D|M)$ is the Bayesian likelihood with $P(D|M) = P(D|\overrightarrow{\theta}_M)P(\overrightarrow{\theta}_M)$. $P(D|\overrightarrow{\theta}_M)$ is the mixture likelihood (3) of the data evaluated at the *maximum aposterior paramters* $\overrightarrow{\theta}_M$. $P(\overrightarrow{\theta}_M)$ is a conjugate prior over the model parameters. Due to the independence assumptions $P(\overrightarrow{\theta}_M)$ decomposes into a product distribution of conjugate priors over π and the individual $\theta_{X_j|g_{j}{jr}}$. For discrete distributions the Dirichlet distribution and for Gaussians a Normal Inverse-Gamma prior was used. The second term needed to evaluate the model posterior is the prior over the model structure P(M) which is given by $P(M) \propto P(K)P(G)$ with $P(K) \propto \gamma^K$ and $P(G) \propto \prod_{j=1}^{p} \alpha^{Z_j}$. $\gamma < 1$ and $\alpha < 1$ are hyper parameter which act as a regularization of the structure learning by introducing a bias towards less complex models into the posterior. Here, α and γ were chosen as weak priors by the heuristic introduced in (Georgi and Schliep (2006)) with a $\delta = 0.05$. Since exhaustive evaluation of all possible structures is infeasible, the structure learning is carried out by a straightforward greedy procedure starting from the full structure matrix (again refer to (Georgi and Schliep (2006)) for details).

3 Results

We applied the CSI mixture model based clustering to the genotype and phenotype data separately, as well as to the fused data set. For each data set we trained models with 1 to 10 components and model selection was performed using the *Normalized Entropy Criterion* (NEC) (C. Biernacki (1999)).

3.1 Genotype clustering

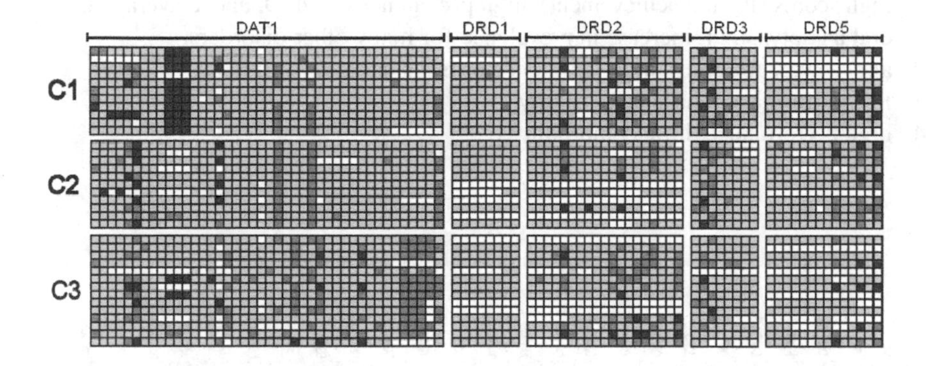

Fig. 2. *VG2* plot (http://pga.gs.washington.edu/VG2.html) of three clusters out of the 7 component genotype clustering. The color code is as follows: rare homozygous is shown in dark grey , heterozygous in medium grey, common homozygous in light gray and missing values in white. It can be seen that the clustering captures strong and distinctive patterns within the genotypes. The plot for the full clustering can be obtained from *http://algorithmics.molgen.mpg.de/pymix/genoclust.html.*

The model selection on the genotype data set indicated 7 components to be optimal. Three example clusters out of this clustering of the genotypes are visualized in Fig. 2. The plot for the full clustering is available from our homepage at *http://algorithmics.molgen.mpg.de/pymix/genoclust.html.* In the figure the rare homozygous alleles are shown in dark grey, the heterozygous alleles are shown in medium grey, the common homozygous alleles in light grey and missing values in white. It can be seen that the clustering recovered strong and distinctive patterns within the genotypes data. When contrasting the clustering with the *linkage disequilibrium* (LD) found between the loci in the data set one can see a strong agreement between high LD loci and loci which are informative for cluster discrimination according to the CSI structure. An interesting observation was that out of the 92 features 71 were found to be uninformative in the CSI structure. In other words only features that carried strong discriminative information with respect for the clustering were influencing the result.

3.2 Phenotype clustering

For the phenotype data the NEC model selection indicated two and four component to be good choices, with the score for two being slightly better. The clusters for the two component model could readily be identified as a high performance and a low performance cluster with respect to the IQ (BD, VOC) and achievement (READING, MATH, SPELLING) features. In fact, the diagnosis features did not contribute strongly to the clustering and most were selected to be uninformative in the CSI structure. When considering the four component clustering a more interesting picture arose. The distinctive features of the four clusters can be summarized as

1. high scores (IQ and achievement), high prevalence of ODD, above average general anxiety, slight increase in prevalence for many other disorders,
2. above average scores, high prevalence of transient and chronic tics,
3. low performance, little comorbidity,
4. high performance, little comorbidity.

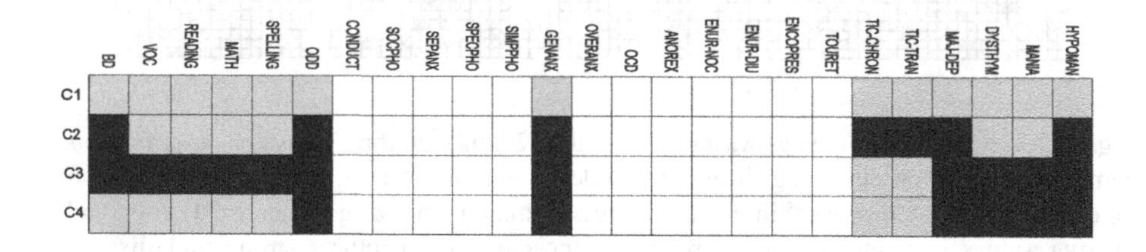

Fig. 3. CSI structure matrix for the four component phenotype clustering. Identical colors within each column denote shared use of parameters. Uninformative features are depicted in white.

The CSI structure matrix for this clustering is shown in Fig. 3. Identical colors within each column of the matrix denote a shared set of parameters. For instance one can see that cluster 1 has a unique set of parameters for the feature *Oppositional Defiancy Disorder* (ODD) and general anxiety (GENANX) while the other clusters share parameters. This indicates that these two features are distinguishing the cluster from the rest of the data set. The same is true for the transient (TIC-TRAN) and chronic tics (TIC-CHRON) features in cluster 2. Moreover one can immediately see that cluster 3 is characterized by distinct parameters for the IQ and achievement features. Finally, one can also consider which features are discriminating different clusters. For instance clusters 3 and 4 share parameters for all features but the IQ and achievement features.

3.3 Joined clustering

The NEC model selection for the fused data set yielded two clusters to be optimal with four being second best. The analysis of the clustering showed that the a small number of genotype features dominated the clustering and that in particular all the phenotype features were selected to be uninformative. Moreover one could observe that the genotype patterns found were more noisy and less distinctive within clusters. From these observations we conclude that phenotypes covered in the data set do not carry meaningful information about the genotypes and vice versa.

4 Discussion

The clustering of geno- and phenotype data separately yielded interesting partitions of the data. For the former the clustering captured strong patterns of LD within the clusters. For the latter we found sub groups of differing levels of IQ and achievement as well as differing degrees of comorbidity. For the fused data set the analysis revealed that there were no strong correlations between the two sources of data. While a positive result in this aspect would have been more interesting, the analysis was exploratory in nature. In particular, while the dopamine pathway is known to be relevant for ADHD, there was no guarantee that the specific genotypes in the data would account for any of the represented phenotypes. As for the CSI mixture method, we showed that it is well suited for the analysis of complex biological data sets. The interpretation of the CSI matrix as a high level overview of the discriminative information of each feature allows for an effortless assessment which features are of relevance to specifically characterize a cluster. This greatly facilitates the analysis of a clustering result for data sets with a large number of features.

5 Acknowledgements

We would like to thank Robert Moyzis and James Swanson (both UC Irvine) for making available the genotype and phenotype data respectively and the German Academic Exchange Service (DAAD) and Martin Vingron for providing funding for this work.

References

Y. BARASH and N. FRIEDMAN (2002): Context-specific Bayesian clustering for gene expression data. *J Comput Biol, 9,169–91.*

C. BIERNACKI, G. CELEUX and G. GOVAERT (1999): An improvement of the NEC criterion for assessing the number of clusters in a mixture model. *Non-Linear Anal., 20,267–272.*

E. H. Jr. COOK, M. A. STEIN, M. D. KRASOWSKI, N. J. COX, D. M. OLKON, J. E. KIEFFER and B. L. LEVENTHAL (1995): Association of attention-deficit disorder and the dopamine transporter gene. *Am. J. Hum. Genet., 56,993–998.*

A. DEMPSTER and N. LAIRD and D. RUBIN. (1977): Maximum likelihood from incomplete data via the EM algorithm. *Journal of the Royal Statistical Society, Series B, 1–38.*

N. FRIEDMAN (1998): The Bayesian Structural EM Algorithm. *Proceedings of the Fourteenth Conference on Uncertainty in Artificial Intelligence,129–138.*

B. GEORGI and A. SCHLIEP (2006): Context-specific Independence Mixture Modeling for Positional Weight Matrices. *Bioinformatics, 22, 166–73.*

M. GILL, G. DALY, S. HERON, Z. HAWI and M. FITGERALD (1997): Confirmation of association between attention deficit hyperactivity disorder and a dopamine transporter polymorphism. *Molec. Psychiat, 2, 311–313.*

F. C. LUFT (2000): Can complex genetic diseases be solved ? *J Mol Med, 78, 469–71.*

G.J. MCLACHLAN and D. PEEL (2000): *Finite Mixture Models*. John Wiley & Sons

J. SWANSON, J. OOSTERLAAN, M. MURIAS, S. SCHUCK, P. FLODMAN, M. A. SPENCE, M. WASDELL,Y. DING, H. C. CHI, M. SMITH, M. MANN, C. CARLSON, J. L. KENNEDY, J. A. SERGEANT, P. LEUNG, Y. P. ZHANG,A. SADEH, C. CHEN, C. K. WHALEN, K. A. BABB, R. MOYZIS and M. I. POSNER (2000b): Attention deficit/hyperactivity disorder children with a 7-repeat allele of the dopamine receptor D4 gene have extreme behavior but normal performance on critical neuropsychological tests of attention. *Proc Natl Acad Sci U S A, 97,4754–4759.*

J. SWANSON, P. FLODMAN, J. L. KENNEDY, M. A. SPENCE, R. MOYZIS, S. SCHUCK, M. MURIAS, J. MORIARITY, C. BARR, M. SMITH and M. POSNER (2000a): Dopamine genes and ADHD. *Neurosci Biobehav Rev, 24, 21–25.*

T. J. WOODRUFF, D. A. AXELRAD, A. D. KYLE, O. NWEKE, G. G. MILLER and B. J. HURLEY (2004): Trends in environmentally related childhood illnesses. *Pediatrics, 113, 1133–40.*

The Noise Component in Model-based Cluster Analysis

Christian Hennig[1] and Pietro Coretto[2]

[1] Department of Statistical Science, University College London,
Gower St, London WC1E 6BT, United Kingdom
chrish@stats.ucl.ac.uk

[2] Dipartimento di Scienze Economiche e Statistiche Universita degli Studi di Salerno
84084 Fisciano - SA - Italy
pcoretto@unisa.it

Abstract. The so-called noise-component has been introduced by Banfield and Raftery (1993) to improve the robustness of cluster analysis based on the normal mixture model. The idea is to add a uniform distribution over the convex hull of the data as an additional mixture component. While this yields good results in many practical applications, there are some problems with the original proposal: 1) As shown by Hennig (2004), the method is not breakdown-robust. 2) The original approach doesn't define a proper ML estimator, and doesn't have satisfactory asymptotic properties.

We discuss two alternatives. The first one consists of replacing the uniform distribution by a fixed constant, modelling an improper uniform distribution that doesn't depend on the data. This can be proven to be more robust, though the choice of the involved tuning constant is tricky. The second alternative is to approximate the ML-estimator of a mixture of normals with a uniform distribution more precisely than it is done by the "convex hull" approach. The approaches are compared by simulations and for a real data example.

1 Introduction

Maximum Likelihood (ML)-estimation of a mixture of normal distributions is a widely used technique for cluster analysis (see, e.g., Fraley and Raftery (1998)). Banfield and Raftery (1993) introduced the term "model-based cluster analysis" for such methods.

In the present paper we are concerned with an idea for improving the robustness of these estimators against outliers and points not belonging to any cluster. For the sake of simplicity, we only deal with one-dimensional data here, but the theoretical results carry over easily to multivariate models. See Section 6 for a discussion of computational issues in the multivariate case.

Observations $x_1, \ldots, x_n$ are modelled as i.i.d. according to the density

$$f_\eta(x) = \sum_{j=1}^{s} \pi_j \varphi_{a_j,\sigma_j^2}(x), \tag{1}$$

where $\eta = (s, a_1, \ldots, a_s, \sigma_1, \ldots, \sigma_s, \pi_1, \ldots, \pi_s)$ is the parameter vector, the number of components $s \in \boldsymbol{N}$ may be known or unknown, (a_j, σ_j) pairwise distinct, $a_j \in \boldsymbol{R}$, $\sigma_j > 0$, $\pi_j > 0$, $j = 1, \ldots, s$, $\sum_{j=1}^{s} \pi_j = 1$ and φ_{a,σ^2} is the density of the normal distribution with mean a and variance σ^2. Estimators of the parameters are denoted by hats.

There is a problem with the ML-estimation of η. If $\hat{a}_j = x_i$ for some i, a mixture component j and $\hat{\sigma}_j \to 0$, the likelihood converges to infinity and the ML-estimator is not properly defined. This has to be prevented by a restriction. $\sigma_j \geq c_0 > 0\ \forall j$ for a given c_0 or

$$\frac{\sigma_i}{\sigma_j} \geq c_0 > 0,\ i, j = 1, \ldots, s, \tag{2}$$

ensure a well-defined ML-estimator (up to label switching of the components). In the present paper we use (2), see Hathaway (1985) for theoretical background.

Having estimated the parameter vector η by ML for given s, the points can be classified by assigning them to the mixture component for which the estimated a posteriori probability p_{ij} that x_i has been generated by the mixture component j is maximized:

$$cl(x_i) = \arg\max_j \ p_{ij},$$

$$p_{ij} = \frac{\hat{\pi}_j \varphi_{\hat{a}_j,\hat{\sigma}_j}(x_i)}{\sum_{k=1}^{s} \hat{\pi}_k \varphi_{\hat{a}_k,\hat{\sigma}_k}(x_i)}. \tag{3}$$

In cluster analysis, the mixture components are interpreted as clusters, though this is somewhat controversial, because mixtures of more than one not well separated normal distributions may be unimodal and could look quite homogeneous.

It is possible to estimate the number of mixture components s by the Bayesian Information Criterion BIC (Schwarz (1978)), which is done for example by the add-on package "mclust" (Fraley and Raftery (1998)) for the statistical software systems R and SPLUS. In the present paper we don't treat the estimation of s. Note that robustness for fixed s is important as well if s is estimated, because the higher s, the more problematic the computation of the ML-estimator, and therefore it is important to have good robust solutions for small s.

Figure 1 illustrates the behaviour of the ML-estimator for normal mixtures in the presence of outliers. The addition of one extreme point to a data set generated from a normal mixture with three mixture components has the effect that the ML estimator joins two of the original components and fits the outlier alone by the third component. Note that the solution depends on the choice of c_0 in (2), because the mixture component to fix the outlier is estimated to have minimum possible variance.

Various approaches to deal with outliers are suggested in the literature about mixture models (note that all of the methods introduced below work for the data in Figure 1 in the sense that the outlier on the right side doesn't affect the classification

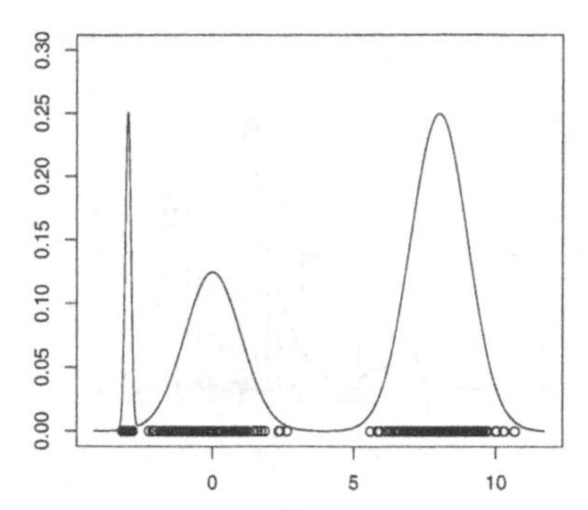

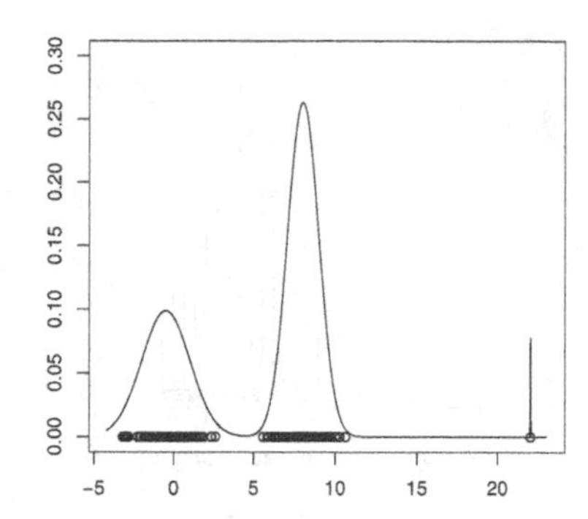

Fig. 1. Left side: artificial data generated from a mixture of three normals with normal mixture ML-fit. Right side: same data with one outlier added at 22 and ML-fit with $c_0 = 0.01$.

of the points on the left side, provided that not too unreasonable tuning constants are chosen where needed). Banfield and Raftery (1993) suggested to add a uniform distribution over the convex hull (i.e., the range for one-dimensional data) to the normal mixture:

$$f_\eta(x) = \sum_{j=1}^{s} \pi_j \varphi_{a_j,\sigma_j^2}(x) + \pi_0 \frac{1(x \in [x_{min}, x_{max}])}{x_{max} - x_{min}}, \tag{4}$$

$\sum_{j=0}^{s} \pi_j = 1$, $\pi_0 \geq 0$, x_{max} and x_{min} denote the maximum and minimum of the data. The uniform component is called the "noise component". The parameters π_j, a_j and σ_j can again be estimated by ML ("BR-noise" in the following").

As an alternative, McLachlan and Peel (2000) suggest to replace the normal densities in (1) by the location/scale family defined by t_ν-distributions (ν could be fixed or estimated). Other families of distributions yielding more robust ML-estimators than the normal could be chosen as well, such as Huber's least favourable distributions as suggested for mixtures by Campbell (1984).

A further idea is to optimize the log-likelihood of (1) for a trimmed set of points, as has already been proposed for the k-means clustering criterion (Cuesta-Albertos, Gordaliza and Matran (1997)).

Conceptually, the noise component approach is very appealing. t-mixtures formally assign all outliers to mixture components modelling clusters. This is not appropriate in most situations from a subject-matter perspective, because the idea of an outlier is that it is essentially different from the main bulk of the data, which in the mixture setup means that it doesn't belong to any cluster. McLachlan and Peel (2000) are aware of this and suggest to classify points in the tail areas of the t-distributions as not belonging to the clusters, but mathematically the outliers are still treated as generated by the mixture components modelling the clusters.

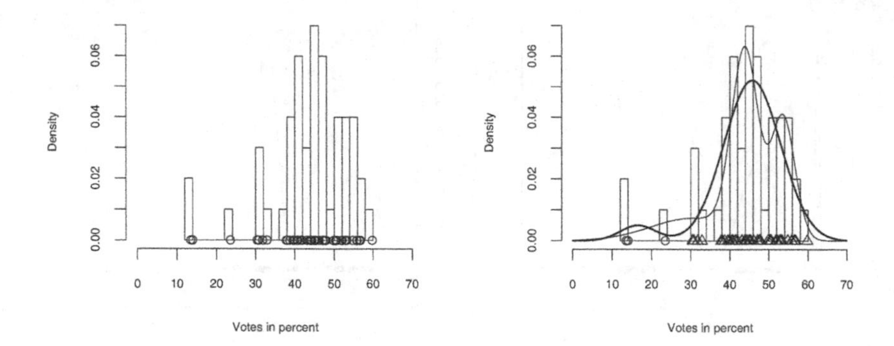

Fig. 2. Left side: votes for the republican candidate in the 50 states of the USA 1968. Right side: fit by mixture of two (thick line) and three (thin line) normals. The symbols indicate the classification by two normals.

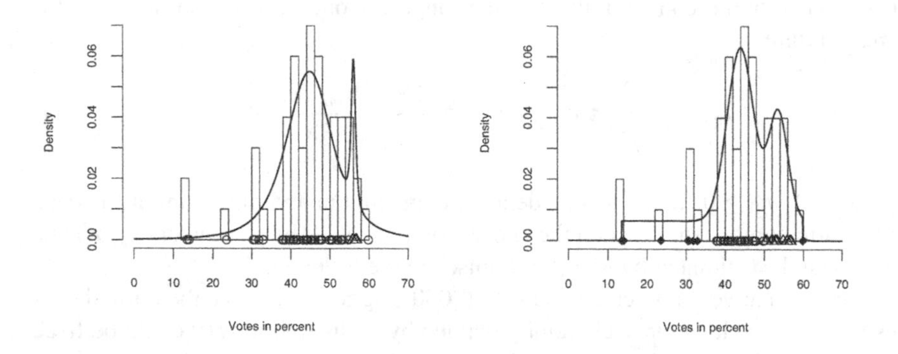

Fig. 3. Left side: votes data fitted by a mixture of two t_3-distributions. Right side: fit by mixture of two normals and BR-noise. The symbols indicate the classifications.

On the other hand, the trimming approach makes a crisp distinction between trimmed outliers and "normal" non-outliers, while in reality it is often unclear whether points on the borderline of clusters should be classified as outliers or members of the clusters. The smoother mixture approach via estimated a posteriori probabilities by analogy to (3) applied to (4) seems to be more appropriate in such situations, while still implying a conceptual distiction between normal clusters and the outlier generating uniform distribution.

As an illustration, consider the dataset shown on the left side of Figure 2 giving the votes in percent for the republican candidate in the 1968 election in the USA (taken from the add-on package "cluster" for R). The main bulk of the data can be roughly separated into two normally looking clusters and there are several states on the left that look atypical. However, it is not so clear where the main bulk ends and states begin to be "outlying", neither is it clear whether the state with the best result for the republican candidate should be considered an outlier. On the right side you see ML-fits by normal mixtures. For $s = 2$ (thick line), one mixture component is taken to fit just three outliers on the left, obscuring the fact that two normals would yield a much more convincing fit for the vast majority of the higher election results. The mixture of three normals (thin line) does a much better job, although it joins several points on the left as a third "cluster" that don't have very much in common and don't look very "normal".

The t_3-mixture ML runs into problems on this dataset. For $s = 2$, it yields a spurious mixture component fitting just four packed points (Figure 3, left side). According to the BIC, this solution is better than the one with $s = 3$, which is similar two the normal mixture with $s = 3$. On the right side of Figure 3 the fit with the noise component approach can be seen, which is similar to three normals in terms of point classification, but provides a useful distinction between normal "clusters" and uniform "outliers".

Another conceptual remark concerns the interpretation of the results. It makes a crucial difference whether a mixture is fitted for the sake of density estimation or for the sake of clustering. If the main interest is in cluster analysis, it is of major importance to interpret the classification and the distinction between "cluster" and "outlier" can be very useful. In such a situation the uniform distribution for the noise component is not chosen because we really believe that the outliers are uniformly distributed, but to mimic the situation that there is no prior information where outliers could be and what could be their distributional shape. The uniform distribution can then be interpreted as "informationless" in a subjective Bayesian fashion.

However, if the main interest is density estimation, it is much more important to come up with an estimator with a reasonable shape of the density. The discontinuities of the uniform may then be judged as unsatisfactory and a mixture of three or even four normals may be preferred. In the present paper we focus on the cluster analytical interpretation.

In Section 2, some theoretical shortcomings of the original noise component approach are highlighted and two alternatives are proposed, namely replacing the uniform distribution over the range of the data by am improper uniform distribution and estimating the range of the uniform component by ML.

In Section 3, theoretical properties of the different noise component approaches are discussed. In Section 4, the computation of the estimators using the EM-algorithm is treated and some simulation results are given in Section 5. The paper is concluded in Section 6. Note that the theory and simulations in this paper are an overview of more detailed results in Pietro Coretto's forthcoming PhD thesis. Proofs and detailed simulation results will be published elsewhere.

2 Two variations on the noise component

2.1 The improper noise component

Hennig (2004) has derived a robustness theory for mixture estimators based on the finite sample addition breakdown point by Donoho and Huber (1983). This breakdown point is defined, in general, as the smallest proportion of points that has to be added to a dataset in order to make the estimation arbitrarily bad, which is usually defined by at least one estimated parameter converging to infinity under a sequence of a fixed number of added points. In the mixture setup, Hennig (2004) defined breakdown as $a_j \to \infty$, $\sigma_j^2 \to \infty$, or $\pi_j \to 0$ for at least one of $j = 1, \ldots, s$. Under (4), the uniform component is not regarded as interesting on its own, but as a helpful device, and its parameters are not included in the breakdown point definition. However, Hennig (2004) showed that for fixed s the breakdown point not only for the normal mixture-ML, but also for the t-mixture-ML and BR-noise is the smallest possible; all these methods can be driven to breakdown by adding a single data point. Note, however, that a point has to be a very extreme outlier for the noise component and t-mixtures to cause trouble, while it's much easier to drive conventional normal mxtures to breakdown.

The main robustness problem with the noise component is that the range of the uniform distribution is determined by the most extreme points, and therefore it depends strongly on where the outliers are.

A better breakdown behaviour (under some conditions on the dataset, i.e., the components have to be well separated in some sense) has been shown by Hennig (2004) for a variant in which the noise component is replaced by an improper uniform density k over the whole real line:

$$f_\eta(x) = \sum_{j=1}^{s} \pi_j \varphi_{a_j,\sigma_j^2}(x) + \pi_0 k. \qquad (5)$$

k has to be chosen in advance, and the other parameters can then be fitted by "pseudo ML" ("pseudo" because (5) does not define a proper density and therefore not a proper likelihood). There are several possibilities to determine k:

- a priori by subject matter considerations, deciding about the maximum density value for which points cannot be considered anymore to lie in a "cluster",
- exploratory, by trying several values and choosing the one yielding the most convincing solution,
- estimating k from the data. This is a difficult task, because k is not defined by a proper probability model. Interpreting the improper noise as a technical device to fit a good normal mixture for most points, we propose the following technique:
  1. Fit (5) for several values of k.
  2. For every k, perform classification according to (3) and remove all points classified as noise.
  3. Fit a simple normal mixture on the remaining (non-noise) points.

4. Choose the k that minimizes the Kolmogorow distance between the empirical distribution of the non-noise points and the fit in step 3. Note that this only works if all candidate values for k are small enough that a certain minimum portion of the data points (50%, say) is classifed as non-noise.

From a statistical point of view, estimating k is certainly most attractive, but theoretically it is difficult to analyze. Particularly, it requires a new robustness theory because the results of Hennig (2004) assume that k is chosen independently of the data. The result for the voting data is shown on the left side of Figure 4. k is lower than for BR-noise, so that the "borderline points" contribute more to the estimation of the normal mixture. The classification is the same. More improvement could be seen if there was a further much more extreme outlier in the dataset, for example a negative number caused by a typo. This would affect the range of the data strongly, but the improper noise approach would still yield the same classification. Some alternative techniques to estimate k are discussed in Coretto and Hennig (2007).

2.2 Maximum likelihood with uniform

A further problem of BR-noise is that the model (4) is data dependent, and its ML estimator is not ML for any data independent model, particularly not for the following one:

$$f_\eta(x) = \sum_{j=1}^{s} \pi_j \varphi_{a_j,\sigma_j^2}(x) + \pi_0 u_{b_1,b_2}(x), \tag{6}$$

where u_{b_1,b_2} is the density of a uniform distribution on the interval $[b_1, b_2]$. This may come as a surprise, because the range of the data is ML for a single uniform distribution, but if it is mixed with some normals, the range of the data is not ML anymore for b_1 and b_2, because f_η is nonzero outside $[b_1, b_2]$. For example, BR-noise doesn't deliver the ML solution for the voting data, which is shown on the right side of Figure 4. In order to prevent the likelihood from converging to infinity for $b_2 - b_1 \to 0$, the restriction (2) has to be extended to $\sigma_0 = \frac{b_2 - b_1}{\sqrt{12}}$, the standard deviation of the uniform.

Taking the ML-estimator for (6) is an obvious alternative ("ML-uniform"). For the voting data the ML solution to fit the uniform component only on the left side seems reasonable. The largest election result is now assigned to one of the normal clusters, to the center of which it is much closer than the outliers on the left to the other normal cluster.

3 Some theory

Here is a very rough overview on some theoretical results which will be published elsewhere in detail:

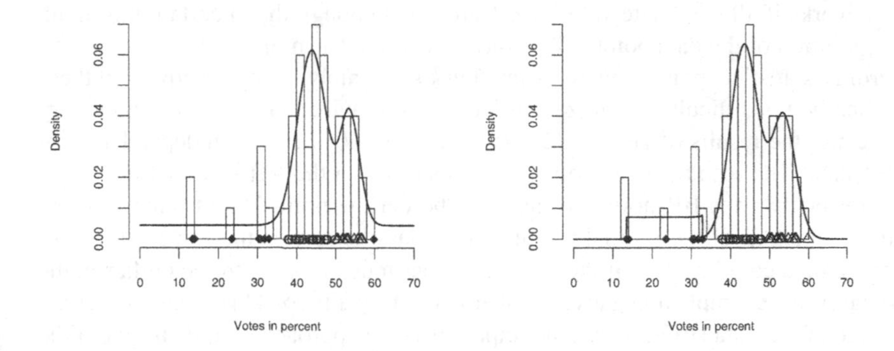

Fig. 4. Left side: votes data fitted by (5) with $s = 2$ and estimated k. Right side: fit by ML for (6), $s = 2$. The symbols indicate the classifications.

Identifiability. All parameters in model (6) are identifiable. This is not surprising because the uniform can be located by the discontinuities in the density (defined as the derivative of the cdf), and mixtures of normals are identifiable. The result involves a new definition of identifiability for mixtures of different families of distributions, see Coretto and Hennig (2006).

Asymptotics. Note that the results below concern parameters, but asymptotic results concerning classification can be derived in a straightforward way from the asymptotic behaviour of the parameter estimators.

BR-noise. $n \to \infty \Rightarrow 1/(x_{max} - x_{min}) \to 0$ whenever $s > 0$. This means that asymptotically the uniform density is estimated to be zero (no points are classified as noise), even if the true underlying model is (6) including a uniform.

ML-uniform. This is consistent for model (6) under (2) including the standard deviation of the uniform. However, at least the estimation of b_1 and b_2 is not asymptotically normal because the uniform distribution doesn't fulfill the conditions for asymptotic normality of ML-estimators.

Improper noise. Unfortunately, even if the density value of the uniform distribution in (6) is known to be k, the improper noise approach doesn't deliver a consistent estimate for the normal parameters in (6). Its asymptotics concerning the canonical parameters estimated by (5), i.e., the value of its "population version", is currently investigated.

Robustness. Unfortunately, ML-uniform is not robust according to the breakdown definition given by Hennig (2004). It can be driven to breakdown by two extreme points in the same way BR-noise can be driven to breakdown by one extreme point, because if two outliers are added on both sides of the original dataset, BR-noise becomes ML for (6).

The improper noise approach with estimated k is robust against the addition of extreme outliers under a sensible initial range of k. Its precise robustness properties still have to be investigated.

4 The EM-algorithm

Nowadays, the ML-estimator for mixtures is often computed by the EM-algorithm, which is shown in various settings to increase the likelihood in every iteration, see Redner and Walker (1984). The principle is as follows:

Start with some initial parameter values which may be obtained by an initial partition of the data. Then iterate the E-step and the M-step until convergence.
E-step: compute the posterior probabilities (3), their analogues for the model under study, respectively, given the current parameter values.
M-step: compute component-wise ML-estimators for the parameters from weighted data, where the weights are given by the E-step.

For given k, the improper noise estimator can be computed precisely in the same way. The proof in Redner and Walker (1984) carries over even though the estimator is only pseudo-ML, because given the data, the improper noise component can be replaced by a proper uniform distribution over some set containing all data points with a density value of k.

For ML-uniform it has to be taken into account that the ML-estimator for a single uniform distribution is always the range of the data. This means for the EM-algorithm that whatever initial interval I is chosen for $[b_1, b_2]$, the uniform mixture component is estimated as the uniform over the range of the data contained in I in the M-step. Particularly, if $I = [x_{min}, x_{max}]$, the EM-estimator yields Banfield and Raftery's noise component as ML-estimator, which is indeed a local optimum of the likelihood in this sense. Therefore, unfortunately, the EM-algorithm is not informative about the parameters of the uniform.

A reasonable approximation of ML-uniform can only be obtained by starting the EM-algorithm several times, either initializing the uniform by all pairs of data points, or, if this is computationally not feasible, by choosing an initial grid of data points from which all pairs of points are used. This could be for example x_{min}, x_{max}, and all empirical $0.1q$-quantiles for $q = 1, \ldots, 9$, or the range of the data could be partitioned into a number of equally long intervals and the data points closest to the interval borders could be chosen. The solution maximizing the likelihood can then be taken.

5 Simulations

Simulations have been carried out to compare the two new proposals ML-uniform and improper noise with BR-noise and ML for t_ν-mixtures. The latter has been carried out with estimated degrees of freedom ν and classification of points as "outliers/noise" in the tail areas of the estimated t-components, according to Chapter 7 of McLachlan and Peel (2000). The ML-uniform has been computed based on a grid of points as explained in Section 4.

Data sets have been generated with $n = 50$, $n = 200$ and $n = 500$, and several statistics have been recorded. The precise simulation results will be published elsewhere. In the present paper we focus on the average misclassification percentages for the datasets with $n = 200$. Data have been simulated from four different parameter choices of the model (6), which are illustrated in Figure 5. For every model, 70 repetitions have been run.

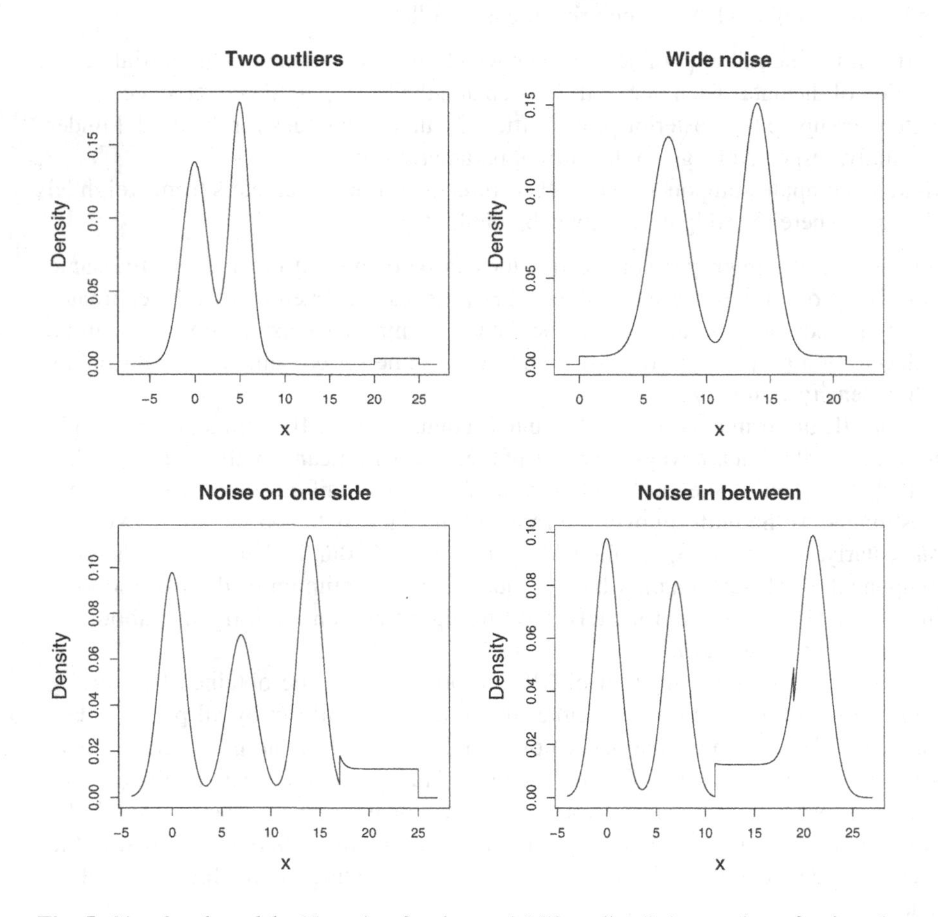

Fig. 5. Simulated models. Note that for the model "2 outliers" the number of points drawn from the uniform component has been fixed to 2.

The misclassification results are given in Table 1. BR-noise yielded the best performance for the "wide noise" model. This is not surprising, because in this model it's very likely that the most extreme points on both sides are generated by the uniform. With two extreme outliers on one side, it was also optimal. However, it per-

Table 1. Average misclassification percentages for $n = 200$

Model/method	BR-noise	t-mixture	improper noise	ML-uniform
Two outliers	2.7	7.3	3.9	3.3
Wide noise	8.0	9.6	8.4	9.3
Noise on one side	10.6	8.3	3.6	5.3
Noise in between	8.8	8.7	5.5	7.3

formed much worse in the two models that generated 10% noise at particular places ("noise on one side" and "noise in between"). The improper noise approach generally performed very well, almost always better than uniform-ML (which was the best method for two of the models for $n = 500$). The t-mixtures-ML didn't perform very well, but this is at least partly due to the fact that all simulated models were of the "normal mixture plus uniform"-type. We will also carry out simulations from t-mixtures in the future.

6 Conclusion

To deal with noise and outliers in cluster analysis, two new methods have been proposed, which are variants of Banfield and Raftery's (1993) noise component, namely the use of an improper density to model the noise and an ML-estimator for a mixture model including a uniform component. Both methods have some theoretical advantages over BR-noise. Simulations showed a good performance particularly for the improper noise component with estimated density value. We find the principle to model outliers and noise by an additional (proper or improper) uniform component appealing, particularly for cluster analysis applications. It allows a smooth classification of points as "noise" or as belonging to a cluster.

Of course it is desirable to apply the ideas to multivariate data as well. This is possible in a straightforward way for the improper noise approach where k is fixed in advance by subject matter considerations. Our proposal to estimate k may work as well for moderate dimensionality, but this is still under investigation.

The ML-uniform approach is problematic in the multivariate setup because of the large number of potentially reasonable support sets for the uniform distribution. In principle it could be applied by assuming the support of the uniform component as rectangular and parallel to the coordinate axes defined by the variables in the data. The ML solution could then be approximated by the best of several hyperrectangles defined by pairs of data points. It remains to see whether this leads to useful clusterings.

References

BANFIELD, J. D. and RAFTERY, A. E. (1993): Model-Based Gaussian and Non-Gaussian Clustering. *Biometrics, 49, 803–821.*

CAMPBELL, N. A. (1984): Mixture models and atypical values. *Mathematical Geology, 16, 465–477.*

CORETTO P. and HENNIG C. (2006): Identifiability for mixtures of distributions from a location-scale family with uniforms. DISES Working Papers No. 3.186, University of Salerno.

CORETTO P. and HENNIG C. (2007): Choice of the improper density in robust improper ML for finite normal mixtures. Submitted.

CUESTA-ALBERTOS, J. A., GORDALIZA, A. and MATRAN, C. (1997): Trimmed k-means: An Attempt to Robustify Quantizers. *Annals of Statistics, 25, 553–576.*

DONOHO, D. L. and HUBER, P. J. (1983): The notion of breakdown point. In P. J. Bickel, K. Doksum, and J. L. Hodges jr. (Eds.): *A Festschrift for Erich L. Lehmann*, Wadsworth, Belmont, CA, 157–184.

FRALEY, C. and RAFTERY, A. E. (1998): How Many Clusters? Which Clustering Method? Answers Via Model Based Cluster Analysis. *Computer Journal, 41, 578–588.*

HATHAWAY, R. J. (1985): A constrained formulation of maximum-likelihood estimates for normal mixture distributions. *Annals of Statistics, 13, 795–800.*

HENNIG, C. (2004): Breakdown points for maximum likelihood-estimators of location-scale mixtures. *Annals of Statistics, 32, 1313–1340.*

MCLACHLAN, G. J. and PEEL, D. (2000): *Finite Mixture Models*, Wiley, New York.

REDNER, R. A. and WALKER, H. F. (1984): Mixture densities, maximum likelihood and the EM algorithm, *SIAM Review, 26, 195–239.*

SCHWARZ, G. (1978): Estimating the dimension of a model, *Annals of Statistics, 6, 461–464.*

An Artificial Life Approach for Semi-supervised Learning

Lutz Herrmann and Alfred Ultsch

Databionics Research Group, Philipps-University Marburg, Germany
{lherrmann,ultsch}@informatik.uni-marburg.de

Abstract. An approach for the integration of supervising information into unsupervised clustering is presented (semi supervised learning). The underlying unsupervised clustering algorithm is based on swarm technologies from the field of Artificial Life systems. Its basic elements are autonomous agents called Databots. Their unsupervised movement patterns correspond to structural features of a high dimensional data set. Supervising information can be easily incorporated in such a system through the implementation of special movement strategies. These strategies realize given constraints or cluster information. The system has been tested on fundamental clustering problems. It outperforms constrained k-means.

1 Introduction

For traditional cluster analysis there is usally a large supply of unlabeled data but little background information about classes. To generate a complete labeling of data can be expensive. Instead, background information might be available as small amount of preclassified input samples that can help to guide the cluster analysis. Consequently, integration of background information into clustering and classification techniques has recently become focus of interest. See Zhu (2006) for an overview.

Retrieval of previously unknown cluster structures, in the sense of multi-mode densities, from unclassified and classified data is called *semi-supervised clustering*. In contrast to semi-supervised classification, semi-supervised clustering methods are not limited to the class labels given in the preclassified input samples. New classes might be discovered, given classes are merged or might be purged.

A particularly promising approach to unsupervised cluster analysis are systems that possess the ability of emergence through self-organization (Ultsch (2007)). This means that systems consisting of a huge number of interacting entities may produce a new, observable pattern on a higher level. Such patterns are said to *emerge* from the self-organizing entities. A biological example for emergence through self-organization is the formation of swarms, e.g. bee swarms or ant colonies.

An example of such nature-inspired information processing techniques is clustering with simulated ants. The ACLUSTER system of Ramos and Abraham (2003) is inspired by ant colonies clustering corpses. It consists of a low-dimensional grid that only carries pheromone intensities. A set of simulated ants moves on the grid's nodes. The ants are used to cluster data objects that are located on the grid. An ant might pick up a data object and drop it later on. Ants are more likely to drop an object on a node whose neighbourhood has similar data objects rather than on nodes with dissimilar objects. Ants move according to pheromone trails on the grid.

In this paper we describe a novel approach for semi-supervised clustering that is based on our unsupervised learning artificial life system (see Ultsch (2000)). The main idea is that a large number of autonomous agents show collective behaviour patterns that correspond to structural features of a high dimensional training set. This approach turns out to be inherently prepared to incorporate additional information from partially labeled data.

2 Artificial life

The artifical life system (ALife) is used to cluster a finite high-dimensional training set $X \subset \mathbb{R}^n$. It consists of a low-dimensional grid $I \subset \mathbb{N}^2$ and a set B of so-called Databots. A Databot carries an input sample of training set X and moves on the grid. Formally, a Databot $i \in B$ is denoted as a triple $(x_i, m(x_i), S_i)$ whereas $x_i \in X$ is the input sample, $m(x_i) \in I$ is the Databot's location on the grid and S_i is a set of movement programs, so-called *strategies*. Later on, mapping of data onto the low-dimensional grid is used for visualization of distance and density structure as described in section 4.

A strategy $s \in S_i$ is a function that assigns probabilites to available directions of movement (north, east, et cetera). The Databot's new location $m'(x_i)$ is chosen at

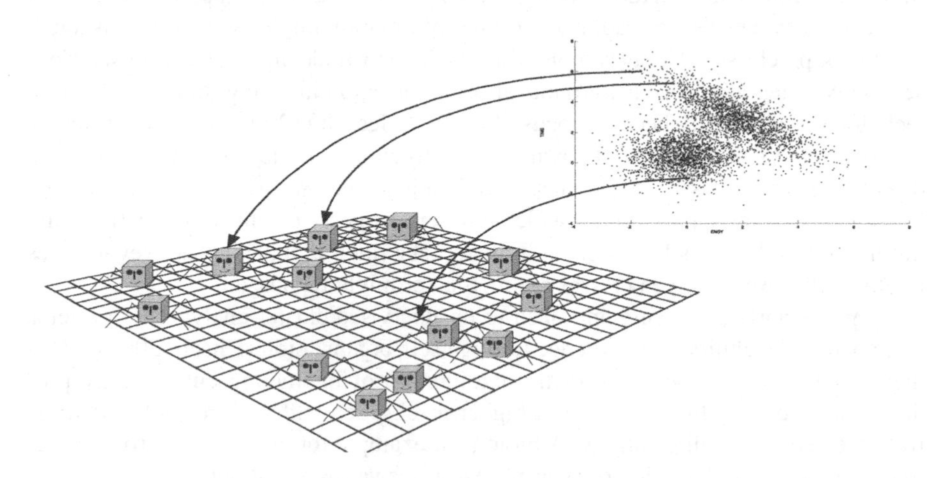

Fig. 1. ALife system: Databots carry high-dimensional data objects while moving on the grid, nearby objects are to be mapped on nearby nodes of the low-dimensional grid

random according to the strategies' probabilites. Several strategies are combined into a single one by weighted averaging of probabilities. Probabilities of movements are to be chosen such that a Databot is more likely to move towards Databots carrying similar input samples than towards Databots with dissimilar input samples. This aims at creation of a sufficiently topography preserving projection $m : X \to I$ (see figure 1). For an overview on strategies see Ultsch (2000).

A generalized view on strategies for topography preservation is given below. For each Databot $(x_i, m(x_i), S_i) \in B$ there is a set of bots F_i (*friends*) it should move towards. Here, the strategy for topography preservation is denoted with s_F. Canonically, F_i is chosen to be the Databots carrying the $k \in \mathbb{N}$ most similar input samples with respect to x_i according to a given dissimilarity measure $d : X \times X \to \mathbb{R}_0^+$, e.g. the euclidean metric on cardinal scaled spaces. Strategy s_F assigns probabilites to all directions of movements such that $m(x_i)$ is more likely to be moved towards $\frac{1}{|F_i|}\sum_{j \in F_i} m(x_j)$ than to any other node on the grid. This can easily be achieved, for example, by vectorial addition of distances for every direction of movement. Additionally, a set of Databots F_i' with the most dissimilar input samples with respect to x_i might inversely be used such that $m(x_i)$ is moved away from its *foes*. A showclass example for s_F is given in figure 2. In analogy to self-organizing maps (Kohonen (1982)), the size of set F_i is decreasing over time. This means that Databots adapt a global ordering before they adapt to local orderings.
Strategies are combined by weighted averaging, i.e. probability of movement towards direction $D \in \{north, east, ...\}$ is $p(D) = \sum_{s \in S_i} w_s s(D) / \sum_{s \in S_i} w_s$ with $w_s \in [0, 1]$ being the weight of strategy s. Linear combination of probabilities is to be preferred over multiplicative because of its compensation. Several combinations of strategies have intensely been tested. It turned out that for obtaining good results a small

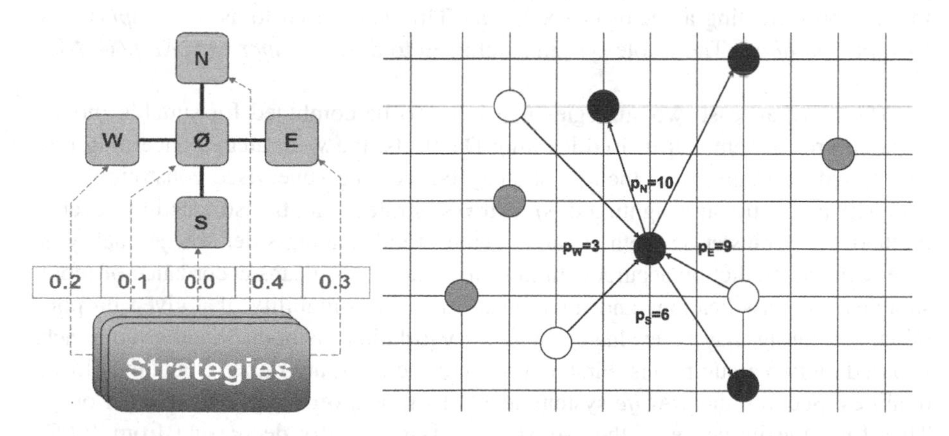

Fig. 2. Strategies for Databots' movements: (**a**) probabilities for directed movements (**b**) set of friends (black) and foes (white), counters resulting from vectorial addition of distances are later on normalized to obtain probabilities, e.g. p_N consists of black northern distances and white southern distances

amount[1] of *random walk* is necessary. This strategy assigns equal probabilities to all available directions in order to overcome local optima by the help of randomness.

3 Semi-supervised artificial life

As described in section 2, the ALife system produces a vector projection for clustering purposes using a movement strategy s_F depending on set F_i. Choice of bots in $F_i \subset B$ is derived from the input samples' similarities with respect to x_i. This is subsumed as *unsupervised constraints* because F_i arises from unlabeled data only.

Background information about cluster memberships is given as pairwise constraints stating that two input samples $x_i, x_j \in X$ belong to the same class (must-link) or different classes (cannot-link). For each input sample x_i this results in two sets: $ML_i \subset X$ denotes the samples that are known to belong to the same class whereas $CL_i \subset X$ contains all samples from different classes. ML_i and CL_i remain empty for unclassified input samples. For each x_i, vector projection $m : X \to I$ has to reflect this by mapping $m(x_i)$ nearby $m(ML_i)$ and far from $m(CL_i)$. This is subsumed as *supervised constraints* because they arise from preclassifications.

The s_F paradigm for satisfaction of unsupervised constraints and how to combine them has already been described in section 2. Same method is applied for satisfaction of supervised constraints. This means that an additional strategy s_{ML} is introduced for Databots carrying preclassified input samples. For such a Databot $(x_i, m(x_i), S_i)$ the set of friends is simply defined as $F_i = ML_i$. According to that strategy, $m(x_i)$ is more likely to be moved towards $\frac{1}{|ML_i|}\sum_{j \in ML_i} m(x_j)$ than to any other node on the grid. This strategy s_{ML} is added to other available strategies. Thus, integration of supervised and unsupervised learning tasks is realized on basis of movement strategies for Databots creating a vector projection m. This is referred to as *semi-supervised learning Databots*. The whole system is referred to as *semi-supervised ALife* (ssALife).

There are at least two strategies that have to be combined for suitable movement control of semi-supervised learning Databots: the s_F strategy concerning unsupervised constraints and the s_{ML} strategy concerning supervised constraints. An adequate proportional weighting of s_F and s_{ML} strategy can be estimated by several methods: Any clustering method can be understood as a classifier whose quality is assessable as prediction accuracy. In this case, accuracy means accordance of input samples' preclassifications and final clustering. The suitability of a given proportional weighting may be evaluated by cross validation methods. Another approach is based on two assumptions. First, cluster memberships are rather global than local qualities. Second, the *ssALife* system adapts to global orderings before local ones. Therefore, the influence of the s_{ML} strategy is constantly decreasing from 100% down to 0 over the training process. The latter method was applied in the current realization of the ssALife system.

---

[1] usually with an absolute weight of 5% up to 10%

4 Semi-Supervised artificial life for cluster analysis

Since ssALife is not an inherent clustering but vector projection method, its visualization capabilities are enhanced using structure maps and the U-Matrix method.

A structure map enhances the regular grid of the ALife system such that each node $i \in I$ contains a high-dimensional codebook vector $m_i \in \mathbb{R}^n$. Structure maps are used for vector projection and quantization purposes, i.e. arbitrary input samples $x \in \mathbb{R}^n$ are assigned to nodes with bestmatching codebook vectors $bm(x) = \arg\min_{i \in I} d(x, m_i)$ with d being the dissimilarity measure from section 2. For a meaningful projection the codebook vectors are to be arranged in a topography preserving manner. This means that neighbouring nodes i, j usually have got codebook vectors m_i, m_j that are neighbouring in the input space. A popular method to achieve that is the Emergent Self-organizing Map (see Ultsch (2003)). In this context, projected input samples $m(x_i), \forall x_i \in X$ from our ssALife system are used for structure map creation. A high-dimensional interpolation based on the self-organizing map's learning technique determines the codebook vectors (Kohonen (1982)).

The U-Matrix (see figure 3 for illustration) is the canonical display of structure maps. The local distance structure is displayed on each grid node as a height value creating a 3D landscape of the high dimensional data space. Clusters are represented as valleys whereas mountain ranges depict cluster boundaries. See Ultsch (2003) for an overview.

Contrairy to common belief, visualizations of structure maps are not clustering algorithms. Segmentation of U-Matrix landscapes into clusters has to be done separately. The U*C clustering algorithm uses an entropy-based heuristic in order to automatically determine the correct number of clusters (Ultsch and Herrmann (2006)). By the help of the watershed-transformation, a structure map decomposes into several coherent regions called basins. Basins are merged in order to form clusters if they share a highly dense region on the structure map. Therefore, U*C combines distance and density information for cluster analysis.

5 Experimental settings and results

In order to evaluate the clustering and self-organizing abilities of ssALife, its clustering performance was measured. The main idea is to use data sets on which the input samples' true classification is known in beforehand. Clustering accuracy can be evaluated as fraction of correctly classified input samples. The ssALife is tested against the well known *constrained k-means* (COPK-Means) from Wagstaff et al. (2001). For each data set, both algorithms got 10% of input samples with the true classification. The remaining samples are presented as unlabeled data.

The data comes from the fundamental clustering problem suite (FCPS). This is a collection of data sets for testing clustering algorithms. Each data set represents a certain problem that arbitrary clustering algorithms shall be able to handle when facing real world data sets. For example, "Chainlink", "Atom" and "Target" contain spatial clusters of linear not separable, i.e. twined, structure. "Lsun",

"EngyTime" and "Wingnut" consist of density defined clusters. For details see `http://www.mathematik.uni-marburg.de/~databionics`.

Comparative results can be seen in table 1. The ssALife method clearly outperforms COPK-Means. COPK-Means suffers from its inability to recognize more complex cluster shapes. As an example, the so-called *EngyTime* data set is shown in figure 3.

Table 1. Percental clustering accuracy: ssALife outperforms COPK-Means, accuracy estimated on fully classified original data over fifty runs with random initialization

data set	COPK-Means	ssALife with U*C
Atom	71	100
Chainlink	65.7	100
Hepta	100	100
Lsun	96.4	100
Target	55.2	100
Tetra	100	100
TwoDiamonds	100	100
Wingnut	93.4	100
EngyTime	90	96.3

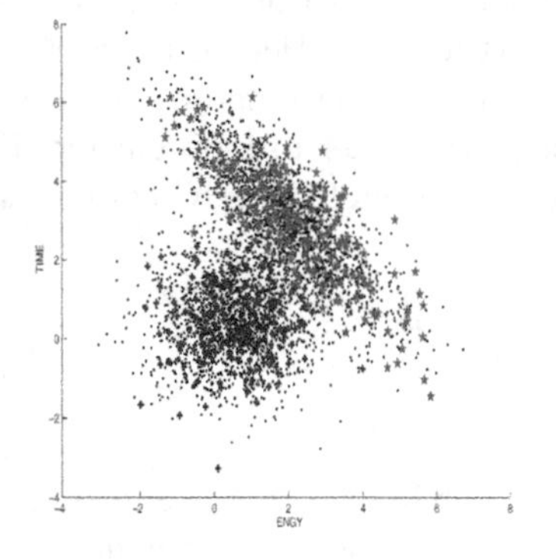

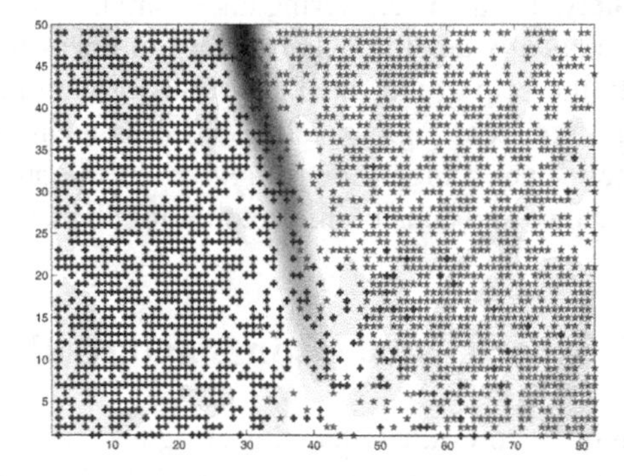

Fig. 3. Density defined clustering problem *EngyTime*: **(a)** partially labeled data **(b)** ssALife produced U-Matrix, clearly visible decision boundary, fully labeled data

6 Discussion

In this work we described a first approach of semi-supervised cluster analysis using autonomous agents called Databots. To our knowledge, this is the first approach that aims for the realization of semi-supervised learning paradigms on basis of a swarm clustering algorithm.

The ssALife system and Ramos' ACLUSTER differ in two ways. First, Databots can be seen as a bijective mapping of input samples onto locations whereas simulated ants have no permanent connection to the data. This facilitates the integration of additional data-related features into the swarm entities. Furthermore, there is no global exchange about topographic information in ACLUSTER, which may lead to discontinuous projections of clusters, i.e. projection errors.

Most popular approaches for semi-supervised learning can be distinguished in two groups (Belkin et al. (2006)). The *manifold assumption* states that input samples with equal class labels are located on manifolds or subspaces, respectively, of the input space (Belkin et al. (2006), Bilenko et al. (2004)). Recovery of such manifolds is accomplished by optimization of an objective function, e.g. for adaption of metrics. The *cluster assumption* states that input samples in the same cluster are likely to have the same class label (Wagstaff et al. (2001), Bilenko et al. (2004)). Again, recovery of such clusters is accomplished by optimization of an objective function. Such objective functions consist of terms for unsupervised cluster retrieval and a loss term that punishes supervised constraint violations. Obviously, the obtainable clustering solutions are predetermined by the inherent cluster shape assumption of the chosen objective function. For example, k-means like clustering algorithms and Mahalanobis like metric adaptions, too, assume linear separable clusters of spherical shape and well-behaved density structure. In contrast to that, the *ssALife* method comes up with a simple yet powerful learning procedure based on movement programs for autonomous agents. This enables a unification of supervised and unsupervised learning tasks without the need for a main objective function. Except for the used dissimilarity measure, the ssALife system does not rely on such objective functions and reaches maximal accuracy on FCPS.

7 Summary

In this paper, cluster analysis is presented on basis of a vector projection problem. Supervised und unsupervised learning of a suitable projection means to incorporate information from topography and preclassifications of input samples. In order to solve this, a very simple yet powerful enhancement of our ALife system was introduced. So-called Databots move the input samples' projection points on a grid-shaped output space. Databots' movements are chosen according to so-called strategies. The unifying framework for supervised and unsupervised learning is simply based on defining an additional strategy that can incorporate preclassifications into the self-organization process.

From this self-organizing process a non-linear display of the data's spatial structure emerges. The display is used for automatic cluster analysis. The proposed method *ssALife* outperforms a simple yet popular algorithm for semi-supervised cluster analysis.

References

BELKIN, M., SINDHWANI, V., NIYOGI, P. (2006): The Geometric Basis of Semi-Supervised Learning. In: O. Chapelle, B. Scholkopf, and A. Zien (Eds.): *Semi-Supervised Learning*. MIT Press, 35-54.

BILENKO, M., BASU, S., MOONEY, R.J. (2004): Integrating Constraints and Metric Learning in Semi-Supervised Clustering. In: *Proc. 21st International Conference on Machine Learning (ICML 2004)*. Banff, Canada, 81-88.

KOHONEN, T. (1982): Self-organized formation of topologically correct feature maps. In: *Biological Cybernetics (43)*. 59-69.

RAMOS, V., ABRAHAM, A. (2003): Swarms on Continuous Data. In: *Proc. Congress on Evolutionary Computation*. IEEE Press, Australia, 1370-1375.

ULTSCH, A. (2000): Visualization and Classification with Artificial Life. In: *Proceedings Conf. Int. Fed. of Classification Societies (ifcs)*. Namur, Belgium.

ULTSCH, A. (2003): Maps for the Visualization of high-dimensional Data Spaces. In: *Proceedings Workshop on Self-Organizing Maps (WSOM 2003)*. Kyushu, Japan, 225-230.

ULTSCH, A., HERRMANN, L. (2006): Automatic Clustering with U*C. Technical Report, Dept. of Mathematics and Computer Science, University of Marburg.

ULTSCH, A. (2007): Emergence in Self-Organizing Feature Maps. In: *Proc. Workshop on Self-Organizing Maps (WSOM 2007)*. Bielefeld, Germany, to appear.

WAGSTAFF, K., CARDIE, C., ROGERS, S., SCHROEDL, S. (2001): Constrained K-means Clustering with Background Knowledge. In: *Proc. 18th International Conf. on Machine Learning*. Morgan Kaufmann, San Francisco, CA, 577-584.

ZHU, X. (2006): Semi-Supervised Learning Literature Survey. Computer Sciences TR 1530. University of Wisconsin, Madison.

Hard and Soft Euclidean Consensus Partitions

Kurt Hornik and Walter Böhm

Department of Statistics and Mathematics
Wirtschaftsuniversität Wien, A-1090 Wien, Austria
{Kurt.Hornik, Walter.Boehm}@wu-wien.ac.at

Abstract. Euclidean partition dissimilarity $d(P,\tilde{P})$ (Dimitriadou et al., 2002) is defined as the square root of the minimal sum of squared differences of the class membership values of the partitions P and $\tilde{P}$, with the minimum taken over all matchings between the classes of the partitions. We first discuss some theoretical properties of this dissimilarity measure. Then, we look at the Euclidean consensus problem for partition ensembles, i.e., the problem to find a hard or soft partition P with a given number of classes which minimizes the (possibly weighted) sum $\sum_b w_b d(P_b,P)^2$ of squared Euclidean dissimilarities d between P and the elements P_b of the ensemble. This is an NP-hard problem, and related to consensus problems studied in Gordon and Vichi (2001). We present an efficient "Alternating Optimization" (AO) heuristic for finding P, which iterates between optimally rematching classes for fixed memberships, and optimizing class memberships for fixed matchings. An implementation of such AO algorithms for consensus partitions is available in the R extension package **clue**. We illustrate this algorithm on two data sets (the popular Rosenberg-Kim kinship terms data and a macroeconomic one) employed by Gordon & Vichi.

1 Introduction

Over the years, a huge number of dissimilarity measures for (hard) partitions has been suggested. Day (1981), building on work by Boorman and Arabie (1972), identifies two leading groups of such measures. *Supervaluation metrics* are derived from supervaluations on the lattice of partitions. *Minimum cost flow (MCF) metrics* are given by the minimum weighted number of admissible transformations required to transform one partition into another.

One such MCF metric is the R-metric of Rubin (1967), defined as the "minimal number of augmentations and removals of single objects" needed to transform one partition into another. This equals twice the Boorman-Arabie A (single element moves) distance, and is also called *transfer distance* in Charon et al. (2006) and *partition-distance* in Gusfield (2002). It can be computed by solving the Linear Sum Assignment Problem (LSAP)

$$\min_{W \in \mathcal{W}_A} \sum\nolimits_{k,l} w_{kl} |C_k \Delta \tilde{C}_l|$$

where $\mathcal{W}_A$ is the set of all matrices $W = [w_{kl}]$ with non-negative elements and row all column sums all one, and the $\{C_k\}$ and $\{\tilde{C}_l\}$ denote the classes of the first and second partition P and $\tilde{P}$, respectively. The LSAP can be solved efficiently in polynomial time using primal-dual algorithms such as the so-called Hungarian method, see e.g. Papadimitriou and Steiglitz (1982).

For possibly *soft* partitions, as e.g. obtained by fuzzy or model-based mixture clustering, the theory of dissimilarities is far less developed. To fix notations and terminology, let n be the number of objects to be classified. A (possibly soft) *partition* P assigns to each object i and class k a non-negative number m_{ik} quantifying the "belongingness" or membership of the object to the class, such that $\sum_k m_{ik} = 1$. We can gather the m_{ik} into the *membership* (matrix) $M = M(P) = [m_{ik}]$ of the partition. In general, M is a stochastic matrix; for hard partitions, it is a binary matrix. Note that M is unique up to permutations of its columns. We refer to the number of non-zero columns of M as the number of classes of the partition, and write $\mathcal{P}_\nu$ and $\mathcal{P}_\nu^H$ for the space of all (possibly soft) partitions with ν classes, and all hard partitions with ν classes, respectively.

In what follows, it will often be convenient to bring memberships to "a common number of classes" (i.e., columns) by adding trailing zero columns as needed. Formally, we can work on the space $\mathcal{P}$ of all stochastic matrices with n rows and infinitely many columns, with the normalization that non-zero columns are the leading ones.

For two hard partitions with memberships M and $\tilde{M}$, we have $|C_k \Delta \tilde{C}_l| = \sum_i |m_{ik} - \tilde{m}_{il}|^p$ for all $p \geq 1$, as $|u|^p = |u|$ if $u \in \{-1,0,1\}$. This strongly suggests to generalize the R-metric to possibly soft partitions via dissimilarities defined as the p-th root of

$$\min_{W \in \mathcal{W}_A} \sum_{k,l} w_{kl} \sum_i |m_{ik} - \tilde{m}_{il}|^p$$

Using $p = 2$ gives *Euclidean dissimilarity* d (Dimitriadou et al. (2002)). Identifying the optimal assignment with its corresponding map π ("permutation" in the possibly augmented case) of the classes of the first to those of the second partition (i.e., $\pi(k) = l$ iff $w_{kl} = 1$ iff C_k is matched with $\tilde{C}_l$), we can use $\sum_{k,l} w_{kl} \sum_i |m_{ik} - \tilde{m}_{il}|^p = \sum_i \sum_k |m_{ik} - \tilde{m}_{i,\pi(k)}|^p$ to obtain

$$d(M, \tilde{M}) = \min_{\Pi} \|M - \tilde{M}\Pi\|_F$$

where the minimum is taken over all permutation matrices Π and $\|M\|_F = (\sum_{i,k} m_{ik}^2)^{1/2}$ is the Frobenius norm. See Hornik (2005b) for details.

For $p = 1$, we get *Manhattan dissimilarity* (Hornik, 2005a). For general p and $W = [w_{kl}]$ constrained to have given row sums α_k and column sums β_l (not necessarily all identical as for the assignment case), we get the Mallows-type distances introduced in Zhou et al. (2005), and motivated from formulations of the Monge-Kantorovich optimal mass transfer problem.

Gordon and Vichi (2001, Model 1) introduce a dissimilarity measure also based on squared distances between optimally matched columns of the membership matrices, but ignoring the "unmatched" columns. This will result in discontinuities (with respect to the natural topology on $\mathcal{P}$) for sequences of membership matrices for which at least one column converges to zero.

In Section 2, we give some theoretical results related to Euclidean partition dissimilarity, and present a heuristic for solving the Euclidean consensus problem for partition ensembles. Section 3 investigates soft Euclidean consensus partitions for two data sets employed in Gordon and Vichi (2001), the popular Rosenberg-Kim kinship terms data and a macroeconomic one.

2 Theory

2.1 Maximal Euclidean dissimilarity

Charon et al. (2006) provide closed-form expressions for the maximal R-metric (transfer distance) between hard partitions with ν and $\tilde{\nu}$ classes, which readily yield

$$\mu_{\nu,\tilde{\nu}} = \max_{M \in \mathcal{P}_\nu^H, \tilde{M} \in \mathcal{P}_{\tilde{\nu}}^H} d(M, \tilde{M}) = \sqrt{n - c_{\min}(\nu, \tilde{\nu})},$$

with the minimum concordance $c_{\min}$ given in Theorem 2 of Charon et al. (2006). One can show (Hornik, 2007b) that the maxima of the Euclidean dissimilarity between (possibly soft) partitions can always be attained at the "boundary", i.e., for hard partitions, such that

$$\max_{M \in \mathcal{P}_\nu, \tilde{M} \in \mathcal{P}_{\tilde{\nu}}} d(M, \tilde{M}) = \max_{M \in \mathcal{P}_\nu^H, \tilde{M} \in \mathcal{P}_{\tilde{\nu}}^H} d(M, \tilde{M}) = \mu_{\nu,\tilde{\nu}}$$

E.g., if $\nu \leq \tilde{\nu}$ and $(\nu - 1)\tilde{\nu} < n$, then $\mu_{\nu,\tilde{\nu}} = (n - \lceil n/\tilde{\nu} \rceil)^{1/2}$. Note that the dissimilarities between soft partitions are "typically" much smaller than for hard ones.

2.2 The Euclidean consensus problem

Aggregating ensembles of clusterings into a consensus clustering by minimizing average dissimilarity has a long history, with key contributions including Mirkin (1974), Barthélemy and Monjardet (1981, 1988), and Wakabayashi (1998). More generally, *clusterwise* aggregation of ensembles of relations (thus containing equivalence relations, i.e., partitions, as a special case) was introduced by Gaul and Schader (1988).

Given an ensemble (profile) of partitions $P_1, \ldots, P_B$ of the same n objects and weights $w_1, \ldots, w_B$ summing to one, a *soft Euclidean consensus partition* (generalized *mean* partition) is defined as a partition which minimizes

$$\sum_{b=1}^{B} w_b d(P, P_b)^2$$

over $\mathcal{P}_\nu$ for given ν. Similarly, a *hard Euclidean consensus partition* minimizes the criterion function over $\mathcal{P}_\nu^H$. Equivalently, one needs to find

$$\min_M \sum_b w_b \min_{\Pi_b} \|M - M_b \Pi_b\|_F^2 = \min_M \min_{\Pi_1,\dots,\Pi_b} \sum_b w_b \|M - M_b \Pi_b\|_F^2$$

over all suitable M and permutation matrices $\Pi_1, \dots, \Pi_B$.

Soft Euclidean consensus partitions can be characterized as follows (see Hornik (2005b)). For fixed $\Pi_1, \dots, \Pi_B$, $\sum_b w_b \|M - M_b \Pi_b\|_F^2 = \|M - \bar{M}\|_F^2 + \sum_b w_b \|M_b \Pi_b\|_F^2 - \|\bar{M}\|_F^2$ where $\bar{M} = \sum_b w_b M_b \Pi_b$ is the weighted mean of the (suitably matched) memberships. If $\bar{M}$ is feasible for M (such that $\nu \geq \max(\nu_1, \dots, \nu_B)$), the overall minimum sought is found by

$$\max_{\Pi_1,\dots,\Pi_B} \|\bar{M}\|_F^2 = \max_{\Pi_1,\dots,\Pi_B} \sum_{k=1}^{\nu} c_{\pi_1(k),\dots,\pi_B(k)},$$

for a suitable B-dimensional cost array c. This is an instance of the Multi-dimensional Assignment Problem (MAP), which is known to be NP-hard.

For hard partitions M and fixed $\Pi_1, \dots, \Pi_B$, $\sum_b w_b \|M - M_b \Pi_b\|_F^2 = \|M\|_F^2 - 2 \sum_b w_b \operatorname{trace}(M' M_b \Pi_b) + \sum_b w_b \|M_b \Pi_b\|_F^2 = \text{const} - 2 \operatorname{trace}(M' \bar{M})$. As $\operatorname{trace}(M' \bar{M}) = \sum_i \sum_k m_{ik} \bar{m}_{ik}$, if again $\nu \geq \max(\nu_1, \dots, \nu_B)$, this can be maximized by choosing, for each row i, $m_{ik} = 1$ for the first k such that $\bar{m}_{ik}$ is maximal for the i-th row of $\bar{M}$. I.e., the optimal M is given by a closest hard partition $H(\bar{M})$ of $\bar{M}$ ("winner-takes-all weighted voting").

Inserting the optimal M yields that the optimal permutations are found by solving

$$\max_{\Pi_1,\dots,\Pi_B} \sum_i \left(\max_{1 \leq k \leq \nu} \sum_b w_b m_{i,\pi_b(k)} \right)$$

which looks "similar" to, if not worse than, the MAP for the soft case.

In both cases, we find that determining Euclidean consensus partitions by simultaneous optimization over the memberships M and permutations $\Pi_1, \dots, \Pi_B$ leads to very hard combinatorial optimization problems, for which solutions by exhaustive search are only possible for very "small" instances. Hornik and Böhm (2007) introduce an "Alternating Optimization" (AO) algorithm based on the natural idea to alternate between minimizing the criterion function $\sum_b w_b \|M - M_b \Pi_b\|_F^2$ over the permutation for fixed M, and over M for fixed permutations. The first amounts to solving B (independent) linear sum assignment problems, the latter to computing suitable approximations to the weighted mean $\bar{M} = \sum_b w_b M_b \Pi_b$ (see above for the case where $\nu \geq \max(\nu_1, \dots, \nu_B)$; otherwise, one needs to "project" or constrain to the space of all M with only ν leading non-zero columns). If every update reduces the criterion function, converge to a fixed point is ensured (it is currently unknown whether these are necessarily local minima of the criterion function). These AO algorithms, which are implemented as methods `"SE"` (default) and `"HE"` of function `cl_consensus` of package **clue** (Hornik, 2007a), provide efficient heuristics for finding the global optimum, provided that the best solution found in "sufficiently many" replications with random starting values is employed.

Table 1. Memberships for the soft Euclidean consensus partition with $\nu = 3$ classes for the Gordon-Vichi macroeconomic ensemble.

Country				Country			
Argentina	0.618	0.374	0.008	Norway	0.082	0.912	0.006
Bolivia	0.666	0.056	0.278	Portugal	0.488	0.452	0.060
Canada	0.018	0.980	0.002	South Africa	0.626	0.366	0.008
Chile	0.632	0.356	0.012	Spain	0.314	0.658	0.028
Egypt	0.750	0.070	0.180	Sudan	0.566	0.088	0.346
France	0.012	0.988	0.000	Sweden	0.050	0.944	0.006
Greece	0.736	0.194	0.070	U.K.	0.112	0.872	0.016
India	0.542	0.076	0.382	U.S.A.	0.062	0.930	0.008
Indonesia	0.616	0.144	0.240	Uruguay	0.680	0.310	0.010
Italy	0.044	0.950	0.006	Venezuela	0.600	0.390	0.010
Japan	0.134	0.846	0.020				

3 Applications

3.1 Gordon-Vichi macroeconomic ensemble

Gordon and Vichi (2001, Table 1) provide soft partitions of 21 countries based on macroeconomic data for the years 1975, 1980, 1985, 1990, and 1995. These partitions were obtained using fuzzy c-means on measurements of variables such as annual per capita gross domestic product (GDP) and the percentage of GDP provided by agriculture. The 1980 and 1990 partitions have 3 classes, the remaining ones two.

Table 1 shows the memberships of the soft Euclidean consensus partition for $\nu = 3$ based on 1000 replications of the AO algorithm. It can be verified by exhaustive search (which is feasible as there are at most $6^5 = 7776$ possible permutation sequences) that this is indeed the optimal solution. Interestingly, one can see that the maximal membership values are never attained in the third column, such that the corresponding closest hard partition (which is also the hard Euclidean consensus partition) has only 2 classes. One might hypothesize that there is a bias towards 2-class partitions as these form the majority (3 out of 5) of the data set, and that 3-class consensus partitions could be obtained by suitably "up-sampling" the 3-class partitions, i.e., increasing their weights w_b. Table 2 indicates how a third consensus class is formed when giving the 3-class partitions w times the weight of the 2-class ones (all these countries are in class 1 for the unweighted consensus partition): The order in which countries join this third class (of the least developed countries) agrees very well with the "sureness" of their classification in the unweighted consensus, as measured by their *margins*, i.e., the difference between the largest and second largest membership values for the respective objects.

3.2 Rosenberg-Kim Kinship terms data

Rosenberg and Kim (1975) describe an experiment where perceived similarities of the kinship terms were obtained from six different "sorting" experiments. In one of

Table 2. Formation of a third class in the Euclidean consensus partitions for the Gordon-Vichi macroeconomic ensemble as a function of the weight ratio w between 3- and 2-class partitions in the ensemble.

w	
1.5	India
2.0	India, Sudan
3.0	India, Sudan
4.5	India, Sudan, Bolivia, Indonesia
10.0	India, Sudan, Bolivia, Indonesia
12.5	India, Sudan, Bolivia, Indonesia, Egypt
∞	India, Sudan, Bolivia, Indonesia, Egypt

these, 85 female undergraduates at Rutgers University were asked to sort 15 English terms into classes "on the basis of some aspect of meaning". There are at least three "axes" for classification: gender, generation, and direct versus indirect lineage. The Euclidean consensus partitions with $\nu = 3$ classes put grandparents and grandchildren in one class and all indirect kins into another one. For $\nu = 4$, {brother, sister} are separated from {father, mother, daughter, son}. Table 3 shows the memberships for a soft Euclidean consensus partition for $\nu = 5$ based on 1000 replications of the AO algorithm.

Table 3. Memberships for the 5-class soft Euclidean consensus partition for the Rosenberg-Kim kinship terms data.

grandfather	0.000	0.024	0.012	0.965	0.000
grandmother	0.005	0.134	0.016	0.840	0.005
granddaughter	0.113	0.242	0.054	0.466	0.125
grandson	0.134	0.111	0.052	0.581	0.122
brother	0.612	0.282	0.024	0.082	0.000
sister	0.579	0.391	0.026	0.002	0.002
father	0.099	0.546	0.122	0.158	0.075
mother	0.089	0.654	0.136	0.054	0.066
daughter	0.000	1.000	0.000	0.000	0.000
son	0.031	0.842	0.007	0.113	0.007
nephew	0.012	0.047	0.424	0.071	0.447
niece	0.000	0.129	0.435	0.000	0.435
cousin	0.080	0.056	0.656	0.033	0.174
aunt	0.000	0.071	0.929	0.000	0.000
uncle	0.000	0.000	0.882	0.071	0.047

Figure 1 indicates the classes and margins for the 5-class solutions. We see that the memberships of 'niece' are tied between columns 3 and 5, and that the margin of 'nephew' is only very small (0.02), suggesting the 4-class solution as the optimal Euclidean consensus representation of the ensemble.

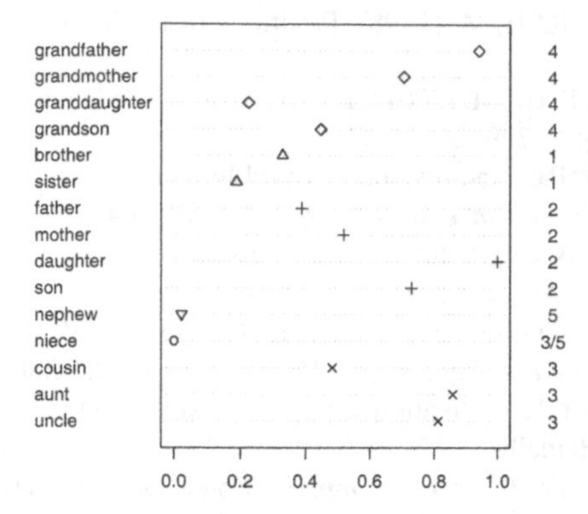

Fig. 1. Classes (incicated by plot symbol and class id) and margins (differences between the largest and second largest membership values) for the 5-class soft Euclidean consensus partition for the Rosenberg-Kim kinship terms data.

Quite interestingly, none of these consensus partitions split according to gender, even though there are such partitions in the data. To take the natural heterogeneity in the data into account, one could try to partition them (perform clusterwise aggregation, Gaul and Schader (1988)), resulting in *meta-partitions* (Gordon and Vichi (1998)) of the underlying objects. Function `cl_pclust` in package **clue** provides an AO heuristic for soft prototype-based partitioning of classifications, allowing in particular to obtain soft or hard meta-partitions with soft or hard Euclidean consensus partitions as prototypes.

References

BARTHÉLEMY, J.P. and MONJARDET, B. (1981): The median procedure in cluster analysis and social choice theory. *Mathematical Social Sciences*, 1, 235–267.

BARTHÉLEMY, J.P. and MONJARDET, B. (1988): The median procedure in data analysis: new results and open problems. In: H. H. Bock, editor, *Classification and related methods of data analysis*. North-Holland, Amsterdam, 309–316.

BOORMAN, S. A. and ARABIE, P. (1972): Structural measures and the method of sorting. In R. N. Shepard, A. K. Romney and S. B. Nerlove, editors, *Multidimensional Scaling: Theory and Applications in the Behavioral Sciences, 1: Theory*. Seminar Press, New York, 225–249.

CHARON, I., DENOEUD, L., GUENOCHE, A. and HUDRY, O. (2006): Maximum transfer distance between partitions. *Journal of Classification*, 23(1), 103–121.

DAY, W. H. E. (1981): The complexity of computing metric distances between partitions. *Mathematical Social Sciences*, 1, 269–287.

DIMITRIADOU, E., WEINGESSEL, A. and HORNIK, K. (2002): A combination scheme for fuzzy clustering. *International Journal of Pattern Recognition and Artificial Intelligence*, 16(7), 901–912.

GAUL, W. and SCHADER, M. (1988): Clusterwise aggregation of relations. *Applied Stochastic Models and Data Analysis*, 4, 273–282.

GORDON, A. D. and VICHI, M. (1998): Partitions of partitions. *Journal of Classification*, 15, 265–285.

GORDON, A. D. and VICHI, M. (2001): Fuzzy partition models for fitting a set of partitions. *Psychometrika*, 66(2), 229–248.

GUSFIELD, D. (2002): Partition-distance: A problem and class of perfect graphs arising in clustering. *Information Processing Letters*, 82, 159–164.

HORNIK, K. (2005a): A CLUE for CLUster Ensembles. *Journal of Statistical Software*, 14(12). URL `http://www.jstatsoft.org/v14/i12/`.

HORNIK, K. (2005b): Cluster ensembles. In C. Weihs and W. Gaul, editors, *Classification – The Ubiquitous Challenge*. Proceedings of the 28th Annual Conference of the Gesellschaft für Klassifikation e.V., University of Dortmund, March 9–11, 2004. Springer-Verlag, Heidelberg, 65–72.

HORNIK, K. (2007a): *clue: Cluster Ensembles*. R package version 0.3-12.

HORNIK, K. (2007b): On maximal euclidean partition dissimilarity. Under preparation.

HORNIK, K. and BÖHM, W. (2007): Alternating optimization algorithms for Euclidean and Manhattan consensus partitions. Under preparation.

MIRKIN, B.G. (1974): The problem of approximation in space of relations and qualitative data analysis. Automatika y Telemechanika, translated in: *Information and Remote Control*, 35, 1424–1438.

PAPADIMITRIOU, C. and STEIGLITZ, K. (1982): *Combinatorial Optimization: Algorithms and Complexity*. Prentice Hall, Englewood Cliffs.

ROSENBERG, S. (1982): The method of sorting in multivariate research with applications selected from cognitive psychology and person perception. In N. Hirschberg and L. G. Humphreys, editors, *Multivariate Applications in the Social Sciences*. Erlbaum, Hillsdale, New Jersey, 117–142.

ROSENBERG, S. and KIM, M. P. (1975): The method of sorting as a data-gathering procedure in multivariate research. *Multivariate Behavioral Research*, 10, 489–502.

RUBIN, J. (1967): Optimal classification into groups: An approach for solving the taxonomy problem. *Journal of Theoretical Biology*, 15, 103–144.

WAKABAYASHI, Y. (1998): The complexity of computing median relations. *Resenhas do Instituto de Mathematica ed Estadistica*, Universidade de Sao Paolo, 3/3, 323–349.

ZHOU, D., LI, J. and ZHA, H. (2005): A new Mallows distance based metric for comparing clusterings. In *ICML '05: Proceedings of the 22nd International Conference on Machine Learning*. ISBN 1-59593-180-5. ACM Press, New York, NY, USA, 1028–1035.

Rationale Models for Conceptual Modeling

Sina Lehrmann and Werner Esswein

Dresden University of Technology, Chair of Information Systems, esp. Systems Engineering, 01062 Dresden, Germany
{sina.lehrmann, werner.esswein}@tu-dresden.de

Abstract. In developing information systems conceptual models are used for varied purposes. Since the modeling process is characterized by interpretation and abstracting the situation at hand it is essential to enclose information about the design process the modelers went through. This aspect is often discarded. But the lack of this information hinders the reuse of past knowledge for later, similar problems encountered and supports the repeat of failures.

The design rationale approaches, discussed in the software engineering community since the 1990s, seem to be an effective means to solve these problems. But the semiformal style of the rationale models challenges the retrieval of the relevant information. The paper explores an approach for classifying issues by its responding alternatives as an access to the complex rationale documentation.

1 Subjectivism in the modeling process

Our considerations are based on a moderate constructivistic position. This attitude of mind has significant consequences on the design of the modeling process as well as on the evaluation of the quality of the resulting model. As it is outlined in (Schütte and Rotthowe (1998)) a model is a result of a cognitive process done by a modeler, who is structuring the considered system according to a specific purpose. Because of the differing thought patterns of the stakeholder a consensus about structuring the problem domain as well as about the model representation has to be defined. In this way the modeling process is a consensus oriented one.

The definition of the application domain terms is an accepted starting point for the process of conceptual modeling (cp. Holten (2003), p. 201). Therefore it is fair to assume that no misinterpretation of the applied terminology occurs.

In order to manage the subjectivity in the modeling process and to support the traceability of the conceptualizations done by the model designer, SCHUETTE and ROTTHOWE proposed the Guidelines of Modeling as generic modeling conventions (cp. Schütte and Rotthowe (1998)). In doing so they also considered not only the significant role of the model designer but also the role of the model user. They claim that the model user is only able to interpret the model in a correct way, if he knows the underlying guidelines of the model design (cp. Schütte and Rotthowe (1998), p. 242).

Model designers are facing similar problems in different projects (cp. Fowler (1997)). Owing to a lack of an explicit and maintained knowledge base containing experiences in model construction and model use, similar problems are solved repeatedly at higher costs than they have to be (cp. Hordijk and Wieringa (2006), p. 353).

Due to the subjectivism in the modeling process it is inevitable to externalize the assumptions and objectives the model bases on. The traceability of the model construction is not only relevant for reusing modeling solutions but also for maintaining the model itself. Stakeholder, who were not involved in the modeling process, are not able to interpret the model in the right way. Particularly with regard to fractional changes of the model, the lack of rationale information could have far-reaching consequences like violating assumptions, constraints or tradeoffs.

Argumentation based models of design rationale ought to be suitable for solving these problems (cp. Dutoit et al. (2006)). Based on the literature about Design Rationale approaches in Software Engineering we derive an approach for reusing experiences in conceptual modeling. For this purpose we use the classification of rationale fragments accessing different rationale models resulting from various modeling projects.

2 The design rationale approach

According to the latest level of knowledge in software engineering issue models which represent the justification for a design in a semiformal manner are the most promising approach to solve the problems described above (cp. Dutoit et al. (2006)). They could be used for structuring the rationale in a more systematic way than text documentations do. In addition, implementing a knowledge base containing the rationales of past modeling projects could improve the efficiency of future modeling processes as well as the quality of the outcoming artifacts.

VAN DER VEN ET AL. identified a general process for creating rationale, which most of the approaches have in common (cp. van der Ven et al. (2006), p. 333).

After the problems are identified and described in problem statements they are evaluated one by one. Alternative solutions are created, evaluated and weighted for their suitability of solving the problem at hand. After an informed decision is made, it is documented along with its justification in a rationale document.

Various approaches for capturing design rationale have been evolved. Most of them are basing on very similar concepts and are more or less restrictive. For our concerns we have chosen the QOC notation, because it is quite expressive and deals directly with evaluation of artifact features (cp. Dutoit et al. (2006), p. 13).

2.1 The QOC-Notation

The Questions, Options, and Criteria (QOC) notation is used for the design space analysis, which ' [...] creates an explicit representation of a structured space of design alternatives and the considerations for choosing among them [...] ” (MacLean et al. (1991), p. 203).

QOC is a semiformal node-and-link diagram. Though it provides a formal structure, the statements within any of the nodes are informal and unrestricted. MACLEAN ET AL. define the three basic concepts, questions, options, and criteria. These concepts and their relations are depicted in Figure 1.

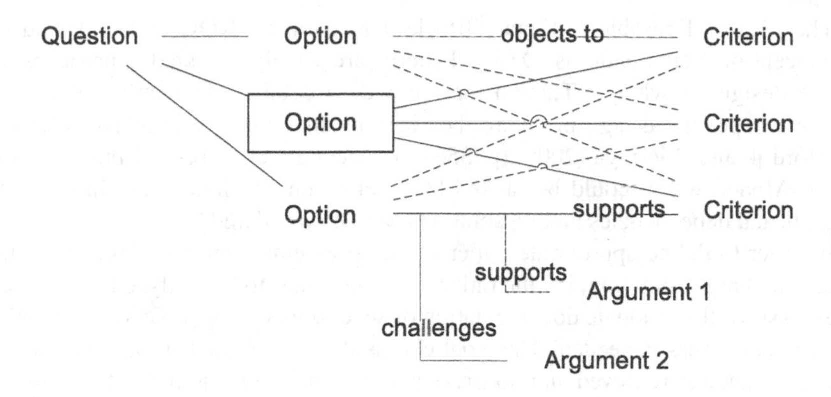

Fig. 1. QOC notation

Questions represent key issues of design decisions not having trivial solutions. They are means for structuring the design space of an artifact. Options are alternative solutions responding to a question. ” [...] Criteria represent the desirable properties of the artifact and requirements that it must satisfy [...] ” (MacLean et al. (1991), p. 208). Because they state the objectives of the design in a clear and structured manner, they form the basis of evaluation, weighting and selection of a design solution. The labeled link between an option and a criterion displays the assessment whether an option satisfy a criterion. In doing so tradeoffs are made explicit and the discussion about choosing among the options turns focus to the purpose the design is made for.

The presented design space analysis is an argumentation based approach. On this account all of the QOC elements could be supported or challenged by arguments. These arguments could play an important role for the evolution of the organizational knowledge base. In the case of reusing design solution the validity of the arguments the primary design decision was based on has to be proven.

One objection to the utility of rationale models is that they are very complex and hardly to manage without any tool support (cp. MacLean et al. (1991), p. 216). Due to the complexity of the rationale models it is necessary to provide an effective retrieval mechanism. Otherwise this kind of documentation seems to be useless for a managed organizational memory.

2.2 Reuse of rationale documentation

Since the capturing of design rationale takes considerable effort, the benefit from using the resulting models has to exceed the costs of their construction.

HORDIJK and WIERINGA propose Reusable Rationale Blocks for reusing design knowledge in order to improve quality and efficiency of design choices (cp. Hordijk and Wieringa (2006)). For achieving this goal they use generalized pieces of decision rationale.

The idea of Reusable Rationale Blocks bases on the QOC approach and on the concept of design patterns. Design Patterns are widely accepted approaches for reusing design knowledge. Though they provide a detailed description of a solution for a repeating design problem, they lack evaluations of alternative solutions (cp. Hordijk and Wieringa (2006), p. 356). But they are appropriate options within a QOC-Model, which could be ranked by a set of quality indicators. In this way tradeoffs and dependencies among solutions can be considered.

In order to define appropriate patterns and to assemble an experience base the documented argumentation, i.e. the rationale models, has to be analyzed. To support the analysis of the rationale documentation of several modeling projects an effective and efficient access is needed. This goal claims that all relevant information to the problem at hand is retrieved and no irrelevant information is element of the answer set. Precision and recall are accepted measures for assessing the achievement of this objective.

The classification scheme presented in the next section could be regarded as an intermediate stage for editing the rationale information of project specific documentations to generate generic rationale information like the described Reusable Rationale Blocks.

3 Classification of rationale fragments

The QOC notation is more restrictive than most of the other approaches and deals directly with the evaluation of artifact features. These are premises for classifying the options of divers rationale models as a systematic entry to the rationale documentation.

To depict our idea we use FOWLERS Analysis Pattern (cp. Fowler (1997)). He discusses different alternatives for modeling derivatives.

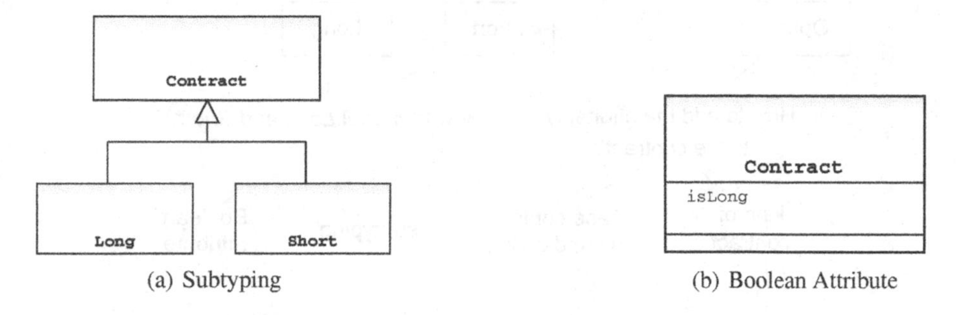

(a) Subtyping (b) Boolean Attribute

Fig. 2. Alternative Modeling of Long and Short

Figure 2 shows two different models of a contract and the distinction between Long and Short. In the first model subtyping is used for this purpose whereas the second one uses the Boolean attribute *isLong*. FOWLER states that both alternatives are equivalent in conceptual modeling (cp. Fowler (1997), p. 177).

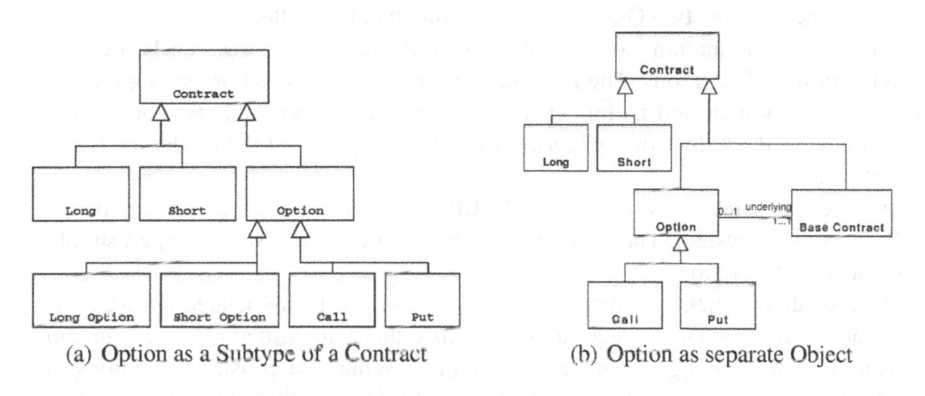

(a) Option as a Subtype of a Contract (b) Option as separate Object

Fig. 3. Different Structures of the Optionality of a Contract

For modeling the concept *Option* FOWLER presents two alternatives depicted in Figure 3 (cp. Fowler (1997), pp. 200ff.). In the first model the optionality of a contract is represented by subtyping. In this way an option is a t'"[...] kind of contract with additional properties and some variant behavior [...]t'" (Fowler (1997), p. 204). The second model differentiates between an option and its underlying base contract. Even FOWLER can give only little advice for choosing among these alternative modeling solutions.

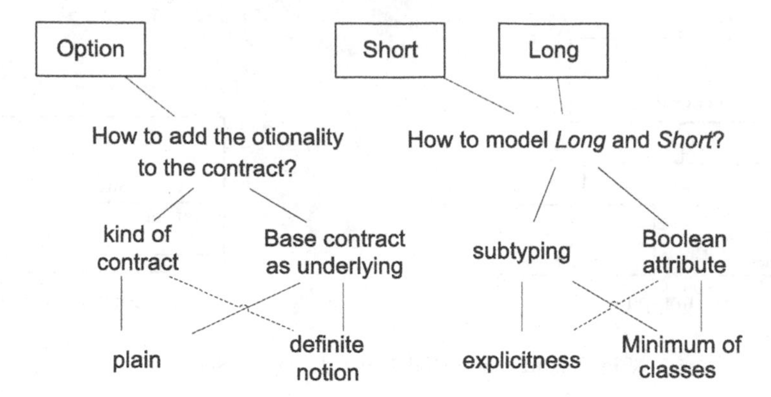

Fig. 4. Example for a Design Space Analysis

For this purpose we analyzed the rationale for the modeling alternatives presented by FOWLER. Figure 4 shows an extract of the rationale model using QOC. The represented discussion bases on the assumption that there has been a decision to include the information objects *Option*, *Long* and *Short* in the model. From these decisions, there follow two Questions concerning the divers alternatives.

On closer examination two different kinds of modeling issues can be derived from the provided solutions. The first one are problem solutions concerning the use of modeling grammar and its influence on the resulting model quality. For solving these problems the knowledge, experiences and assumptions of the modeling expert are decisive.

As a second kind of issues we can identify questions concerning the structuring of the considered system. The expertise and the instinct of the domain expert should dominate this discussion.

A rationale fragment contains at least a question and its associated options, criteria, and arguments. One single question deals either with structuring the problem domain or with applying the modeling grammar. While the considered options in the QOC model can be identified by means of the formal structure, the statements within the nodes are facing the common problems of information retrieval. If we can presume a defined terminology both of the application domain and of the modeling grammar a classification of the Options can identify Questions concerning similar design problems discussed in several rationale models. The resulting classification can be used as a starting point for the analysis of the archived rationale documentation in order to accumulate and aggregate the specific project experiences.

To exemplify our thoughts Figure 5 depicts a possible classification of rationale fragments. The two main branches, problem domain and modeling grammar, categorize the rationale information according to the experiences of the domain expert and the modeling expert respectively.

The differentiation between these two kinds of modeling issues is also reflected in the two principles of the Guidelines of Modeling, construction adequacy and language suitability (cp. Schütte and Rotthowe (1998), p. 246). Just these principles

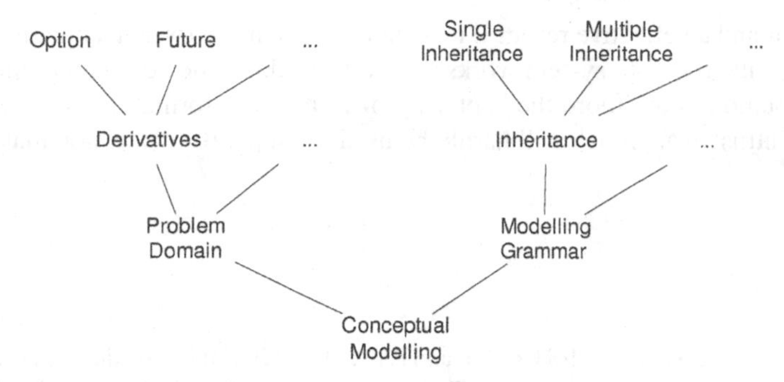

Fig. 5. Classification of Rationale Fragments

reveal that information modeling is characterized by various decision problems. So the choice of the information objects, relevant for the modeling problem, determines the appropriateness of the resulting model. Furthermore an agreement about the application of certain modeling techniques has to be settled.

The branch referring to the usability and utility of the modeling grammar deserves closer attention. Rationale documentations concerning these kinds of issues are not only useful for the model designer and user, but they are also invaluable as feedback information for an incremental knowledge base for the designers of the modeling method.

Experiences in the method use, i.e. usage of the modeling grammar, are discovered as an essential resource for the method engineering process (cp. Rossi et al. (2004)). ROSSI ET AL. stress these kind of information as a complementary part of the method rationale documentation. They define the method construction rationale and the method use rationale as a coherent unit of rationale information.

4 Conclusion

The paper suggests that a classification of design rationale fragments can support the analysis and reuse of modeling experiences resulting in an explicit and systematic structured organizational memory.

Owing to the subjectivism in the modeling process the application of an argumentation based design rationale approach could assist the reasoning in design decisions and the reflection of the resulting model. Furthermore Reusable Rationale Blocks are valuable assets for estimating the quality of the prospective conceptual model.

The semiformality of the complex rationale models challenges the retrieval of documented discussions relevant to a specific modeling problem. The paper presents an approach for classifying issues by its responding alternatives as a systematic entry in the rationale models as a starting point for the analysis of modeling experiences.

What is needed now is empirical research on the impact of design rationale modeling on the resulting conceptual model. An appropriate notation has to be elaborated. This is not a trivial mission because of the tradeoff between a flexible modeling grammar and an effective retrieval mechanism. The more formal a notation is the more precise the retrieval system works. The other side of the coin is that the more formal a notation is the more the capturing of rationale information is interfering. But a high intrusive approach will hardly be used for supporting decision making on the fly.

References

DUTOIT, A.H., McCALL, R., MISTRIK, I. and PAECH, B. (2006): Rationale Management in Software Engineering: Concepts and Techniques. In: A.H. Dutoit, R. McCall, I. Mistrík and B. Paech (Eds.): *Rationale Management in Software Engineering*. Springer, Berlin, 1–48.

FOWLER, M. (1997): *Analysis Patterns: Reusable Object Models*, Addison-Wesley, Menlo Park.

HOLTEN, R. (2003): *Integration von Informationssystemen. Theorie und Anwendung im Supply Chain Management.* Habilitationsschrift, Westfälische Wilhelms-Universität Münster.

HORDIJK, W. and WIERINGA, R. (2006): Reusable Rationale Blocks: Improving Quality and Efficiency of Design Choices. In: A.H. Dutoit, R. McCall, I. Mistrík and B. Paech (Eds.): *Rationale Management in Software Engineering*. Springer, Berlin, 353–370.

MACLEAN, A., YOUNG, R.M., BELLOTTI, V.M.E. and MORAN, T.P. (1991):Questions, Options and Criteria: Elements of Design Space Analysis. *Human-Computer Interaction, 6(1991) 3/4, 201–250.*

ROSSI, M., RAMESH, B., LYYTINEN, K. and TOLVANEN, J.-P. (2004): Managing Evolutionary Method Engineering by Method Rationale. *Journal of the Association for Information Systems, 5(2004) 9, 356–391.*

SCHÜTTE, R. (1999): Architectures for Evaluating the Quality of Information Models - a Meta and an Object Level Comparison. In: J. Akoka, M. Bouzeghoub, I. Comyn-Wattiau and E. Métais (Eds.): *Conceptual Modeling - ER '99, 18th International Conference on Conceptual Modeling, Paris, France, November, 15-18, 1999, Proceedings*. Springer, Berlin, 490–505.

SCHÜTTE, R. and ROTTHOWE, T. (1998): The Guidelines of Modeling - An Approach to Enhance the Quality in Information Models. In: T.W. Ling, S. Ram and M.L. Lee (Eds.): *Conceptual Modeling - ERt'98, 17th International Conference on Conceptual Modeling, Singapore, November 16-19, 1998, Proceedings*. Springer, Berlin, 240–254.

VAN DER VEN, J.S., JANSEN, A.G.J., NIJHUIS, J.A.G. and BOSCH, J. (2006): Design Decisions: The Bridge between Rationale and Architecture. In: A.H. Dutoit, R. McCall, I. Mistrík and B. Paech (Eds.): *Rationale Management in Software Engineering*. Springer, Berlin, 329–348.

Measures of Dispersion and Cluster-Trees for Categorical Data

Ulrich Müller-Funk

ERCIS, Leonardo-Campus 3, 48149 Münster, Germany
mueller-funk@ercis.de

Abstract. A clustering algorithm, in essence, is characterized by two features (1) the way in which the heterogeneity within resp. between clusters is measured (objective function) (2) the steps in which the splitting resp. fusioning proceeds. For categorical data there are no "standard indices" formalizing the first aspect. Instead, a number of ad hoc concepts have been used in cluster analysis, labelled "similarity", "information", "impurity" and the like. To clarify matters, we start out from a set of axioms summarizing our conception of "dispersion" for categorical attributes. To no surprise, it turns out, that some well-known measures, including the Gini index and the entropy, qualify as measures of dispersion. We try to indicate, how these measures can be used in unsupervised classification problems as well. Due to its simple analytic form, the Gini index allows for a dispersion-decomposition formula that can be made the starting point for a CART-like cluster tree. Trees are favoured because of i) factor selection and ii) communicability.

1 Motivation

Most data sets in business administration show attributes of mixed type i.e. numerical and categorical ones. The classical text-book advice to cluster data of this kind can be summarized as follows

a) Measure (dis-)similarities among attribute vectors separately on the basis of either kind of attributes and unite both the resulting numbers in a (possibly weighted) sum.
b) In order to deal with the categorical attributes, encode them in a suitable (binary) way and look for coincidences all over the resulting vectors. Condense your findings with the help of one of the numerous existing matching coefficients.

(cf. Fahrmeir et al. (1996), p. 453). This advice, however, is bad policy for at least two reasons. Treating both parts of the attribute vectors separately amounts to saying that both groups of variables are independent—which only can be claimed in exceptional cases. By looking for bit-wise coincidences, as in step two, one completely looses contact with the individual attributes. This feature, too, is statistically undesirable. For that reason it seems to be less harmful to categorize numerical quantities and to deal with all variables simultaneously—but to avoid matching coefficients and the like. During the last decade roughly a dozen agglomerative or partitioning cluster algorithms for categorical data have been proposed, quite a few based on the concept of entropy. Examples include "COOLCAT" (Barbará et al. (2002)) or "LIMBO" (Andritsos et al. (2004)). These approaches, no doubt, have their merits. For various reasons, however, it would be advantageous to rely on a divisive, tree-structured technique that

a) supports the choice of relevant factors,
b) helps to identify the resulting clusters and renders the device communicable to practitioners.

In other words, we favour some unsupervised analogue to CART or CHAID.

That type of procedure, furthermore, facilitates the use of prior information on the attribute level as it will be seen in Section 3. Within that context comparisons of attributes should not be based any longer on similarity-measures but on quantities that allow for a model-equivalent and accordingly, can be related to the underlying probability source. For that purpose we shall work out the concept of "dispersion" in Section 2 and discuss starting points for cluster algorithms in Section 3. The material in Section 2 may bewilder some readers as it seems that "somebody should have written down something like that long time ago". Despite some efforts, however, no source in the literature could be spotted.

There is another important aspect that has to be addressed. Categorical data is typically organized in form of tables or cubes. Obviously, the number of cells exponentially increases with the number of factors taken into consideration. This, in turn, will result in many empty or sparsely populated cells and render the analysis obsolete. In order to circumvent this difficulty, some form of "sequential sub-cube clustering" is needed (and will be reported elsewhere).

2 Measures of dispersion

What is a meaningful splitting criterion? There are essentially three answers provided in the literature, "impurity", "information" and "distance". The axiomization of impurity is somewhat scanty. Every symmetric functional of a probability vector qualifies as a measure of impurity iff it is minimal (zero) in the deterministic case and takes its maximum value at the uniform distribution (cf. Breiman et al. (1984), p. 24). That concept is not very specific and it hardly gives way to an interpretation in terms of "intra-class-density" or "inter-class-sparsity". Information, on the other hand, can be made precise by means of axioms that uniquely characterize the Shannon entropy (cf. Rényi (1971), p. 442). The reading of those axioms in the realm of classification and clustering is disputable. Another approach to splitting is based on probability metrics measuring the dissimilarity of stochastic vectors representing different classes. Various types of divergences figure prominently in that context (cf. Teboulle et al. (2006), for instance). That approach, no doubt, is conceptually sound but suffers from a technical drawback in the present context. Divergences are defined in terms of likelihood ratios and, accordingly, are hardly able to distinguish among (exactly or approximately) orthogonal probabilities. Orthogonality among cluster-representatives, however, is considered to be a desirable feature.

A time-tested road to clustering for objects represented by quantitative attribute-vectors is based on functions of the covariance matrix (e.g. determinant or trace). It is near at hand to mimic those approaches in the presence of qualitative characteristics. However, there seems to be no source in the literature that systematically specifies a notion like "variability", "volatility", "diversity", "dispersion" etc. for categorical variates. In order to make this conception precise, we consider functionals D,

$$D : \mathcal{P} \to [0, \infty[$$

where $\mathcal{P}$ denotes the class of all finite stochastic vectors, i.e. $\mathcal{P}$ is the union of the sets $\mathcal{P}_K$ comprising all probability vectors of length $K \geq 2$. D, of course, will be subject to further requirements:

(PI) "invariance w.r.t. permutations (relabellings)"

$$D(p_{\sigma_1}, \ldots, p_{\sigma_K}) = D(p_1, \ldots, p_K)$$

for all $p = (p_1, \ldots, p_K) \in \mathcal{P}_K$ and all permutations σ.

(MD) "dispersion is minimal in the deterministic case"

$$D(p) = 0 \quad \text{iff } p \text{ is an unit vector.}$$

(MA) "D is monotone w.r.t. to majorization"

$$p <_m q \to D(p) \geq D(q) \quad p, q \in \mathcal{P}_K \ .$$

In particular, D takes its maximum at the uniform distribution (cf. Tong (1980), p. 102ff for the definition of $<_m$ and some basics).

(SC) "splitting cells increases dispersion"

$$D(p_1, \ldots, p_{k-1}, r, s, p_{k+1}, \ldots, p_K) \geq D(p_1, \ldots, p_{k-1}, p_k, p_{k+1}, \ldots, p_K) \ .$$

where $p \in \mathcal{P}_K, 0 \leq r, s$ and $r + s = p_k$.

(MP) "mixing probabilities increases dispersion"

$$D((1-r)p + rq) \geq (1-r)D(p) + rD(q)$$

for $0 < r < 1$ and $p, q \in \mathcal{P}_K$. In addition to concavity we assume D to be continuous on all of $\mathcal{P}_K, K \geq 2$.

(EC) "consistency w.r. to empty cells"

$$D(p_1,\ldots,p_K,0) \leq D(p_1,\ldots,p_K)$$

for all $p \in \mathcal{P}_K$.

Definition 1. *A functional D satisfying (PI), (MD), (MA), (SC), (MP), (EC) is called a (categorical) measure of dispersion.*

Some comments on this definition.

1. The majorization ordering seems to be a "natural" choice and it guarantees that D is also a measure of impurity. "$<_m$" could be replaced, however, by an ordering expressing concentration around the mode. The restriction to unimodal probabilities (frequencies) and the dependency on a measure of location to be specified in advance, is somewhat undesirable.
2. In an earlier draft, (EC) was formulated with "=" instead of "$\leq$". Some helpful remarks made by C. Henning and A. Ultsch lead to this modification. It allows for measures that relate dispersion to the length of the stochastic vector. This might be meaningful in tree-building in order to prevent a preferential treatment of attributes exhibiting many levels. Such an index, for instance, could take on the form

$$p \in int(\mathcal{P}_K) \Rightarrow D(p) = w_K \sum_k g(p_k),$$

   where g is some "suitable" function (see below) and w_K are some discounting weights.
3. In case of ordinal variates, it makes sense to restrict the class of permutations in (PI).

For the sake of convenience (and "w.l.o.g") the axioms above were formulated by means of a linearly ordered indexing set. With two-way (or higher order) tables, multiple indices $k = (i, j) \in K = I \times J$ are more convenient. The marginal resp. conditional distributions associated with probabilities (or empirical frequencies) (p_{ij}) on a $I \times J$-table are denoted as usual, e.g.

$$p_1^{(1)} = p_{i.} \text{ or } p^{2|1}(j|i) = p_{ij}/p_{i.} \;.$$

The next assertion parallels the well-known formula "$\sigma^2(Y) \geq \mathbb{E}(\sigma^2(Y|X))$".

Proposition 1. *Let D be a measure of dispersion and (p_{ij}) probabilities on a two-way-table. Then,*

$$D(p^{(2)}) \geq \sum_i D(p^{2|1}(\cdot|i))\, p_i^{(1)} \;.$$

Proof. Consequence of (MP) □

The proposition implies that any measure of dispersion induces a predictive measure of association A_D,

$$A_D(2|1) = 1 - \frac{\sum_i D(p^{2|1}(\cdot|i))p_i^{(1)}}{D(p^{(2)})} = \sum_i A_D(2|1=i)p_i^{(1)} ,$$

where $A_D(2|1=i)$ represents the conditional predictive strength of level i. For $D(p) = 1 - p_{max}$, A_D ist closely related to Goodman-Kruskal's lambda. The measures A_D can be employed, for instance, to construct association rules.

In what follows we shall restrict our attention to functionals D of the form

$$D_g(p) = \sum_i g(p_i)$$

where g is a continuous, concave function on $[0,1]$, $g(0) = g(1) = 0$ and $g(t) > 0$ for $0 < t < 1$.

Examples.

i.) $g(t) = t(1-t) \Rightarrow D_g(p) = 1 - \sum_i p_i^2 = \sum\sum_{i \neq j} p_i p_j = trace(\Sigma)$,

$$\Sigma = diag(p) - pp^T$$

i.e. D is the Gini-index resp. the generalized variance. More general Beta densities could be employed as well.

ii.) $g(t) = -t \log t \Rightarrow D_g(p) = -\sum_i p_i \log p_i$

i.e. D is the Shannon entropy.

Proposition 2. *a) D_g is a measure of dispersion.*
b) If g is strictly concave, then D_g takes its unique maximum on $\mathcal{P}_K$ at the uniform distribution $u_K = K^{-1}(1,\ldots,1)$.

Proof. (PI), (MD) and (MP) are immediate consequences of the definition. (MA) follows from a well-known lemma by Schur (cf. Tong (1980), Lemma 6.2.1). In order to see (SC), just write $r = \alpha p_K, s = (1-\alpha)p_K$ and employ concavity. □

Obviously, $D_g(p)$ can efficiently be estimated from an multinomial i.i.d. sample $X^{(1)},\ldots,X^{(N)}$, $\hat{p}_N = N^{-1}\sum_n X^{(n)}$. In fact, $D_g(\hat{p}_N)$ is the strongly consistent ML-estimator of $D_g(p)$ and distributional aspects can be settled with the help of the Δ-method.

Proposition 3. *Let p be an interior point of $\mathcal{P}_K$.*

a) *If $p \not\equiv u_K$, then*

$$\mathcal{L}_p\left(\sqrt{N}\left(D_g(\hat{p}_N) - D_g(p)\right)\right) \to \mathcal{N}(0, \Gamma_g)$$

where $\Gamma_g = (\ldots, g'(p_k), \ldots)\Sigma(\ldots, g'(p_k), \ldots)^T$.

b) *If $p = u_K$, then*

$$\mathcal{L}_u\left(N\left(D_g(\hat{p}_N) - D_g(p)\right)\right) \to \mathcal{L}\left(\sum_j \lambda_j Y_j^2\right) ,$$

where $Y_1, \ldots, Y_K$ is a sample of standard-normal variates and where $\lambda_1, \ldots, \lambda_K$ denote the eigenvalues of $\Sigma^{1/2} H \Sigma^{1/2}$, $H = diag(\ldots, u''(p_k), \ldots)$.

Proof.

a) is a direct consequence of Witting and Müller-Funk (1995), Satz 5.107 b), p. 107 ("Delta method").
b) follows from their Satz 5.127, p. 134. □

The limiting distribution in b) must be worked out for every g seperately. For the Gini index D_G this becomes

$$\mathcal{L}_u\left(N\left(D_G(\hat{p}_N) - 1 + \frac{1}{K}\right)\right) \to \mathcal{L}\left(\frac{K-1}{K}\sum_{i=1}^{K-1} Y_i^2\right) .$$

3 Segmentation

Again, we start out from a sample of categorical (multinomial) attribute vectors, $X^{(1)}, \ldots, X^{(N)}$. In general, a clustering corresponds to a partition of the objects $\{1, \ldots, N\}$. With categorical data we shall demand, that vectors contributing to the same cell should always be united in the same cluster. With that convention, a clustering now corresponds to a partitioning of the cells, i.e. is related to the attributes. That makes it easy to formulate further constraints on the attribute-level. For instance, it can be required in the segmentation process that cells pertaining to some ordinal factor only come along in intervals within a cluster. As already indicated, we are mainly interested in building up cluster-trees on the basis of some measure D_g.

Now let $\hat{p}(m)$ be the average of all observations in cluster $C(m)$. According to our convention, these cluster-representatives become orthogonal. If $C(m)$ is further decomposed into two subclusters $C_L(m)$ and $C_R(m)$, then

$$D_g\left(\hat{p}(m)\right) - \hat{p}_L(m) D_g\left(\hat{p}_L(m)\right) - \hat{p}_R(m) D_g\left(\hat{p}_R(m)\right) \geq 0$$

is the gain in dispersion within clusters and is to to be maximized. A look at the corresponding formula characterizing a CART-tree (cf. Breiman et al (1984), p. 25), reveals that in the absence of the information in labels, a posteriori probabilities are merely replaced by "centroids". Matters become even more transparent in case of the Gini-index D_G. This is due to the identity

$$D_G(\alpha p + \beta q) = \alpha^2 D_G(p) + \beta^2 D_G(q) + 2\alpha\beta(1 - p^T q)$$

where $p, q \in \mathcal{P}_K, \alpha, \beta \geq 0, \alpha + \beta = 1$, resulting in the general decomposition-formula:

$$\begin{aligned} D_G(\hat{p}_N) &= \sum_l \hat{\pi}_N^2(l) D_G(\hat{p}_N(l)) + \sum\sum_{l \neq m} \hat{\pi}_N(l)\hat{\pi}_N(m)\left(1 - \hat{p}_N^T(l)\hat{p}_N(m)\right) \\ &= D_G(within) + D_G(between) \end{aligned}$$

where $\hat{\pi}_N(m)$ denotes the proportion of observations in cluster $C(m)$. With our convention $\hat{\pi}_N(m) = \hat{p}_N(m)$ and the decomposition formula becomes

$$D_G(\hat{p}_N) = \sum_m \hat{p}_N^2(m) D_G(\hat{p}_N(m)) + \sum\sum_{l \neq m} \hat{p}_N(l)\,\hat{p}_N(m) \ .$$

Here, the quantity to be maximized simply becomes $\hat{p}_L(m) \cdot \hat{p}_R(m)$. Accordingly, a cluster is divided into subclasses of approximately the same size if no further prior information (restriction) is added. This solution to the clustering problem of course, is rather blunt and undesirable for most applications. It provokes the question, however whether related measures (like the entropy) really produce partitions that allow for a better statistical interpretation. It remains to see, moreover, how well the approach performs if restrictions, prior probabilities or label-information is provided.

There is a promising alternative route based on the predictive measures of association introduced earlier. At each node the best predictor-attribute is selected. Attribute cells with a low conditional predictive power are merged into one and the node is branched out accordingly. The procedure stops if predictive power falls short a prescribed critical value. The whole device, in fact, can be interpreted as some form of non-linear factor analysis. It will be part of forthcoming work.

References

ANDRITSOS, P., TSAPARAS, P., MILLER, R.J. and SEVCIK, K.C. (2004): LIMBO: Scalable clustering of categorical data. In: E. Bertino, S. Christodoulakis, D. Plexousakis, V. Christophides, M. Koubarakis, K. Böhm and E. Ferrari (Eds.): *Advances in Database Technology—EDBT 2004*. Springer, Berlin, 123–146.

BARBARA, D., LI, Y. and COUTO, J. (2002): COOLCAT: An entropy-based algorithm for categorical clustering. *Proceedings of the Eleventh International Conference on Information and Knowledge Management, 582–589.*

BREIMAN, L., FRIEDMAN, J.H., OLSHEN, R.A. and STONE, C.J. (1984): *Classification and Regression Trees.* CRC Press, Florida.

FAHRMEIR, L., HAMERLE, A. and TUTZ, G. (1996): *Multivariate statistische Methoden.* de Gruyter, Berlin.

RENYI, A. (1971): *Wahrscheinlichkeitsrechnung. Mit einem Anhang über Informationstheorie.* VEB Deutscher Verlag der Wissenschaften, Berlin.

TEBOULLE, M., BERKHIN, P., DHILLON, I., GUAN, Y. and KOGAN, J. (2006): Clustering with entropy-like k means algorithms. In: J. Kogan, C. Nicholas, and M. Teboulle (Eds.): *Grouping Multidimensional Data: Recent Advances in Clustering.* Springer Verlag, New York, 127–160.

TONG, Y.L. (1980): Probability inequalities in multivariate distributions. In: Z.W. Birnbaum and E. Lukacs (Eds.): *Probability and Mathematical Statistics.* Academic Press, New York.

WITTING, H., MÜLLER-FUNK, U. (1995): *Mathematische Statistik II – Asymptotische Statistik: Parametrische Modelle und nicht-parametrische Funktionale.* Teubner, Stuttgart.

Information Integration of Partially Labeled Data

Steffen Rendle and Lars Schmidt-Thieme

Information Systems and Machine Learning Lab, University of Hildesheim
{srendle, schmidt-thieme}@ismll.uni-hildesheim.de

Abstract. A central task when integrating data from different sources is to detect identical items. For example, price comparison websites have to identify offers for identical products. This task is known, among others, as record linkage, object identification, or duplicate detection.

In this work, we examine problem settings where some relations between items are given in advance – for example by EAN article codes in an e-commerce scenario or by manually labeled parts. To represent and solve these problems we bring in ideas of semi-supervised and constrained clustering in terms of pairwise must-link and cannot-link constraints. We show that extending object identification by pairwise constraints results in an expressive framework that subsumes many variants of the integration problem like traditional object identification, matching, iterative problems or an active learning setting.

For solving these integration tasks, we propose an extension to current object identification models that assures consistent solutions to problems with constraints. Our evaluation shows that additionally taking the labeled data into account dramatically increases the quality of state-of-the-art object identification systems.

1 Introduction

When information collected from many sources should be integrated, different objects may refer to the same underlying entity. Object identification aims at identifying such equivalent objects. A typical scenario is a price comparison system where offers from different shops are collected and identical products have to be found. Decisions about identities are based on noisy attributes like product names or brands. Moreover, often some parts of the data provide some kind of label that can additionally be used. For example some offers might be labeled by a European Article Number (EAN) or an International Standard Book Number (ISBN). In this work we investigate problem settings where such information is provided on some parts of the data. We will present three different kinds of knowledge that restricts the set of consistent solutions. For solving these constrained object identification problems we extend the generic object identification model by a collective decision model that is guided by both constraints and similarities.

2 Related work

Object identification (e.g. Neiling 2005) is also known as record linkage (e.g. Winkler 1999) and duplicate detection (e.g. Bilenko and Mooney 2003). State-of-the-art methods use an adaptive approach and learn a similarity measure that is used for predicting the equivalence relation (e.g. Cohen and Richman 2002). In contrast, our approach also takes labels in terms of constraints into account.

Using pairwise constraints for guiding decisions is studied in the community of semi-supervised or constrained clustering – e.g. Basu et al. (2004). However, the problem setting in object identification differs from this scenario because in semi-supervised clustering typically a small number of classes is considered and often it is assumed that the number of classes is known in advance. Moreover, semi-supervised clustering does not use expensive pairwise models that are common in object identification.

3 Four problem classes

In the classical object identification problem $C_{classic}$ a set of objects X should be grouped into equivalence classes E_X. In an adaptive setting, a second set Y of objects is available where the perfect equivalence relation E_Y is known. It is assumed that X and Y are disjoint and share no classes – i.e. $E_X \cap E_Y = \emptyset$.

In real world problems often there is no such clear separation between labeled and unlabeled data. Instead only the objects of some subset Y of X are labeled. We call this problem setting the iterative problem C_{iter} where (X, Y, E_Y) is given with $X \supseteq Y$ and $Y^2 \supseteq E_Y$. Obviously, consistent solutions E_X have to satisfy $E_X \cap Y^2 = E_Y$. Examples of applications for iterative problems are the integration of offers from different sources where some offers are labeled by a unique identifier like an EAN or ISBN, and iterative integration tasks where an already integrated set of objects is extended by new objects.

The third problem setting deals with integrating data from n sources, where each source is assumed to contain no duplicates at all. This is called the class of matching problems C_{match}. Here the problem is given by $\mathcal{X} = \{X_1, \ldots, X_n\}$ with $X_i \cap X_j = \emptyset$ and the set of consistent equivalence relations $\mathcal{E}$ is restricted to relations E on X with $E \cap X_i^2 = \{(x,x)|x \in X_i\}$. Traditional record linkage often deals with matching problems of two data sets ($n = 2$).

At last, there is the class of pairwise constrained problems C_{constr}. Here each problem is defined by (X, R_{ml}, R_{cl}) where the set of objects X is constrained by a must-link R_{ml} and a cannot-link relation R_{cl}. Consistent solutions are restricted to equivalence releations E with $E \cap R_{cl} = \emptyset$ and $E \supseteq R_{ml}$. Obviously, R_{cl} is symmetric and irreflexive whereas R_{ml} has to be an equivalence relation. In all, pairwise constrained problems differ from iterative problems by labeling relations instead of labeling objects. The constrained problem class can better describe local informations like two offers are the same/ different. Such information can for example be provided by a human expert in an active learning setting.

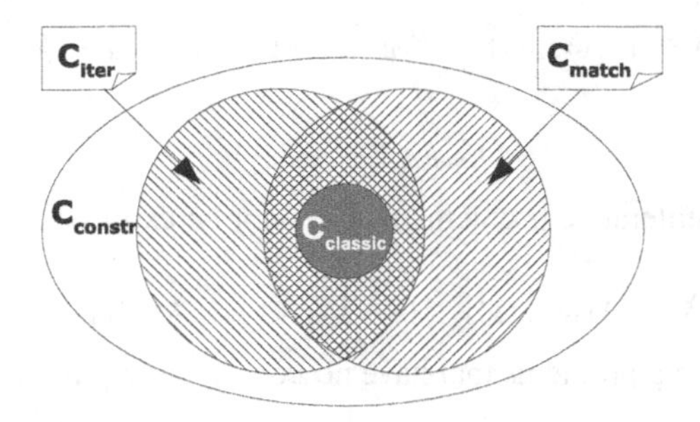

Fig. 1. Relations between problem classes: $C_{classic} \subset C_{iter} \subset C_{constr}$ and $C_{classic} \subset C_{match} \subset C_{constr}$.

We will show, that the presented problem classes form a hierarchy $C_{classic} \subset C_{iter} \subset C_{constr}$ and that $C_{classic} \subset C_{match} \subset C_{constr}$ but neither $C_{match} \subseteq C_{iter}$ nor $C_{iter} \subseteq C_{match}$ (see Figure 1). First of all, it is easy to see that $C_{classic} \subseteq C_{iter}$ because any problem $X \in C_{classic}$ corresponds to an iterative problem without labeled data ($Y = \emptyset$). Also $C_{classic} \subseteq C_{match}$ because an arbitrary problem $X \in C_{classic}$ can be transformed to a matching problem by considering each object as its own dataset: $X_1 = \{x_1\}, \ldots, X_n = \{x_n\}$. On the other hand, $C_{iter} \not\subseteq C_{classic}$ and $C_{match} \not\subseteq C_{classic}$, because $C_{classic}$ is not able to formulate any restriction on the set of possible solutions $\mathcal{E}$ as the other classes can do. This shows that:

$$C_{classic} \subset C_{match}, \quad C_{classic} \subset C_{iter} \tag{1}$$

Next we will show that $C_{iter} \subset C_{constr}$. First of all, any iterative problem (X, Y, E_Y) can be transformed to a constrained problem (X, R_{ml}, R_{cl}) by setting $R_{ml} \leftarrow \{(y_1, y_2) | y_1 \equiv_{E_Y} y_2\}$ and $R_{cl} \leftarrow \{(y_1, y_2) | y_1 \not\equiv_{E_Y} y_2\}$. On the other hand, there are problems $(X, R_{ml}, R_{cl}) \in C_{constr}$ that cannot be expressed as an iterative problem, e.g.:

$$X = \{x_1, x_2, x_3, x_4\}, \quad R_{ml} = \{(x_1, x_2), (x_3, x_4)\}, \quad R_{cl} = \emptyset$$

If one tries to express this as an iterative problem, one would assign to the pair (x_1, x_2) the label l_1 and to (x_3, x_4) the label l_2. But one has to decide whether or not $l_1 = l_2$. If $l_1 = l_2$, then the corresponding constrained problem would include the constraint $(x_2, x_3) \in R_{ml}$, which differs from the original problem. Otherwise, if $l_1 \neq l_2$, this would imply $(x_2, x_3) \in R_{cl}$, which again is a different problem. Therefore:

$$C_{iter} \subset C_{constr} \tag{2}$$

Furthermore, $C_{match} \subseteq C_{constr}$ because any matching problem $X_1, \ldots, X_n$ can be expressed as a constrained problem with:

$$X = \bigcup_{i=1}^{n} X_i, \quad R_{cl} = \{(x, y) | x, y \in X_i \wedge x \neq y\}, \quad R_{ml} = \emptyset$$

There are constrained problems that cannot be translated into a matching problem. E.g.:

$$X = \{x_1, x_2, x_3\}, \quad R_{ml} = \{(x_1, x_2)\}, \quad R_{cl} = \emptyset$$

Thus:

$$C_{match} \subset C_{constr} \tag{3}$$

At last, there are iterative problems that cannot be expressed as matching problems, e.g.:

$$X = \{x_1, x_2, x_3\}, \quad Y = \{x_1, x_2\}, \quad x_1 \equiv_{E_Y} x_2$$

And there are matching problems that have no corresponding iterative problem, e.g.:

$$X_1 = \{x_1, x_2\}, \quad X_2 = \{y_1, y_2\}$$

Therefore:

$$C_{match} \not\subseteq C_{iter}, \quad C_{iter} \not\subseteq C_{match} \tag{4}$$

In all we have shown that C_{constr} is the most expressive class and subsumes all the other classes.

4 Method

Object Identification is generally done by three core components (Rendle and Schmidt-Thieme (2006)):

1. *Pairwise Feature Extraction* with a function $f : X^2 \to \mathbb{R}^n$.
2. *Probabilistic Pairwise Decision Model* specifying probabilities for equivalences $P[x \equiv y]$.
3. *Collective Decision Model* generating an equivalence relation E over X.

The task of feature extraction is to generate a feature vector from the attribute descriptions of any two objects. Mostly, heuristic similarity functions like TFIDF-Cosine-Similarity or Levenshtein distance are used. The probabilistic pairwise decision model combines several of these heuristic functions to a single domain specific similarity function (see Table 1). For this model probabilistic classifiers like SVMs, decision trees, logic regression, etc. can be used. By combining many heuristic functions over several attributes, no time-consuming function selection and fine-tuning has to be performed by a domain-expert. Instead, the model automatically learns which similarity function is important for a specific problem. Cohen and Richman (2002) as well as Bilenko and Mooney (2003) have shown that this approach is successful. The collective decision model generates an equivalence relation over X by using $sim(x,y) := P[x \equiv y]$ as learned similarity measure. Often, clustering is used for this task (e.g. Cohen and Richman (2002)).

Table 1. Example of feature extraction and prediction of pairwise equivalence $P[x_i \equiv x_j]$ for three digital cameras.

Object	Brand	Product Name	Price
x_1	Hewlett Packard	Photosmart 435 Digital Camera	118.99
x_2	HP	HP Photosmart 435 16MB memory	110.00
x_3	Canon	Canon EOS 300D black 18-55 Camera	786.00

Object Pair	TFIDF-Cos. Sim. (Product Name)	FirstNumberEqual (Product Name)	Rel. Difference (Price)	Feature Vector	$P[x_i \equiv x_j]$
(x_1, x_2)	0.6	1	0.076	(0.6, 1, 0.076)	0.8
(x_1, x_3)	0.1	0	0.849	(0.1, 0, 0.849)	0.2
(x_2, x_3)	0.0	0	0.860	(0.0, 0, 0.860)	0.1

4.1 Collective decision model with constraints

The constrained problem easily fits into the generic model above by extending the collective decision model by constraints. As this stage might be solved by clustering algorithms in the classical problem, we propose to solve the constrained problem by a constraint-based clustering algorithm. To enforce the constraint satisfaction we suggest a constrained hierarchical agglomerative clustering (HAC) algorithm. Instead of a dendrogram the algorithm builds a partition where each cluster should contain equivalent objects. Because in an object identification task the number of equivalence classes is almost never known, we suggest model selection by a (learned) threshold θ on the similarity of two clusters in order to stop the merging process. A simplified representation of our constrained HAC algorithm is shown in Algorithm 1. The algorithm initially creates a new cluster for each object (line 2) and afterwards merges clusters that contain objects constrained by a mustlink (line 3-7). Then the most similar clusters, that are not constrained by a cannotlink, are merged until the threshold θ is reached.

From a theoretical point of view this task might be solved by an arbitrary, probabilistic HAC algorithm using a special initialization of the similarity matrix and minor changes in the update step of the matrix. For satisfaction of the constraints R_{ml} and R_{cl}, one initializes the similarity matrix for $X = \{x_1, \ldots, x_n\}$ in the following way:

$$A^0_{j,k} = \begin{cases} +\infty, & \text{if } (x_j, x_k) \in R_{ml} \\ -\infty, & \text{if } (x_j, x_k) \in R_{cl} \\ P[x_j \equiv x_k] & \text{otherwise} \end{cases}$$

As usual, in each iteration the two clusters with the highest similarity are merged. After merging cluster c_l with c_m the dimension of the square matrix A reduces by one – both in columns and rows. For ensuring constraint satisfaction, the similarities between $c_l \cup c_m$ to all the other clusters have to be recomputed:

$$A^{t+1}_{n,i} = \begin{cases} +\infty, & \text{if } A^t_{l,i} = +\infty \vee A^t_{m,i} = +\infty \\ -\infty, & \text{if } A^t_{l,i} = -\infty \vee A^t_{m,i} = -\infty \\ sim(c_l \cup c_m, c_i) & \text{otherwise} \end{cases}$$

For calculating the similarity *sim* between clusters, standard linkage techniques like single-, complete- or average-linkage can be used.

Algorithm 1 Constrained HAC Algorithm

1: **procedure** CLUSTERHAC(X, R_{ml}, R_{cl})
2: $P \leftarrow \{\{x\} | x \in X\}$

3: **for all** $(x,y) \in R_{ml}$ **do**
4: $c_1 \leftarrow c$ *where* $c \in P \wedge x \in c$
5: $c_2 \leftarrow c$ *where* $c \in P \wedge y \in c$
6: $P \leftarrow (P \setminus \{c_1, c_2\}) \cup \{c_1 \cup c_2\}$
7: **end for**

8: **repeat**
9: $(c_1, c_2) \leftarrow \underset{c_1,c_2 \in P \wedge (c_1 \times c_2) \cap R_{cl} = \emptyset}{\text{argmax}} sim(c_1, c_2)$
10: **if** $sim(c_1, c_2) \geq \theta$ **then**
11: $P \leftarrow (P \setminus \{c_1, c_2\}) \cup \{c_1 \cup c_2\}$
12: **end if**
13: **until** $sim(c_1, c_2) < \theta$

14: **return** P
15: **end procedure**

4.2 Algorithmic optimizations

Real-world object identification problems often have a huge number of objects. An implementation of the proposed constrained HAC algorithm has to consider several optimization aspects. First of all, the cluster similarities should be computed by dynamic programming. So the similarities between clusters have to be collected just once and afterward can be inferred by the similarities, that are already given in the similarity-matrix:

$$sim_{sl}(c_1 \cup c_2, c_3) = \max\{sim_{sl}(c_1, c_3), sim_{sl}(c_2, c_3)\} \qquad \textit{single-linkage}$$
$$sim_{cl}(c_1 \cup c_2, c_3) = \min\{sim_{cl}(c_1, c_3), sim_{cl}(c_2, c_3)\} \qquad \textit{complete-linkage}$$
$$sim_{al}(c_1 \cup c_2, c_3) = \frac{|c_1| \cdot sim_{al}(c_1, c_3) + |c_2| \cdot sim_{al}(c_2, c_3)}{|c_1| + |c_2|} \qquad \textit{average-linkage}$$

Second, a blocker should reduce the number of pairs that have to be taken into account for merging. Blockers like the canopy blocker (McCallum et al. (2000))

Table 2. Comparison of F-Measure quality of a constrained to a classical method with different linkage techniques. For each data set and each method the best linkage technique is marked bold.

Data Set	Method	Single Linkage	Complete Linkage	Average Linkage
Cora	classic/constrained	0.70/0.92	0.74/0.71	**0.89/0.93**
DVD player	classic/constrained	**0.87**/0.94	0.79/0.73	0.86/**0.95**
Camera	classic/constrained	0.65/**0.86**	0.60/0.45	**0.67**/0.81

reduce the amount of pairs very efficiently, so even large data sets can be handled. At last, pruning should be applied to eliminate cluster pairs with similarity below θ_{prune}. These optimizations can be implemented by storing a list of cluster-distance-pairs which is initialized with the pruned candidate pairs of the blocker.

5 Evaluation

In our evaluation study we examine if additionally guiding the collective decision model by constraints improves the quality. Therefore we compare constrained and unconstrained versions of the same object identification model on different data sets. As data sets we use the bibliographic Cora dataset that is provided by McCallum et al. (2000) and is widely used for evaluating object identification models (e.g. Cohen et al. (2002) and Bilenko et al. (2003)), and two product data sets of a price comparison system.

We set up an iterative problem by labeling N% of the objects with their true class label. For feature extraction of the Cora model we use TFIDF-Cosine-Similarity, Levenshtein distance and Jaccard distance for every attribute. The model for the product datasets uses TFIDF-Cosine-Similarity, the difference between prices and some domain-specific comparison functions. The pairwise decision model is chosen to be a Support Vector Machine. In the collective decision model we run our constrained HAC algorithm against an unconstrained ('classic') one. In each case, we run three different linkage methods: single-, complete- and average-linkage. We report the average F-Measure quality of four runs for each of the linkage techniques and for constrained and unconstrained clustering. The F-Measure quality is taken on all pairs that are unknown in advance – i.e. pairs that do not link two labeled objects.

$$F\text{-}Measure = \frac{2 \cdot Recall \cdot Precision}{Recall + Precision}$$

$$Recall = \frac{TP}{TP + FN}, \quad Precision = \frac{TP}{TP + FP}$$

Table 2 shows the results of the first experiment where $N = 25\%$ of the objects for Cora and $N = 50\%$ for the product datasets provide labels. As one can see, the best constrained method always clearly outperforms the best classical method. When switching from the best classical to the best constrained method, the relative error reduces by 36% for Cora, 62% for DVD-Player and 58% for Camera. An informal

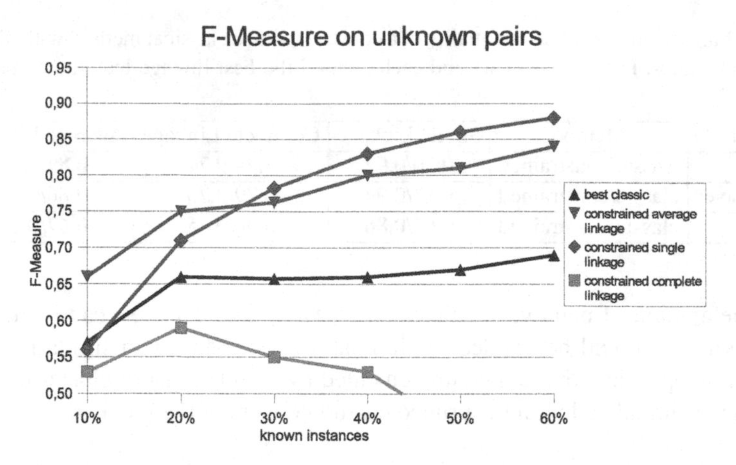

Fig. 2. F-Measure on Camera dataset for varying proportions of labeled objects.

significance test shows that in this experiment the best constrained method is better than the best classic one.

In a second experiment (see Figure 2) we increased the amount of labeled data from $N = 10\%$ to $N = 60\%$ and report results for the Camera dataset for the best classical method and the three constrained linkage techniques. The figure shows that the best classical method does not improve much beyond more than 20% labeled data. In contrast, when using the constrained single- or average-linkage technique the quality on non-labeled parts improves always with more labeled data. When few constraints are available average-linkage tends to be better than single-linkage whereas single-linkage is superior in the case of many constraints. The reason are the cannot-links that prevent single-linkage from merging false pairs. The bad performance of constrained complete-linkage can be explained by must-link constraints that might result in diverse clusters (Algorithm 1, line 3-7). For any diverse cluster, complete-linkage can not find any cluster with similarity greater than θ and so after the initial step, diverse clusters are not merged any more (Algorithm 1, line 8-13).

6 Conclusion

We have formulated three problem classes that encode knowledge and restrict the space of consistent solutions. For solving problems of the most expressive class C_{constr}, that subsumes all the other classes, we have proposed a constrained object identification model. Therefore the generic object identification model was extended in the collective decision stage to ensure constraint satisfaction. We proposed a HAC algorithm with different linkage techniques that is guided by both a learned similarity measure and constraints. Our evaluation has shown, that this method with single- or average-linkage is effective and using constraints in the collective stage clearly outperforms non-constrained state-of-the-art methods.

References

BASU, S. and BILENKO, M. and MOONEY, R. J. (2004): A Probabilistic Framework for Semi-Supervised Clustering. In: *Proceedings of the 10th International Conference on Knowledge Discovery and Data Mining (KDD-2004).*

BILENKO, M. and MOONEY, R. J. (2003): Adaptive Duplicate Detection Using Learnable String Similarity Measures. In: *Proceedings of the 9th International Conference on Knowledge Discovery and Data Mining (KDD-2004).*

COHEN, W. W. and RICHMAN, J. (2002): Learning to Match and Cluster Large High-Dimensional Data Sets for Data Integration. In: *Proceedings of the 8th International Conference on Knowledge Discovery and Data Mining (KDD-2002).*

MCCALLUM, A. K., NIGAM K. and UNGAR L. (2000): Efficient Clustering of High-Dimensional Data Sets with Application to Reference Matching. In: *Proceedings of the 6th International Conference On Knowledge Discovery and Data Mining (KDD-2000).*

NEILING, M. (2005): Identification of Real-World Objects in Multiple Databases. In: *Proceedings of GfKl Conference 2005.*

RENDLE, S. and SCHMIDT-THIEME, L. (2006): Object Identification with Constraints. In: *Proceedings of 6th IEEE International Conference on Data Mining (ICDM-2006).*

WINKLER W. E. (1999): *The State of Record Linkage and Current Research Problems.* Technical report, Statistical Research Division, U.S. Census Bureau.

Part III

Multidimensional Data Analysis

Data Mining of an On-line Survey - A Market Research Application

Karmele Fernández-Aguirre[1], María I. Landaluce[2], Ana Martín[1]* and Juan I. Modroño[1]

[1] Universidad del País Vasco (EHU/UPV), Spain
karmele.fernandez@ehu.es
[2] Universidad de Burgos (UBU), Spain

Abstract. In this work we apply several data mining techniques that give us deep insight into knowledge extraction from a marketing survey addressed to the potential buyers of an university gift shop. The techniques are classified as symmetrical and non-symmetrical. An advocation for such combination is given as conclusion.

1 Introduction

When a large dataset is obtained from a survey including a large number of questions it is necessary to extract the information and the relationships inherent to the data in an ordered and effective way. The data is usually a mixture of subsets of quantitative, categorical (closed questions) and frecuency (open-ended) questions.

In this work we analyze data extracted from an on-line survey by means of different and complementary methods divided in two categories: symmetrical and non-symmetrical. The former will be some factor method complemented with classification, whereas the latter will comprise some sort of regression models. After presenting data and objectives (section 2) we outline methodology and results (section 3) and finally give some conclusions (section 4).

2 Data and objectives

The University of the Basque Country (UPV/EHU), as part of a large project which main aim is revamping its corporate image, is about launching a corporate shop (also considered as a gift or souvenir shop). In order to better know its potential buyers and the potential success of it, it has set up an online survey to collect information on its acceptability.

* Authors gratefully acknowledge financial support from Grupo de Investigación Consolidado DEC UPV/EHU GIU06/53.

Such on-line survey is addressed to the members of the research and teaching staff, administrative staff and the students of the university. Its main objectives are to evaluate buying propensity about the corporate products, identify potential buyers' and non-buyers' profiles, know desirable characteristics of the products and obtain a function to be named and considered as a "propensity to buy".

Table 1 contains the sampling technical characteristics. The access to filling in the survey was possible only by invitation and there was a period of one month for doing so. The number of invitations or sample size was fixed per strata and chosen in order to get a maximum error of 2% of the variability range of the responses for a 95% confidence level. The sampling was thus proportionallly random and the results were encouraging, with a global response rate of around 40%, though not equally distributed.

Table 1. Technical characteristics of the on-line survey.

	Students	Admin. Staff	Research & Teaching
Population	48995	1128	3982
Sample size	2289	768	1499
Response (%)	547 (23.9)	444 (57.81)	754 (50.30)
Sampling error	0.042	0.036	0.032
Confidence level	0.95	0.95	0.95

The most relevant questions included in the sample were: a question over general satisfaction about being a member of the university (5 point scale), a binary question on general interest about buying the corporate articles, 26 questions on the valuation (from 1 to 4) of the same number of products (shown in a photo), valuation (from 1 to 7) of 8 proposed desirable characteristics of products (sober, traditional, stylish, modern, practical, artistic, daring and original) and personal information (gender, age, post and campus - up to three possible -). We were particularly interested in getting information on preferences on the products so we intentionally dropped the middle point in product valuation questions. These questions are those which we analyze by means of both non-symmetrical and symmetrical methods. We have made this distinction in order to differentiate between methods that assume some sort of causality or relationship direction in the variables (i.e., regression methods) and those who don't (as factor methods).

3 Methodology and results

3.1 Symmetrical methods: Exploratory multivariate techniques

Depending upon which kind of variables are to be considered as active we can consider a Principal Components (PCA) or a Multiple Correspondence Analysis (MCA), see e.g., Greenacre (1984), Lebart et al. (1984), Lebart (1994).

PCA of continuous variables and classification

We first consider as active variables the scores given to the question of the desirable characteristics of products (original, sober, ...), which are measured in a 7 point scale and may arguably be considered as near-continuous variables. The variables regarding personal characteristics as gender or age are considered as supplementary variables, as well as the variables reflecting satisfaction with the institution and the interest in buying.

The first factor is a size factor which distinguishes between persons who select higher scores for all or most such characteristics from those who select lower values. Those who give higher marks are also people who manifest a greater satisfaction, interest in buying and are over 44 years old. The positive side of the second factor corresponds to higher scores given to sober, traditional, stylish and artistic and to respondents over 44, teaching-research staff and men and the negative side corresponds to higher scores given to daring, original and modern. Finally, the third factor locates individuals scoring high the term practical, who are mostly students and under 30.

After performing a hierarchical clustering on the PCA first 5 axes, using the generalized Ward criterion, this results in three clusters. The first one (46%) corresponds exactly to those on the positive side of the first factor (over 44, fully satisfied, with buying interest, high scores to all characteristics). The second one (31%) to individuals who rank high the characteristics of original, daring, modern and practical and who are students, under 30, neither satisfied or dissatisfied and who do not manifest buying interest. This is a group who might be attracted to the first group, composed of feasible buyers, by improving the characteristics of the products in the way they consider important. The last cluster (23%) give low scores to most of the characteristics and manifest no interest in buying and are also indifferent to the institution. This group seems a difficult one to reach to.

This first analysis provides three main directions of variability by means of a PCA. The clustering over the main factors helps to group individuals into homogeneous families where each cluster represents a market segment with different characteristics and reachable through different marketing strategies or perhaps products not considered here.

MCA of categorical variables and classification

As a second factor method, we choose the categorical variables referring to valuation of the 26 articles (after seeing a displayed photo) in a scale 1-4 as the active variables of a MCA. As supplementary variables we choose the products characteristics, the satisfaction variable, the intention to buy and the individuals' personal data.

Figure 1 shows the projection of the active categories on the MCA main plane. It shows how the first factor represents a global propensity to buy, roughly ordering categories from left to right with respect to their probability to buy, from lower to higher. The plane shows a typical Guttman effect with the second factor reflecting differences between extreme and centered opinions.

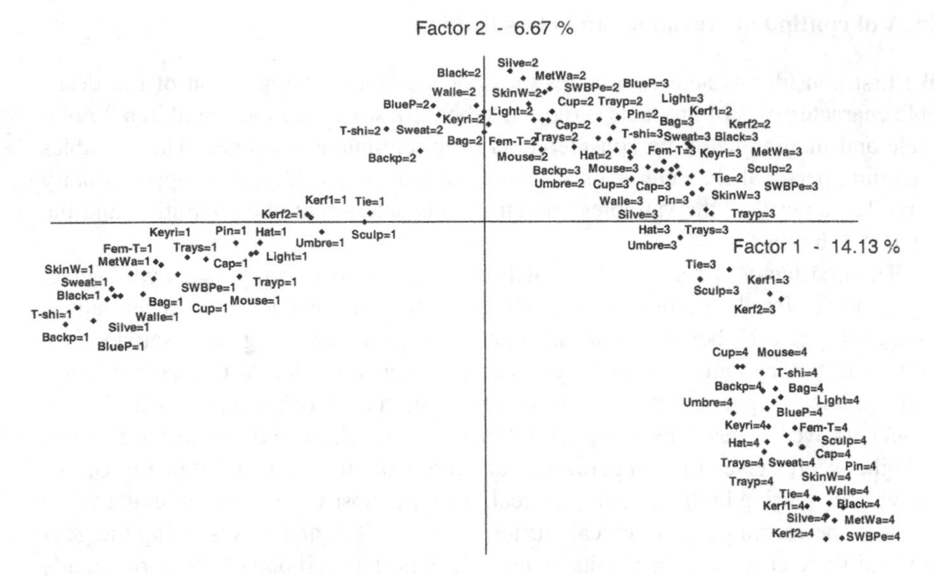

Fig. 1. MCA: active categories on plane (1,2).

With respect to the projections of the supplementary categories, it is shown in Figure 2 that the first factor is positively related to the satisfaction with the institution and the declared propensity to buy. This shows the relationship of these variables with the overall propensity to buy individually the 26 products.

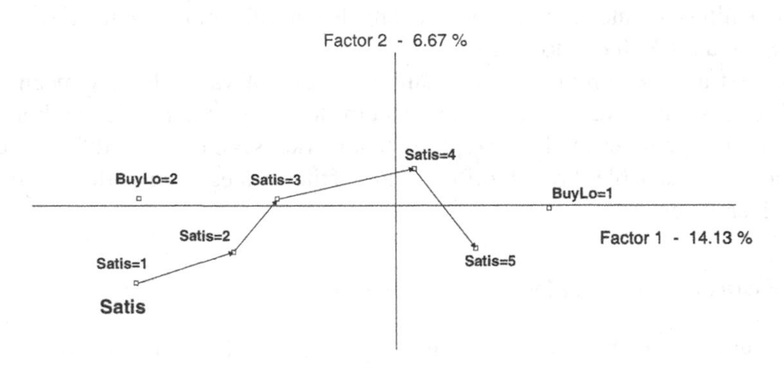

Fig. 2. MCA: supplementary categories on plane (1,2).

A mixed classification in three steps is carried out on 8 MCA first principal axes. This process starts by choosing a partition in 10 clusters with random initial centers and then update those centers calculating the centroids of the groups of individuals nearest to the centers (K-means algorithm); the process is repeated until the clusters are stable. We reduce further the number of clusters by means of a hierarchical algorithm (generalized Ward's method) and refine the resulting partition with a consolidation step with re-assignment (testing moving centers with convergence achieved in 7 iterations). This results in a partition of 6 classes with an inter inertia over total inertia ratio of 55.62%. The positions of the final centers on the plane are given in Figure 3, and are following the pattern set by the active categories on this same plane.

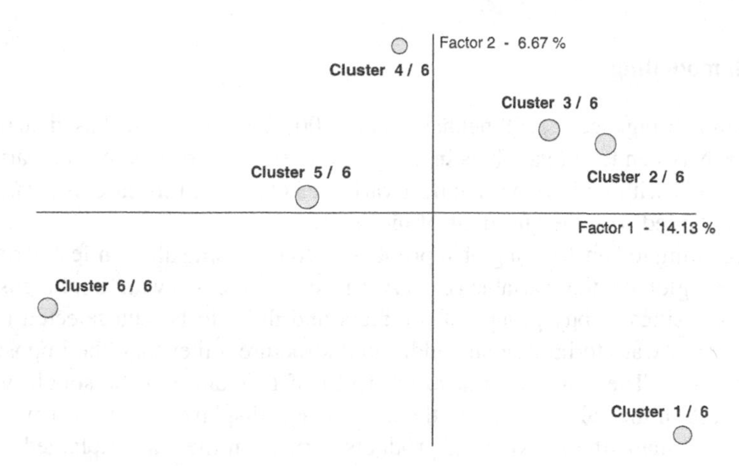

Fig. 3. Classification on MCA factors. Clusters centers and relative sizes represented by circle diameters.

The partition description is as follows. Cluster 1 (15.73%) contains those who would prior buy, say is very likely to buy for many products, are over 44, fully satisfied, females, members of the teaching and research staff, give high scores to stylish and traditional. Cluster 2 (17.91%) is formed by those who are likely to buy, over 44, would prior buy and rank highly stylish, traditional and sober. In cluster 3 (17.74%) predominate those who say it is unlikely to buy sober and stylish products (metallic) but it is likely to buy original, modern and practical products (textiles and bags). Cluster 4 (12.80%) groups individuals unlikely to buy anything with low scores for stylish products. Cluster 5 (18.66%) is composed of individuals very unlikely to buy, aged between 18 and 22, students, from Gipuzkoa campus, neither satisfied or dissatisfied and with low scores on traditional, sober or stylish. Finally, on cluster 6 (17.16%) are those who are very unlikely to buy, between 30 and 44, males and with low marks for all characteristics of the products.

This MCA confirms the tight relationship between the interest to buy articles featuring the logo (before visualization), the degree of satisfaction about the institution and the scores given to the proposed desirable characteristics of the products. The clustering process shows marketing implications on the buyers' and non-buyers' personal characteristics and on which articles are perceived as stylish, traditional and sober and which ones as modern, original and practical. Furthermore, the parabolic path apperaring in Figure 1 is similar to those shown in Figures 2 and 3, reinforcing its interpretation as an indicator of the propensity to buy the displayed products.

3.2 Non-symmetrical methods: regression related techniques

In this section we consider methods where one variable is chosen to be depending on others. In this work, the variable of interest is the probability, or propension, to buy and is exactly our choice for the endogeneous variable.

PLS path modelling

PLS path modelling (see, *e.g.*, Tenenhaus et al. (2005)) is a technique based on the relationships between latent variables in a regression framework where such variables are constructed with underlying manifest variables (MV). In this case, the variables are those obtained with the questions of the survey.

We are going to construct a global propensity to buy using all manifest variables, resulting in a global latent variable (LV). At the same time, we want unidimensional partial propensities to buy groups of products and these to be autoselected by the data, we do not want to impose any additional structure, other than the imposed by the model itself. These will also have the form of LVs and will be sought with a previous PCA of the valuations of all the 26 products displayed in the survey.

Table 2 contains the 8 groups of products formed in the way explained above. These groupings originate directly 8 partial LVs, using mode B.

Table 2. Groups of products to be considered as LV.

label	LV	products
umbh	ξ_1	umbrella, hat
tie	ξ_2	tie, kerchief no.1, kerchief no.2
textiles	ξ_3	T-shirt, T-shirt-V, sweater, cap
bag	ξ_4	plastic tray, leather tray, backpack, bag, cup
wat	ξ_5	leather-strapped watch, metallic-strapped watch, wallet
mous	ξ_6	keyring, lighter, mousepad
scul	ξ_7	pin, sculpture
pens	ξ_8	blue pen, black pen, silver pen, silver pen in wooden case

Selecting all products valuations, we construct the global propensity to buy using mode A. Finally, we formulate the external model $\xi = \sum_{j=1}^{8} \beta_j \xi_j + \nu$.

Figure 4 shows the path model specified. The numbers are correlations and show relatively high values between the partial LVs and the global one. We can also see the pairwise correlations between individual MVs and the LVs.

The actual estimates of the external model parameters are given in equation (1). These show higher values for textiles, bags and pens products groups, which are those with a higher acceptability among the respondents.

$$E(\xi) = 0.0865 * \text{umbh} + 0.1335 * \text{tie} + 0.2041 * \text{textiles} + 0.2114 * \text{bag} \\ + 0.1791 * \text{wat} + 0.1292 * \text{mous} + 0.0881 * \text{scul} + 0.2322 * \text{pens} \quad (1)$$

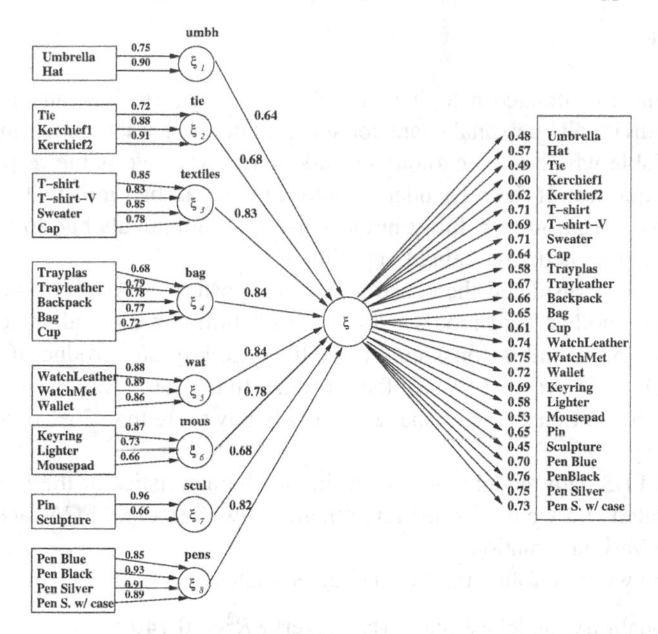

Fig. 4. PLS path diagram for products to be sold at the university shop.

In order to get a potential buyers' characterization (similar to the projection of supplementary variables in a factor analysis), we perform a regression on the desirable characteristics of the products and the respondents' personal characteristics. This is actually a Principal Components Regression (PCR), since the desirable characteristics are highly correlated, selecting 2 main components out of the 7 original variables.

$$\begin{aligned} E(\xi) = & -0.85 + 0.07 * \text{F1 (orig., daring, practical, artistic, modern)} \\ & +0.11 * \text{F2 (traditional, sober, stylish)} - 0.25 * \text{male} \\ & +0.15 * \text{satisfied} + 0.26 * \text{very satisfied} + 0.07 * \text{age(+44)} \\ & +0.06 * \text{teaching-research staff} - 0.10 * \text{higher education} \\ & +1.18 * \text{overall propensity to buy a logo product} \\ & +0.14 * \text{campus: Araba} + 0.12 * \text{campus: Bizkaia} \end{aligned}$$

$$R^2 = 0.4848$$

All parameters whose estimates are shown are significant at the 5% level, both using bootstrap confidence intervals and usual t-test statistics. These estimates show how those individuals most satisfied with the university are more likely to buy, along with women. It is also so for those who have a prior intention to buy, members of teaching and research staff, older age and those proceeding from the campuses of Bizkaia and Araba from over those from Gipuzkoa. With respect to product characteristics, those marking as more important the terms traditional, sober and stylish are more likely to buy than individuals giving more importance to aspects as modern, practical and so on.

Logit models

Finally, we have calculated a logit regression (see, *e.g.*, Hosmer and Lemeshow (2000)) on individuals' personal characteristics, products characteristics and the satisfaction variable where the dichotomous endogeneus variable is the response (yes or no) to the question if the respondent would, in general, buy university corporate products. This is a prior probability in the sense that individuals had to respond to that question before actually seeing the products.

We have also considered the construction of a posterior probability to buy and then estimated another logit model with this probability as the endogeneous variable. Thus, an individual is considered to be likely to buy one product if he or she scores 3 (likely) or 4 (very likely) for that product. In the same way, an individual is considered to buy articles if he or she would likely buy more than 25% of all articles (at least 7 articles).

As in the PLS path model case, the desirable characteristics of the products are highly correlated and we have substituted them by two principal PCA factors (after performing a Varimax rotation).

We end up with the following two model estimates:

1. Prior probability model estimates (Nagelkerke $R^2 = 0.140$):

$$
\begin{aligned}
X'\beta = & -0.510 + 0.267 * \text{teach./res.} + 0.307 * \text{Bizkaia} + 0.398 * \text{age over 44} \\
& +0.797 * \text{satisfied} + 1.160 * \text{very satisfied} + \\
& +0.220 * \text{F1 (innovative+practical)} + 0.272 * \text{F2 (classic)}
\end{aligned}
$$

2. Posterior estimates (Nagelkerke $R^2 = 0.502$):

$$
\begin{aligned}
X'\beta = & -1.298 + 0.537 * \text{student} + 0.584 * \text{teach./res.} - 0.794 * \text{male} \\
& +0.367 * \text{satisfied} + 0.710 * \text{very satisfied} + 0.339 * \text{F2 (classic)} \\
& +2.979 * \text{buying initial interest}
\end{aligned}
$$

The prior probability model yields very similar results to those from the PLS path model and the factor analyses performed in the previous subsection. The posterior probability model yields, with a better fit, results not so similar, what can be due to the particular construction of the endogeneous variable. That construction is sensitive but also subjective and it can only be considered as a help to better know the structure of the data.

4 Conclusions

Each different technique used shows specific, though related, conclusions given its different objectives. The symmetrical methods (PCA, MCA) combined with Cluster Analysis help to learn what is contained in the data, including relationships and classifications of similar individuals. On the other hand, non-symmetrical methods as

PLS or Logit regressions allow for modelling individuals' global and partial (group) behaviour using inference tools to select a better model with a good fit to the data.

The methods exposed above extract consistently some facts from this particular data. The gift shop potential buyers' general characteristics become clear (satisfied with the institution, members of the teaching-research staff, women...). At the same time, it is also clear the general characteristics of the articles shown (traditional, ...) and the sort of characteristics of possible successful articles not covered in current product line (practical, original or modern). It seems that a better, more modern, design is needed to reach other market segments.

The marketing implications obtained have been somewhat conditioned upon the actual articles displayed with photographs in the on-line questionnaire. It has been observed that many have been perceived as stylish and traditional (generally of a metallic aspect) and of little appeal for the young. As a general issue, this work recommends the promotion of articles with the characteristics mentioned above and, particularly, belonging to the groups of textiles, bags and desktop articles which would yield a better acceptance for this target public in the opening university gift shop.

All in one, it can be said that these data mining techniques yield useful directions for the university marketing policy, regarding the corporate shop. The combination of techniques, though never fully exhaustive, reinforces the confidence on the results as it is improbable to having missed important patterns in the data.

References

GREENACRE, M. (1984): *Theory and Applications of Correspondence Analysis*. (Academic Press, London)

HOSMER, D. R. and LEMESHOW, S. (2000): *Applied Logistic Regression*. 2nd Edition, Wiley & Sons Inc, USA

LEBART, L. (1994): *Complementary use of correspondence analysis and cluster analysis*. In: Greenacre, M.J. and Blasius, J. (Eds.): *Correspondence Analysis in the Social Sciences*.

LEBART, L., MORINEAU, A. and WARWICK, K. (1984): *Multivariate Descriptive Statistical Analysis*. (Wiley, NewYork)

TENENHAUS, M., E. VINZI, V., CHATELIN, Y.M. and LAURO, C. (2005): *PLS path modeling*. Computational Statistics & Data Analysis, 48, 159–205.

Nonlinear Constrained Principal Component Analysis in the Quality Control Framework

Michele Gallo[1] and Luigi D'Ambra[2]

[1] Department of Social Science, University of Naples - L'Orientale
Largo S.Giovanni Maggiore 30, 80134 Naples, Italy
mgallo@unior.it,

[2] Department of Mathematics and Statistics, University of Naples - Federico II
Via Cinthia 26, 80126 Naples, Italy
dambra@unina.it

Abstract. Many problems in industrial quality control involve n measurements on p process variables $X_{n,p}$. Generally, we need to know how the quality characteristics of a product behavior as process variables change. Nevertheless, there may be two problems: the linear hypothesis is not always respected and q quality variables $Y_{n,q}$ are not measured frequently because of high costs. B-spline transformation remove nonlinear hypothesis while principal component analysis with linear constraints (CPCA) onto subspace spanned by column X matrix. Linking $Y_{n,q}$ and $X_{n,p}$ variables gives us information on the $Y_{n,q}$ without expensive measurements and off-line analysis. Finally, there are few uncorrelated latent variables which contain the information about the $Y_{n,q}$ and may be monitored by multivariate control charts. The purpose of this paper is to show how the conjoint employment of different statistical methods, such as B-splines, Constrained PCA and multivariate control charts allow a better control on product or service quality by monitoring directly the process variables. The proposed approach is illustrated by the discussion of a real problem in an industrial process.

1 Introduction

Frequently firms have to define how to select the process parameters which mostly influence the quality characteristics of a product. The selection of the "optimal" combination of parameters and the choice of statistical methods to solve this problem could be no simple question. In this paper, we have proposed some statistical techniques to determinate the "best" technology for pasta production.

Quality characteristics of pasta, tested in laboratory, can be divided in two clusters: "colour-appeal" and "taste". Customers prefer clear and amber pasta without red vein. Besides, the pasta must be characterise by "al dente" stage in case of overcooking or undercooking (Abecassis et al., 1992).

In this paper, we suggest a nonlinear approach to select the "best" technology for pasta production, spaghetti about 0.04 in (diameter), and choose process parameters was to monitor.

In the first step, we define the different setting of the manufacturing process which can be used. To obtain an optimal setting, it is necessary to consider three process parameters: temperature (***T***), drying time (***DT***), damp (***D***). Forty-five tests have been running with different combinations of process parameters. At the same time, quality characteristics have been measured by six variables: viscosity on a 1-9 category scale (***V***), judgement on taste in case of overcooking (***N1***) and undercooking (***N2***) on a 0-9 category scale, homogeneity of red (***A***), yellow (***B***), brown (***100-L***). In the second step, we define every new relation between response variables ($Y_{45,6}$) and process variables ($X_{45,7}$) by-means of multivariate statistical methods such as Constrained Principal Component Analysis (CPCA - D'Ambra and Lauro, 1982). In the third step, since CPCA analysis shows a horseshoes effect in data set, we propose a B-spline transformation in data before interpreting results. In the last step, we define, by means a Shewhart charts, the "optimal" combination of process parameters which produces the "best" pasta.

The use of traditional control charts to monitor the process variables instead of the response ones is a good solution for many reasons. First, the process variables are measured much more frequently, usually in the order of seconds or minutes as compared to hours for the response variables. Second, process variables are generally measured in a more precise way than response variables. Third, CPCA components are always independent even when single variables are correlated.

The aim of this paper is to show how the CPCA methods can be used in case of nonlinear data and the employment of techniques like Multivariate Principal Component Charts (MPCC - MacGregor and Kourti, (1995)) can aid in the interpretation of results. The paper is organised as follows. In Section 2 CPCA method is applied to pasta data. A horseshoes effect is present on raw data. Different approaches to solve this problem is given in Section 3, in particular, B-spline transformation on X data is applied. In Section 4 the results of CPCA on B-spline transformed data are tested by a stability analysis. A first interpretation of CPCA results is given in Section 5.

2 Constrained principal component analysis

Let $X_{n,p}$ and $Y_{n,q}$ be the raw data matrices associated with two sets of quantitative variables observed on the same experimental units. Furthermore let Q and D be symmetric and positive definite matrices of qth-order and nth-order respectively. In the remainder of the paper, we will consider X and Y standardised data matrices, hence $Q = 1$. The CPCA (D'Ambra and Lauro, 1982) aim is to analyse the structure of the explained variability of the Y data set given the process variables X. Let

$$P_X = X(X'D_X X)^{-1}X' \tag{1}$$

be the D-orthogonal projector onto the space spanned by the columns of X CPCA consists in carrying out a PCA on the matrix

$$\widehat{Y} = P_X Y \tag{2}$$

Figure 1 shows a scatter plot of the first two principal components of the statistical study $(\widehat{Y}, D, I)$. It explain nearly all the data variability (87.80%) but in this representation the second axis is a special arched function of the first axis. CPCA creates a serious artifact called the horseshoes effect. This is a problem because CPCA perform better when the 45 experimental tests have a monotonic distributions along gradients (i.e. either increase or decrease but not both). To resolves horseshoes problem and gives more interpretable results, nonlinear transformation of data can be used (Gifi, 1990).

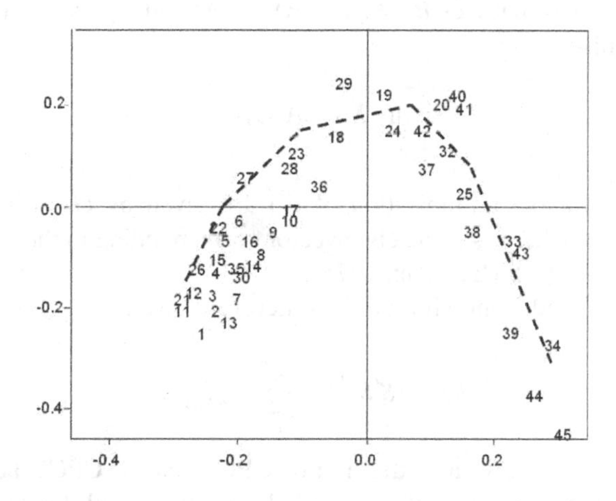

Fig. 1. Plot of the first and second Constrained Principal Component.

3 Nonlinear Constrained Principal Component Analysis

B-spline approach (Durand, 1993) allows a greater flexibility in the adjustment of dependence between the X and Y sets of variables.

Let $S_j(x_j)B_j$ be the transformation of x_j-column, $j = 1, \ldots, p$, $S_j(n,k)$ the B-basis spline with a priori fixed order and knots (De Boor, 1978; Eubank, 1988), $B_j(k,q)$ is the matrix of coefficient.

Similarly we can write S as:

$$S_{(n, \sum k)} = [S_1(x_1)| \ldots |S_p(x_p)] \tag{3}$$

and B as:

$$B_{(\sum k, q)} = \begin{bmatrix} B_1 \\ \vdots \\ B_p \end{bmatrix} \tag{4}$$

Consider the following multivariate model

$$X = S\widehat{B} + \widehat{E} \tag{5}$$

In order to estimate the $\widehat{B}$ we will minimise the trace $\widehat{E}'\widehat{E}$. Then

$$\min_{B} \|X - S\widehat{B}\|^2 \tag{6}$$

Consider the class of reduced-rank regression for the multivariate linear model with $rank(\widehat{B}) = r \leq [min(\Sigma k, q)]$ (Izenman, 1975). With such condition there will exist two (non-unique) matrices $\widehat{B}_r = \widehat{A}_r \widehat{G}_r$ where $\widehat{A}_r$ and $\widehat{G}_r$ are both of rank r. So we have to minimise

$$\min_{\widehat{A}_r \widehat{G}_r} \|X - S\widehat{A}_r \widehat{G}_r\|^2 \tag{7}$$

The solutions for the minimisation of (7) are given by $\widehat{G}_r = [v'_1 \ldots v'_r]$, $\widehat{A}_r = (S'S)^{-1}S'X[v_1 \ldots v_r]$ where v_k is the eigenvector corresponding to the k largest eigenvalue λ_k of $Y'S(S'S)^{-1}S'Y$ (Izenman, 1975).

The regression coefficient with rank r is therefore given by

$$\widehat{B}_r = (S'S)^1 S'X[\sum_{k=1}^{r} v_k v'_k] \tag{8}$$

This solution is linked to an extension of CPCA, called CPCC-additive spline, concerning a PCA of the image of y_j onto B-basis spline with knots chosen in the range of each y_j, $j = 1, \ldots, p$. In this case we have $\widehat{Y}^* = P_s Y = S(S'S)^{-1}S'Y$ and we carrying out PCA of the statistical study $(\widehat{Y}^*, D, I)$.

A second approach (Durand, 1993) searches a matrix transformation C of X and a matrix R to minimise the distance between the scalar product operators $YY'D$ and $CRC'D$:

$$\min_{C,R} \|YY'D - CRC'D\|^2 \tag{9}$$

with $C = S\widehat{B}$ and where S is the B-spline matrix with a priori fixed order and knots. The minimum can be attained by an approximate solution based on an alternate iterative procedure.

A more recent method is the two-stage approach to engine mapping by using B-spline basis functions at the second stage to describe the effects of one or more factors (splined factors) and low-order monomials to represent the main effects and interactions of the remaining (nonsplined) factors (Grove et al., 2004).

In this paper we have used the first approach. The first principal component of the statistical study $\widehat{Y}^*, D, I$ explains the 81.10% of the total variation of the matrix Y. The 96.80% of the total variance is explained by the first two principal components. A stability analysis can be performed to evaluate the goodness of the results.

4 Stability analysis

Daudin (1988) suggests the study of stability by bootstrap. The basic idea of bootstrap is to generate many new matrices starting from the raw data. Any new matrix is obtained by a random replacement of the original rows. Applying bootstrap on $\widehat{Y}^*$, we generate m new matrices ${}^l\widehat{Y}^*$ where $l = 1,\ldots,m$. Let λ_i and $\widehat{\mu}_i$ be the ith eigenvalue and the associated eigenvector of the correlation matrix of $\widehat{Y}^*$ and ${}^l\widehat{\mu}_f$ the fth eigenvector of the correlation matrix of ${}^l\widehat{Y}^*$. Furthermore let

$$ {}^l\rho_{if} = cos({}^l\widehat{\mu}_f,\widehat{\mu}_i) = \sqrt{\lambda_i}\frac{{}^l\widehat{\mu}'_f,\widehat{\mu}_i}{({}^l\widehat{\mu}'_f,\widehat{\mu}_f)^{1/2}} \tag{10} $$

and

$$ {}^l\gamma_k = \Sigma_{i<k}\Sigma_{f<k}{}^l\rho_{if}^2 \tag{11} $$

where $i,f = 1,\ldots,k$, and k is the number of the examined eigenvalues.

Plotting ${}^l\gamma$ respect x-axis and j versus y-axis , thc orientation of the first two eigenvectors seems to be stable (Figure 2). In fact, it is not considerably modified over the 250 replications.

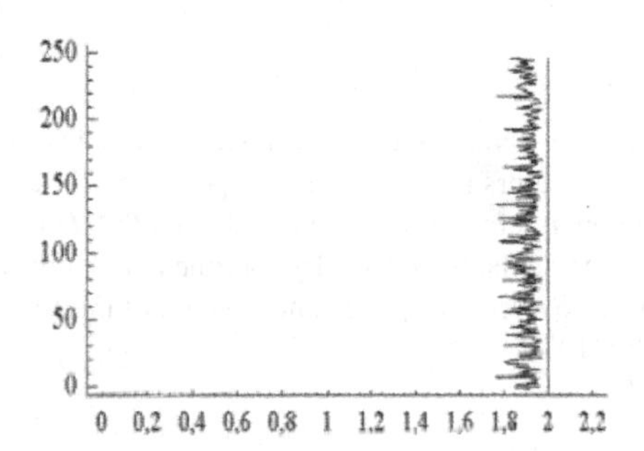

Fig. 2. The stability representation for the first two eigenvectors.

The stability of the components can be confirmed by the following quantity

$$ \mathrm{MSE}(k) = \frac{1}{m}\Sigma_{i<k}\Sigma_{f<k}({}^l\gamma_k - k)^2. \tag{12} $$

If MSE(k) is near to zero, the examined k components are stable. Here for $k = 2$ and $m = 250$ the result is 0.000084.

5 Results and interpretation

Figure 3 shows the representation of 45 samples of the first two principal components, where the percent of the total variability explained is 96.80. This percentage allows a good description of data structure and stability analysis indicates that this data structure could be considered stable. Furthermore, the dotted lines show that B-spline transformation have smoothed raw data and the problem of nonlinearity would seem to be eliminated.

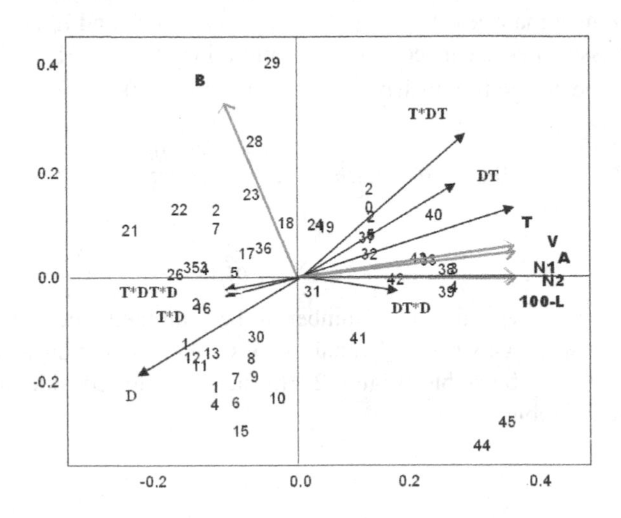

Fig. 3. Plot of the first and second Nonlinear Constrained Principal Component; the points are the 45 different tests and the vectors are variables: temperature (***T***); drying time (***DT***); damp (***D***); interaction between temperature, drying time and damp (***T*DT*D***); temperature and drying time(***T*DT***); temperature and damp (***T*D***); drying time and damp (***DT*D***); viscosity (***V***); judgement on taste in case of overcooking (***N1***) and undercooking (***N2***); homogeneity of red (***A***), yellow (***B***), brown (***100-L***).

The first axis of representation could be called "taste" as it is characterised by contributions of viscosity and judgement on taste together with contributions of red and brown colour. All these variables are positively correlated with "taste" (about 0.97). The second axis could be called "colour-appeal", as the yellow colour contributes to this axis with 98%. The process variables which have a positive influence on "taste" are temperature, drying time, their interaction and the interaction between drying time and damp. On the contrary, all the other variables have a negative influence. The second axis is characterised mostly by the fact that drying time and damp are each at the opposite side of the other, this contrast influences the homogeneity of yellow.

The PCA of statistical study $(\widehat{Y}^*, D, I)$ indicates only which process variables influence the quality characteristics of products. The direction where to look for the best combination of process parameters (Abecassis et al., 1992) is along the diagonal **D-DT** (Figure 3). This information could be not sufficient clear because along the diagonal **D-DT** there are a lot of different combinations of process parameters. In this case, a graphic display, such as Shewhart charts, could give some information about the optimal combination to choose for the production of pasta.

The scores could be projected onto Shewhart charts where Central Line (CL), Upper Decision Line (UDL) and Lower Decision Line (LDL) are 0, 0+3 and 0-3 respectively. In this paper, these "Multivariate Principal Component Charts" (MPCC) are used for the first principal component (Figure 4.a), and the second one (Figure 4.b) or both, according to marketing decisions, that is maximise "taste" or "colour-appeal" or choose the optimal mix of "taste" and "colour-appeal".

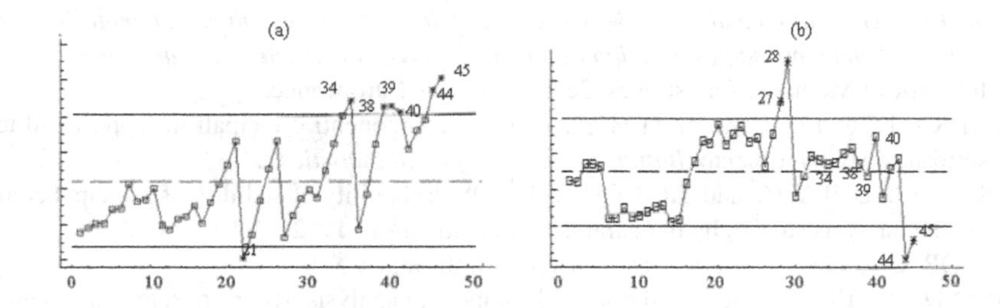

Fig. 4. MPCC for the first (a) and the second (b) Nonlinear Constrained Principal Component.

In Figure 4.a, the experimental tests 34, 38, 39, 40, 44 and 45 could suggest that temperature must be higher than 100°C to give the best value for "taste". Figure 4.b shows that the best value for "colour-appeal" is obtained in correspondence of temperature 90°C, drying time 2.5 or 5, and damp 5.5. The "optimal" mixture of "taste" and "colour-appeal" is obtained in correspondence of the maximum value taken in Figure 4.b, by the experimental tests which are out of the UDL in Figure 4.a. The experimental test 40 could be represents the "optimal" combination of parameters in term of "taste" and "colour appeal".

6 Concluding remarks

Today the advent of on-line process computer system have totally changed the nature of the data that are available. The use of multivariate statistical methods is necessary to treat the problems associated with these large volumes of messy data. We can use all the information contained in data, to improve the quality of products and processes. Multivariate analysis as Constrained Principal Component Analysis could be employed to determine the relationships between the quality characteristic of products with the process parameters. In this way, we can select the best technology to get a quality product and/or to monitor the quality characteristics of product by process variables.

In many situation it is reasonable attend to the presence of anomalies observations, in these cases principal components are influenced and may not capture the variation of the regular observations. Therefore, data reduction based on PCA becomes unreliable. When outliers are present in the data, to obtain a more accurate estimates at noncontaminated data sets and more robust estimates at contaminated data a method for robust principal component analysis could be used (Hubert et al., 2005).

References

ABECASSIS, J., DURAND, J.F., MEOT, J.M., and VASSEUR, P. (1992): *Etude non linéaire de l'influence des conditions de séchage sur la qualité des pâtes alimentaires par l'analysis en composantes principales par rapport à des variables instrumentales.* Agro-Industrie et Methodes Statistiques, 3èmes Journees Europeennes.

D'AMBRA, L., and LAURO, N. (1982): Analisi in componenti principali in rapporto ad un sottospazio di riferimento. *Italian Journal of Applied Statistics*, 1.

DAUDIN, J.J., DUBY, C., and TRECOURT, P. (1988): Stability of stability of principal component analysis studied by bootstrap method. *Statistics* , 19, 2.

DE BOOR, C. (1978): *A practical guide to splines.* Springer, N.Y.

DURAND, J.F. (1993): Generalized principal component analysis with respect to instrumental variables via univariate spline trasformations. *Computational Statistics Data Analysis*, 16, 423-440.

EUBANK, R.L. (1988): *Smoothing splines and non parametric regression.* Markel Dekker and Bosel, N.Y.

GIFI, A. (1990): *Nonlinear Multivariate Analysis.* Wiley, Chichester, England.

GROVE, D.M., WOODS, D.C., and LEWIS, S.M. (2004): Multifactor B-Spline Mixed Models in Designed Experiments for the Engine Mapping Problem. *Journal of Quality Technology*, 36, 4, 380-391.

IZERNMAN, A.J. (1975): Reduced-rank regression for the multivariate linear model. *Journal of Multivariate Analysis*, 5.

HUBERT, M., ROUSSEEUW, P.J., and BRANDEN, K.V. (2005): ROBPCA: A New Approach to Robust Principal Component Analysis. *Technometrics*, 47, 1.

MACGREGOR, J.F. and KOURTI, T. (1995): Statistical Process Control of Multivariate Processes. Control Engineering Practice, 3, 3, 403-414.

Non Parametric Control Chart by Multivariate Additive Partial Least Squares via Spline

Rosaria Lombardo[1], Amalia Vanacore[2] and Jean-Francçois Durand[3]

[1] Faculty of Economics, Second University of Naples, Italy
rosaria.lombardo@unina2.it
[2] Faculty of Engineering, University of Naples "Federico II", Italy
amalia.vanacore@unina.it
[3] Faculty of Maths, University of Montpelier II, France
jfd@ensam.inra.fr

Abstract. Statistical process control (SPC) chart is aimed at monitoring a process over time in order to detect any special event that may occur and find assignable causes for it. Controlling both product quality variables and process variables is a complex problem. Multivariate methods permit to treat all the data simultaneously extracting information on the "directionality" of the process variation. Highlighting the dependence relationships between process variables and product quality variables, we propose the construction of a non-parametric chart, based on Multivariate Additive Partial Least Squares Splines; proper control limits are built by applying the Bootstrap approach.

1 Introduction

The multivariate nature of product quality (response or output variables) and process characteristics (predictors or input variables) highlights the limits of any analysis based exclusively on descriptive and univariate statistics. On the other hand, the possibility for process managers of extracting knowledge from large databases, opens the way to analyze the multivariate dependence relationships between quality product and process variables via predictive and regressive techniques like PLS (Tenenhaus, 1998; Wold, 1966) and its generalizations (Durand, 2001; Lombardo *et al.*, 2007). In this paper, the application of a multivariate control chart based on a generalization of PLS-T^2 chart (Kourti and MacGregor, 1996) is proposed in order to analyze the in-control process and monitoring it over time. Furthermore, in order to face the problem of the unknown distribution of the statistic to be charted, a non-parametric approach is applied for the selection of the control limits. Distribution-free or non-parametric control charts have been proposed in literature to overcome the problems related to the lack of normality in process data. An overview in literature on univariate non-parametric control charts is given by Chakraborti *et al.* (2001). The principles on which non-parametric control charts rest can be generalized to multivariate settings. In particular, the bootstrap approach to estimate control limits (Wu and Wang, 1997; Jones and Woodall, 1998; Liu and Tang, 1996) has been followed.

2 Multivariate control charts based on projection methods

A standard multivariate quality control problem occurs when an observed vector of measurements on quality characteristics exhibits a significant shift from a set of target (or standard) values. The first attempt to face the problem of multivariate process control is due to Hotelling (1947) who introduced the well-known T^2 chart based on variance-covariance matrix. Successively, different approaches to take into account the multivariate nature of the problem were proposed (Woodall, Ncube, 1985; Lowry *et al.*, 1992; Jackson, 1991; Liu, 1995; Kourti and MacGregor, 1996, MacGregor, 1997). In particular, we focus on the approach based on PLS components proposed by Kourti and MacGregor (1996), in order to monitor over time the dependence structure between a set of process variables and one or more product quality variables (Hawkins, 1991). The PLS approach proves to be effective in presence of a low-ratio of observations to variables and in case of multicollinearity among the predictors, but a major limit of this approach is that it assumes a linear dependence structure. Generally, linearity assumption in a model is reasonable as first research step, but in practice relationships between the process variables and the product quality variables are often non-linear and in order to study the dependence structure it could be much more appropriate the use of non-linear models (PLS via Spline, i.e. PLSS; Durand, 2001) as proposed by Vanacore and Lombardo (2005). The PLSS-T^2 chart allows to handle non-linear dependence relationships in data structure, missing values and outliers, but it presents two major drawbacks: 1) it does not take into account the possible effect of interactions between process variables; 2) it requires testing normality assumption on the component scores, even when original data are multinormal (in fact, in case of spline, i.e. non linear transformations of original process variables, the multinormality assumption cannot be guaranteed anymore). To overcome these drawbacks we present non-parametric Multivariate Additive PLS Spline-T^2 chart based on Multivariate Additive PLSS (MAPLSS, Lombardo *et al.*, 2007) briefly described in sub-section 2.2.

2.1 Review of MAPLSS

MAPLSS is just the application of linear PLS regression of the response (matrix $\mathbf{Y}$ of dimension n,q) on linear combinations of the transformed predictors (matrix $\mathbf{X}$ of dimension n,p) and their interactions. The predictors and bivariate interactions are transformed via a set of $K = d+1+m$ (d is the spline degree and m is the knot number) basis functions, called B-splines $B_l(.)$, so as to represent any spline as a linear combination

$$s(x,\beta) = \sum_{l=1}^{K} \beta_l B_l(x),$$

where $\beta = (\beta_1, .., \beta_K)$ is the vector of spline coefficients computed via regression of $y \in I\!R$ on the $B_l(.)$ The centered coding matrix or design matrix including interactions becomes

$$\mathbf{B} = [\underbrace{\ldots \mathbf{B}^i \ldots}_{i \in K_1} | \underbrace{\ldots \mathbf{B}^{k,l} \ldots}_{(k,l) \in K_2}], \tag{1}$$

where K_1 and K_2 are index sets for single variables and bivariate interactions, respectively. In a generic form, the MAPLSS model, for the response j, can be written as

$$\hat{y}^j(A) = \sum_{l \varepsilon L} \hat{\beta}^j_l(A)\mathbf{B}^l, \tag{2}$$

where A is the space dimension parameter and L is the index set pointing out the predictors as well as the bivariate interactions retained by MAPLSS. It is thus a purely additive model that depends on A which in turn depends on the spline parameters (i.e. degree, number and location of knots).
Increasing the order of interaction in MAPLSS implies expanding the dimension of the design matrix **B**. MAPLSS constructs a sequence of centered and uncorrelated predictors, i.e. the MAPLSS (latent) components $(\mathbf{t}^1, ..., \mathbf{t}^A)$. We now briefly describe the MAPLSS building-model stage. In the first phase we do not consider interactions in the design matrix. This phase consists of the following steps

step 1 Denote $\mathbf{B}_0 = \mathbf{B}$ and $\mathbf{Y}_0 = \mathbf{Y}$ the design and response data matrices, respectively. Define $\mathbf{t}^1 = \mathbf{B}_0\mathbf{w}^1$ and $\mathbf{u}^1 = \mathbf{Y}_0\mathbf{c}^1$ as the first MAPLSS components, where the weighting unit vectors $\mathbf{w}^1$ and $\mathbf{c}^1$ are computed by maximizing the covariance between linear compromises of the transformed predictors and response variables, $cov(\mathbf{t}_1, \mathbf{u}_1)$.

step k Compute the generic MAPLSS component

$$\mathbf{t}^k = \mathbf{B}_{k-1}\mathbf{w}^k\mathbf{u}^k = \mathbf{Y}_{k-1}\mathbf{c}^k. \tag{3}$$

Update the new matrices $\mathbf{B}_k$ and $\mathbf{Y}_k$ as the residuals of the least-squares regressions on the components previously computed using the orthogonal projection operator $\mathbf{P}_{\mathbf{t}^k}$ on $\mathbf{t}^k$, that is $\mathbf{P}_{\mathbf{t}^k} = \mathbf{t}^k\mathbf{t}^{k\prime}/\|\mathbf{t}^k\|^2$, we write

$$\mathbf{B}_k = \mathbf{B}_{k-1} - \mathbf{P}_{\mathbf{t}^k}\mathbf{B}_{k-1} \tag{4}$$

$$\mathbf{Y}_k = \mathbf{Y}_{k-1} - \mathbf{P}_{\mathbf{t}^k}\mathbf{Y}_{k-1}. \tag{5}$$

Final Step The algorithm stops on the base of the A number of components defined by PRESS criterion.

In the second phase of the MAPLSS building-model stage, we individually evaluate all possible interactions. The rule for accepting a candidate bivariate interaction is based on the gain in fit (R^2) and prediction (GCV criterion) compared to that of the model with main effects only. Then, the selected interactions are ordered in decreasing value for consideration to adding them step-by-step to the main effects model. At

the end, in the final phase we include in the design matrix **B** the selected interactions and repeat the algorithm from *step 1* to the *final step*.
A simple way to illustrate the contribution of predictors to response variables, consists of ordering the predictors with respect to their decreasing influence on the response $\hat{y}^j(A)$, using as a criterion, the range of the $s_i(\mathbf{x}^i, \hat{\beta}_\mathbf{i}^\mathbf{j}(\mathbf{A}))$ values of the transformed sample $\mathbf{x}^i$ (see figure 3). One can also use the same criterion to prune the model, by eliminating the predictors and/or the interactions of low influence so as to obtain a more parsimonious model.

2.2 MAPLSS-T^2 chart

Based on a generalization of PLS chart, taking into account not only the original process variables, but also their bivariate interactions, in this paper, we discuss the applicability of a new chart called MAPLSS-T^2 chart. Following the procedure used for the construction of multivariate control charts based on projection methods like PCA-T^2 chart(Jackson, 1991), PLS-T^2 chart (Kourti and MacGregor, 1996) and PLSS-T^2 chart (Vanacore and Lombardo, 2005), the MAPLSS-T^2 chart is based on the first A components. The MAPLSS-T^2 chart is an effective monitoring tool: it incorporates the variability structure underlying process data and quality product data extracting information on the directionality of the process variation. The scores of each new observation are monitored by the MAPLSS-T^2 control chart based on the following statistic

$$T_A^2 = \sum_{a=1}^{A} \frac{(t_a^2)}{\lambda_a} \tag{6}$$

where λ_a and $\mathbf{t}_a$ for $a = 1,...,A$ are the eigenvalues and the component scores, respectively, of the previously defined covariance matrix. The control limits of the MAPLSS-T^2 chart are based on the percentiles q_α (for $\alpha \leq 10\%$) of the empirical distributions, F_N, of MAPLSS component scores, computed on a large number N of bootstrap samples

$$\alpha = P(T_A^2 \leq q_\alpha | F_N). \tag{7}$$

Multivariate control charts can detect an unusual event but do not provide a reason for it. Following the diagnostic approach proposed by Kourti and MacGregor (1996) and using some new tools, we can investigate observations falling out of the limits through

(1) bar plots of standardized out-of control scores ($t_a/\sqrt{\lambda_a}$ for $a = 1,...,A$), to focus on the most important dimensions;
(2) bar plot of the contributions of the process variables on the dimensions identified as the most important ones, to evaluate how each process variable involved in the calculation of that score contributes to it;
(3) bar plot of the contributions of the process variables on product variables (measured by the spline range) to evaluate the importance of process variables.

3 Application: monitoring the painting process of hot-rolled aluminium foils

In this section we illustrate the usefulness of MAPLSS-T^2 chart and the related diagnosis tools by applying them to monitor a real manufacturing process. We focus on the modeling phase of statistical process control. The data refer to a manufacturing firm of Naples, specialized in hot-rolling of aluminium foils. The manufacturing process consists in simultaneously painting the lower and upper surfaces of an aluminium foil. The process starts by setting the aluminium roll on the unwinding swift. The aluminium foil, pulled by the draught rein that manages the crossing speed, reaches the painting station where it is uniformly painted on both surfaces by deflector rolls. The paint drying and polymerization is realized in a flotation oven consisting of 6 distinct modules (each module is characterized by a specific temperature and can be gradually boosted and independently tuned up).
The process stops by rewinding the aluminium roll. The key product quality characteristics are the uniformity and stability of the alumium painting. Both of them depend on the Peak Metal Temperature, *PMT* , reached during the polymerization. By managing the temperatures of the stay of the aluminium foil in the oven, one can influence the *PMT* . Thus *PMT* has been selected as the only quality product variable, whereas the temperatures characterizing the six modules (*T1,T2,T3,T4,T5,T6*) and the post-combustion temperature (*Tpost*) have been selected as process variables. The MAPLSS-T^2 control chart is built on an historical data set of $n = 100$ independent unit samples. The computational strategy consists in performing at first the MAPLSS regression (see Table 1) using low degree and knot number (degree=1, knots=1), deciding the dimension space A by Cross Validation (we get $A = 3$ with $PRESS = 0.15$). Using the balance between the goodness of fit (R^2) and thriftness ($PRESS$), we select only one interaction among the candidates, the resulting best one is *T4*T5*. Afterwards we extract $N = 500$ Bootstrap samples and perform MAPLSS

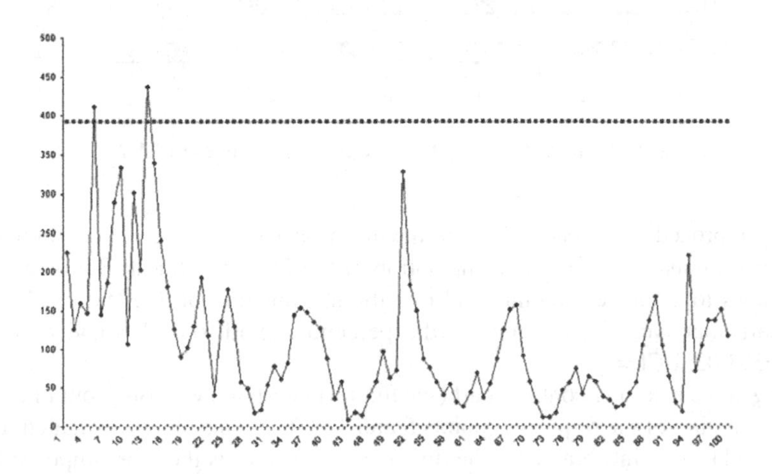

Fig. 1. MAPLSS$-T^2$ control chart.

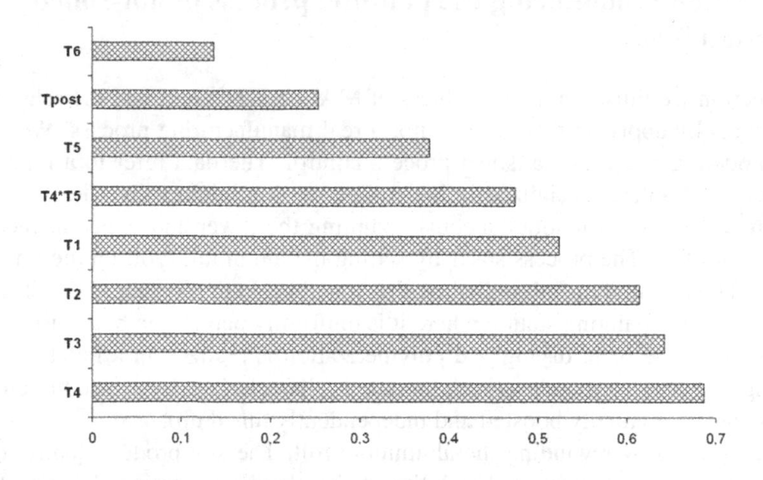

Fig. 2. Bar plot of contributions of process variables to the second dimension

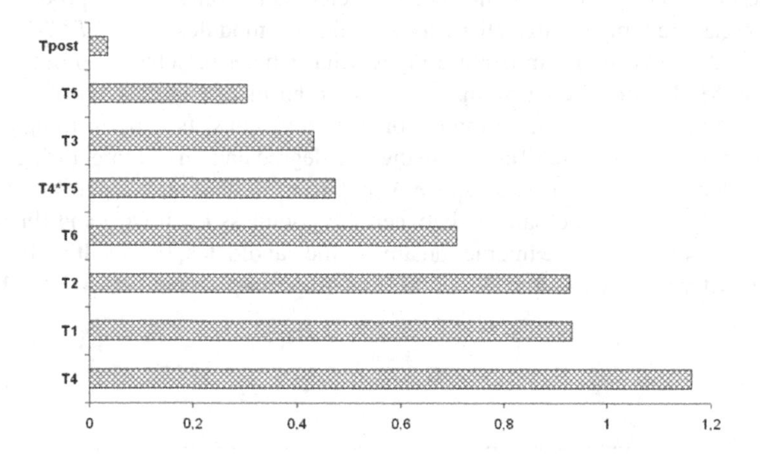

Fig. 3. Bar plot of contributions of process variables to *PMT* .

regression procedure on each of them, having properly fixed the model parameters (degree=1, knots=1, A=3). The computation of the T^2 scores for all Bootstrap samples allows to estimate the empirical distribution function of T^2. We fix the control chart upper and lower limits at the percentiles with $\alpha = 1\%$ and $\alpha = 99\%$ (UCL=393.03, LCL=2.81)
Looking at the resulting control chart (see figure 1) we note two points out of control at the beginning of the sequence (points 5 and 13). They must be investigated, using bar plots (1) for points 5 and 13, the dimension 2 results as the most important one for both out of control points. The bar plot (2) of process variables which contribute

Table 1. MAPLSS results: R^2 according to the dimension

Dimension A	R^2	%cum.
1	0.74	74%
2	0.16	89.6
3	0.03	92.3

to dimension 2 (figure 2) highlights that the most important process variables are the temperature in zone 4 (*T4*) zone 3 (*T3*), zone 2 (*T2*), zone 1 (*T1*), the interaction between temperatures in zone 4 and zone 5 (*T4*T5*), ecc. In particular, *T4* has a strong effect on dimension 2 as well as on the quality product variable (Fig. 3). In Fig. 3 we read in decreasing order the most important predictors on *PMT*, a part from *T4*, the other important process variables are *T1*, *T2*, *T6*, *T4*T5*, and so on. It is interesting to observe that the interaction between *T4*T5* is more important than the simple process variable given by *T5* (Fig. 2 and 3). After the diagnosis analysis, the causes for observed out of control points have been detected. In fact the expert technologist suggests that the out-of-control signals are the consequence of a 'transition phenomenon' due to a calibration problem in the feedback of the automatic loop (i.e. the methane valve opens when temperature is naturally rising). Having identified and removed the causes for the out of control signals, the modeling phase of the MAPLSS-T^2 chart requires that the control limits should be recomputed excluding the out of control points. The modeling phase ends when all points are inside the control limits.

4 Conclusion

In this paper a powerful non-parametric multivariate process control chart has been proposed for monitoring a manifacturing process. By simultaneously monitoring process and product variables, MAPLSS-T^2 chart quickly detects and diagnoses unusual events that may occur during the process. The proposed non-parametric control chart allows to handle collinear variables, missing values, outliers and interactions between variables, without imposing any distributional assumption. Further developments of this work could be related to the construction of a chart of the Squared Prediction Error (SPE; Kourti and MacGregor, 1996) on MAPLSS model, in order to monitor any change in the covariance structure and verify that the process conditions during the monitoring stage are not different with respect to the time the in control MAPLSS model was developed.

References

CHAKRABORTI, S., VAN DER LAAN, P. and BAKIR, S.T. (2001): Nonparametric control chart: an overview and some results. *Journal of Quality Technology, 33, pp. 304-315.*

DURAND, J.F. (2001): Local Polynomial additive regression through PLS and splines: PLSS. *Chemiometric and Intelligent Laboratory systems, 58, pp. 235-246.*

HAWKINS, D.M. (1991): Multivariate Quality Control based on regression-adjusted variables.*Technometrics, 33, pp. 61-75.*

HOTELLING, H. (1947): *Multivariate Quality Control, in Techniques of Statistical Analysis.* Eds. Eisenhart, Hastay and Wallis, MacGraw Hill, New York.

JACKSON, J. E. (1991): *A User Guide to Principal Component.* Wiley, New York.

JONES, A. L. and WOODALL, W.H. (1998): The performance of Bootstrap Cotrol Charts. *Journal of Quality Technology, 30, pp. 362-375.*

KOURTI, T. and MACGREGOR, J. F. (1996): Multivariate SPC Methods for Process and Product Monitoring. *Journal of Quality Technology, 28, pp. 409-428.*

LIU, Y.R. (1995): Control Charts for Multivariate Processes. *Journal of the American Statistical Association, 90, pp. 1380-1387.*

LIU, Y.R. and TANG, J. (1996): Control Charts for Dependent and Independent Measurement based on Bootstrap Methods.*Journal of the American Statistical Association, 91, pp. 1694-1700.*

LOMBARDO, R., DURAND, J.F., DE VEAUX, R. (2007): Multivariate Additive Partial Least Squares via Splines. *Submitted.*

LOWRY, C. A., WOODALL, W. H., CHAMP, C. W., RIGDON, S. E. (1992): A multivariate EWMA control chart. *Technometrics, 34, pp. 46-53.*

MACGREGOR, J.F. (1997): Using On-line Process Data to Improve Quality: Challenges for Statisticians.*International Statistical Review, 65, pp. 309-323.*

NOMIKOS, P. and MACGREGOR, J.F. (1995): Multivariate SPC Charts for Monitoring Batch Processes. *Technometrics, 37, pp. 41-59.*

TENENHAUS, M. (1998): *La Règression PLS, Thèorie et Pratique.* Eds. Technip, Paris.

VANACORE, A. and LOMBARDO, R. (2005): Multivariate Statistical Control Charts by non-linear Partial Least Squares.*Proceedings of Cladag 2005 conference, pp. 525-528.*

WOLD, H. (1966): Estimation of principal components and related models by iterative least squares. *In Multivariate Analysis, (Eds.) P.R. Krishnaiah, New York: Academic Press, pp. 391-420.*

WOODALL, W. H. and NCUBE, M. M. (1985): Multivariate CUSUM quality control procedures. *Technometrics, 27, pp. 285-292.*

Simple Non Symmetrical Correspondence Analysis

Antonello D'Ambra[1], Pietro Amenta[1] and Valentin Rousson[2]

[1] University of Sannio, Benevento, Italy
{andambra, amenta}@unisannio.it
[2] Department of Biostatistics, University of Zürich, Switzerland
rousson@ifspm.unizh.ch

Abstract. Simple Component Analysis (SCA) was introduced by Rousson and Gasser (2004) as an alternative to Principal Component Analysis (PCA). The goal of SCA is to find the "optimal simple system" of components for a given data set, which may be slightly correlated and suboptimal compared to PCA but which is easier to interpret.

Aim of the present paper paper is to consider an extension of SCA to categorical data. In particular, we consider a simple version of the Non Symmetrical Correspondence Analysis (D'Ambra and Lauro, 1989). This latter approach can be seen as a centered PCA on the column profile matrix with suitable metrics enabling to describe the association in two way contingency table in cases where one categorical variable is supposed to be the explanatory variable and the other the response.

1 Introduction

It is well known that Principal Component Analysis (PCA) is optimal in at least two ways: principal components extract a maximum of the variability of the original variables and they are uncorrelated. The former ensures that a minimum of "total information" will be missed when looking at the first few principal components. The latter warrants that the extracted information will be organized in an optimal way: we may look at one principal component after the other, separately, without taking into account the rest.

Unfortunately, principal components often lack interpretability. They define some abstract scores which often are not meaningful, or not well interpretable in practice. The same remark applies to all methods based on PCA.

Simple Component Analysis (SCA) was introduced by Rousson and Gasser (2004) as an alternative to Principal Component Analysis. The goal of SCA was to find the "optimal simple system" of components for a given data set. A component was considered to be simple if the number of possibles values for its loadings was restricted to three (a positive one, zero and a negative one). Optimality of a syztem of components was defined as in Gervini and Rousson (2004). At the end, the optimal simple system defined by SCA may be slightly correlated and suboptimal compared to PCA but will be easier to interpret. Thus, SCA may represent a worth alternative to PCA if the loss of optimality remains modest.

Aim of the present paper is to consider an extension of SCA to categorical data. In particular, we consider a simple version of the Non Symmetrical Correspondence Analysis (D'Ambra and Lauro, 1989). This latter approach can be seen as a PCA performed on the column profile matrix with the same weighting system of Correspondence Analysis but in a different metric.

Advantages of the method are illustrated with a well known data set.

2 Non symmetrical correspondence analysis

In many fields, the researcher is interested to study the relationship between two or more variables. When the variables are collected in a contingency table, classical statistic tools like correspondence analysis (CA) are applied in order to measure and visualize the strength of the association.

The CA is based on the decomposition of the index ϕ^2 of Pearson, which is a symmetric measure of association. This approach however is no longer appropriate when one has to study a two way contingency table where one categorical variable is supposed to be the explanatory variable and the other the response. To overcome this problem, D'Ambra and Lauro (1989) introduced the Non Symmetrical Correspondence Analysis (NSCA). This approach decomposes the numerator of the Goodman-Kruskal τ (1954), which is an asymmetric measure of association in a contingency table.

Given two categorical variables I and J, the goal of NSCA is to evaluate the influence of categories of the explanatory variable J on the distribution of the reponse I.

Let $\mathbf{N} = (n_{ij})$ and $\mathbf{P} = \frac{N}{n} = (p_{ij}) = (\frac{n_{ij}}{n})$ be the absolute and relative two-way contingency table of dimension $I \times J$ where I and J also denote the number of categories of the response and the explanatory variable, respectively, based on n individuals. Let $p_{i.} = \sum_{j=1}^{J} p_{ij}$ and $p_{.j} = \sum_{i=1}^{I} p_{ij}$ be the column and row marginals, respectively, and let $\mathbf{D}_j = diag(p_{.j})$.

Finally, let

$$\Pi = (\pi_{ij}) = (\frac{p_{ij}}{p_{.j}} - p_{i.})$$

be the matrix describing the conditional distribution of I given J. This matrix contains information on the I conditional distributions $\frac{p_{ij}}{p_{.j}}$ adjusted to the row marginal $p_{i.}$, and is hence a weighted average of the column profiles.

From a geometrical point of view, the purpose of NSCA is to evaluate in the space $\mathbb{R}^I$ the spread of the cloud of points defined by Π around its centroid according to an appropriate weighting system. A global measure of dispersion is given by the inertia

$$In = \tau_{num} = \sum_{i=1}^{I} \sum_{j=1}^{J} p_{.j} \left(\frac{p_{ij}}{p_{.j}} - p_{i.} \right)^2 .$$

NSCA looks for the orthonormal basis which accounts for the largest part of inertia to visualize the dependence structure between J and I in a lower dimensional space. Solutions are given by the eigen-analysis of the variance covariance matrix $\mathbf{S} = \Pi \mathbf{D}_j \Pi'$ whose general term (i,i') is given by

$$s_{ii'} = \sum_{j=1}^{J} p_{.j} \left(\frac{p_{ij}}{p_{.j}} - p_{i.} \right) \left(\frac{p_{i'j}}{p_{.j}} - p_{i'.} \right)$$

where $p_{i.}$ denotes the centre of gravity of the ith row of $\mathbf{P}$. This is achieved also by the generalized singular value decomposition of $\Pi = \sum_{m=1}^{M*} \lambda_m \mathbf{a_m b_m}'$ with $M* \leq M = min[(I,J)-1]$ and where the scalar λ_m is the singular value (we shall note $\Delta = diag(\lambda_m)$), $\mathbf{a}_m$ and $\mathbf{b}_m$ are orthonormal singular vectors in an unweighted and weighted metric, respectively, such that $\mathbf{a}'_m \mathbf{a}_m = 1$, $\mathbf{a}'_m \mathbf{a}_{m'} = 0$ and $\mathbf{b}'_m \mathbf{D}_j \mathbf{b}_m = 1$, $\mathbf{b}'_m \mathbf{D}_j \mathbf{b}_{m'} = 0$ for $m \neq m'$.

In the previous decomposition, the numerator of the Goodman and Kruskal τ (1954) can be decomposed as $\tau_{num} = \sum_{m=1}^{M*} \lambda_m^2$.

The factorial row and column coordinates are given by $\psi_m = \sqrt{\lambda_m} \mathbf{a}_m$ and $\varphi_m = \sqrt{\lambda_m} \mathbf{D}_j^{-1/2} \mathbf{b}_m$, respectively. Finally, factorial coordinates can be also obtained from the transition formulae:

$$\varphi_m = \left(\frac{1}{\sqrt{\lambda_m}} \right) \sum_i \left(\frac{p_{ij}}{p_{.j}} - p_{i.} \right) \psi_m$$

$$\psi_m = \left(\frac{1}{\sqrt{\lambda_m}} \right) \sum_j p_{.j} \left(\frac{p_{ij}}{p_{.j}} - p_{i.} \right) \varphi_m$$

See D'Ambra and Lauro (1989) for further details and remarks.

3 Simple non symmetrical correspondence analysis

It is possible to show that NSCA corresponds to a PCA of the profile matrix Π with suitable row and column metrics. This is equivalent (Tenenhaus and Young, 1985) to study the statistical triplet $(\Pi, \mathbf{I}, \mathbf{D}_j)$ where the identity matrix $\mathbf{I}$ denotes the metric and $\mathbf{D}_j$ the weighting system. Thus, like all PCA-based methods, the components produced by NSCA are optimal but may lack interpretability, as recalled in the Introduction.

In the similar way as SCA was introduced as an alternative to PCA, we shall now introduce a technique called Simple NSCA as an alternative to NSCA. For this, we shall use similar concepts and algorithms as in SCA. Note that while one makes the distinction between simple block-components and simple difference-components in SCA, we shall here consider only difference components (i.e. components with both positive and negative loadings), since NSCA does not produce block-components (i.e. components where all loadings share the same sign). Thus, we shall consider simple components with loadings proportional to vectors with only three different values (a negative value, zero, and a positive value), the sum of the loadings being zero for each component (defining hence proper contrasts of categories).

The goal of Simple NSCA is to find the optimal system of components among the simple ones, where optimality is calculated according to Gervini and Rousson (2004).

The percentage of extracted variability $\mathbf{V}(\mathbf{L})$ accounted by a system $\mathbf{L}$ of $m = min(I,J) - 1$ components is given by

$$\mathbf{V}(\mathbf{L}) = \frac{\mathbf{l}_1'\mathbf{S}\mathbf{l}_1}{tr(\Delta)} + \frac{1}{tr(\Delta)} \sum_{k=2}^{m} \mathbf{l}_k' \left(\mathbf{S} - \mathbf{S}\mathbf{L}_{(k-1)}(\mathbf{L}_{(k-1)}'\mathbf{S}\mathbf{L}_{(k-1)})^{-1}\mathbf{L}_{(k-1)}'S\right)\mathbf{l}_k,$$

where $\mathbf{l}_k$ is the kth column of $\mathbf{L}$, and where $\mathbf{L}_{(k-1)}$ is the $m \times (k-1)$ matrix containing the first $(k-1)$ columns of $\mathbf{L}$.

Whereas the numerator of the first term of this sum is equal to the variance of the first component, the numerator of the kth term can be interpreted as the variance of the part of the kth component which is not explained by (which is independent from) the previous $(k-1)$ components. Thus, correlations are "penalized" by this criterion which is hence uniquely maximized by PCA, i.e. by taking $\mathbf{L} = \mathbf{E}_m$, the matrix of the first m eigenvectors of $\mathbf{S}$ (Gervini and Rousson, 2004). The optimality of a system $\mathbf{L}$ is then calculated as $\mathbf{V}(\mathbf{L})/\mathbf{V}(\mathbf{E}_m)$.

In our sequential algorithms below, the kth simple component is obtained by regressing the original row/column categories on the previous $k-1$ simple components already in the system, by computing the first eigenvector of the residual variance hence obtained, and by shrinking this eigenvector towards the simple difference component which maximizes optimality. Here are two algorithms providing simple components for the rows and the columns.

Simple solutions for the rows

1. Let $\mathbf{S} = \Pi\mathbf{D}_j\Pi'$, let $\mathbf{L}$ be an empty matrix and let $\hat{\mathbf{S}} = \mathbf{S}$.
2. Let $\mathbf{a} = (a_1, \ldots, a_I)'$ be the first eigenvector of $\hat{\mathbf{S}}$.
3. For each cut-off value among $g = \{0, |a_1|, \ldots, |a_I|\}$, consider the shrunken vector $\mathbf{b}(g) = \{b_1(g), ..., b_I(g)\}'$ with elements $b_k(g) = \text{sign}(a_k)$ if $|a_k| > g$ and $b_k(g) = 0$ otherwise (for $k = 1, \ldots, I$). Update and normalize it such that $\sum b_k(g) = 0$ and $\sum b_k^2(g) = 1$.
4. Include into the system the difference component $\mathbf{b}(g)$ which maximizes $\mathbf{b}(g)'\hat{\mathbf{S}}\mathbf{b}(g)$ (i.e. add the column $\mathbf{b}(g)$ to the matrix of loadings $\mathbf{L}$).
5. If the maximum number of components is attained stop. Otherwise let $\hat{\mathbf{S}} = \mathbf{S} - \mathbf{S}\mathbf{L}(\mathbf{L}'\mathbf{S}\mathbf{L})^{-1}\mathbf{L}'\mathbf{S}$ and go back to step 2.

Simple solutions for the columns

1. Let $\mathbf{S} = \mathbf{D}_j^{1/2}\Pi'\Pi\mathbf{D}_j^{1/2}$, let $\mathbf{L}$ be an empty matrix and let $\hat{\mathbf{S}} = \mathbf{S}$.
2. Let $\mathbf{a} = (a_1, \ldots, a_J)'$ be the first eigenvector of $\hat{\mathbf{S}}$.

3. For each cut-off value among $g = \{0, |a_1|, \ldots, |a_J|\}$, consider the shrunken vector $\mathbf{b}(g) = \{b_1(g), \ldots, b_I(g)\}'$ with elements $b_k(g) = \text{sign}(a_k)$ if $|a_k| > g$ and $b_k(g) = 0$ otherwise (for $k = 1, \ldots, J$). Update and normalize it such that $\sum b_k(g) = 0$ and $\sum b_k^2(g) = 1$.
4. Include into the system the difference component $\mathbf{b}(g)$ which maximizes $\mathbf{b}(g)'\hat{\mathbf{S}}\mathbf{b}(g)$ (i.e. add the column $\mathbf{b}(g)$ to the matrix of loadings $\mathbf{L}$).
5. If the maximum number of components is attained, let $\mathbf{L} = \mathbf{D}_j^{-1/2}\mathbf{L}$ and stop. Otherwise let $\hat{\mathbf{S}} = \mathbf{S} - \mathbf{SL}(\mathbf{L}'\mathbf{SL})^{-1}\mathbf{L}'\mathbf{S}$ and go back to step 2.

4 Father's and son's occupations data

To illustrate the technique of Simple NSCA, we applied it to the well known Father's and Son's Occupations. This data set (Perrin, 1904) was collected to study whether and how the professional occupation of some man depends on the occupation of his father. Occupations of 1550 men were cross-classified according to father's and son's occupation reparted into 14 occupations.

The conclusion of the study was that such a dependence existed. Two measures of predicability, the Goodman-Kruskal's τ (1954) and the Light and Margolin's $C = (n-1)(I-1)\tau$ (1971), have been computed. Note that the C-statistic can be used to formally test for association, being asymptotically chi-squared distributed with $(I-1)(J-1)$ degrees of freedom under the hypothesis of no association (Light and Margolin, 1971).

The overall increase in predicability of a man's occupation when knowing the occupation of his father was equal to 14% ($\tau = 0.14$; $C = 2880.8$; $df = 169$, $p_{value}\langle 0.0001$).

According to the NSCA decomposition of the numerator of τ ($\tau_{num} = \sum_{k=1}^{M} \lambda_k^2 = 0.1288$), we have for the first two axes $\lambda_1 = 0.24$ and $\lambda_2 = 0.16$, which are the weights of the axes in the joint plot of Figure 1. The first axis accounts for $100 \times (0.24)^2/0.1288 = 43.7\%$ of the dependence between the two variables while the second one represents 20.7%. Therefore Figure 1 accounts for 64.4% of the total inertia.

Unfortunately, the two-dimensional NSCA solution (Figure 1) does not give a clear description of the dependence of the two variables as well as of the association between rows and columns. Thus, NSCA is difficult to interpret and a simple solution has been calculated according to Simple NSCA.

From Table 1, one can see that the first component defined by Simple NSCA for the rows contrasts son's occupation "Art" versus the group of occupations {Army, Divinity, Law, Medicine, Politics & Court and Scholarship & Science}. This simple component explains 42.5% of the variance compared to 43.7% for optimal solution above. Thus, the first simple row solution is 42.5%/43.7%=97.4% optimal. One can conclude that the influence of father's occupation on son's occupation mainly contrasts these two groups of occupation. The second simple row solution provided by Simple NSCA contrasts son's occupation "Divinity" versus the group of occupations {Army and Politics & Court}.

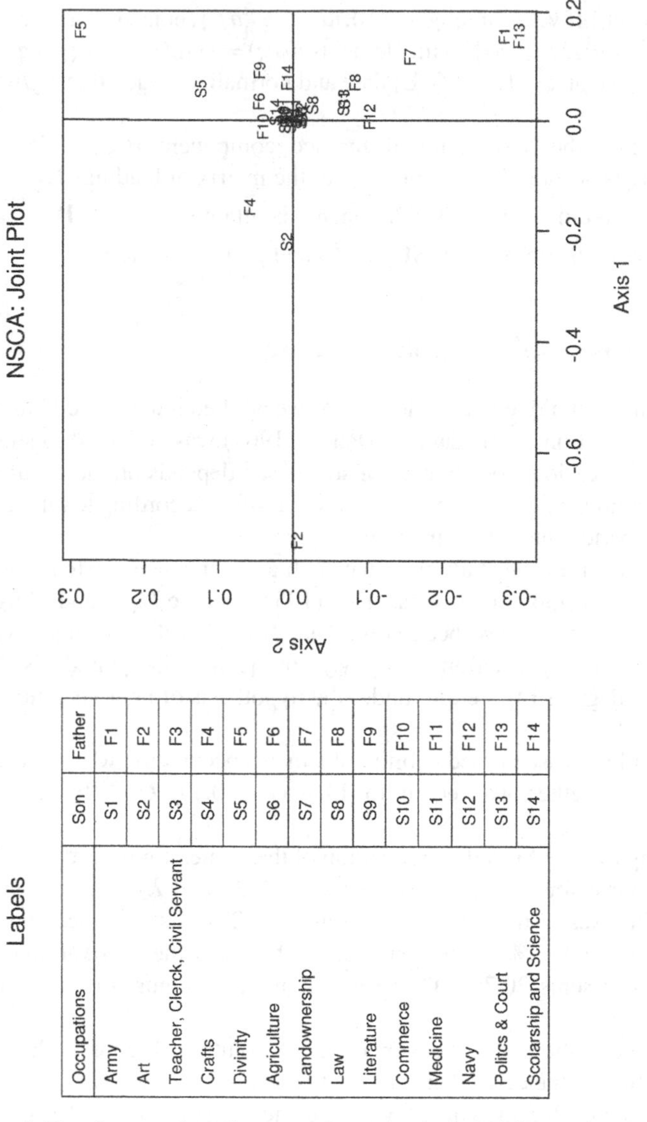

Fig. 1. Non Symmetrical Correspondence Analysis (NSCA): Joint plot.

The same table also contains the Simple NSCA solution for the columns. The first simple column solution contrasts father's occupation "Art" versus "Divinity", and is 81.9% optimal. The second simple column solution contrast groups of father's occupations {Army, Landownership, Law and Politics & Court} versus {Art and Divinity} with an optimality value of 90.4%. Similarly, further simple constrats can be defined for both the rows and the columns (see Table 1 for the first 5 solutions).

Table 1. Simple NSCA solutions for the first five axes.

	SON (row)					FATHER (column)				
	Axis1	Axis2	Axis3	Axis4	Axis5	Axis1	Axis2	Axis3	Axis4	Axis5
Army	0,15	-0,41	-0,44	-0,37	-0,50	0,00	-0,89	-1,20	3,21	0,00
Art	-0,93	0,00	0,00	0,00	0,00	-2,04	1,77	-1,20	0,00	0,00
TCCS	0,00	0,00	0,00	0,00	0,00	0,00	0,00	0,00	0,00	0,00
Crafts	0,00	0,00	0,00	0,00	0,00	0,00	0,00	0,86	0,00	0,00
Divinity	0,15	0,82	-0,44	0,00	0,00	2,04	1,77	-1,20	0,00	0,00
Agricolture	0,00	0,00	0,00	0,00	0,00	0,00	0,00	0,86	0,00	0,00
Landownership	0,00	0,00	0,00	0,00	0,00	0,00	-0,89	-1,20	0,00	0,00
Law	0,15	0,00	0,33	0,55	-0,50	0,00	-0,89	0,86	-1,61	-2,65
Literature	0,00	0,00	0,33	0,00	0,00	0,00	0,00	0,86	0,00	0,00
Commerce	0,00	0,00	0,33	0,00	0,00	0,00	0,00	0,86	0,00	0,00
Medicine	0,15	0,00	0,00	-0,37	0,50	0,00	0,00	0,86	0,00	2,65
Navy	0,00	0,00	0,00	0,00	0,00	0,00	0,00	0,00	0,00	0,00
POLCOURT	0,15	-0,41	-0,44	0,55	0,50	0,00	-0,89	-1,20	-1,61	0,00
SCSCIENCE	0,15	0,00	0,33	-0,37	0,00	0,00	0,00	0,86	0,00	0,00
Explained variance (%)										
Optimal solution	43,70	64,40	75,30	83,00	89,20	43,70	64,40	75,30	83,00	89,20
Simple solution	42,50	62,20	72,30	79,70	85,70	35,80	58,20	68,50	75,10	80,30
Optimality	97,40	96,60	96,10	96,10	96,10	81,90	90,40	91,00	90,50	90,00

Note: TCCS, POLCOURT and SCSCIENCE stand for "Teacher, Clerck and Civil Servant", "Politics & Court" and "Scolarship & Science", respectively.

To better summarize and visualize the relationship between father's and son's occupation, it is helpful to plot the solutions for rows and columns for each axis on a same graphic (Figure 2). One can see that the first Simple NSCA solution highlights the fact that a son has the tendency to choose the same occupation as his father if this occupation is "Art", while father's occupation "Divinity" is linked with a son's occupation within {Army, Divinity, Law, Medicine, Politics & Court and Scholarship & Science}. Similarly, one can try to interpret the second Simple NSCA solution.

In summary, Simple NSCA provides a clearcut picture of the situation, the optimality of the first two axes being in this example of more than 95% (for the rows) and 90% (for the columns). Thus, the price to pay for simplicity is about 5% (for the rows) and 10% (for the columns), which is not much. In this sense, Simple NSCA may be a worth alternative to NSCA.

5 Conclusions

In general, all PCA-based methods are tuned to condense information in an optimal way. However, they define some abstract scores which often are not meaningful or not well interpretable in practice. This was also the case in our example above for

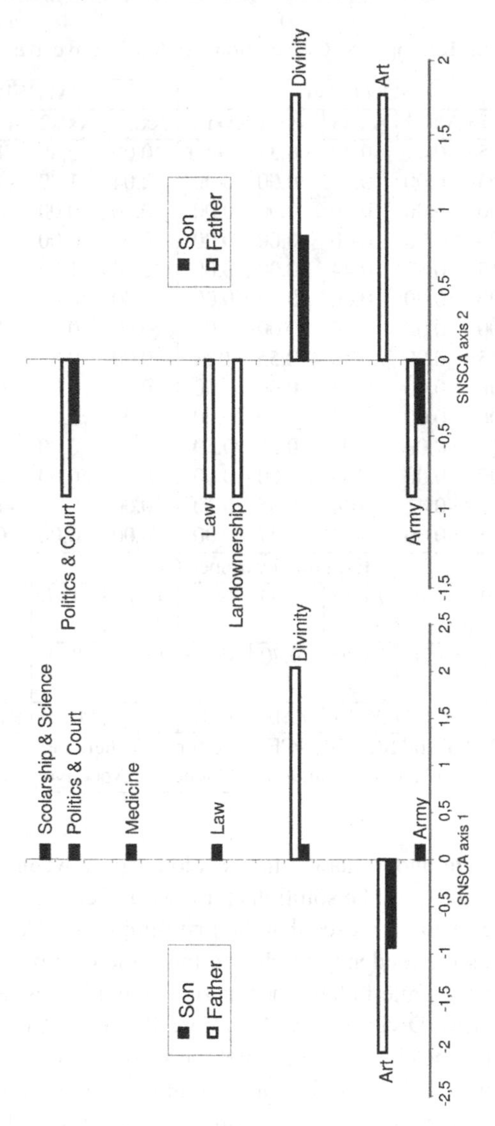

Fig. 2. Summary of Simple NSCA solutions for the axes 1 and 2.

NSCA. To enhance interpretability, Simple NSCA focus on simplicity and seeks for "optimal simple components", as illustrated in our example. It provides a clear-cut interpretation of the association between rows and columns, the price to pay for simplicity being relatively low. In this sense, Simple NSCA may be a worth alternative to NSCA. Extensions of this approach for the Classical Correspondence Analysis and for ordinal variables are under investigation.

References

D 'AMBRA, L. and LAURO, N.C. (1989): Non symmetrical analysis of three-way contingency tables. In: R. Coppi and S. Bolasco (Eds.): *Multiway Data Analysis.* North Holland, 301–314.

LIGHT, R. J. and MARGOLIN, B. H. (1971): An analysis of variance for categorical data. *Journal of the American Statistical Association, 66, 534–544.*

GERVINI, D. and ROUSSON, V. (2004): Criteria for evaluating dimension-reducing components for multivariate data. *The American Statistician, 58, 72–76.*

GOODMAN, L. A. and KRUSKAL, W. H. (1954): Measures of association for cross-classifications. *Journal of the American Statistical Association, 49, 732–7644.*

PERRIN, E. (1904): On the Contingency Between Occupation in the Case of Fathers and Sons. *Biometrika, 3, 4, 467–469.*

ROUSSON, V. and GASSER, Th. (2004): Simple component analysis. *Applied Statistics, 53, 539–555.*

TENENHAUS, M. and YOUNG, F.W., (1985): An analysis and synthesis of multiple correspondence analysis, optimal scaling, dual scaling and other methods for quantifying categorical multivariate data. *Psychometrika, 50, 90–104.*

Factorial Analysis of a Set of Contingency Tables

Amaya Zárraga and Beatriz Goitisolo

Departamento de Economía Aplicada III, UPV/EHU, Bilbao, Spain
{amaya.zarraga, beatriz.goitisolo}@ehu.es

Abstract. The aim of this work is to present a method of joint factorial analysis of several contingency tables. This method that we have called Simultaneous Analysis (SA), is especially appropriate to analyze frequency tables whose row margins are different, for example when the tables are from different samples or different time points. Furthermore, SA may be applied to the joint analysis of more than two data tables in which rows refer to the same entities, but columns may be different.

SA allows us to maintain the structure of each table in the overall analysis by centering each table internally with its margins, as is done in Correspondence Analysis (CA) and provides a joint description of the different structures contained within each table. Besides jointly studying the intrastructure of the tables, SA permits an overall comparison of the similarities and differences between the tables.

1 Introduction

The need of jointly analyzing several contingency tables has produced several factorial methods.

Some of the proposed methods consist in the analysis of the table obtained as sum of the separated contingency tables and/or the analysis of the table obtained as juxtaposition of the initial tables (Cazes (1980) and (1981)) and the Intra Analysis (Escofier (1983)). Nevertheless, in Zárraga and Goitisolo (2002) it is shown that there are situations where none of these methods permits an analysis of the similarities among rows that mantains the similarity in the analyses of the separated tables.

The aim of this work is to present a factorial method for the joint analysis of several contingency tables that allows, in a similar way to correspondence analysis, the study of the similarity among the set of rows, of columns and the relations between both sets.

Also cite the non symmetrical analysis (D' Ambra and Lauro (1984) and Lauro and D' Ambra (1989)) and more recently the Multiple Factor Analysis for Contingency Tables (Pagès and Bécue-Bertaut (2006)).

2 Methodology

Let $\mathbf{T} = \{1,\ldots,t,\ldots,T\}$ be the set of contingency tables to be analyzed. Each of them classifies the answers of $n_{..t}$ individuals with respect to two categorical variables. All the tables have one of the variables in common, in this case the row variable with categories $\mathbf{I} = \{1,\ldots,i,\ldots,I\}$. The other variable of each contingency table can be different or the same variable observed at different time points or in different subsamples. On concatenating all these contingency tables, a joint set of columns $\mathbf{J} = \{1,\ldots,j,\ldots,J\}$ is obtained. The element n_{ijt} corresponds to the total number of individuals who choose simultaneously the categories $i \in \mathbf{I}$ of the first variable and $j \in \mathbf{J}_t$ of the second variable, for table $t \in \mathbf{T}$. Sums are denoted in the usual way, for example, $n_{i.t} = \sum_{j\in\mathbf{J}_t} n_{ijt}$, and n denotes the grand total of all T tables.

In order to maintain the internal structure of each table t, SA begins by obtaining the relative frequencies of each table as usually done in CA: $p^t_{ij} = n_{ijt}/n_{..t}$ so that $\sum_{i\in\mathbf{I}}\sum_{j\in\mathbf{J}_t} p^t_{ij} = 1$ for each table t. It is important to keep in mind that these relative frequencies are different from those obtained when calculating the relative frequency for the whole matrix: $p_{ijt} = n_{ijt}/n$.

The method that we propose is carried out in three stages.

2.1 Stage one: CA of each contingency table

Since in SA it is important for each table to maintain its own structure, the first stage carries out a classical CA of each of the T contingency tables. These separate analyses also allow us to check for the existence of structures common to the different tables. From these analyses it is possible to obtain the weighting used in the next stage.

CA on the t-th contingency table can be carried out by calculating the singular value decomposition (SVD) of the matrix X^t, whose general term is:

$$\sqrt{p^t_{i.}}\left(\frac{p^t_{ij}-p^t_{i.}\,p^t_{.j}}{p^t_{i.}\,p^t_{.j}}\right)\sqrt{p^t_{.j}}$$

Let $\mathbf{D}^t_r$ and $\mathbf{D}^t_c$ be the diagonal matrices whose diagonal entries are respectively the marginal row frequencies $p^t_{i.}$ and column frequencies $p^t_{.j}$. From the SVD of each table X^t we retain the first squared singular value (or eigenvalue, or principal inertia), denoted by λ^t_1.

2.2 Stage two: analysis of intrastructure

In the second stage, in order to balance the influence of each table in the joint analysis, measured by the inertia, and to prevent this joint analysis from being dominated by a particular table, SA will include a weighting on each table, α_t. With this aim, in SA, $\alpha_t = 1/\lambda^t_1$, where λ^t_1 denotes the first eigenvalue (square of first singular value) of the separate CA of table t (stage one). This weight is similar to the one used in Multiple Factor Analysis (MFA) (Escofier and Pagès (1988)).

As a result, SA proceeds by performing a principal component analysis (PCA) of the matrix X, $X = \left[\sqrt{\alpha_1}X^1 \ \dots \ \sqrt{\alpha_t}X^t \dots \sqrt{\alpha_T}X^T\right]$

The PCA results are also obtained using the SVD of X, giving singular values $\sqrt{\lambda_s}$ on the s-th dimension and corresponding left and right singular vectors $\mathbf{u}_s$ and $\mathbf{v}_s$.

We calculate projections on the s-th axis of the columns as principal coordinates $\mathbf{g}_s$, $\mathbf{g}_s = \lambda_s \mathbf{D}_c^{-1/2} \ \mathbf{v}_s$ where $\mathbf{D}_c$ $(J \times J)$, is a diagonal matrix of all the column masses, that is all the $\mathbf{D}_c^t$.

One of the aims of the joint analysis of several data tables is to compare them through the points corresponding to the same row in the different tables. These points will be called *partial rows* and denoted by i^t.

The projection on the s-th axis of each partial row is denoted by f_{is}^t and the vector of projections of all the partial rows for table t is denoted by $\mathbf{f}_s^t$, $\mathbf{f}_s^t = (\mathbf{D}_r^t)^{-1/2} \ [0 \dots \sqrt{\alpha_t}X^t \dots 0] \ \mathbf{v}_s$

Especially when the number of tables is large, comparison of partial rows is complicated. Therefore each partial row will be compared with the (overall) row, projected as $\mathbf{f}_s = (\mathbf{D}_w)^{-1} \ \left[\sqrt{\alpha_1}X^1 \ \dots \ \sqrt{\alpha_t}X^t \ \dots \sqrt{\alpha_T}X^T\right] \ \mathbf{v}_s = (\mathbf{D}_w)^{-1} \ X \ \mathbf{v}_s$ where $\mathbf{D}_w$ is the diagonal matrix whose general term is $\sum_{t \in \mathbf{T}} \sqrt{p_{i.}^t}$. The choice of this matrix $\mathbf{D}_w$ allows us to expand the projections of the (overall) rows to keep them inside the corresponding set of projections of partial rows, and is appropriate when the partial rows have different weights in the tables. With this weighting the projections of the overall and partial rows are related as follows:

$$f_{is} = \sum_{t \in \mathbf{T}} \frac{\sqrt{p_{i.}^t}}{\sum_{t \in \mathbf{T}} \sqrt{p_{i.}^t}} \ f_{is}^t$$

So the projection of a row is a weighted average of the projections of partial rows. It is closer to those partial rows that are more similar to the overall row in terms of the relation expressed by the axis and have a greater weight than the rest of the partial rows. The dispersal of the projections of the partial rows with regard to the projection of their (overall) row indicates discrepancies between the same row in the different tables.

Notice that if $p_{i.}^t$ is equal in all the tables then $\mathbf{f}_s = (1/T) \sum_{t \in \mathbf{T}} \mathbf{f}_s^t$, that is the overall row is projected as the average of the projections of the partial rows.

Interpretation rules for simultaneous analysis

In SA the transition relations between projections of different points create a simultaneous representation that provides more detailed knowledge of the matter being studied.

Relation between f_{is}^t and g_{js}: The projection of a partial row on axis s depends on the projections of the columns:

$$f_{is}^t = \frac{\sqrt{\alpha_t}}{\sqrt{\lambda_s}} \sum_{j \in \mathbf{J}_t} \frac{p_{ij}^t}{p_{i.}^t} \ g_{js}$$

Except for the factor $\sqrt{\alpha_t/\lambda_s}$, the projection of a partial row on axis s, is, as in CA, the centroid of the projections of the columns of table t.

Relation between f_{is} and g_{js}: The projection of an overall row on axis s may be expressed in terms of the projections of the columns as follows:

$$f_{is} = \sum_{t\in \mathbf{T}} \sqrt{\alpha_t} \frac{\sqrt{p^t_{i.}}}{\sum_{t\in \mathbf{T}} \sqrt{p^t_{i.}}} \left(\frac{1}{\sqrt{\lambda_s}} \sum_{j\in \mathbf{J}_t} \frac{p^t_{ij}}{p^t_{i.}} g_{js} \right)$$

The projection of the row is therefore, except for the coefficients $\sqrt{\alpha_t/\lambda_s}$, the weighted average of the centroids of the projections of the columns for each table.

Relation between g_{js} and f_{is} or f^t_{is}: The projection on the axis s, of the column j for table t, can be expressed in the following way:

$$g_{js} = \frac{\sqrt{\alpha_t}}{\lambda_s} \left(\sum_{i\in \mathbf{I}} \left(\sum_{t\in \mathbf{T}} \sqrt{p^t_{i.}} \right) \sqrt{p^t_{i.}} \left(\frac{p^t_{ij} - p^t_{i.} p^t_{.j}}{p^t_{i.} p^t_{.j}} \right) f_{is} \right)$$

This expression shows that the projection of a column is placed on the side of the projections of the rows with which it is associated, compared to the hypothesis of independence, and on the opposite side of the projections of those to which it is less associated.

This projection is, according to partial rows:

$$g_{js} = \sqrt{\frac{\alpha_t}{\lambda_s}} \left(\sum_{i\in \mathbf{I}} \sqrt{p^t_{i.}} \left(\frac{p^t_{ij} - p^t_{i.} p^t_{.j}}{p^t_{i.} p^t_{.j}} \right) \left(\sum_{t\in \mathbf{T}} \sqrt{p^t_{i.}} f^t_{is} \right) \right)$$

The same aids to interpretation are available in SA as in standard factorial analysis as regards the contribution of points to principal axes and the quality of display of a point on axis s.

2.3 Stage three: comparison of the tables: interstructure

In order to compare the different tables, SA allows us, to represent each of them by means of a point and to project them on the axes.

The coordinate of table t on axis s, f_{ts}, represents the projected inertia of the table on the axis and, therefore, indicates the importance of the table in the determination of the axis. Thus, $f_{ts} = \sum_{j\in \mathbf{J}_t} p^t_{.j}\, g^2_{js} = \text{Inertia}_s(t)$ where $\text{Inertia}_s(t)$ represents the projected inertia of the sum of columns of the table t on the axis s.

Due to the weighting of the tables chosen by SA, the maximum value of this inertia on the first axis is 1. A value of f_{ts} close to 0 would indicate orthogonality between the first axes of the separate analyses with regard the Simultaneous Analysis. A value of f_{ts} close to 1 would indicate that the axis of the joint analysis is approximately the same as in the separate analysis of each table. So, if all the tables present a coordinate close to the maximum value, 1, on the first factorial axis of the SA, the projected inertia onto it is approximately T, the number of tables, and this confirms that this first direction is accurately depicting the relevant associations of each table.

2.4 Relations between factors of the analyses

In SA it is also possible to calculate the following measurements of the relation between the factors of the different analyses.

Relation between factors of the individual analyses: The correlation coefficient can be used to measure the degree of similarity between the factors of the separate CA of different tables. This is possible when the marginals $p^t_{i.}$ are equal.

When $p^t_{i.}$ are not equal, Cazes (1982) proposes calculating the correlation coefficient between factors, assigning weight to the rows corresponding to the margins of one of the tables. Therefore, these weights, and the correlation coefficient as well, depend on the choice of this reference table. In consequence, we propose to solve this problem of the weight by extending the concept of generalized covariance (Méot and Leclerc (1997)) to that of generalized correlation (Zárraga and Goitisolo (2003)).

The relation between the factors s and s' of the tables t and t' respectively would be calculated as:

$$r\left(\mathbf{f}_{st}, \mathbf{f}_{s't'}\right) = \sum_{i \in \mathbf{I}} \frac{f_{ist}}{\sqrt{\lambda^t_s}} \sqrt{p^t_{i.}} \sqrt{p^{t'}_{i.}} \frac{f_{is't'}}{\sqrt{\lambda^{t'}_{s'}}}$$

where f_{ist} and $f_{is't'}$ are the projections on the axes s and s' of the separate CA of the tables t and t' respectively and where λ^t_s and $\lambda^{t'}_{s'}$ are the inertias associated with these axes. This measurement allows us to verify whether the factors of the separate analyses are similar and check the possible rotations that occur.

Relation between factors of the SA and factors of the separate analyses: Likewise, it is possible to calculate for each factor s of the SA, the relation with each of the factors s' of the separate analyses of the different tables:

$$r\left(\mathbf{f}_{s't}, \mathbf{f}_{s}\right) = \sum_{i \in \mathbf{I}} \frac{f_{is't}}{\sqrt{\lambda^t_{s'}}} \sqrt{p^t_{i.}} \left(\sum_{t \in \mathbf{T}} \sqrt{p^t_{i.}}\right) \frac{f_{is}}{\sqrt{\lambda_s}}$$

If all the tables of frequencies analysed have the same row weights this measurement is reduced to:

$$r\left(\mathbf{f}_{s't}, \mathbf{f}_{s}\right) = \sum_{i \in \mathbf{I}} \frac{p^t_{i.}\, f_{is't}\, f_{is}}{\sqrt{\sum_{i \in \mathbf{I}} p^t_{i.} \left(f_{is't}\right)^2} \sqrt{\sum_{i \in \mathbf{I}} p^t_{i.} \left(f_{is}\right)^2}}$$

that is, the classical correlation coefficient between the factors of the separate analyses and the factors of SA.

3 Application

In this section we apply SA to the data taken from an on-line survey drawn up by the Spanish Ministry of Education and Science, from January to March 2006, to Spanish students who participate in the Erasmus program in European universities.

This application presents a comparative study for Spanish students, according to gender, of the relationships between the countries that they choose as destination to carry out the university interchange in the Erasmus program and the scientific fields in which they are studying.

The 15 countries that they choose as destination are Austria, Belgium, Czech Republic, Denmark, Finland, France, Germany, Ireland, Italy, Netherlands, Norway, Poland, Portugal, Sweden and United Kingdom. The scientific fields in which they are studying are: Social and Legal Sciences, Engineering and Technology, Humanities, Health Science and Experimental Science.

Therefore, we have two data tables whose rows (countries) and columns (scientific fields) correspond to the same modalities but refer to two different sets of individuals, depending on their gender. In these tables both the marginals and the grand-totals are different. This fact suggests analyzing the tables by SA since the results of applying other methods can be affected by the above mentioned differences (Zárraga and Goitisolo (2002)).

The first factorial plane of SA (figure 1) explains nearly 60% of total inertia. In the plane we observe that male and female students of Humanities Area, Health Science and specially Engineering and Technology have a similar behavior in the choice of the country of destination to realize their studies, whereas students of Social and Legal Sciences and of Experimental Science choose different countries as destiny depending on their gender.

The plane shows that students of Humanities Area, both male and female, choose the United Kingdom as destiny country, followed by Ireland. The countries chosen as destiny for students of both gender of Engineering and Technology are mainly Austria, Sweden and Denmark. Finally, the males and females students of Health Science Area prefer Portugal and Finland.

The students of Experimental Science Area select different countries to realize the interchange depending on their gender. While male students go mainly to Portugal and Netherlands, females go to Norway.

Also students of Social and Legal Sciences Area have a different behavior. The Netherlands and Ireland are selected as destiny country by males and females but males also go to Belgium, the United Kingdom and Italy while females do it to Norway and Sweden.

The projection of partial rows of each table, joined by segments, allows us to appreciate the differences between males and females in each destiny country. We will only remark some of them.

For example, United Kingdom is a country to which males and females students go in a greater proportion among the students of Humanities. Nevertheless males also choose United Kingdom to carry out Social and Legal studies whereas females do not.

Male and female students that come to Portugal agree in selecting this country over the average for Health degrees. But, males also go to Portugal to study Experimental Science while females prefer this country for studies of Engineering and Technology.

Spanish students who go to Finland share the selection of this country over the rest of the countries to study in the areas of Health and Engineering but there are more females in the former area and males in the last one.

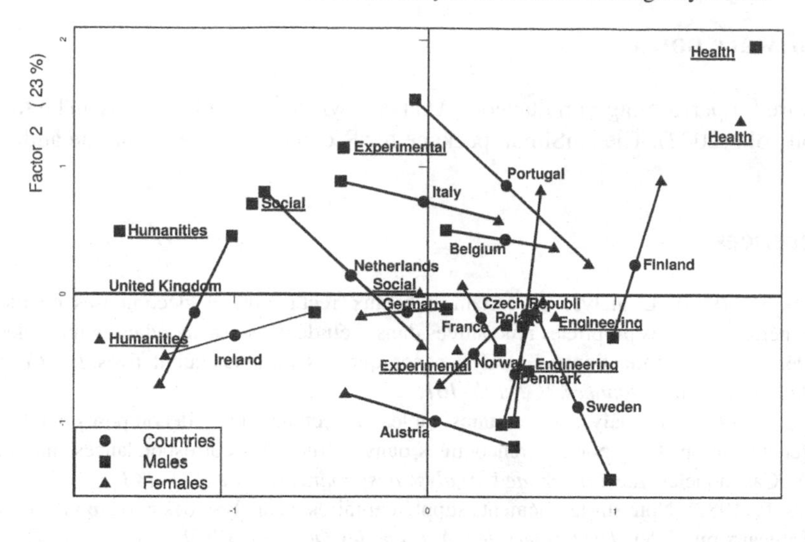

Fig. 1. Projection of columns, overall rows and partial rows

In the other hand, not big differences between males and females are found in Germany, France, Belgium and Norway as it is indicate by the close projections of overall and partial rows.

As conclusion of this application we can say that Simultaneous Analysis allows us to show the common structure inside each table as well as the differences in the structure of both tables. A more extensive application to the joint study of the inter and intra-structure of a bigger number of contingency tables can be found in Zárraga and Goitisolo (2006).

4 Discussion

The joint study of several data tables has given rise to an extensive list of factorial methods, some of which have been gathered by Cazes (2004), for both quantitative and categorical data tables. In the correspondence analysis (CA) approach Cazes shows the similarity between some methods in the case of proportional row margins and shows the problem that arises in a joint analysis when the row margins are different or not proportional.

Comments on the appropriateness of SA and a comparison with different methods, especially with Multiple Factor Analysis for Contingency Tables (Pagès and Bécue-Bertaut (2006)), in the cases where row margins are equal, proportional and not proportional between the tables can be found in Zárraga and Goitisolo (2006).

5 Software notes

Software for performing Simultaneous Analysis, written in S-Plus 2000 can be found in Goitisolo (2002). The AnSimult package for R can be obtained from the authors.

References

CAZES, P. (1980): L' analyse de certains tableaux rectangulaires décomposés en blocs: généralisation des propriétes rencontrées dans l' étude des correspondances multiples. I. Définitions et applications à l' analyse canonique des variables qualitatives. *Les Cahiers de l' Analyse des Données, V, 2, 145–161.*

CAZES, P.(1981): L' analyse de certains tableaux rectangulaires décomposés en blocs: généralisation des propriétes rencontrées dans l' étude des correspondances multiples. IV. Cas modèles. *Les Cahiers de l' Analyse des Données, VI, 2, 135–143.*

CAZES, P. (1982): Note sur les éléments supplémentaires en analyse des correspondances II. Tableaux multiples. *Les Cahiers de l' Analyse des Données, VII, 133–154.*

CAZES, P. (2004): Quelques methodes d' analyse factorielle d' une serie de tableaux de données. *La Revue de Modulad, 31, 1–31.*

D' AMBRA, L. and LAURO, N. (1989): Non symetrical analysis of three-way contingency tables. *Multiway Data Analysis, 301–315.*

ESCOFIER, B. (1983): Généralisation de l' analyse des correspondances à la comparaison de tableaux de fréquence. *INRIA, Mai, 207, 1–33.*

ESCOFIER, B and PAGÈS, J. (1988 (1998, 3e édition)): *Analyses Factorielles Simples et Multiples. Objetifs, méthodes et interprétation.* Dunod, París.

GOITISOLO, B. (2002): *El análisis simultáneo. Propuesta y aplicación de un nuevo método de análisis factorial de tablas de contingencia.* Phd Thesis. Basque Country University Press. Bilbao. Spain.

LAURO, N. and D' AMBRA, L. (1984): L' Analyse non symétrique des correspondances. *Data Analysis and Informatics, III, 433–446.*

MÉOT, A. and LECLERC, B. (1997): Voisinages a priori et analyses factorielles: Illustration dans le cas de proximités géographiques. *Revue de Statistique Appliquée, XLV, 25–44.*

PAGÈS, J. and BÉCUE-BERTAUT, M. (2006): Multiple Factor Analysis for Contingency Tables. In: M. Greenacre and J.Blasius (Eds.): *Multiple Correspondence Analysis and Related Methods*. Chapman & Hall/CRC, 299–326.

ZÁRRAGA, A. and GOITISOLO, B. (2002): Méthode factorielle pour l'analyse simultanée de tableaux de contingence. *Revue de Statistique Appliquée L(2), 47-70.*

ZÁRRAGA, A. and GOITISOLO, B. (2003): Étude de la structure Inter-tableaux à travers l'Analyse Simultanée. *Revue de Statistique Appliquée LI(3), 39-60.*

ZÁRRAGA, A. and GOITISOLO, B. (2006): Simultaneous Analysis: A Joint Study of Several Contingency Tables with Different Margins. In: M. Greenacre and J.Blasius (Eds.): *Multiple Correspondence Analysis and Related Methods*. Chapman & Hall/CRC, 327–350.

Part IV

Analysis of Complex Data

Graph Mining: Repository vs. Canonical Form

Christian Borgelt and Mathias Fiedler

European Center for Soft Computing
c/ Gonzalo Gutiérrez Quirós s/n, 33600 Mieres, Spain
christian.borgelt@softcomputing.es
mail@mathias-fiedler.info

Abstract. In frequent subgraph mining one tries to find all subgraphs that occur with a user-specified minimum frequency in a given graph database. The basic approach is to grow subgraphs, adding an edge and maybe a node in each step, to count the number of database graphs containing them, and to eliminate infrequent subgraphs. The predominant method to avoid redundant search (the same subgraph can be grown in several ways) is to define a canonical form that uniquely identifies a graph up to automorphisms. The obvious alternative, a repository of processed subgraphs, has received fairly little attention yet. However, if the repository is laid out as a hash table with a carefully designed hash function, this approach is competitive with canonical form pruning. In experiments we conducted, the repository-based approach could sometimes outperform canonical form pruning by 15%.

1 Introduction

Frequent subgraph mining consists in the task to find all subgraphs that occur with a user-specified minimum frequency in a given database of (attributed) graphs. Since this problem appears in applications in biochemistry, web mining, and program flow analysis, it has attracted a lot of attention, and several algorithms were proposed to tackle it. Some of them rely on principles from inductive logic programming and describe graphs by logical expressions (Finn *et al.* 1998). However, the vast majority transfers techniques developed originally for frequent item set mining. Examples include MolFea (Kramer *et al.* 2001), FSG (Kuramochi and Karypis 2001), MoSS/MoFa (Borgelt and Berthold 2002), gSpan (Yan and Han 2002), Closegraph (Yan and Han 2003), FFSM (Huan *et al.* 2003), and Gaston (Nijssen and Kok 2004). A related, but slightly different approach is used in Subdue (Cook and Holder 2000).

The basic idea of these approaches is to grow subgraphs into the graphs of the database, adding an edge and maybe a node (if it is not already in the subgraph) in each step, to count the number of graphs containing each grown subgraph, and to eliminate infrequent subgraphs. All found frequent subgraphs are reported (or often only the subset of so-called *closed* subgraphs).

While in frequent item set mining it is trivial to ensure that each item set is checked only once, it is a core problem in frequent subgraph mining how to avoid redundant search. The reason is that the same subgraph can be grown in several ways, namely by adding the same nodes and edges in different orders. Although multiple tests of the same subgraph do not invalidate the result of a subgraph mining algorithm, they can be devastating for its execution time.

One of the most elegant ways to avoid redundant search is to define a canonical description of a (sub)graph. Combined with a specific way of growing the subgraphs, such a canonical description can be used to check whether a given subgraph has been considered in the search before. For example, Borgelt (2006) studied a family of such canonical forms, which comprises the special forms used in gSpan (Yan and Han 2002) and Closegraph (Yan and Han 2003) as well as the one underlying MoSS/MoFa (Borgelt and Berthold 2002).

However, canonical form pruning is not the only way to avoid redundant search. A simpler and much more straightforward approach is a repository of already processed subgraphs, against which each grown subgraph is checked. Nevertheless this approach is rarely used, has actually not even been properly investigated yet. To our knowledge only two existing algorithms use a repository, namely MoSS/MoFa, which prunes with a canonical form by default, but offers the optional use of a repository, and Gaston (Nijssen and Kok 2004), in which a repository is used in the final phase for general graphs, since Gaston's canonical form is restricted to trees. In order to close this gap, this paper examines repository-based pruning and compares it to canonical form pruning. Surprisingly enough, a repository-based approach is highly competitive and could sometimes outperform canonical form pruning by 15%.

2 Canonical form pruning

The core idea underlying a canonical form is to construct a code word that uniquely identifies a graph up to automorphisms. The characters of this code word describe the connection structure of the graph. If the graph is attributed (labeled), they also comprise information about edge and node attributes. While it is straightforward to capture the attribute information, it is less obvious how to describe the connection structure. For this, the nodes of the graph must be numbered (more generally: endowed with unique labels), because we need to specify the source and the destination node of an edge. Unfortunately, different ways of numbering the nodes of a graph yield different code words, because they lead to different descriptions of an edge (simply because the indices of source and destination node differ). In addition, the edges can be listed in different orders. Different possible solutions to these two problems give rise to different canonical forms (see Borgelt (2006) for details).

However, given a (systematic) way of numbering the nodes of a graph and a sorting criterion for the edges, a canonical description is derived as follows: each numbering of the nodes yields a code word, which is the concatenation of the sorted edge descriptions. The resulting code words are sorted lexicographically. The lexicographically smallest code word is the canonical description. (It should be noted that the graph can be reconstructed from this code word.)

Canonical code words are used in the search as follows: the process of growing subgraphs is associated with a way of building code words for them. Most naturally, the code word of a subgraph is obtained by simply concatenating the descriptions of its edges in the order in which they are added in the search. Since each possible subgraph needs to be checked only once, we may choose to process it only in the node of the search tree, in which its code word (as constructed by the search) is the canonical code word. Otherwise the subgraph (and thus the search tree rooted at it) is pruned.

It follows that we cannot use just any possible canonical form. If extended code words are built by appending the next edge description to the code word of the current subgraph, then the canonical form must have the so-called *prefix property*: any prefix of a canonical code word must be a canonical code word itself. Since we plan to extend only graphs in canonical form, the prefix property is needed to ensure that all possible subgraphs can be reached in the search. A simple way to ensure that a canonical form has the prefix property is to confine oneself to spanning tree numberings of the nodes of a graph.

In a straightforward algorithm (the code words of) all possible extensions of a subgraph are created and checked for canonical form. Extensions in canonical form are processed further, the rest is discarded. However, canonical forms also give rise to restrictions of the extensions of a subgraph, because for certain extensions one can see immediately that they lead to a non-minimal code word. For the two most important canonical forms, namely those that are based on a breadth-first (MoSS/Mofa) and a depth-first spanning tree numbering (gSpan/Closegraph), these are (for details see Borgelt (2006)):

- *maximum source extensions*
  Only nodes having an index no less than the maximum source of an edge may be extended (the source of an edge is the node with the smaller index).
- *rightmost path extensions*
  Only the nodes on the rightmost path of the spanning tree used for numbering the nodes may be extended (children of a node are sorted by index).

While reasons of space prevent us from reviewing details, restricted extensions are important to mention here. The reason is that they can be exploited for the repository approach as well, because they are an inexpensive way of avoiding most of the redundancy imminent in the search. (Note, however, that they cannot rule out all redundancy, as there are no perfect "simple rules".)

3 Repository of processed subgraphs

A repository of processed subgraphs is the most straightforward way of avoiding redundant search. Every encountered frequent subgraph is stored in a data structure, which allows us to check quickly whether a given subgraph is contained in it or not. Whenever a new subgraph is created, this data structure is accessed and if it contains the subgraph, we know that it has already been processed and thus can be discarded.

Only subgraphs that are not contained in the repository are extended and, of course, inserted into the repository.

There are two main issues one has to address when designing such a data structure. In the first place, we have to make sure that each subgraph is stored using a minimal amount of memory, because the number of processed subgraphs is usually huge. (This consideration may be one of the main reasons why a subgraph repository is so rarely used.) Secondly, we have to make the containment test as fast as possible, since it will be carried out frequently.

In order to achieve the first objective, we exploit that we only want to store graphs that appear in at least one graph of the database (which usually resides in memory anyway). Therefore we can store a subgraph by listing the edges of one embedding (that is, one occurrence of the subgraph in a graph of the database). Note that it suffices to list the edges, since the search is usually restricted to connected subgraphs and thus the edges also identify all nodes.[1]

It is pleasing to observe that this way of storing a subgraph can also make it easier to check whether a given subgraph is equivalent to it (isomorphism test). The rationale is to fix an order of the database graphs and to create the embeddings of all subgraphs in this order. Then we do not store an arbitrary embedding, but one into the first database graph it is contained in. For a new subgraph, for which we want to know whether it is in the repository, we can then check whether the first database graph containing it coincides with the one underlying the stored embedding. If it does not, we already know that the subgraphs (the new one and the stored one to which it is compared) cannot be equivalent, since equivalent subgraphs have the same embeddings.

However, if the database graphs coincide, we carry out the actual isomorphism test by also relying on the embeddings. We mark the embedding that is stored in the repository (that is, its edges) in the containing database graph. Then we traverse all embeddings of the new subgraph into the same graph[2] and check whether for any of them all edges are marked. If such an embedding exists, the two subgraphs (the new one and the stored one) must be equivalent, otherwise they differ. Obviously, this isomorphism test is linear in the number of edges and thus very efficient. It should be kept in mind, though, that it can be costly if a subgraph possesses a large number of embeddings into the same graph, because in the worst case (that is, if the two subgraphs are not isomorphic) all of these embeddings have to be checked. However, our experiments showed that this is an unlikely case, since especially larger subgraphs most of the time possess only a single embedding per database graph.

Even though an isomorphism test of the described form is fairly efficient, one should try to avoid it. Apart from the obvious checks whether the number of nodes and edges, the support in the graph database and the number of embeddings coin-

[1] The only exception are subgraphs consisting of a single node. Fortunately, such subgraphs need not be stored, since they cannot be created in more than one way, thus making it unnecessary to check whether they have been processed before.

[2] This is straightforward in our implementation, since in order to facilitate and accelerate forming extensions, we keep a list of all embeddings of a subgraph.

cide (naturally these must all be equal for isomorphic subgraphs), we employ a hash function that is computed from local graph properties. The basic idea is to combine the node and edge attributes and the node degrees, hoping that this allows us to distinguish non-isomorphic subgraphs. In particular, we combine for each edge the edge attribute and the attribute and degree of the two incident nodes into a number. For each node we compute a number from the node attribute, the node degree, the attributes of its incident edges and the attributes of the other nodes these edges are incident to. These numbers (one for each node and one for each edge) are then combined with the total numbers of nodes and edges to yield a hash code.[3]

The computed hash code is used in the standard way to build a hash table, thus making it possible to restrict the isomorphism test to (a subset of) the subgraphs in one hash bin (a subset, because some collisions can be resolved by comparing the support etc., see above). By carefully tuning the parameters of the hash function we tried to minimize the number of collisions.

4 Comparison

Considering how canonical form pruning and repository-based pruning work, we can make the following observations, which already give hints w.r.t. their relative performance (and which we use to explain our experimental findings):

Canonical form pruning has the advantage that we only have to carry out one test (for canonical form) in order to determine whether a subgraph needs to be processed or not (even though this test can be expensive). It has the disadvantage that it is most costly for the subgraphs that are in canonical form (and thus have to be processed), because for these subgraphs all possibilities to construct a code word have to be tried. For non-canonical code words the test usually terminates earlier, since it can often construct fairly quickly a prefix that is smaller than the code word of the subgraph to test.

Repository-based pruning has the advantage that it often allows to decide very quickly that a subgraph has not been processed yet (for example, if a hash bin is empty). Together with comparing the numbers of nodes and edges, the support etc., this suggests that a repository-based approach is fastest for subgraphs that actually have to be processed. Only if these simple tests fail (as for equivalent subgraphs), we have to carry out the isomorphism test.

As a consequence, we expect repository-based pruning to perform well if the number of subgraphs to be processed is large compared to the number of subgraphs to be discarded (as the repository is usually faster for the former).

[3] A technical remark: we do not only combine these numbers by summing them and computing their bitwise exclusive or, but also apply bitwise shifts of varying width in order to cover the full range of values of (32 bit) integer numbers.

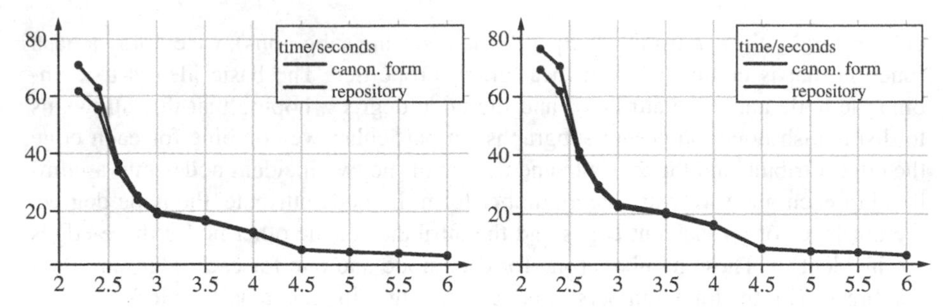

Fig. 1. Experimental results on the IC93 data set, search time vs. minimum support in percent. Left: maximum source extensions, right: rightmost path extensions.

5 Experiments

In order to test our repository-based pruning experimentally, we implemented it as part of the MoSS program[4], which is written in Java. As a test dataset (to which we confine ourselves here due to limits of space) we used a subset of the *Index Chemicus* from 1993. The results we obtained with different restricted extensions (maximum source and rightmost path, see Section 2) are shown in Figures 1 to 3. The horizontal axis shows the minimal support in percent.

Figure 1 shows the execution times in seconds. The upper graph refers to canonical form pruning, the lower to repository-based pruning. The times do not differ much, but diverge for lower support values, reaching 15% advantage for the repository-based approach together with maximum source extensions.

Figure 2 shows the numbers of subgraphs considered in the search and provides a basis for explanations of the observed behavior. The graphs refer (from top to bottom) to the number of generated subgraphs, the number checked for duplicates, the number of processed subgraphs, and the number of (discarded) duplicates (difference between the two preceding curves).

Note that about half of the work is done by minimum support pruning (which discards all subgraphs that do not appear in the user-specified minimum number of database graphs), as it is responsible for the difference between the two top curves. The subgraphs discarded in this way may be unique or not—we need not care, since they do not qualify anyway.

Canonical form or repository-based pruning only serve the purpose to get rid of the subgraphs between the two middle curves. That the gap between them is fairly small compared to their vertical location indicates the high quality of restricted extensions: most redundancy is already removed by them and only fairly few redundant subgraphs still need to be detected. (Note that the gap is smaller for maximum source extensions, which is the main reason for the usually lower execution times achieved by this approach).

---

[4] MoSS is available for download under the Gnu Lesser (Library) General Public License at http://www.borgelt.net/moss.html.

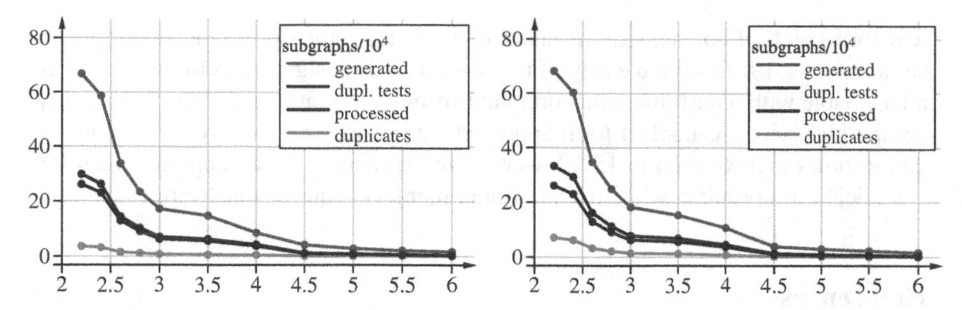

Fig. 2. Experimental results on the IC93 data set, numbers of subgraphs used in the search. Left: maximum source extensions, right: rightmost path extensions.

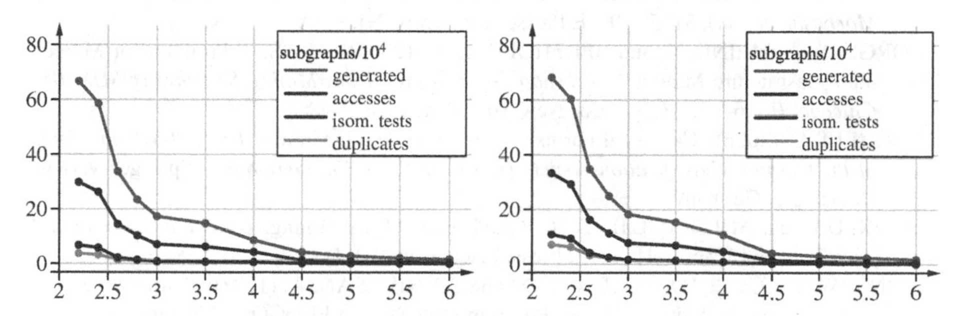

Fig. 3. Experimental results on the IC93 data set, performance of repository-based pruning. Left: maximum source extensions, right: rightmost path extensions.

Figure 3 finally shows the performance of repository-based pruning (mainly the effectiveness of the hash function). All curves are the same as in the preceding figure, with the exception of the third curve from the top, which shows the number of isomorphism tests. Subgraphs in the gap between this curve and the one above it have to be processed and are identified as such without any isomorphism test. Only subgraphs in the (small) gap between this curve and the bottom curve (the number of actual duplicates) have to be identified and discarded with the help of isomorphism tests.

Note that for a perfect hash function (which maps only equivalent subgraphs to the same value) the two bottom curves would coincide. Note also that a canonical form can be seen as a perfect hash function (with a range of values that does not fit into an integer), since it uniquely identifies a graph.

6 Summary

In this paper we investigated the widely neglected possibility to avoid redundant search in frequent subgraph mining with a repository of already encountered subgraphs. Even though it may be less elegant than the more popular approach of canonical forms and, of course, requires additional memory for storing the subgraphs, it should not be dismissed too easily. If the repository is designed carefully, namely as a hash table with a hash function computed from local graph properties, it is highly competitive with a canonical form approach. In our experiments we observed execution times that were up to 15% lower for the repository-based approach than for canonical form pruning, while the additional memory requirements were bearable.

References

BORGELT, C., and BERTHOLD, M.R. (2002): Mining Molecular Fragments: Finding Relevant Substructures of Molecules. *Proc. IEEE Int. Conf. on Data Mining (ICDM 2002, Maebashi, Japan)*, 51–58. IEEE Press, Piscataway, NJ, USA

BORGELT, C., MEINL, T., and BERTHOLD, M.R. (2005): MoSS: A Program for Molecular Substructure Mining. *Workshop Open Source Data Mining Software (OSDM'05, Chicago, IL)*, 6–15. ACM Press, New York, NY, USA

BORGELT, C. (2006): Canonical Forms for Frequent Graph Mining. *Proc. 30th Ann. Conf. of the German Classification Society (GfKl 2006, Berlin, Germany).* Springer-Verlag, Heidelberg, Germany

COOK, D.J., and HOLDER, L.B. (2000) Graph-Based Data Mining. *IEEE Trans. on Intelligent Systems* 15(2):32–41. IEEE Press, Piscataway, NJ, USA

FINN, P.W., MUGGLETON, S., PAGE, D., and SRINIVASAN, A. (1998): Pharmacore Discovery Using the Inductive Logic Programming System PROGOL. *Machine Learning*, 30(2-3):241–270. Kluwer, Amsterdam, Netherlands

HUAN, J., WANG, W., and PRINS, J. (2003): Efficient Mining of Frequent Subgraphs in the Presence of Isomorphism. *Proc. 3rd IEEE Int. Conf. on Data Mining (ICDM 2003, Melbourne, FL)*, 549–552. IEEE Press, Piscataway, NJ, USA

INDEX CHEMICUS — Subset from 1993. Institute of Scientific Information, Inc. (ISI). Thomson Scientific, Philadelphia, PA, USA 1993
http://www.thomsonscientific.com/products/indexchemicus/

KRAMER, S., DE RAEDT, L., and HELMA, C. (2001): Molecular Feature Mining in HIV Data. *Proc. 7th ACM SIGKDD Int. Conf. on Knowledge Discovery and Data Mining (KDD 2001, San Francisco, CA)*, 136–143. ACM Press, New York, NY, USA

KURAMOCHI, M., and KARYPIS, G. (2001): Frequent Subgraph Discovery. *Proc. 1st IEEE Int. Conf. on Data Mining (ICDM 2001, San Jose, CA)*, 313–320. IEEE Press, Piscataway, NJ, USA

NIJSSEN, S., and KOK, J.N. (2004): A Quickstart in Frequent Structure Mining Can Make a Difference. *Proc. 10th ACM SIGKDD Int. Conf. on Knowledge Discovery and Data Mining (KDD2004, Seattle, WA)*, 647–652. ACM Press, New York, NY, USA

YAN, X., and HAN, J. (2002): gSpan: Graph-Based Substructure Pattern Mining. *Proc. 2nd IEEE Int. Conf. on Data Mining (ICDM 2003, Maebashi, Japan)*, 721–724. IEEE Press, Piscataway, NJ, USA

YAN, X., and HAN, J. (2003): Closegraph: Mining Closed Frequent Graph Patterns. *Proc. 9th ACM SIGKDD Int. Conf. on Knowledge Discovery and Data Mining (KDD 2003, Washington, DC)*, 286–295. ACM Press, New York, NY, USA

Classification and Retrieval of Ancient Watermarks

Gerd Brunner and Hans Burkhardt

Institute for Pattern Recognition and Image Processing, Computer Science Faculty,
University of Freiburg, Georges-Koehler-Allee 052, 79110 Freiburg, Germany
{gbrunner, Hans.Burkhardt}@informatik.uni-freiburg.de

Abstract. Watermarks in papers have been in use since 1282 in Medieval Europe. Watermarks can be understood much in the sense of being an ancient form of a copyright signature. The interest of the International Association of Paper Historians (IPH) lies specifically in the categorical determination of similar ancient watermark signatures.

The highly complex structure of watermarks can be regarded as a strong and discriminative property. Therefore we introduce edge-based features that are incorporated for retrieval and classification. The feature extraction method is capable of representing the global structure of the watermarks, as well as local perceptual groups and their connectivity. The advantage of the method is its invariance against changes in illumination and similarity transformations.

The classification results have been obtained with leave-one out tests and a support vector machine (SVM) with an intersection kernel. The best retrieval results have been received with thc histogram intersection similarity measure. For the 14 class problem we obtain a true positive rate of more than 87%, that is better than any earlier attempt.

1 Introduction

Ancient watermarks served as a mark for the paper mill that made the sheet. Hence, they served as a unique identifier and as a quality label. Nowadays, scientists from the International Association of Paper Historians (IPH) try to identify unique watermarks in order to get known the evolution of commercial and cultural exchanges between cities in the Middle Ages (IHP 1998). The work is tedious since there are approximately 600.000 known watermarks and their number is steadily growing.

In this paper we present a structure-based feature approach in order to automatically retrieve and classify ancient watermarks. In the following we show that structure is a well suited feature to discriminate ancient watermarks.

Next, we present relevant work that is followed by a section on the actual feature computation. In the second part of this article we show the most important results. We summarize our contribution with a discussion of the results and a final conclusion.

1.1 Related work

To date, there have been attempts to classify and retrieve watermark images, both by textual- and content-based approaches. Textual approaches have been developed by Del Marmol (1987) and Briquet (1923). As a matter of fact, pure textual classification systems can be error prone. Watermark labels and or textual descriptions might be very old, erroneous or just not detailed enough. Therefore, more recent attempts have been undertaken in order to focus on the *real* content of watermark images. In Rauber et al. (1997) the authors used a 16-bin large circular histogram computed around the center of gravity of each watermark image. In addition, eight directional filters were applied to each image and used as a feature vector. The algorithms were tested on a small watermark database consisting of 120 images, split up into 12 different classes. The system achieved a probability of 86% that the first retrieved image belongs to the same class as the query image. A different approach was taken by the authors in Riley and Eakins (2002) who used three sets of various global moment features and three sets of component-based features. The latter set of features consists of several shape descriptors which are extracted from various image regions.

In the following we will show that the structure of watermarks can be most efficiently represented by features taken from a set of straight line segments. Therefore, we will extract sets of segments and compute features from them on different scales.

2 Feature extraction

The geometric structure of watermarks is a strong descriptor. Therefore, we compute a hierarchy of structural features, namely global and local ones. The former ones depict a holistic scene representation and the latter ones take local perceptual groups and their connectivity into account. As mentioned earlier we represent the structure of the watermarks by straight line segments. In order to extract the line segments we have adopted the algorithms of Pope and Lowe (1994) and Kovesi (2002). In the first step we create an edge map with the Canny detector. Next, the algorithm scans through the binary edge map, where the neighborhood of every edge pixel is investigated in order to form line segments. The final segments serve as a ground truth for the further feature computation.

Global Features

Let $\mathbf{L} = \{\mathbf{l}_i \,|\, i = 1, 2, ..., N\}$, be a set of line segments obtained from a watermark image. Then, we compute geometric properties of $\mathbf{L}$ such as the angles of all segments between each other, the relative lengths of every segment and the relative Euclidean distance between all segment mid-points.

In detail, the angle between two segments $\mathbf{l}_i$ and $\mathbf{l}_j$ is defined as:

$$cos(\theta_{ij}) = \frac{\mathbf{l}_i \cdot \mathbf{l}_j}{||\mathbf{l}_i \cdot \mathbf{l}_j||_2}, \quad (1)$$

with $||\cdot||_2$ being the $L2-Norm$. The angle is in the range of $[-\pi,\pi]$. The relative length of a segment $\mathbf{l}_i$ can be written as:

$$len(\mathbf{l}_i) = \frac{\sqrt{(x_i^e - x_i^b)^2 + (y_i^e - y_i^b)^2}}{\sqrt{(x_{max} - x_0)^2 + (y_{max} - y_0)^2}}, \tag{2}$$

where x_i^b, x_i^e, y_i^b and y_i^e denote the coordinates of the segment's begin and end points. The denominator is a scaling factor in respect to the longest possible line segment[1] with (x_0, y_0) and (x_{max}, y_{max}) as the begin and end point coordinates. The Euclidean distance between the mid-points p_i^c and p_j^c of the segments $\mathbf{l}_i$ and $\mathbf{l}_j$ is defined as

$$dist^c(\mathbf{l}_i, \mathbf{l}_j) = \frac{\sqrt{(x_j^c - x_i^c)^2 + (y_j^c - y_i^c)^2}}{\sqrt{(x_{max} - x_0)^2 + (y_{max} - y_0)^2}}, \tag{3}$$

with x_i^c, x_j^c, y_i^c and y_j^c as the coordinates of the segment mid-points. The denominator fulfills the same scaling purpose as the one in Equation 2. Thus, the relative length of a segment and the relative distance between two segments is limited to the range $[0,1]$. The relative representation ensures invariance under isotropic scaling.

Now, that the three basic properties of a set of line segments are computed, we can incorporate this information into Euclidean distance matrices (EDM). An EDM is a two-dimensional array consisting of distances taken from a set of entities, that can be coordinates or points from a feature space. Thus, an EDM incorporates distance knowledge. For our feature computation, EDMs are used in order to represent the relative geometric connectivity for a set of straight line segments. Specifically, we define three EDMs: one based on segment angles $\mathbf{E}^{ang}$ (see Equation 1) a second one based on relative segment lengths $\mathbf{E}^{len}$ (see Equation 2) and a third one based on relative distances between segments $\mathbf{E}^{dist}$ (see Equation 3). The matrix of $\mathbf{E}^{ang}$ can be written as:

$$\mathbf{E}^{ang} = \begin{bmatrix} e_{11}^{ang} & e_{12}^{ang} & \cdots & e_{1n}^{ang} \\ e_{21}^{ang} & e_{22}^{ang} & \cdots & e_{2n}^{ang} \\ \vdots & \vdots & \ddots & \vdots \\ e_{n1}^{ang} & e_{n2}^{ang} & \cdots & e_{nn}^{ang} \end{bmatrix}, \tag{4}$$

and each element is computed according to

$$e_{ij}^{ang} = ||\theta_i - \theta_j||, \tag{5}$$

where the values of θ_i and θ_j are in the range of $[-\pi,\pi]$. The angles are taken between the line segments i and j. $\mathbf{E}^{len}$ and $\mathbf{E}^{dist}$ can be represented in a similar fashion.

Next, we compute three histograms from the previously created EDMs. The histograming step is necessary since the size of the EDMs can differ, i.e. the number of line segments is not the same for each watermark.ming step is necessary since the size of EDMs can differ, i.e. the number of line segments is not the same for each

[1] The longest possible line segment is as long as the diagonal of the image.

watermark. The three histograms can be understood as a holistic representation of a set of segments. The final concatenation of the three histograms resembles a global feature and is invariant against similarity transformations.

Local features

The previously developed global features encode a complete watermark. However, local structural information plays an important role, too. Watermarks commonly exhibit certain local regularities in their structure. In order to tackle this problem we introduce local features that are based on perceptual groups of line segments.

Therefore, we define subsets of line segments from every watermark which are unique, eminent structural entities with well defined relations: **Parallelity**, **Perpendicularity**, **Diagonality** ($\frac{\pi}{4}, \frac{3\pi}{4}$). These groups are formed according to angular relations between segments and will be used in order to compute geometric relations between their members.

The four subsets reflect line segments with certain relations. In fact, we will extract similar features as we did in the global case. Following that methodology, we can compute three EDMs: $\mathbf{E}_*^{ang}$, $\mathbf{E}_*^{len}$ and $\mathbf{E}_*^{dist}$, for each of the four extracted sets of segments. Note that the $*$ is a placeholder for the four sets. Specifically, we define the angles between two segments, the relative segment lengths and the relative distance between two segments according to Equations 1, 2 and 3 for every subset of line segments.

Then we create three histograms for every subset of line segments. The histograms represent geometric relations of perceptual segment subsets. Since three histograms have been formed for every set, we obtain 12 histograms in total. The final set of local feature vectors is obtained by concatenation of all 12 histograms.

Feature representation

In our experiments we have empirically determined the best resolution for the histograms. For the angle based histograms[2] we have incorporated 36 bins, that corresponds to a 10° resolution with respect to angles. The resolutions for every length based histogram[3] is 15 bins, which results in a robust and compact feature. The final feature vector is obtained by the concatenation of all global and local feature histograms.

3 Results

3.1 Data description

The Swiss Paper Museum in Basel provided us a subset of their digital watermark database. The database used in the subsequent experiments consists of about 1800

[2] Histograms that are computed from the following EDMs: $\mathbf{E}^{ang}$ (global features) and $\mathbf{E}_*^{ang}$ (local features).

[3] Histograms that are computed from the following EDMs: $\mathbf{E}^{len}$, $\mathbf{E}^{dist}$ (global features) and $\mathbf{E}_*^{len}$, $\mathbf{E}_*^{dist}$ (local features).

images, split up into 14 classes. : *Eagle*, *Anchor1*, *Anchor2*, *Coat of Arm*, *Circle*, *Bell*, *Heart*, *Column*, *Hand*, *Sun*, *Bull Head*, *Flower*, *Cup* and *Other objects*. The class memberships are according to the Briquet catalog (Briquet 1923). Figure 1 shows scanned sample watermark images. A detailed description of the scanning setup can be found in Rauber (1998). In fact, the watermarks are digitized from the original sources. Specifically, each ancient document was scanned three times (front, back and by transparency) in order to obtain a high quality *digital copy*, where the last scan contains all necessary information (Rauber 1998). A semi-automatic method, that is describe in (Rauber 1998), delivers the *final* images. The method incorporates a global contrast, contour enhancement and grey-level inversion. Figure 2 shows sample images after the method was applied.

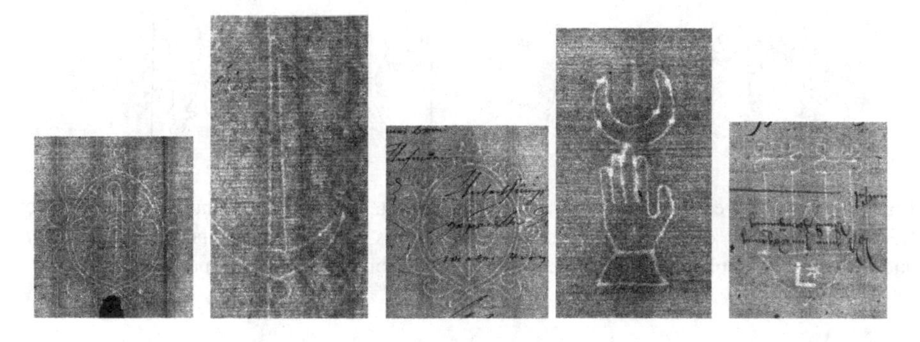

Fig. 1. Samples of scanned ancient watermark images (courtesy Swiss Paper Museum, Basel).

3.2 Ancient Watermark Retrieval

For retrieval we have computed the features offline for all watermarks. At retrieval time, only the feature vector for the query watermark has to be computed. The retrieval results are obtained with the histogram intersection similarity measure.

Figure 3 shows a set of 10 watermark images. The first image is the query, the second one is the identical match, indicated by the 1 above the image. The subsequent images are sorted in decreasing similarity, as it is indicated by the numbers above each image. It is interesting to observe that most of the retrieved anchors show the same orientation. A closer look at the query image reveals that it is featured with a tiny cross atop and with *cusp-like* structures at the outer endings[4]. The retrieved images clearly show that both of these small scale structures are present in all of the displayed images. In Figure 4 we can see another retrieval result. Table 1 shows the averaged class-wise precision and recall at $N/2$, where N is the number of class

[4] Note, that the class *Anchor1* possesses a large intra-class variation of shapes, i.e. many anchors have no crosses or show very different endings.

Fig. 2. Sample filigrees from the watermark database after enhancement and binarization (see Rauber 1998). Each of the two rows shows watermarks from the same class, namely *Heart* and *Eagle*. The samples show the large intra-class variability of the watermark database.

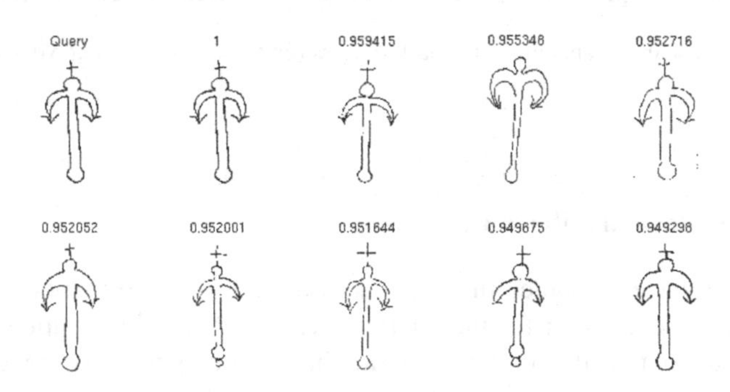

Fig. 3. Retrieval result obtained with our structure-based features from the class *Anchor1* of the watermark database.

Table 1. Averaged precision and recall at $N/2$ for the watermark database.

Classes	1	2	3	4	5	6	7	8	9	10	11	12	13	14
N	322	115	139	71	91	44	197	126	99	33	14	31	17	416
$P(N/2)$	.492	.243	.214	.144	.109	.244	.173	.097	.442	.068	.190	.802	.556	.283
$R(N/2)$	.528	.139	.302	.197	.088	.182	.152	.191	.263	.061	.143	.710	.352	.361

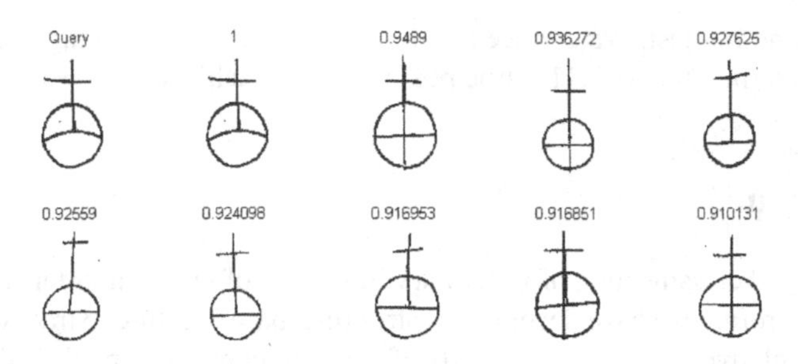

Fig. 4. Retrieval result of the class *Circle* from the watermark database, under the usage of global and local structural features.

members. Due to place limitations the watermark classes have been assigned a number[5], where one refers to the class *Eagle* and 14 to the class *Other objects*. However, we do observe some classes of worse performance. That is to a large extent due to the high intra-class variation of the database. Figure 2 shows the large intra-class variation for two sample classes. Since CBIR performs a similarity ranking some class members can be less similar to a certain query (from the same class) then images from other classes. Visual inspections have shown that this argumentation holds for the classes *Eagle* and *Coat of Arm*. The reason is that eagle motives are very common in heraldry, i.e. about half of the members of the class *Coat of Arms* have some kind of eagle embedded on a shield or armorial bearings. Similar observations hold for some other classes.

3.3 Ancient Watermark Classification

In the previous section we have retrieved watermark images. Now we want to learn the feature distribution of every class in the feature space. Therefore, the classification of the watermark images is treated as a learning problem. The classification results are obtained with leave-one out tests and SVMs under the usage of different kernel. Specifically, we have obtained the best results with the intersection kernel and a cost parameter $C = 2^{20}$. We have used the same features as for the retrieval task. The feature vectors have been normalized according to zero mean and unit variance. Table 2 shows the class-wise true and false positive rates which have been obtained

Table 2. Class-wise true positive (TP) and false positive (FP) rates for the watermark database.

Classes	1	2	3	4	5	6	7	8	9	10	11	12	13	14	Total
TP	.919	.870	.871	.465	.758	.773	.817	.865	.919	.546	.571	1.00	.824	.995	.874
FP	.037	.001	.019	.012	.011	.003	.025	.008	.002	.004	.001	0	0	.008	.125

[5] The class names are listed in Section 3.1.

with a leave-one-out test. We can see that for most of the classes a high recognition rate is achieved. In total, a 87.41% true positive rate is achieved.

4 Conclusion

The retrieval and classification of watermark images is of great importance for paper historians. Therefore we have developed a structure-based feature extraction method that encodes relative spatial arrangements of line segments. The method determines relations on global and local scales. The results show that structure is a powerful descriptor for the current problem. The retrieval results show that the proposed features work very well.

Next, we have performed a classification of the watermark images. A support vector machine with intersection kernel was able to successfully learn the characteristics of every class. A classification rate (true positive rate) of more than 87% is an indicator of a good performance. In future work, we would like to apply the structural features to a larger database of watermarks and investigate partial matching as well.

References

BRIQUET, C. M. (1923): Les filigranes, Dictionnaire historique des marques de papier des leur apparition vers 1282 jusqu'en 1600. *Tome I B IV, Deuxieme edition.* Verlag Von Karl W. Hiersemann, Leipzig.

DEL MARMOL, F. (1987): Dictionnaire des filigranes classes en groupes alphabetique et chronologiques. *Namur: J. Godenne, 1900. -XIV, 192.*

IHP (1998): International Standard for the Registration of Watermarks. *International Association of Paper Historians (IHP).* Isbn 0250-8338.

KOVESI, P. D. (2002): Edges Are Not Just Steps. *Proceedings of the Fifth Asian Conference on Computer Vision, Melbourne, 822–827.*

POPE, A. R. and LOWE, D. G. (1994): Vista: A Software Environment for Computer Vision Research. *CVPR, 768-772.*

RAUBER, C. (1998): Acquisition, archivage et recherche de documents accessibles par le contenu: Application à la gestion d'une base de données d'images de filigranes. *Ph.D. Dissertation No. 2988.* University of Geneva, Switzerland.

RAUBER, C. and PUN, T. and TSCHUDIN, P. (1997): Retrieval of images from a library of watermarks for ancient paper identification. *EVA 97, Elekt. Bildverarbeitung und Kunst, Kultur, Historie.* Berlin, Germany.

RILEY, K. J. and EAKINS, J. P. (2002): Content-Based Retrieval of Historical Watermark Images: I-tracings. *Image and Video Retrieval, International Conference, CIVR. LNCS 2383, 253-261, Springer.*

Segmentation and Classification of Hyper-Spectral Skin Data

Hannes Kazianka[1], Raimund Leitner[2] and Jürgen Pilz[1]

[1] University of Klagenfurt, Institute of Statistics, Alpen-Adria-Universität Klagenfurt, Universitätsstraße 65-67, 9020 Klagenfurt, Austria
{hannes.kazianka, juergen.pilz} @uni-klu.ac.at

[2] CTR Carinthian Tech Research AG, Europastraıe 4/1, 9524 Villach, Austria
Raimund.Leitner@ctr.at

Abstract. Supervised classification methods require reliable and consistent training sets. In image analysis, where class labels are often assigned to the entire image, the manual generation of pixel-accurate class labels is tedious and time consuming. We present an independent component analysis (ICA)-based method to generate these pixel-accurate class labels with minimal user interaction. The algorithm is applied to the detection of skin cancer in hyper-spectral images. Using this approach it is possible to remove artifacts caused by sub-optimal image acquisition. We report on the classification results obtained for the hyper-spectral skin cancer data set with 300 images using support vector machines (SVM) and model-based discriminant analysis (MclustDA, MDA).

1 Introduction

Hyper-spectral images consist of several, up to hundred, images acquired at different - mostly narrow band and contiguous - wavelengths. Thus, a hyper-spectral image contains pixels represented as multidimensional vectors with elements indicating the reflectivity at a specific wavelength. For a contiguous set of narrow band wavelengths these vectors correspond to spectra in the physical meaning and are equal to spectra measured with e.g. spectrometers.

Supervised classification of hyper-spectral images requires a reliable and consistent training set. In many applications labels are assigned to the full image instead of to each individual pixel even if instances of all the classes occur in the image. To obtain a reliable training set it may be necessary to label the images on a pixel by pixel basis. Manually generating pixel-accurate class labels requires a lot of effort; cluster-based automatic segmentation is often sensitive to measurement errors and illumination problems. In the following we present a labelling strategy for hyper-spectral skin cancer data that uses PCA, ICA and K-Means clustering. For the classification of unknown images, we compare support vector machines and model-based discriminant analysis.

Section 2 describes the methods that are used for the labelling approach. The classification algorithms are discussed in Section 3. In Section 4 we present the segmentation and classification results obtained for the skin cancer data set and Section 5 is devoted to discussions and conclusions.

2 Labelling

Hyper-spectral data are highly correlated and contain noise which adversely affects classification and clustering algorithms. As the dimensionality of the data equals the number of spectral bands, using the full spectral information leads to computational complexity. To overcome the curse of dimensionality we use PCA to reduce the dimensions of the data, and inherently also unwanted noise. Since different features of the image may have equal score values for the same principal component, an additional feature extraction step is proposed. ICA makes it possible to detect acquisition artifacts like saturated pixels and inhomogeneous illumination. Those effects can be significantly reduced in the spectral information giving rise to an improved segmentation.

2.1 Principal Component Analysis (PCA)

PCA is a standard method for dimension reduction and can be performed by singular value decomposition. The algorithm gives uncorrelated principal components. We assume that those principal components that correspond to very low eigenvalues contribute only to noise. As a rule of thumb, we chose to retain at least 95% of the variability which led to selecting 6-12 components.

2.2 Independent Component Analysis (ICA)

ICA is a powerful statistical tool to determine hidden factors of multivariate data. The ICA model assumes that the observed data, x, can be expressed as a linear mixture of statistically independent components, s. The model can be written as

$$x = As$$

where the unknown matrix A is called the mixing matrix. Defining W as the unmixing matrix we can calculate s as

$$s = Wx.$$

As we have already done a dimension reduction, we can assume that noise is negligible and A is square which implies $W = A^{-1}$. This significantly simplifies the estimation of A and s. Providing that no more than one independent component has Gaussian distribution, the model can be uniquely estimated up to scalar multipliers. There exists a variety of different algorithms for fitting the ICA model. In our work we focused on the two most popular implementations which are based on maximisation of non-Gaussianity and minimisation of mutual information respectively: FastICA and FlexICA.

FastICA

The FastICA algorithm developed by Hyvärinen et al. (2002) uses negentropy, $J(y)$, as a measure of Gaussianity. Since negentropy is zero for Gaussian variables and always nonnegative one has to maximise negentropy in order to maximise non-Gaussianity. To avoid computation problems the algorithm uses an approximation of negentropy: If G denotes a nonquadratic function and we want to estimate one independent component s we can approximate

$$J(y) \approx [E\{G(y)\} - E\{G(\nu)\}]^2,$$

where ν is a standardised Gaussian variable and y is an estimate of s. We adopt to use $G(y) = \log\cosh y$ since this has been shown to be a good choice. Maximisation directly leads to a fixed-point iteration algorithm that is $20-50$ times faster than other ICA implementations. To estimate several independent components a deflationary orthogonalisation method is used.

FlexICA

Mutual information is a natural measure of information that members of a set of random variables have on the others. Choi et al. (2000) proposed an ICA algorithm that attempts to minimise this quantity. All independent components are estimated simultaneously using a natural gradient learning rule with the assumption that the source signals have the generalized Gaussian distribution with density

$$q_i(y_i) = \frac{r_i}{2\sigma_i\Gamma(1/r_i)} \exp\left(-\frac{1}{r_i}\left|\frac{y_i}{\sigma_i}\right|^{r_i}\right).$$

Here r_i denotes the Gaussian exponent which is chosen in a flexible way depending on the kurtosis of the y_i.

2.3 Two-Stage K-Means clustering

From a statistical point of view it may be inappropriate to use K-means clustering since K-means cannot use all the higher order information that ICA provides. There are several approaches that avoid using K-means, for example Shah et al. (2005) proposed the ICA mixture model (ICAMM). However, for large images this algorithm fails to converge. We developed a 2-stage K-means clustering strategy that works particularly well with skin data. The choice of 5 resp. 3 clusters for the K-means algorithm has been determined empirically for the skin cancer data set.

1. Drop ICs that contain a high amount of noise or correspond to artifacts.
2. Perform K-means clustering with 5 clusters.
3. Those clusters that correspond to healthy skin are taken together into one cluster. This cluster is labelled as *skin*.
4. Perform a second run of K-means clustering on the remaining clusters (inflamed skin, lesion, etc.). This time use 3 clusters. Label the clusters that correspond to the mole and melanoma centre as *mole* and *melanoma*. The remaining clusters are considered to be 'regions of uncertainty'.

3 Classification

This section describes the classification methods that have been investigated. The preprocessing steps for the training data are the same as in the segmentation task: Dimension reduction using PCA and feature extraction performed by ICA. Using the Bayesian Information Criterion (BIC), the data were reduced to 6 dimensions.

3.1 Mixture Discriminant Analysis (MDA)

MDA assumes that each class j can be modelled as a mixture of R_j subclasses. The subclasses have a multivariate Gaussian distribution with mean vector μ_{jr}, $r = 1,\ldots,R_j$, and covariance matrix Σ, which is the same for all classes. Hence, the mixture model for class j has the density

$$m_j(x) = |2\pi\Sigma|^{-\frac{1}{2}} \sum_{r=1}^{R_j} \pi_{jr} \exp\left\{ -\frac{(x-\mu_{jr})\,\Sigma^{-1}\,(x-\mu_{jr})}{2} \right\},$$

where π_{jr} denote the mixing probabilities for the j-th subclass, $\sum_{r=1}^{R_j} \pi_{jr} = 1$. The parameters $\theta = (\mu_{jr}, \Sigma, \pi_{jr})$ can be estimated using an EM-algorithm or, as Hastie et al. (2001) suggest, using optimal scoring. It is also possible to use flexible discriminant analysis (FDA) or penalized discriminant analysis (PDA) in combination with MDA. The major drawback of this classification approach is that, similar to LDA which is also described in Hastie et al. (2001), the covariance matrix is fixed for all classes and the number of subclasses for each class has to be set in advance.

3.2 Model-based Discriminant Analysis (MclustDA)

MclustDA, proposed by Fraley et al. (2002), extends MDA in a way that the covariance in each class is parameterized using the eigenvalue decomposition

$$\Sigma_r = \lambda_r D_r A_r D_r^T, \quad r = 1,\ldots,R_j.$$

The volume of the component is controlled by λ_r, A_r defines the shape and D_r is responsible for the orientation. The model selection is done using the BIC and the maximum likelihood estimation is performed by an EM-algorithm.

3.3 Support Vector Machines (SVM)

The aim of support vector machines is to find a hyperplane that optimally separates two classes in a high-dimensional feature space induced by a Mercer kernel $K(\mathbf{x},\mathbf{z})$. In the L^2-norm case the Lagrangian dual problem is to find λ^* that solves the following convex optimization problem:

$$\max_{\lambda} \sum_{i=1}^{m} \lambda_i - \frac{1}{2} \sum_{i=1}^{m} \sum_{j=1}^{m} \lambda_i \lambda_j y_i y_j \left(K(\mathbf{x}_i, \mathbf{x}_j) + \frac{1}{C} \delta_{ij} \right) \text{ s.t.} \sum_{i=1}^{m} \lambda_i y_i = 0,\ \lambda_i \geq 0,$$

where $\mathbf{x}_i$ are training points belonging to classes y_i. The cost parameter C and the kernel function have to be chosen to suit to the problem. It is also possible to use different cost parameters for unbalanced data as was suggested by Veropoulos et al. (1999).

Although SVMs were originally designed as binary classifiers, there exists a variety of methods to extend them to $k > 2$ classes. In our work we focused on *one-against-all* and *one-against-one* SVMs. The one-against-all formulation trains each class against all remaining classes resulting in k binary SVMs. The one-against-one formulation uses $\frac{k(k-1)}{2}$ SVMs, each separating one class from one another.

4 Results

A set of 310 hyper-spectral images (512×512 pixels and 300 spectral bands) of malign and benign lesions were taken in clinical studies at the Medical University Graz, Austria. They are classified as *melanoma* or *mole* by human experts on the basis of a histological examination. However, in our survey we distinguish between three classes, *melanoma, mole* and *skin*, since all these classes typically occur in the images. The segmentation task is especially difficult in this application: We have to take into account that *melanoma* typically occurs in combination with *mole*. To reduce the number of outliers in the training set we define a 'region of uncertainty' as a transition region between the kernels of *mole* and *melanoma* and between the lesion and the skin.

4.1 Training

Figures 1(b) and 1(c) display the first step of the K-Means strategy described in Section 2.3. The original image displayed in Figure 1(a) shows a mole that is located in the middle of a hand. For PCA-transformed data, as in Figure 1(b), the algorithm performs poorly and the classes do not correspond to lesion, *mole* and *skin* regions (left and bottom). Even the lesion is in the same class together with an illumination problem. If the data is also transformed using ICA, as in Figure 1(c), the lesion is already identified and there exists a second class in the form of a ring around the lesion which is the desired 'region of uncertainty'. The other classes correspond to wrinkles on the hand.

Figure 1(d) shows the second K-Means step for the PCA transformed data. Although the second K-Means step makes it possible to separate the lesion from the illumination problem it can be seen that the class that should correspond to the kernel of the mole is too large. Instances from other classes are present in the kernel. The second K-Means step with the ICA preprocessed data is shown in Figure 1(e). Not only the kernel is reliably detected but there also exists a transition region consisting of two classes. One class contains the border of the lesion. The second class separates the kernel from the remaining part of the mole.

We believe that the FastICA algorithm is the most appropriate ICA implementation

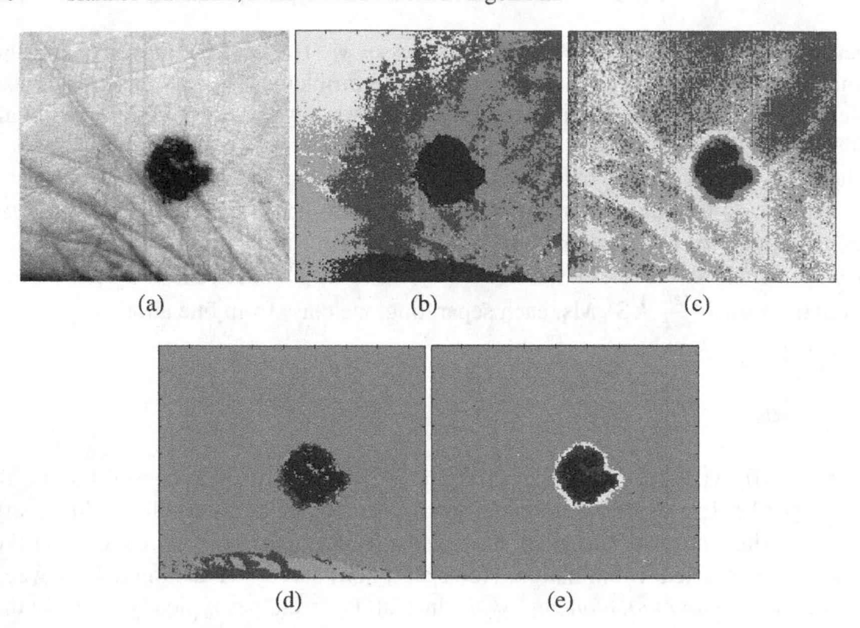

Fig. 1. The two iteration steps of the K-Means approach for both PCA ((b) and (d)) and ICA ((c) and (e)) are displayed together with the original image (a). The different gray levels indicate the cluster the pixel has been assigned to.

for this segmentation task. The segmentation quality for both methods is very similar, however the FastICA algorithm is faster and more stable.
To generate a training set of 12.000 pixel spectra per class we labelled 60 *mole* images and 17 *melanoma* images using our labelling approach. The pixels in the training set are chosen randomly from the segmented images.

4.2 Classification

In Table 1 we present the classification results obtained for the different classifiers described in Section 3. As a test set we use 57 *melanoma* and 253 *mole* images. We use the output of the LDA classifier as a benchmark.
LDA turns out to be the worst classifier for the recognition of *moles*. Nearly one half of the *mole* images are misclassified as *melanoma*. On the other hand LDA yields excellent results for the classification of *melanoma*, giving rise to the presumption that there is a large bias towards the *melanoma* class. With MDA we use three subclasses in each class. Although both MDA and LDA keep the covariance fixed, MDA models the data as mixture of Gaussians leading to a significantly higher recognition rate compared to LDA. Using FDA or PDA in combination with MDA does not improve the results. MclustDA performs best among these classifiers. Notice however, that BIC overestimates the number of subclasses in each class which is between 14 and 21. For all classes the model with varying shape, varying volume and varying

Table 1. Recognition rates obtained for the different classifiers

Pre-Proc.	Class	MDA	MclustDA	LDA
FlexICA	Mole	84.5%	86.5%	56.1%
	Melanoma	89.4%	89.4%	98.2%
FastICA	Mole	84.5%	87.7%	56.1%
	Melanoma	89.4%	89.4%	98.2%
Pre-Proc.	Class	OAA-SVM	OAO-SVM	unbalanced SVM
FlexICA	Mole	72.7%	69.9%	87.7%
	Melanoma	92.9%	94.7%	89.4%
FastICA	Mole	71.5%	69.9%	87.3%
	Melanoma	92.9%	94.7%	89.4%

orientation of the mixture components is chosen. This extra flexibility makes it possible to outperform MDA even though only half of the training points could be used due to memory limitations. Another significant advantage of MclustDA is its speed, taking around 20 seconds for a full image.

Since misclassification of *melanoma* into the *mole* class is less favourable than misclassification of *mole* into the *melanoma* class, we clearly have unbalanced data in the skin cancer problem. According to Veropoulos et al. (1999) we can choose $C_{melanoma} > C_{mole} = C_{skin}$. We obtain the best results using the polynomial kernel of degree three with $C_{melanoma} = 0.5$ and $C_{mole} = C_{skin} = 0.1$. This method is clearly superior when compared with the other SVM approaches. For the one-against-all (OAA-SVM) and the one-against-one (OAO-SVM) formulation we use Gaussian kernels with $C = 2$ and $\sigma = 20$. A drawback of all the SVM classifiers, however, is that training takes 20 hours (Centrino Duo 2.17GHz, 2GB RAM) and classification of a full image takes more than 2 minutes.

We discovered that different ICA implementations have no significant impact on the quality of the classification output. FlexICA performs slightly better for the unbalanced SVM and one-against-all-SVM. FastICA gives better results for MclustDA. For all other classifiers the performances are equal.

5 Conclusion

The combination of PCA and ICA makes it possible to detect both artifacts and the lesion in hyper-spectral skin cancer data. The algorithm projects the corresponding features on different independent components; dropping the independent components that correspond to the artifacts and applying a 2-stage K-Means clustering leads to a reliable segmentation of the images. It is interesting to note that for the *mole* images in our study there is always one single independent component that carries the information about the whole lesion. This suggests very simple segmentation in the case where the skin is healthy: keep the single independent component that contains the desired information and perform the K-Means steps. For *melanoma*

images the spectral information about the lesion is contained in at least two independent components, leading to reliable separation of the melanoma kernel from the mole kernel.
Unbalanced SVM and MclustDA yield equally good classification results, however, because of its computational performance MclustDA is the best classifier for the skin cancer data in terms of overall accuracy.
The presented segmentation and classification approach does not use any spatial information. In future research Markov random fields and contextual classifiers could be used to take into account the spatial context.
In a possible application, where the physician is assisted by system which pre-screens patients, we have to take care of high sensitivity which is typically accompanied with a loss in specificity. Preliminary experiments showed that a sensitivity of 95% is possible at the cost of 20% false-positives.

References

ABE, S. (2005): *Support Vector Machines for Pattern Classification.* Springer, London.

CHOI, S., CICHOCKI, A. and AMARI, S. (2000): Flexible Independent Component Analysis. *Journal of VLSI Signal Processing, 26(1/2), 25-38.*

FRALEY, C. and RAFTERY, A. (2002): Model-Based Clustering, Discriminant Analysis, and Density Estimation. *Journal of the American Statistical Association, 97, 611–631.*

HASTIE, T., TIBSHIRANI, R. and FRIEDMAN, J. (2001): *The Elements of Statistical Learning.* Springer, New York.

HYVÄRINEN, A., KARHUNEN, J. and OJA, E. (2001): *Independent Component Analysis.* Wiley, New York.

SHAH, C., ARORA, M. and VARSHNEY, P. (2004): Unsupervised classification of hyperspectral data: an ICA mixture model based approach. *International Journal of Remote Sensing, 25, 481–487.*

VEROPOULOS, K., CAMPBELL, C. and CRISTIANI, N. (1999): Controlling the Sensitivity of Support Vector Machines. *Proceedings of the Sixteenth International Joint Conference on Artificatial Intelligence, Workshop ML3, 55–60.*

FSMTree: An Efficient Algorithm for Mining Frequent Temporal Patterns

Steffen Kempe[1], Jochen Hipp[1] and Rudolf Kruse[2]

[1] DaimlerChrysler AG, Group Research, 89081 Ulm, Germany
{Steffen.Kempe, Jochen.Hipp}@daimlerchrysler.com
[2] Dept. of Knowledge Processing and Language Engineering,
University of Magdeburg, 39106 Magdeburg, Germany
Kruse@iws.cs.uni-magdeburg.de

Abstract. Research in the field of knowledge discovery from temporal data recently focused on a new type of data: interval sequences. In contrast to event sequences interval sequences contain labeled events with a temporal extension. Mining frequent temporal patterns from interval sequences proved to be a valuable tool for generating knowledge in the automotive business. In this paper we propose a new algorithm for mining frequent temporal patterns from interval sequences: *FSMTree*. *FSMTree* uses a prefix tree data structure to efficiently organize all finite state machines and therefore dramatically reduces execution times. We demonstrate the algorithm's performance on field data from the automotive business.

1 Introduction

Mining sequences from temporal data is a well known data mining task which gained much attention in the past (e.g. Agrawal and Srikant (1995), Mannila et al. (1997), or Pei et al. (2001)). In all these approaches, the temporal data is considered to consist of events. Each event has a label and a timestamp. In the following, however, we focus on temporal data where an event has a temporal extension. These temporally extended events are called temporal intervals. Each temporal interval can be described by a triplet (b,e,l) where b and e denote the beginning and the end of the interval and l its label.

At DaimlerChrysler we are interested in mining interval sequences in order to further extend the knowledge about our products. Thus, in our domain one interval sequence may describe the history of one vehicle. The configuration of a vehicle, e.g. whether it is an estate car or a limousine, can be described by temporal intervals. The build date is the beginning and the current day is the end of such a temporal interval. Other temporal intervals may describe stopovers in a garage or the installation of additional equipment. Hence, mining these interval sequences might help us in tasks like quality monitoring or improving customer satisfaction.

2 Foundations and related work

As mentioned above we represent a temporal interval as a triplet (b,e,l).

Definition 1. *(Temporal Interval) Given a set of labels $l \in L$, we say the triplet $(b,e,l) \in \mathbb{R} \times \mathbb{R} \times L$ is a temporal interval, if $b \leq e$. The set of all temporal intervals over L is denoted by I.*

Definition 2. *(Interval Sequence) Given a sequence of temporal intervals, we say $(b_1,e_1,l_1),(b_2,e_2,l_2),\ldots,(b_n,e_n,l_n) \in I$ is an interval sequence, if*

$$\forall (b_i,e_i,l_i),(b_j,e_j,l_j) \in I, i \neq j : b_i \leq b_j \wedge e_i \geq b_j \Rightarrow l_i \neq l_j \tag{1}$$

$$\begin{gathered} \forall (b_i,e_i,l_i),(b_j,e_j,l_j) \in I, i < j : \\ (b_i < b_j) \vee (b_i = b_j \wedge e_i < e_j) \vee (b_i = b_j \wedge e_i = e_j \wedge l_i < l_j) \end{gathered} \tag{2}$$

hold. A given set of interval sequences is denoted by $\mathbb{S}$.

Equation 1 above is referred to as the *maximality assumption* (Höppner (2002)). The maximality assumption guarantees that each temporal interval A is maximal, in the sense that there is no other temporal interval in the sequence sharing a time with A and carrying the same label. Equation 2 requires that an interval sequence has to be ordered by the beginning (primary), end (secondary) and label (tertiary, lexicographically) of its temporal intervals.

Without temporal extension there are only two possible relations. One event is before (or after as the inverse relation) the other or they coincide. Due to the temporal extension of temporal intervals the possible relations between two intervals become more complex. There are 7 possible relations (or 13 if one includes inverse relations). These interval relations have been described in Allen (1983) and are depicted in Figure 1. Each relation of Figure 1 is a temporal pattern on its own that consists of two temporal intervals. Patterns with more than two temporal intervals are straightforward. One just needs to know which interval relation exists between each pair of labels. Using the set of Allen's interval relations $\mathbb{I}$, a temporal pattern is defined by:

Definition 3. *(Temporal Pattern) A pair $P = (s,R)$, where $s : 1,\ldots,n \rightarrow L$ and $R \in \mathbb{I}^{n \times n}$, $n \in \mathbb{N}$, is called a "temporal pattern of size n" or "n-pattern".*

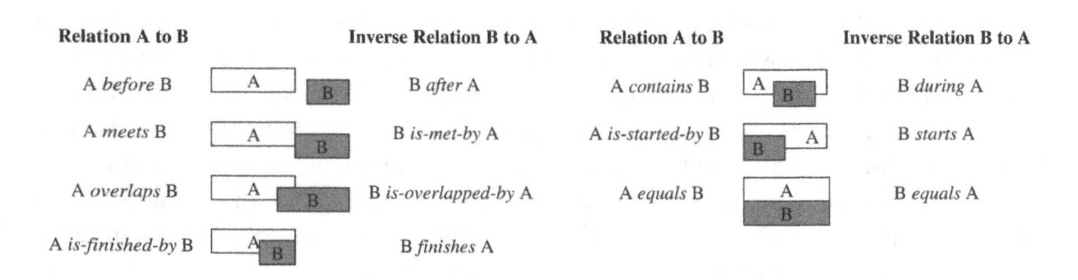

Fig. 1. Allen's Interval Relations

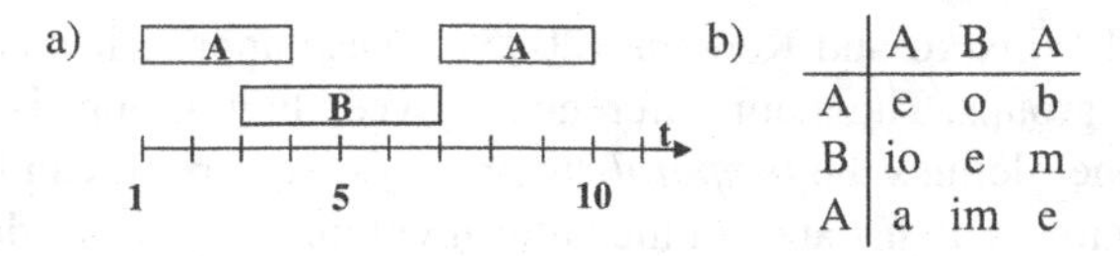

Fig. 2. a) Example of an interval sequence: (1,4,A), (3,7,B), (7,10,A) b) Example of a temporal pattern (e stands for *equals*, o for *overlaps*, b for *before*, m for *meets*, io for *is-overlapped-by*, etc.)

Figure 2.a shows an example of an interval sequence. The corresponding temporal pattern is given in Figure 2.b.

Note that a temporal pattern is not necessarily valid in the sense that it must be possible to construct an interval sequence for which the pattern holds true. On the other hand, if a temporal pattern holds true for an interval sequence we consider this sequence as an instance of the pattern.

Definition 4. *(Instance) An interval sequence* $S = (b_i, e_i, l_i)_{1 \le i \le n}$ *conforms to a n-pattern* $P = (s, R)$, *if* $\forall i, j : s(i) = l_i \wedge s(j) = l_j \wedge R[i, j] = \mathrm{ir}([b_i, e_i], [b_j, e_j])$ *with function* ir *returning the relation between two given intervals. We say that the interval sequence S is an instance of temporal pattern P. We say that an interval sequence S′* contains *an instance of P if* $S \subseteq S'$, *i.e. S is a subsequence of* S'.

Obviously a temporal pattern can only be valid if its labels have the same order as their corresponding temporal intervals have in an instance of the pattern. Next, we define the support of a temporal pattern.

Definition 5. *(Minimal Occurrence) For a given interval sequence S a time interval (time window)* $[b, e]$ *is called a minimal occurrence of the k-pattern* P $(k \ge 2)$, *if (1.) the time interval* $[b, e]$ *of S contains an instance of* P, *and (2.) there is no proper subinterval* $[b', e']$ *of* $[b, e]$ *which also contains an instance of* P. *For a given interval sequence S a time interval* $[b, e]$ *is called a minimal occurrence of the* 1*-pattern* P, *if (1.) the temporal interval* (b, e, l) *is contained in S, and (2.)* l *is the label in* P.

Definition 6. *(Support) The support of a temporal pattern P for a given set of interval sequences* $\mathbb{S}$ *is given by the number of minimal occurrences of P in* $\mathbb{S}$: $Sup_{\mathbb{S}}(P) = |\{[b, e] : [b, e] \text{ is a minimal occurrence of } P \text{ in } S \wedge S \in \mathbb{S}\}|$.

As an illustration consider the pattern *A before A* in the example of Figure 2.a. The time window $[1, 11]$ is not a minimal occurrence as the pattern is also visible e.g. in its subwindow $[2, 9]$. Also the time window $[5, 8]$ is not a minimal occurrence. It does not contain an instance of the pattern. The only minimal occurrence is $[4, 7]$ as the end of the first and the beginning of the second *A* are just inside the time window.

The mining task is to find all temporal patterns in a set of interval sequences which satisfy a defined minimum support threshold. Note that this task is closely related to frequent itemset mining, e.g. Agrawal et al. (1993).

Previous investigations on discovering frequent patterns from sequences of temporal intervals include the work of Höppner (2002), Kam and Fu (2000), Papapetrou et al. (2005), and Winarko and Roddick (2005). These approaches can be divided into two different groups. The main difference between both groups is the definition of support. Höppner defines the *temporal support of a pattern*. It can be interpreted as the probability to see an instance of the pattern within the time window if the time window is randomly placed on the interval sequence. All other approaches count the number of instances for each pattern. The pattern counter is incremented once for each sequence that contains the pattern. If an interval sequence contains multiple instances of a pattern then these additional instances will not further increment the counter.

For our application neither of the support definitions turned out to be satisfying. Höppner's temporal support of a pattern is hard to interpret in our domain, as it is generally not related to the number of instances of this pattern in the data. Also neglecting multiple instances of a pattern within one interval sequence is inapplicable when mining the repair history of vehicles. Therefore we extended the approach of minimal occurrences in Mannila (1997) to the demands of temporal intervals. In contrast to previous approaches, our support definition allows (1.) to count the number of pattern instances, (2.) to handle multiple instances of a pattern within one interval sequence, and (3.) to apply time constraints on a pattern instance.

3 Algorithms FSMSet and FSMTree

In Kempe and Hipp (2006) we presented *FSMSet*, an algorithm to find all frequent patterns within a set of interval sequences $\mathbb{S}$. The main idea is to generate all frequent temporal patterns by applying the Apriori scheme of candidate generation and support evaluation. Therefore *FSMSet* consists of two steps: generation of candidate sets and support evaluation of these candidates. These two steps are alternately repeated until no more candidates are generated. The Apriori scheme starts with the frequent 1-patterns and then successively derives all k-candidates from the set of all frequent (k-1)-patterns.

In this paper we will focus on the support evaluation of the candidate patterns, as it is the most time consuming part of the algorithm. *FSMSet* uses finite state machines which subsequently take the temporal intervals of an interval sequence as input to find all instances of a candidate pattern.

It is straightforward to derive a finite state machine from a temporal pattern. For each label in the temporal pattern a state is generated. The finite state machine starts in an initial state. The next state is reached if we input a temporal interval that contains the same label as the first label of the temporal pattern. From now on the next states can only be reached if the shown temporal interval carries the same label as the state and its interval relation to all previously accepted temporal intervals is the same as specified in the temporal pattern. If the finite state machine reaches its last state it also reaches its final accepting state. Consequently the temporal intervals that have been accepted by the state machine are an instance of the temporal pattern.

The minimal time window in which this pattern instance is visible can be derived from the temporal intervals which have been accepted by the state machine. We

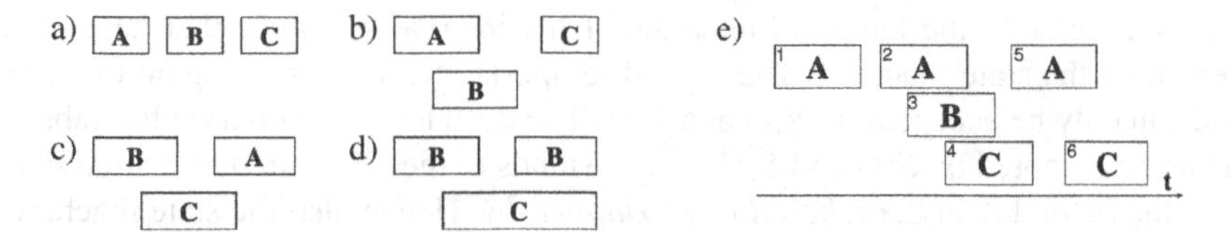

Fig. 3. a) – d) four candidate patterns of size 3 e) an interval sequence

Table 1. Set of state machines of *FSMSet* for the example of Figure 3. Each column shows the new state machines that have been added by *FSMSet*.

	1	2	3	4	5	6
$S_a()$	$S_a(1)$	$S_a(2)$	$S_c(3)$	$S_c(3,4)$	$S_a(5)$	$\mathbf{S_a(1,3,6)}$
$S_b()$	$S_b(1)$	$S_b(2)$	$S_d(3)$	$S_d(3,4)$	$S_b(5)$	$\mathbf{S_b(2,3,6)}$
$S_c()$			$S_a(1,3)$		$\mathbf{S_c(3,4,5)}$	
$S_d()$			$S_b(2,3)$			

know that the time window contains an instance but we do not know whether it is a minimal occurrence. Therefore *FSMSet* applies a two step approach. First it will find all instances of a pattern using state machines. Then it prunes all time windows which are not minimal occurrences.

To find all instances of a pattern in an interval sequence *FSMSet* is maintaining a set of finite state machines. At first, the set only contains the state machine that is derived from the candidate pattern. Subsequently, each temporal interval from the interval sequence is shown to every state machine in the set. If a state machine can accept the temporal interval, a copy of the state machine is added to the set. The temporal interval is shown only to one of these two state machines. Hence, there will always be a copy of the initial state machine in the set trying to find a new instance of the pattern. In this way *FSMSet* also can handle situations in which single state machines do not suffice. Consider the pattern *A meets B* and the interval sequence (1, 2, A), (3, 4, A), (4, 5, B). Without using look ahead a single finite state machine would accept the first temporal interval (1, 2, A). This state machine is stuck as it cannot reach its final state because there is no temporal interval which *is-met-by* (1, 2, A). Hence the pattern instance (3, 4, A), (4, 5, B) could not be found by a single state machine. Here this is not a problem because there is a copy of the first state machine which will find the pattern instance.

Figure 3 and Table 1 give an example of *FSMSet*'s support evaluation. There are four candidate patterns (Figure 3.a – 3.d) for which the support has to be evaluated on the given interval sequence in Figure 3.e.

At first, a state machine is derived for each candidate pattern. The first column in Table 1 corresponds to this initialization (state machines S_a – S_d). Afterwards each temporal interval of the sequence is used as input for the state machines. The first temporal interval has label *A* and can only be accepted by the state machines $S_a()$ and $S_b()$. Thus the new state machines $S_a(1)$ and $S_b(1)$ are added. The numbers in brackets refer to the temporal intervals of the interval sequence that have been accepted by the state machine. The second temporal interval carries again the label A and can only be accepted by $S_a()$ and $S_b()$. The third temporal interval has label B and can be accepted by $S_c()$ and $S_d()$. It also stands to the first A in the relation *after* and to the second A in the relation *is-overlapped-by*. Hence also the state machines $S_a(1)$ and $S_b(2)$ can accept this interval. Table 1 shows all new state machines for each temporal interval of the interval sequence. For this example the approach of *FSMSet* needs 19 state machines to find all three instances of the candidate patterns.

A closer examination of the state machines in Table 1 reveals that many state machines show a similar behavior. E.g. both state machines S_c and S_d accept exactly the same temporal intervals until the fourth iteration of *FSMSet*. Only the fifth temporal interval cannot be accepted by S_d. The reason is that both state machines share the common subpattern *B overlaps C* as their first part (i.e. a common prefix pattern). Only after this prefix pattern is processed their behavior can differ. Thus we can minimize the algorithmic costs of *FSMSet* by combining all state machines that share a common prefix. Combining all state machines of Figure 3 in a single data structure leads to the prefix tree in Figure 4. Each path of the tree is a state machine. But now different state machines can share states, if their candidate patterns share a common pattern prefix. By using the new data structure we derive a new algorithm for the support evaluation of candidate patterns — *FSMTree*.

Instead of maintaining a list of state machines *FSMTree* maintains a list of nodes from the prefix tree. In the first step the list only contains the root node of the tree. Afterwards all temporal intervals of the interval sequence are processed subsequently. Each time a node of the set can accept the current temporal interval its corresponding child node is added to the set. Table 2 shows the new nodes that are added in each step if we apply the prefix tree of Figure 4 to the example of Figure 3. Obviously the algorithmic overhead is reduced significantly. Instead of 19 state machines *FSMTree* only needs 11 nodes to find all pattern instances.

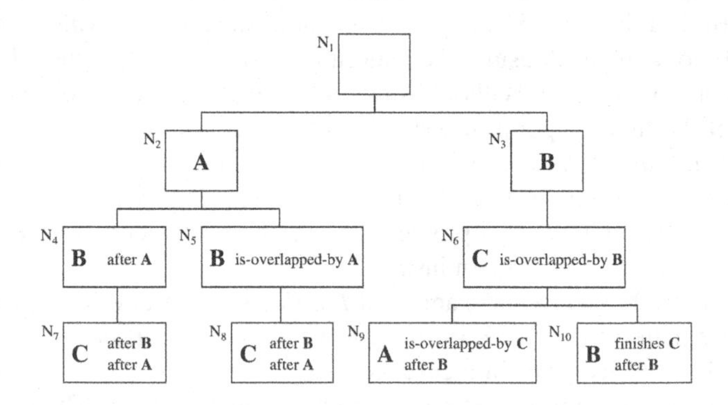

Fig. 4. FSMTree: prefix tree of state machines based on the candidates of Figure 3

Table 2. Set of nodes of *FSMTree* for the example of Figure 3. Each column gives the new nodes that have been added by *FSMTree*.

	1	2	3	4	5	6
$N_1()$	$N_2(1)$	$N_2(2)$	$N_3(3)$	$N_6(3,4)$	$N_2(5)$	$\mathbf{N_7(1,3,6)}$
			$N_4(1,3)$		$\mathbf{N_9(3,4,5)}$	$\mathbf{N_8(2,3,6)}$
			$N_5(2,3)$			

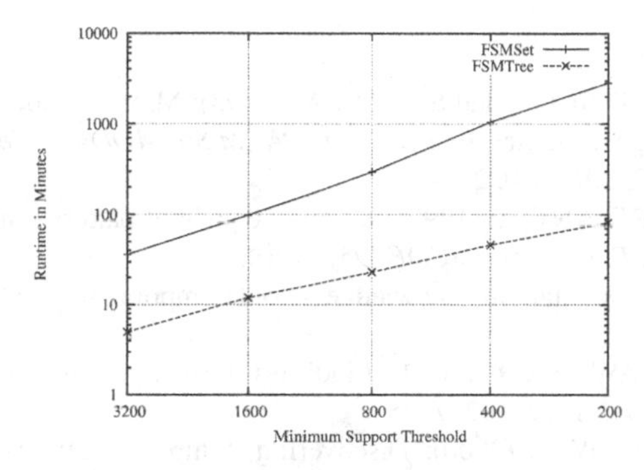

Fig. 5. Runtimes of *FSMSet* and *FSMTree* for different support thresholds.

4 Performance evaluation and conclusions

In order to evaluate the performance of FSMTree in a real application scenario we employed a dataset from our domain. This dataset contains information about the history of 101 250 vehicles. There is one sequence for each vehicle. Each sequence comprises between 14 and 48 temporal intervals. In total, there are 345 different labels and about 1.4 million temporal intervals in the dataset.

We performed 5 different experiments varying the minimum support threshold from 3 200 down to 200. For each experiment we measured the runtimes of *FSMSet* and *FSMTree*. The algorithms are implemented in Java and all experiments were carried out on a SUN Fire X2100 running at 2.2 GHz.

Figure 5 shows that *FSMTree* clearly outperforms *FSMSet*. In the first experiment *FSMTree* reduced the runtime from 36 to 5 minutes. The difference between *FSMSet* and *FSMTree* even grows as the minimum support threshold gets lower. For the last experiment *FSMSet* needed two days while it took *FSMTree* only 81 minutes. The reason for *FSMTree*'s huge runtime advantage at low support threshold is that as the support threshold decreases the number of frequent patterns increases. Consequently the number of candidate patterns increases too. The number of candidates is the same for *FSMSet* and *FSMTree* but *FSMTree* combines all patterns with common prefix patterns. If there are more candidate patterns the chance for common prefixes increases. Therefore *FSMTree*'s ability to reduce the runtime will increase (compared to *FSMSet*) as the support threshold gets lower.

In this paper we presented *FSMTree*: a new algorithm for mining frequent temporal patterns from interval sequences. *FSMTree* is based on the Apriori approach of candidate generation and support evaluation. For each candidate pattern a finite state machine is derived to parse the input data for instances of this pattern. *FSMTree* uses a prefixtree-like data structure to efficiently organize all finite state machines. In our application of mining the repair history of vehicles *FSMTree* was able to dramatically reduce execution times.

References

AGRAWAL, R., IMIELINSKI, T. and SWAMI, A. (1993): Mining association rules between sets of items in large databases. In: *Proc. of the ACM SIGMOD Int. Conf. on Management of Data (ACM SIGMOD '93).* 207–216.

AGRAWAL, R. and SRIKANT, R. (1995): Mining sequential patterns. In: *Proc. of the 11th Int. Conf. on Data Engineering (ICDE '95).* 3–14.

ALLEN, J. F. (1983): Maintaining knowledge about temporal intervals. *Commun. ACM*, 26(11):832–843.

HÖPPNER, F. and KLAWONN, F. (2002): Finding informative rules in interval sequences. *Intelligent Data Analysis*, 6(3):237–255.

KAM, P.-S. and FU, A. W.-C. (2000): Discovering Temporal Patterns for Interval-Based Events. In: *Data Warehousing and Knowledge Discovery, 2nd Int. Conf., DaWaK 2000.* Springer, 317–326.

KEMPE, S. and HIPP, J. (2006): Mining Sequences of Temporal Intervals. In: *10th Europ. Conf. on Principles and Practice of Knowledge Discovery in Databases* Springer, Berlin-Heidelberg, 569–576.

MANNILA, H., TOIVONNEN, H. and VERKAMO, I. (1997): Discovery of frequent episodes in event sequences. *Data Mining and Knowl. Discovery*, 1(3):259–289.

PAPAPETROU, P., KOLLIOS, G., SCLAROFF, S. and GUNOPULOS, D. (2005): Discovering frequent arrangements of temporal intervals. In: *5th IEEE Int. Conf. on Data Mining (ICDM '05).* 354–361.

PEI, J., HAN, J., MORTAZAVI, B., PINTO, H., CHEN, Q., DAYAL, U. and HSU, M. (2001): Prefixspan: Mining sequential patterns by prefix-projected growth. In: *Proc. of the 17th Int. Conf. on Data Engineering (ICDE '01).* 215–224.

WINARKO, E. and RODDICK, J. F. (2005): Discovering Richer Temporal Association Rules from Interval-Based Data. In: *Data Warehousing and Knowledge Discovery, 7th Int. Conf., DaWaK 2005.* Springer, Berlin-Heidelberg, 315–325.

A Matlab Toolbox for Music Information Retrieval

Olivier Lartillot, Petri Toiviainen and Tuomas Eerola

University of Jyväskylä, PL 35(M), FI-40014, Finland
lartillo@campus.jyu.fi

Abstract. We present *MIRToolbox*, an integrated set of functions written in Matlab, dedicated to the extraction from audio files of musical features related, among others, to timbre, tonality, rhythm or form. The objective is to offer a state of the art of computational approaches in the area of Music Information Retrieval (MIR). The design is based on a modular framework: the different algorithms are decomposed into stages, formalized using a minimal set of elementary mechanisms, and integrating different variants proposed by alternative approaches – including new strategies we have developed –, that users can select and parametrize. These functions can adapt to a large area of objects as input.

This paper offers an overview of the set of features that can be extracted with *MIRToolbox*, illustrated with the description of three particular musical features. The toolbox also includes functions for statistical analysis, segmentation and clustering.

One of our main motivations for the development of the toolbox is to facilitate investigation of the relation between musical features and music-induced emotion. Preliminary results show that the variance in emotion ratings can be explained by a small set of acoustic features.

1 Motivation and approach

MIRToolbox is a *Matlab* toolbox dedicated to the extraction of musically-related features in audio recordings. It has been designed in particular with the objective of enabling the computation of a large range of features from databases of audio files, that can be applied to statistical analyses.

We chose to base the design of the toolbox on *Matlab* computing environment, as it offers good visualisation capabilities and gives access to a large variety of other toolboxes. In particular, the *MIRToolbox* makes use of functions available in public-domain toolboxes such as the *Auditory Toolbox* (Slaney, 1998), *NetLab* (Nabney, 2002), or *SOMtoolbox* (Vesanto, 1999). It appeared that such computational framework, because of its general objectives, could be useful to the research community in Music Information Retrieval (MIR), but also for teaching. For that reason, a particular attention has been paid concerning the ease of use of the toolbox. The functions are called using a simple and adaptive syntax. More expert users can specify a large range of options and parameters.

The different musical features extracted from the audio files are highly interdependent: in particular, as can be seen in figure 1, some features are based on same initial computations. In order to improve the computational efficiency, it is important to avoid redundant computations of these common components. Each of these intermediary components, and the final musical features, are therefore considered as *building blocks* that can be freely articulated one with each other. Besides, in keeping with the objective of optimal ease of use of the toolbox, each building block has been conceived in a way that it can adapt to the type of input data.

2 Feature extraction

Figure 1 shows an overview of the main features considered in the toolbox. All the different processes start from the audio signal (on the left) and form a chain of operations developed horizontally rightwise. The vertical disposition of the processes indicates an increasing order of complexity of the operations, from simplest computation (top) to more detailed auditory modelling (bottom). Each musical feature is related to the different broad musical dimensions traditionally defined in music theory. In bold are highlighted features related to pitch, to tonality (chromagram, key strength and key Self-Organising Map, or SOM) and to dynamics (Root Mean Square, or RMS, energy). In bold italics are indicated features related to rhythm: namely tempo, pulse clarity and fluctuation. In simple italics are highlighted a large set of features that can be associated to timbre. Among them, all the operators in grey italics can be in fact applied to many others different representations: for instance, statistical moments such as centroid, kurtosis, etc., can be applied to either spectra, envelopes, but also to any histogram based on any given feature.

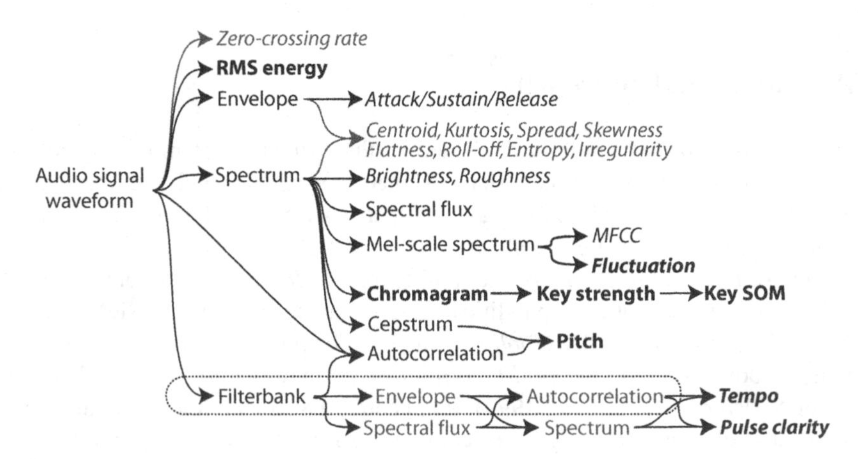

Fig. 1. Overview of the musical features that can be extracted with *MIRToolbox*.

2.1 Example: Timbre analysis

One common way of describing timbre is based on MFCCs (Rabiner and Juang, 1993; Slaney, 1998). MFCCs, providing a measure of spectral shape, has been found to be a good predictor of timbral similarity. Figure 2 shows the diagram of operations. First, the audio sequence is described in the spectral domain, using an FFT. The spectrum is converted from the frequency domain to the Mel-scale domain: the frequencies are rearranged into 40 frequency bands called Mel-bands. The envelope of the Mel-scale spectrum is described through a Discrete Cosine Transform. The values obtained through this transform are the MFCCs. Usually only a restricted number of them (for instance the 13 first ones) are selected. The computation can be carried in a window sliding through the audio signal, resulting in a series of MFCC vectors, one for each successive frame, that can be represented column-wise in a matrix. Figure 2 shows an example of such matrix. The MFCCs are generally applied to distance computation between frames, and therefore to segmentation tasks.

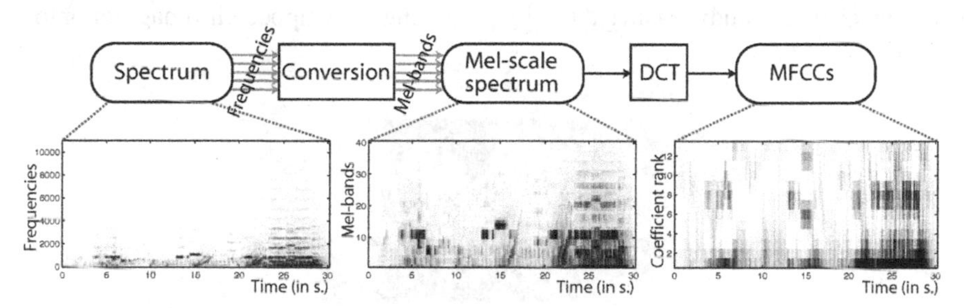

Fig. 2. Successive steps for the computation of MFCCs, illustrated with the analysis of an audio excerpt decomposed into frames.

2.2 Example: Tonality analysis

The spectrum is converted from the frequency domain to the pitch domain by applying a log-frequency transformation. The distribution of the energy along the pitches is called the *chromagram*. The chromagram is then wrapped, by fusing the pitches belonging to same pitch classes. The wrapped chromagram shows therefore a distribution of the energy with respect to the twelve possible pitch classes. Krumhansl and Schmuckler (Krumhansl, 1990) proposed a method for estimating the tonality of a musical piece (or an extract thereof) by computing the cross-correlation of its pitch class distribution with the distribution associated to each possible tonality. These distributions have been established through listening experiments (Krumhansl and Kessler, 1982). The most prevalent tonality is considered to be the tonality candidate with highest correlation (or *key strength*). This method was originally designed for the analysis of symbolic representations of music but has been extended to audio analysis through an adaptation of the pitch class distribution to the chromagram representation (Gomez, 2006). Figure 3 displays the successive steps of this approach.

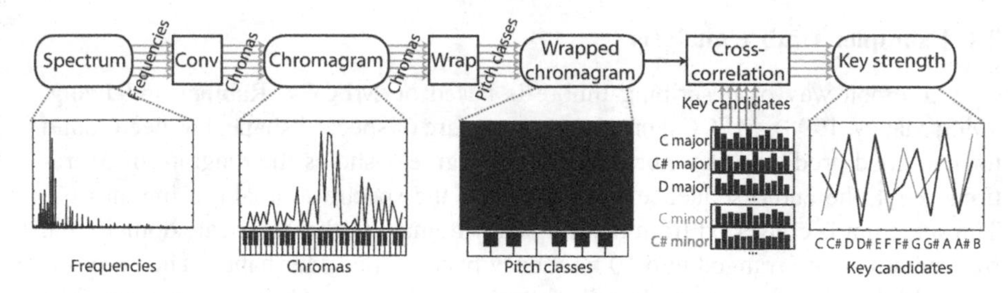

Fig. 3. Successive steps for the calculation of chromagram and estimation of key strengths, illustrated with the analysis of an audio excerpt, this time not decomposed into frames.

A richer representation of the tonality estimation can be drawn with the help of a self-organizing map (SOM), trained by the 24 tonal profiles (Toiviainen and Krumhansl, 2003). The configurations of the 24 classes after the training on the SOM corresponds to studies in music theory. The estimation of the tonality of the musical piece under study is carried out by projecting its wrapped chromagram onto the SOM.

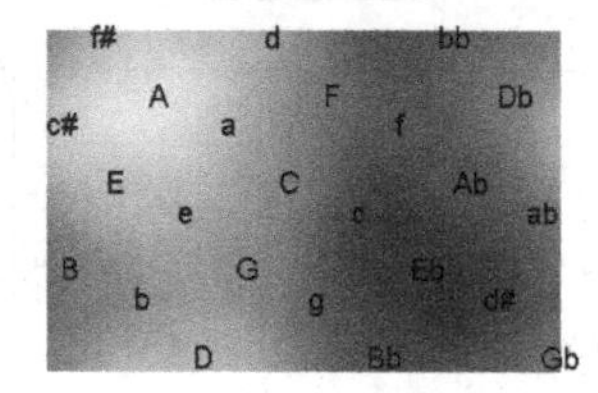

Fig. 4. Activity pattern of a self-organizing map representing the tonal configuration of the first two seconds of Mozart *Sonata in A major*, K 331. High activity is represented by bright nuances.

2.3 Example: Rhythm analysis

One common way of estimating the rhythmic pulsation, described in figure 5, is based on auditory modelling (Tzanetakis and Cook, 1999). The audio signal is first decomposed into auditory channels using a bank of filters. The envelope of each channel is extracted. As pulsation is generally related to increase of energy only, the envelopes are differentiated, half-wave rectified, before being finally summed together again. This gives a precise description of the variation of energy produced by each note event from the different auditory channels.

After this onset detection, the periodicity is estimated through autocorrelation. However, if the tempo variates throughout the piece, an autocorrelation of the whole sequence will not show clear periodicities. In such cases it is better to compute the autocorrelation on a moving window. This yields a periodogram that highlights the different periodicities, as shown in figure 5. In order to focus on the periodicities that are more perceptible, the periodogram is filtered using a resonance curve (Toiviainen and Snyder, 2003), after which the best tempos are estimated through peak picking, and the results are converted into beat per minutes. Due to the difficulty of choosing among the possible multiples of the tempo, several candidates (three for instance) may be selected for each frame, and a histogram of all the candidates for all the frames, called periodicity histogram, can be drawn.

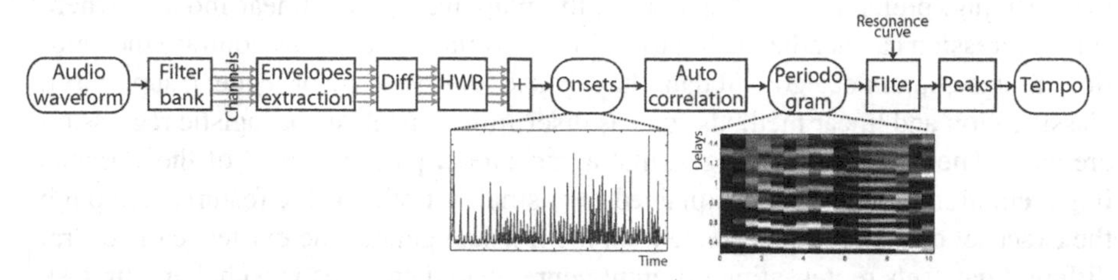

Fig. 5. Successive steps for the estimation of tempo illustrated with the analysis of an audio excerpt. In the periodogram, high autocorrelation values are represented by bright nuances.

3 Data analysis

The toolbox includes diverse tools for data analysis, such as a peak extractor, and functions that compute histograms, entropy, zero-crossing rates, irregularity or various statistical descriptors (centroid, spread, skewness, kurtosis, flatness) on data of various types, such as spectrum, envelope or histogram. The peak picker can accept any data returned by any other function of the *MIRtoolbox* and can adapt to the different kinds of data of any number of dimensions. In the graphical representation of the results, the peaks are automatically located on the corresponding curves (for 1D data) or bit-map images (for 2D data). We have designed a new strategy of peak selection, based on a notion of *contrast*, discarding peaks that are not sufficiently contrastive (based on a certain threshold) with the neighbouring peaks. This adaptive filtering strategy hence adapts to the local particularities of the curves. Its articulation with other more conventional thresholding strategies leads to an efficient peak picking module that can be applied throughout the *MIRtoolbox*.

More elaborate tools have also been implemented that can carry out higher-level analyses and transformations. In particular, audio files can be automatically segmented into a series of homogeneous sections, through the estimation of temporal discontinuities in timbral features (Foote and Cooper, 2003). The resulting segments can then be clustered into classes, suggesting a formal analysis of the musical piece. Supervised classification of musical samples can also be performed, using techniques such as K-Nearest Neighbours or Gaussian Mixture Models. The results of feature extraction processes can be stored as text files of various format, such as the ARFF format that can be exported in the Weka machine learning environment (Witten and Frank, 2005).

4 Application to the study of music and emotion

The toolbox is conceived in the context of a project investigating the interrelationships between perceived emotions and acoustic features. In a first study, musical features of musical materials collected form a large number of recent empirical studies in music and emotions (15 so far) are systematically reanalysed. For emotion rating based on interval scales – using emotion dimensions such as emotional valence (liking, preference) and activity –, the mapping applies linear models, where ridge regression can handle the highly collinear variables. If on the contrary the emotion ratings contain categorical data (happy, sad, angry, scary or tender), supervised classification and linear methods such as discriminant analysis or logistic regression are used. The early results suggest that a substantial part (50-70%) of the variance in the emotion ratings can be explained by a small set of acoustic features, although the exact set of features is dependent on the musical genre. The existence of several different data sets representing different genres and data types is a challenging task for the selection of the appropriate statistical measures.

A second study focuses on musical timbre. Listeners' rating of 110 short instrument sounds (of constant pitch height, loudness and duration, but varying timbre) on four bipolar scales (valence, energy arousal, tension arousal, and preference) were correlated with the acoustic features. We found that positively valenced sounds tend to be dark in sound colour (see figure 6). Energy arousal, on the other hand, is more related to high spectral flux and high brightness ($R^2 = .62$). The tension arousal ratings ($R^2 = .70$) are a mixture of high brightness, and roughness and inharmonicity. These observations extend the earlier findings relating to timbre and emotions (Scherer and Oshinsky, 1977; Juslin, 1997).

The emotional connotations induced by instrument sounds alone are consistent across listeners and can be meaningfully connected to acoustic descriptors of timbre. Certain aspects of these features are already known from the expressive speech literature (e.g., brightness or high-frequency energy, Juslin and Laukka, 2003), but musical sounds have distinctive features such as inharmonicity and roughness which are reflected in emotion ratings. According to our view, subtle nuances in music-induced emotions can only be studied from the audio representation using tools that extract acoustic features in a fashion that is relevant for the perceptual processing of sounds. Further work is necessary for establishing the effectiveness and reliability of higher-level features – such as rhythmic patterns, tonality and harmony – in terms of their correspondence with the listener observations.

Furthermore, both the *MIDItoolbox* and the multivariate techniques that were used to extract meaningful results out of the audio descriptors in the examples are well suited to other, similar tasks in content analysis of musical audio. For example, genre, instrument and artist recognition are such application areas where the extracted features and multivariate statistics are used (see Tzanetakis and Cook, 2002). Applications such as these are especially important commercially, as ever-increasing volumes of recordings need to be automatically indexed.

Following our first Matlab toolbox, called *MIDItoolbox* (Eerola and Toiviainen, 2004), dedicated to the analysis of symbolic representations of music, the *MIRtool-*

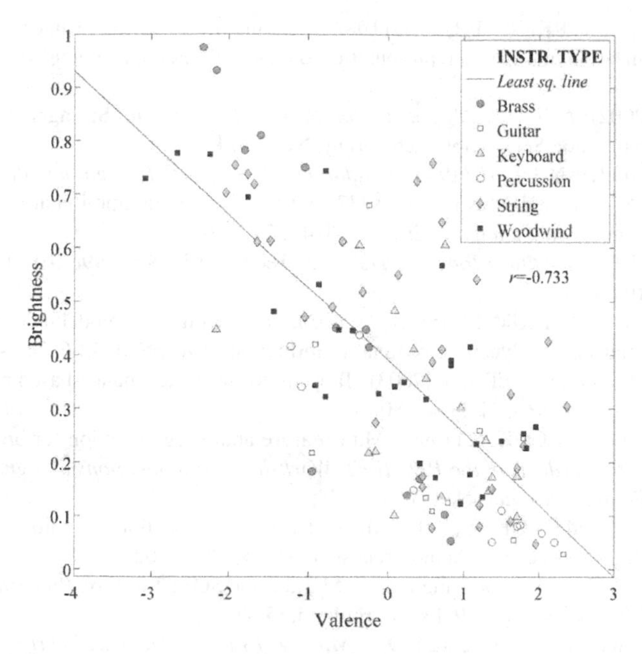

Fig. 6. Correlation between listener rating of valence and acoustic brightness of short instrument sounds ($r = -.733$).

box will be available for free, from September 2007 on, at the following address: `http://users.jyu.fi/~lartillo/mirtoolbox`

This work has been supported by the European Commission (NEST project "Tuning the Brain for Music", code 028570).

References

EEROLA, T. and TOIVIAINEN, P. (2004): MIR in Matlab: The MIDI Toolbox. *Proceedings of 5th International Conference on Music Information Retrieval*, 22–27, Barcelona.

FOOTE, J. and COOPER, M. (2003): Media segmentation using self-similarity decomposition. In *Proceedings of SPIE Storage and Retrieval for Multimedia Databases*, 5021, 167-75.

GOMEZ, E. (2006): Tonal description of polyphonic audio for music content processing. *INFORMS Journal on Computing*, 18-3, 294–304.

JUSLIN, P. N. (1997): Emotional communication in music performance: A functionalist perspective and some data. *Music Perception*, 14, 383–418.

JUSLIN, P. N. and LAUKKA, P. (2003): Communication of emotions in vocal expression and music performance: Different channels, same code? Psychological Bulletin (129), 770-814.

KRUMHANSL, C. (1990): *Cognitive Foundations of Musical Pitch.* Oxford University Press, New York.

KRUMHANSL, C. and KESSLER, E. J. (1982): Tracing the dynamic changes in perceived tonal organization in a spatial representation of musical keys. *Psychological Review*, 89, 334–368.

NABNEY, I. (2002): *NETLAB: Algorithms for Pattern Recognition.* Springer Advances In Pattern Recognition Series, Springer-Verlag, New-York.

RABINER, L. and JUANG, B. H. (1993): *Fundamentals of Speech Recognition.* Prentice-Hall.

SCHERER, K. R. and OSHINSKY J. S. (1977): Cue utilization in emotion attribution from auditory stimuli. *Motivation and Emotion*, 1-4, 331–346.

SLANEY, M. (1998): *Auditory Toolbox Version 2.* Technical Report 1998-010, Interval Research Corporation.

TOIVIAINEN, P. and KRUMHANSL, C. (2003): Measuring and modeling real-time responses to music: The dynamics of tonality induction, *Perception*, 32-6, 741–766.

TOIVIAINEN, P. and SNYDER J. S. (2003): Tapping to Bach: Resonance-based modeling of pulse. *Music Perception*, 21(1), 43–80.

TZANETAKIS, G and COOK, P. (1999): Multifeature audio segmentation for browsing and annotation. *Proceedings of the 1999 IEEE Workshop on Applications of Signal Processing to Audio and Acoustics.* New-York.

TZANETAKIS, G. and COOK, P. (2002): Musical genre classification of audio signals. IEEE Transactions on Speech and Audio Processing, 10(5), 293Ð302.

VESANTO, J. (1999): Self-organizing map in Matlab: the SOM Toolbox. *Proceedings of the Matlab DSP Conference 1999.* Espoo, Finland,35–40.

WITTEN, I. H. and FRANK, E. (2005): *Data Mining: Practical Machine Learning Tools and Techniques*, 2nd Edition. Morgan Kaufmann, San Francisco, 2005.

A Probabilistic Relational Model for Characterizing Situations in Dynamic Multi-Agent Systems

Daniel Meyer-Delius[1], Christian Plagemann[1], Georg von Wichert[2], Wendelin Feiten[2], Gisbert Lawitzky[2] and Wolfram Burgard[1]

[1] Department for Computer Science, University of Freiburg, Germany
{meyerdel,plagem,burgard}@informatik.uni-freiburg.de
[2] Information and Communications, Siemens Corporate Technology, Germany
{georg.wichert,wendelin.feiten,gisbert.lawitzky}@siemens.com

Abstract. Artificial systems with a high degree of autonomy require reliable semantic information about the context they operate in. State interpretation, however, is a difficult task. Interpretations may depend on a history of states and there may be more than one valid interpretation. We propose a model for spatio-temporal situations using hidden Markov models based on relational state descriptions, which are extracted from the estimated state of an underlying dynamic system. Our model covers concurrent situations, scenarios with multiple agents, and situations of varying durations. To evaluate the practical usefulness of our model, we apply it to the concrete task of online traffic analysis.

1 Introduction

It is a fundamental ability for an autonomous agent to continuously monitor and understand its internal states as well as the state of the environment. This ability allows the agent to make informed decisions in the future, to avoid risks, and to resolve ambiguities. Consider, for example, a driver assistance application that notifies the driver when a dangerous situation is developing, or a surveillance system at an airport that recognizes suspicious behaviors. Such applications do not only have to be aware of the current state, but also have to be able to interpret it in order to act rationally.

State interpretation, however, is not an easy task as one has to also consider the spatio-temporal context, in which the current state is embedded. Intuitively, the agent has to understand the *situation* that is developing. The goals of this work are to formally define the concept of *situation* and to develop a sound probabilistic framework for modeling and recognizing situations.

Related work includes Anderson *et al.* (2004) who propose relational Markov models with fully observable states. Fern and Givan (2004) describe an inference technique for sequences of hidden relational states. The hidden states must be inferred from observations. Their approach is based on logical constraints and uncertainties are not handled probabilistically. Kersting *et al.* (2006) propose logical hidden Markov models where the probabilistic framework of hidden Markov models is integrated with a logical representation of the states. The states of our proposed situation models are represented by conjunctions of logical atoms instead of single atoms and we present a filtering technique based on a relational, non-parametric probabilistic representation of the observations.

2 Framework for modeling and recognizing situations

Dynamic and uncertain systems can in general be described using dynamic Baysian networks (DBNs) (Dean and Kanazawa (1989)). DBNs consist of a set of random variables that describe the system at each point in time t. The state of the system at time t is denoted by x_t and z_t represents the observations. Furthermore, DBNs contain the conditional probability distributions that describe how the random variables are related.

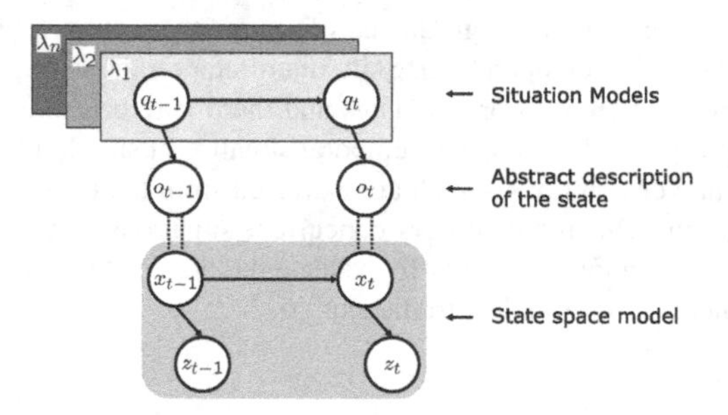

Fig. 1. Overview of the framework. At each time step t, the state x_t of the system is estimated from the observations z_t. A relational description o_t of the estimated state is generated and evaluated against the different situation models $\lambda_1, \ldots, \lambda_n$.

Intuitively, a situation is an interpretation associated to some states of the system. In principle, situations could be represented in such a DBN model by introducing additional latent situation variables and by defining their influence on the rest of the system. Since this would lead to an explosion of network complexity already for moderately sized models, we introduce a relational abstraction layer between the system DBN used for estimating the state of the system, and the situation models used to recognize the situations associated to the state of the system. In this framework, we sequentially estimate the system state x_t from the observations z_t in the DBN model using the Bayes filtering scheme. In a second step within each time step, we transform the state estimate x_t to a relational state description o_t, which is then used to recognize instances of the different situation models. Figure 1 visualizes the structure of our proposed framework for situation recognition.

3 Modeling situations

Based on the DBN model of the system outlined in the previous section, a situation can be described as a sequence of states with a meaningful interpretation. Since in general we are dealing with continuous state variables, it would be impractical or even impossible to reason about states, and state sequences directly in that space. Instead, we use an abstract representation of the states, and define situations as sequences of these abstract states.

3.1 Relational state representation

For the abstract representation of the state of the system, relational logic will be used. In relational logic, an *atom* $\mathtt{r}(\mathtt{t}_1,\ldots,\mathtt{t}_n)$ is an n-tuple of terms $\mathtt{t}_i$ with a relation symbol `r`. A *term* can be either a variable `R` or a constant `c`. Relations can be defined over the state variables or over features that can be directly extracted from them. Table 1 illustrates possible relations defined over the *distance* and *bearing* state variables in a traffic scenario.

Table 1. Example distance and bearing relations for a traffic scenario.

Relation	Distances
equal(R,R′)	[0 m, 1 m)
close(R,R′)	[1 m, 5 m)
medium(R,R′)	[5 m, 15 m)
far(R,R′)	[15 m, ∞)

Relation	Bearing angles
in_front_of(R,R′)	[315°, 45°)
right(R,R′)	[45°, 135°)
behind(R,R′)	[135°, 225°)
left(R,R′)	[225°, 315°)

An *abstract state* is a conjunction of logical atoms (see also Cocora *et al.* (2006)). Consider for example the abstract state $q \equiv \mathtt{far}(\mathtt{R},\mathtt{R}'),\mathtt{behind}(\mathtt{R},\mathtt{R}')$, which represents all states in which a car is *far* and *behind* another car.

3.2 Situation models

Hidden Markov models (HMMs) (Rabiner (1989)) are used to describe the admissible sequences of states that correspond to a given situation. HMMs are temporal probabilistic models for analyzing and modeling sequential data. In our framework we use HMMs whose states correspond to conjunctions of relational atoms, that is, abstract states as described in the previous section. The state transition probabilities of the HMM specify the allowed transitions between these abstract states. In this way, HMMs specify a probability distribution over sequences of abstract states.

To illustrate how HMMs and abstract states can be used to describe situations, consider a *passing* maneuver like the one depicted in Figure 2. Here, a reference car is passed by a faster car on the left hand side. The maneuver could be coarsely described in three steps: (1) the passing car is behind the reference car, (2) it is left of it, (3) and it is in front. Using, for example, the bearing relations presented in Table 1,

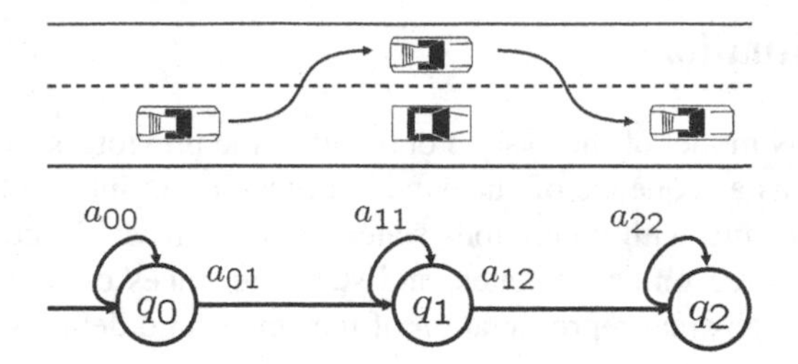

Fig. 2. *passing* maneuver and corresponding HMM.

an HMM that describes this sequences could have three states, one for each step of the maneuver: $q_0 = \texttt{behind}(\texttt{R},\texttt{R}')$, $q_1 = \texttt{left}(\texttt{R},\texttt{R}')$, and $q_2 = \texttt{in\_front\_of}(\texttt{R},\texttt{R}')$. The transition model of this HMM is depicted in Figure 2. It defines the allowed transitions between the states. Observe how the HMM specifies that when in the second state (q_1), that is, when the passing car is left of the reference car, it can only remain left (q_1) or move in front of the reference car (q_2). It is not allowed to move behind it again (q_0). Such a sequence would not be a valid *passing* situation according to our description.

A situation HMM consists of a tuple $\lambda = (Q, A, \pi)$, where $Q = \{q_0, \dots, q_N\}$ represents a finite set of N states, which are in turn abstract states as described in the previous section, $A = \{a_{ij}\}$ is the state transition matrix where each entry a_{ij} represents the probability of a transition from state q_i to state q_j, and $\pi = \{\pi_i\}$ is the initial state distribution, where π_i represents the probability of state q_i being the initial state. Additionally, just as for the DBNs, there is also an observation model. In our case, this observation model is the same for every situation HMM, and will be described in detail in Section 4.1.

4 Recognizing situations

The idea behind our approach to situation recognition is to instantiate at each time step new candidate situation HMMs and to track these over time. A situation HMM can be instantiated if it assigns a positive probability to the current state of the system. Thus, at each time step t, the algorithm keeps track of a set of active situation hypotheses, based on a sequence of relational descriptions.

The general algorithm for situation recognition and tracking is as follows. At every time step t,

1. Estimate the current state of the system x_t (see Section 2).
2. Generate relational representation o_t from x_t: From the estimated state of the system x_t, a conjunction o_t of grounded relational atoms with an associated probability is generated (see next section).
3. Update all instantiated situation HMMs according to o_t: Bayes filtering is used to update the internal state of the instantiated situation HMMs.

4. Instantiate all non-redundant situation HMMs consistent with o_t: Based on o_t, all situation HMMs are grounded, that is, the variables in the abstract states of the HMM are replaced by the constant terms present in o_t. If a grounded HMM assigns a non-zero probability to the current relational description o_t, the situation HMM can be instantiated. However, we must first check that no other situation of the same type and with the same grounding has an overlapping internal state. If this is the case, we keep the oldest instance since it provides a more accurate explanation for the observed sequence.

4.1 Representing uncertainty at the relational level

At each time step t, our algorithm estimates the state x_t of the system. The estimated state is usually represented through a probability distribution which assigns a probability to each possible hypothesis about the true state. In order to be able to use the situation HMMs to recognize situation instances, we need to represent the estimated state of the system as a grounded abstract state using relational logic.

To convert the uncertainties related to the estimated state x_t into appropriate uncertainties at the relational level, we assign to each relation the probability mass associated to the interval of the state space that it represents. The resulting distribution is thus a histogram that assigns to each relation a single cumulative probability. Such a histogram can be thought of as a piecewise constant approximation of the continuous density. The relational description o_t of the estimated state of the system x_t at time t is then a grounded abstract state where each relation has an associated probability.

The probability $P(o_t|q_i)$ of observing o_t while being in a grounded abstract state q_i is computed as the product of the matching terms in o_t and q_i. In this way, the observation probabilities needed to estimate the internal state of the situation HMMs and the likelihood of a given sequence of observations $O_{1:t} = (o_1, \ldots, o_t)$ can be computed.

4.2 Situation model selection using Bayes factors

The algorithm for recognizing situations keeps track of a set of active situation hypothesis at each time step t. We propose to decide between models at a given time t using Bayes factors for comparing two competing situation HMMs that explain the given observation sequence. Bayes factors (Kass and Raftery (1995)) provide a way of evaluating evidence in favor of a probabilistic model as opposed to another one. The Bayes factor $B_{1,2}$ for two competing models λ_1 and λ_2 is computed as

$$B_{12} = \frac{P(\lambda_1 | O_{t_1:t_1+n_1})}{P(\lambda_2 | O_{t_2:t_2+n_2})} = \frac{P(O_{t_1:t_1+n_1} | \lambda_1) P(\lambda_1)}{P(O_{t_2:t_2+n_2} | \lambda_2) P(\lambda_2)}, \tag{1}$$

that is, the ratio between the likelihood of the models being compared given the data. The Bayes factor can be interpreted as evidence provided by the data in favor of a model as opposed to another one (Jeffreys (1961)).

In order to use the Bayes factor as evaluation criterion, the observation sequence $O_{t:t+n}$ which the models in Equation 1 are conditioned on, must be the same for the two models being compared. This is, however, not always the case, since situation can be instantiated at any point in time. To solve this problem we propose a solution used for sequence alignment in bio-informatics (Durbin et al. (1998)) and extend the situation model using a separate *world* model to account for the missing part of the observation sequence. This *world* model in our case is defined analogously to the bigram models that are learn from the corpora in the field of natural language processing (Manning and Schütze (1999)). By using the extended situation model, we can use Bayes factors to evaluate two situation models even if they where instantiated at different points in time.

5 Evaluation

Our framework was implemented and tested in a traffic scenario using a simulated 3D environment. *TORCS - The Open Racing Car Simulator* (Espié and Guionneau) was used as simulation environment. The scenario consisted of several autonomous vehicles with simple driving behaviors and one reference vehicle controlled by a human operator. Random noise was added to the pose of the vehicles to simulate uncertainty at the state estimation level. The goal of the experiments is to demonstrate that our framework can be used to model and successfully recognize different situations in dynamic multi-agent environments. Concretely, three different situations relative to a reference car where considered:

1. The *passing* situation corresponds to the reference car being passed by another car. The passing car approaches the reference car from behind, it passes it on the left, and finally ends up in front of it.
2. The *aborted passing* situation is similar to the *passing* situation, but the reference car is never fully overtaken. The passing car approaches the reference car from behind, it slows down before being abeam, and ends up behind it again.
3. The *follow* situation corresponds to the reference car being followed from behind by another car at a short distance and at the same velocity.

The structure and parameters of the corresponding situation HMMs where defined manually. The relations considered for these experiments where defined over the relative distance, position, and velocity of the cars.

Figure 3 (left) plots the likelihood of an observation sequence corresponding to a passing maneuver. During this maneuver, the passing car approaches the reference car from behind. Once at close distance, it maintains the distance for a couple of seconds. It then accelerates and passes the reference car on the left to finally end up in front of it. It can be observed in the figure how the algorithm correctly instantiated the different situation HMMs and tracked the different instances during the execution of the maneuver. For example, the *passing* and *aborted passing* situations where instantiated simultaneously from the start, since both situation HMMs initially

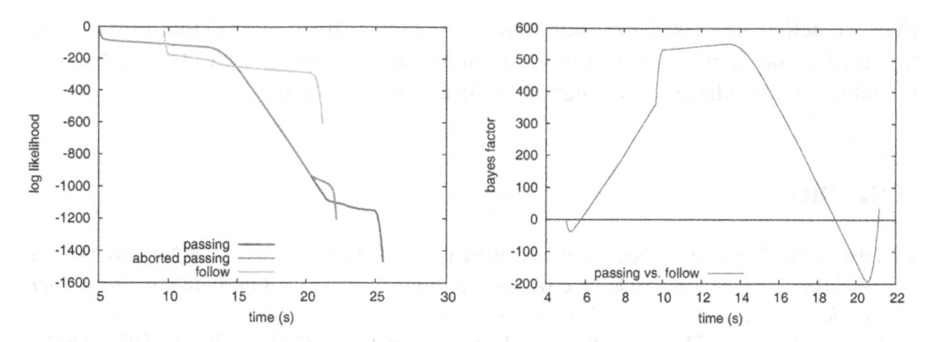

Fig. 3. (Left) Likelihood of the observation sequence for a passing maneuver according to the different situation models, and (right) Bayes factor in favor of the *passing* situation model against the other situation models.

describe the same sequence of observations. The *follow* situation HMM was instantiated, as expected, at the point where both cars where close enough and their relative velocity was almost zero. Observe too that at this point, the likelihood according to the *passing* and *aborted passing* situation HMMs starts to decrease rapidly, since these two models do not expect both cars to drive at the same speed. As the passing vehicle starts changing to the left lane, the HMM for the *follow* situation stops providing an explanation for the observation sequence and, accordingly, the likelihood starts to decrease rapidly until it becomes almost zero. At this point the instance of the situation is not tracked anymore and is removed from the active situation set. This happens since the *follow* situation HMM does not expect the vehicle to speed up and change lanes.

The Bayes factor in favor of the *passing* situation model compared against the *follow* situation model is depicted in Figure 3 (right). A positive Bayes factor value indicates that there is evidence in favor of the *passing* situation model. Observe that up to the point where the *follow* situation is actually instantiated the Bayes factor keeps increasing rapidly. At the time where both cars are equally fast, the evidence in favor of the *passing* situation model starts decreasing until it becomes negative. At this point there is evidence against the *passing* situation model, that is, there is evidence in favor of the *follow* situation. Finally, as the passing vehicle starts changing to the left lane the evidence in favor of the *passing* situation model starts increasing again. Figure 3 (right) shows how Bayes factors can be used to make decisions between competing situation models.

6 Conclusions and further work

We presented a general framework for modeling and recognizing situations. Our approach uses a relational description of the state space and hidden Markov models to represent situations. An algorithm was presented to recognize and track situations in an online fashion. The Bayes factor was proposed as evaluation criterion between two competing models. Using our framework, many meaningful situations can be modeled. Experiments demonstrate that our framework is capable of tracking multiple situation hypotheses in a dynamic multi-agent environment.

References

ANDERSON, C. R., DOMINGOS, P. and WELD, D. A. (2002): Relational Markov models and their application to adaptive web navigation. *Proc. of the International Conference on Knowledge Discovery and Data Mining (KDD).*

COCORA, A., KERSTING, K., PLAGEMANN, C. and BURGARD, W. and DE RAEDT, L. (2006): Learning Relational Navigation Policies. *Proc. of the IEEE/RSJ International Conference on Intelligent Robots and Systems (IROS).*

COLLETT, T., MACDONALD, B. and GERKEY, B. (2005): Player 2.0: Toward a Practical Robot Programming Framework. *In: Proceedings of the Australasian Conference on Robotics and Automation (ACRA 2005).*

DEAN, T. and KANAZAWA, K. (1989): A Model for Reasoning about Persistence and Causation. *Computational Intelligence, 5(3):142-150.*

DURBIN, R., EDDY, S., KROGH, A. and MITCHISON, G. (1998):*Biological Sequence Analysis.* Cambridge University Press.

FERN, A. and GIVAN, R. (2004): Relational sequential inference with reliable observations. *Proc. of the International Conference on Machine Learning.*

JEFFREYS, H. (1961): *Theory of Probability (3rd ed.).* Oxford University Press.

KASS, R. and RAFTERY, E. (1995): Bayes Factors. *Journal of the American Statistical Association, 90(430):773-795.*

KERSTING, K., DE RAEDT, L. and RAIKO, T. (2006): Logical Hidden Markov Models. *Journal of Artificial Intelligence Research.*

MANNING, C.D. and SCHÜTZE, H. (1999): *Foundations of Statistical Natural Language Processing.* The MIT Press.

RABINER, L. (1989): A tutorial on hidden Markov models and selected applications in speech recognition. *Proceedings of the IEEE, 77(2):257–286.*

ESPIÉ, E. and GUIONNEAU, C. *TORCS - The Open Racing Car Simulator.* http://torcs.sourceforge.net

Applying the Q_n Estimator Online

Robin Nunkesser[1], Karen Schettlinger[2] and Roland Fried[2]

[1] Department of Computer Science, Univ. Dortmund, 44221 Dortmund, Germany
Robin.Nunkesser@udo.edu

[2] Department of Statistics, Univ. Dortmund, 44221 Dortmund, Germany
{schettlinger, fried}@statistik.uni-dortmund.de

Abstract. Reliable automatic methods are needed for statistical online monitoring of noisy time series. Application of a robust scale estimator allows to use adaptive thresholds for the detection of outliers and level shifts. We propose a fast update algorithm for the Q_n estimator and show by simulations that it leads to more powerful tests than other highly robust scale estimators.

1 Introduction

Reliable online analysis of high frequency time series is an important requirement for real-time decision support. For example, automatic alarm systems currently used in intensive care produce a high rate of false alarms due to measurement artifacts, patient movements, or transient fluctuations around the chosen alarm limit. Preprocessing the data by extracting the underlying level (the signal) and variability of the monitored physiological time series, such as heart rate or blood pressure can improve the false alarm rate. Additionally, it is necessary to detect relevant changes in the extracted signal since they might point at serious changes in the patient's condition.

The high number of artifacts observed in many time series requires the application of *robust* methods which are able to withstand some largely deviating values. However, many robust methods are computationally too demanding for real time application if efficient algorithms are not available.

Gather and Fried (2003) recommend Rousseeuw and Croux's (1993) Q_n estimator to measure the variability of the noise in robust signal extraction. The Q_n possesses a breakdown point of 50%, i.e. it can resist up to almost 50% large outliers without becoming extremely biased. Additionally, its Gaussian efficiency is 82% in large samples, which is much higher than that of other robust scale estimators: for example, the asymptotic efficiency of the median absolute deviation about the median (MAD) is only 36%. However, in an online application to moving time windows the MAD can be updated in $O(\log n)$ time (Bernholt et al. (2006)), while the fastest algorithm known so far for the Q_n needs $O(n \log n)$ time (Croux and Rousseeuw (1992)), where n is the width of the time window.

In this paper, we construct an update algorithm for the Q_n estimator which, in practice, is substantially faster than the offline algorithm and implies an advantage for online application. The algorithm is easy to implement and can also be used to compute the Hodges-Lehmann location estimator (HL) online. Additionally, we show by simulation that the Q_n leads to resistant rules for shift detection which have higher power than rules using other highly robust scale estimators. This better power can be explained by the well-known high efficiency of the Q_n for estimation of the variability.

Section 2 presents the update algorithm for the Q_n. Section 3 describes a comparative study of rules for level shift detection which apply a robust scale estimator for fixing the thresholds. Section 4 draws some conclusions.

2 An update algorithm for the Q_n and the HL estimator

For data $x_1,\ldots,x_n$, $x_i \in \mathbb{R}$ and $k = \binom{\lfloor n/2 \rfloor + 1}{2}$, $\lfloor a \rfloor$ denoting the largest integer not larger than a, the Q_n scale estimator is defined as

$$\hat{\sigma}^{(Q)} = c_n^{(Q)} \left\{ |x_i - x_j|, 1 \leq i < j \leq n \right\}_{(k)} ,$$

corresponding to approximately the first quartile of all pairwise differences. Here, $c_n^{(Q)}$ denotes a finite sample correction factor for achieving unbiasedness for the estimation of the standard deviation σ at Gaussian samples of size n. For online analysis of a time series $x_1,\ldots,x_N$, we can apply the Q_n to a moving time window $x_{t-n+1},\ldots,x_t$ of width $n < N$, always adding the incoming observation x_{t+1} and deleting the oldest observation x_{t-n+1} when moving the time window from t to $t+1$. Addition of x_{t+1} and deletion of x_{t-n+1} is called an *update* in the following.

It is possible to compute the Q_n as well as the HL estimator of n observations with an algorithm by Johnson and Mizoguchi (1978) in running time $O(n \log n)$, which has been proved to be optimal for offline calculation. An optimal online update algorithm therefore needs at least $O(\log n)$ time for insertion or deletion, respectively, since otherwise we could construct an algorithm faster than $O(n \log n)$ for calculating the Q_n from scratch. The $O(\log n)$ time bound was achieved for $k = 1$ by Bespamyatnikh (1998). For larger k - as needed for the computation of Q_n or the HL estimator - the problem gets more difficult and to our knowledge there is no online algorithm, yet. Following an idea of Smid (1991), we use a buffer of possible solutions to get an online algorithm for general k, because it is easy to implement and achieves a good running time in practice. Theoretically, the worst case amortized time per update may not be better than the offline algorithm, because $k = O(n^2)$ in our case. However, we can show that our algorithm runs substantially faster for many data sets.

Lemma 1. *It is possible to compute the Q_n and the HL estimator by computing the kth order statistic in a multiset of form $X + Y = \{x_i + y_j \mid x_i \in X \text{ and } y_j \in Y\}$.*

Proof. For $X = \{x_1,\ldots,x_n\}$, $k' = \binom{\lfloor n/2 \rfloor + 1}{2}$, and $k = k' + n + \binom{n}{2}$ we may compute the Q_n in the following way:

$$c_n^{(Q)}\{|x_i - x_j|, 1 \le i < j \le n\}_{(k')} = c_n^{(Q)}\{x_{(i)} - x_{(n-j+1)}, 1 \le i, j \le n\}_{(k)} \ .$$

Therefore me may compute the Q_n by computing the kth order statistic in $X + (-X)$.

To compute the HL estimator $\hat{\mu} = \text{median}\{(x_i + x_j)/2, 1 \le i \le j \le n\}$, we only need to compute the median element in $X/2 + X/2$ following the convention that in multisets of form $X + X$ exactly one of $x_i + x_j$ and $x_j + x_i$ appears for each i and j. □

To compute the kth order statistic in a multiset of form $X + Y$, we use the algorithm of Johnson and Mizoguchi (1978). Due to Lemma 1, we only consider the online version of this algorithm in the following.

2.1 Online algorithm

To understand the online algorithm it is helpful to look at some properties of the offline algorithm. It is convenient to visualize the algorithm working on a partially sorted matrix $B = (b_{ij})$ with $b_{ij} = x_{(i)} + y_{(j)}$, although B is, of course, never constructed. The algorithm utilizes, that $x_{(i)} + y_{(j)} \le x_{(i)} + y_{(\ell)}$ and $x_{(j)} + y_{(i)} \le x_{(\ell)} + y_{(i)}$ for $j \le \ell$. In consecutive steps, a matrix element is selected, regions in the matrix are determined to be certainly smaller or certainly greater than this element, and parts of the matrix are excluded from further consideration according to a case differentiation. As soon as less than n elements remain for consideration, they are sorted and the sought-after element is returned. The algorithm may easily be extended to compute a *buffer* $\mathcal{B}$ of size s of matrix elements $b_{(k-\lfloor (s-1)/2 \rfloor):n^2}, \ldots, b_{(k+\lfloor s/2 \rfloor):n^2}$.

To achieve a better computation time in online application, we use balanced trees, more precisely indexed AVL-trees, as the main data structure. Inserting, deleting, finding and determining the rank of an element needs $O(\log n)$ time in this data structure. We additionally use two pointers for each element in a balanced tree. In detail, we store X, Y, and $\mathcal{B}$ in separate balanced trees and let the pointers of an element $b_{ij} = x_{(i)} + y_{(j)} \in \mathcal{B}$ point to $x_{(i)} \in X$ and $y_{(j)} \in Y$, respectively. The first and second pointer of an element $x_{(i)} \in X$ points to the smallest and greatest element such that $b_{ij} \in \mathcal{B}$ for $1 \le j \le n$. The pointers for an element $y_{(j)} \in Y$ are defined analogously.

Insertion and deletion of data points into the buffer $\mathcal{B}$ correspond to the insertion and deletion of matrix rows or columns in B. We only consider insertions into and deletions from X in the following, because they are similar to insertions into and deletions from Y.

Deletion of element x_{del}

1. Search in X for x_{del} and determine its rank i and the elements b_s and b_g pointed at.
2. Determine $y_{(j)}$ and $y_{(\ell)}$ with the help of the pointers such that $b_s = x_{(i)} + y_{(j)}$ and $b_g = x_{(i)} + y_{(\ell)}$.
3. Find all elements $b_m = x_{(i)} + y_{(m)} \in \mathcal{B}$ with $j \le m \le \ell$.
4. Delete these elements b_m from $\mathcal{B}$, delete x_{del} from X, and update the pointers accordingly.

5. Compute the new position of the kth element in $\mathcal{B}$.

Insertion of element x_{ins}

1. Determine the smallest element b_s and the greatest element b_g in $\mathcal{B}$.
2. Determine with a binary search the smallest j such that $x_{\text{ins}} + y_{(j)} \geq b_s$ and the greatest ℓ such that $x_{\text{ins}} + y_{(\ell)} \leq b_g$.
3. Compute all elements $b_m = x_{\text{ins}} + y_{(m)}$ with $j \leq m \leq \ell$.
4. Insert these elements b_m into $\mathcal{B}$, insert x_{ins} into X and update pointers to and from the inserted elements accordingly.
5. Compute the new position of the kth element in $\mathcal{B}$.

It is easy to see, that we need a maximum of $O(|\text{deleted elements}| \log n)$ and $O(|\text{inserted elements}| \log n)$ time for deletion and insertion, respectively. After deletion and insertion we determine the new position of the kth element in $\mathcal{B}$ and return the new solution or recompute $\mathcal{B}$ with the offline algorithm if the kth element is not in $\mathcal{B}$ any more. We may also introduce bounds on the size of $\mathcal{B}$ in order to maintain linear size and to recompute $\mathcal{B}$ if these bounds are violated.

For the running time we have to consider the number of elements in the buffer that depend on the inserted or deleted element and the amount the kth element may move in the buffer.

Theorem 1. *For a constant signal with stationary noise, the expected amortized time per update is $O(\log n)$.*

Proof. In a constant signal with stationary noise, data points are *exchangeable* in the sense that the rank of each data point in the set of all data points is equiprobable. Assume w.l.o.g. that we only insert into and delete from X. Consider for each rank i of an element in X the number of buffer elements depending on it, i.e. $\left|\{i \mid b_{ij} \in \mathcal{B}\}\right|$. With $O(n)$ elements in $\mathcal{B}$ and equiprobable ranks of the observations inserted into or deleted from X, the expected number of buffer elements depending on an observation is $O(1)$. Thus, the expected number of buffer elements to delete or insert during an update step is also $O(1)$ and the expected time we spend for the update is $O(\log n)$.

To calculate the amortized running time, we have to consider the number of times $\mathcal{B}$ has to be recomputed. With equiprobable ranks, the expected amount the kth element moves in the buffer for a deletion and a subsequent insertion is 0. Thus, the expected time the buffer has to be recomputed is also 0 and consequently, the expected amortized time per update is $O(\log n)$. □

2.2 Running time simulations

To show the good performance of the algorithm in practice, we conducted some running time simulations for online computation of the Q_n. The first data set for the simulations suits the conditions of Theorem 1, i.e. it consists of a constant signal with standard normal noise and an additional 10% outliers of size 8. The second data set is the same in the first third of the time period, before an upward shift of size 8 and a linear upward trend in the second third and another downward shift of size 8

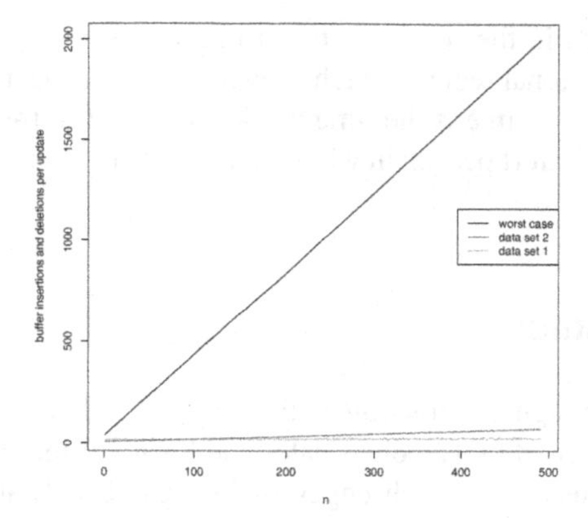

Fig. 1. Insertions and deletions needed for an update with growing window size n.

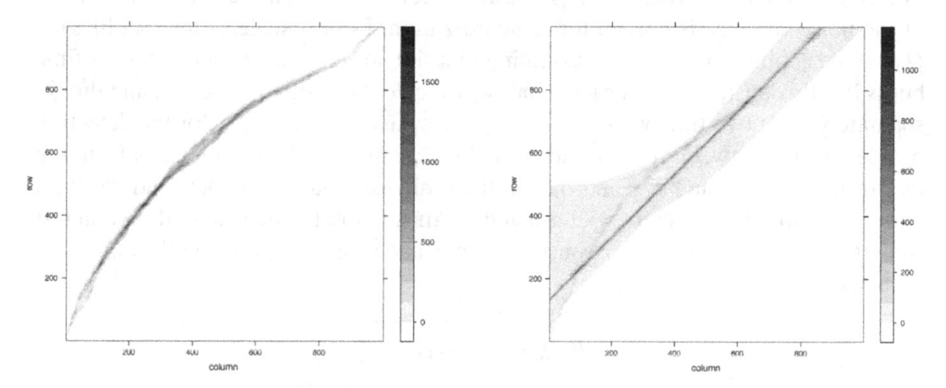

Fig. 2. Positions of $\mathcal{B}$ in the matrix B for data set 1 (left) and 2 (right).

and a linear downward trend in the final third occur. The reason to look at this data set is to analyze situations with shifts, trends and trend changes, because these are not covered by Theorem 1.

We analyzed the average number of buffer insertions and deletions needed for an update when performing $3n$ updates of windows of size n with $10 \leq n \leq 500$. Recall, that the insertions and deletions directly determine the running time. A variable number of updates assures similar conditions for all window widths. Additionally, we analyzed the position of $\mathcal{B}$ over time visualized in the matrix B when performing 3000 updates with a window of size 1000.

We see in Figure 1 that the number of buffer insertions and deletions for the first data set seems to be constant as expected, apart from a slight increase caused by the 10% outliers. The second data set causes a stronger increase, but is still far from the theoretical worst case of $4n$ insertions and deletions.

Considering Figure 2 we gain some insight into the observed number of update steps. For the first data set, elements of $\mathcal{B}$ are restricted to a small region in the matrix B. This region is recovered for the first third of the second data set in the right-hand side figure. The trends in the second data set cause $\mathcal{B}$ to be in an additional, even more concentrated diagonal region, which is even better for the algorithm. The cause for the increased running time is the time it takes to adapt to trend changes. After a trend change there is a short period, in which parts of $\mathcal{B}$ are situated in a wider region of the matrix B.

3 Comparative study

An important task in signal extraction is the fast and reliable detection of abrupt level shifts. Comparison of two medians calculated from different windows has been suggested for the detection of such edges in images (Bovik and Munson (1986), Hwang and Haddad (1994)). This approach has been found to give good results also in signal processing (Fried (2007)). Similar as for the two-sample t-test, an estimate of the noise variance is needed for standardization. Robust scale estimators like the Q_n can be applied for this task. Assuming that the noise variance can vary over time but is locally constant within each window, we calculate both the median and the Q_n separately from two time windows $y_{t-h+1},\ldots,y_t$ and $y_{t+1},\ldots,y_{t+k}$ for the detection of a level shift between times t and $t+1$. Let $\tilde{\mu}_{t-}$ and $\tilde{\mu}_{t+}$ be the medians from the two time windows, and $\hat{\sigma}_{t-}$ and $\hat{\sigma}_{t+}$ be the scale estimate for the left and the right window of possibly different widths h and k. An asymptotically standard normal test statistic in case of a (locally) constant signal and Gaussian noise with a constant variance is

$$\frac{\tilde{\mu}_{t+}-\tilde{\mu}_{t-}}{\sqrt{0.5\pi(\hat{\sigma}_{t-}^2/h+\hat{\sigma}_{t+}^2/k)}}$$

Critical values for small sample sizes can be derived by simulation.

Figure 3 compares the efficiencies of the Q_n, the median absolute deviation about the median (MAD) and the interquartile range (IQR) measured as the percentage variance of the empirical standard deviation as a function of the sample size n, derived from 200000 simulation runs for each n. Obviously, the Q_n is much more efficient than the other, 'classical' robust scale estimators.

The higher efficiency of the Q_n is an intuitive explanation for median comparisons standardized by the Q_n having higher power than those standardized by the MAD or the IQR if the windows are not very short. The power functions depicted in Figure 3 for the case $h=k=15$ have been derived from shifts of several heights $\delta=0,1,\ldots,6$ overlaid by standard Gaussian noise, using 10000 simulation runs each. The two-sample t-test, which is included for the reason of comparison, offers under Gaussian assumptions higher power than all the median comparisons, of course. However, Figure 3 shows that its power can drop down to zero because of a single outlier, even if the shift is huge. To see this, a shift of fixed size 10σ was generated, and a single outlier of increasing size into the opposite direction of the shift inserted briefly after the shift. The median comparisons are not affected by a single outlier even if windows as short as $h=k=7$ are used.

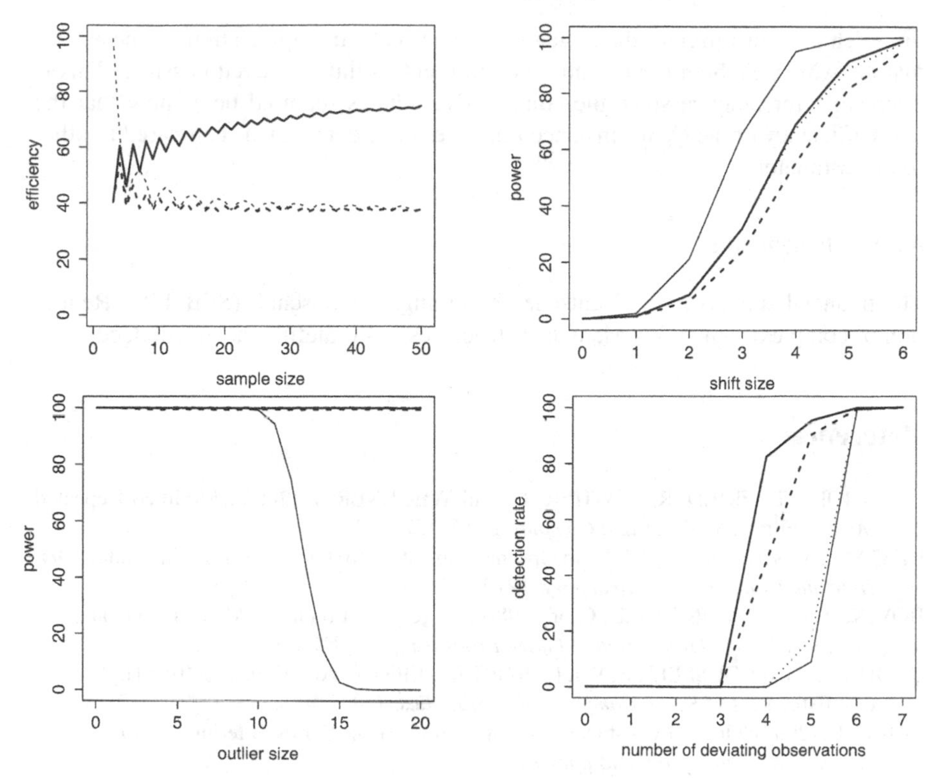

Fig. 3. Gaussian efficiencies (top left), power of shift detection (top right), power for a 10σ-shift in case of an outlier of increasing size (bottom left), and detection rate in case of an increasing number of deviating observations (bottom right): Q_n (solid), *MAD* (dashed), *IQR* (dotted), and S_n (dashed-dot). The two-sample t-test (thin solid) is included for the reason of comparison.

As a final exercise, we treat shift detection in case of an increasing number of deviating observations in the right-hand window. Since a few outliers should neither mask a shift nor cause false detection when the signal is constant, we would like a test to resist the deviating observations until more than half of the observations are shifted, and to detect a shift from then on. Figure 3 shows the detection rates calculated as the percentage of cases in which a shift was detected for $h = k = 7$. Median comparisons with the Q_n behave as desired, while a few outliers can mask a shift when using the *IQR* for standardization, similar as for the t-test. This can be explained by the *IQR* having a smaller breakdown point than the Q_n and the *MAD*.

4 Conclusions

The proposed new update algorithm for calculation of the Q_n scale estimator or the Hodges-Lehmann location estimator in a moving time window shows good running time behavior in different data situations. The real time application of these estimators, which are both robust and quite efficient, is thus rendered possible. This is interesting for practice since the comparative studies reported here show that the good efficiency of the Q_n for instance improves edge detection as compared to other robust estimators.

Acknowledgements

The financial support of the Deutsche Forschungsgemeinschaft (SFB 475, "Reduction of complexity in multivariate data structures") is gratefully acknowledged.

References

BERNHOLT, T., FRIED, R., GATHER, U. and WEGENER, I. (2006): Modified Repeated Median Filters. *Statistics and Computing, 16, 177–192.*

BESPAMYATNIKH, S. N. (1998): An Optimal Algorithm for Closest-Pair Maintenance. *Discrete and Computational Geometry, 19 (2), 175–195.*

BOVIK, A. C. and MUNSON, D. C. Jr. (1986): Edge Detection using Median Comparisons. *Computer Vision, Graphics, and Image Processing, 33, 377–389.*

CROUX, C.t'and ROUSSEEUW, P. J. (1992): Time-Efficient Algorithms for Two Highly Robust Estimators of Scale. *Computational Statistics, 1, 411–428.*

FRIED, R. (2007): On the Robust Detection of Edges in Time Series Filtering. *Computational Statistics & Data Analysis, to appear.*

GATHER, U. and FRIED, R. (2003): Robust Estimation of Scale for Local Linear Temporal Trends. *Tatra Mountains Mathematical Publications, 26, 87–101.*

HWANG, H. and HADDAD, R. A. (1994): Multilevel Nonlinear Filters for Edge Detection and Noise Suppression. *IEEE Trans. Signal Processing, 42, 249–258.*

JOHNSON, D. B. and MIZOGUCHI, T. (1978): Selecting the kth Element in $X+Y$ and $X_1 + X_2 + \ldots + X_m$. *SIAM Journal on Computing, 7 (2), 147–153.*

ROUSSEEUW, P.J. and CROUX, C. (1993): Alternatives to the Median Absolute Deviation. *Journal of the American Statistical Association, 88, 1273–1283.*

SMID, M. (1991): Maintaining the Minimal Distance of a Point Set in Less than Linear Time. *Algorithms Review, 2, 33–44.*

A Comparative Study on Polyphonic Musical Time Series Using MCMC Methods

Katrin Sommer and Claus Weihs

Lehrstuhl für Computergestützte Statistik,
Universität Dortmund, 44221 Dortmund, Germany
sommer@statistik.uni-dortmund.de

Abstract. A general harmonic model for pitch tracking of polyphonic musical time series will be introduced. Based on a model of Davy and Godsill (2002) the fundamental frequencies of polyphonic sound are estimated simultaneously. For an improvement of these results a preprocessing step was be implemented to build an extended polyphonic model.

All methods are applied on real audio data from the McGill University Master Samples (Opolko and Wapnick (1987)).

1 Introduction

The automatic transcription of musical time series data is a wide research domain. There are many methods for the pitch tracking of monophonic sound (e.g. Weihs and Ligges (2006)). More difficult is the distinction of polyphonic sound because of the properties of the time series of musical sound.

In this research paper we describe a general harmonic model for polyphonic musical time series data, based on a model of Davy and Godsill (2002). After transforming this model to an hierarchical bayes model the fundamental frequencies of this data can be estimated with MCMC methods.

Then we consider a preprocessing step to improve the results. For this, we introduce the design of an alphabet of artificial tones.

After that we apply the polyphonic model to real audio data from the McGill University Master Samples (Opolko and Wapnick (1987)). We demonstrate the building of an alphabet on real audio data and present the results of utilising such an alphabet. Further, we show first results of combining the preprocessing step and the MCMC methods. Finally the results are discussed and an outlook to future work is given.

2 Polyphonic model

In this section the harmonic polyphonic model will be introduced and its components will be illustrated. The model is based on the model of Davy and Godsill (2002) and has the following structure:

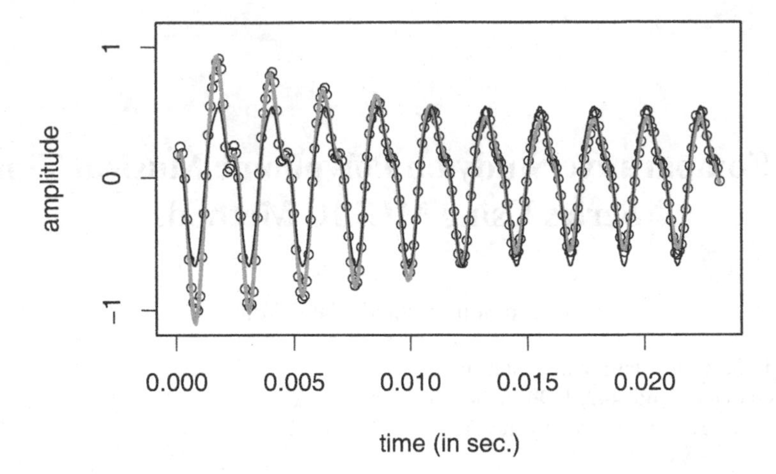

Fig. 1. Illustration of the modelling with basis functions. Modelling time-variant amplitudes of a real audio signal

$$y_t = \sum_{k=1}^{K} \sum_{h=1}^{H} \sum_{i=0}^{I} \phi_{t,i} \left\{ a_{k,h,i} \cos\left(2\pi h f_k / f_s t\right) + b_{k,h,i} \sin\left(2\pi h f_k / f_s t\right) \right\} + \varepsilon_t,$$

The number of observations of the audio signal y_t is $T, t \in \{0, \ldots, T-1\}$. Each signal is normalized to $[-1, 1]$ since the absolute overall loudness of different recordings is not relevant. The signal y_t is made up of K tones each composed out of harmonics from H_k partial tones. In this paper the number of tones K is assumed to be known. The first partial of the k-th tone is the fundamental frequency f_k, the other $H_k - 1$ partials are called overtones. Further, f_s is the sampling rate.

To reduce the number of parameters to be estimated, the amplitudes $a_{k,h,t}$ and $b_{k,h,t}$ of the $k-$th tone and the h-th partial tone at each timepoint t are modelled with $I+1$ basis functions. The basis functions $\phi_{t,i}$ are equally spaced hanning windows with 50% overlap:

$$\phi_{t,i} := \cos^2\left[\pi(t - i\Delta)/(2\Delta)\right] \mathbf{1}_{[(i-1)\Delta,(i+1)\Delta]}(t), \Delta = (T-1)/I.$$

So the $a_{k,h,i}$ and $b_{k,h,i}$ are the amplitudes of the k-th tone, the h-th partial tone and the i-th basis function. Finally, ε_t is the model error.

Figure 1 shows the necessity of using basis functions and thus modelling time-variant amplitudes. In the figure the points are the observations of the real signal. The assumption of constant amplitudes over time cannot depict the higher amplitudes at the beginning of the tone (black line). Modelling with time-variant amplitudes (grey line) leads to better results.

The model can be written as a hierarchical bayes model. The estimation of the parameters results from stochastic search for the best coefficients in a given region with different prior distributions. The region and the probabilities are specified by distributions. This leads to the implementation of MCMC methods (Gilks et al. (1996)).

For the sampling of the fundamental frequency f_k variants of the Metropolis-Hastings-Algorithm are used where the candidate frequencies are generated in different ways.

In the first variant the candidate for the fundamental frequency is sampled from a uniform distribution in the range of the possible frequencies. In the second variant the new candidate for the fundamental frequency is the half or the double frequency of the actual fundamental frequency. In the third variant a random walk is used which allows small changes of the fundamental frequency f_k to get a more precise result.

For the determination of the number of partial tones H_k a reversible jump MCMC was implemented. In each iteration of the MCMC-computation one of these algorithms is chosen with a distinct probability.

The parameters of the amplitude $a_{k,h,i}$ and $b_{k,h,i}$ are computed conditional on the fundamental frequency f_k and the number of partial tones H_k.

There is no full generation of the posterior distributions due to the computational burden. Instead we use a stopping criterion to stop the iterations if the slope of the model error is no longer significant (Sommer and Weihs (2006)).

3 Extended polyphonic model

An extented polyphonic model with an additional preprocessing step to the MCMC-algorithms will be established in this section. The results of this step could be the starting values for the MCMC algorithm in order to improve the results.

For this purpose we constructed an alphabet of artificial tones. These artificial tones are compared with the audio data to be analysed. The artificial tones are composed by evaluating the periodograms of seven time intervals with 512 observations of a real audio signal with 50% overlap. So a time interval of 2048 observations is regarded. At a sampling rate of 11 025 Hz a time interval of 0.186 seconds is observed.

These seven periodograms are averaged to a mean periodogram. For better comparability all values in this periodogram are set to zero which are smaller than one percent of the maximum peak. All artificial tones together form the alphabet.

In figure 2 (upper part) a periodogram of a c4 (262 Hz) played by an electric guitar can be seen. The lower part of figure 2 shows the small values of the periodogram. The horizontal line reflects the value of one percent of the maximum value of the periodogram. All values below this line are set to zero in the alphabet.

To determine the correct notes, every combination of two artificial tones of the alphabet is matched to the periodogram of the real audio signal. The modified periodograms of the two artificial tones are summed up to one periodograms. These periodograms are compared with the audio signal. The notes corresponding to the two artificial tones which cause minimal error are considered as estimates for the true notes. Finally, voting over ten time intervals leads to the estimation of the fundamental frequencies.

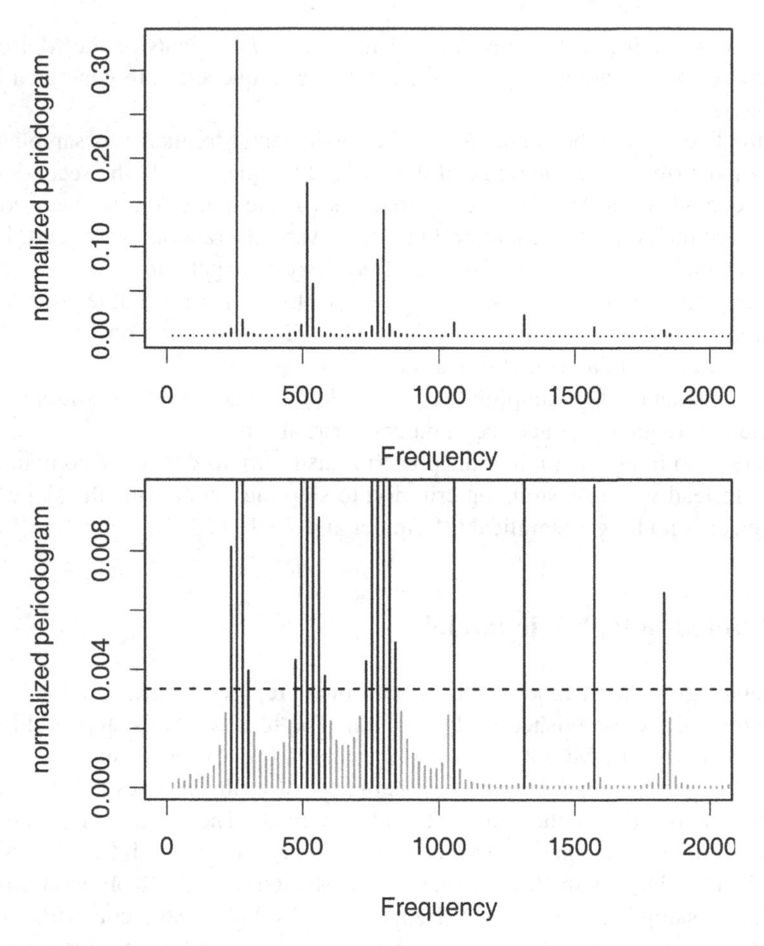

Fig. 2. Periodogram of note c4 played with an electric guitar. Original (upper part) and zoomed in with cut-off line (lower part)

4 Results

In this section results of estimating the fundamental frequencies of real audio data will be figured out. First, the data used in our studies will be introduced. Then first results are shown. Further the construction of an alphabet will be reconsidered and then the results based on this alphabet are depicted. Finally additional results are shown.

4.1 Data

The data used for our monophonic and polyphonic studies are real audio data from the McGill University Master Samples (Opolko and Wapnick (1987)). We chose 5 instruments (electric guitar, piano, violin, flute and trumpet) each with 5 notes (262,

Table 1. 1 if both notes were correctly identified, 0 otherwise. The left hand table requires the exact note to be estimated, the right table also counts octaves of the note as correct.

	instrument				
notes	flu	guit	pian	trum	viol
c4–c4	1	1	1	1	1
c4–e4	0	1	0	0	1
c4–g4	0	0	0	0	0
c4–a4	1	1	1	0	0
c4–c5	1	1	1	1	1

	instrument				
notes	flu	guit	pian	trum	viol
c4–c4	1	1	1	1	1
c4–e4	0	1	0	1	1
c4–g4	1	1	1	1	1
c4–a4	1	1	1	0	0
c4–c5	1	1	1	1	1

330, 390, 440 and 523 Hz) out of two groups of instruments, string instruments and wind instruments. The three string instruments are played in different ways, namely picked, struck and bowed. The two wind instruments are a woodwind instrument and a brass instrument.

For polyphonic data we superimposed the oszillations of two tones. The first tone was a c4 (262 Hz) played by the piano. This tone was combined with each instrument–tone combination we used. So we had 25 datasets each normalized to $[-1, 1]$. The pitches of the tones were tracked over ten time intervals of $T = 512$ observations with 50% overlap at a sampling rate of 11 025 Hz. The number of observations in one time interval is a tradeoff between the computational burden and the quality of the estimate. The estimate of the notes is the result of voting over the ten time intervals. The estimated notes are the two notes which occur in the ten time intervals most often.

4.2 First results with polyphonic model

The first step in our analysis was to consider how good the model works and if the pitch of a tone is estimated exactly. For this purpose we made a first study with monophonic data. The results of the study with monophonic time series data were very promising. In most cases the correct note was estimated and the deviations from the correct fundamental frequencies were minor (Sommer and Weihs (2006)).

The results of the estimation of polyphonic time series data are not as promising as the results with monophonic time series data. There are many notes which are not estimated correctly. The left side of Table 1 shows 1 if both notes were estimated correctly and 0 otherwise. In 15 of the 25 experiments both notes were estimated correctly. Counting octaves of the notes as correct increases the number of correct estimates to 21 (see the right hand side of Table 1). It can be seen that all notes of the combination c4–g4 are estimated incorrectly, but they are correct by counting the octaves of the right notes as correct (Sommer and Weihs (2007)).

Analysing the data over 20, 30 and 50 time intervals results in the same outcomes. So it seems to be adequate to examine 10 time intervals. In longer interval series new correctly estimated notes could not be determined.

Table 2. 1 if both notes AND instruments are correctly recognized after voting and 1* if both notes are estimated correcty, but not the instrument (left), including octaves of the correct notes (right). In 22 (left) and 23 (right) cases both notes are estimated correctly, in 18 cases for both tones the correct instrument is recognized.

	instrument				
notes	flu	guit	pian	trum	viol
c4–c4	0	0	1*	1*	1
c4–e4	1	1	1	1	1*
c4–g4	1	1	1	1	1
c4–a4	1	1	1	1	1
c4–c5	1	1	0	1*	1

	instrument				
notes	flu	guit	pian	trum	viol
c4–c4	0	0	1*	1*	1
c4–e4	1	1	1	1	1*
c4–g4	1	1	1	1	1
c4–a4	1	1	1	1	1
c4–c5	1	1	1*	1*	1

4.3 Results with extended polyphonic model

In a first study with an alphabet of artificial tones we used 30 notes from g3 (196 Hz) to c6 (1 047 Hz) of the same five instruments as for the studies in section 4.2. The choice of this range is restricted by the availability of the data of the McGill University Master Samples. The mean periodogram is computed out of seven periodograms each with $T = 512$ observations with 50% overlap at a sampling rate of 11 025 Hz. The first 1000 observations of a note were not considered for this periodogram in order to omit the attack of an instrument. Overall there are 150 artificial notes in the alphabet.

With this alphabet 11 325 pairwise comparisons of two artificial tones with the audio signal have to be computed. The results of the estimates of the same 25 note-combinations used in the previous study can be seen in table 2. The left hand side of the table shows that in 22 of 25 cases the fundamental frequency of both notes is estimated correctly. If octaves of the correct notes are counted as correct this number increases to 23 (right hand side of table 2).

Further, the entries in table 2 are annotated with a star if the instruments are not recognized correctly. This means that only in 18 of 22 cases (18 of 23) the instruments of both notes are identified correctly. Moreover, it can be seen that the cases where the notes are estimated incorrectly occur only in the first and last rows of the tables. So the correct estimation of the notes seems to be a problem if both notes are the same or one is the octave of the other.

4.4 Further results

Using these estimated notes as starting values for the MCMC algorithm in order to estimate the fundamental frequencies more precisely does not lead to an improvement of the results of the preprossing. To the contrary, the results are comparable to the results without this preprocessing step. In most of the cases the estimated notes are the octave of the correct notes. Often, the MCMC algorithm leads to an estimate $\hat{H} = 1$ of the number of partial tones. This often meant that only the octave of the fundamental frequency is found and neither the fundamental itself nor any other overtones.

A solution to this problem is the limitation of the possible range of frequencies. Restricting the frequency to the same range which the alphabet is covering and forcing the number of partial tones to be greater than 1 yields 20 respectively 24 correct estimations of both notes. A further improvement can be achieved by applying two chains in the MCMC algorithm. Starting values for both chains are equal, namely the results of the preprocessing. For each time interval the chain with the minimal model error is chosen. Voting over the ten time intervals results in 22 respectively 25 correct estimates. There are no more incorrectly estimated notes, in the worst case octaves of the correct notes. Also, voting is based on many more correct notes in the individual time intervals of 512 observations than in our previous studies, i.e. now typically five or six estimates are correct in contrast to three before.

5 Conclusion

In this paper a pitch tracking model for polyphonic musical time series data has been introduced. The unknown parameters are estimated with an MCMC algorithm as a stochastic optimization procedure. Because of the unfavorable results in a first study with polyphonic data the polyphonic model was extended and a preprocessing step was implemented. The application of an alphabet of artificial notes leads to promising results. The combination of the preprocessing and the MCMC algorithm is even more encouraging after the limitation of the frequency range.

Further work will extend the alphabet by using more artificial tones and considering attack, sustain and release, the different phases of a realisation of a note. An additional aim is the construction of a complete alphabet on the whole audio data of the McGill Universitiy Master Samples.

Acknowledgements

This work has been supported by the Graduiertenkolleg "Statistical Modelling" of the German Research Foundation (DFG).

References

DAVY, M. and GODSILL, S. J. (2002): Bayesian Harmonic Models for Musical Pitch Estimation and Analysis. *Technical Report 431*, Cambridge University Engineering Department.

GILKS, W. R., RICHARDSON, S. and SPIEGELHALTER D. J. (1996): *Markov Chain Monte Carlo in Practice*, Chapman & Hall.

OPOLKO, F. and WAPNICK, J. (1987): McGill University Master Samples [Compact disc]: Montreal, Quebec: McGill University.

SOMMER K. and WEIHS C. (2006): Using MCMC as a stochastic optimization procedure for music time series. In: V. Batagelj, H.H. Bock, A. Ferligoj, and A. Ziberna (Eds.): *Data Science and Classifiction* , Springer, Heidelberg, 307–314.

SOMMER K. and WEIHS C. (2007): Using MCMC as a stochastic optimization procedure for monophonic and polyphonic sound. In: R. Decker and H. Lenz (Eds.): *Advances in Data Analysis*, Springer, Heidelberg, 645–652.

WEIHS, C. and LIGGES, U. (2006): Parameter Optimization in Automatic Transcription of Music. In: Spiliopoulou, M., Kruse, R., Nürnberger, A., Borgelt, C. and Gaul, W. (eds.): *From Data and Information Analysis to Knowledge Engineering*. Springer, Berlin, 740 – 747.

Collective Classification for Labeling of Places and Objects in 2D and 3D Range Data

Rudolph Triebel[1], Óscar Martínez Mozos[2] and Wolfram Burgard[2]

[1] Autonomous Systems Lab, ETH Zürich, Switzerland
rudolph.triebel@mavt.ethz.ch
[2] Department of Computer Science, University of Freiburg, Germany
{omartine,burgard}@informatik.uni-freiburg.de

Abstract. In this paper, we present an algorithm to identify types of places and objects from 2D and 3D laser range data obtained in indoor environments. Our approach is a combination of a collective classification method based on associative Markov networks together with an instance-based feature extraction using nearest neighbor. Additionally, we show how to select the best features needed to represent the objects and places, reducing the time needed for the learning and inference steps while maintaining high classification rates. Experimental results in real data demonstrate the effectiveness of our approach in indoor environments.

1 Introduction

One key application in mobile robotics is the creation of geometric maps using data gathered with range sensors in indoor environments. These maps are usually used for navigation and represent free and occupied spaces. However, whenever the robots are designed to interact with humans, it seems necessary to extend these representations of the environment to improve the human-robot communication. In this work, we present an approach to extend indoor laser-based maps with semantic terms like "corridor", "room", "chair", "table", etc, used to annotate different places and objects in 2D or 3D maps. We introduce the instance-based associative Markov network (iAMN), which is an extension of associative Markov networks together with instance-based nearest neighbor methods. The approach follows the concept of collective classification in the sense that the labeling of a data point in the space is partly influenced by the labeling of its neighboring points. iAMNs classify the points in a map using a set of features representing these points. In this work, we show how to choose these features in the different cases of 2D and 3D laser scans. Experimental results obtained in simulation and with real robots demonstrate the effectiveness of our approach in various indoor environments.

2 Related work

Several authors have considered the problem of adding semantic information to 2D maps. Koenig and Simmons (1998) apply a pre-programmed routine to detect doorways. Althaus and Christensen (2003) use sonar data to detect corridors and doorways. Moreover, Friedman et al. (2007) introduce Voronoi random fields as a technique for mapping the topological structure of indoor environments. Finally, Martinez Mozos et al. (2005) use AdaBoost to create a semantic classifier to classify free cells in occupancy maps.

Also the problem of recognizing objects from 3D data has been studied intensively. Osada et al. (2001) propose a 3D object recognition technique based on shape distributions. Additionally, Huber et al. (2004) present an approach for parts-based object recognition. Boykov and Huttenlocher (1999) propose an object recognition method based on Markov random fields. Finally, Anguelov et al. (2005) present an associative Markov network approach to classify 3D range data. This paper is based on our previous work (Triebel et al. (2007)) which introduces the instance-based associative Markov networks.

3 Collective classification

In most standard spatial classification methods, the label of a data point only depends on its local features but not on the labeling of nearby data points. However, in practice one often observes a statistical dependence of the labeling associated to neighboring data points. Methods that use the information of the neighborhood are denoted as *collective classification* techniques. In this work, we use a collective classifier based on associative Markov networks (AMNs) (Taskar et al. (2004)), which is improved with an instance-based nearest-neighbor (NN) approach.

3.1 Associative Markov networks

An associative Markov network is an undirected graph in which the nodes are represented by N random variables $y_1, \ldots, y_N$. In our case, these random variables are discrete and correspond to the semantic label of each of the data points $\mathbf{p}_1, \ldots, \mathbf{p}_N$, each represented by a vector $\mathbf{x_i} \in \mathbb{R}^L$ of local features. Additionally, edges have associated a vector $\mathbf{x}_{ij}$ of features representing the relationship between the corresponding nodes. Each node y_i has an associated non-negative potential $\varphi(\mathbf{x}_i, y_i)$. Similarly, each edge (y_i, y_j) has a non-negative potential $\psi(\mathbf{x}_{ij}, y_i, y_j)$ assigned to it. The node potentials reflect the fact that for a given feature vector $\mathbf{x}_i$ some labels are more likely to be assigned to $\mathbf{p}_i$ than others, whereas the edge potentials encode the interactions of the labels of neighboring nodes given the edge features $\mathbf{x}_{ij}$. Whenever the potential of a node or edge is high for a given label y_i or a label pair (y_i, y_j), the conditional probability of these labels given the features is also high. The conditional probability that is represented by the network is expressed as:

$$P(\mathbf{y} \mid \mathbf{x}) = \frac{1}{Z} \prod_{i=1}^{N} \varphi(\mathbf{x}_i, y_i) \prod_{(ij) \in E} \psi(\mathbf{x}_{ij}, y_i, y_j), \tag{1}$$

where the partition function $Z = \sum_{\mathbf{y}'} \prod_{i=1}^{N} \varphi(\mathbf{x}_i, y_i') \prod_{(ij)\in E} \psi(\mathbf{x}_{ij}, y_i', y_j')$.

The potentials can be defined using the log-linear model proposed by Taskar et al. (2004). However, we use a modification of this model in which a weight vector $\mathbf{w}^k \in \mathbb{R}^{d_n}$ is introduced for each class label $k = 1, \ldots, K$. Additionally, a different weight vector $\mathbf{w}_e^{k,l}$, with $k = y_i$ and $l = y_j$ is assigned to each edge. The potentials are then defined as:

$$\log \varphi(\mathbf{x}_i, y_i) = \sum_{k=1}^{K} (\mathbf{w}_n^k \cdot \mathbf{x}_i) y_i^k \tag{2}$$

$$\log \psi(\mathbf{x}_{ij}, y_i, y_j) = \sum_{k=1}^{K} \sum_{l=1}^{K} (\mathbf{w}_e^{k,l} \cdot \mathbf{x}_{ij}) y_i^k y_j^l, \tag{3}$$

where y_i^k is an indicator variable which is 1 if point $\mathbf{p}_i$ has label k and 0, otherwise. In a further refinement step in our model, we introduce the constraints $\mathbf{w}_e^{k,l} = \mathbf{0}$ for $k \not\equiv l$ and $\mathbf{w}_e^{k,k} \geq \mathbf{0}$. This results in $\psi(\mathbf{x}_{ij}, k, l) = 1$ for $k \not\equiv l$ and $\psi(\mathbf{x}_{ij}, k, k) = \lambda_{ij}^k$, where $\lambda_{ij}^k \geq 1$. The idea here is that edges between nodes with different labels are penalized over edges between equally labeled nodes.

If we reformulate Equation 1 as the conditional probability $P_{\mathbf{w}}(\mathbf{y} \mid \mathbf{x})$, where the parameters ω are expressed by the weight vectors $\mathbf{w} = (\mathbf{w}_n, \mathbf{w}_e)$, and plugging in Equations (2) and (3), we then obtain that $\log P_{\mathbf{w}}(\mathbf{y} \mid \mathbf{x})$ equals

$$\sum_{i=1}^{N} \sum_{k=1}^{K} (\mathbf{w}_n^k \cdot \mathbf{x}_i) y_i^k + \sum_{(ij)\in E} \sum_{k=1}^{K} (\mathbf{w}_e^{k,k} \cdot \mathbf{x}_{ij}) y_i^k y_j^k - \log Z_{\mathbf{w}}(\mathbf{x}). \tag{4}$$

In the learning step we try to maximize $P_{\mathbf{w}}(\mathbf{y} \mid \mathbf{x})$ by maximizing the margin between the optimal labeling $\hat{\mathbf{y}}$ and any other labeling $\mathbf{y}$ (Taskar et at. (2004)). This margin is defined by:

$$\log P_{\omega}(\hat{\mathbf{y}} \mid \mathbf{x}) - \log P_{\omega}(\mathbf{y} \mid \mathbf{x}). \tag{5}$$

The inference in the unlabeled data points is done by finding the labels $\mathbf{y}$ that maximize $\log P_{\mathbf{w}}(\mathbf{y} \mid \mathbf{x})$. We refer to Triebel et al. (2007) for more details.

3.2 Instance-based AMNs

The main drawback of the AMN classifier explained previously, which is based on the log-linear model, is that it separates the classes linearly. This assumes that the features are separable by hyper-planes, which is not justified in all applications. This does not hold for instance-based classifiers such as the nearest-neighbor (NN), in which a query data point $\tilde{\mathbf{p}}$ is assigned to the label that corresponds to the training data point $\mathbf{p}$ whose features $\mathbf{x}$ are closest to the features $\tilde{\mathbf{x}}$ of $\tilde{\mathbf{p}}$. In the learning step, the NN classifier simply stores the entire training data set and does not compute a reduced set of training parameters.

To combine the advantage of instance-based NN classification with the AMN approach, we convert the feature vector $\tilde{\mathbf{x}}$ of the query point $\tilde{\mathbf{p}}$ using the transform $\tau : \mathbb{R}^L \rightarrow \mathbb{R}^K$: $\tau(\tilde{\mathbf{x}}) = (d(\tilde{\mathbf{x}}, \hat{\mathbf{x}}_1), \ldots, d(\tilde{\mathbf{x}}, \hat{\mathbf{x}}_{\mathbf{K}}))$, where K is the number of classes and $\hat{\mathbf{x}}_{\mathbf{k}}$ denotes the training example with label k closest to $\tilde{\mathbf{x}}$. The transformed features are more easily separable by hyperplanes. Additionally, the N nearest neighbors can be used in the transform function.

4 Feature extraction in 2D maps

In this paper, indoor environments are represented by two dimensional occupancy grid maps (Moravec (1988)). The unoccupied cells of a grid map form an 8-connected graph which is used as the input to the iAMN. Each cell is represented by a set of single-valued geometrical features calculated from the 360° laser scan in that particular cell as shown by Martínez Mozos et al. (2005).

Three dimensional scenes are presented by point clouds which are extracted with a laser scan. For each 3D point we computed *spin images* (Johnson (1997)) with a size of 5×10 bins. The spherical neighborhood for computing the spin images had a radius between 10 and 15cm, depending on the resolution of the input data.

5 Feature selection

One of the problems when classifying points represented by range data consists in selecting the size L of the features vectors $\mathbf{x}$. The number of possible features that can be used to represent each data point is usually very large and can easily be in the order of hundreds. This problem is known as *curse of dimensionality*. There are at least two reasons to try to reduce the size of the feature vector. The most obvious one is the computational complexity, which in our case, is also the more critical. We have to learn an inference in networks with thousands of nodes. Another reason is that although some features may carry a good classification when treated separately, maybe there is a little gain if they are combined together if they have a high mutual correlation (Theodoridis and Koutroumbas (2006)).

In our approach, the size of the feature vector for 2D data points is of the order of hundreds. The idea is to reduce the size of the feature vectors when used with the iAMN and at the same time try to maintain their class discriminatory information. To do this we apply a scalar feature selection procedure which uses a class separability criterion and incorporates correlation information. As separability criterion C, we use the Fisher's discrimination ratio (FDR) extended to the multi-class case (Theodoridis and Koutroumbas (2006)). For a scalar feature f and K classes $\{w_1, \ldots, w_K\}$, $C(f)$ can be defined as:

$$C(f) = FDR_f = \sum_{i}^{K} \sum_{j \neq i}^{K} \frac{(\mu_i - \mu_j)^2}{\sigma_i + \sigma_j}, \tag{6}$$

where the subscripts i, j refer to the mean and variance of the classes w_i and w_j respectively. Additionally, the cross-correlation coefficient between any two features f and g given T training examples is defined as:

$$\rho_{fg} = \frac{\sum_{t=1}^{T} x_{tf} x_{tg}}{\sqrt{\sum_{t=1}^{T} x_{tf}^2 \sum_{t=1}^{T} x_{tg}^2}}, \tag{7}$$

where x_{tf} denotes the value of the feature f in the training example t. Finally, the selection of the best L features involves the following steps:

- Select the first feature f_1 as $f_1 = \text{argmax}_f C(f)$.
- Select the second feature f_2 as:

$$f_2 = \underset{f \neq f_1}{\text{argmax}} \left\{ \alpha_1 C(f) - \alpha_2 |\rho_{f_1 f}| \right\},$$

  where α_1 and α_2 are weighting factors.
- Select f_l, $l = 1, \ldots, L$, such that:

$$f_l = \underset{f \neq f_r}{\text{argmax}} \left\{ \alpha_1 C(f) - \frac{\alpha_2}{l-1} \sum_{r-1}^{l} |\rho_{f_r f}| \right\}, \quad r = 1, 2, \ldots, l-1$$

6 Experiments

The approach described above has been implemented and tested in several 2D maps and 3D scenes. The goal of the experiment is to show the effectiveness of the iAMN in different indoor range data.

6.1 Classification of places in 2D maps

This experiment was carried out using the occupancy grid map of the building 79 at the University of Freiburg. For efficiency reasons we used a grid resolution of $20cm$, which lead us to a graph of 8088 nodes. The map was divided into two parts, the left one used for learning, and the right one used for classification purposes (Figure 1). For each cell we calculate 203 geometrical features. This number was reduced to 30 applying the feature selection of Section 5. The right image of Figure 1 shows the resulting classification with a success rate of 97.6%.

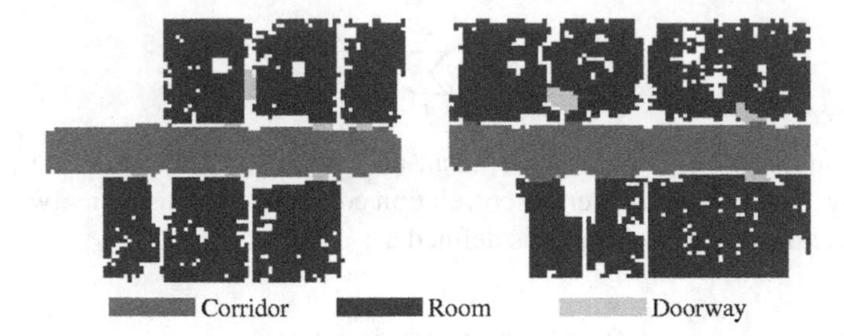

Fig. 1. The left image depicts the training map of building 79 at the University of Freiburg. The right image shows the resulting classified map using an iAMN with 30 selected features.

6.2 Classification of objects in 3D scenes

In this experiment we classify 3D scans of objects that appear in a laboratory of the building 79 of the University of Freiburg. The laboratory contain tables, chairs, monitors and ventilators. For each object class, an iAMN is trained with 3D range scans each containing just one object of this class (apart from tables, which may have screens standing on top of them). Figure 2 shows three example training objects. A complete laboratory in the building 79 of the University of Freiburg was later scanned with a 3D laser. In this 3D scene all the objects appear together and the scene is used as a test set. The resulting classification is shown in Figure 3. In this experiment 76.0% of the 3D points where classified correctly.

6.3 Comparison with previous approaches

In this section we compare our results with the ones obtained using other approaches for place and object classification. First, we compare the classification of the 2D map when using a classifier based on AdaBoost as shown by Martinez Mozos et al. (2005). In this case we obtained a classification rate of 92.1%, in contrast with the 97.6% obtained using iAMNs. We believe that the reason for this improvement is the neighboring relation between classes, which is ignored when using the AdaBoost approach. In a second experiment, we compare the resulting classification of the 3D scene with the one obtained when using AMN and NN. As we can see in Table 1, iAMNs perform better than the other approaches. A posterior statistical analysis using the t-student test indicates that the improvement is significant at the 0.05 level. We additionally realized different experiments in which we used the 3D scans of isolated objects for training and test purposes. The results are shown in Table 1 and they confirm that iAMN outperform the other methods.

7 Conclusions

In this paper we propose a semantic classification algorithm that combines associative Markov networks with an instance-based approach based on nearest neighbor.

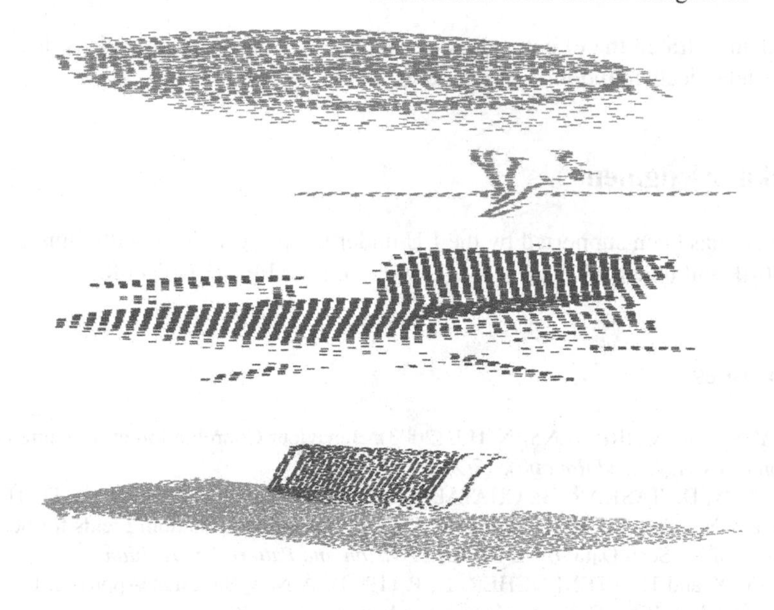

Fig. 2. 3D scans of isolated objects used for training: a ventilator, a chair and a table with a monitor on top.

Fig. 3. Classification of a complete 3D range scan obtained in a laboratory at the University of Freiburg.

Table 1. Classification results in 3D data

Data set	NN	AMN	iAMN
Complete scene	63%	62%	76%
Isolated objects	81%	72%	89%

Furthermore, we show how this method can be used to classify points described by features extracted from 2D and 3D laser scans. Additionally, we present an approach to reduce the number of features needed to represent each data point, while maintaining their class discriminatory information. Experiments carried out in 2D and 3D maps demonstrated the effectiveness of our approach for semantic classification of places and objects in indoor environments.

8 Acknowledgment

This work has been supported by the EU under the project CoSy with number FP6-004250-IP and under the project BACS with number FP6-IST-027140.

References

ALTHAUS, P. and CHRISTENSEN, H.I. (2003): Behaviour Coordination in Structured Environments. *Advanced Robotics, 17(7), 657–674.*

ANGUELOV, D., TASKAR, B., CHATALBASHEV, V., KOLLER, D., GUPTA, D., HEITZ, G. and NG, A. (2005): Discriminative Learning of Markov Random Fields for Segmentation of 3D Scan Data. *IEEE Computer Vision and Pattern Recognition.*

BOYKOV, Y. and HUTTENLOCHER. D. P. (1999): A New Bayesian Approach to Object Recognition. *IEEE Computer Vision and Pattern Recognition.*

FRIEDMAN, S., PASULA, S. and FOX, D. (2007): Voronoi Random Fields: Extracting the Topological Structure of Indoor Environments via Place Labeling. *International Joint Conference on Artificial Intelligence.*

HUBER, D., KAPURIA, A., DONAMUKKALA, R. R. and HEBERT, M. (2004): Parts-Based 3D Object Classification. *IEEE Computer Vision and Pattern Recognition.*

JOHNSON, A. (1997): *Spin-Images: A Representation for 3-D Surface Matching*. PhD thesis, Robotics Institute, Carnegie Mellon University, Pittsburgh, PA.

KOENIG, S. and SIMMONS, R. (1998): Xavier: A Robot Navigation Architecture Based on Partially Observable Markov Decision Process Models. In: Kortenkamp, D. and Bonasso, R. and Murphy, R. (Eds). *Artificial Intelligence Based Mobile Robotics: Case Studies of Successful Robot Systems*. MIT-Press, 91–122.

MARTINEZ MOZOS, O., STACHNISS, C. and BURGARD, W. (2005): Supervised Learning of Places from Range Data using Adaboost. *IEEE International Conference on Robotics & Automation.*

MORAVEC, H. P. (1988): Sensor Fusion in Certainty Grids for Mobile Robots. *AI Magazine, 61–74.*

OSADA, R., FUNKHOUSER, T., CHAZELLE, B. and DOBKIN, D. (2001): Matching 3D Models with Shape Distributions. *Shape Modeling International 154–166.*

TASKAR, B., CHATALBASHEV, V. and KOLLER, D. (2004): Learning Associative Markov Networks.*International Conference on Machine Learning.*

THEODORIDIS, S. and KOUTROUMBAS, K. (2006): *Pattern Recognition.* Academic Press, 3rd Edition, 2006.

TRIEBEL, R., SCHMIDT, R., MARTINEZ MOZOS, O. and BURGARD, W. (2007): Instace-based AMN Classification for Improved Object Recognition in 2D and 3D Laser Range Data. *International Joint Conference on Artificial Intelligence*

Lag or Error? - Detecting the Nature of Spatial Correlation

Mario Larch[1] and Janette Walde[2]

[1] ifo Institute for Economic Research at the University of Munich,
Poschingerstrasse 5, 81679 Munich, Germany
larch@ifo.de

[2] Department of Statistics, University of Innsbruck, Faculty of Economics and Statistics,
Universitaetsstrasse 15, 6020 Innsbruck, Austria
janette.walde@uibk.ac.at

Abstract. Theory often suggests spatial correlations without being explicit about the exact form. Hence, econometric tests are used for model choice. So far, mainly Lagrange Multiplier tests based on ordinary least squares residuals are employed to decide whether and in which form spatial correlation is present in Cliff-Ord type spatial models. In this paper, the model selection is based both on likelihood ratio and Wald tests using estimates for the general model and information criteria. The results of the conducted large Monte Carlo study suggest that Wald tests on the spatial parameters after estimation of the general model are the most reliable approach to reveal the nature of spatial correlation.

1 Introduction

Theoretical considerations frequently suggest proximity and/or similarity between observational units as important determinant. Econometric models trying to capture the proximity and/or similarity are referred to as 'spatial models'. Spatial models are nowadays employed widely. Spatial correlation can have numerous reasons, e.g. interaction between cross-sectional units could be due to environmental circumstances, network externalities, market interdependencies, strategic effects such as tax setting behavior and vote seeking behavior, contagion problems, population and employment growth, or the determinants of welfare expenditures. For a state-of-the-art overview, see the book by Anselin, Florax and Rey (2004). A Google Scholar search with the words 'spatial correlation cliff ord' lead to 1,770 hits. This kind of spatial models capture the proximity between observational units either by introducing a spatially lagged (endogenous or exogenous) variable or by modeling spatial correlation in the error term. In either way it is necessary to specify a weighting scheme which specifies the proximity or similarity. A common example for the former is the inverse distance between the capitals, whereas for the latter the membership in regional trade groups or the common language are examples.

In most cases theory is silent about the explicit functional form of the spatial interaction. In many applications modeling either a spatial lag in the endogenous variable and/or a lag in the error term cooperates with the theory. Including both, a spatially lagged endogenous variable and spatial correlation in the error term, may therefore be useful in order to obtain white noise errors and valid hypothesis tests for the regression parameters. The spatial autoregressive model with spatial autoregressive disturbances is then an obvious model to start with. However, this general model is so far not considered as the starting point for model selection/specification.

For the choice of the econometric model, there are basically two different approaches that can be employed: the 'bottom-up' or the 'top-down' approach. In the spatial econometric literature the classical specification search approach has been predominant, which is the 'specific to general' or 'bottom-up' approach. First a model without spatially lagged variables is estimated. Afterwards, Lagrange Multiplier (LM) tests for the spatial error model or the spatial lag model using ordinary least squares (OLS) residuals are employed to decide whether spatial correlation is present or not. If the null hypothesis of a test for a spatial autoregressive process is rejected, a spatial variant is calculated (see Florax et al. (2003)). Florax and de Graaff (2004) suggest to rely on the ad hoc decision rule that whichever test statistic is greater and significantly different form zero, points to the right alternative. Note, however, that LM tests for the spatial error and the spatial lag model exhibit power against both alternatives.

The second approach is a 'general to specific' or 'top-down' approach put forward by Hendry (1979), and in spatial econometrics by Florax et al. (2003). The 'top-down' approach starts with a very general model that allows for spatial correlation among various variables. A sequence of specification tests progressively simplifies the model. We propose to use the 'top-down' approach with the spatial autoregressive model with spatial autoregressive disturbances as the general model. The appropriateness of this approach is shown in a large Monte Carlo study, using maximum likelihood (ML) and generalized method of moments (GMM) estimators.

2 Model and test statistics

We describe the estimation approaches for the spatial autoregressive model with spatial autoregressive disturbances (henceforth short SARAR(1,1)), i.e. in our case the most general model. The estimation procedure for the other models are then easily obtained by implying the restriction $\rho = 0$ for the spatial error model (abbreviated by SARAR(0,1)) and $\lambda = 0$ for the spatial lag model (denoted subsequently by SARAR(1,0)). We restrict ourself to these classes of model choice and do not consider other possible functional forms or misspecifications (see for an analysis of misspecification resulting form an improper weighting matrix Dubin (2003) and for misspecification concerning the functional form McMillen (2003)).

The data generating process (DGP) for the SARAR(1,1) model considered in our study is given by:

$$y = \rho W y + X\beta + u, \qquad u = \lambda W u + \varepsilon, \tag{1}$$

where y is the $n \times 1$ vector of the dependent variable, n is the sample size, X is the $n \times k$ matrix of the independent variables, k is the number of independent variables, β is the $k \times 1$ vector of coefficients, W is a given $n \times n$ weighting matrix, ρ is the coefficient of the spatially lagged dependent variable, λ is the spatial error correlation coefficient, and ε is the $n \times 1$ disturbance term. The disturbances ε_i $(i = 1,...,n)$ are assumed to be $i.i.d.(0, \sigma^2)$ with finite second and fourth moments. Further we assume that all diagonal elements of the row normalized weighting matrix W are zero, the absolute values of ρ and λ are less than 1, and thus the matrices $(I - \rho W)$ and $(I - \rho W)$ are nonsingular.

2.1 Estimation approaches

We use two different approaches to estimate our models: (i) Maximum Likelihood, and (ii) GMM. For the maximum likelihood estimator two of the first order conditions are employed to get the concentrated log-likelihood function $LL_c = Fkt(\rho, \lambda; X, y)$. This is a non-linear function in the two parameters ρ and λ (Anselin (1988a)). The standard errors of all the estimators are obtained via the information matrix.

The second approach is based on generalized method of moments (GMM). The GMM estimator is a two-stage least squares procedure that uses additional moment conditions to estimate the spatial parameters. To account for the endogeneity of $\mathbf{Wy}$, all independent variables as well as the once and twice spatially lagged independent variables ($[\mathbf{X}, \mathbf{WX}, \mathbf{W}^2\mathbf{X}]$) serve as instruments as recommended by Kelejian and Prucha (1999). Kelejian and Prucha (1999) proposed a three-step procedure. In the first step a consistent estimate for the residuals is obtained by two-stage least squares (2SLS). These residuals are used in the moment conditions to estimate the spatial correlation coefficient of the error term. In the final step, a Cochrane-Orcutt type transformation is applied and the parameters are estimated by 2SLS on the transformed values. Lee (2003) proved that these instruments do not lead to asymptotically efficient parameter estimates. He suggests to use $\mathbf{H} = (\mathbf{I} - \lambda\mathbf{W})[\mathbf{X}, \mathbf{W}(\mathbf{I} - \rho\mathbf{W})^{-1}\mathbf{X}\hat{\beta}]$ as instruments, where $\hat{\beta}$ are the estimates from the first-step regression. We apply these optimal instruments by replacing ρ and λ with their estimates from the first step. The standard errors for the regression coefficients and the coefficient of the spatially lagged dependent variable are readily obtained from the last stage regression. However, in order to obtain the standard error for the spatial error parameter, we have to apply the estimator suggested in Kelejian and Prucha (2006).

2.2 Applied tests for model selection

First the 'specific to general' approaches are described. These tests start from the most simple model and turn to more complicated ones if the test statistic rejects the simple model. In the applied framework the most simple model is one without spatial lag and spatial error, i.e. OLS regression.

Available tests are mainly LM tests, which only rely on the estimates of the model unter the null hypothesis. Basically the LM tests suggested by Anselin et al. (1996) are implemented. As the LM tests are based on OLS resiudals, $\hat{\mathbf{u}}$ denotes the estimated residuals from the OLS regression, and $\hat{\sigma}^2 = (1/n)\hat{\mathbf{u}}'\hat{\mathbf{u}}$. Further, we have to distinguish whether we assume the second spatial parameter to be zero or not. The following definitions will simplify the expressions: $T = tr((\mathbf{W}' + \mathbf{W})\mathbf{W})$, $\mathbf{M} = \mathbf{I} - \mathbf{X}(\mathbf{X}'\mathbf{X})^{-1}\mathbf{X}'$, $\hat{J}_{\rho\beta} = \frac{1}{n\hat{\sigma}^2}[(\mathbf{W}\mathbf{X}\hat{\beta})'\mathbf{M}(\mathbf{W}\mathbf{X}\hat{\beta}) + T\hat{\sigma}^2]$. Now the following tests can be conducted:

$$\text{Model:} \quad y = X\beta + u, \text{ assumption: } \rho = 0, H_0 : \lambda = 0 \tag{2}$$

$$LM_\lambda = \frac{(\hat{u}'W\hat{u}/\hat{\sigma}^2)^2}{T}.$$

$$\text{Model:} \quad y = X\beta + u, \text{ assumption: } \lambda = 0, H_0 : \rho = 0 \tag{3}$$

$$LM_\rho = \frac{(\hat{u}'Wy/\hat{\sigma}^2)^2}{n\hat{J}_{\rho\beta}}.$$

$$\text{Model:} \quad y = \rho Wy + X\beta + u, H_0 : \lambda = 0 \tag{4}$$

$$LM_\lambda^* = \frac{[\hat{u}'W\hat{u}/\hat{\sigma}^2 - T(n\hat{J}_{\rho\beta})^{-1}\hat{u}'Wy/\hat{\sigma}^2]^2}{T[1 - T(n\hat{J}_{\rho\beta})]^{-1}}.$$

$$\text{Model:} \quad y = X\beta + u, u = \lambda Wu + \varepsilon, H_0 : \rho = 0 \tag{5}$$

$$LM_\rho^* = \frac{[\hat{u}'Wy/\hat{\sigma}^2 - \hat{u}'W\hat{u}/\hat{\sigma}^2]^2}{n\hat{J}_{\rho\beta} - T}.$$

LM tests for ρ and λ in the case of spatial correlation in the error term or in the dependent variable respectively, which are assumed to be estimated, were derived by Anselin (1988b):

$$LM_\lambda^A = \frac{[\hat{u}'W_2\hat{u}/\hat{\sigma}^2]^2}{T_{22} - (T_{21A})^2 v\hat{a}r(\hat{\rho})}, \tag{6}$$

$$LM_\rho^A = \frac{[\hat{u}'B'BW_1y]^2}{H_{rho} - H_{\theta\rho} v\hat{a}r(\hat{\theta}) H'_{\theta\rho}}, \tag{7}$$

where $T_{21A} = tr[W_2W_1A^{-1} + W_2'W_1A^{-1}]$, $A = I - \hat{\rho}W_1$, $\theta' = (\beta'\lambda\sigma 12)'$, $B = I - \hat{\lambda}W_2$, $H_\rho = trW^2 + tr(BWB^{-1})'(BWB^{-1}) + \frac{1}{\sigma^2}(BWX\beta)' \; (BWX\beta)$ and $H'_{\theta\rho} = \begin{pmatrix} \frac{1}{\sigma^2}(BX)'BWX\beta \\ tr(WB^{-1})'BWB^{-1} + trWWB^{-1} \\ 0 \end{pmatrix}$, with $v\hat{a}r(\tilde{\theta})$ as the estimated variance matrix for the parameter vector θ in the null model.

Besides the described LM test we calculate likelihood ratio (LR) tests. Therefore one needs to calculate both the restricted and the unrestricted model, i.e. $LR = -2(LL_r - LL_{ur})$, where LL_{ur} (LL_r) denotes the value of the maximized log-likelihood of the unrestricted (restricted) model.

Third, we calculate Wald tests for both the MLE and the GMM approach. The SARAR(1,1) model has to be estimated in order to test against more sparse variants. Hence, these tests are in the vein of a 'general to specific' methodology. Given the estimates for the general model, we can conduct the Wald test for ρ and λ: $W_\rho = \hat{\rho}/\hat{\sigma}_\rho$ and $W_\lambda = \hat{\lambda}/\hat{\sigma}_\lambda$, where $\hat{\rho}$ and $\hat{\lambda}$ are the estimates of the general model under consideration, and $\hat{\sigma}_\rho$ and $\hat{\sigma}_\lambda$ are the estimated standard errors thereof. Note that with the estimates of the SARAR(1,1) model we can conduct a test for joint significance of ρ and λ for both, the MLE and the GMM estimators.

Fourth, widely used information criteria are implemented in order to obtain the true DGP. The Akaike information criterion (AIC), the bias corrected Akaike criterion (AIC_c), and the Schwartz information criterion (BIC) are calculated (e.g., Belitz and Lang (2006)).

3 Monte Carlo study

All test evaluations are done using a sample size of 400. The regression coefficient vector β is set to be $(1,1)$. The independent variable is drawn randomly from the uniform distribution between zero and twenty. The remainder noise is normally distributed with mean zero and variance one. For each setting of the true DGP 1000 Monte Carlo data sets are calculated which leads to a 95% confidence interval for the nominal significance level of $5\% \pm 1.35\%$.

Two different weighting schemes are employed. The units are ordered regularly in a square grid of size $\sqrt{n} \times \sqrt{n}$. The first weighting matrix uses the Moore (Queen, e.g., Anselin (1988b)) neighborhood with radius $r = 1$. After row normalizing the matrix, the weighting matrix $\mathbf{W}$ is obtained, and denoted henceforth as $\mathbf{W}_1$. As second weighting matrix ($\mathbf{W}_2$) the distance d_{ij} between observation units i and j is computed and the elements of the weighting matrix are calculated as $1/d_{ij}$ if $i \not\equiv j$. The diagonal elements are set to zero. In order to limit the neighboring influence, additionally the elements of the weighting matrix are set to zero if the distance is greater than 7.1 (which corresponds to a radius of 5). Hence, the weighting matrix based on the Moore neighborhood ($\mathbf{W}_1$) is sparser and demonstrates less spatial connectivity than the one based on the distance ($\mathbf{W}_2$).

In order to obtain the power function the true spatial correlation parameters are varied in the following way $(\rho,\lambda) = (0,0.5),(0.05,0.5),(0.1,0.5),(0.15,0.5),(0.2,0.5),(0.5,0),(0.5,0.05),(0.5,0.1),(0.5,0.15),(0.5,0.2)$.

4 Results

Let us first analyze the experiments with SARAR(1,1) as true DGP. In order to obtain the size and the power of the Wald test the spatial parameter λ (ρ) is fixed at the value of 0.5. The second spatial parameter ρ (λ) is varied from 0 to 0.2. The actual size of the Wald test with the null hypothesis $H_0: \rho = 0$ ($H_0: \lambda = 0$) is not significantly different from the nominal size of 5%. The joint hypothesis test supports the alternative hypothesis with 100% as well as the Wald test for the corresponding second spatial parameter $\lambda(\rho)$. Hence, the correct more parsimonious model under the null hypothesis is detected accordingly.

When the null hypothesis is tested whether the spatial correlation parameter ρ is zero in the presence of spatial correlation in the error term the Wald test has a very good power although the power is higher with the sparser weighting matrix. The latter characteristic is a general feature for all conducted tests. The power for the spatial error parameter in the presence of a non-zero spatial lag parameter is lower. However, the power of the Wald test in this circumstances is (much) greater than the power achievable by using Lagrange Multiplier tests. In Figure 1c the best performing LM test is plotted, i.e. LM^A. All LM tests relying on OLS residuals fail seriously to detect the true DGP.

Comparable to the performance of the Wald test based on MLE estimates is the Wald test based on GMM estimates but only in detecting the significant lag parameter in the presence of a significant spatial error parameter. In the reverse case the Wald test using GMM estimates is much worse.

As a further model selection approach the performance of information criteria is analyzed. The performance of the classical Akaike information criterion and the bias corrected AIC are almost identical. In Figure 1d the share of cases in which AIC/AIC$_c$ identifies the correct DGP is plotted on the y-axis. All information criteria fail in more than 15% of the cases to identify the correct more parsimonious model, i.e. SARAR(1,0) or SARAR(0,1) instead of SARAR(1,1). However, in the remaining experiments ($\rho = 0.05, ..., 0.2$ or $\lambda = .05, ..., 0.2$) AIC/AIC$_c$ is comparable to the performance of the Wald test. BIC performs better than AIC/AIC$_c$ to detect SARAR(1,0) or SARAR(0,1) instead of SARAR(1,1) but much worse in the remaining experiments.

In order to be able to propose a general procedure for model selection the approach must also be suitable if the true DGP is SARAR(1,0) or SARAR(0,1). In this case the Wald test based on the general model has again the appropriate size and a very good power. Further the sensitivity on different weighting matrices is less severe. However, the power is smallest for the test with the null hypothesis $H_0 : \lambda = 0$ and with distance as weighting scheme $\mathbf{W}_2$. The Wald test using GMM estimates is again comparable when testing for the spatial lag parameter but worse when testing for the spatial error parameter.

Not significantly different from the power function of the Wald test based on the general model are both LM statistics based on OLS residuals. However, in this case LM^A fails to identify the correct DGP.

The Wald test outperforms the information criteria regarding the identification of SARAR(1,0) or SARAR(0,1). If OLS is the DGP, the correct model is chosen only about two thirds of the time by AIC/AIC$_c$ but comparably often to Wald by BIC. If SARAR(1,0) is the data generating process all information criteria perform poorer than the Wald test independent of the underlying weighting scheme. If the SARAR(0,1) is the data generating process, BIC is worse than the Wald test, and AIC/AIC$_c$ has a slightly higher performance for small values of the spatial parameter but is outperformed by the Wald test for higher values of the spatial parameters. For the sake of completeness it is noted that no valid model selection can be conducted using likelihood ratio tests.

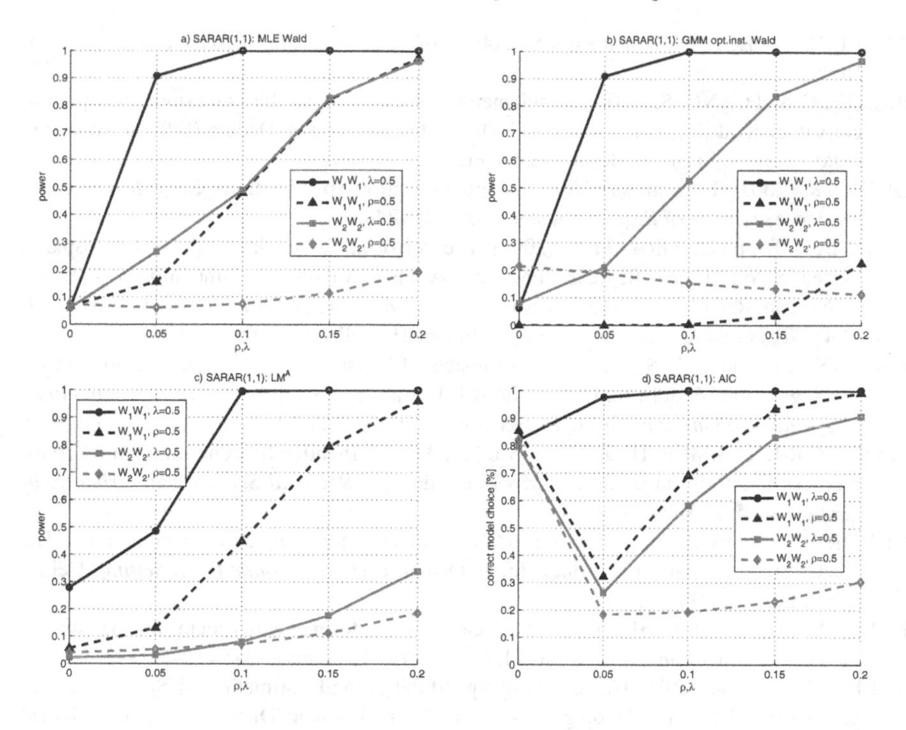

Fig. 1. a) Power of the Wald test based on the general model and MLE estimates. b) Power of the Wald test based on the general model and GMM estimates. c) Power of the Lagrange Multiplier test using LM^A as test statistic. d) Correct model choice of the better performing information criterion (AIC/AIC$_c$).

To conclude, we find that the 'general to specific' approach is the most suitable procedure to identify the correct data generating process (DGP) regarding Cliff-Ord type spatial models. Independent whether the true DGP is a SARAR(1,1), SARAR(1,0), SARAR(0,1), or just a regression model without any spatial correlation, the general model should be estimated and the Wald tests conducted. The chance to identify the true DGP is than higher compared to the alternative model choice criteria based on the LM tests, LR tests or on information criteria like AIC, AIC$_c$ or BIC.

References

ANSELIN, L. (1988a): Lagrange Multiplier Test Diagnostics for Spatial Dependence and Spatial Heterogeneity. *Geographical Analysis, 20, 1–17.*

ANSELIN, L. (1988b): *Spatial Econometrics: Methods and Models.* Kluwer Academic Publishers, Boston.

ANSELIN, L., BERA, A., FLORAX, R. and YOON, M. (1996): Simple Diagnostic Tests for Spatial Dependence, *Regional Science and Urban Economics, 26, 77–104.*

ANSELIN, L., FLORAX, R. and REY, S. (2004): *Advances in Spatial Econometrics*. Springer, Berlin.

BELITZ, C. and LANG, S. (2006), Simultaneous selection of variables and smoothing parameters in geoadditive regression models. In H.-J. Lenz, and R. Decker (Eds.):*Advances in Data Analysis*. Springer, Berlin-Heidelberg, forthcoming.

DUBIN, R. (2003): Robustness of Spatial Autocorrelation Specifications: Some Monte Carlo Evidence, *Journal of Regional Science, 43, 221–248.*

FLORAX, R.J., and DE GRAAFF, T. (2004): The Performance of Diagnostic Tests for Spatial Dependence in Linear Regression Models: A Meta-Analysis of Simulation Studies. In: L. Anselin, R.J. Florax, and S.J. Rey (Eds.): *Advances in Spatial Econometrics - Methodology, Tools and Applications*. Springer, Berlin-Heidelberg, 29-65.

FLORAX, R.J., and REY, S.J. (1995): The Impacts of Misspecified Spatial Interaction in Linear Regression Models. In: L. Anselin, R.J. Florax, and S.J. Rey (Eds.): *New Directions in Spatial Econometrics*. Springer, Berlin-Heidelberg, 111-135.

FLORAX, R.J., FOLMER, H., and REY, S.J. (2003): Specification Searches in Spatial Econometrics: The Relevance of HendryŠs Methodology, *Regional Science and Urban Economics, 33, 557–579.*

HENDRY, D. (1979): Predictive Failure and Econometric Modelling in Macroeconomics: The Transactions Demand for Money. In: P. Ormerod (Ed.): *Economic Modelling*. Heinemann, London, 217-242.

KELEJIAN, H., and PRUCHA, I. (1999): A Generalized Moments Estimator for the Autoregressive Parameter in a Spatial Model, *International Economic Review, 40, 509–533.*

KELEJIAN, H., and PRUCHA, I. (2006): Specification and Estimation of Spatial Autoregressive Models with Autoregressive and Heteroskedastic Disturbances, unpublished manuscript, University of Maryland.

LEE, L. (2003): Best Spatial Two-Stage Least Squares Estimators for a Spatial Autoregressive Model with Autoregressive Disturbances, *Econometric Reviews, 22, 307–335.*

MCMILLEN, D. (2003): Spatial Autocorrelation or Model Misspecification?, *International Regional Science Review, 26, 208–217.*

Part V

Exploratory Data Analysis and Tools for Data Analysis

Urban Data Mining Using Emergent SOM

Martin Behnisch[1] and Alfred Ultsch[2]

[1] Institute of industrial Building Production, University of Karlsruhe (TH), Englerstraße 7, D-76128 Karlsruhe, Germany
Martin.Behnisch@email.de
[2] Data Bionics Research Group
Philipps-University Marburg, D-35032 Marburg, Germany
ultsch@informatik.uni-marburg.de

Abstract. The term of Urban Data-Mining is defined to describe a methodological approach that discovers logical or mathematical and partly complex descriptions of urban patterns and regularities inside the data. The concept of data mining in connection with knowledge discovery techniques plays an important role for the empirical examination of high dimensional data in the field of urban research. The procedures on the basis of knowledge discovery systems are currently not exactly scrutinised for a meaningful integration into the regional and urban planning and development process. In this study ESOM is used to examine communities in Germany. The data deals with the question of dynamic processes (e.g. shrinking and growing of cities). In the future it might be possible to establish an instrument that defines objective criteria for the benchmark process about urban phenomena. The use of GIS supplements the process of knowledge conversion and communication.

1 Introduction

Comparisons of cities and typological grouping processes are methodical instruments to develop statistical scales and criteria about urban phenomena. Harris started in 1943, who ranked US cities according to industrial specialization data; many of the other studies that followed added occupational data to the classification models. Later on, in the 1970s, classification studies were geared to measuring social outcomes and shifted more towards the goals of public policy. Forst (1974) presents an investigation of german cities by using social and economic variables. In Great Britain, Craig (1985) employed a cluster analysis technique to classify 459 local authority districts, based on the 1981 Census of Population. Hill et al. (1998) classified US cities by using the city's population characteristics. Most of the mentioned classification studies use economic, social, and demographic variables as a basis for their classifications which are usually calculated by hierarchical algorithms (e.g. WARD, K-Means). Geospatial objects are analysed by Demsar (2006). These former approaches of city classification are summarized in Behnisch (2007).

The purpose of this article is to find groups (clusters) of communities with the same dynamic characteristics in Germany (e.g. shrinking and growing of cities).

The Application of Emergent Self Organizing Maps (ESOM) and the corresponding U*C-Algorithm is proposed for the task of City Classification. The term of Urban Data Mining (Behnisch, 2007) is defined to describe a methodological approach that discovers logical or mathematical and partly complex descriptions of urban patterns and regularities inside the data. The result can suggests a general typology and can lead to the development of prediction models using subgroups instead of the total population.

2 Inspection and transformation of data

Four variables were selected for the classification analysis. The variables characterise a city's dynamic behaviour. The data was created by the German BBR (Federal Office for Building and Regional Planning) and refers to the statistics of inhabitants (V_1), migration (V_2), employment (V_3) and mobility (V_4). The dynamic processes are characterised by positive or negative percentage quotations between the year 1999 and 2003. The inspection of data includes the visualisation in form of histograms, QQ-Plots, PDE-Plots (Ultsch, 2003) and Box-Plots. The authors decided to use transformation measurements such as ladder of power to take into account restrictions of statistics (Hand et al., 2001 or Ripley, 1996). Figure 1 and Figure 2 show an example for the distribution of variables. As a result of pre-processing the authors find a mixture of two distributions with decision boundary zero in each of the four variables. All variables are transformed by using $\text{Slog}(x) = \text{sign}(x) \cdot \log(|x| + 1)$.

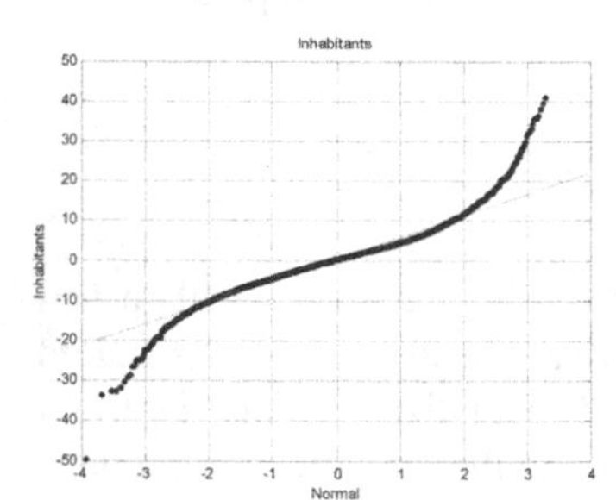

Fig. 1. QQ-Plot(inhabitants)

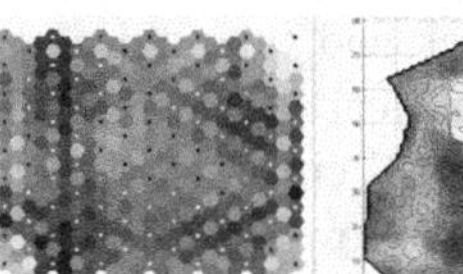
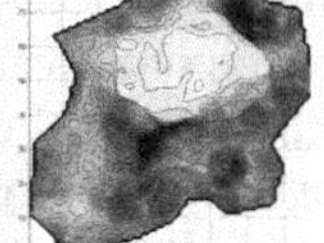

Fig. 2. PDE-Plot(Sloginhabitants)

The first hypothesis to the distribution of each variable is a bimodal distribution of lognormal distributed data (Data > 0: skewed right, Data < 0: skewed left).

The result of the detailed examination is summarized in Table 1. The data follows a lognormal distribution. Decision boundaries will be used to form a basis for a manual classification process and support the interpretation of results.

Pertaining to the classification approach (e.g. U*-Matrix and subsequent U*C-Algorithm) and according to the Euclidian Distance the data need to be standardized. Figure 3 shows Scatter-Plots of the transformed variables.

Table 1. Examination of the four distributions

Variable	Slog(Data)	Decision Boundaries	Size of Classes
inhabitants	bimodal distribution	C1: Data ≤ 0	[5820], 46,82%
		C2: Data > 0	[6610], 53,18%
migration	bimodal distribution	C1: Data ≤ 0	[4974], 40,02%
		C2: Data > 0	[7456], 59,98%
employment	bimodal distribution	C1: Data ≤ 0	[7492], 60,27%
		C2: Data > 0	[4938], 39,73%
mobility	multimodal distribution	C1: Data ≤ 0	[2551], 20,52%
		C2: $0 <$ Data < 50	[9317], 74,96%
		C3: Data ≥ 50	[562], 4,52%

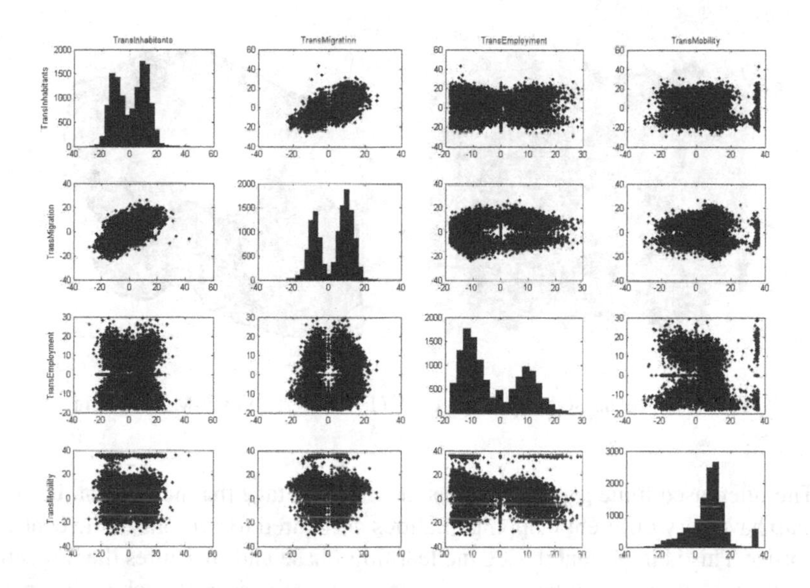

Fig. 3. Scatter-Plots of transformed variables

3 Method

In the field of urban planning and regional science data are usually multidimensional, spatially correlated and especially heterogeneous. These properties make classical data mining algorithms often inappropriate for this data, as their basic assumptions cease to be valid. The power of self-organization allows the emergence of structure in data and supports visualization, clustering and labelling concerning a combined distance and density based approach. To visualize high-dimensional data, a projection from the high dimensional space onto two dimensions is needed. This projection onto a grid of neurons is called SOM map. There are two different SOM usages. The first are SOM, introduced by Kohonen (1982). Neurons are identified with clusters in the data space (k-means SOM) and there are very few neurons. The second are

SOM where the map space is regarded as a tool for the visualization of the otherwise high dimensional data space. These SOM consist of thousands or tens of thousand neurons. Such SOM allow the emergence of intrinsic structural features of the data space and therefore they are called Emergent SOM (Ultsch, 1999). The map of an ESOM preserves the neighbourhood relationships of the high dimensional data and the weight vectors of the neurons are thought as sampling point of the data. The U-Matrix has become the canonical tool for the display of the distance structures of the input data on ESOM. The P-Matrix takes density information into account. The combination of U-Matrix and P-Matrix leads to the U*Matrix. On this U*-Matrix a cluster structure in the data set can be detected directly. Compare the examples in Figure 4 using the same data to see in an appropriate way, whether there are cluster structures.

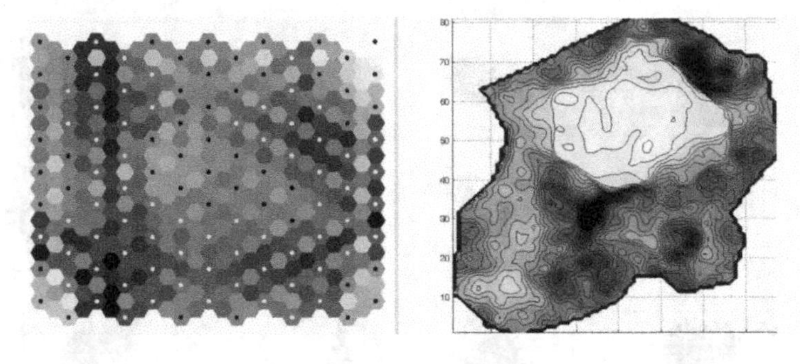

Fig. 4. K-Means-SOM by Kaski et al. (1999), left and U*-Matrix, right

The often used finite grid as map has the disadvantage that neurons at the rim of the map have very different mapping qualities compared to neurons in the centre vs. the border. This is important during the learning phase and structures the projection. In many applications important clusters appear in the corner of such a planar map. Using ESOM methods for clustering has the advantage of a nonlinear disentanglement of complex structures.

The clustering of the ESOM can be performed at two different levels. The Bestmatch Visualization can be used to mark data points that represents a neuron with a defined characteristic. Bestmatches and thus corresponding data points can be manually grouped into several clusters. Not all points need to be labelled, outliers are usually easily detected and can be removed. Secondly the neurons can be clustered by using a clustering algorithm, called U*C, which is based on grid projections and uses distance and density information (Ultsch (2005)). In most times an aggregation process of objects is necessary to build up a meaningful classification. Assigning a name to a cluster is one of the most important processes in order to define the meaning of a cluster. The interpretation is based on the attribute values. Moreover it is possible to integrate techniques of Knowledge Discovery to understand the structure in a complementary form and support the finding of an appropriate cluster denomination. Examples are the symbolic algorithms such as SIG* or U-Know (Ultsch (2007)) which lead to significant properties for each cluster and a fundamental knowledge based description.

4 Results

A first classification is based on the dichotomic characteristics of the four variables. 24 Classes are detected by using the decision boundaries (Variable > 0 or Variable < 0). The further aggregation leads to the five classes of Table 2. The classed are content adressed to the approved pressure factors for urban dynamic development (population and employment). The purpose of such a wise classification was to sharpen characteristics and to find a special label.

Table 2. Classes of Urban Dynamic Phenomena

Label	Inhabitants	Migration	Employment
Shrinking of Inhabitants and Employment	low	low	low
Shrinking but influx	low	high	low
Growing of Employment	low		high
Growing of Inhabitants	high		low
Growing of Inhabitants and Employment	high		high

An ESOM with 50x82 neurons is trained with the pre-processed data to proof the defined structure. The corresponding U*-Map delivers a geographical landscape of the input data on to a projected map (imaginary axis). The cluster boundaries are expressed by mountains that means the value of height defines the distance between different objects which is displayed on the z-Axis. A valley describes similar objects, characterized by small U-heights on the U*-Matrix. Data points found in coherent regions are assigned to one cluster. All local regions lying in the same cluster have the same spatial properties.

The U*-Map (Island View) can be seen in Figure 5 in connection to the U*-Matrix of Figure 6 including the clustering results of U*C-Algorithm with 11 classes. The existing clusters are described by the U-Know Algorithm and the symbolic description is comparable to the dichotomic properties. The interpretation of the clustering results leads finally to the same five main classes realized by the content-based aggregation. It is remarkable that the structure of the first classification can be recognized by using later Emergent SOM.

Figure 7 determines the five main cluster solution and displays the spatial structure of the classified objects. It is obvious to see that growing processes can be found in the southern and western part of Germany and shrinking processes can be localized in the eastern part. Shrinking processes also exist in areas of traditional coal and steel industry.

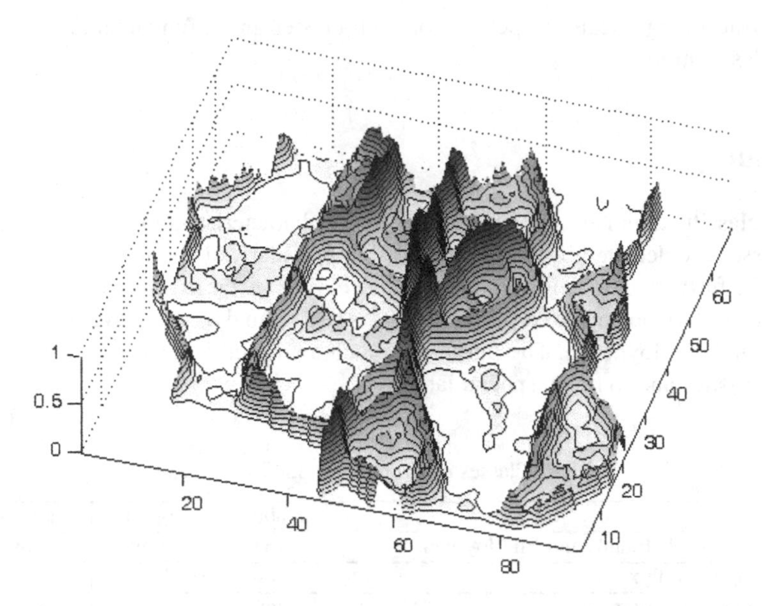

Fig. 5. U*-Map (Island View)

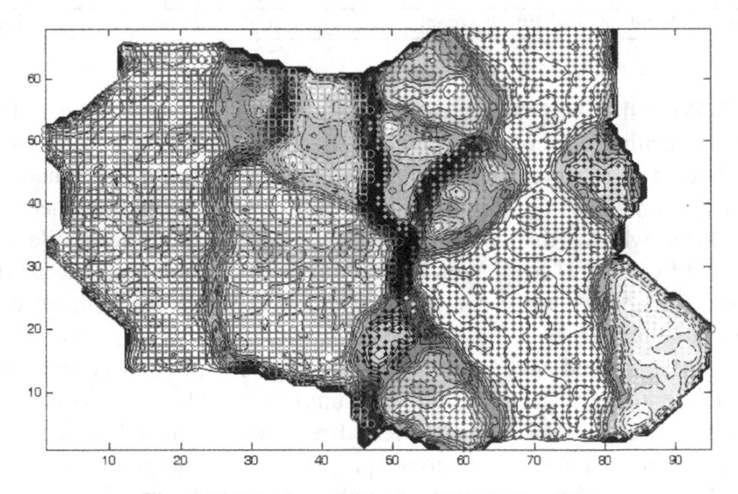

Fig. 6. U*Matrix and Result of U*-C-Algorithm

5 Conclusion

The authors present a classification approach in connection with geospatial data. The central issue of the grouping processes are the shrinking and growing phenomena in Germany. First the authors examine the pool of data and show the importance for the investigation of distributions according to the dichotomic properties. Afterwards it is shown that the use of Emergent SOMs is an appropriate method for clustering and

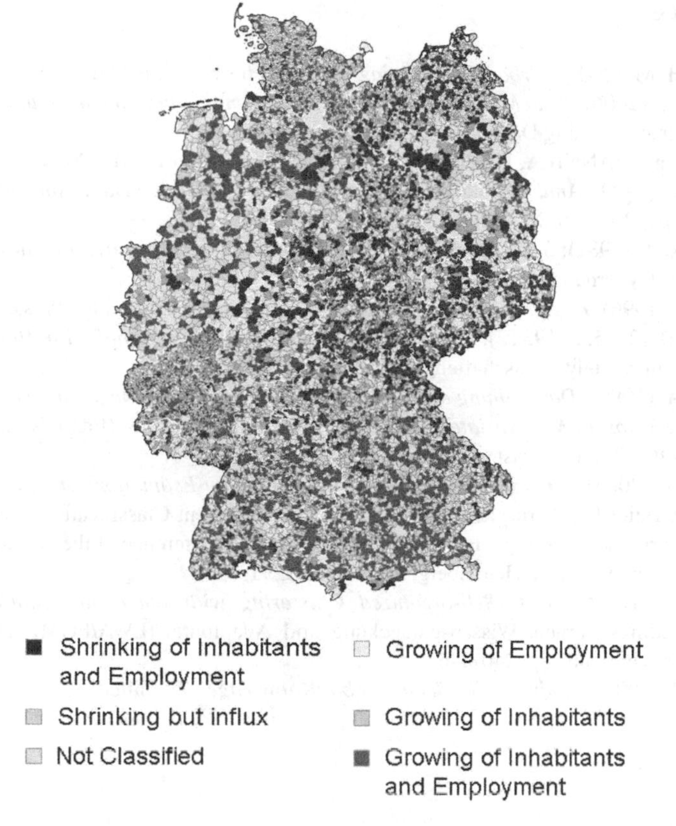

Fig. 7. Localisation of Shrinking and Growing Municipalities in Germany

classification. The advantage is to visualize the structure of data and later on to define a number of feasible cluster using U*C-algorithm or manual bestmatch grouping processes. The application of existing visual methods especially U*-Matrix shows that it is possible to detect meaningful classes among a large amount of geospatial objects. For example typical hierarchical algorithm would fail to examine 12430 objects. As such, the authors believe that the presented procedure of the wise classification and the ESOM approach complements the former proposals for city classification. It is expected that in the future the concept of data mining in connection with knowledge discovery techniques will get an increasing importance for the urban research and planning processes (Streich, 2005). Such approaches might lead to a benchmark system for regional policy or other strategical institutions. To get more data for a deeper empirical examination it is necessary to conduct field investigation in selected areas.

References

BEHNISCH, M. (2007): *Urban Data Mining.* Doctoral thesis, Karlsruhe (TH).

DEMSAR, U. (2006): *Data Mining of geospatial data: combining visual and automatic methods.* Urban Planning Department, KTH Stockholm.

HAND, D. and MANNILA, H. (2001): *Principles of Data Mining.* MIT Press.

KASKI et al. (1999): *Analysis and Visualization of Gene Expression Data using Self Organizing Maps*, Proc NSIP.

KOHONEN, T. (1982): *Self-Organizing formation of topologically correct feature maps.* Biological Cybernetics 43, 59-69.

RIPLEY, B. (1996): *Pattern Recognition and Neural Networks.* Cambridge Press.

STREICH, B. (2005, S. 193 ff.): *Stadtplanung in der Wissensgesellschaft - Ein Handbuch.* VS Verlag für Sozialwissenschaften, Wiesbaden.

ULTSCH, A. (1999): *Data Mining and Knowledge Discovery with Emergent Self Organizing Feature Maps for Multivariate Time Series*, In: Oja, E., Kaski, S. (Eds.): Kohonen Maps, pp. 33-46, Elsevier, Amsterdam.

ULTSCH, A. (2003): *Pareto Density Estimation: A Density Estimation for Knowledge Discovery*, Baier D., Wernecke K.D. (Eds), In Innovations in Classification, Data Science, and Information Systems - Proceedings 27th Annual Conference of the German Classification Society, Berlin, Heidelberg, Springer, pp. 91-100

ULTSCH, A. (2005): *U*C Self-organized Clustering with Emergent Feature Map*, In Prooceedings Lernen, Wissensentdeckung und Adaptivität (LWA/FGML 2005), Saarbrücken, Germany, pp. 240-246

ULTSCH, A (2007): *Mining for Understandable Knowledge.* Submitted.

KNIME: The Konstanz Information Miner

Michael R. Berthold, Nicolas Cebron, Fabian Dill, Thomas R. Gabriel, Tobias Kötter, Thorsten Meinl, Peter Ohl, Christoph Sieb, Kilian Thiel and Bernd Wiswedel

ALTANA Chair for Bioinformatics and Information Mining,
Department of Computer and Information Science, University of Konstanz,
Box M712, 78457 Konstanz, Germany
contact@knime.org

Abstract. The Konstanz Information Miner is a modular environment, which enables easy visual assembly and interactive execution of a data pipeline. It is designed as a teaching, research and collaboration platform, which enables simple integration of new algorithms and tools as well as data manipulation or visualization methods in the form of new modules or nodes. In this paper we describe some of the design aspects of the underlying architecture and briefly sketch how new nodes can be incorporated.

1 Overview

The need for modular data analysis environments has increased dramatically over the past years. In order to make use of the vast variety of data analysis methods around, it is essential that such an environment is easy and intuitive to use, allows for quick and interactive changes to the analysis process and enables the user to visually explore the results. To meet these challenges data pipelining environments have gathered incredible momentum over the past years. Some of today's well-established (but unfortunately also commercial) data pipelining tools are InforSense KDE (InforSense Ltd.), Insightful Miner (Insightful Corporation), and Pipeline Pilot (SciTegic). These environments allow the user to visually assemble and adapt the analysis flow from standardized building blocks, which are then connected through pipes carrying data or models. An additional advantage of these systems is the intuitive, graphical way to document what has been done.

KNIME, the Konstanz Information Miner provides such a pipelining environment. Figure 1 shows a screenshot of an example analysis flow. In the center, a flow is reading in data from two sources and processes it in several, parallel analysis flows, consisting of preprocessing, modeling, and visualization nodes. On the left a repository of nodes is shown. From this large variety of nodes, one can select data sources, data preprocessing steps, model building algorithms, as well as visualization tools and drag them onto the workbench, where they can be connected to other nodes. The

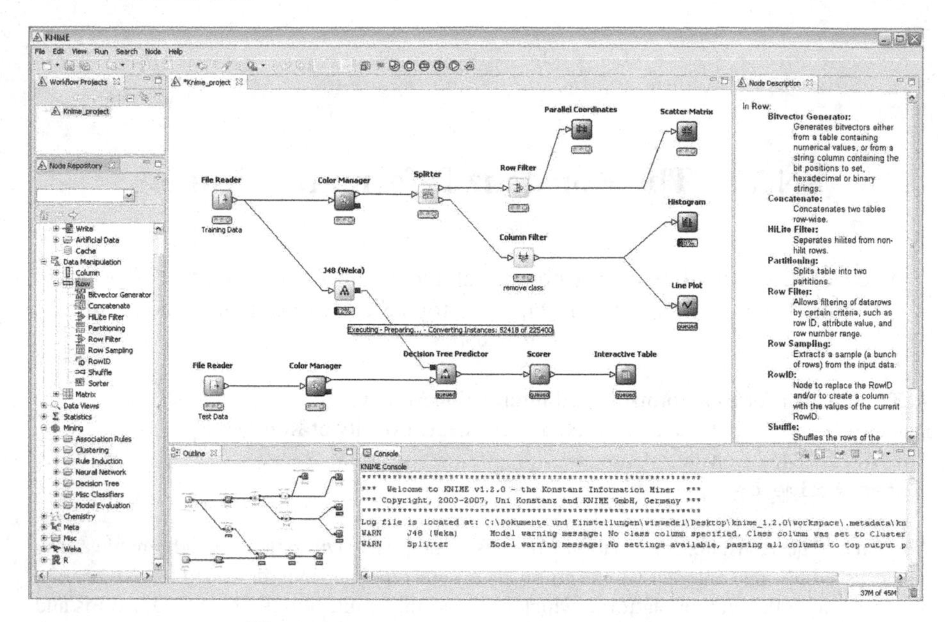

Fig. 1. An example analysis flow inside KNIME.

ability to have all views interact graphically (*visual brushing*) creates a powerful environment to visually explore the data sets at hand. KNIME is written in Java and its graphical workflow editor is implemented as an Eclipse (Eclipse Foundation (2007)) plug-in. It is easy to extend through an open API and a data abstraction framework, which allows for new nodes to be quickly added in a well-defined way.

In this paper we describe some of the internals of KNIME in more detail. More information as well as downloads can be found at http://www.knime.org.

2 Architecture

The architecture of KNIME was designed with three main principles in mind.

- Visual, interactive framework: Data flows should be combined by simple drag&drop from a variety of processing units. Customized applications can be modeled through individual data pipelines.
- Modularity: Processing units and data containers should not depend on each other in order to enable easy distribution of computation and allow for independent development of different algorithms. Data types are encapsulated, that is, no types are predefined, new types can easily be added bringing along type specific renderers and comparators. New types can be declared compatible to existing types.
- Easy expandability: It should be easy to add new processing nodes or views and distribute them through a simple plugin mechanism without the need for complicated install/deinstall procedures.

In order to achieve this, a data analysis process consists of a pipeline of nodes, connected by edges that transport either data or models. Each node processes the arriving data and/or model(s) and produces results on its outputs when requested. Figure 2 schematically illustrates this process. The type of processing ranges from basic data operations such as filtering or merging to simple statistical functions, such as computations of mean, standard deviation or linear regression coefficients to computation intensive data modeling operators (clustering, decision trees, neural networks, to name just a few). In addition, most of the modeling nodes allow for an interactive exploration of their results through accompanying views. In the following we will briefly describe the underlying schemata of data, node, workflow management and how the interactive views communicate.

2.1 Data structures

All data flowing between nodes is wrapped within a class called `DataTable`, which holds meta-information concerning the type of its columns in addition to the actual data. The data can be accessed by iterating over instances of `DataRow`. Each row contains a unique identifier (or primary key) and a specific number of `DataCell` objects, which hold the actual data. The reason to avoid access by Row ID or index is scalability, that is, the desire to be able to process large amounts of data and therefore not be forced to keep all of the rows in memory for fast random access. KNIME employs a powerful caching strategy which moves parts of a data table to the hard drive if it becomes too large. Figure 3 shows a UML diagram of the main underlying data structure.

2.2 Nodes

Nodes in KNIME are the most general processing units and usually resemble one node in the visual workflow representation. The class `Node` wraps all functionality and

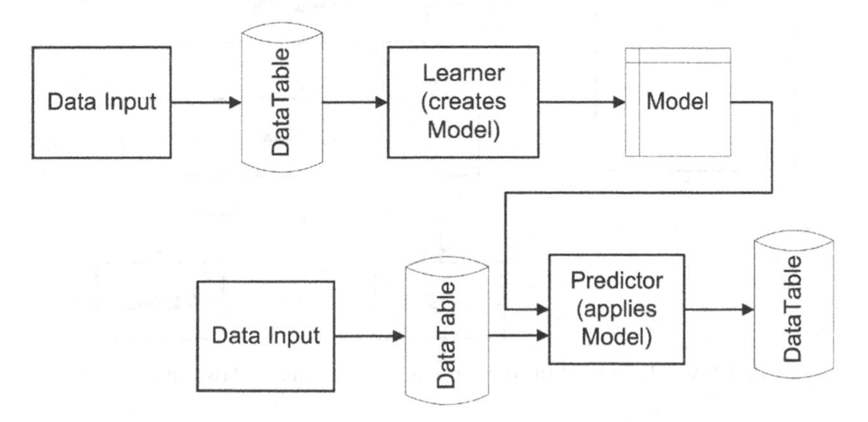

Fig. 2. A schematic for the flow of data and models in a KNIME workflow.

makes use of user defined implementations of a `NodeModel`, possibly a `NodeDialog`, and one or more `NodeView` instances if appropriate. Neither dialog nor view must be implemented if no user settings or views are needed. This schema follows the well-known Model-View-Controller design pattern. In addition, for the input and output connections, each node has a number of `Inport` and `Outport` instances, which can either transport data or models. Figure 4 shows a UML diagram of this structure.

2.3 Workflow management

Workflows in KNIME are essentially graphs connecting nodes, or more formally, a direct acyclic graph (DAG). The `WorkflowManager` allows to insert new nodes and to add directed edges (connections) between two nodes. It also keeps track of the status of nodes (configured, executed, ...) and returns, on demand, a pool of executable nodes. This way the surrounding framework can freely distribute the workload among a couple of parallel threads or – in the future – even a distributed cluster of servers. Thanks to the underlying graph structure, the workflow manager is able to determine all nodes required to be executed along the paths leading to the node the user actually wants to execute.

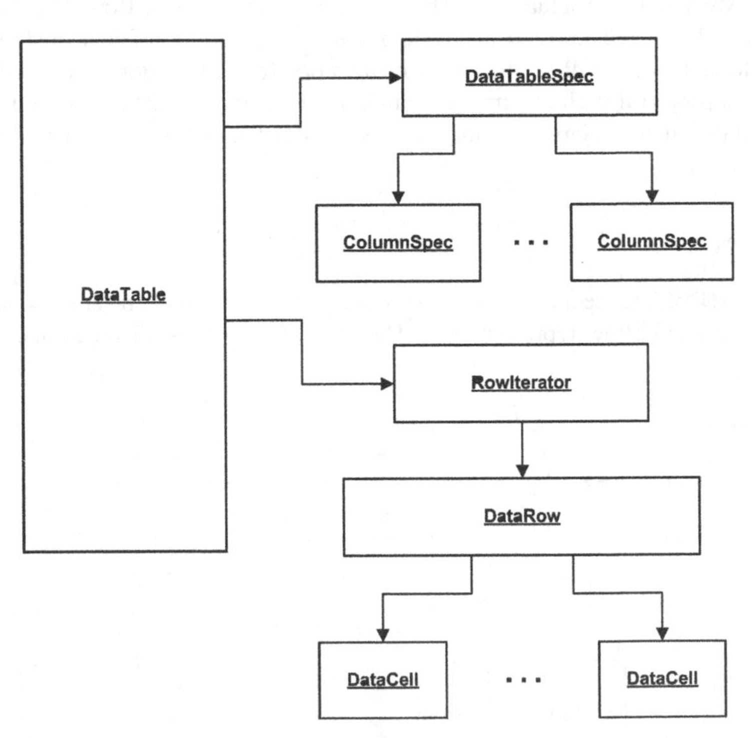

Fig. 3. A UML diagram of the data structure and the main classes it relies on.

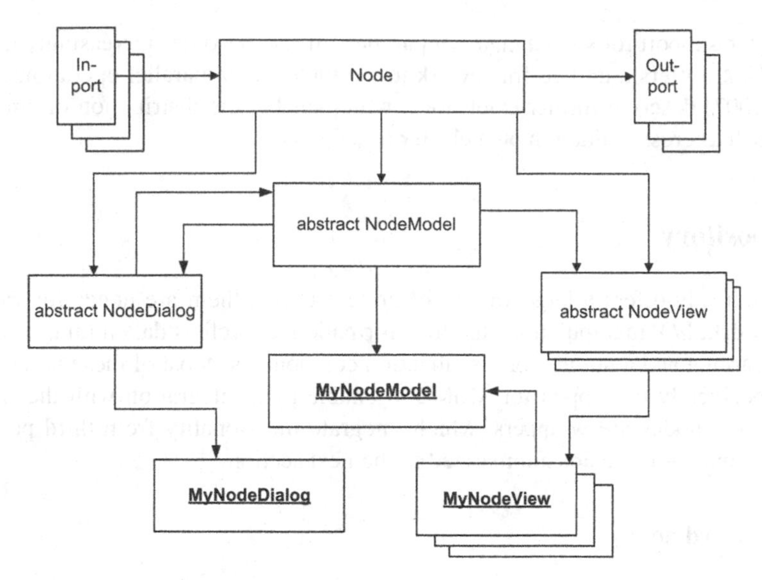

Fig. 4. A UML diagram of the Node and the main classes it relies on.

2.4 Views and interactive brushing

Each Node can have an arbitrary number of views associated with it. Through receiving events from a `HiLiteHandler` (and sending events to it) it is possible to mark selected points in such a view to enable visual brushing. Views can range from simple table views to more complex views on the underlying data (e. g. scatterplots, parallel coordinates) or the generated model (e. g. decision trees, rules).

2.5 Meta nodes

So-called *Meta Nodes* wrap a sub workflow into an encapsulating special node. This provides a series of advantages such as enabling the user to design much larger, more complex workflows and the encapsulation of specific actions. To this end some customized meta nodes are available, which allow for a repeated execution of the enclosed sub workflow, offering the ability to model more complex scenarios such as cross-validation, bagging and boosting, ensemble learning etc. Due to the modularity of KNIME, these techniques can then be applied virtually to any (learning) algorithm available in the repository.

Additionally, the concept of Meta Nodes helps to assign dedicated servers to this subflow or export the wrapped flow to other users as a predefined module.

2.6 Distributed processing

Due to the modular architecture it is easy to designate specific nodes to be run on separate machines. But to accommodate the increasing availability of multi-core machines, the support for shared memory parallelism also becomes increasingly important. KNIME offers a unified framework to parallelize data-parallel operations. Sieb et al. (2007) describe further extensions, which enable the distribution of complex tasks such as cross validation on a cluster or a GRID.

3 Repository

KNIME already offers a large variety of nodes, among them are nodes for various types of data I/O, manipulation, and transformation, as well as data mining and machine learning and a number of visualization components. Most of these nodes have been specifically developed for KNIME to enable tight integration with the framework; other nodes are wrappers, which integrate functionality from third party libraries. Some of these are summarized in the next section.

3.1 Standard nodes

- Data I/O: generic file reader, and reader for the attribute-relation file format (ARFF), database connector, CSV and ARFF writer, Excel spreadsheet writer
- Data manipulation: row and column filtering, data partitioning and sampling, sorting or random shuffling, data joiner and merger
- Data transformation: missing value replacer, matrix transposer, binners, nominal value generators
- Mining algorithms: clustering (k-means, sota, fuzzy c-means), decision tree, (fuzzy) rule induction, regression, subgroup and association rule mining, neural networks (probabilistic neural networks and multi-layer-perceptrons)
- Visualization: scatter plot, histogram, parallel coordinates, multidimensional scaling, rule plotters
- Misc: scripting nodes

3.2 External tools

KNIME integrates functionality of different open source projects that essentially cover all major areas of data analysis such as WEKA (Witten and Frank (2005)) for machine learning and data mining, the R environment (R Development core team (2007)) for statistical computations and graphics, and JFreeChart (Gilbert (2005)) for visualization.

- WEKA: essentially all algorithm implementations, for instance support vector machines, Bayes networks and Bayes classifier, decision tree learners
- R-project: console node to interactively execute R commands, basic R plotting node
- JFreeChart: various line, pie and histogram charts

The integration of these tools not only enriches the functionality available in KNIME but has also proven to be helpful to overcome compatibility limitations when the aim is on using these different libraries in a shared setup.

4 Extending KNIME

KNIME already includes plug-ins to incorporate existing data analysis tools. It is usually straightforward to create wrappers for external tools without having to modify these executables themselves. Adding new nodes to KNIME, also for native new operations, is easy. For this, one needs to extend three abstract classes:

- `NodeModel`: this class is responsible for the main computations. It requires to overwrite three main methods: `configure()`, `execute()`, and `reset()`. The first takes the meta information of the input tables and creates the definition of the output specification. The `execute` function performs the actual creation of the output data or models, and `reset` discards all intermediate results.
- `NodeDialog`: this class is used to specify the dialog that enables the user to adjust individual settings that affect the node's execution. A standardized set of `DefaultDialogComponent` objects allows the node developer to quickly create dialogs when only a few standard settings are needed.
- `NodeView`: this class can be extended multiple times to allow for different views onto the underlying model. Each view is automatically registered with a `HiLiteHandler` which sends events when other views have hilited points and allows to launch events in case points have been hilit inside this view.

In addition to the three model, dialog, and view classes the programmer also needs to provide a `NodeFactory`, which serves to create new instances of the above classes. The factory also provides names and other details such as the number of available views or a flag indicating absence or presence of a dialog.

A wizard integrated in the Eclipse-based development environment enables convenient generation of all required class bodies for a new node.

5 Conclusion

KNIME, the Konstanz Information Miner offers a modular framework, which provides a graphical workbench for visual assembly and interactive execution of data pipelines. It features a powerful and intuitive user interface, enables easy integration of new modules or nodes, and allows for interactive exploration of analysis results or trained models. In conjunction with the integration of powerful libraries such as the WEKA data mining toolkit and the R-statistics software, it constitutes a feature rich platform for various data analysis tasks.

KNIME is an open source project available at http://www.knime.org. The current release version 1.2.1 (as of 14 May 2007) has numerous improvements over the first public version released in July 2006. KNIME is actively maintained by a group of about 10 people and has more than 6 000 downloads so far. It is free for non-profit and academic use.

References

INFORSENSE LTD.: InforSense KDE. http://www.inforsense.com/kde.html.

INSIGHTFUL CORPORATION: Insightful Miner. http://www.insightful.com/products/iminer/default.asp.

SCITEGIC: Pipeline Pilot. http://www.scitegic.com/products/overview/.

ECLIPSE FOUNDATION (2007):*Eclipse 3.2 Documentation*. http://www.eclipse.org.

GILBERT, D. (2005): *JFreeChart Developer Guide*. Object Refinery Limited, Berkeley, California. http://www.jfree.org/jfreechart.

R DEVELOPMENT CORE TEAM (2007): *R: A language and environment for statistical computing*. R Foundation for Statistical Computing, Vienna, Austria. ISBN 3-900051-07-0. http://www.R-project.org.

SIEB C., MEINL T., and BERTHOLD, M. R. (2007): Parallel and distributed data pipelining with KNIME. *Mediterranean Journal of Computers and Networks, Special Issue on Data Mining Applications on Supercomputing and Grid Environments*. To appear.

WITTEN, I. H. and FRANK, E (2005): *Data Mining: Practical machine learning tools and techniques*. Morgan Kaufmann, San Francisco. http://www.cs.waikato.ac.nz/~ml/weka/index.html.

A Pattern Based Data Mining Approach

Boris Delibašić[1], Kathrin Kirchner[2] and Johannes Ruhland[2]

[1] University of Belgrade, Faculty of Organizational Sciences, Center for Business Decision-Making, 11000 Belgrade, Serbia
boris.delibasic@fon.bg.ac.yu
[2] Friedrich-Schiller-University,
Department of Business Information Systems, 07737 Jena, Germany
{k.kirchner, j.ruhland}@wiwi.uni-jena.de

Abstract. Most data mining systems follow a data flow and toolbox paradigm. While this modular approach delivers ultimate flexibility, it gives the user almost no guidance on the issue of choosing an efficient combination of algorithms in the current problem context. In the field of Software Engineering the Pattern Based development process has empirically proven its high potential. Patterns provide a broad and generic framework for the solution process in its entirety and are based on equally broad characteristics of the problem. Details of the individual steps are filled in at later stages. Basic research on pattern based thinking has provided us with a list of generally applicable and proven patterns. User interaction in a pattern based approach to data mining will be divided into two steps: (1) choosing a pattern from a generic list based an a handful of characteristics of the problem and later (2) filling in data mining algorithms for the subtasks.

1 Current situation in data mining

The current situation in the data mining area is characterized by a plethora of algorithms and variants. The well known WEKA collection (Witten and Frank (2005))implements approx. 100 different algorithms. However, there is little guidance in selecting and using the appropriate algorithm for the problem at hand as each algorithm may also have its very specific strengths and weaknesses.

As Figure 1 shows for large German companies, the most signifact problems in data mining are application issues and the management of the process as a whole and not the lack of algorithms (Hippner, Merzenich and Stolz (2002)). Standardizing the process as proposed by Fayyad et.al (1996) and later refined into the CRISP-DM model (Chapman et.al. (2000)) has resulted in a well established phase model with preprocessing, mining and postprocessing steps, but has failed to give hints for chosing a proper sequence of processing tools or avoidance of pitfalls.

Design has always elements of integrated and modular solutions. Integrated solutions provide us with simplicity, but the lack of adaptability. Modular solutions give us the ability to have greater influence on our solution, but ask for more knowledge

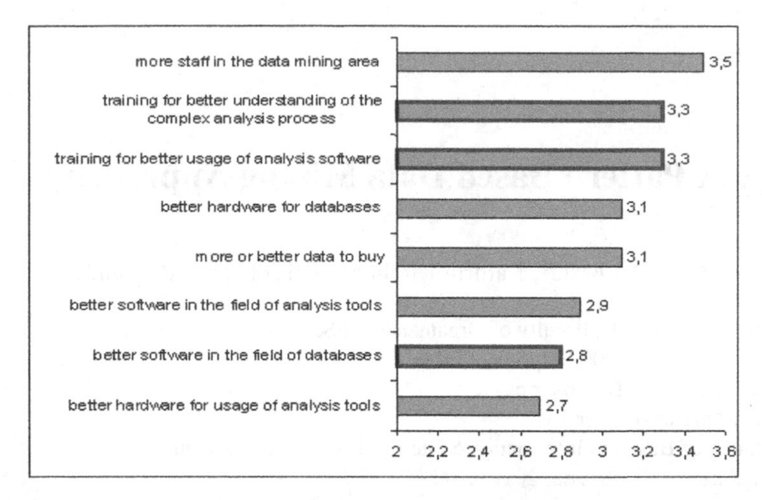

Fig. 1. Proposals for improvement of current data mining projects (46 questionnaires, average scores, 1 = no improvement, 5 = highly improvable)

and human attendance. In reality all solutions are between full modularity and full integrality (Eckert and Clarkson (2005)). We believe that for solving problems in the data mining area, it is more appropriate to use a modular solution, than an integrated one.

Patterns are meant to be experience packages that give a broad outline on how to solve specific aspects of complex problems. Complete solutions are built through chaining and nesting of patterns. Thus they go beyond the pure structuring goal. They have proven their potential in diverse fields of science.

2 Introduction to patterns

Patterns are already very popular in software design as the well known GOF-patterns for Object Oriented Design exemplify. (Gamma et.al. (1995)). Patterns we envisage are, however, applicable to a much wider context. With the development of pattern theories in various areas (architecture, IS, tele-communications, organization) it seems that also the problems of adaptability and maintenance of DM algorithms can be solved using patterns.

The protagonist of the pattern movement, Cristopher Alexander defines a pattern as a three-part rule that expresses the relation between a certain context, a problem and a solution. It is at the same time a thing that happens in the world (empirics) and a rule that tells us, how to create that thing (process rule) and when to create it (context specificity). It is at the same time a process, a description of a thing that is alive, and a process that generates that thing (Alexander (1979)). Alexander's work was concentrated in identifying patterns in architecture covering a broad range from urban planning to details of interior design. The patterns are shells, which allows various realizations, all of which will solve the problem.

Fig. 2. Small public square pattern

We shall illustrate the essence and power of C. Alexander style patterns by two examples. On Figure 2 a pattern named *Small public squares* is presented. Such squares enable people in large cities to gather, communicate and develop a community feeling. The core of the pattern is to make such squares not too large, lest they will be deserted and look strange to people.

Another example is shown on Figure 3. The pattern *Entrance transition* advocates and enables a smooth transition between the outdoor and indoor space in a house. People do not like instant transition. It makes them feel uncomfortable, and the house ugly.

Fig. 3. Entrance transition pattern

Alexander (2002b) says:

1. Patterns contain life.
2. Patterns support each other: the life and existence of one pattern influences the life and existence of another pattern.
3. Patterns are built of patterns, this way their composition can be explained.
4. The whole (the space in which patterns are implemented to) gets its life depending on the density and intensity of the patterns inside the whole.

We want to provide the user with the abilty to make data mining (DM) solutions by nesting and pipelining of patterns. In that way, the user will concentrate on the problems he wants to solve through the deployment of some key patterns. He may then nest patterns deep enough to get the job done at the data processing level. Current DM algorithms and DM process paradigms don't provide users with such an ability, as they are typically based on the data flow diagrams approach principle. A standard problem solution in the SPSS Clementine system is shown on Figure 4; it is a documentation of a chosen solution rather than a solution guide.

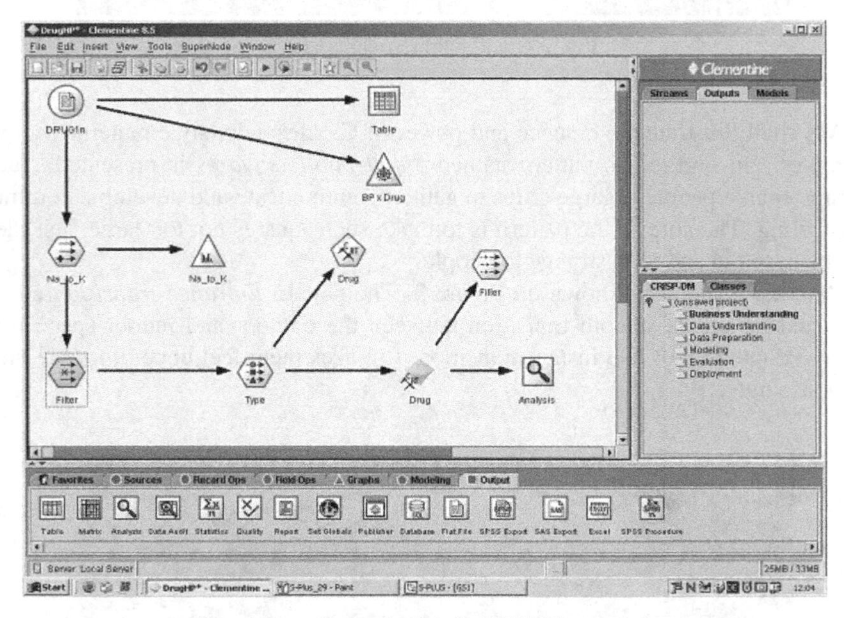

Fig. 4. Data flow principle in SPSS Clementine

3 Some data mining patterns

We have already developed some archetypical DM patterns. For their formal representation the J.O. Coplien Pattern Formalization Form has been used (Coplien and Zhao (2005), Coplien (1996), p 8). This form consists of the following elements: *Context*, *Problem*, *Forces*, *Solutions* and *Resulting context*.

A pattern is applicable within a *Context* (description of the world) and creates a *Resulting Context*, as the application of the Pattern will change the state space.

Problem describes what produces the uncomfortable feeling in a certain situation. *Forces* are keys for pattern understanding. Each force will yield a quality critereon for any solution, and as forces can be (and generally are) conflicting, the relative importance of forces will drive a good solution into certain areas of the solution space, hence their name. In many contexts, for instance, the relative importance of the conflicting forces of economic, time and quality considerations will render a particular solution a good or a bad compromise.

When a problem, forces as problem descriptors, are well understood, then a *solution* is most often easily evaluated. Understanding of a problem is crucial for finding a solution. Patterns are functions that transform problems in a certain context into solutions. Patterns are familiar and popular concepts, because they systematize repeatedly occuring solutions in nature. The solution, the pattern itself, resolves forces in a problem and provides a good solution. On the other hand, a pattern is always a compromise, it is not easy to recognize. Because it is a compromise it resolves some forces, but may add to the context space new ones.

A pattern is best recognized through solving and generalizing real problems. The quality and applicability of patterns may change over time as new forces gain relevance or new solutions become available. The process of recognizing and deploying patterns is continuous. For example, house building changed very much when concrete was invented.

3.1 The Condense Pattern: a popular DM pattern

The pattern is shown in the Coplien form is

1. *Context*: The collection of data is completed.
2. *Problem*: Data matrix is too large for efficient handling.
3. *selected Forces*: Efficiency of DM algorithms depends upon the number of cases and variables. Irrelevant cases and variables will hamper learning capabilities of DM algorithm. Leaving out a case or a variable may lead to errors and delete special, but important cases.
4. *Solution*: Condense the data matrix
5. *Resulting context*: manageable data matrix with some information loss

The *Condense* pattern is a typical preprocessing pattern that has found diverse applications, for example on variables (by – for example – calculating a score, choosing a representative variable or through clustering of variables), on cases (through sampling, clustering with subsequent use of centers only, etc) or in transformation of continuous variables (e.g. through equal width binning, equal frequency binning).

3.2 The Divide et Impera Pattern

A second pattern which is widely used in data mining is *Divide et Impera*. It can also be described in the Coplien pattern form:

1. *Context*: A data mining problem is too large/complicated to be solved in one step.
2. *Problem*: Structuring of the task

3. *Forces*: It is not possible to subdivide the problem, there are many strongly interrelated facets influencing the problem. The sheer combination of subproblem solutions may be grossly suboptimal. Subproblems may have very different relevance for the global problem. Complexity of a generated subproblem may be grossly out of proportion to its relevance. *Solution*: Divide the problem into subproblems that are more easily solved (and quite often structurally similar to the original one) and build the solution to the complete problem as a combination.
4. *Resulting context*: a set of smaller problems, more palatable to solution
   - It is possible that the problem structure is bad or the effort has not been reduced in sum.
   - The effort has not been reduced in sum.

The *Divide et Impera* pattern can be used for problem structuring where the problem is too complex to solve it in one step. It is found as a typical meta heuristic in many algorithms such as decision trees or divisive clustering. Other application areas, which also vouch for its broad applicability, are segmented marketing (if an across-the-board marketing strategy is not feasible, try to form homogeneous segments and cater to their needs), or the division of labor within divisional organizations.

3.3 More patterns in data mining

We have already identified a lot of other patterns in the field of data mining. Some of them are:

- Combine voting(with boosting, bagging, stacking, etc. as corresponding algorithms),
- Training / Retraining (supervised mining, etc.),
- Solution analysis,
- Categorization and
- Normalization

This list is in no way closed. Every area of human interest has its characteristic patterns. However, there is not an infinite number of patterns, but always a limited one. Collecting them and making them available for users gives the users the possibility to model the DM process, but also to understand the DM process through patterns.

4 Summary and outlook

Pattern based data-mining offers some attractive features

1. The algorithm creators and the algortihm users have different interests and different need. These sides often don't understand each others needs and, quite often, do not need to know about specific details relevant to the other side. A pattern is something that is understandable to all people.

2. Science converges. Concepts in one area of science is applicable in another area. Patterns support these processes. This potential is comparable to the promises of Systems Theory.
3. Decision for a specific algorithm can be postponed to later stages. A solution path as a whole will be sketched through patterns and algorithms need only be filled in immediately prior to processing. Using differnet algorithms in places will not invalidate the solution path, creating "late binding" at the algorithm level.

Current Data Mining applications occasionally provide the user with first traces of pattern based DM. Figure 5 shows the example of Bagging of Classifiers within the TANAGRA project and its graphical user interface (Rakotomalala (2004)). Bagging cannot be described with a pure data flow paradigm, rather a nesting of a classifier pattern within the bagging pattern is needed. This nested structure will then be pipelined with pre- and postprocessing patterns.

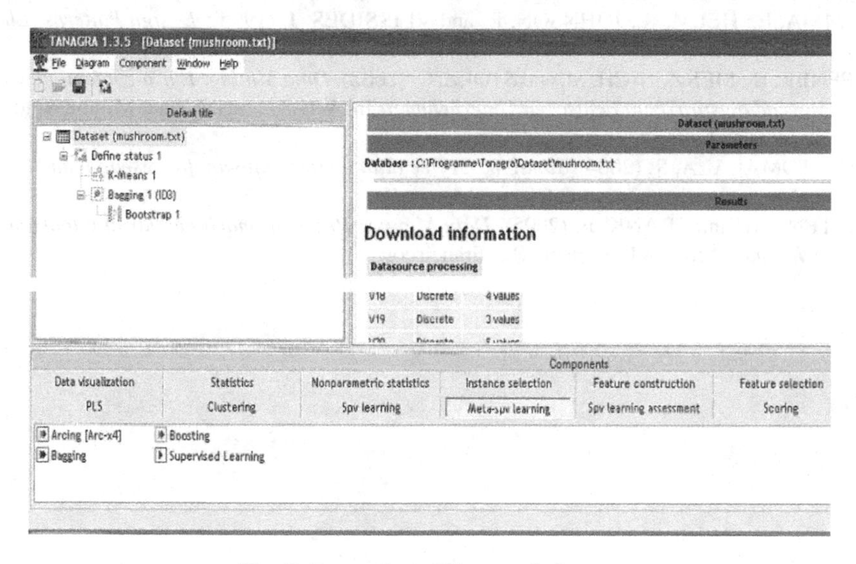

Fig. 5. Screenshot of Tanagra Software

Further steps in our project are to

- collect a list of patterns which are useful in the whole knowledge discovery process and data mining (list will be open-ended).
- integrate these patterns into data mining software to help design ad-hoc algorithms, choose an existing one or have guidance in the data mining process.
- develop a software prototype with our pattern and make experiments with users: how it works and what are the benefits.

References

ALEXANDER, C. (1979): *The Timeless Way of Building*, Oxford University Press.

ALEXANDER, C. (2002a): *The Nature of Order Book 1: The Phenomenon of Life*, The Center for Environmental Structure, Berkeley, California.

ALEXANDER, C. (2002b): *The Nature of Order Book 2: The Process of Creating Life*, The Center for Environmental Structure, Berkeley, California.

CHAPMAN, P., CLINTON, J., KERBER, R., KHABAZA, T., REINARTZ, T., SHEARER, C. and WIRTH, R. (2000): *CRISP-DM 1.0. Step-by-step data mining guide*, www.crisp-dm.org.

COPLIEN, J.O.(1996): *Software Patterns*, SIGS Books & Multimedia.

COPLIEN, J.O. and ZHAO, L. (2005): *Toward a General Formal Foundation of Design - Symmetry and Broken Symmetry*, Brussels: VUB Press.

ECKERT, C. and CLARKSON, J. (2005): *Design Process Improvement: a review of current practice*, Springer Verlag London.

FAYYAD, U.M., PIATETSKY-SHAPIRO, G. and UTHURUSAMY, R. (Ed.) (1996): *Advances in Knowledge Discovery and Data Mining*, MIT Press.

GAMMA, E., HELM, R., JOHNSON, R. and VLISSIDES, J. (1995): *Design Patterns. Elements of Reusable Object-Oriented Software*, Addison-Wesley.

HIPPNER, H., MERZENICH, M. and STOLZ, C. (2002): *Data Mining: Einsatzpotentiale und Anwendungspraxis in deutschen Unternehmen*, In: WILDE, K.D.: Data Mining Studie, absatzwirtschaft.

RAKOTOMALALA, R. (2004): *Tanagra – A free data mining software for research and education*, www.eric.univ-lyon2.fr/~rico/tanagra/.

WITTEN, I.H. and FRANK, E. (2005): *Data Mining: Practical machine learning tools and techniques*, Morgan Kaufmann, San Francisco.

A Framework for Statistical Entity Identification in R

Michaela Denk

EC3 – E-Commerce Competence Center,
Donau-City-Str. 1, 1220 Vienna, Austria
michaela.denk@ec3.at

Abstract. Entity identification deals with matching records from different datasets or within one dataset that represent the same real-world entity when unique identifiers are not available. Enabling data integration at record level as well as the detection of duplicates, entity identification plays a major role in data preprocessing, especially concerning data quality. This paper presents a framework for statistical entity identification in particular focusing on probabilistic record linkage and string matching and its implementation in R. According to the stages of the entity identification process, the framework is structured into seven core components: data preparation, candidate selection, comparison, scoring, classification, decision, and evaluation. Samples of real-world CRM datasets serve as illustrative examples.

1 Introduction

Ensuring data quality is a crucial challenge in statistical data management aiming at improved usability and reliability of the data. Entity identification deals with matching records from different datasets or within a single dataset that represent the same real-world entity and, thus, enables data integration at record level as well as the detection of duplicates. Both can be regarded as a means of improving data quality, the former by completing datasets through adding supplementary variables, replacing missing or invalid values, and appending records for additional real-world entities, the latter by resolving data inconsistencies. Unless sophisticated methods are applied, data integration is also a potential source of 'dirty' data: duplicate or incomplete records might be introduced. Besides its contribution to data quality, entity identification is regarded as a means of increasing the efficiency of the usage of available data as well. This is of particular interest in official statistics, where the reduction of the responder burden is a prevailing issue. In general, applications necessitating statistical entity identification (SEI) are found in diverse fields such as data mining, customer relationship management (CRM), bioinformatics, criminal investigations, and official statistics. Various frameworks for entity identification have been proposed (see for example Denk (2006) or Neiling (2004) for an overview), most of them concentrating on particular stages of the process, such the author's

SAS implementation of a metadata framework for record linkage procedures (Denk (2002)). Moreover, commercial as well as 'governmental' software (especially from national statistical institutes) is available (for a survey cf. Herzog et al. (2007) or Gill (2001)).

Based on the insights gained from the EU FP5 research project DIECOFIS (Denk et al. (2004 & 2005)) in the context of the integration of enterprise data sources, a framework for statistical entity identification has been designed (Denk (2006)) and implemented in the free software environment for statistical computing R (R Development Core Team (2006)). Section 2 provides an overview of the underlying methodological framework, Section 3 introduces its implementation. The functionality of the framework components is discussed and illustrated by means of demo samples of real-world CRM data. Section 4 concludes with a short summary and an outlook on future work.

2 Methodological framework

Statistical entity identification aims at finding a classification rule assigning each pair of records from the original dataset(s) to the set of links (identical entities or duplicates) or the set of non-links (distinct entities), respectively. Frequently, a third class is introduced containing undetermined record pairs (possible links/duplicates) for which the final linkage status can only be set by using supplementary information (usually obtained via clerical review). The process of deriving such a classification rule can be structured into seven stages (Denk (2006)). In the initial *data preparation* stage, *matching variables* are defined and undergo various transformations to become suitable for the usage in the ensuing processing stages. In particular, string variables have to be preprocessed to become comparable among datasets (Winkler (1994)). In the *candidate selection* or *filtering* stage, candidate record pairs with a higher likelihood of representing identical real-world entities are selected (Baxter et al. (2003)), since a detailed comparison, scoring, and classification of all possible record pairs from the cross product of the original datasets is extremely time-consuming (if accomplishable at all). In the third stage, the *comparison* or *profiling* stage, similarity profiles are determined which consist of compliance measures of the records in a candidate pair with respect to the specified matching variables, in which the treatment of string variables (Navarro (2001)) and missing values is the most challenging (Neiling (2004)). Based on the similarity patterns, the *scoring* stage estimates matching scores for the candidate record pairs. In general, matching scores are defined as ratios of the conditional probabilities of observing a particular similarity pattern provided that the record pair is a true match or non-match respectively, or as the binary or natural logarithm thereof (Fellegi and Sunter (1969)). The conditional probabilities are estimated via the classical EM algorithm (Dempster et al. (1977)) or one of the problem-specific EM variants (Winkler (1994)). In the ensuing *classification* stage classification rules are determined. Especially in the record linkage setting, rules are based on prespecified error levels for erroneous links and non-links through two score thresholds that can be directly obtained from the estimated conditional probabilities (Fellegi and Sunter (1969)) or via comparable training data with known true matching status (Belin and Rubin (1995)). In the *decision* stage, examined record pairs are finally assigned to the set of links or non-links and inconsistent values of linked records with respect to common variables are resolved. If 1:n or 1:1 assignment of records is targeted, the m:n assignment resulting from the classification stage has to be refined (Jaro (1989)). The seventh and final stage focuses on the *evaluation* of the entity identification process. Training data (e.g. from previous studies or from a sample for which the true matching status has been determined) are required to provide sound estimates of quality measures. A contingency table of the true versus the estimated matching status is used as a basis for the calculation of misclassification rates and other overall quality criteria, such as precision and recall, which can be visualized for varying score thresholds.

3 Implementation

The SEI framework is structured according to the seven stages of the statistical entity identification process. For each stage there is one component, i.e. one function, that establishes an interface to the lower-level functions which implement the respective methods. The outcome of each stage is a list containing the processed data and protocols of the completed processing stages. Table 1 provides an overview of the functionality of the components and the spectrum of available methods. Methods not yet implemented are *italicised.*

3.1 Sample data

As an illustrative example, samples of real-life CRM datasets are used originating from a register of casino customers and their visits (approx. 150,000) and a survey on customer satisfaction. Common (and thus potential matching) variables are first and last name, sex, age group, country, region, and five variables related to previous visits and the playing behaviour of the customers (visit1, visit2, visit3, visit4, and lottery). The demo datasets correspond to a sample of 100 survey records for which the visitor ID is also known and 100 register entries from which 70 match the survey sample and the remaining 30 were drawn at random. I.e., the true matching status of all 10,000 record pairs is known. The data snippet shows a small subset of the first dataset.

```
       fname    sex agegroup country visit1 visit2 ...
711    GERALD   m   41-50    Austria 1      1
13     PAOLO    m   41-50    Italy   1      1
164988 WEIFENG  m   19-30    other   0      1
```

3.2 Data preparation

`preparation(data, variable, method, label, ...)` provides an interface to the `phoncode()` function from the STRINGMATCH toolbox (Denk (2007)) as well as

Table 1. Component Functionality and Methodological Range

Component	Functionality	Methods
Preparation	parsing	*address and name parsing in different languages*
	standardisation	dictionary provided by the user *integrated dictionaries*
	phonetic coding	American Soundex, Original Russel Soundex NYSIIS, ONCA, Daitch-Mokotoff, Koelner Phonetik, Reth-Schek-Phonetik *(Double) Metaphone, Phonex, Phonet, Henry*
Filtering	single-pass	cross product / no selection, blocking, sorted neighbourhood, string ranking *hybrid*
	multi-pass	*sequence of single-pass*
Comparison	universal	binary, frequency-based
	metric variables	tolerance intervals, (absolute distance)^p, Canberra
	string variables: phonetic coding	see above
	string variables: token-based	Jaccard, n -gram, maximal match, longest common subsequence, TF-IDF
	string variables: edit distances	*Damerau-Levenstein, Hamming, Needleman-Wunsch, Monge-Elkan, Smith-Waterman*
	string variables: Jaro algorithms	Jaro, Jaro-Winkler Jaro-McLaughlin, Jaro-Lynch
Scoring	binary outcomes	two-class EM *two-class EM interactions, three-class EM*
	frequency based	*Fellegi-Sunter, two-class EM frequency based*
	similarities	*two-class EM approximate*
	any	*logistic regression*
Classification	no training data	Fellegi-Sunter empiric, Fellegi-Sunter pattern
	training data	*Belin-Rubin*
Decision	assignment	greedy *LSAP*
	review	*possible links, inconsistent values*
Evaluation	confusion matrix	absolute, relative
	quality measures	false match rate Fellegi-Sunter & Belin-Rubin, false non-match rate Fellegi-Sunter & Belin-Rubin, accuracy, precision, recall, f-measure, specificity, unclassified pairs
	plots	*varying classification rules*

the functions `standardise()` and `parse()`. By this means, `preparation()` phonetically codes, standardizes, or parses the `variable`(s) in data frame `data` according to the specified `method`(s) (default: American Soundex (`'asoundex'`)) and appends the resulting variable(s) with the defined `label`(s) to the `data`. The default label is composed of the specified variables and methods. At the moment, a selection of popular phonetic coding algorithms and standardization with user-provided dictionaries are implemented, whereas parsing is not yet supported. The ellipsis indicates additional method-specific arguments, e.g., the dictionary according to which standardisation should be carried out. The following code chunk illustrates the usage of the function.

```
> preparation(data=d1, variable='lname',
method='asoundex')

       ... lname        ... asoundex.lname
115256 ... WESTERHEIDE  ... W236
200001 ... BESTEWEIDE   ... B233
200002 ... WESTERWELLE  ... W236
```

3.3 Candidate selection

`candidates(data1, data2, method, selvars1, selvars2, key1, key2, ...)` provides an interface to the functions `crossproduct()`, `blocking()`, `sortedneighbour()`, and `stringranking()`. Candidate record pairs from data frames `data1` and `data2` are created and filtered according to the specified `method` (default: `'blocking'`). In case of a deduplication scenario, data2 does not have to be specified. `selvars1` and `selvars2` specify the variables that the filtering is based on. The ellipsis indicates additional method-specific arguments, e.g. the extent `k` of the neighbourhood for sorted neighbourhood filtering or the string similarity measure to be used for string ranking. The following examples illustrate the usage of the function. In contrast to the full cross product of the datasets with 10,000 record pairs, sorted neighbourhood by region, age group, and sex reduces the list of candidate pairs to 1,024, and blocking by Soundex code of last name retains only 83 candidates.

```
> candidates(data1=d1.prep, data2=d2.prep,
method='blocking',selvars1='asoundex.lname')
> candidates(data1=d1.prep, data2=d2.prep,
method='sorted', selvars1=c
('region','agegroup','sex'), k=10)
```

3.4 Comparison

`comparison(data, matchvar1, matchvar2, method, label, ...)` makes use of the `stringsim()` function from the STRINGMATCH toolbox (Denk (2007)) as well as the functions `simplecomp()` for simple (dis-)agreement and `metcomp()` for similarities of metric variables. `comparison()` computes the similarity profiles for the candidate pairs in data frame `data` with respect to the specified matching variable(s) `matchvar1`, `matchvar2` according to the selected `method` and appends the resulting variable(s) with the defined `label`(s) to `data`. The ellipsis indicates additional method-specific arguments, e.g. different types of weights for Jaro or edit distance algorithms. In the current implementation, missing values are not specially treated.

```
> comparison(data=d12, matchvar1=c('fname.d1',
 'lname.d1','visit1.d1'), matchvar1=c('fname.d1',
'lname.d1','visit1.d1'),
method=c('jaro','asoundex','simple'))
```

```
  fname.d1  fname.d2 ... jaro.fname c.asound.lname simple.visit1
1 GERALD    SELJAMI  ... 0.53175    0.00000        0.00000
2 PAOLO     SELJAMI  ... 0.39524    0.00000        0.00000
3 WEIFENG   SELJAMI  ... 0.42857    0.00000        1.00000
```

3.5 Scoring

`scoring(data, profile, method, label, wtype, ...)` estimates matching scores for the candidate pairs in data frame `data` from the specified similarity `profile` according to the selected `method` and appends the resulting variable with the defined `label` to the `data`. `wtype` indicates the score to be computed, e.g. `'LR'` for likelihood ratio (default). The ellipsis indicates additional method-specific arguments, for example the maximum number of iterations for the EM algorithm. The following example illustrates the usage of the function. The output is shown together with the output of `classification()` and `decision()` in section 3.7.

```
> scoring(data=d12, profile=31:39, method='EM01',
 wtype='LR')
```

3.6 Classification

`classification(data, scorevar, method, mu, lambda, label, ...)` determines a classification rule for the candidate pairs in data frame `data` according to the selected `method` (default: empirical Fellegi-Sunter) based on prespecified error levels `mu` and `lambda` and the matching score in `scorevar`. The estimated matching status is appended to the `data` as a variable with the defined `label`. The ellipsis indicates additional method-specific arguments, for instance a data frame holding the `training` data and the position or label of the true matching status `trainstatus`. The following example illustrates the usage of the function. The result is shown in section 3.7.

```
> classification(data=d12, scorevar='score.EM01',
 method='FSemp')
```

3.7 Decision

`decision(data, keys, scorevar, classvar, atype, method, label, ...)` provides an interface to the function `assignment()` that enables 1:1, 1:n/n:1 and particular m:n assignments of the examined records. Eventually, features supporting the review of undetermined record pairs and inconsistent values in linked pairs are intended. decision() comes to a final decision concerning the matching status of the record pairs in data frame data based on the preliminary classification in classvar, the matching score scorevar, and the specified method (default: 'greedy'). keys specifies the positions or labels of the key variables referring to the records from the original data frames. atype specifies the target type of assignment (default: '1:1'). A variable with the defined label is appended to the data. The ellipsis indicates additional method-specific arguments not yet determined. The following example illustrates the usage of the function. In this case, 60 pairs first classified as links as well as all 112 possible links were transferred to the class of non-links.

```
> decision(data=d12, keys=1:2, scorevar='score.EM01',
classvar='class.FSemp', atype='1:1', method='greedy')

  fname.d1  fname.d2 ... score.EM01 class.FSemp 1:1.greedy
1 GERALD    SELJAMI  ... 6.848e-03  L           N
2 PAOLO     SELJAMI  ... 1.709e-04  P           N
3 WEIFENG   SELJAMI  ... 1.709e-05  P           N
```

3.8 Evaluation

evaluation(data, true, estimated, basis, plot, xaxes, yaxes, ...) computes the confusion matrix and various quality measures, e.g. false match and non-match rates, recall, precision, for the given data frame data containing the candidate record pairs with the estimated and true matching status. basis discerns whether the confusion matrix and quality measures should be based on the number of 'pairs' (default) or the number of 'records'. plot is a flag indicating whether a plot of two quality measures xaxes and yaxes, typically precision and recall, should be created (default: FALSE). The ellipsis indicates additional method-specific arguments not yet determined. The following example illustrates the usage of the function.

```
> evaluation(data=d12, true='true',
 estimated='1:1.greedy')
```

4 Conclusion and future work

The SEI framework introduced in this paper poses a considerable step towards statistical entity identification in R. It consists of seven components according to the stages of the entity identification process, viz. the preparation of matching variables, the selection of candidate record pairs, the creation of similarity patterns, the estimation of matching scores, the (preliminary) classification of record pairs into links, non-links, and possible links, the final decision on the classification and on inconsistent values in linked records, and the evaluation of the results. The projected and current range of functionality of the framework were presented. Future work consists in the explicit provision for missing values in the framework as well as the implementation of additional algorithms for the most components. The main focus is on further scoring and classification algorithms that significantly contribute to the completion of the framework which will finally be provided as an R package.

References

BAXTER, R., CHRISTEN, P., and CHURCHES, T. (2003): A Comparison of Fast Blocking Methods for Record Linkage. In: *Proc. 1st Workshop on Data Cleaning, Record Linkage, and Object Consolidation, 9th ACM SIGKDD.* Washington, D.C., August 2003.

BELIN, T.R. and RUBIN, D.B. (1995): A Method for Calibrating False-Match Rates in Record Linkage. *J. American Statistical Association, 90, 694–707.*

DEMPSTER, A.P., LAIRD, N.M. and RUBIN, D.B. (1977): Maximum Likelihood from Incomplete Data via the EM-Algorithm. *J. Royal Statistical Society (B), 39, 1–38.*

DENK, M. (2002): Statistical Data Combination: A Metadata Framework for Record Linkage Procedures. Doctoral thesis, Dept. of Statistics, University of Vienna.

DENK, M. (2006): A Framework for Statistical Entity Identification to Enhance Data Quality. Report wp6dBiz14_br1. (EC3, Vienna, Austria). Submitted.

DENK, M. (2007): The StringMatch Toolbox: Determining String Compliance in R. In: *Proc. IASC 07 – Statistics for Data Mining, Learning and Knowledge Extraction.* Aveiro, Portugal, August 2007. Accepted.

DENK, M., FROESCHL, K.A., HACKL, P. and RAINER, N. (Eds.) (2004): *Special Issue on Data Integration and Record Matching, Austrian J. Statistics, 33.*

DENK, M., HACKL, P. and RAINER, N. (2005): String Matching Techniques: An Empirical Assessment Based on Statistics Austria's Business Register. *Austrian J. Statistics, 34(3), 235–250.*

FELLEGI, I.P. and SUNTER, A.B. (1969): A Theory for Record Linkage. *J. American Statistical Association, 64, 1183–1210.*

GILL, L.E. (2001): Methods for automatic record matching and linking in their use in National Statistics. GSS Methodology Series, NSMS25, ONS UK.

HERZOG, T.N., SCHEUREN, F.J. and WINKLER, W.E. (2007): *Data Quality and Record Linkage Techniques.* Springer, New York.

JARO, M.A. (1989): Advances in Record-Linkage Methodology as Applied to Matching the 1985 Census of Tampa, Florida. *Journal of the American Statistical Association, 84, 414–420.*

NAVARRO, G. (2001): A guided tour to approximate string matching. *ACM Computing Surveys, 33(1), 31–88.*

NEILING, M. (2004): Identifizierung von Realwelt-Objekten in multiplen Datenbanken. Doctoral thesis, TU Cottbus. In German.

R DEVELOPMENT CORE TEAM (2006): *R: A language and environment for statistical computing.* R Foundation for Statistical Computing, Vienna, Austria.

WINKLER, W.E. (1994): Advanced Methods for Record Linkage. In: *Proc. Section on Survey Research Methods.* American Statistical Association, 467–472.

Combining Several SOM Approaches in Data Mining: Application to ADSL Customer Behaviours Analysis

Francoise Fessant, Vincent Lemaire, Fabrice Clérot

R&D France Telecom, 22307 Lannion, France
{francoise.fessant,vincent.lemaire,fabrice.clerot}@orange-ftgroup.com

Abstract. The very rapid adoption of new applications by some segments of the ADSL customers may have a strong impact on the quality of service delivered to all customers. This makes the segmentation of ADSL customers according to their network usage a critical step both for a better understanding of the market and for the prediction and dimensioning of the network. Relying on a "bandwidth only" perspective to characterize network customer behaviour does not allow the discovery of usage patterns in terms of applications. In this paper, we shall describe how data mining techniques applied to network measurement data can help to extract some qualitative and quantitative knowledge.

1 Introduction

Broadband access for home users and small or medium business and especially ADSL (Asymmetric Digital Subscriber Line) access is of vital importance for telecommunication companies, since it allows them to leverage their copper infrastructure so as to offer new value-added broadband services to their customers. The market for broadband access has several strong characteristics:

- there is a strong competition between the various actors,
- although the market is now very rapidly increasing, customer retention is important because of high acquisition costs,
- new applications or services may be picked up very fast by some segments of the customers and the behaviour of these applications or services may have a very strong impact on the quality of service delivered to all customers (and not only those using these new applications or services).

Two well-known examples of new applications or services with possibly very demanding requirements in term of bandwidth are peer-to-peer file exchange systems and audio or video streaming.

The above characteristics explain the importance of an accurate understanding of the customer behaviour and a better knowledge of the usage of broadband access. The notion of "usage" is slowly shifting from a "bandwidth only" perspective to a much broader perspective which involves the discovery of usage patterns in terms of applications or services. The knowledge of such patterns is expected to give a much better understanding of the market and to help anticipate the adoption of new services or applications by some segments and allow the deployment of new resources before the new usage effects hit all the customers.

Usage patterns are most often inferred from polls and interviews which allow an in-depth understanding but are difficult to perform routinely, suffer from the small size of the sampled population and cannot easily be extended to the whole population or correlated with measurements (Anderson et al. (2002)). "Bandwidth only" measurements are performed routinely on a very large scale by telecommunication companies (Clement et al. (2002)) but do not allow much insight into the usage patterns since the volumes generated by different applications can span many orders of magnitude.

In this paper, we report another approach to the discovery of broadband customers' usage patterns by directly mining network measurement data. After a description of the data used in the study and their acquisition process, we explain the main steps of the data mining process and we illustrate the ability of our approach to give an accurate insight in terms of usages patterns of applications or services while being highly scalable and deployable. We focus on two aspects of customers' usages: usage of types of applications and customers' daily traffic; these analyses suppose to observe the data at several levels of detail.

2 Network measurements and data description

2.1 Probes measurements

The network measurements are performed on ADSL customer traffic by means of a proprietary network probe working at the SDH (Synchronous Digital Hierarchy) level between the Broadband Access Server (BAS) and the Digital Subscriber Line Access Multiplexer (DSLAM). This on-line probe allows to read and store all the relevant fields of the ATM (Asynchronous Transfer Mode) cells and of the IP/TCP headers. From now, 9 probes equip the network; they observe about 18000 customers non-stop (a probe can observe about 2000 customers on a physical link). Once the probe is in place, data collection is performed automatically. A detailed description of the probe architecture can be found in (Francois (2002)).

2.2 Data description

For the study reported here, we gathered one month of data, on one site, for about two thousand customers. The data give the volumes of data exchanged in the upstream and downstream directions of twelve types of applications (web, peer-to-peer, ftp, news, mail, db, control, games, streaming, chat, others and unknown) sampled for each 6 minutes window for each customer. Most of the types of applications correspond to a group of well-known TCP ports, except the last two which relate to some well known but “obscure" ports (others) or dynamic ones (unknown). Since much of peer-to-peer traffic uses dynamic ports, peer-to-peer applications are recognized from a list of application names by scanning the payloads at the application level and not by relying on the well-known ports only. This is done transparently for the customers; no other use is made of such data than statistical analysis.

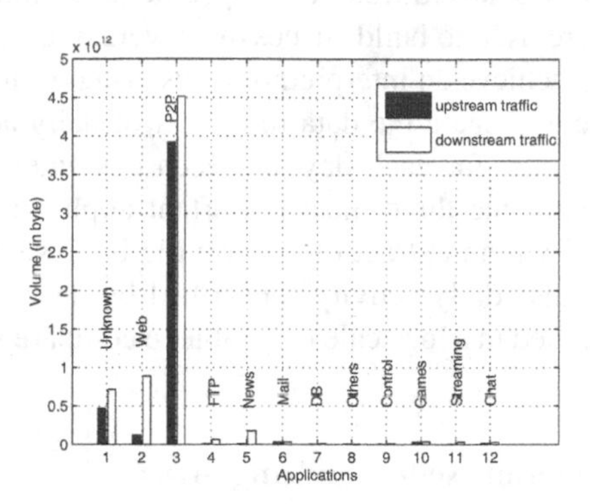

Fig. 1. Volume of the traffic on the applications

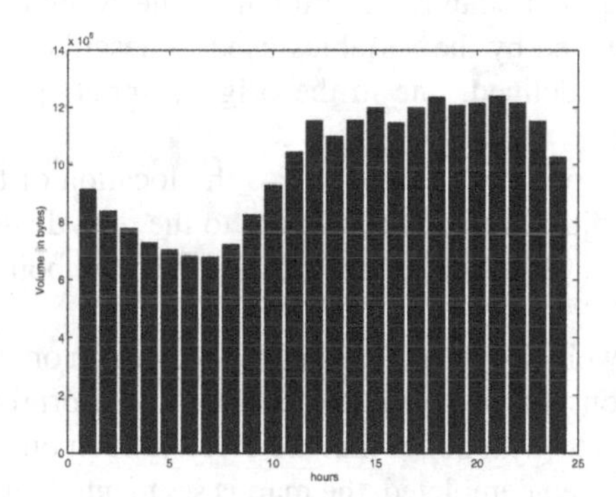

Fig. 2. Average hourly volume

Figure 1 plots the distribution of the total monthly traffic on the applications (all days and customers included) for one site in September 2003 (the volumes are given in bytes). About 90 percent of the traffic is due to peer-to-peer, web and unknown applications and all the monitored sites show a similar distribution. Figure 2 plots the average hourly volume for the same month and the same site, irrespective of the applications. We can observe that the night traffic remains significant.

3 Customer segmentation

3.1 Motivation

The motivation of this study is a better understanding of the customers' daily traffic on the applications. We try to answer the question: **who is doing what and when?**

To achieve this task we have developed a specific data mining process based on Kohonen maps. They are used to build successive layers of abstraction starting from low level traffic data to achieve an interpretable clustering of the customers.

For one month, we aggregate the data into a set of daily activity profiles given by the total hourly volume, for each day and each customer, on each application (we confined ourselves to the three most important applications in volume: peer-to-peer, web and unknown; an extract of the log file is presented Figure 3). In the following, "usage" means "daily activity" described by hourly volumes. The daily activity profiles are recoded in a log scale to be able to compare volumes with various orders of magnitude.

3.2 Data segmentation using self-organizing maps

We choose to cluster our data with a Self Organizing Map (SOM) which is an excellent tool for data survey because it has prominent visualization properties. A SOM is a set of nodes organized into a 2-dimensional[1] grid (the map). Each node has fixed coordinates in the map and adaptive coordinates (the weights) in the input space. The input space is spanned by the variables used to describe the observations. Two Euclidian distances are defined, one in the original input space and one in the 2-dimensional space.

The self-organizing process slightly moves the location of the nodes in the data definition space -i.e. adjusts weights according to the data distribution. This weight adjustment is performed while taking into account the neighbouring relation between nodes in the map.

The SOM has the well-known ability that the projection on the map preserves the proximities: observations that are close to each other in the original multidimensional input space are associated with nodes that are close to each other on the map.

After learning has been completed, the map is segmented into clusters, each cluster being formed of nodes with similar behaviour, with a hierarchical agglomerative clustering algorithm. This segmentation simplifies the quantitative analysis of the map (Vesanto and Alhoniemi (2000), Lemaire and Clérot (2005)). For a complete description of the SOM properties and some applications, see (Kohonen (2001)) and (Oja and Kaski (1999)).

3.3 An approach in several steps for the segmentation of customers

We have developed a multi-level exploratory data analysis approach based on SOM. Our approach is organized in five steps (see Figure 6):

[1] All the SOMs in this article are square maps with hexagonal neighborhoods.

• In a first step, we analyze each application separately. We cluster the set of all the daily activity profiles (irrespective of the customers) by application. For example, if we are interested in a classification of web down daily traffic, we only select the relevant lines in the log file (Figure 3) and we cluster the set of all the daily activity profiles for the application. We obtained a map with a limited number of clusters (Figure 4): the typical days for the application. We proceed in the same way for all the other applications.

As a result we end up, for each application, with a set of "typical application days" profiles which allow us to understand how the customers are globally using their broadband access along the day, for this application. Such "typical application days" form the basis of all subsequent analysis and interpretations.

client	day	application	volume
client 1	day 1	unknown-up	volume-day-unknown-up-11
client 1	day 1	P2P-up	volume-day-P2P-up-11
client 1	day 2	unknown-up	volume-day-unknown-up-12
...	...	...	...
client 2	day 1	web-down	volume-day-web-down-21
client 2	day 3	unknown-up	volume-day-unknown-up-23
client 2	day 3	web-up	volume-day-web-up-23
client 2	day 3	web-down	volume-day-web-down-23
client 2	day 5	P2P-down	volume-day-P2P-down-25
...	...	...	...

Fig. 3. log file : each application volume (last column) is a curve similar to the one plotted Figure 2

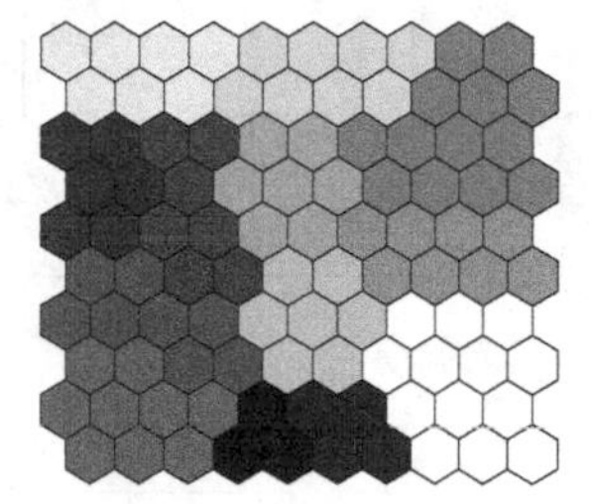

Fig. 4. Typical Web-down days

• In a second step we gather the results of previous segmentations to form a global daily activity profile: for one given day, the initial traffic profile for an application is replaced by a vector with as many dimensions as segments of typical days obtained previously for this application.

The profile is attributed to its cluster; all the components are set to zero except the one associated with the represented segment (Figure 5). This component is set to one. We do the same for the other applications. The binary profiles are then concatenated to form the global daily activity profile (the applications are correlated at this level for the day).

• In a third step, we cluster the set of all these daily activity profiles (irrespective of the customers). As a result we end up with a limited number of "typical day" profiles which summarize the daily activity profiles. They show how the three applications are simultaneously used in a day.

• In a fourth step, we turn to individual customers described by their own set of daily profiles. Each daily profile of a customer is attributed to its "typical day" cluster and we characterize this customer by a profile which gives the proportion of days spent in each "typical day" for the month.

• In a fifth step, we cluster the customers as described by the above activity profiles and end up with "typical customers". This last clustering allows to link cus-

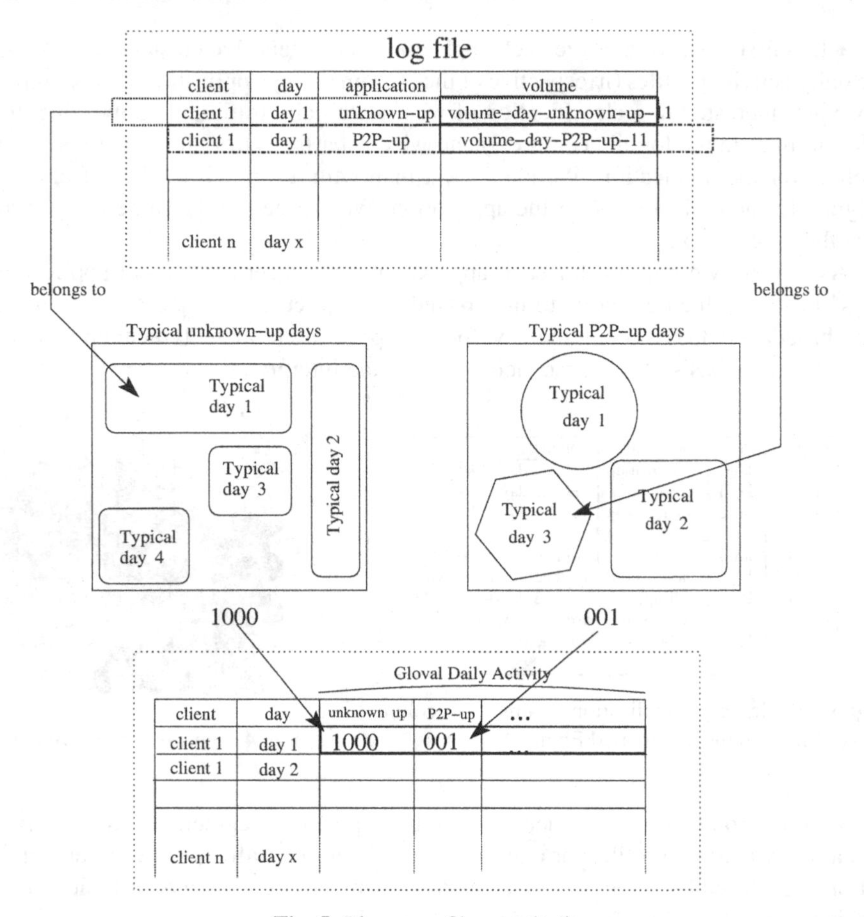

Fig. 5. Binary profile constitution

tomers to daily activity on applications.

The process (Figure 6) exploits the hierarchical structure of the data: a customer is defined by his days and a day is defined by its hourly traffic volume on the applications. At the end of each stage, an interpretation step allows to incrementally extract knowledge from the analysis results. The unique visualization ability of the self organizing map model makes the analysis quite natural and easy to interpret. More details about such kind of approach on another application can be found in (Clérot and Fessant (2003)).

3.4 Clustering results

We experiment with the site of Fontenay in September 2003. All the segmentations are performed with dedicated SOMs (experiments have been done with the SOM Toolbox package for matlab (Vesanto et al. (2000)).

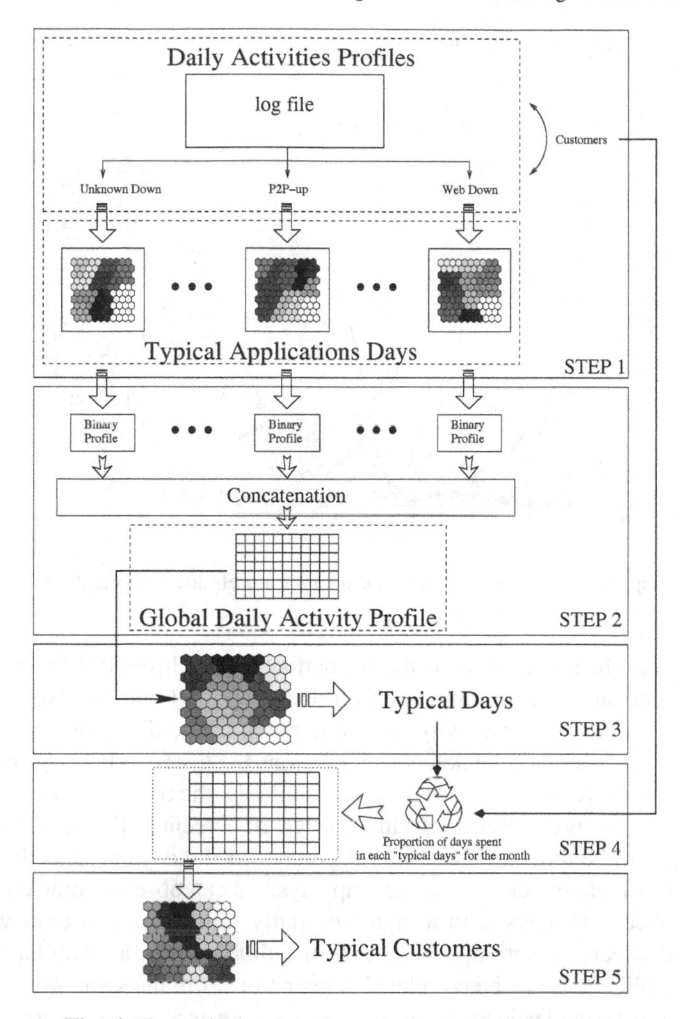

Fig. 6. The multi-level exploratory data analysis approach.

The first step leads to the formation of 9 to 13 clusters of "typical application days" profiles, depending on the application. Their behaviours can be summarized into inactive days, days with a mean or high activity on some limited time periods (early or late evening, noon for instance), and days with a very high activity on a long time segment (working hours, afternoon or night).

Figure 7 illustrates the result of the first step for one application: it shows the mean hourly volume profiles of the 13 clusters revealed after the clustering for the web down application (the mean profiles are computed by the mean of all the observations that have been classified in the cluster; the hourly volumes are plotted in natural statistics). The other applications can be described similarly.

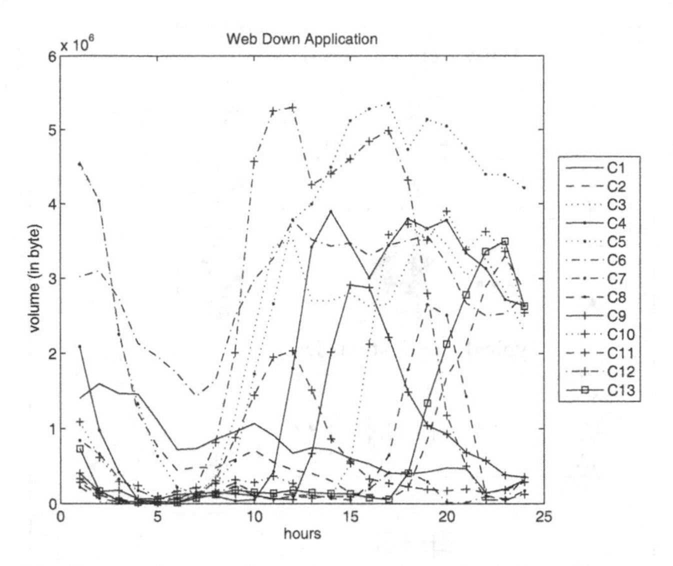

Fig. 7. Mean daily volumes of clusters for web down application

The second clustering leads to the formation of 14 clusters of "typical days". Their behaviours are different in terms of traffic time periods and intensity. The main characteristics are a similar activity in up and down traffic directions and a similar usage of the peer-to-peer and unknown applications in clusters. The usage of the web application can be quite different in intensity. Globally, the time periods of traffic are very similar for the three applications in a cluster. 10 percent of the days show a high daily activity on the three applications, 25 percent of the days are inactive days. If we project the other applications on the map days, we can observe some correlations between applications: days with a high web daily traffic are also days with high mail, ftp and streaming activities and the traffic time periods are similar. The chat and games applications can be correlated to peer-to-peer in the same way.

The last clustering leads to the formation of 12 clusters of customers which can be characterized by the preponderance of a limited number of typical days.

Figure 8 illustrates the characteristic behaviour of one "typical customer" (cluster 6) which groups 5 percent of the very active customers on all the applications (with a high activity all along the day, 7 days out of 10 and very little days with no activity). We plot the mean profile of the cluster (computed by the mean of all the customers classified in the cluster (up left, in black). We also give the mean profile computed on all the observations (bottom left, in grey), for comparison.

The profile can be discussed according to its variations against the mean profile in order to reveal its specific characteristics. The visual inspection of the left part of Figure 8 shows that the mean customer associated with the cluster is mainly active on "typical day 12" for 78 percent of the month. The contributions of the other "typical days" are low and are lower than the global mean. Typical day 12 corresponds to very active days. The mean profile of "typical day 12" is shown in the right top part

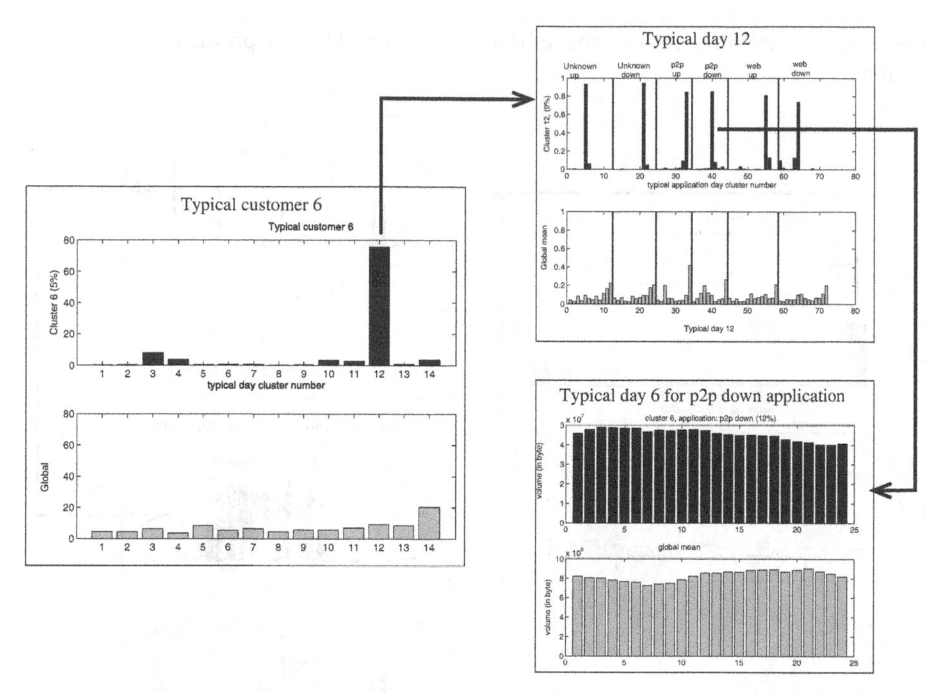

Fig. 8. Profile of one cluster of customers (up left) and mean profile (bottom) and profiles of associated typical days and typical application days

of the figure in black. The day profile is formed by the aggregation of the individual application clustering results (a line delimits the set of descriptors for each application). We also give the mean profile computed on all the observations (bottom, in grey).

Typical day 12 is characterized by a preponderant typical application day on each application (from 70 percent to 90 percent for each). These typical application days correspond to high daily activities.

For example, we plot the mean profile of "typical day 6" for the peer-to-peer down application in the same figure (right bottom; in black the hourly profile of the typical day for the application and in grey the global average hourly profile; the volumes are given in bytes). These days show a very high activity all along the day and even at night for the application (12 percent of the days). Figure 8 schematizes and synthesizes the complete customer segmentation process.

Our step-by-step approach aims at striking a practical balance between the faithful representation of the data and the interpretative power of the resulting clustering. The segmentation results can be exploited at several levels according to the level of details expected. The customer level gives an overall view on the customer behaviours. The analysis also allows a detailed insight into the daily cycles of the customers in the segments. The approach is highly scalable and deployable and clustering technique used allows easy interpretations. All the other segments of customers can be discussed similarly in terms of daily profiles and hourly profiles on the applications.

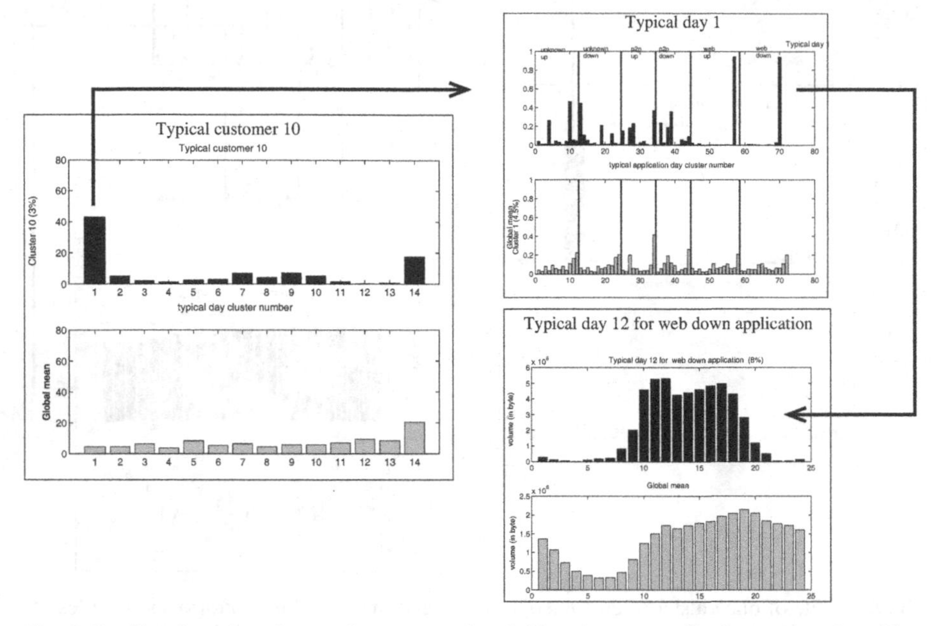

Fig. 9. Profile of another cluster of customers (top left) and mean profile (bottom) and profiles of associated typical days and typical application days

We have identified segments of customers with a high or very high activity all along the day on the three applications (24 percent of the customers), others segments of customers with very little activity (27 percent of the customers) and segments of customers with activity on some limited time periods on one or two applications, for example, a segment of customers with overall a low activity mainly restricted to working hours on web applications. This segment is detailed in Figure 9.

The mean customer associated with cluster 10 (3 percent of the customers) is mainly active on "typical day 1" for 42 percent of the month. The contributions on the other "typical days" are close to the global mean. Typical day 1 (4.5 percent of the days) is characterized by a preponderant typical application day on web application only (both in up and down directions); no specific typical day appears for the two other applications. The characteristic web days are working days with a high daily web activity on the segment 10h-19h.

Figure 10 depicts the organization of the 12 clusters on the map (each of the clusters is identified by a number and a colour). The topological ordering inherent to the SOM algorithm is such that clusters with close behaviours lie close on the map and it is possible to visualize how the behaviour evolves in a smooth manner from one place of the map to another. The map is globally organized along an axis going from the north east (cluster 12) to the south west (cluster 6), from low activity to high activity on all the applications, non-stop all over the day.

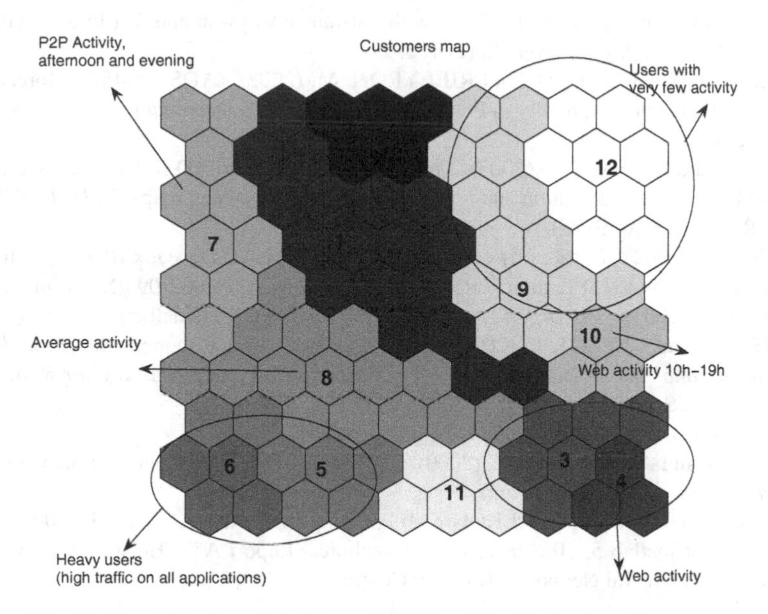

Fig. 10. Interpretation of the learned SOM and its 12 clusters of customers

4 Conclusion

In this paper, we have shown how the mining of network measurement data can reveal the usage patterns of ADSL customers. A specific scheme of exploratory data analysis has been presented to give lightings on the usages of applications and daily traffic profiles. Our data-mining approach, based on the analysis and the interpretation of Kohonen self-organizing maps, allows us to define accurate and easily interpretable profiles of the customers. These profiles exhibit very heterogeneous behaviours ranging from a large majority of customers with a low usage of the applications to a small minority with a very high usage.

The knowledge gathered about the customers is not only qualitative; we are also able to quantify the population associated to each profile, the volumes consumed on the applications or the daily cycle.

Our methodologies are continuously in development in order to improve our knowledge of customer's behaviours.

References

ANDERSON, B., GALE, C., JONES, M., and McWILLIAMS, A. (2002). Domesticating broadband-what consumers really do with flat-rate, always-on and fast Internet connections. *BT Technology Journal*, 20(1):103–114.

CLEMENT, H., LAUTARD, D., and RIBEYRON, M. (2002). ADSL traffic: a forecasting model and the present reality in France. In *WTC (World Telecommunications Congress)*, Paris, France.

CLEROT, C. and FESSANT, F. (2003). From IP port numbers to ADSL customer segmentation: knowledge aggregation and representation using Kohonen maps. In *DATAMINING IV*, Rio de Janeiro, Brazil.

FRANCOIS, J. (2002). Otarie: observation du traffic d'accès des réseaux IP en exploitation. France Télécom R&D Technical Report FT.R&D /DAC-DT/2002-094/NGN (in French).

KOHONEN, T. (2001). *Self-Organizing Maps*. Springer-Verlag, Heidelberg.

LEMAIRE, V. and CLEROT, F. (2005) The many faces of a Kohonen Map,. *Studies in computational Intelligence (SCI) 4, 1-13 (Classification and Clustering for Knowledge Discovery)*. Springer.

OJA, E. and KASKI, S. (1999). *Kohonen maps*. Elsevier.

VESANTO, J. and ALHONIEMI, E. (2000). Clustering of the self organizing map. In *IEEE Transactions of Neural Networks*.

VESANTO, J., HIMBERG, J., ALHONIEMI, E., and PARHANKANGAS, J. (2000). Som toolbox for matlab 5. Technical Report Technical Report A57, Helsinki University of Technology, Neural Networks Research Centre.

On the Analysis of Irregular Stock Market Trading Behavior

Markus Franke, Bettina Hoser and Jan Schröder

Information Services and Electronic Markets
Universität Karlsruhe (TH), Germany
{franke, hoser, schroeder}@iism.uni-karlsruhe.de

Abstract. In this paper, we analyze the trading behavior of users in an experimental stock market with a special emphasis on irregularities within the set of regular trading operations. To this end the market is represented as a graph of traders that are connected by their transactions. Our analysis is executed from two perspectives: On a micro scale view fraudulent transactions between traders are introduced and described in terms of the patterns they typically produce in the market's graph representation. On a macro scale, we use a spectral clustering method based on the eigensystem of complex Hermitian adjacency matrices to characterize the trading behavior of the traders and thus characterize the market. Thereby, we can show the gap between the formal definition of the market and the actual behavior within the market where deviations from the allowed trading behavior can be made visible. These questions are for instance relevant with respect to the forecast efficiency of experimental stock markets since manipulations tend to decrease the precision of the market's results. To demonstrate this we show some results of the analysis of a political stock market that was set up for the 2006 state parliament elections in Baden-Wuerttemberg, Germany.

1 Introduction

Stock markets do not only attract the *good* traders but also the ones who try to manipulate the market. The approaches used by malign traders differ with respect to the design of the market, but altogether tend to bias its outcome. In this contribution, we present basic behavior patterns that are characteristic of irregular trading activities and discuss an approach for their detection. We concentrate on patterns that tend to appear in prediction markets. In the first section we adopt a micro scale perspective, describing the traders' individual motivation for malicious actions and deriving the characteristics of two basic patterns. The second and third section approach a market's transaction records from a broader (macro) view. The market data is analyzed by means of a clustering method on the results of a certain type of eigensystem analysis, finding a reliable way of discovering the patterns sought.

2 Irregular trading behavior in a market

There are several incentives to act in a fraudulent way which result in the basic patterns price manipulation and circular trading. In this introductory section, we will show these basic patterns that constitute the micro scale view of the market activities. They can be made visible when the money or share flows in the market are used to generate a graph of traders and flows as shown in section 4.2.

Price manipulation are for instance motivated by idealistic reasons: Members of parties that may or may not take the hurdle of five percent introduced by German electoral laws have an incentive to set a price slightly above 5% in order to signal that every vote given for this party counts. This in turn is expected to motivate electors to vote who have not yet decided whether to vote at all (see Franke et al. (2006) and Hansen et al. (2004)). On the other hand, opponents may be induced to lower the prices for a rivaling party in order to discourage the voters of this party. These cases are quite easily detectable, since traders without such a bias in their trading behavior should have an approximately balanced ratio between buy and sell actions – this includes the number of offers and transactions as well as trading volumes. Manipulators, on the other hand, have a highly imbalanced ratio, since they become either a sink (when increasing the price) or a source (when decreasing the price) of shares of the respective party. Thus, these traders can be found by calculating the ratios for each trader and each share and setting a cutoff.

The other basic micro pattern, circular trading, is egoistically motivated; its objective is to increase the trader's endowment by transferring money (either in monetary units or in undervalued shares) from one or several satellite accounts that are also controlled by the fraudulent trader to a central account. In its most extreme form, the pattern leads to many accounts with a balance close to zero that have only traded with one other account in a circular pattern: shares are sold by the satellite to the central account at the lower end of the spread and then bought back at the higher end of the spread, resulting in a net flow of money from the satellite to the central account. Often this pattern is preluded by a widening of the spread by buying from and selling to the traders whose offers form the spread boundary in order to increase the leverage of each transaction between the fraudulent accounts. We have seen cases where the order book was completely emptied prior to the money transfer. This pattern is only present in markets where the cost of opening an account lies below the benefit of doing so, i.e. the initial endowment given to the trader.

While the most extreme form is easily detectable, we need a criterion for the more subtle forms. In terms of the flows between traders, circular trading implies that transferring a similar number of shares in each direction, the amounts of money exchanged differ significantly. In other words, there is a nontrivial net flow of money to the fraudulent account from one or several accounts to the central, fraudulent one. The problem lies here in the definition of net flow. Optimally, it should be calculated as the deviation from the "true" price at the time of the offer or trade times the number of shares. Unfortunately, the true price is only known at the close of the market. As a remedy, the current market price could be used. However, as we have seen, the market price may be manipulated and thus is quite unreliable, especially during the periods in which fraud occurs. The other, favorable approach is to use the volume of the trades, i.e. the number of shares times the price, as a substitute. For subsequent transactions with equal numbers of shares, the net flow is equivalent to the difference in the volumes, for other types of transactions, this is at least an approximation that facilitates the detection of circular trading.

3 Analysis of trading behavior with complex valued Eigensystem analysis

To analyze the market on a macro scale, we use an eigensystem analysis method. The method is fully described in Geyer-Schulz and Hoser (2005). In the next two sections we will give a short introduction to the technique and the necessary results with respect to the following analysis.

3.1 Spectral analysis of Hermitian adjacency matrices

The eigensystem analysis described in Geyer-Schulz and Hoser (2005) results in a full set of eigenvalues (spectrum) Λ with $\lambda_1, \lambda_2, \ldots, \lambda_l$ and their corresponding eigenvectors $\mathbf{X}$ with $\mathbf{x}_1, \mathbf{x}_2, \ldots, \mathbf{x}_l$ where the properties of the flow representation guarantee that the matrix becomes Hermitian and thus the eigenvalues are real while the components of the eigenvectors can be complex. This eigensystem represents a full orthonormal system where Λ and $\mathbf{X}$ can be written in the Fourier sum representation $\sum_{k=1}^{l} \lambda_k P_k = H$ with $P_k = \mathbf{x}_k \mathbf{x}_k^*$; H denotes the linear transformation $H = A_{\mathbb{C}} \cdot e^{-i\frac{\pi}{4}}$ with $A_{\mathbb{C}} = A + i \cdot A^t$ and A the real valued adjacency matrix of the graph. The projectors P_k are computed as the complex outer product of $\mathbf{x}_k$ and represent a substructure of the graph. We identify the relevant projectors by their covered data variance which can be calculated from the eigenvalues since the overall data variance is given as $\sum_{k=1}^{l} \lambda_k^2$. We detect the most central vertex in the graph by its absolute value $|\mathbf{x}_{max,m}|$ of the eigenvector component corresponding to the largest eigenvalue $|\lambda_{max}|$. This also holds for the most central vertices in each substructure identified by the projectors P_k.

3.2 Clustering within the eigensystem

Given the eigensystem as introduced in the last section we take the set of positive eigenvalues Λ^+ with $\lambda_1^+, \lambda_2^+, \ldots, \lambda_t^+$ and their corresponding eigenvectors $\mathbf{X}^+$ with $\mathbf{x}_1^+, \mathbf{x}_2^+, \ldots, \mathbf{x}_t^+$ and build the matrix $R_{n \times t} = \left(\lambda_1^+ \mathbf{x}_1^+ | \lambda_2^+ \mathbf{x}_2^+ | \ldots | \lambda_t^+ \mathbf{x}_t^+\right)$. With this matrix and its complex conjugate we build the matrix $S_{n \times n} = R * R^*$ as the scalar product matrix. Since we work in Hilbert space, distances are defined by the following scalar products: $||\mathbf{x} - \mathbf{y}||^2 = \langle \mathbf{x} - \mathbf{y} \mid \mathbf{x} - \mathbf{y} \rangle = ||\mathbf{x}||^2 + ||\mathbf{y}||^2 - 2Re(\langle \mathbf{x} \mid \mathbf{y} \rangle)$. Distances become minimal if the real part of the scalar product becomes maximal. Within this matrix S we find the clusters p_k by assigning the vertices of the network to the cluster such that a vertex i belongs to a cluster p_k if $Re(S_{i,p_k}) = max_j Re(S_{i,j})$. As at least one of the eigenvalues of Λ has to be negative due to $\sum_{k=1}^{l} \lambda_k = 0$, the minimum number of clusters is at least one, at most $l-1$ for the analyzed network. For details to this approach see Hoser and Schröder (2007).

4 Analysis of the dataset

When analyzing an actual market to discover fraudulent traders, the basic patterns introduced in section 2 reflect these traders' behavior (or part of it) within the market. To describe the actions taken by the traders we use the eigensystem analysis together with the spectral clustering method described in section 3. In order to demonstrate the use of this powerful method, we transform the transaction data of the market into a network as detailed in section 4.2. Eigensystem analysis is advantageous for the analysis here as it takes into account not only the relations from one node to the next, but computes the status of one node recursively from the information on the status of all other nodes within the network and is therefore referred to as a centrality measure (for the idea see Brin and Page (1998)).

4.1 Description of the dataset

We analyze a dataset generated by the political stock market system PSM used for the prediction of the 2006 state elections in Baden-Wuerttemberg, Germany. The traders were mainly readers of rather serious and politically balanced newspapers all over the election region. The market ran from January, 31st 2006 until election day on March, 26th 2006 for about twelve weeks and was stopped with the closing time of the polling stations at 18:00 CET when the first official information on the voters' decision is allowed to be released. More detailed data on the market is given in Table 1.

Table 1. Statistical Data on the 2006 state parliament elections in Baden-Wuerttemberg in Germany

Number of traders (at least one sell or buy transaction)	306 traders
Number of traders (at least one sell transaction)	190 traders
Number of traders (at least one buy transaction)	291 traders
Number of transactions	10786 transactions
Number of shares	7 shares
Avg. volume per trade	214.6 shares
Avg. money flow per trade	2462.1 monetary units
Money flow in total	26556378 monetary units
Share flow in total	2314197 shares

Traders in the market are given 100.000 monetary units (MU) as initial endowment. The market itself ran a continuous double auction market mechanism where offers by traders are executed immediately if they match. For each share an order

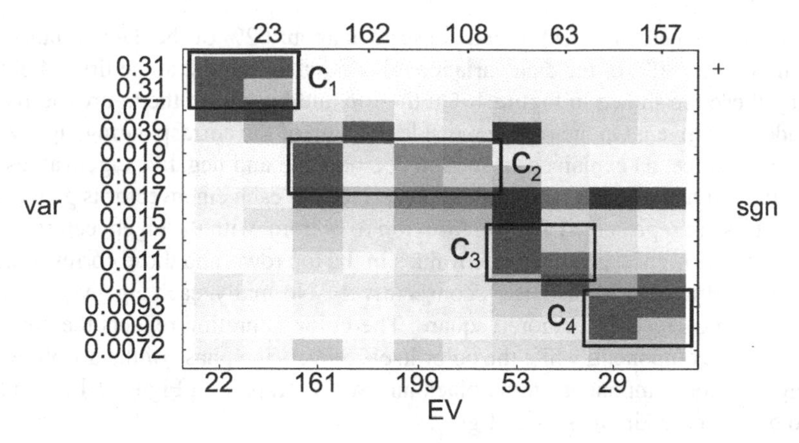

Fig. 1. Eigenvectors of the traders within the most prominent clusters

book is provided by the system where buy and sell offers are added and subsequently removed in the case of matching or withdrawal.

4.2 Generating the network

In markets with a central market instance the traders usually communicate only with this central instance; trades are executed against "the market", and a direct communication between traders does not take place. This results in an anonymous two-mode network perspective where a trader has no information on his counterparts in the offers as well as in the transaction partners. As the idea of the fraudulent action in the circular trading pattern from section 2 essentially deals with the knowledge of the counterpart trader we build the trader-to-trader network where the set of nodes consists of the traders that appeared in the transaction records. These traders have issued at least one offer that was matched and executed by the market mechanism. The edges of the network are set as the monetary flow between each pair of traders in the network (price times number of shares).

4.3 Results of the analysis

Within an agile market with random and normal distributed matching between the acting traders, *good* traders should appear in this analysis with a relatively balanced money flow as argued in section 2. Acting in fraudulent patterns, on the other hand, leads to a bias of these flows regarding the fraudulent trader, his counterparts and their connected traders.

Applying eigensystem analysis to the complex valued adjacency matrix as defined in section 3 reveals the patterns of trading behavior within the data set. The spectrum of the market shows symmetry since the largest and smallest eigenvalues have the same absolute value, but different signs. A symmetric spectrum points towards a star-structured graph. The variances given in Figure 1 reveal that the first pattern (first two eigenvalues) already describes about 62% of the data variance. To reach more than 80% of the data variance it is sufficient to look at the first 14 eigenvectors; these are shown in Figure 1. On the top and bottom of the figure the IDs of the traders are given. On the right hand side the sign of the corresponding eigenvalue is depicted, since, as explained in section 3.1 positive and negative eigenvalues exist. On the left hand side the covered data variance for each eigenvector is given. The eigenvectors are represented as rows from top to bottom, with the eigenvectors corresponding to the highest absolute eigenvalues in the top rows, and those corresponding to lower absolute eigenvalues listed consecutively. Normally, each eigenvector component is represented as a colored square. The color saturation reflects the absolute value for this component, while the color itself reflects the phase of the absolute valued eigenvector component. In the black and white graphic in Figure 1 both values had to be combined in the shade of grey.

As can clearly be seen, there are four blocks c_1 - c_4 in this figure. The block c_1 in the upper left hand corner shows that traders with IDs 1847 und 1969 had an almost balanced trading communication between them, and the volume was large. The second block c_2 in the middle of the figure represents the trading behavior of the group of traders with IDs 1922, 1775, 1898 and 1858. Here it can be stated that the connection between 1922, 1898 and 1858 is quite strong, and the trading behavior was nearly balanced between 1922 and 1858, while the behavior between 1922 and 1898 has a stronger outbound direction from 1922 to 1858. Between the first and second block the eigenvectors 3 and 4 describe normal trading behavior as defined by the market. The third block c_3 shows the traders with IDs 1924 and 1948. These again show a nearly balanced behavior, as do the traders with IDs 1816 and 1826 in the lower right hand corner of the figure.

These results were compared to the trading data in the data base. The result is given in Figure 2. The setup is similar to Figure 1 and it can easily be verified that the trading behavior is consistent with the results from the eigensystem analysis. Whenever the eigensystem analysis revealed a nearly balanced trading behavior this holds true even if the absolute values of transactions are different, since the order of magnitude stays approximately the same. The important aspect lies in the difference between the values, as it shows the transfer of money from one trader to the other.

It can thus be seen that the eigensystem reveals overall information about the trading behavior on the market, when transformed into a trader to trader network. On the other hand an analysis of each trader and his or her trading behavior towards other traders can be detected at the same time. Since the method used is an eigenanalysis the absolute value of each eigenvector component is similar to the eigenvector centrality used e.g. by Google (Brin and Page (1998)) to define relevant actors in a graph. Our approach though allows a decomposition of the market in the distinguishable trading patterns respectively subgroups of traders.

To visualize and illustrate the results of the eigensystem analysis as a graph, we have taken the respective subgraph which shows the relevant actors as found by the eigensystem analysis, embedded into the network of all their trading counterparts in Figure 3. As can be seen the relevant actors really have many connections within

	1847	1969	1922	1775	1898	1858	1924	1948	1816	1826
1847		761860							9	
1969	704600									
1922				10056	191060	90392		22800		
1775			5145		20000		6628			2632
1898			92400			198			247	
1858	2698		90035		4407				300	
1924				2				123470		
1948	2151						92310			
1816		3				518		2628		109360
1826									108560	

Fig. 2. Reduced adjacency matrix entries for the traders within the most prominent clusters among themselves

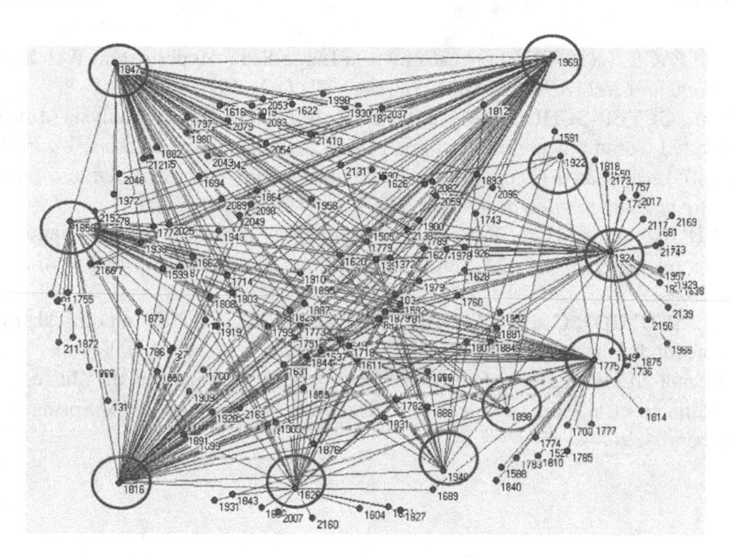

Fig. 3. Unweighted subgraph of the traders within the most prominent clusters to all related traders

the market and even amongst each other, which again validates the results of the eigensystem analysis.

5 Conclusion

As manipulation within electronic trading systems is limited to behavioral aspects and the usual amount of data is quite high, irregular acting is likely to remain hidden within the mass of data when using naïve fraud analysis techniques. Also structural effects of networks blur a clear view. We found that a recursive network analysis approach, facilitated by a trader-to-trader network supports the discovery of irregular patterns. Especially by means of the chosen network those traders are followed who try to use the network in their own favor and thus break the anonymity assumed by the market system.

Further research will focus on the analysis of the mix of several patterns, the detection of plain patterns in very noisy trading data as well as the weight functions for the edges within the network transaction graph. On the side of the analysis technique, comparison of traditional stock market measurements and the measures that arise from the approach of analyzing the behavioral aspects in electronic trading systems in a network analysis context are of special interest.

Acknowledgment. We gratefully acknowledge funding of the two projects SESAM and STOCCER by the Federal Ministry of Education and Research (BMBF).

References

BRIN, S. and PAGE, L.(1998): The Anatomy of a Large-Scale Hypertextual Web Search Engine. *Computer Networks and ISDN Systems, 30 (1–7), 107–117.*

FRANKE, M., GEYER-SCHULZ, A. and HOSER, B. (2006): On the Analysis of Asymmetric Directed Communication Structures in Electronic Election Markets. In: F. Billari et al. (Eds.):*Agent-Based Computational Modelling. Applications in demography, social, economic and environmental sciences.* Physica, Heidelberg, 37–59.

GEYER-SCHULZ, A. and HOSER, B. (2005): Eigenspectralanalysis of Hermitian Adjacency Matrices for the Analysis of Group Substructures. *Journal of Mathematical Sociology, 29(4), 265–294.*

HANSEN, J., SCHMIDT, C. and STROBEL, M. (2004): Manipulation in political stock markets - preconditions and evidence. *Applied Economics Letters, 11, 459–463.*

HOSER, B. and SCHROEDER, J. (2007): Automatic Determination of Clusters. In: K.-H. Waldmann et al. (Eds.): *Operations Research Proceedings 2006.* Springer, Berlin-Heidelberg, 439-444.

A Procedure to Estimate Relations in a Balanced Scorecard

Veit Köppen[1], Henner Graubitz[2], Hans-K. Arndt[2] and Hans-J. Lenz[1]

[1] Institut für Produktion, Wirtschaftsinformatik und Operations Research
Freie Universität Berlin, Germany
{koeppen, hjlenz}@wiwiss.fu-berlin.de

[2] Arbeitsgruppe Wirtschaftsinformatik - Managementinformationssysteme
Otto-von-Guericke-Universität Magdeburg, Germany
{graubitz, arndt}@iti.cs.uni-magdeburg.de

Abstract. A Balanced Scorecard is more than a business model because it moves performance measurement to performance management. It consists of performance indicators which are inter-related. Some relations are hard to find, like soft skills. We propose a procedure to fully specify these relations. Three types of relationships are considered. For the function types inverse functions exist. Each equation can be solved uniquely for variables at the right hand side. By generating noisy data in a Monte Carlo simulation, we can specify function type and estimate the related parameters. An example illustrates our procedure and the corresponding results.

1 Related work

Indicator systems are appropriate instruments to define business targets and to measure management indicators together. Such a system should not be just a system of hard indicators; it should be used as a system with control in which one can bring hard indicators and management visions together.

In the beginning of the 90's Johnson and Kaplan (1987) published the idea how to bring a company's strategy and used indicators together. This system, also known as Balanced Scorecards (BSC), is developed until now.

The relationships between those indicators are hard to find. According to Marr (2004), companies understand better their business if they visualise relations between available indicators. However, some indicators influence each other in cause and effect relations which increases the validity of these indicators. Unusually, compared to a study of Ittner et al (2003) and Marr (2004) 46% of questioned companies do not or are not able to visualise cause-and-effect relations of indicators.

Several approaches try to solve the existing shortcomings.

A possible way to model fuzzy relations in a BSC is described in Nissen (2006). Nevertheless, this leads to restrictions in the variable domains.

Blumenberg et al (2006) concentrate on Bayesian Belief Networks (BBN) and try to predict value chain figures and enhanced corporate learning. The weakness of this prediction method is that it does not contain any loops which BSCs may contain. Loops within BSCs must be removed if BBN are used to predict causes and effects in BSCs.

Banker et al (2004) suggest calculating trade-offs between indicators. The weakness of this solution is that they concentrate on one financial and three nonfinancial performance indicators and try to derive management decisions.

A totally different way of predicting relations in BSCs is the usage of system dynamics. System Dynamics is usually used to simulate complex dynamic systems (Forrester (1961)). Various publications exist of how to combine these indicators with dynamics systems to predict economic scenarios in a company, e.g. Akkermans et al (2002). In contrast to these approaches we concentrate on existing performance indicators and try to predict relationships between these indicators instead of predicting economic scenarios. It is similar to the methods of system identification. In contrast, our approach calculates in a more flexible way all models within the described model classes (see section 3).

2 Balanced scorecards

"If you can't measure it, you can't manage it" (Kaplan and Norton (1996), p. 21). With this sentence the BSC inventors Kaplan and Norton made a statement which describes a common problem in the industry: you can not manage a company if you don't have performance indicators to manage and control your company.Kaplan and Norton presented the BSC – a management tool for bringing the current state of the business and the strategy of the company together. It is a result of previous indicator systems. Nevertheless, a BSC is more than a business system (Friedag & Schmidt 2004). Kaplan & Norton (2004) emphasise this in their further development of Strategy Maps.

However, what are these performance indicators and how can you measure it. PreiSSner (2002) divides the functionality of indicators into four topics: operationalisation ("indicators should be able to reach your goal"), animation ("a frequent measurement gives you the possibility to recognise important changes"), demand ("it can be used as control input") and control ("it can be used to control the actual value"). Nonetheless, we understand an indicator as defined in (Lachnit 1979).

But before a decision is made which indicator is added to the BSC and the corresponding perspective the importance of the indicator has to be evaluated. Kaplan & Norton divide indicators additionally into hard and soft, short and long-term objectives. They also consider cause and effect relations. The three main aspects are: 1. All indicators that do not make sense are not worthwhile being included into a BSC; 2. While building a BSC, a company should differentiate between performance and result indicators; 3. All non-monetary values should influence monetary values. Based on these indicators we are now able to build up a complete system of indicators which turns into or influences each other and seeks a measurement for one of the following four perspectives: (1) Financial Perspective to reflect the financial performance like the return on investment; (2) Customer Perspective to summarize all indicators of the customer/company relationships; (3) Business Process Perspective to give an overview about key business processes; (4) Learning and Growth Perspective which measures the company's learning curve.

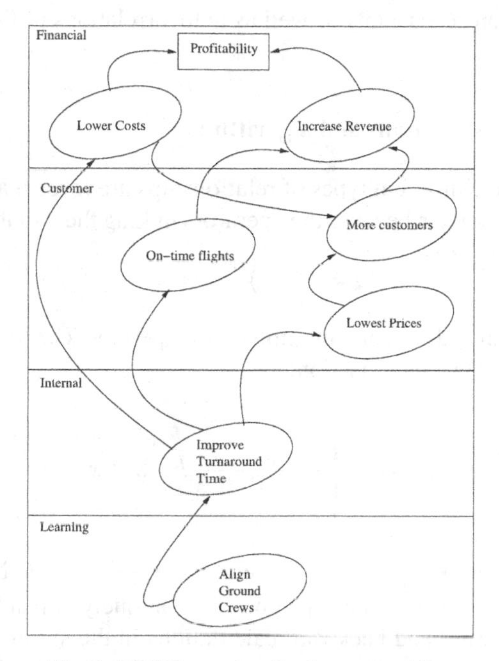

Fig. 1. BSC Example of a domestic airline

By splitting a company into four different views the management of a company gets the chance of a quick overview. The management can focus on its strategic goal and is able to react in time. They are able to connect qualitative performance indicators with one or all business indicators. Moreover the construction of an adequate equation system might be impossible.

Nevertheless the relations between indicators should be elaborated and an approximation of the relations of these indicators should be considered. In this case multivariate density estimation is an appropriate tool for modeling the relations of the business. Figure 1 shows a simple BSC of an airline company. Profitability is the main figure of interest but additionally seven more variables are useful for managing the company. Each arc visualizes the cause and effect relations. This example is taken from "The Balanced Scorecard Institute"[1].

[1] www.balancedscorecard.org

3 Model

To quantify the relationships in a given data set different methods for parameter estimation are used. Measurement errors within the data set are allowed, but these errors are assumed to have a mean value of zero. For each indicator within the data set no missing data is assumed. To quantify the relationships correctly it is further assumed that intermediate results are included in the data set. Otherwise the relationships will not be covered. Heteroscedasticity as well as autocorrelations of the data is not considered.

3.1 Relationships, estimations and algorithm

In our procedure three different types of relationships are investigated. The first two function types are unknown because the operators linking the variables are unknown:

$$z = f(x,y) = x \otimes y \tag{1}$$

where $\otimes$ represent an addition or a multiplication operator. The third type includes a parametric type of real valued function:

$$y = f_\theta(x) = \begin{cases} p & x \leq a \\ \frac{c}{1+e^{-d\cdot(x-g)}} + h & a < x \leq b \\ q & x > b \end{cases} \tag{2}$$

with $\theta = (abcdgh)$ and $p = \frac{c}{1+e^{-d\cdot(a-g)}} + h$ and $q = \frac{c}{1+e^{-d\cdot(b-g)}} + h$. Note, that all three function types are assumed to be separable, i.e. uniquely solvable for x or y in 1 and x in 2. Thus forward and backward calculations in the system of indicators are possible. As a data set is tested independently with respect to the described function types a Ŝidàk correction has to be applied (cf. Abdi (2007)).

Additive relationships between three indicators ($Y = X_1 + X_2$) are detected via multiple regression. The model is:

$$Y = \beta_0 + \beta_1 \cdot X_1 + \beta_2 \cdot X_2 + u \tag{3}$$

where $u \sim N(0, \sigma^2)$. The relationship is accepted if level of significance of all explanatory variables is high and $\beta_0 = 0$, $\beta_1 = 1$ and $\beta_2 = 1$. The multiplicative relationship $Y = X_1 \cdot X_2$ is detected by the regression model:

$$Y = \beta_0 + \beta_1 \cdot Z + u \text{ with } Z = X_1 \cdot X_2, u \sim N(0, \sigma^2). \tag{4}$$

The relationship is accepted if the level of significance of the explanatory variable is high and $\beta_0 = 0$ and $\beta_1 = 1$. The nonlinear relationship between two indicators according to equation 2 is detected by parameter estimation based on nonlinear regression:

$$Y = \frac{c}{1+e^{-d\cdot(X-g)}} + h + u \quad \forall a < x \leq b; u \sim N(0, \sigma^2). \tag{5}$$

In a first step the indicators are extracted from a business database, files or tools like excel spreadsheets. The number of extracted indicators is denoted by n. In the second step all possible relationships have to be evaluated. For the multiple regression scenario $\frac{n!}{3!\cdot(n-3)!}$ cases are relevant. Testing multiplicative relationships demands $\frac{n!}{2\cdot(n-3)!}$ test cases. The nonlinear regression needs to be performed $\frac{n!}{(n-2)!}$ times. All regressions are performed in R. The univariate and the multivariate linear regression are performed with the `lm` function from the R-base stats package. The nonlinear regression is fitted by the `nls` function in the stats package and the level of significance is evaluated. If additionally the estimated parameter values are in given boundaries the relationship is accepted.

The pseudo code of the the complete environment is given in algorithm 3.1.

Algorithm 1 Estimation Procedure

Require: data matrix $data[M_{t\times n}]$ with t observations for n indicators significance level, boundaries for parameter
Ensure: detected relationships between indicators
1: **for** $i = 1$ to $n - 2$ AND $j = i + 1$ to $n - 1$ AND $k = j + 1$ to n **do**
2: estimation by `lm`(data[,i] data[,j] + data[,k])
3: **if** significant AND parameter estimates within boundaries **then**
4: Relationship "Addition" found
5: **end if**
6: **end for**
7: **for** $i = 1$ to n AND $j = 1$ to $n - 1$ AND $k = j + 1$ to n **do**
8: **if** i != j AND i != k **then**
9: set Z := data[,j] · data[,k]
10: estimation by `lm`(data[,i] Z)
11: **if** significant AND parameter estimates within boundaries **then**
12: Relationship "Multiplication" found
13: **end if**
14: **end if**
15: **end for**
16: **for** $i = 1$ to n AND $j = 1$ to n **do**
17: **if** i != j **then**
18: estimation by `nls`(data[,j] c/(1+exp(-d+g*data[,i])) + h)
19: **if** significant **then**
20: "Nonlinear Relationship" found
21: **end if**
22: **end if**
23: **end for**

4 Case study

For our case study we create an artificial model with 16 indicators and 12 relationships, see Fig. 2. It includes typical cases of the real world.

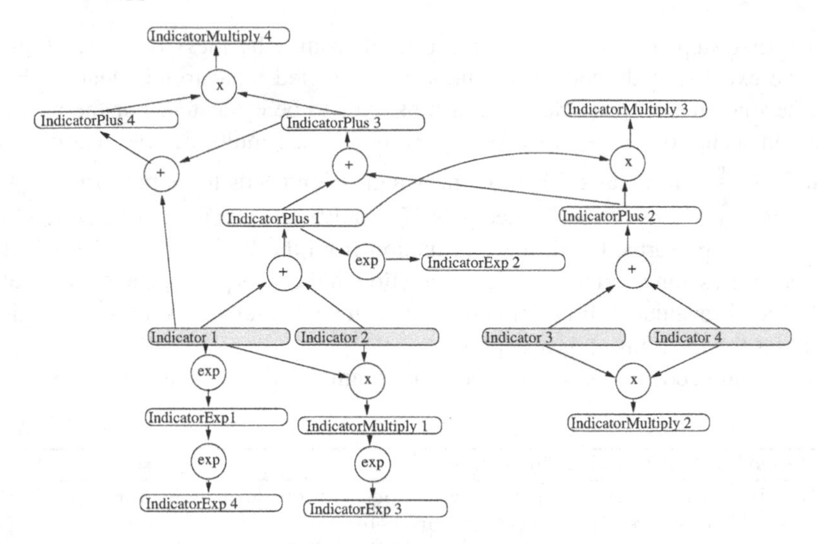

Fig. 2. Artificial Example

Indicators 1-4 are independently and randomly distributed. In Fig. 2 they are displayed in grey and represent the basic input for the simulated BSC system. All other indicators are either functional dependent on two indicators related by an addition or multiplication or functional dependent on an indicator according to equation 2. Some of these indicators effect other quantities or represent leaf nodes in the BSC model graph, cf. Fig. 2. Based on the fact that indicators may not be precisely measured we add noise to some indicators, see Tab. 1. Note, that IndicatorPlus4 has a skewed added noise whereas the remaining added noise is symmetrical.

In our case study we hide all given relationships and try to identify them, cf. section 3.

Table 1. Indicator Distributions and Noise

Indicator	Distribution	Indicator	added Noise	Indicator	Noise
Indicator1	$N(100, 10^2)$	IndicatorPlus1	$N(0,1)$	IndicatorExp1	$N(0,1)$
Indicator2	$N(40, 2^2)$	IndicatorPlus4	$E(1)-1$	IndicatorExp4	$U(-1,1)$
Indicator3	$U(-10,10)$	IndicatorMultiply1	$N(0,1)$		
Indicator4	$E(2)$	IndicatorMultiply4	$U(-1,1)$		

5 Results

The case study runs in three different stages: with 1k, 10k, and 100k randomly distributed data. The results are similar and can be classified into four cases: (1) if a relation exists and it was found (displayed black in Fig. 3), (2) if a relation was found but does not exist (displayed with a pattern in Fig. 3) (error of the second kind), (3) if no relation was found but one exists in the model (displayed white in Fig. 3) (error of the first kind), and (4) if no relation exists and no one was found. Additionally the results have been split according to the operator class (see Tab. 2).

Table 2. Identification Results

Observations	1k			10k			100k		
	+	*	Exp	+	*	Exp	+	*	Exp
(2)	0	3	27	0	5	48	0	2	49
(3)	1	0	3	1	0	3	1	0	3
	560	1680	240	560	1680	240	560	1680	240

Hence, Tab. 2 shows that the results for all experiments are similar for the operators addition and multiplication. For non-linear regression, relationships could not be discovered properly.

The additive relation of IndicatorPlus4 was the only non-detective relation, see observation (3) in Tab. 2. This is caused by the fact that the indicator has an added noise which is skewed. In such a case the identification is not possible.

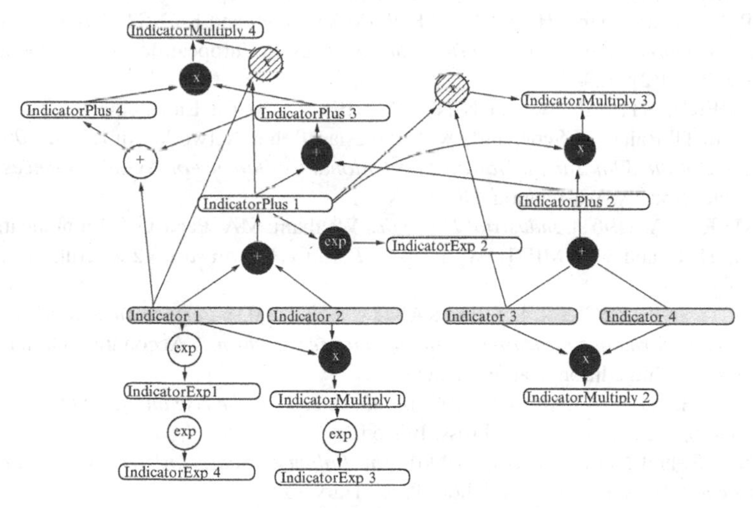

Fig. 3. Results of the Artificial Example for 100k observations

6 Conclusion and outlook

Traditional regression analysis allows estimating the cause and effect dependencies within a profit seeking organization. Univariate and multivariate linear regression exhibit the best results whereas skewed noise in the variables destroys the possibility to detect these relationships.

Non-linear regression has a high error output due to the fact that optimization has to be applied and starting values are not always at hand. The results from the non-linear regression should only be carefully taken into account.

In future work we try to improve our results while removing indicators for which we calculate a nearly 100% secure relationship. Additionally we plan to work on real data which also includes the possibility of missing data for indicators. Research aims at creating a company's BSC with relevant business figures while looking only at a company's indicator system.

References

ABDI, H. (2007): Bonferroni and Sidak corrections for multiple comparisons. In: N.J. Salkind (Ed.): *Encyclopedia of Measurement and Statistics.* Thousand Oaks (CA): Sage: 103–107.

AKKERMANS, H. and VAN OORSCHOT, KIM (2002): *Developing a balanced scorecard with system dynamics* in Proceeding of 2002 International System Dynamics Conference.

BANKER, R. D. and Chang, H. and JANAKIRAMAN, S. N. and KONSTANS, C. (2004): *A balanced scorecard analysis of performance metrics.* in European Journal of Operational Research 154(2): 423–436.

BLUMENBERG, STEFAN A. and HINZ, DANIEL J. (2006): Enhancing the Prognostic Power of IT Balanced Scorecards with Bayesian Belief Networks. In *HICSS '06: Proceedings of the 39th Annual Hawaii International Conference on System Sciences* IEEE Computer Society, Washington, DC, USA

FORRESTER, J. W. (1961). *Industrial Dynamics* Waltham, MA: Pegasus Communications.

FRIEDAG, H.R. and SCHMIDT, W. (2004): *Balanced Scorecard.* 2nd edition. Haufe, Planegg.

ITTNER, C.D. and LARCKER, D.F. and RANDALL, T. (2003): *Performance implications of strategic performance measurement in financial service firms".* Accounting Organization and Society, 2nd edition. Haufe, Planegg.

JOHNSON, T.H. and KAPLAN, R.S. (1987): *Relevance lost: the rise and fall of management accounting* . Harvard Business Press, Boston.

KAPLAN, R.S. and NORTON, D.P. (1996): *The Balanced Scorecard. Translating Strategy Into Action.* Harvard Business School Press, Harvard.

KÖPPEN, V. and LENZ, H.-J. (2006): A comparison between probabilistic and possibilistic models for data validation. In: Rizzi, A. & Vichi, M. (Eds.) *Compstat 2006 Ű Proceedings in Computational Statistics* , Springer, Rome.

LACHNIT, L. (1979): *Systemorientierte Jahresabschlussanalyse.* Betriebswirtschaftlicher Verlag Dr. Th. Gabler KG, Wiesbaden.

MARR, B. (2004): *Business Performance Measurement: Current State of the Art.* Cranfield University, School of Management, Centre for Business Performance.

NISSEN, V. (2006): Modelling Corporate Strategy with the Fuzzy Balanced Scorecard. In: Hüllermeier, E. et al. (Eds.): *Proceedings Symposium on Fuzzy Systems in Computer Science FSCS 2006*: 121– 138, Magdeburg.

PREISSNER, A. (2002): *Balanced Scorecard in Vertrieb und Marketing: Planung und Kontrolle mit Kennzahlen*, 2nd ed. Hanser Verlag, München, Wien

The Application of Taxonomies in the Context of Configurative Reference Modelling

Ralf Knackstedt and Armin Stein

European Research Center for Information Systems
{ralf.knackstedt, armin.stein}@ercis.uni-muenster.de

Abstract. The manual customisation of reference models to suite special purposes is an exhaustive task that has to be accomplished thoroughly to preserve, explicit and extend the inherit intention. This can be facilitated by the usage of automatisms like those being provided by the Configurative Reference Modelling approach. Thus, the reference model has to be enriched by data describing for which scenario a certain element is relevant. By assigning this data to application contexts, it builds a taxonomy. This paper aims to illustrate the advantage of the usage of this taxonomy during three relevant phases of Configurative Reference Modelling, *Project Aim Definition*, *Construction* and *Configuration* of the configurable reference model.

1 Introduction

Reference information models – in this context solely called *reference models* – give recommendations for the structuring of information systems as best or common practices and can be used as a starting basis for the development of application specific information system models. The better the reference models are matched with the special features of individual application contexts, the bigger the benefit of reference model use. Configurable reference models contain rules that describe how different application specific variants are derived. Each of these rules is placed together with a condition and an implication. Each condition describes one application context of the reference model. The respective implication determines the relevant model variant. For describing the application contexts configuration parameters are used. Their specification forms a taxonomy. Based upon a procedure model this paper highlights the usefulness of taxonomies in the context of Configurative Reference Modelling. Thus, the paper is structured as follows: First, the Configurative Reference Modelling approach and its procedure model is being described. Afterwards, the usefulness of the application of taxonomies is being shown during the respective phases. An outlook on future research areas concludes the paper.

2 Configurative Reference Modelling and the application of taxonomies

2.1 Configurative Reference Modelling

Reference models are representations of knowledge recorded by domain experts to be used as guidelines for every day business as well as for further research. Their purpose is to structure and store knowledge and give recommendations like best or common practices. They should be of general validity in terms of being applicable for more than one user (see Schuette (1998); vom Brocke (2003); Fettke, Loos (2004)). Currently 38 of them have been clustered and categorised, spanning domains like logistics, supply chain management, production planing and control or retail (see Braun, Esswein (2006)).

General applicability is a necessary requirement for a model to be characterised as reference model, as it has to grant the possibility to be adopted by more than one user or company. Thus, the reference model has to include information about different business models, different functional areas or different purposes for its usage. A reference model for retail companies might have to cover economic levels like *Retail* or *Wholesale*, trading levels like *Inland trade* or *Foreign trade* as well as functional areas like *Sales*, *Production Planning and Control* or *Human Resource Management*. While this constitutes the general applicability for a certain domain, one special company usually needs just one suitable instance of this reference model, for example *Retail/Inland Trade*, leaving the remaining information dispensable. This yields the problem that the perceived demand of information for each individual will be hardly met. The information delivered – in terms of models of different types which might consist of different element types and hold different element instances – might either be too little or too extensive, hence the addressee will be overburdened on the one hand or insufficiently supplied with information on the other hand. Consequently, a person requiring the model for the purpose of developing the database of a company might not want to be burdened with models of the technique Event-driven Process Chain (EPC), whose purpose is to describe processes, but with Entity Relationship Model (ERM), used to describe data structures. To compensate this in a conventional manner, a complex manual customisation of the reference model is necessary to meet the addressees demand. Another implication is the maintenance of the reference model. Every time changes are committed to the reference model, every instance has to be manually updated as well.

This is where Configurable Reference Models come into operation. The basic idea is to attach parameters to elements of the integrated reference model in advance, defining the contexts to which these elements are relevant (see e. g. Knackstedt (2006)). In reference to the example given above this means that certain elements of the model might just be relevant for one of the economic levels – *retail* or *wholesale* –, or for both of them. The user eventually selects the best suited parameters for his purpose and the respective configured model is generated automatically. This leads to the conclusion that the lifecycle of a configurable reference model can be divided into two parts called *Development* and *Usage* (see Schlagheck (2000)).

The first part – relevant for the reference model developer – consists of the phases *Project Aim Definition, Model Technique Definition, Model Construction* and *Evaluation* for the developer, whereas the second one – relevant for the user – includes the phases *Project Aim Definition, Search and Selection* of existing and suitable reference models and *Model Configuration*. The configured model can be further adapted to satisfy individual needs (see Becker et al. 2004). Several phases can be identified, where the application of taxonomies can be of value, especially *Project Aim Definition* and *Model Construction* (for the developer) and *Model Configuration* (for the user). Fig. 1 gives an overview of the phases, where the ones that will be discussed in detail are solid, the ones actually not relevant are greyed out. The output of both *Development* and *Usage* is printed in italics.

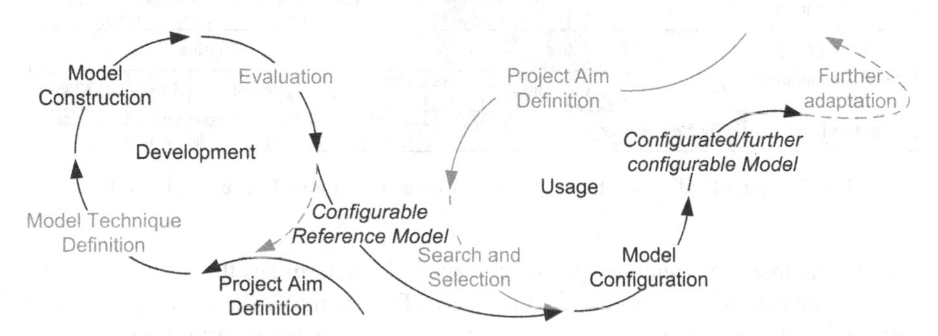

Fig. 1. Development and Usage of Configurable Reference Models

2.2 Project aim definition

During the first phase, *Project Aim Definition*, the developers have to agree on the purpose of the reference model to build. They have to decide for which domain the model should be used, which business models should be supported, which functional areas should be integrated to support the distribution for different perspectives and so on. To structure these parameters, a *morphological box* has become apparent to be applicable. First, all instances for each possible characteristic have to be listed. By shading the relevant parameters for the reference model, the developers commit themselves to one common project aim and reduce the given complexity. Thus, the emerging morphological box constitutes a *taxonomy*, implying the variants included in the integrated configurative reference model (see fig. 2; Mertens, Lohmann (2000)). By generating this taxonomy, the developers get aware of all possible included variants, thus getting a better overview of the *to-be*-state of the model. One special variant of the model will later on be generated by choosing one or a set of the parameters by the user. The choice of parameters should be supported by an underlying ontology that can be used throughout both *Development* and *Usage* (see Knackstedt et al. (2006)). The developers have to decide whether or not dependencies between parameters exist. In some cases, the choice of one

Characteristic	Characteristic form				
Business level	Retailer	Wholesaler			
Retail Business	Third-party business	Pooled payment business	Promotion business	Service business	
Trading Level	Inland trade	Foreign trade			
Horizontal cooperation	Retailers	Wholesalers	Other cooperation		
Vertical Cooperation	Retail and wholesale	Wholesale and industrial companies	Retail and industry companies	Retail, wholesale and Industrial companies	
Contact orientation	Stationary	Itinerant	Mail Order		
Sales contact form	Sales person	Self-service	Catalog	Vending Machine	
Beneficiary	Investment goods trade	Consumer goods trade			
Range extend	Wide and deep range	Wide and shallow range	Narrow and deep range	Narrow and shallow range	
Pricing policy	Active	Passive			
Purchase initiation through	Visit to store	Letter/fax	Telephone (call center)	Internet	Push (e.g., Clubs)
Logistics handling	By the customer (collect)	By the retailer/intermediary (delivery)	Through the internet (for digital products)		

Fig. 2. Example of a morphological box, used as taxonomy. Becker et al. (2001)

specific parameter within one specific characteristic determines the necessity of another parameter within another characteristic. For example, the developers might decide that the choice of `ContactOrientation=MailOrder` determines the choice of `PurchaseInitiationThrough=AND(Internet;Letter/Fax)`.

2.3 Construction

During the *Model Construction* phase, the configurable reference model has to be developed in regards to the decisions made during the preceding phase *Project Aim Definition*. The example in fig. 3 illustrates an EPC regarding the payment of a bill, distinguishing whether the bill originates from a national or an international source. If the origin of the bill is national, it can be paid immediately, otherwise it has to be cross-checked by the international auditing. This scenario can only take place, if both instances of the characteristic `TradingLevel`, namely `InlandTrade` and `ForeignTrade`, are chosen. If all clients of a company are settled abroad or (in the meaning of an exclusive `or`) all of them are inland, the check for the origin is not necessary. The cross-check with the international auditing has only to take place, if the bill comes from abroad. To store this information in the model, the respective parameters are attached to the respective model elements in form of a term and can later be evaluated to *true* or *false*. Only if the equation is evaluated to *true* or if there is no term attached to an element, the respective element may remain in the configured model. Thus, for example, the function *check for origin* stays, if the term `TradingLevel=AND(Foreign;Inland)` is true, which happens if both parameters are selected. If only one is selected, the equation returns *false* and the element will be removed from the model.

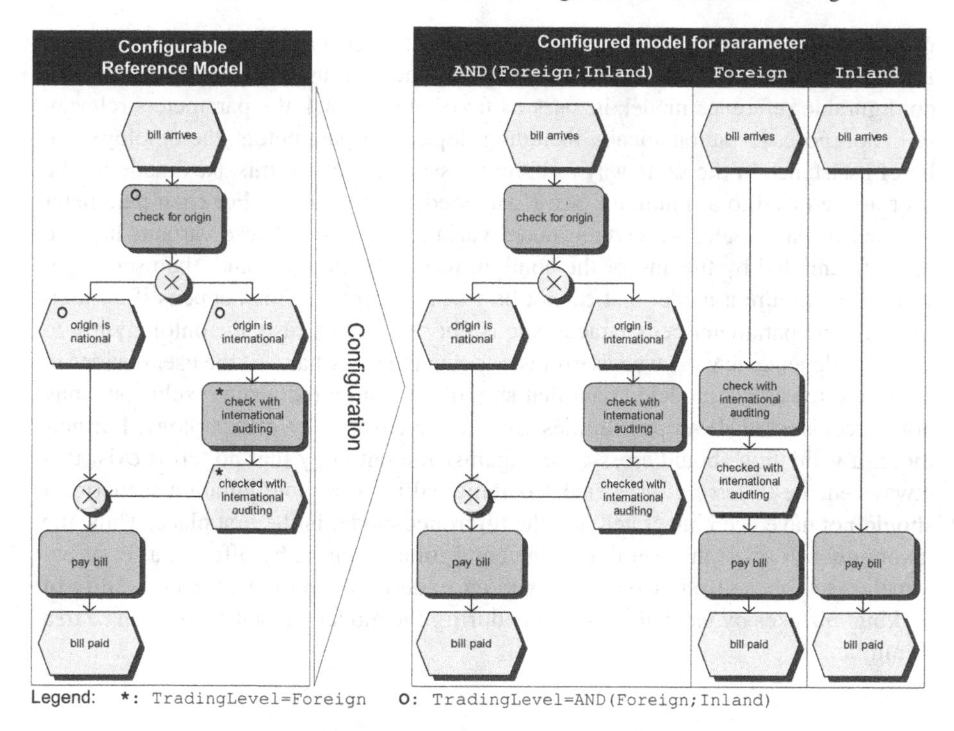

Fig. 3. Annotated parameters to elements, resulting model variants

To specify these terms, which can get complex if many characteristics are used, a term editor application has been developed, which enables the user to attach them to the relevant elements. Here again, the ontology can support the developer by automatically testing for correctness and reasonableness of dependent parameters (see Knackstedt et al. (2006)). Opposite to dependencies, exclusions take into account that under certain circumstances parameters may not be chosen together. This minimises the risk of defective modelling and raises the consistency level of the configurable reference model. In the example given above, if the developer selects `SalesContactForm=VendingMachine`, the parameter `Beneficiary` may not be `InvestmentGoodsTrade`, as investment goods can hardly be bought via a vending machine. Thus, the occurrence of both statements concatenated with a logical `AND` is not allowed. The same fact has to be regarded when evaluating dependencies: If, like stated above, `ContactOrientation=MailOrder` determines the choice of `PurchaseInitiationThrough=AND(Internet;Letter/Fax)`, the same statement may not occur with a preceded `NOT`. Again, the previously generated taxonomy can support the developer by structurising the included variants.

2.4 Configuration

The *Usage* phase of a configurable reference model starts independently from its development. During the *Project Aim Definition* phase the potential user defines the parameters to determine which reference model best meets his needs. He has to search for it during the *Search and Selection* phase. Once the user has selected a certain configurable reference model, he uses its taxonomy to pick the parameters relevant to his purpose. By automatically including dependent parameters, the ontology can be of assistance in the same way as before, assuring that the mistakes made by the user are reduced to a minimum (see Knackstedt et al. (2006)). For each parameter – or set of parameters – a certain model variant is created. These variants have to be differentiated by the aim of the configuration. On the one hand, the user might want to configure a model that cannot be further adapted. This happens if a maximum of one parameter per characteristic is chosen. In this case, the ontology has to consider dependencies as well as exclusions. On the other hand, if the user decides to configure towards a model variant that should be configured again, exclusions may not be considered. Both possibilities have to be covered by the ontology. Furthermore, a validation should cross-check against the ontology that no terms exist that always equate to *false*. If an element is removed in every configuration scenario, it should not have been integrated into the reference model in the first place. Thus, the taxonomy can assist the user during the configuration phase by offering a set of parameters to choose from. Combined with an underlying ontology, the possibility of making mistakes by using the taxonomy during the model adaptation is reduced to a minimum.

3 Conclusion

As well as the ontology, the taxonomy used as a basic element throughout the phases of Configurative Reference Modelling has to meet certain demands. Most importantly, the developers have to carefully select the constituting characteristics and associated parameters. It has to be possible for the user to distinguish between several options, so they can make a clear decision to configure the model towards the variant relevant for his purpose. This means that each parameter has to be understandable and be delimited from the others, which – for example – can be arranged by supplying a manual or guide. Moreover, the parameters may neither be too abstract nor too detailed. The taxonomy can be of use during the three relevant phases. As mentioned before, the user has to be assisted in the usage of the taxonomy by automatically including or excluding parameters as defined by the ontology. Furthermore, only such parameters should be chosen, that have an effect on the model that is comparative to the necessary effort to identify it. Parameters that have no effect at all or are not used should be removed as well, to decreases the complexity for both the developer and the user. If the choice of a parameter results in the removal of only one element and its identification takes a very long time, it should be removed from the taxonomy because of its little effect at high costs. Thus, the way the adaptation process is supported by the taxonomy strongly depends on the associated ontology.

4 Outlook

The resulting effect of the selection of one parameter to configure the model shows its relevance and can be measured either by the quantity or by the importance of the elements that are being removed. Each parameter can be associated with a certain cost that emerges due to the time it takes the user to identify it. Thus, *cheap* parameters are easy to identify and have a huge effect once selected. *Expensive* parameters instead are hard to identify and have little effect on the model. Further research should first try to benchmark, which combinations of parameters of a certain reference model are chosen most often. In doing so, the developer has the chance to concentrate on the evolution of these parts of the reference model. Second, it should be possible to identify *cheap* parameters by either running simulations on reference models, measuring the effect a parameter has – even in combination with other parameters –, or by auditing the behavior of reference model users – which is feasible in a limited way due to the small distribution of configurable reference models. Third, configured models should be rated with costs, so cheap variants can be identified and – the other way round – the responsible parameters can be identified. To sum up, a objective function should be developed, enabling the calculation of the costs for the configuration of a certain model variant in advance by giving the selected parameters as input. It should have the form $C(MV) = \sum_{k=1}^{n} \frac{C(P_k)}{R(P_k)}$ with $C(MV)$ being the cost function of a certain model variant derived from the reference model by using n parameters, $C(P_k)$ being the cost function of a single parameter and $R(P_k)$ being a function weighting the relevance of a single parameter P, which is used for the configuration of the respective model variant. Furthermore, the usefulness of the application of the taxonomy has to be evaluated by empirical studies in every day business. This will be realised for the configuration phase by integrating consultancies into our research and giving them a taxonomy for a certain domain at hand. With the application of supporting software tools, we hope that the adoption process of the reference model can be facilitated.

References

BECKER, J., DELFMANN, P. and KNACKSTEDT, R. (2004): Konstruktion von Referenzmodellierungssprachen – Ein Ordnungsrahmen zur Spezifikation von Adaptionsmechanismen fuer Informationsmodelle. *Wirtschaftsinformatik, 46, 4, 251–264.*

BECKER, J., UHR, W. and VERING, O. (2001): *Retail Information Systems Based on SAP Products.* Springer Verlag, Berlin, Heidelberg, New York.

BRAUN, R. and ESSWEIN, W. (2006): Classification of Reference Models. In: *Advances in Data Analysis: Proceedings of the 30th Annual Conference of The Gesellschaft fuer Klassifikation e.V., Freie Universitaet Berlin, March 8–10, 2006.*

DELFMANN, P., JANIESCH, C., KNACKSTEDT, R., RIEKE, T. and SEIDEL, S. (2006): Towards Tool Support for Configurative Reference Modelling – Experiences from a Meta Modeling Teaching Case. In: *Proceedings of the 2nd Workshop on Meta-Modelling and Ontologies (WoMM 2006). Lecture Notes in Informatics. Karlsruhe, Germany, 61–83.*

FETTKE, P. and LOOS, P. (2004): Referenzmodellierungsforschung. *Wirtschaftsinformatik, 46, 5, 331–340.*

KNACKSTEDT, R. (2006): *Fachkonzeptionelle Referenzmodellierung einer Managementunterstuetzung mit quantiativen und qualitativen Daten. Methodische Konzepte zur Konstruktion und Anwendung*. Logos-Verlag, Berlin.

KNACKSTEDT, R., SEIDEL, S. and JANIESCH, C. (2006): Konfigurative Referenzmodellierung zur Fachkonzeption von Data-Warehouse-Systemen mit dem H2-Toolset. In: J. Schelp, R. Winter, U. Frank, B. Rieger, K. Turowski (Hrsg.): *Integration, Informationslogistik und Architektur. DW2006, 21.–22. Sept. 2006, Friedrichshafen. Lecture Notes in Informatics. Bonn, Germany, 61–81.*

MERTENS, P. and LOHMANN, M. (2000): Branche oder Betriebstyp als Klassifikationskriterien fuer die Standardsoftware der Zukunft? Erste Ueberlegungen, wie kuenftig betriebswirtschaftliche Standardsoftware entstehen koennte. In: F. Bodendorf, M. Grauer (Hrsg.): *Verbundtagung Wirtschaftsinformatik 2000. Shaker Verlag, Aachen, 110–135.*

SCHLAGHECK, B. (2000): *Objektorientierte Referenzmodelle fuer das Prozess- und Projektcontrolling. Grundlagen – Konstruktion – Anwendungsmoeglichkeiten.* Deutscher Universitaets-Verlag, Wiesbaden.

SCHUETTE, R. (1998): *Grundsaetze ordnungsmaessiger Referenzmodellierung. Konstruktion konfigurations- und anpassungsorientierter Modelle.* Deutscher Universitaets-Verlag, Wiesbaden.

VOM BROCKE, J. (2003): *Referenzmodellierung. Gestaltung und Verteilung von Konstruktionsprozessen.* Logos Verlag, Berlin.

Two-Dimensional Centrality of a Social Network

Akinori Okada

Graduate School of Management and Information Sciences
Tama University, 4-1-1 Hijirigaoka Tama-shi, Tokyo 206-0022, Japan
okada@tama.ac.jp

Abstract. A procedure of deriving the centrality in a social network is presented. The procedure uses the characteristic values and the vectors of a matrix of friendship relationships among actors. While the centrality of an actor has been usually derived by the characteristic vector corresponding to the largest characteristic value, the present study uses not only the characteristic vector corresponding to the largest characteristic value but also that corresponding to the second largest characteristic value. Each actor has two centralities. The interpretation of two centralities, and the comparison with the additive clustering are presented.

1 Introduction

When we have a symmetric social network among a set of actors, where the relationship from actors j to k is equal to the relationship from actors k to j, the centrality of each actor who constitutes a social network is very important to find the features and the structure of the social network. The centrality of an actor represents the importance, significance, power, or popularity of the actor to form relationships with the other actors in the social network. Several procedures to derive the centrality of each actor in the social network have been introduced (ex. Hubbell (1965)). Bonacich (1972) introduced a procedure to derive the centrality of an actor by using the characteristic (eigen) vector of a matrix of friendship relationships or friendship choices among a set of actors. The matrix of friendship relationships which is dealt with by these procedures is assumed to be symmetric.

The procedure of Bonacich (1972) is based on the characteristic vector corresponding to the largest characteristic (eigen) value. Each element of the characteristic vector represents the centrality of each actor. The procedure has one good property that the centrality of an actor is defined recursively by the weighted sum of the centralities of all actors, where the weight is the strength of the friendship relationship between the actor and the other actors. The procedure was extended to deal with an asymmetric matrix of friendship relationships (Bonachich (1991)), where (a) the relationship from actors j to k is not same as that from actors k to j or (b) relationships between a set of actors and another set of actors. The first case (a) means the one-mode two-way data, and the second case (b) means the two-mode two-way data. These procedures utilized the characteristic vector which corresponds to the largest characteristic value. Wright and Evitts (1961) also introduced a procedure to derive the centrality of an actor utilizing the characteristic vectors which correspond to more than one (largest) characteristic value. While Wright and Evitts (1961) say the purpose is to derive the centrality, they focus their attention to summarize the relationships among actors just like applying factor analysis to the matrix of friendship relationships.

The purpose of the present study is to introduce a procedure to derive the centrality of each actor of a social network by using the characteristic vectors which correspond to more than one largest characteristic value of the matrix of friendship relationships. Although the present procedure is based on more than one characteristic vectors, the purpose is to derive the centrality of actors but not to summarize relationships among actors in a social network.

2 The procedure

The present procedure deals with a symmetric matrix of friendship relationships. Suppose we are dealing with a social network consisits of n actors. Let A be an $n \times n$ matrix representing friendship relationships among actors in a social network. The (j,k) element of A, a_{jk}, represents the relationship between actor j and k; when actors j and k are friends each other

$$a_{jk} = 1, \tag{1}$$

and when actors j and k are not friends each other

$$a_{jk} = 0. \tag{2}$$

Because the relationships among actors are symmetric, the matrix A is symmetric; $a_{jk} = a_{kj}$.

The characteristic vectors of $n \times n$ matrix A which correspond to two largest characteristic values are derived. Each characteristic value represents the salience of the centrality represented by the corresponding characteristic vector. The jth element of a characteristic vector represents the centrality of actor j along the feature or the aspect represented by the corresponding characteristic vector.

3 The analysis and the result

In the present study, the social network data among 16 families were analyzed (Wasserman and Faust (1994, p. 744, Table B6)). The data show the marital relationships among 16 families. Thus the actor in the present data is the family. The relationships are represented by a 16×16 matrix. Each element represents whether there was a marital tie between two families corresponding to a row and a column

(Wasserman and Faust (1994, p. 62)). The (j,k) element of the matrix is equal to 1, when there is a marital tie between families j and k, and is equal to 0, when there is no marital tie between families j and k. In the present analysis, the unity was embedded in the diagonal elements of the matrix of friendship relationships.

The five largest characteristic values of the 16×16 friendship relationship matrix were 4.233, 3.418, 2.704, 2.007, and 1.930. The corresponding characteristic vectors for the two largest characteristic values are shown in the second and the third columns of Table 1.

Table 1. Characteristic vectors

	Dimension 1	Dimension 2
	Characteristic values	
Actor (Family)	4.233	3.418
1 Acciaiuoli	0.129	0.134
2 Albizzi	0.210	0.300
3 Barbadori	0.179	0.053
4 Bischeri	0.328	-0.260
5 Castellani	0.296	-0.353
6 Ginori	0.094	0.123
7 Guadagni	0.283	0.166
8 Lamberteschi	0.086	0.076
9 Medici	0.383	0.434
10 Pazzi	0.039	0.117
11 Peruzzi	0.339	-0.385
12 Pucci	0.000	0.000
13 Ridolfi	0.301	0.124
14 Salviati	0.137	0.236
15 Strozzi	0.404	-0.382
16 Tornabuoni	0.281	0.285

Two characteristic values are 4.233 and 3.418 each of which represents the relative salience of the centrality over the all 16 actors along the feature or aspect shown by each of the two characteristic vectors. The two centralities represent two different features or aspects, called Dimensions 1 and 2 (see Figure 1), of the importance, significance, power, or popularity of actors. The second column, which represents the characteristic vector corresponding the largest characteristic value, has non-negative elements. These figures show the centrality of the 16 actors along the feature or the aspects of Dimension 1. The larger value shows the larger centrality of an actor. Actor 15 has the largest value 0.404, and has the largest centrality among the 16 actors. Actors 4, 9, 11, and 13 have larger centralities as well. Actor 12 has the smallest value 0.000, and has the smallest centrality among the 16 actors. Actors 6, 8, and 10 also have small centralities.

The third column represents the characteristic vector corresponding to the second largest characteristic value. While the characteristic vector corresponding to the largest characteristic value represented in the second column has all non-negative elements, the characteristic vector corresponding to the second largest characteristic value has negative elements. Actors 2 and 9 have larger positive elements. On the contrary, actors 4, 5, 11, and 15 have substantive negative elements. The meaning and the interpretation of the characteristic vector which corresponds to the second largest characteristic value will be discussed in the next section.

4 Discussion

Two characteristic vectors each corresponding to the largest and the second largest characteristic values represent the centralities of each actor along two different features or aspects of Dimensions 1 and 2. The 16 elements of the first characteristic vector seem to represent the overall (global) centrality or popularity of an actor among the actors in the social network (cf. Scott (1991, pp. 85-89)). For each actor, the number of ties with the other 15 actors were calculated. Each of the 16 figures shows the overall centrality or popularity of the actor among actors in the social network. The correlation coefficient between the elements of the first characteristic vector and these figures were 0.90. This tells that the elements of the first characteristic vector shows the overall centrality or popularity of the actor in the social network. This is the meaning of the feature or the aspect given by the first characteristic vector of Dimension 1.

The jth element of the first characteristic vector shows the strength of actor j in extending or accepting friendship relationships with the other actors in the social network as a whole. The strength of the friendship relationship between actors j and k along Dimension 1 is represented by the product of the jth and the kth elements of the first characteristic vector. Because all elements of the first characteristic vector are non-negative, the product of any two elements of the first characteristic vector is non-negative. The larger the product is, the stronger the tie between two actors is.

The second characteristic vector has the positive (non-negative) and the negative elements as well. Thus, there are three cases of the product of two elements of the second characteristic vector;
(a) the product of two non-negative elements is non negative
(b) the product of two negative elements is positive, and
(c) the product of a positive element and a negative element is negative.
In the case of (a) the interpretation of the element of the second characteristic vector is the same as that of the first characteristic vector. But in the cases of (b) and (c), it is difficult to interpret the meaning of the elements by the same manner as that for case (a). Because the element of the matrix of friendship relationships was defined by Equations (1) and (2), the larger value or the positive value of the product of any two elements of the second characteristic vector shows the larger or positive friendship relationship between two corresponding actors, and the smaller value or the negative value shows the smaller or negative (friendship) relationship between two corresponding actors. The product of two negative elements of the second characteristic vector is positive, and the positive figure shows the positive friendship rela-

tionship between two actors. The product of the positive and the negative elements is negative, and the negative figure shows the negative friendship relationship between two actors.

The features or the aspect represented by the second characteristic vector can be regarded as the local centrality or popularity within a subgroup (cf. Scott (1991, pp.85-89)). As shown in Table 2, some actors have positive and some actors have negative elements on Dimension 2 or the second characteristic vector. We can consider that there are two subgroups of actors; one subgroup consists of actors having positive elements of the second characteristic vector, and another subgroup consists of those having negative elements of the second characteristic vector, and that two subgroups are not friendly. When two actors belong to the same subgroup, the product of the two corresponding elements of the second characteristic vector is positive (cases (a) and (b) above), suggesting the positive friendship relationship between two actors. On the other hand, when two actors belong to two different subgroups, which means that one actor has the positive element and another actor has the negative element, the product of the two corresponding elements of the second characteristic vector is negative (case (c) above), suggesting the negative friendship relationship between two actors.

Table 1 shows that actor 4, 5, 11, and 15 have negative elements on the second characteristic vector. This means that the second characteristic vector suggests two subgroups of actors each consists of;
Subgroup 1: actors 1, 2, 3, 6, 7, 8, 9, 10, (12), 13, 14, and, 16
Subgroup 2: actors 4, 5, 11, and, 15
The two subgroups are graphically shown in Figure 1, where the horizontal dimension (Dimension 1) corresponds to the first characteristic vector, and vertical dimension (Dimension 2) corresponds to the second characteristic vector. Each actor is represented as a point having the coordinate of the corresponding element of the first characteristic vector on Dimension 1 and that of the second characteristic vector on Dimension 2. Figure 1 shows that four members who belong to the second subgroup are located closely each other and are separated from the other 12 actors. This seems to validate the interpretation of the feature or the aspect represented by the second characteristic vector.

The element of the second characteristic vector represents to which subgroup each actor belongs by its sign (positive or negative). The element represents the centrality of an actor among actors within the subgroup to which the actor belongs, because the product of the two elements corresponding to two actors belong to the same subgroup is positive regardless of the sign of the elements. The absolute value of the element of the second characteristic vector tells the local centrality or popularity among actors in the same subgroup to which the actor belongs, and the degree of periphery or unpopularity among actors in another subgroup to which the actor does not belong. The number of ties with actors who are in the same subgroup of that actor is calculated for each actor. The correlation coefficient between the absolute value of the elements of the second characteristic vector and the number of ties within a subgroup was 0.85. This tells that the absolute values of the elements of the second characteristic vector shows the centrality of an actor in each of the two subgroups. Because the correlation coefficient was derived over the two subgroups, the centralities can be compared between subgroups 1 and 2.

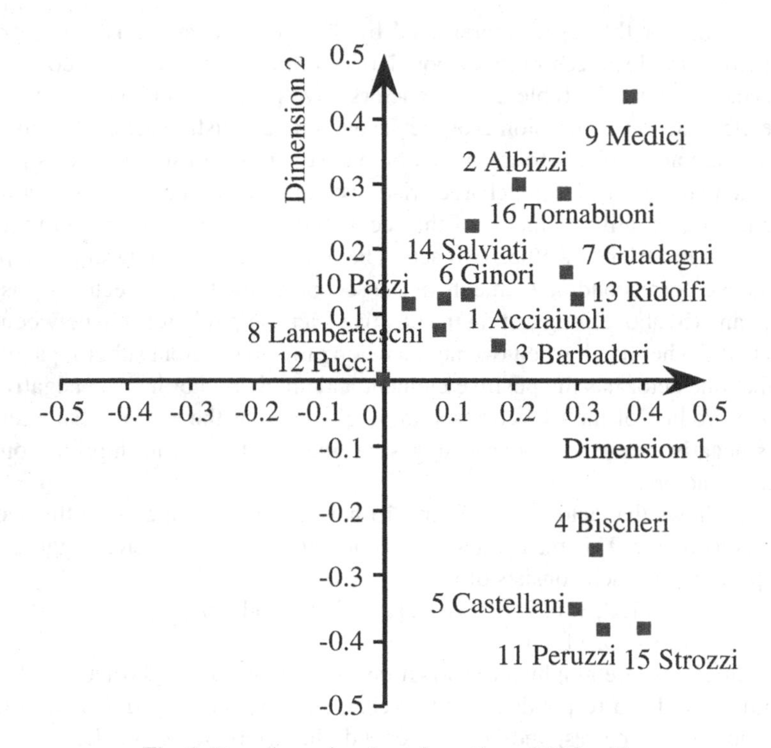

Fig. 1. Two-dimensional configuration of 16 families

The interpretation of the feature or the aspect of the second characteristic vector reminds us of the ADCLUS model (Arabie and Carroll (1980); Arabie, Carroll, and DeSarbo (1987); Shepard and Arabie, (1979)). In the ADCLUS model, each object can belong to more than one cluster, and each cluster has its own weight which shows the salience of that cluster. Table 2 shows the result of the application of ADCLUS to the present friendship relationships data.

Table 2. Result of the ADCLUS analysis

Cluster	Weight	1	2	3	4	5	6	7	8	9	10	11	12	13	14	15	16
Cluster 1	1.88	0	0	0	1	1	0	0	0	0	0	1	0	0	0	1	0
Universal	-0.09	1	1	1	1	1	1	1	1	1	1	1	1	1	1	1	1

In Table 2, the second row represents whether each of the 16 actors belongs to cluster 1 (when the element is 1) or does not belong to cluster 1 (when the element is

0). The third row represents the universal cluster, to which all actors belong, representing the additive constant of the data (Arabie, Carroll, and DeSarbo (1987, p. 58)). As shown in Table 2, actors 4, 5, 11, and 15 belong to cluster 1. These four actors are coincide with those having the negative elements of the second characteristic vector in Table 1.

The result derived by the analysis using ADCLUS and the result derived by using the characteristic values and vectors are very similar. But they have several different points. In the result derived by using ADCLUS, the strength of the friendship relationship between two actors is represented as the sum of two terms; (a) the weight for the universal cluster, and (b) the weight for cluster 1 if the two actors belong to cluster 1. The first term is constant for all combinations of any two actors, and the second term is the weight for the first cluster (when two actors belong to cluster 1) or zero (when one or none of the two actors belong to cluster 1). In using the characteristic vectors, the strength of the friendship relationship between two actors are represented also as the sum of two terms; (a) the product of the two elements of the first characteristic vector, and (b) the product of the two elements of the second characteristic vector. The first and the second terms are not constant for all combinations of two actors but each combination of two actors has its own value, because each actor has its own elements on the first and the second characteristic vectors. The first and the second characteristic vectors are orthogonal, because the matrix of friendship relationships is assumed to be symmetric, and the two characteristic values are different. The correlation coefficient between the first and the second characteristic vectors is zero. The clusters derived by the analysis using ADCLUS does not have the property even if two or more clusters were derived by the analysis.

In the present analysis only one cluster was derived by the analysis using ADCLUS. It seems interesting to compare the result derived by ADCLUS having more than one cluster with the result based on the characteristic vectors corresponding to the third largest and further characteristic values. The comparisons of the present procedure with concepts used in the graph theory seem necessary to thoroughly evaluate the present procedure. The present procedure assumes that the strength of the friendship relationship between actors j and k is represented by the product of the centralities of actors j and k. But the strength of the friendship relationship between two actors is defined as the sum of the centralities of the two actors by using the conjoint measurement (Okada (2003)). Which of the product or the sum of two centralities is more easily understood, or more practical in applications should be examined. The original idea of the centrality has been extended to the asymmetric or rectangular social network (Bonacich (1991, 2001)). The present idea can also be extended rather easily to deal with the asymmetric or the rectangular case as well.

Acknowledgments

The author would like to express his appreciation to Hiroshi Inoue for his helpful suggestions to the present study. The author also wishes to thank two anonymous referees for the valuable reviews which were very helpful to improve the earlier version of the present paper. The present paper was prepared, in part, when the author was at the Rikkyo (St. Paul's) University.

References

ARABIE, P. and CARROLL, J.D. (1980): MAPCLUS: A Mathematical Programming Approach to Fitting the ADCLUS Model. *Psychometrika, 45, 211–235.*

ARABIE, P., CARROLL, J.D., and DeSARBO, W.S. (1987): *Three-Way Scaling and Clustering*. Sage Publications, Newbury Park.

BONACICH, P. (1972): Factoring and Weighting Approaches to Status Scores and Clique Identification. *Journal of Mathematical Sociology, 2, 113–120.*

BONACICH, P. (1991): Simultaneous Group and Individual Centralities. *Social Networks, 13, 155–168.*

BONACICH, P. and LLOYD, P. (2001): Eigenvector-Like Measures of Centrality for Asymmetric Relations. *Social Networks, 23, 191–201.*

HUBBELL, C.H. (1965): An Input-Output Approach to Clique Identification. *Socimetry, 28, 277–299.*

OKADA, A. (2003): Using Additive Conjoint Measurement in Analysis of Social Network Data. In: M. Schwaiger, and O. Opitz (Eds.): *Exploratory Data Analysis in Empirical Research.* Springer, Berlin, 149-156.

SCOTT, J. (1991): *Social Network Analysis: A Handbook.* Sage Publications, London.

SHEPARD, R.N. and ARABIE, P. (1979): Additive Clustering: Representation of Similarities as a Combinations of Discrete Overlapping Properties. *Psychological Review, 86, 87–123.*

WASSERMAN, S. and FAUST, K. (1994): *Social Network Analysis: Methods and Applications*. Cambridge University Press, Cambridge.

WRIGHT, B. and EVITTS, M.S. (1961): Direct Factor Analysis in Sociometry. *Sociometry, 24, 82–98.*

Benchmarking Open-Source Tree Learners in R/RWeka

Michael Schauerhuber[1], Achim Zeileis[1], David Meyer[2], Kurt Hornik[1]

[1] Department of Statistics and Mathematics
Wirtschaftsuniversität Wien
1090 Wien, Austria
[2] Institute for Management Information Systems
Wirtschaftsuniversität Wien
1090 Wien, Austria
{Michael.Schauerhuber, Achim.Zeileis, Kurt.Hornik}@wu-wien.ac.at

Abstract. The two most popular classification tree algorithms in machine learning and statistics — C4.5 and CART — are compared in a benchmark experiment together with two other more recent constant-fit tree learners from the statistics literature (QUEST, conditional inference trees). The study assesses both misclassification error and model complexity on bootstrap replications of 18 different benchmark datasets. It is carried out in the R system for statistical computing, made possible by means of the **RWeka** package which interfaces R to the open-source machine learning toolbox **Weka**. Both algorithms are found to be competitive in terms of misclassification error—with the performance difference clearly varying across data sets. However, C4.5 tends to grow larger and thus more complex trees.

1 Introduction

Due to their intuitive interpretability, tree-based learners are a popular tool in data mining for solving classification and regression problems. Traditionally, practitioners with a machine learning background use the C4.5 algorithm (Quinlan, 1993) while statisticians prefer CART (Breiman, Friedman, Olshen and Stone, 1984). One important reason for this is that free reference implementations have not been easily available within an integrated computing environment. RPart, an open-source implementation of CART, has been available for some time in the S/R package **rpart** (Therneau and Atkinson, 1997) while the open-source implementation J4.8 for C4.5 became available more recently in the **Weka** machine learning package (Witten and Frank, 2005) and is now accessible from within R by means of the **RWeka** package (Hornik, Zeileis, Hothorn and Buchta, 2007). With these software tools available, the algorithms can be easily compared and benchmarked on the same computing platform: the R system for statistical computing (R Development Core Team 2006). The principal concern of this contribution is to provide a neutral and unprejudiced review, especially taking into account classical beliefs (or preconceptions) about performance differences between C4.5 and CART and heuristics for the choice of hyper-parameters. With this in mind, we carry out a benchmark comparison, including different strategies for hyper-parameter tuning as well as two further constant-fit tree models—QUEST (Loh and Shih, 1997) and conditional inference trees (Hothorn, Hornik and Zeileis, 2006). The learners are compared with respect to misclassification error and model complexity on each of 18 different benchmarking data sets by means of simultaneous confidence intervals (adjusted for multiple testing). Across data sets, the performance is aggregated by consensus rankings.

2 Design of the benchmark experiment

The simulation study includes a total of six tree-based methods for classification. All learners were trained and tested in the framework of Hothorn, Leisch, Zeileis and Hornik (2005) based on 500 bootstrap samples for each of 18 data sets. All algorithms are trained on each bootstrap sample and evaluated on the remaining out-of-bag observations. Misclassification rates are used as predictive performance measures, while model complexity requirements of the algorithms under study are measured by the number of estimated parameters (number of splits plus number of leafs). Performance and model complexity distributions are assessed for each algorithm on each of the datasets. In our setting, this results in 108 performance distributions (6 algorithms on 18 data sets), each of size 500. For comparison on each individual data set, simultaneous pairwise confidence intervals (Tukey all-pair comparisons) are used. For aggregating the pairwise dominance relations across data sets, median linear order consensus rankings are employed following Hornik and Meyer (2007). A brief description of the algorithms and their corresponding implementation is given below.

CART/RPart: *Classification and regression trees* (CART, Breiman et al., 1984) is the classical recursive partitioning algorithm which is still the most widely used in the statistics community. Here, we employ the open-source reference implementation of Therneau and Atkinson (1997) provided in the R package **rpart**. For determining the tree size, cost-complexity pruning is typically adopted: either by using a 0- or 1-standard-errors rule. The former chooses the complexity parameter associated with the smallest prediction error in cross-validation (RPart0), whereas the latter chooses the highest complexity parameter which is within 1 standard error of the best solution (RPart1).

C4.5/J4.8: C4.5 (Quinlan, 1993) is the predominantly used decision tree algorithm in the machine learning community. Although source code implementing C4.5 is available in Quinlan (1993), it is not published under an open-source license. Therefore, the Java implementation of C4.5 (revision 8), called J4.8, in **Weka** is the de-facto open-source reference implementation. For determining the tree size, a heuristic confidence threshold C is typically used which is by default set to $C = 0.25$ (as recommended in Witten and Frank, 2005). To evaluate the

Table 1. Artificial [$*$] and non artificial benchmarking data sets

Data set	# of obs.	# of cat. inputs	# of num. inputs
breast cancer	699	9	-
chess	3196	36	-
circle $*$	1000	-	2
credit	690	-	24
heart	303	8	5
hepatitis	155	13	6
house votes 84	435	16	-
ionosphere	351	1	32
liver	345	-	6
Pima Indians diabetes	768	-	8
promotergene	106	57	-
ringnorm $*$	1000	-	20
sonar	208	-	60
spirals $*$	1000	-	2
threenorm $*$	1000	-	20
tictactoe	958	9	-
titanic	2201	3	-
twonorm $*$	1000	-	20

influence of this parameter, we compare the default J4.8 algorithm with a tuned version where C and the minimal leaf size M (default: $M = 2$) are chosen by cross-validation (J4.8(cv)). A full grid search for $C = 0.01, 0.05, 0.1, \ldots, 0.5$ and $M = 2, 3, \ldots, 10, 15, 20$ is used in the cross-validation.

QUEST: *Quick, unbiased and efficient statistical trees* are a class of decision trees suggested by Loh and Shih (1997) in the statistical literature. QUEST popularized the concept of unbiased recursive partitioning, i.e., avoiding the variable selection bias of exhaustive search algorithms (such as CART and C4.5). A binary implementation is available from `http://www.stat.wisc.edu/~loh/quest.html` and interfaced in the R package **LohTools** which is available from the authors upon request.

CTree: *Conditional inference trees* (Hothorn et al., 2006) are a framework of unbiased recursive partitioning based on permutation tests (i.e., conditional inference) and applicable to inputs and outputs measured at arbitrary scale. An open-source implementation is provided in the R package **party**.

The benchmarking datasets shown in Table 1 were taken from the popular UCI repository of machine learning databases (Newman, Hettich, Blake and Merz, 1998) as provided in the R package **mlbench**.

3 Results of the benchmark experiment

3.1 Results on individual datasets: Pairwise confidence intervals

Here, we exemplify—using the well-known Pima Indians diabetes and breast cancer data sets—how the tree algorithms are assessed on a single data set. Simultaneous confidence intervals are computed for all 15 pairwise comparisons of the 6 learners. The resulting dominance relations are used as the input for the aggregation analyses in Section 3.2.

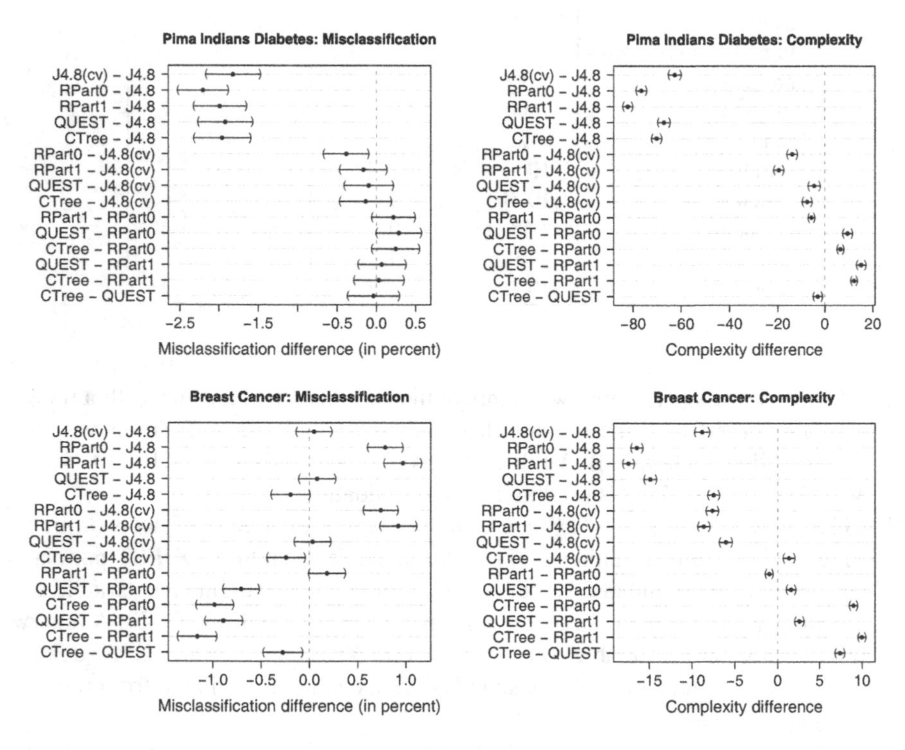

Fig. 1. Simultaneous confidence intervals of pairwise performance differences (left: misclassification, right: complexity) for Pima Indians diabetes (top) and breast cancer (bottom) data.

As can be seen from the performance plots for Pima Indian diabetes in Figure 1, standard J4.8 is outperformed (in terms of misclassification as well as model complexity) by the other tree learners. All other algorithm comparisons indicate equal predictive performances, except for the comparison of RPart0 and J4.8(cv), where the former learner performs slightly better than the latter. On this particular dataset tuning enhances the predictive performance of J4.8, while the misclassification rates of the differently tuned RPart versions are not subject to significant changes. In terms of model complexity J4.8(cv) produces larger trees than the other learners. Looking

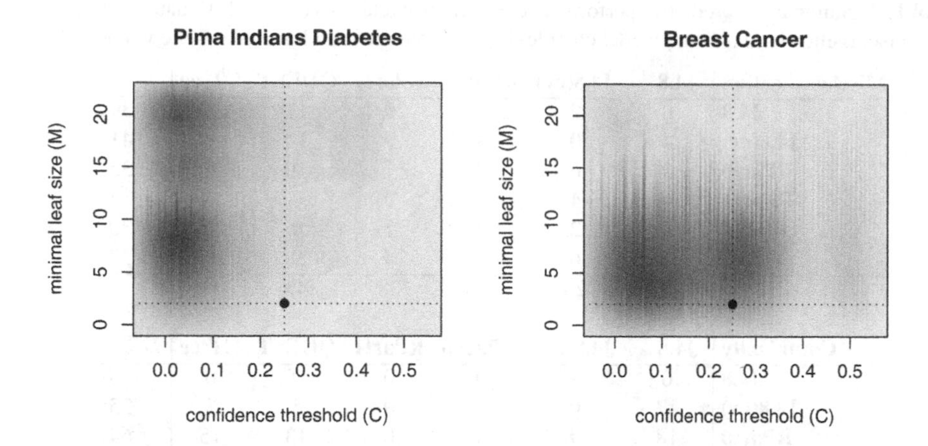

Fig. 2. Distribution of J4.8(cv) parameters obtained through cross validation on Pima Indians diabetes and breast cancer data sets.

at the breast cancer data yields a rather different picture: Both RPart versions are outperformed by J4.8 or its tuned alternative in terms of predictive accuracy. Similar to Pima Indians diabetes, J4.8 and J4.8(cv) tend to build significantly larger trees than RPart. On this dataset, CTree has a slight advantage over all other algorithms except J4.8 in terms of predictive accuracy. For J4.8 as well as RPart, tuning docs not promise to increase predictive accuracy significantly. A closer look at the differing behavior of J4.8(cv) under cross validation for both data sets is provided in Figure 2. In contrast to the breast cancer example, the results based on the Pima Indians diabetes dataset (on which tuning of J4.8 caused a significant performance increase) show a considerable difference in choice of parameters. The multiple inference results gained from all datasets considered in this simulation experiment (just like the results derived from the two datasets above) form the basis on which further aggregation analyses of Section 3.2 are built upon.

3.2 Results across data sets: Consensus Rankings

Having $18 \times 6 = 108$ performance distributions of the 6 different learners applied to 18 bootstrap data settings at hand, aggregation methods can do a great favor to allow for summarizing and comparing algorithmic performance. The underlying dominance relations derived from the multiple testing are summarized by simple sums in Table 2 and by the corresponding median linear order rankings in Table 3. In Table 2, rows refer to winners, while columns denote the losers. For example J4.8 managed to outperform QUEST on 11 datasets and 4 times vice versa, i.e., on the remaining 3 datasets, J4.8 and QUEST perform equally well.

The median linear order for misclassification reported in Table 3 suggests that tuning of J4.8 instead of using the heuristic approach is worth the effort. A similar

Table 2. Summary of predictive performance dominance relations across all 18 datasets based on misclassification rates and model complexity (columns refer to losers, rows are winners).

Misclassification	J4.8	J4.8(cv)	RPart0	RPart1	QUEST	CTree	$\sum$
J4.8	0	2	9	9	11	8	39
J4.8(cv)	4	0	8	9	11	9	41
RPart0	5	6	0	7	10	7	35
RPart1	6	4	1	0	8	6	25
QUEST	4	2	2	5	0	7	20
CTree	7	6	7	8	9	0	37
$\sum$	26	20	27	38	49	37	

Complexity	J4.8	J4.8(cv)	RPart0	RPart1	QUEST	CTree	$\sum$
J4.8	0	1	0	0	2	0	3
J4.8(cv)	17	0	0	0	5	3	25
RPart0	18	18	0	0	13	15	64
RPart1	18	18	16	0	14	15	81
QUEST	15	13	5	4	0	10	47
CTree	18	14	3	2	8	0	45
$\sum$	86	64	24	6	42	43	

Table 3. Median linear order consensus rankings for algorithm performance

	Misclassification	Complexity
1	J4.8(cv)	RPart1
2	J4.8	RPart0
3	RPart0	QUEST
4	CTree	CTree
5	RPart1	J4.8(cv)
6	QUEST	J4.8

conclusion can be made for the RPart versions. Here, the median linear order suggests that the common one standard error rule performs worse. For both cases, the underlying dominance relation figures of Table 2 catch our attention. Regarding the first case, J4.8(cv) only dominates J4.8 in four of six data settings, in which a significant test decision for performance differences could be made. In addition the remaining 12 data settings yield equivalent performances. Therefore superiority of J4.8(cv) above J4.8 is questionable. In contrast the superiority of RPart0 vs. RPart1 seems to be more reliable but still the number of data settings producing tied results is high. A comparison of the figures of CTree and the RPart versions confirms previous findings (Hothorn et al., 2006) that CTree and RPart often perform equally well. The question concerning the dominance relation between J4.8 and RPart cannot be answered easily: Overall, the median linear order suggests that the J4.8 decision tree versions are superior to the RPart tree learners in terms of predictive performance. But still, looking at the underlying relations of the best performing versions of both algorithms

(J4.8(cv) and RPart0) reveals that a confident decision concerning predictive superiority cannot be made. The number of differences in favor of J4.8(cv) is only two and no significant differences are reported on four data settings. A brief look at the complexity ranking (Table 3) and the underlying complexity dominance relations (Table 2, bottom) shows that J4.8 and its tuned version produce more complex trees than the RPart algorithms. While analogous analyses of comparing J4.8 versions to CTree do not indicate confident predictive performance differences, superiority of the J4.8 versions versus QUEST in terms of predictive accuracy is evident.

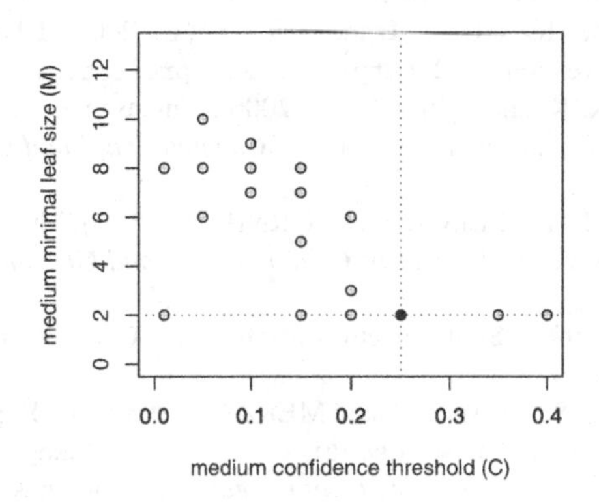

Fig. 3. Medians of the J4.8(cv) tuning parameter distributions for C and M

To aggregate the tuning results from J4.8(cv), Figure 3 depicts the median C and M parameters chosen for each of the 18 parameter distributions. It confirms the finding from the individual breast cancer and Pima Indians diabetes results (see Figure 2) that the parameter chosen by cross-validation can be far off the default values for C and M.

4 Discussion and further work

In this paper, we present results of a medium scale benchmark experiment with a focus on popular open-source tree-based learners available in R. With respect to our two main objectives – performance differences between C4.5 and CART, and heuristic choice of hyper-parameters – we can conclude: (1) The fully cross-validated J4.8(cv) and RPart0 perform better than their heuristic counterparts J4.8 (with fixed hyper-parameters) and RPart1 (employing a 1-standard-error rule). (2) In terms of predictive performance, no support for the claims of (clear) superiority of either algorithm can be found: J4.8(cv) and RPart0 lead to similar misclassification results, however J4.8(cv) tends to grow larger trees. Overall, this suggests that many beliefs or preconceptions about the classical tree algorithms should be (re-)assessed using benchmark studies. Our contribution is only a first step in this direction and further steps will require a larger study with additional datasets and learning algorithms.

References

BREIMAN, L., FRIEDMAN, J., OLSHEN, R. and STONE, C. (1984): *Classification and Regression Trees*. Wadsworth, Belmont, CA, 1984.

HORNIK, K. and MEYER, D. (2007): Deriving Consensus Rankings from Benchmarking Experiments In: *Advances in Data Analysis (Proceedings of the 30th Annual Conference of the Gesellschaft für Klassifikation e.V., March 8–10, 2006, Berlin)*, Decker, R., Lenz, H.-J. (Eds.), Springer-Verlag, 163–170.

HORNIK, K., ZEILEIS, A., HOTHORN, T. and BUCHTA, C. (2007): **RWeka***: An R Interface to* **Weka**. R package version 0.3-2. `http://CRAN.R-project.org/`.

HOTHORN, T., HORNIK, K. and ZEILEIS, A. (2006): Unbiased Recursive Partitioning: A Conditional Inference Framework. *Journal of Computational and Graphical Statistics*, 15(3), 651–674.

HOTHORN, T., LEISCH, F., ZEILEIS, A. and HORNIK, K. (2005): The Design and Analysis of Benchmark Experiments. *Journal of Computational and Graphical Statistics*, 14(3), 675–699.

LOH, W. and SHIH, Y. (1997): Split Selection Methods for Classification Trees. *Statistica Sinica*, 7, 815–840.

NEWMAN, D., HETTICH, S., BLAKE, C. and MERZ C. (1998): UCI Repository of Machine Learning Databases. `http://www.ics.uci.edu/~mlearn/MLRepository.html`.

QUINLAN, J. (1993): *C4.5: Programs for Machine Learning*. Morgan Kaufmann Publishers, Inc., San Mateo, CA.

R DEVELOPMENT CORE TEAM (2006): *R: A Language and Environment for Statistical Computing*. R Foundation for Statistical Computing, Vienna, Austria. ISBN 3-900051-07-0, `http://www.R-project.org/`.

THERNEAU, T. and ATKINSON, E. (1997): An Introduction to Recursive Partitioning Using the **rpart** Routine. *Technical Report*. Section of Biostatistics, Mayo Clinic, Rochester, `http://www.mayo.edu/hsr/techrpt/61.pdf`.

WITTEN, I., and FRANK, E. (2005): *Data Mining: Practical Machine Learning Tools and Techniques*. Morgan Kaufmann, San Francisco, 2nd edition.

From Spelling Correction to Text Cleaning – Using Context Information

Martin Schierle[1], Sascha Schulz[2], Markus Ackermann[3]

[1] DaimlerChrysler AG, Germany
martin.schierle@daimlerchrysler.com
[2] Humboldt-University, Berlin, Germany
Sascha.Schulz@gmail.com
[3] University of Leipzig, Germany
markus.ackermann@informatik.uni-leipzig.de

Abstract. Spelling correction is the task of correcting words in texts. Most of the available spelling correction tools only work on isolated words and compute a list of spelling suggestions ranked by edit-distance, letter-n-gram similarity or comparable measures. Although the probability of the best ranked suggestion being correct in the current context is high, user intervention is usually necessary to choose the most appropriate suggestion (Kukich, 1992).

Based on preliminary work by Sabsch (2006), we developed an efficient context sensitive spelling correction system *dcClean* by combining two approaches: the edit distance based ranking of an open source spelling corrector and neighbour co-occurrence statistics computed from a domain specific corpus. In combination with domain specific replacement and abbreviation lists we are able to significantly improve the correction precision compared to edit distance *or* context based spelling correctors applied on their own.

1 Introduction

In this paper we present a domain specific and context-based text preparation component for processing noisy documents in the automobile quality management. More precisely the task is to clean texts coming from vehicle workshops, being typed in at front desks by service advisors and expanded by technicians. Those texts are always written down under pressure of time, using as few words as possible and as many words as required to describe an issue. Consequently this kind of textual data is extremely noisy, containing common language, technical terms, lots of abbreviations, codes, numbers and misspellings. More than 10% of all terms are unknown with respect to commonly used dictionaries.

In literature basically two approaches are discussed to handle text cleaning: dictionary based algorithms like Aspell work in a semi-automatic way by presenting suggestions for unknown words, which are not in a given dictionary. For all of those words a list of possible corrections is returned and the user has to check the context to choose the appropriate one. This is applicable when supporting users in creating single documents. But for automatically processing large amounts of textual data this is impossible.

Context based spelling correction systems like WinSpell or IBMs csSpell only make use of context information. The algorithms are trained on a certain text corpus to learn probabilities of word occurrences. When analysing a new document for misspellings every word is considered as being suspicious. A context dependent probability for the appearance is calculated and the most likely word is applied. A more detailed introduction into context-based spelling correction can be found in Golding and Roth (1995), Golding and Roth (1999) and Al-Mubaid and Trümper (2001).

In contrast to existing work dealing *either* with context information *or* using dictionaries in our work we combine both approaches to increase efficiency as well as to assure high accuracy. Obviously, text cleaning is a lot more then just spelling correction. Therefore we add technical dictionaries, abbreviation lists, language recognition techniques and word splitting and merging capabilities to our approach.

2 Linguistics and context sensitivity

This section gives a brief introduction into underlying correction techniques. We distinguish two main approaches: linguistic and context based algorithms. Phonetic codes are a linguistic approach to pronunciation and are usually based on initial ideas of Russell and Odell. They developed the *Soundex* code, which maps words with similar pronunciations by assigning numbers to certain groups of letters. Together with the first letter of the word the first three digits result in the Soundex code where the same codes mean same pronunciation. Current developments are the *Metaphone* and the improved *Double Metaphone* (Phillips, 2000) algorithm by Lawrence Philips, the latter one is currently used by the ASpell algorithm. Edit distance based algorithms are a second member of linguistic methods. They calculate word distances by counting letter-wise transformation steps to change one word into another. One of the best known member is the *Levenshtein* algorithm, which uses the three basic operations replacement, insertion and deletion to calculate a distance score for two words.

Context based correction methods usually take advantage of two statistical measures: word frequencies and co-occurrences. The word frequency $\mathrm{f}(w)$ for a word w counts the frequency of its appearance within a given corpus. This is done for every unique word (or token) using the raw text corpus. The result is a list of unique words which can be ordered and normalised with the total number of words. With co-occurrences we refer to a pair of words, which commonly appear in similar contexts. Assuming statistical independence between the occurrence of two words w_1 and w_2, the estimated probability $P_E(w_1 w_2)$ of them occurring in the same context can be easily calculated with:

$$P_E(w_1 w_2) = P(w_1)P(w_2) \tag{1}$$

If the common appearance of two words is significantly higher than expected they can be regarded as co-occurrent. Given a corpus and context related word frequencies the co-occurrence measure can be calculated using pointwise mutual information or likelihood ratios. We utilize pointwise mutual information with a minimum threshold filtering.

$$\mathrm{Cooc}(w_1, w_2) = \log \frac{P(w_1 w_2)}{P(w_1)P(w_2)} \tag{2}$$

Depending on the size of a given context, one can distinguish between three kinds of co-occurrences. Sentence co-occurrences are pairs of words which appear significantly often together within a sentence, neighbour co-occurrences occur significantly often side by side and window-based co-occurrences are normally calculated within a fixed window size. A more detailed introduction into co-occurrences can be found in Manning and Schütze (1999) and Heyer et al. (2006).

3 Framework for text preparation

Before presenting our approach we will discuss requirements which we identified as being important for text preparation. They are based on general considerations but contain some application specific characteristics. In the second part we will explain our framework for a joint text preparation.

3.1 General considerations

Because we focus on automatically processing large amounts of textual documents we have to ensure *fast processing* and *minimal human interaction*. Therefore, after a configuration process the system must be able to clean texts autonomously. The *correction error has to be minimized*, but in contradiction to evaluations found in the literature we propose a very conservative correction strategy. If there is an unknown word it will only be corrected when certain thresholds are reached. As one can see during the evaluation we rather take a loss in recall than inserting an incorrect word. To detect words, which have to be corrected, we rely on *dictionaries*. If a word cannot be found in (general and custom prepared) dictionaries we regard it as suspicious. This is different to pure context based approaches; for instance Golding and Roth (1995) consider every word as suspicious. But this leads to two problems: first the calculation is computational complex and second we imply a new error probability of changing a correct word to an incorrect one. Even if the probability is below 0.1%, regarding 100.000 terms it would result in 100 misleading words.

Our proposed correction strategy can be seen as an interaction between linguistic and context based approaches, extended by word frequency and manually created replacement lists. The latter ones are essential to expand abbreviations, harmonize synonyms, and speed-up replacements of very common spelling errors.

3.2 Cleaning workflow

Automatic text cleaning covers several steps which depend on each other and should obviously be processed in a row. We developed a sequential workflow, consisting of independent modules, which are plugged into IBMs Unstructured Information Management Architecture (UIMA) framework. This framework allows to implement analysis modules for unstructured information like text documents and to plug them into a run-time environment for execution. The UIMA-Framework takes care about resource management, encapsulation of document data and analysis results and even distribution of algorithms on different machines. For processing documents, the

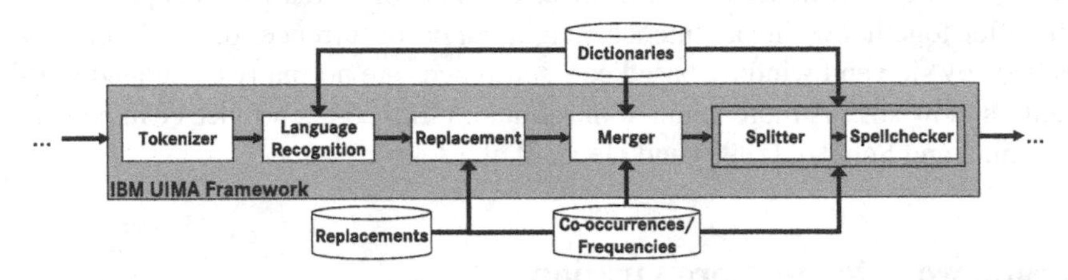

Fig. 1. *dcClean* workflow

UIMA framework uses annotations, the original text content is never changed, all correction tasks are recorded as annotations. The *dcClean* workflow (figure 1) consists of several custom-developed UIMA modules which we will explain in detail according to their sequence.

Tokenizer: The tokenizer is the first module, splitting the text into single tokens. It can recognise regular words, different number pattern (such as dates, mileages, money values), domain dependent codes and abbreviations. The challenge is to handle non-letter characters like slashes or dots. For example a slash between two tokens can indicate a sentence border, an abbreviation (e.g. "r/r" = "right rear") or a number pattern. To handle all different kinds of tokens we parametrize the tokenizer with a set of regular expressions.

Language recognition: Because all of the following modules use resources which might be in different languages, the text language has to be recognised before further processing. Therefore we included the LanI library from the University of Leipzig. Based on characteristic word frequency lists for different languages the library determines the language for a given document by comparing the tokens to the lists. Because we process domain specific documents, the statistical properties are different in comparison to regular language specific data. Thus it is important to include adequate term frequency lists.

Replacements: This module handles all kinds of replacements, because we deal with a lot of general and custom abbreviations and a pure spelling correction cannot handle this. We manually created replacement lists R for abbreviations, synonyms, certain multi-word-terms and misspellings, which are frequent but spelling correction algorithms fail to correct them properly. The replacement module works on word tokens, uses language dependent resources and incorporates context information. This is very important for the replacement of ambiguous abbreviations. If the module finds for example the word "lt", this can mean both "light" and "left". To handle this we look-up the co-occurrence levels of each possible replacement as left neighbour of the succeeding word $\text{Cooc}(w, w_{right})$ and as right neighbour of the preceding word $\text{Cooc}(w, w_{left})$. The result is the replacement w' with the maximum sum of co-occurrence level to the left neighbour word w_{left} and to the right neighbour word w_{right}.

$$\text{CoocSum}(w) = \text{Cooc}(w, w_{left}) + \text{Cooc}(w, w_{right}) \tag{3}$$

$$w' = \underset{w \in R}{\text{argmax}}(\text{CoocSum}(w)) \tag{4}$$

Merging: To merge words, which where split by an unintended blank character, this module sequentially checks two successive words for correct spelling, if one or both words are not contained in the dictionary but the joint word is, the two words get annotated as joint representation.

Splitting and spelling correction: The last module of our workflow treats spelling errors and word splittings. To correct words which are not contained in the dictionary, *dcClean* uses the Java based ASpell implementation Jazzy and incorporates word co-occurrences, word frequencies (both were calculated using a reference corpus), a custom developed weighting schema and a splitting component. If the

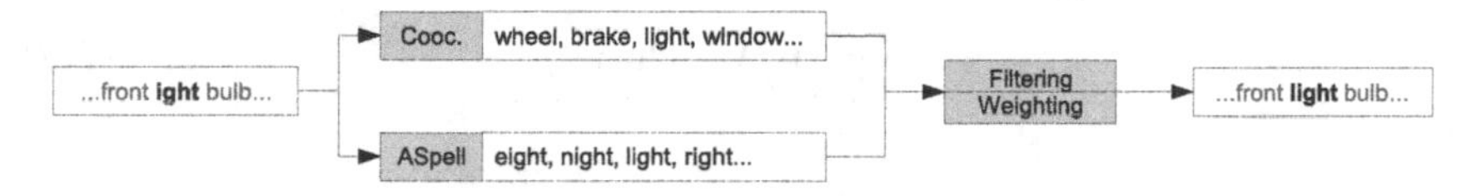

Fig. 2. Spelling correction example

module finds an unknown word w_m, it passes it to the ASpell component. The ASpell algorithm creates suggestions with the use of phonetic codes (Double Metaphone algorithm) and edit distances (Levenshtein distance). Therefore the algorithm creates a set S'_{ASpell} containing all words of dictionary D which have the same phonetic code as the potential misspelling w_m or as one of its variants $v \in V$ with an edit distance of one.

$$V = \{v | \text{Edit}(v, w_m) <= \theta_{edit1}\} \tag{5}$$

$$S'_{ASpell} = \{w | (\text{Phon}(w) = \text{Phon}(w_m) \vee \exists v : \text{Phon}(v) = \text{Phon}(w))\} \tag{6}$$

Then set S'_{ASpell} is filtered according to edit based threshold θ_{edit2}:

$$S_{ASpell} = \{w | w \in S'_{ASpell} \wedge \text{Edit}(w, w_m) < \theta_{edit2}\} \tag{7}$$

A context based set of suggestions S_{Cooc} is generated using co-occurrence statistics. Therefore we use a similar technique as during the replacements: this time we look-up all co-occurrent words as left neighbour of the succeeding word $\text{Cooc}(w, w_{right})$ and as right neighbour of the preceding word $\text{Cooc}(w, w_{left})$ of the misspelling.

$$S'_{Cooc} = \{w | \text{Cooc}(w, w_{left}) > \theta_{Cooc} \vee \text{Cooc}(w, w_{right}) > \theta_{Cooc}\} \tag{8}$$

The co-occurrence levels are summed up and filtered by Levenshtein distance measure to ensure a certain word based similarity.

$$S_{Cooc} = \{w | w \in S'_{Cooc} \wedge \text{Edit}(w, w_m) < \theta_{edit3}\} \quad (9)$$

The third set of suggestions S_{Split} is created using a splitting algorithm. This algorithm provides the capability to split words, which are unintentionally written as one word. Therefore the splitting algorithm creates a set of suggestions S_{Split}, containing all possible splittings of word w into two parts s_1^w and s_2^w with $s_1^w \in D$ and $s_2^w \in D$. To select the best matching correction $\tilde{w}$ the three sets of suggestions S_{ASpell}, S_{Cooc} and S_{Split} are joined

$$S = S_{ASpell} \bigcup S_{Cooc} \bigcup S_{Split} \quad (10)$$

and weighted according to their co-occurrence statistics, or – if there are no significant co-occurrents – according to their frequencies. For weighting the splitting suggestions we use the average frequencies or co-occurrence measures of both words. The correction $\tilde{w}$ is the element with the maximum weight.

$$\text{Weight}(w) = \begin{cases} \text{Cooc}(w), \text{ if } \exists w' \in S : \text{Cooc}(w') > \theta_{Cooc}, \\ \text{f}(w) \quad \text{else} \end{cases} \quad (11)$$

$$\tilde{w} = \underset{w \in S}{\text{argmax}}(\text{Weight}(w)) \quad (12)$$

4 Experimental results

To evaluate the pure spelling correction component of our framework we just consider error types which other spellcheckers can handle as well, which excludes merging, splitting or replacing words. We use a training corpus of one million domain specific documents (500MB of textual data) to calculate word frequencies and co-occurrence statistics. The evaluation is performed on a test set consisting of 679 misspelled terms including their context.

We compared *dcClean* with the dictionary based Jazzy spellchecker based on the ASpell algorithm and IBMs context based spellchecker csSpell. To get comparable results we set the Levenshtein distance threshold for *dcClean* and Jazzy to the same value, the confidence threshold for csSpell is set to 10% (This threshold is based on former experiments.). During the evaluation we counted words which were corrected accurately, words which were corrected, but the correction was mistaken, and words which were not changed at all (As explained in section 3.1 changes of correct words to incorrect ones are not considered, because this can be avoided by the use of dictionaries.).

As can be seen in figure 3 *dcClean* outperforms both spellcheckers using either dictionaries or context statistics. The improvement in relation to Jazzy is due to the fact, that the Aspell algorithm just returns a set of suggestions. For this evaluation we always chose the best suggestion. But sometimes there are several similar ranked suggestions and in a certain number of cases the best result is not the one that fits

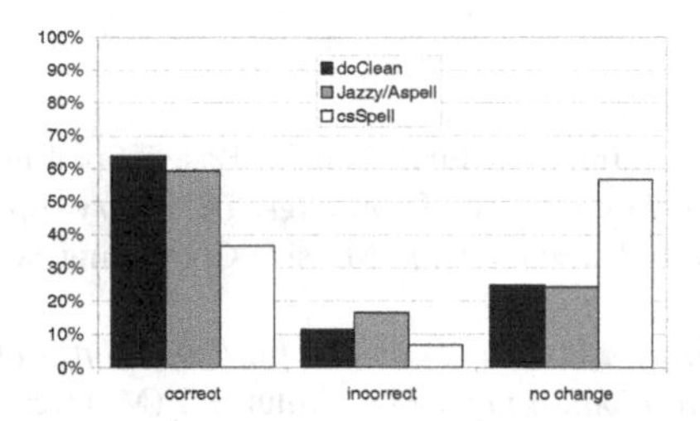

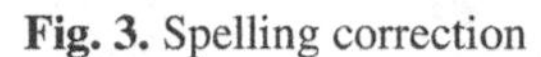
Fig. 3. Spelling correction

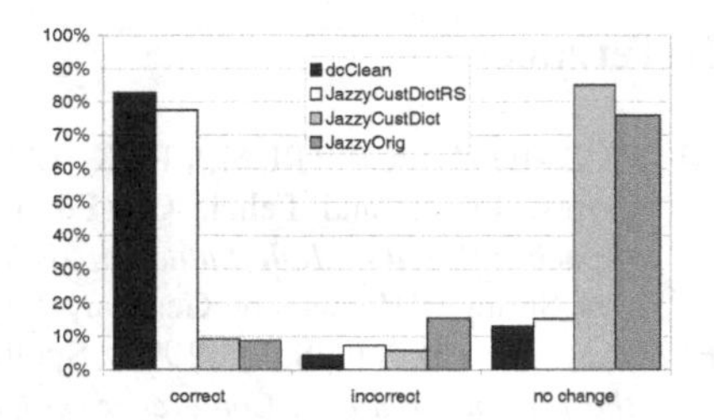

Fig. 4. Entire text cleaning workflow

the context. However, when solely choosing corrections from the context as done by csSpell, even a very low confidence threshold of 10% leads to the fact that most words are not changed at all. A reason for this are our domain specific documents where misspellings are sometimes as frequent as the correct word, or words are continuously misspelled.

To explain the need for a custom text cleaning workflow we show the improvements of our *dcClean* framework in comparison with a pure dictionary based spelling correction. Therefore we set up Jazzy with three different configurations: (1) using a regular English dictionary, (2) using our custom prepared dictionary and (3) using our custom prepared dictionary, splittings and replacements. The test set contains 200 documents with 769 incorrect tokens and 9137 tokens altogether. Figure 4 shows that Jazzy with a regular dictionary performs very poorly. Even with our custom dictionary there are only slight improvements. The inclusion of replacement lists leads to the biggest enhancements. This can be explained by the amount of abbreviations and technical terms used in our data. But *dcClean* with its context sensitivity outperforms even this.

5 Conclusion and future work

In this paper we explained how to establish an entire text cleaning process. We showed how the combination of linguistic and statistic approaches improves not only spelling correction but also the entire cleaning task. The experiments illustrated that our spelling correction component outperformed dictionary or context based approaches and that the whole cleaning workflow performs better than using only spelling correction.

In future work we will try to utilise *dcClean* for more languages than English, which might be especially difficult for languages like German, that have many compound words and a rich morphology. We will also extend our spelling correction component to handle combined spelling errors, like two words which are misspelled and accidentally written as one word. Another important part of our future work will be a more profound analysis of the suggestion weighting algorithm. A combination of frequency, co-occurrence level and Levenshtein distance may allow further improvements.

References

AL-MUBAID, H. and TRÜMPER, K. (2006): Learning to Find Context-Based Spelling Errors, In: E. and Felici. G. (Eds.):*Data Mining and Knowledge Discovery Approaches Based on Rule Induction Techniques. Triantaphyllou.* Massive Computing Series, Springer, Heidelberg, Germany, 597–628.

ELMI, M. A. and EVENS, M. (1998): Spelling correction using context. In: *Proceedings of the 17th international Conference on Computational Linguistics* - Volume 1 (Montreal, Quebec, Canada). Morristown, NJ, 360–364.

GOLDING, A. R (1995): A Bayesian hybrid method for context-sensitive spelling correction. In: *Proceedings of the Third Workshop on Very Large Corpora*, Boston, MA.

GOLDING, A. R., and ROTH, D. (1999): A Winnow based approach to context-sensitive spelling correction. *Machine Learning 34(1-3):107-130. Special Issue on Machine Learning and Natural Language.*

HEYER, G., QUASTHOFF, U. and WITTIG, T. (2006): *Text Mining: Wissensrohstoff Text – Konzepte, Algorithmen, Ergebnisse.* W3L Verlag, Herdecke, Bochum.

KUKICH, K. (1992). *Techniques for Automatically Correcting Words in Text. ACM Comput. Surv. 4:377–439.*

MANNING, C. and SCHÜTZE, H. (1999): *Foundations of Statistical Natural Language Processing.* The M.I.T. Press, Cambridge (Mass.) and London. 151–187.

PHILIPS, L. SABSCH, R. (2006): *Kontextsensitive und domänenspezifische Rechtschreibkorrektur durch Einsatz von Wortassoziationen.* Diplomarbeit, Universität Leipzig.

Root Cause Analysis for Quality Management

Christian Manuel Strobel[1] and Tomas Hrycej[2]

[1] University of Karlsruhe (TH), Germany
mstrobel@statistik.uni-karlsruhe.de
[2] Formerly DaimlerChrysler AG, Germany
tomas_hrycej@yahoo.de

Abstract. In industrial practice, quality management for manufacturing processes is often based on *process capability indices* (PCI) like C_p,C_{pm},C_{pk} and C_{pmk}. These indices measure the behavior of a process incorporating its statistical variability and location and provide a unitless quality measure. Unfortunately, PCIs are not able to identify those factors, having the major impact on quality as they are only based on measurement results and do not consider the explaining process parameters. In this paper an Operational Research approach, based on Branch and Bound is derived, which combines both, the numerical measurements and the nominal process factors. This combined approach allows to identify the main source for minor or superior quality of a manufacturing process.

1 Introduction

The quality of a manufacturing process can be seen as the ability to manufacture a certain product within its specification limits U, L and as close as possible to its target value T, describing the point, where its quality is optimal. In literature, numerous process capability indices have been proposed in order to provide a unitless quality measures to determine the performance of a process, relating the preset specification limits to the actual behavior (Kotz and Johnson (2002)). This behavior can be described by the process variation and process location. In order to state future quality of a manufacturing process based on the past performance, the process is supposed to be stable or in control. This means, that both, process mean and process variation has to be, on the long run, in between pre-defined limits. A common technique to monitor this are control charts, one of the tools, provided by Statistical Process Control.

The basic idea for the most common indices is to assume, that the considered manufacturing process follows a normal distribution and the distance between the upper and lower specification limit should equal 6σ. The commonly recognized "basic" PCIs C_p,C_{pm},C_{pk} and C_{pmk} can be summarized by a superstructure, which was introduced in Vännman (1995) and is referred to in literature as $C_p(u,v)$:

$$C_p(u,v) = \frac{d - u|\mu - M|}{3\sqrt{\sigma^2 + v(\mu - T)^2}} \tag{1}$$

where σ is the process standard deviation, μ the process mean, $d = (U - L)/2$ tolerance width, $m = (U + L)/2$ the mid-point between the two specification limits and T the target value. The "basic" PCIs can be obtained by setting u and v to:

$$\begin{aligned} C_p &\equiv C_p(0,0); \; C_{pk} \equiv C_p(1,0) \\ C_{pm} &\equiv C_p(0,1); \; C_{pmk} \equiv C_p(1,1) \end{aligned} \tag{2}$$

Estimators for the indices can be obtained by substituting μ by the sample mean $\overline{X} = \sum_{i=1}^{n} X_i/n$ and σ by the sample variance $S^2 = \sum_{i=1}^{n}(X_i - \overline{X})^2/(n-1)$. They provide stable and reliable point estimators for processes following a normal distribution. However, in practice, this requirement is hardly met, thus the basic PCIs as defined in (1) are not appropriated for process with non-normal distributions. What is really needed are indices which do not depend on any kind of distribution in order to be useful for measuring quality of a process.

$$C_p'(u,v) = \frac{d - u|m - M|}{3\sqrt{[\frac{F_{99.865} - F_{0.135}}{6}]^2 + v(m - T)^2}} \tag{3}$$

In Pearn and Chen (1997) a generalization of the PCIs superstructure (1) is introduced, in order to cover those cases, where the underlying data does not follow a Gaussian distribution. The authors replaced the process standard deviation σ by the 99.865 and 0.135 quantiles of the empiric distribution function and μ by the median of the process. The idea behind this substitution is, that the difference between the quantiles $F_{99.865}$ and $F_{0.135}$ again equals 6σ or $C_p'(u,v) = 1$, assuming the special case, that the process follows a gaussian distribution. The special PCIs C_p',C_{pm}',C_{pk}' and C_{pk}' can be obtained by applying u and v as in (2).

Assuming that the following assumptions hold, a class of non-parametric indices and a particular specimen thereof can be introduced: every manufacturing process is defined by two distinct sets. Let Y be the set of influence variables (process parameters or process factors) and X the corresponding goal variables or measurements results, then a class of process indices can be defined as:

Definition 1. *Let X and Y describe a manufacturing process. Furthermore, let $f(x,y)$ be the empirical density of the underlying process and $w(x)$ a kernel function. Then*

$$Q := \frac{\int_x \int_y w(x) f(x,y) dy dx}{\int_x \int_y f(x,y) dy dx} \tag{4}$$

defines a class of empirical process indices.

Obviously, if $w(x) = x$ or $w(x) = x^2$ we obtain the first and resp. the second moment of the process, as $\int_x \int_y f(x,y) dy dx \equiv 1$. But, to measure the quality of a process, we are interested in the relationship of the designed specification limits and the process behavior. A possibility is to chose the kernel function $w(x)$ in such way, that it becomes a function of the designed limits U and L.

Definition 2. *Let X, Y and $f(x,y)$ be defined as in definition 1. Let U,L be specification limits. The Empirical Capability Index (E_{ci}) is defined as:*

$$E_{ci} = \frac{\int_x \int_y \mathbf{1}_{(L \leq x \leq U)} f(x,y) dy dx}{\int_x \int_y f(x,y) dy dx} \tag{5}$$

The E_{ci} measures the percentage of data points which are in between the specification limits U and L. Therefore, it is more sensitive to outliners compared to the common, non-parametric indices. A disadvantage is, that for processes, having all data points within the specification limits, the E_{ci} always equals one, and therefore does not provide a comparable quality measure. To avoid this, the specification limits U and L have to be modified, in order to get "further into the sample", by linking them to the behavior of the process.

However, after measuring the quality of a process, one might be interested if there are subsets of influence variables values, such that the quality of a process becomes better, if constraining the process only to this parameters. In the following section a non-parametric, numerical approach for identify those parameters is derived and an algorithm, which efficiently solves this problem is presented.

2 Root Cause Analysis

In literature a common technique to identify significant discrete parameters having an impact on numeric variables like measurement results, is the Analysis of Variance (ANOVA). As a limiting factor, techniques of the Variance Analysis are only useful, if the problem is of lower dimension. Additionally these variables should be well balanced or have a simple structure. Another constraint is the assumption, that the analyzed data has to follow a multivariate Gaussian distribution. In most applications these requirements are hardly ever met. The distribution of the parameters describing the measure variable is in general not Gaussian and of higher dimension. Also the combinations of the cross product of the parameters are non-uniformly and sparely populated nor have a simple dependence structure. Therefore, the method of Variance Analysis is not applicable. What is really needed, is a more general approach to identify the variables, responsible for minor or superior quality.

A process can be defined as a set of influence variables (i.e. process parameters) $Y = (Y^1, \ldots, Y^n)$ consisting of values $Y^i = y^i_1, \ldots, y^i_{m_i}$ and a set of corresponding goal variables (i.e. measurement results) $X = (X_1, \ldots, X_n)$. If constraining the influence variable values to a subset $\overline{Y} \subseteq Y$, $\overline{Y}$ defines a *sub-process* of the original process Y. The support of a sub-process $\overline{Y}$ can be written as $N(X|\overline{Y}) := \int_x \int_{y \in \overline{Y}} f(x,y) dy dx$ and consequently, a conditional PCI is defined as $Q(X|\overline{Y})$. Any of the indices defined in the previous section can be used, whereby the value of the respective index is calculated on the conditional subset $\overline{X} \subseteq X$.

In order to determine those parameters having the greatest impact on quality, an optimal sub-process, consisting of optimal influence combinations, has to be identified. A naive approach would be, to maximize $Q(X|\overline{Y})$ over all sub-processes $\overline{Y} \in Y$.

Unfortunately, in general this yields a sub-process, which would only have a limited support $(N(X|Y^*)\langle n)$. A better approach is to think in economic terms and to weighten the factors responsible for minor quality, which we want to remove, by the costs of eliminating them. In practise this is not feasible, as to track the actual costs is too expensive. But it is likely, that rare factors, which are responsible for lower quality are "cheaper" to remove than frequent influences. In other words, sub-processes with high support are preferable.

Often the available sample set for process optimization is small, having numerous influence variables but only few measurement results. By limiting ourselves only to combinations of variables, we might get too small sub-process (having low support). Therefore, we extend the possible solutions to combinations of variables and their values - the search space for optimal sub-processes is spanned by the powerset of the influence parameters $\mathbb{P}(Y)$. The two sided problem, to find the parameter set combining on the one hand a optimal quality measure and on the other hand a maximal support, can be summarized by the the following optimization problem:

Definition 3.

$$(P) = \begin{cases} N(X|\underline{\bar{Y}}) \rightarrow max \\ \underline{Q}(X|\underline{Y}) \geq q_{min} \\ \underline{Y} \in \mathbb{P}(Y) \end{cases}$$

The solution of the optimization problem is the subset of process parameters with maximal support among those processes, having a better quality than the given threshold q_{min}. Often, q_{min} is set to the common values for process capability of 1.33 or 1.67.

Due to the nature of the application domain, the investigated parameters are discrete which inhibits an analytical solution but allows the use of *Branch and Bound* techniques. In the following we derive an algorithm which solves the optimization problem (3) by avoiding the evaluation of the exponential amount of possible combinations, spanned by the cross product of the influence parameters. In order to achieve this, a efficient cutting rule is derived in the next section.

Branch and bound algorithm

To efficiently store and access the necessary information and to apply Branch and Bound techniques, a multitree was chosen as representing data structure. Each node of the multitree represents a possible combination of the influence parameters (sub-process) and is build out of the combination of the parents influence set and a new influence variable and its value(s). Fig. 1 depicts the data structure, whereby each nodes stands for all elements of the powerset of the considered variable.

To find the optimal solution to the optimization problem (3), a depth-first search is applied to traverse the tree using a Branch and Bound principle. The idea, to branch and bound the traverse of the tree is based on the following thoughts: by descending a branch of the tree, the number of constraints is increasing, as new influence variables are added and therefore the sub-process support decreases (compare Fig. 1). Thus, if a node has a support lower than a given minimum support, there is no possibility

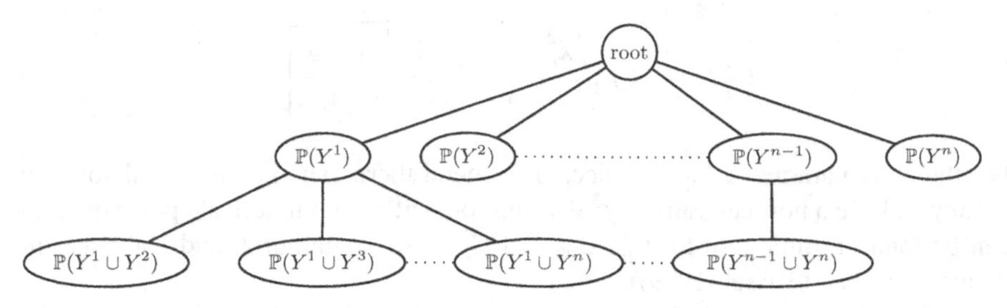

Fig. 1. Organization of the used multitree data structure

to find a node (sub-process) with a higher support in the branch below. This reduces the time to find the optimal solution significantly, as a good portion of the tree to traverse, can be omitted.

Algorithm 1 Branch & Bound algorithm for process optimization

1: **procedure** TRAVERSETREE($\bar{Y}$)
2: $\quad \mathcal{Y} := \{\text{sub-nodes of } \bar{Y}\}$
3: $\quad$ **for all** $y \in \mathcal{Y}$ **do**
4: $\quad\quad$ **if** $N(X|y) > n_{max}$ and $Q(X|y) \geq q_{min}$ **then**
5: $\quad\quad\quad n_{max} = N(X|y)$
6: $\quad\quad$ **end if**
7: $\quad\quad$ **if** $N(X|y) > n_{max}$ and $Q(X|y) < q_{min}$ **then**
8: $\quad\quad\quad$ `TraverseTree`(y)
9: $\quad\quad$ **end if**
10: $\quad$ **end for**
11: **end procedure**

In many real world applications, the influence domain is mixed, consisting of discrete data and numerical variables. To enable a joint evaluation of both influence types, the numerical data is transformed into nominal data by mapping the continuous data onto pre-set quantiles. In most our applications, we chose 10%, 20%, 80% and 90% quantile, as they performed the best.

Verification

The optimum of the problem (3) can only be defined in statistical terms, as in practice the sample sets are small and the quality measures are only point estimators. Therefore, confidence intervals have to be used in order to get a more valid statement of the real value of the considered PCI. In the special case, where the underlying data follows a normal distribution, it is straight forward to construct a confidence interval. As the distribution of $\frac{C_p}{\hat{C}_p}$ ($\hat{C}_p$ denotes the estimator of C_p) is known, a $(1-\alpha)\%$ confidence interval for C_p is given by

$$C(X) = \left[\hat{C}_p\sqrt{\frac{\chi^2_{n-1;\frac{\alpha}{2}}}{n-1}}, \hat{C}_p\sqrt{\frac{\chi^2_{n-1;1-\frac{\alpha}{2}}}{n-1}}\right] \tag{6}$$

For the other parametric basic indices, in general there exits no analytical solution as they all have a non-centralized χ^2 distribution. Different numerical approximation can be found in literature for C_{pm},C_{pk} and C_{pmk} (see Balamurali and Kalyanasundaram (2002) and Bissel (1989)).

If there is no possibility to make an assumption about the distribution of the data, computer based, statistical methods as the Bootstrap method are used to calculate a confidence intervals. In Balamurali and Kalyanasundaram (2002), the authors present three different methods for calculating confidence intervals and a simulation study. As result, the method called BCa-Method outperformed the other two methods, and therefore is used in our applications for assigning confidence intervals for the non-parametric basic PCIs, as described in (3). For the Empirical Capability Index E_{ci} a simulation study showed that the Bootstrap-Standard-Method, as defined in Balamurali and Kalyanasundaram (2002), performed the best. A (1-α)% confidence interval for the E_{ci} can be obtained by

$$C(X) = \left[\hat{E}_{ci} - \Phi^{-1}(1-\alpha)\sigma_B, \hat{E}_{ci} + \Phi^{-1}(1-\alpha)\sigma_B\right] \tag{7}$$

where $\hat{E}_{ci}$ denotes an estimator for E_{ci}, σ_B the Bootstrap standard deviation and Φ^{-1} the inverse standard normal.

As the results of the introduced algorithm are based on sample sets, it is important to verify the soundness of the founded solutions. Therefore, the sample set to analyze is to be randomly divided into two disjoint sets: training and test set. A set of possible optimal sub-process is generated, by applying the describe algorithm and the referenced Bootstrap-methods to calculate confidence intervals. In a second step, the root cause analysis algorithm is applied to the test set. The final output is a verified sub-process.

3 Computational results

A proof on concept was performed using data of a foundry plant and engine manufacturing in the premium automotive industry. The 32 analyzed sample sets comprised measurement results describing geometric characteristics like the position of drill holes or surface texture of the produced products and the corresponding influence sets. The data sets consist of 4 to 14 different values, specifying for example a particular machine number or a workers name. An additional data set, recording the results of a cylinder twist measurement having 76 influence variables, was used to evaluated the algorithm for numerical parameter sets. Each of the analyzed data sets has at least 500 and at most 1000 measurement results.

The evaluation was performed for the non-parametric C_p and the empirical capability index E_{ci} using the describe Branch and Bound principle. Additionally a

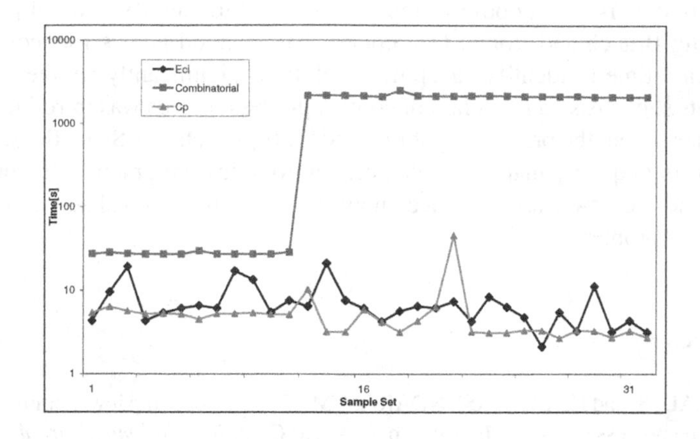

Fig. 2. Computational time for combinatorial search vs. Branch and Bound

combinatorial search for the optimal solution was carried out to demonstrate the efficiency of our approach. The reduction of computational time, using the Branch and Bound principle, amounted to two orders of magnitude in comparison to the combinatorial search as can be seen in Fig. 2. In average, the Branch and Bound method outperformed the combinatorial search by the factor of 230. For the latter it took in average 23 minutes to evaluating the available data sets. However, using Branch and Bound reduced the computing time in average to only 5.7 seconds for the non-parametric C_p and to 7.2 seconds using the E_{ci}. The search for an optimal solution was performed to depth of 4, which means, that all sub-process have no more than 4 different influence variables. A higher depth level did not yield any other results, as the support of the sub-processes diminishes with increasing number of influence variables. Obviously, the computational time for finding the optimal sub-process increases with the number of influence variables and their values. This fact explains the significant jump of the combinatorial computing time, as the first 12 sample sets are made up of only 4 influence variables, whereas the others consist of up to 17 different influence variables.

As the number of influence parameters of the numerical data set where, compared to the other data sets, significantly larger, it took, about 2 minutes to find the optimal solution. The combinatorial search was not performed, as 76 influence variables each with 4 values would have take too long.

4 Conclusion

In this paper we have presented a root cause analysis algorithm for process optimization, with the goal to identify those process parameters having a server impact on the quality of a manufacturing process. The basic idea was to transform the search for those quality drivers into a optimization problem and to identify optimal parameter subsets using Branch and Bound techniques. This method allows for reducing the computational time to identifying optimal solutions significantly, as the computational results show. Also a new class of convex process indices was introduced and a particular specimen, the process capability index, E_{ci} is defined. Since the search for quality drivers in quality management is crucial to industrial practice, the presented algorithm and the new class of indices may be useful for a broad scope of quality and reliability problems.

References

BALAMURALI S. and KALYANASUNDARAM M. (2002): Bootstrap lower confidence limits for the process capability indices Cp, Cpk and Cpm. *International Journal of Quality & Reliability Management, 19, 1088–1097.*

BISSELL A. (1990): How Reliable is Your Capability Index? *Applied Statistics, 39, 331–340* .

KOTZ, S. and JOHNSON, N. (2002): Process Capability Indices – A Review, 1992 2000. *Journal of quality technology, 34, 2–53.*

PEARN, W. and CHEN. K. (1997): Capability indices for non-normal distributions with an application in electrolytic capacitor manufacturing. *Microelectronics Reliability, 37, 1853–1858.*

VÄNNMANN, K. (1995): A Unified Approach to Capability Indices. *Statistica Sinica, 5, 805–820* .

Finding New Technological Ideas and Inventions with Text Mining and Technique Philosophy

Dirk Thorleuchter

Fraunhofer INT, Appelsgarten 2, 53879 Euskirchen, Germany
Dirk.Thorleuchter@int.fraunhofer.de

Abstract. Text mining refers generally to the process of deriving high quality information from unstructured texts. Unstructured texts come in many shapes and sizes. It may be stored in research papers, articles in technical periodicals, reports, documents, web pages etc. Here we introduce a new approach for finding textual patterns representing new technological ideas and inventions in unstructured technological texts.

This text mining approach follows the statements of technique philosophy. Therefore a technological idea or invention represents not only a new mean, but a new purpose and mean combination. By systematic identification of the purposes, means and purpose-mean combinations in unstructured technological texts compared to specialized reference collections, a (semi-) automatic finding of ideas and inventions can be realized. Characteristics that are used to measure the quality of these patterns found in technological texts are comprehensibility and novelty to humans and usefulness for an application.

1 Introduction

The planning of technological and scientific research and development (R&D-) programs is a very demanding task, e.g. in the R&D-program of the German ministry of defense there are at least over 1000 different R&D-projects running simultaneously. They all refer to about 100 different technologies in the context of security and defense. There is always a lot of change in these programs - a lot of projects starting new and a lot of projects running out. One task of our research group is finding new R&D-areas for this program. New ideas or new inventions are a basis for a new R&D-area. That means for planning new R&D-areas it is necessary to identify a lot of new technological ideas and inventions from the scientific community (Ripke et al. (1972)). Up to now, the identification of new ideas and inventions in unstructured texts is done manually (that means by humans) without the support of text mining. Therefore in this paper we will describe the theoretical background of the text mining approach to discover (semi-) automatically textual patterns representing new ideas and inventions in unstructured technological texts.

Hotho (2004) describes the characteristics that are used to measure the quality of these textual patterns extracted by knowledge discovery tasks. The characteristics are comprehensibility and novelty to the users and usefulness for a task. In this paper the users are program planers or researchers and the task is to find ideas and inventions which can be used as basis for new R&D-areas.

It is known from the cognition research that analysis and evaluation of textual information requires the knowledge of a context (Strube (2003)). The selection of the context depends on the users and the tasks. Referring to our users and our task, we have on one hand textual information about world wide existing technological R&D-projects (furthermore this is called "raw information"). This information contains a lot of new technological ideas and inventions. New means, that ideas and inventions are unknown to the user (Ipsen (2002)). On the other hand we have descriptions about own R&D-projects. This represents our knowledge base and furthermore this is called "context information". Ideas and inventions in the context information are already known to the user.

To create a text mining approach for finding ideas and inventions inside the raw information we have to create a common structure for raw and context information first. This is necessary for the comparison between raw and context information e.g. to distinguish new (that means unknown) ideas and inventions from known ideas and inventions.

In short we have to do 2 steps: 1. Create a common structure for raw and context information as a basis for the text mining approach. 2. Create a text mining approach for finding new, comprehensible and useful ideas and inventions inside the raw information. Below we describe step 1 and 2 in detail.

2 A common structure for raw and context information

In order to perform knowledge discovery tasks (e.g. finding ideas and inventions) it is required that raw information and context information have to be structured and formatted in a common way as described above. In general the structure should be rich enough to allow for interesting knowledge discovery operations and it should be simple enough to allow an automatically converting of all kind of textual information in a reasonable cost as described by Feldman et al.(1995).

Raw information is stored in research papers, articles in technical periodicals, reports, documents, databases, web pages etc. That means raw information contains a lot of different structures and formats. Normally context information also contains different structures and formats. Converting all structures and formats to a common structure and format for raw and context information by keeping all structure information available costs plenty of work. Therefore our structure approach is to convert all information into plain text format. That means firstly we destroy all existing structures and secondly build up a new common structure for raw and context information.

The new structure should refer to the relationship between terms or term-combinations (Kamphusmann (2002)). In this paper we realize this by creating sets of domain specific terms which occur in the context of a term or a combination of terms. For the structure formulation we define the term unit as word.

First we create a set of domain specific terms.

Definition 1. *Let (a text) $T = [\omega_1, .., \omega_n]$ be a list of terms (words) ω_i in order of appearance and let $n \in N$ be the number of terms in T and $i \in [1, .., n]$. Let $\Sigma = \{\tilde{\omega}_1, .., \tilde{\omega}_m\}$ be a set of domain specific stop terms (Lustig (1986)) and let $m \in N$ be the number of terms in Σ. Ω - the set of domain specific terms in text T - is defined as the relative complement $T without \Sigma$. Therefore:*

$$\Omega = T \backslash \Sigma \tag{1}$$

For each $\omega_i \in \Omega$ we create a set of domain specific terms which occur in the context of term ω_i.

Definition 2. *Let $l \in N$ be a context length of term ω_i that means the maximum distance between ω_i and a term ω_j in text T. Let the distance be the number of terms (words) which occur between ω_i and ω_j including the term ω_j and let $j \in [1, .., n]$. Φ_i is defined as a set of those domain specific terms which occur in an l-length context of term ω_i in text T:*

$$\Phi_i = \{\omega_j | (\omega_j \in \Omega) \wedge (|i - j| \leq l) \wedge (\omega_i \not\equiv \omega_j)\} \tag{2}$$

For each combination of terms in Φ_i we create a set of domain specific terms which occur in the context of this combination of terms.

Definition 3. *Let $\delta_p \in \Omega$ be a term in a list of terms with number $p \in [1, .., \mu]$. Let $\delta_1, .., \delta_\mu$ be a list of terms - in further this will be called term-combination - with $\delta_p \not\equiv \delta_q \forall p \not\equiv q \in [1, .., \mu]$ that occurs together in an l-length context of term δ_1 in text T. Let $\mu \in N$ be the number of terms in the term-combination $\delta_1, .., \delta_\mu$. $\Xi^T_{\delta_1,..,\delta_\mu}$ is defined as the set of domain specific terms which occur together with the term combination $\delta_1, .., \delta_\mu$ in an l-length context of term δ_1 in text T:*

$$\Xi^T_{\delta_1,..,\delta_\mu} = \bigcup \Phi_i \backslash \bigcup_{p=2}^{\mu} \delta_p \Bigg| \delta_1 = \omega_i \wedge \bigcup_{p=2}^{\mu} \delta_p \subset \Phi_i \tag{3}$$

In the Figure 1 an example for the relationships in set $\Xi^T_{\delta_1,..,\delta_\mu}$ is presented. The term-combination (sensor, infrared, uncooled) has a relationship to the term-combination (focal, array, plane) because uncooled infrared sensors can be built by using the focal plane array technology.

The text T could be a) the textual raw information or b) the textual context information. As result we get in case of a) $\Xi^{raw}_{\delta_1,..,\delta_\mu}$ and in case of b)$\Xi^{context}_{\delta_1,..,\delta_\mu}$.

Definition 4. *To identify terms or term-combinations in the raw information which also occur in the context information - that means the terms or term-combinations are known to the user - we define $\Xi^{known}_{\delta_1,..,\delta_\mu}$ as the set of terms which occur in $\Xi^{raw}_{\delta_1,..,\delta_\mu}$ and $\Xi^{context}_{\delta_1,..,\delta_\mu}$:*

$$\Xi^{known}_{\delta_1,..,\delta_\mu} = \Xi^{raw}_{\delta_1,..,\delta_\mu} \cap \Xi^{context}_{\delta_1,..,\delta_\mu} \tag{4}$$

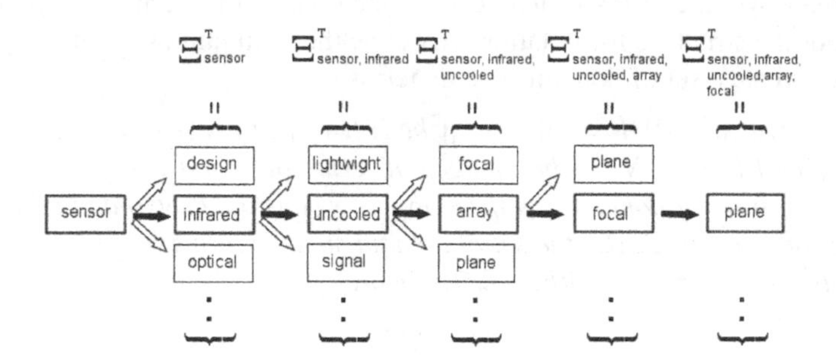

Fig. 1. Example for the relationships in $\Xi^T_{\delta_1,..,\delta_\mu}$: Uncooled infrared sensors can be build by using the focal plane array technology.

3 Relevant aspects for the text mining approach from technique philosophy

The text mining approach follows the statements of technique philosophy (Rohpohl (1996)). Below we describe some relevant aspects of the statements and some specific conclusions for our text mining approach.

a) A technological idea or invention represents not only a new mean, but a new purpose and mean combination. That means to find an idea or invention it is necessary to identify a mean and an appertaining purpose in the raw information. Appertaining means that purpose and mean shall occur together in an l-length context. Therefore for our text mining approach we firstly want to identify a mean and secondly we want to identify an appertaining purpose or vice versa.
b) Purposes and means can be exchanged. That means a purpose can become a mean in a specific context and vice versa. Example: A raw material (mean) is used to create an intermediate product (purpose). The intermediate product (mean) is then used to produce a product (purpose). In this example the intermediate product changes from purpose to mean because of the different context. Therefore for our text mining approach it is possible to identify textual patterns representing means or purposes. But it is not possible to distinguish between means and purposes without the knowledge of the specific context.
c) A purpose or a mean is represented by a technical term or by several technical terms. Therefore purposes or means can be represented by a combination of domain specific terms (e.g. $\delta_1,..,\delta_\mu$) which occur together in an l-length context. The purpose-mean combination is a combination of 2 term-combinations and it also occurs in an l-length context as described in 3 a). For the formulation a term-combination $\delta_1,..,\delta_\mu$ represents a mean (a purpose) only if $\Xi^{raw}_{\delta_1,..,\delta_\mu} \neq \oslash$, which means there are further domain-specific terms representing a purpose (a mean) which occur in an l-length context together with the term-combination $\delta_1,..,\delta_\mu$

in the raw information.

d) To find an idea or invention that is really new to the user, the purpose-mean combination must be unknown to the user. That means a mean and an appertaining purpose in the raw information must not occur as mean and as appertaining purpose in the context information. For the formulation the term-combination $\delta_1,..,\delta_\mu$ from 3 c) represents a mean (a purpose) in a new idea or invention only if $\Xi^{known}_{\delta_1,..,\delta_\mu} = \oslash$, which means there are no further domain-specific terms which occur in an l-length context together with the term-combination $\delta_1,..,\delta_\mu$ in the raw and in the context information.

e) To find an idea or invention that is comprehensible to the user, either the purpose or the mean must be known to the user. That means one part (a purpose or a mean) of the new idea or invention is known to the user and the other part is unknown. The user understand the known part because it is also a part of a known idea or invention that occurs in the context information and therefore he gets an access to the new idea or invention in the raw information.
That means the terms representing either the purpose or the mean in the raw information must occur as purpose or mean in the context information. For the formulation the term-combination $\delta_1,..,\delta_\mu$ from 3 d) represents a mean (a purpose) in a comprehensible idea or invention only if $\Xi^{context}_{\delta_1,..,\delta_\mu} \neq \oslash$, which means $\delta_1,..,\delta_\mu$ is known to the user and there are further domain-specific terms representing a purpose (a mean) which occur in an l-length context together with the term-combination $\delta_1,..,\delta_\mu$ in the context information.

f) Normally an idea or an invention is useful for a specific task. Transferring an idea or an invention to a different task makes it sometimes necessary that the idea or invention has to be changed to become useful for the new task. To change an idea or invention you have to change either the purpose or the mean. That is because the known term-combination $\delta_1,..,\delta_\mu$ from 3 e) must not be changed, otherwise it will become unknown to the user and then the idea or invention is not comprehensible to the user as described in 3 e).

g) After some evaluation we get the experience that for finding ideas and inventions the number of known terms (e.g. representing a mean) and the number of unknown terms (e.g. representing the appertaining purposes) shall be well balanced. Example: one unknown term among many known terms often indicates that an old idea got a new name. Therefore the unknown term is probably not a mean or a purpose. That means the probability that $\delta_1,..,\delta_\mu$ is a mean or a purpose increases when μ is close to the cardinality of $\Xi^{raw}_{\delta_1,..,\delta_\mu}$.

h) There are often domain specific stop terms (like better, higher, quicker, integrated, minimized etc.) which occur with ideas and inventions. They point to a changing purpose or a changing mean and can be indicators for ideas and inventions.

i) An identified new idea or invention can be a basis for further new ideas and inventions. That means all ideas and inventions that are similar to the identified new idea and invention are also possible new ideas and inventions.

4 A text mining approach for finding new ideas and inventions

In this paper we want to create a text mining approach by applying point 3 a) to 3 g). Further we want to prove the feasibility of our text mining approach.

Firstly we want to identify a mean and secondly we want to identify an appertaining purpose below as described in 3 a). The other case - firstly identify a purpose and secondly identify an appertaining mean - is trivial because of the purpose-mean dualism described in 3 b).

Definition 5. *We define $p(\Xi^{raw}_{\delta_1,..,\delta_\mu})$ as the probability that the term-combination $\delta_1,..,\delta_\mu$ in the raw information is a mean. That means whether μ is close to the cardinality of $\Xi^{raw}_{\delta_1,..,\delta_\mu}$ or not as described in 3 g):*

$$p(\Xi^{raw}_{\delta_1,..,\delta_\mu}) = \begin{cases} \frac{\left|\Xi^{raw}_{\delta_1,..,\delta_\mu}\right|}{\mu} & \mu > \left|\Xi^{raw}_{\delta_1,..,\delta_\mu}\right| \\ \frac{\mu}{\left|\Xi^{raw}_{\delta_1,..,\delta_\mu}\right|} & \mu \leq \left|\Xi^{raw}_{\delta_1,..,\delta_\mu}\right| \end{cases} \quad (5)$$

The user determines a minimum probability p_{min}. For the text mining approach the term-combinations $\delta_1,..,\delta_\mu$ are means only if

a) $\Xi^{raw}_{\delta_1,..,\delta_\mu} \not\equiv \oslash$ as described in 3 c),

b) $\Xi^{known}_{\delta_1,..,\delta_\mu} = \oslash$ as described in 3 d) to get a new idea or invention,

c) $\Xi^{context}_{\delta_1,..,\delta_\mu} \not\equiv \oslash$ as described in 3 e) to get a comprehensible idea or invention and

d) $p(\Xi^{raw}_{\delta_1,..,\delta_\mu}) \geq p_{min}$ as described in 3 g).

For each of these term-combinations we collect all appertaining purposes (that means the combinations of all further terms) which occur in an l-length context together with $\delta_1,..,\delta_\mu$ in the raw information.

We present each $\delta_1,..,\delta_\mu$ as a known mean and all appertaining unknown purposes to the user. The user selects the suited purposes for his task or he combines some purposes to a new purpose. That means he changes the purpose to become useful for his task as described in 3 f). Additionally it is possible that the user changes known means to known purposes and appertaining purposes to appertaining means as described in 3 b) because at this point the user gets the knowledge of the specific context.

With this selection the user gets the purpose-mean combination that means he gets an idea or invention. This idea or invention is novel to him because of 3 d) and it is comprehensible to him because of 3 e). Further it is useful for his application because the user selects the suited purposes for his task.

5 Evaluation and outlook

We have done a first evaluation with a text about R&D-projects from the USA as raw information (Fenner et al. (2006)), a text about own R&D-projects as context information (Thorleuchter (2007)), a stop word list created for the raw information and the parameter values $l = 8$ and $p_{min} = 50\%$. The aim is to find new, comprehensible and useful ideas and inventions in the raw information. According to human experts the number of these relevant elements - the so-called "ground truth" for the evaluation - is eighteen. That means eighteen ideas or inventions can be used as basis for new R&D-areas. With the text mining approach we extracted about fifty patterns (retrieved elements) from the raw information. The patterns have been evaluated by the experts. Thirteen patterns are new, comprehensible and useful ideas or inventions that means thirteen from fifty patterns are relevant elements. Five new, comprehensible and useful ideas or inventions are not found by the text mining approach. Therefore, as result we get a precision value of about 26% and a recall value of about 72%. This is not representative because of the small number of relevant elements but we think this is above chance and it is sufficient to prove the feasibility of the approach.

For future work firstly we will enlarge the stop word list to a general stop word list for technological texts and optimize the parameters concerning the precision and recall value. Secondly we will enlarge the text mining approach with further thoughts e.g. the two thoughts described in 3 h) and 3 i). The aim of this work shall be to get better results for the precision and recall value. Thirdly we will implement the text mining approach to a web based application. That will help the users to find new, comprehensible and useful ideas and inventions with this text mining approach. Additionally with this application it will be easier for us to do a representative evaluation.

6 Acknowledge

This work was supported by the German Ministry of Defense. We thank Joachim Schulze for his constructive technical comments and Jörg Fenner for helping collect the raw and context information and evaluate the text mining approach.

References

FELDMAN, R. and DAGAN, I. (1995): Kdt - knowledge discovery in texts. In: *Proceedings of the First International Conference on Knowledge Discovery (KDD)*. Montreal, 112–113.

FENNER, J. and THORLEUCHTER, D. (2006): Strukturen und Themengebiete der mittelstandsorientierten Forschungsprogramme in den USA. Fraunhofer INT's edition, Euskirchen, 2.

HOTHO, A. (2004): *Clustern mit Hintergrundwissen*. Univ. Diss., Karlsruhe, 29.

IPSEN, C. (2002): *F&E-Programmplanung bei variabler Entwicklungsdauer*. Verlag Dr. Kovac, Hamburg, 10.

KAMPHUSMANN, T. (2002): *Text-Mining*. Symposion Publishing, Düsseldorf, 28.

LUSTIG, G. (1986): *Automatische Indexierung zwischen Forschung und Anwendung*. Georg Olms Verlag, Hildesheim, 92.

RIPKE, M. and STÖBER, G. (1972): Probleme und Methoden der Identifizierung potentieller Objekte der Forschungsförderung. In: H. Paschen and H. Krauch (Eds.): *Methoden und Probleme der Forschungs- und Entwicklungsplanung*. Oldenbourg, München, 47.

ROHPOHL, G. (1996): Das Ende der Natur. In: L. Schäfer and E. Sträker (Eds.): *Naturauffassungen in Philosophie, Wissenschaft und Technik*. Bd. 4, Freiburg, München, 151.

STRUBE, G. (2003): Menschliche Informationsverarbeitung. In: G. Görz, C.-R. Rollinger and J. Schneeberger (Eds.): *Handbuch der Künstlichen Intelligenz*. 4. Auflage, Oldenbourg, München, 23–28.

THORLEUCHTER, D. (2007): Überblick über F&T-Vorhaben und ihre Ansprechpartner im Bereich BMVg. Fraunhofer Publica, Euskirchen, 2–88.

Investigating Classifier Learning Behavior with Experiment Databases

Joaquin Vanschoren and Hendrik Blockeel

Computer Science Dept., K.U.Leuven,
Celestijnenlaan 200A, 3001 Leuven, Belgium

Abstract. Experimental assessment of the performance of classification algorithms is an important aspect of their development and application on real-world problems. To facilitate this analysis, large numbers of such experiments can be stored in an organized manner and in complete detail in an *experiment database*. Such databases serve as a detailed log of previously performed experiments and a repository of verifiable learning experiments that can be reused by different researchers. We present an existing database containing 250,000 runs of classifier learning systems, and show how it can be queried and mined to answer a wide range of questions on learning behavior. We believe such databases may become a valuable resource for classification researchers and practitioners alike.

1 Introduction

Supervised classification is the task of learning from a set of classified training examples $(x, c(x))$, where $x \subset X$ (the instance space) and $c(x) \in C$ (a finite set of classes), a classifier function $f : X \rightarrow C$ such that f approximates c (the target function) over X. Most of the existing algorithms for learning f are heuristic in nature, and try to (quickly) approach c by making some assumptions that may or may not hold for the given data. They assume c to be part of some designated set of functions (the hypothesis space), deem some functions more likely than others, and strictly consider consistency with the observed training examples (not with X as a whole). While there is theory relating such heuristics to finding c, in many cases this relationship is not so clear, and the utility of a certain algorithm needs to be evaluated empirically.

As in other empirical sciences, experiments should be performed and described in such a way that they are easily verifiable by other researchers. However, given the fact that the exact algorithm implementation used, its chosen parameter settings, the used datasets and the experimental methodology all influence the outcome of an experiment, it is practically not self-evident to completely describe such experiments. Furthermore, there exist complex interactions between data properties, parameter settings and the performance of learning algorithms. Hence, to thoroughly study these interactions and to assess the generality of observed trends, we need a sufficiently large sample of experiments, covering many different conditions, organized in a way that makes their results easily accessible and interpretable.

For these reasons, Blockeel (2006) proposed the use of *experiment databases*: databases describing a large number of learning experiments in complete detail, serving as a detailed log of previously performed experiments and an (online available) repository of learning experiments that can be reused by different researchers. Blockeel and Vanschoren (2007) provide a detailed account of the advantages and disadvantages of experiment databases, and give guidelines for designing them. As a proof of the concept, they present a concrete implementation that contains a full description of the experimental conditions and results of 250,000 runs of classifier learning systems, together with a few examples of its use and results that were obtained from it.

In this paper we provide a more detailed discussion of how this database can be used in practice to store the results of many learning experiments and to obtain a clear picture of the performance of the involved algorithms and the effects of parameter settings and dataset characteristics. We believe that this discussion may be of interest to anyone who may want to use this database for their own purposes, or set up a similar databases for their own research.

We describe the structure of the database in Sect. 2 and the experiments in Sect. 3. In Sect. 4 we illustrate the power of this database by showing how SQL queries and data mining techniques can be used to investigate classifier learning behavior. Section 5 concludes.

2 A database for classification experiments

To efficiently store and allow queries about all aspects of previously performed classification experiments, the relationships between the involved learning algorithms, datasets, experimental procedures and results are captured in the database structure, shown in Fig. 1. Since many of these aspects are parameterized, we use *instantiations* to uniquely describe them. As such, an `Experiment` (central in the figure) consists of instantiations of the used learner, dataset and evaluation method.

First, a `Learner instantiation` points to a learning algorithm (`Learner`), which is described by the algorithm name, version number, a url where it can be downloaded, and some generally known or calculated properties (Van Someren (2001), Kalousis & Hilario (2000)), like the used approach (e.g. neural networks) or how susceptible it is to noise. Then, if an algorithm is parameterized, the parameter settings used in each learner instantiation (one of which is set as default) are stored in table `Learner_parval`. Because algorithms have different numbers and kinds of parameters, we store each parameter value assignment in a different row (in Fig. 1 only two are shown). A `Learner_parameter` is described by the learner it belongs to, its name and a specification of sensible or suggested values, to facilitate experimentation.

Secondly, the used `Dataset`, which can be instantiated with a randomization of the order of its attributes or examples (e.g. for incremental learners), is recorded by

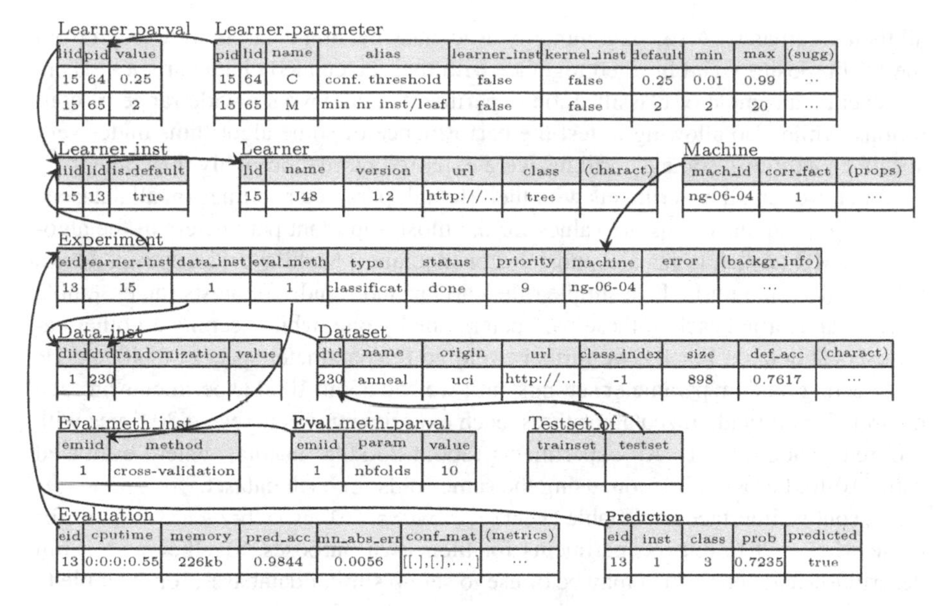

Fig. 1. A simplified schema of the experiment database.

its name, download url(s), the index of the class attribute and some information on its origin (e.g. to which repository it belongs or how it was generated artificially). In order to investigate whether the performance of an algorithm is linked to certain kinds of data, a large set of dataset characterization metrics is stored, most of which are described in Peng et al. (2002). These can be useful to help gain insight into an algorithm's behavior and, conversely, assess a learner's suitability for handling new learning problems[1].

Finally, we must store an evaluation of the experiments. The evaluation method (e.g. cross-validation) is stored together with its parameters (e.g. the number of folds). If a dataset is divided into a training set and a test set, this is defined in table `Testset_of`. The result of the evaluation of each experiment is described in table `Evaluation` by a wide range of evaluation metrics for classification, including the contingency tables. To compare cpu times, a factor describing the relative speed of the used `Machine` is stored as part of the machine description. The last table in Fig. 1 stores the (probabilities of the) predictions returned by each experiment, which may be used to calculate new performance measures without rerunning the experiments.

3 The experiments

To populate the database with experiments, we selected 54 classification algorithms from the WEKA platform (Witten and Frank (2005)) and inserted them together with

[1] New data and algorithm characterizations can be added at any time by adding more columns and calculating the characterizations for all datasets or algorithms.

all their parameters. Also, 86 commonly used classification datasets were taken from the UCI repository and inserted together with their calculated characteristics. Then, to generate a sample of classification experiments that covers a wide range of conditions, while also allowing to test the performance of some algorithms under very specific conditions, some algorithms were explored more thoroughly than others.

First, we ran all experiments with their default parameter settings on all datasets. Secondly, we defined sensible values for the most important parameters of the algorithms SMO (which trains a support vector machine), MultilayerPerceptron, J48 (a C4.5 implementation), 1R (a simple rule learner) and Random Forests (an ensemble learner) and varied each of these parameters one by one, while keeping all other parameters at default. Finally, we further explored the parameter spaces of J48 and 1R by selecting random parameter settings until we had about 1000 experiments on each dataset. For all randomized algorithms, each experiment was repeated 20 times with different random seeds. All experiments (about 250,000 in total) where evaluated using 10-fold cross-validation, using the same folds for each dataset.

An online interface is available at `http://www.cs.kuleuven.be/~dtai/expdb/` for those who want to reuse experiments for their own purposes, together with a full description and code which may be of use to set up similar databases, for example to store, analyse and publish the results of large benchmark studies.

4 Using the database

We will now illustrate how easy it is to use this experiment database to investigate a wide range of questions on the behavior of learning algorithms by simply writing the right queries and interpreting the results, or by applying data mining algorithms to model more complex interactions.

4.1 Comparing different algorithms

A first question may be "How do all algorithms in this database compare on a specific dataset D?" To investigate this, we query for the learning algorithm name and evaluation result (e.g. predictive accuracy), linked to all experiments on (an instance of) dataset D, which yields the following query:

```
SELECT l.name, v.pred_acc
FROM experiment e, learner_inst li, learner l, data_inst di,
dataset d, evaluation v
WHERE v.eid = e.eid and e.learner_inst = li.liid and li.lid = l.lid
and e.data_inst = di.diid and di.did = d.did and d.name='D'
```

We can now interpret the returned results, e.g. by drawing a scatterplot. For dataset `monks-problems-2` (a near-parity problem), this yields Fig. 2, giving a clear overview of how each algorithm performs and (for those algorithms whose parameters where varied) how much variance is caused by different parameter settings. Only a few algorithms surpass default accuracy (67%) and while some cover a wide

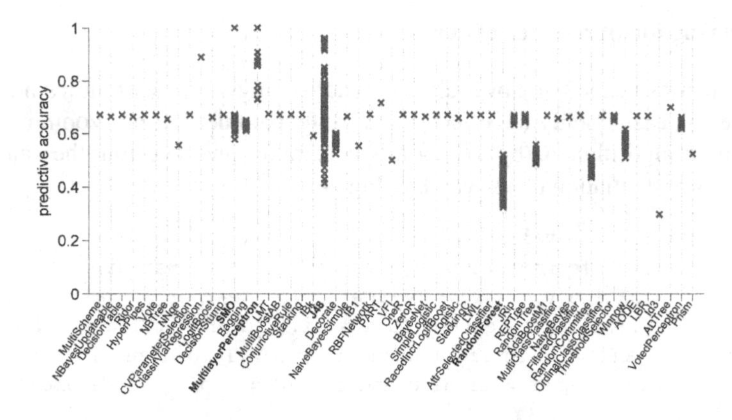

Fig. 2. Algorithm performance comparison on the `monks-problems-2_test` dataset.

spectrum (like J48), others jump to 100% accuracy for certain parameter settings (SMO with higher-order polynomial kernels and MultilayerPerceptron when enough hidden nodes are used).

We can also compare two algorithms A1 and A2 on *all* datasets by joining their performance results (with default settings) on each dataset, and plotting them against each other, as shown in Fig. 3. Moreover, querying also allows us to use aggregates and to order results, e.g. to directly build rankings of all algorithms by their *average* error over all datasets, using default parameters:

```
SELECT l.name, avg(v.mn_abs_err) AS avg_err FROM experiment e,
learner l, learner_inst li, evaluation v WHERE v.eid = e.eid and
e.learner_inst = li.liid and li.lid = l.lid and li.default = true
GROUP BY l.name ORDER BY avg_err asc
```

Similar questions can be answered in the same vein. With small adjustments, we can query for the variance,... of each algorithm's error (over all or a single dataset), study how much error rankings differ from one dataset to another, or study how parameter optimization affects these rankings.

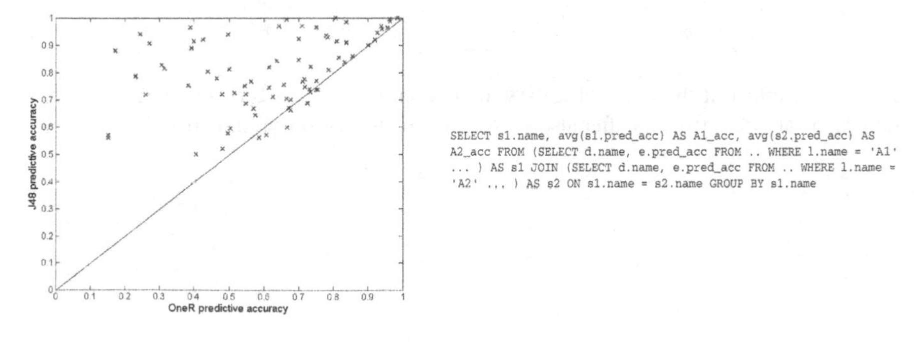

J48 predictive accuracy
OneR predictive accuracy
0
0.1
0.2
0.3
0.4
0.5
0.6
0.7
0.8
0.9
1
SELECT s1.name, avg(s1.pred_acc) AS A1_acc, avg(s2.pred_acc) AS
A2_acc FROM (SELECT d.name, e.pred_acc FROM .. WHERE l.name = 'A1'
...) AS s1 JOIN (SELECT d.name, e.pred_acc FROM .. WHERE l.name =
'A2' ...) AS s2 ON s1.name = s2.name GROUP BY s1.name

Fig. 3. Comparing relative performance of J48 and OneR with a single query.

4.2 Querying for parameter effects

Previous queries generalized over all parameter settings. Yet, starting from our first query, we can easily study the effect of a specific parameter P by "zooming in" on the results of algorithm A (by adding this constraint) and selecting the value of P linked to (an instantiation of) A, yielding Fig. 4a:

```
SELECT v.pred_acc, lv.value
FROM experiment e, learner_inst li, learner l, data_inst di,
dataset d, evaluation v, learner_parameter lp, learner_parval lv
WHERE v.eid = e.eid and e.learner_inst = li.liid and li.lid = l.lid
and l.name='A' and lv.liid=li.liid and lv.pid = lp.pid and lp.
name='P' and e.data_inst = di.diid and di.did = d.did and d.name='D'
```

Sometimes the effect of a parameter P may be dependent on the value of another parameter. Such a parameter P2 can however be controlled (e.g. by demanding its value to be larger than V) by extending the previous query with a constraint requiring that the learner instances additionally are amongst those where parameter P2 obeys those constraints.

```
WHERE ... and lv.liid IN
(SELECT lv.liid FROM learner_parval lv, learner_parameter lp
WHERE lv.pid = lp.pid and lp.name='P2' and lv.value>V)
```

Launching and visualizing such queries yield results such as in Fig. 4, clearly showing the effect of the selected parameter and the variation caused by other parameters. As such, it is immediately obvious how general an observed trend is: all constraints are explicitly mentioned in the query.

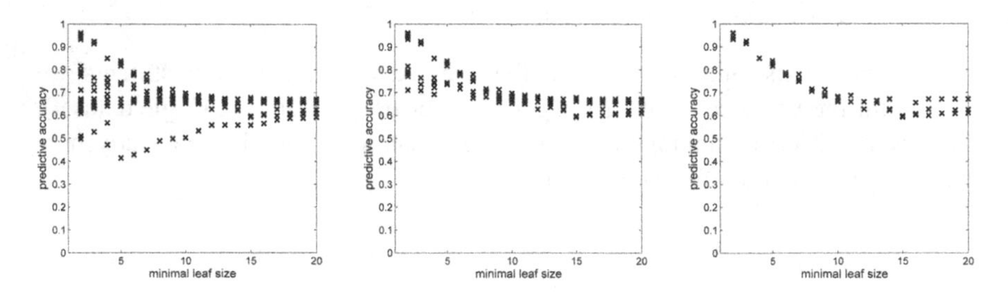

Fig. 4. The effect of the minimal leafsize of J48 on `monks-problems-2_test` (a), after requiring binary trees (b), and after also suppressing reduced error pruning (c)

4.3 Querying for the effect of dataset properties

It also becomes easy to investigate the interactions between data properties and learning algorithms. For instance, we can use our experiments to study the effect of a dataset's size on the performance of algorithm A[2]:

```
SELECT v.pred_acc, d.nr_examples
FROM experiment e, learner_inst li, learner l, data_inst di,
dataset d, evaluation v
WHERE v.eid = e.eid and e.learner_inst = li.liid and li.lid = l.lid
and l.name='A' and e.data_inst = di.diid and di.did = d.did
```

4.4 Applying data mining techniques to the experiment database

There can be very complex interactions between parameter settings, dataset characteristics and the resulting performance of learning algorithms. However, since a large number of experimental results are available for each algorithm, we can apply data mining algorithms to model those interactions.

For instance, to automatically learn which of J48's parameters have the greatest impact on its performance on `monks-problems-2_test` (see Fig. 4), we queried for the available parameter settings and corresponding results. We discretized the performance with thresholds on 67% (default accuracy) and 85%, and we used J48 to generate a (meta-)decision tree that, given the used parameter settings, predicts in which interval the accuracy lies. The resulting tree (with 97.3% accuracy) is shown in Fig. 5. It clearly shows which are the most important parameters to tune, and how they affect J48's performance.

Likewise, we can study for which dataset characteristics one algorithm greatly outperforms another. Starting from the query in Fig. 3, we additionally queried for a wide range of data characteristics and discretized the performance gain of J48 over 1R in three classes: "draw", "win_J48" (4% to 20% gain), and "large_win_J48" (20% to 70% gain). The tree returned by J48 on this meta-dataset is shown in Fig. 6, and clearly shows for which kinds of datasets J48 has a clear advantage over OneR.

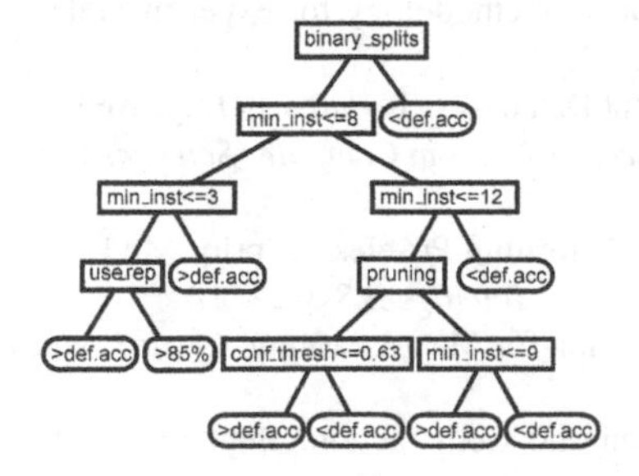

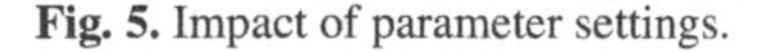
Fig. 5. Impact of parameter settings.

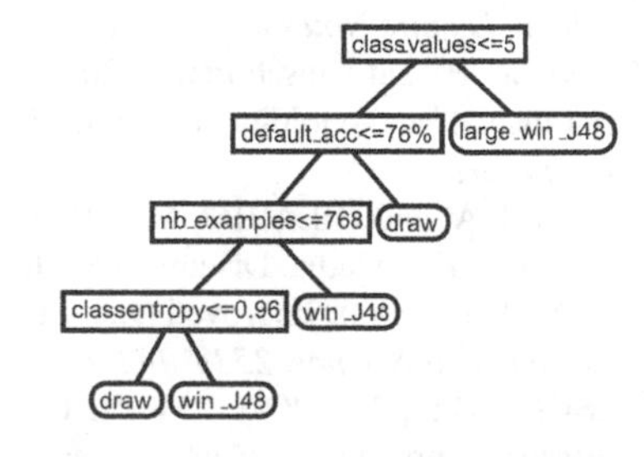

Fig. 6. Impact of dataset properties.

[2] To control the value of additional dataset properties, simply add these constraints to the list: `WHERE ... and d.nr_attributes>5`.

4.5 On user-friendliness

The above SQL queries are relatively complicated. Part of this is however a consequence of the relatively complex structure of the database. A good user interface, including a graphical query tool and an integrated visualization tool, would greatly improve the usability of the database.

5 Conclusions

We have presented an experiment database for classification, providing a well-structured repository of fully described classification experiments, thus allowing them to be easily verified, reused and related to theoretical properties of algorithms and datasets. We show how easy it is to investigate a wide range of questions on the behavior of these learning algorithms by simply writing the right queries and interpreting the results, or by applying data mining algorithms to model more complex interactions. The database is available online and can be used to gain new insights into classifier learning and to validate and refine existing results. We believe this database and underlying software may become a valuable resource for research in classification and, more broadly, machine learning and data analysis.

Acknowledgements

We thank Anneleen Van Assche and Celine Vens for their useful comments and help building meta-decision trees and Anton Dries for implementing the dataset characterizations. Hendrik Blockeel is Postdoctoral Fellow of the Fund for Scientific Research - Flanders (Belgium) (FWO-Vlaanderen), and this research is further supported by GOA 2003/08 "Inductive Knowledge Bases".

References

BLOCKEEL, H. (2006): Experiment databases: A novel methodology for experimental research. *Lecture Notes in Computer Science, 3933, 72-85.*

BLOCKEEL, H. and Vanschoren J. (2007): Experiment Databases: Towards an Improved Experimental Methodology in Machine Learning. *Lecture Notes in Computer Science, 4702, to appear.*

KALOUSIS, A. and HILARIO, M. (2000): Building Algorithm Profiles for prior Model Selection in Knowledge Discovery Systems. *Engineering Intelligent Syst., 8(2).*

PENG, Y. et al. (2002): Improved Dataset Characterisation for Meta-Learning. *Lecture Notes in Computer Science, 2534, 141-152.*

VAN SOMEREN, M. (2001): Model Class Selection and Construction: Beyond the Procrustean Approach to Machine Learning Applications. *Lecture Notes in Computer Science, 2049, 196-217.*

WITTEN, I.H. and FRANK, E. (2005): *Data Mining: Practical Machine Learning Tools and Techniques (2nd edition).* Morgan Kaufmann.

Part VI

Marketing and Management Science

Conjoint Analysis for Complex Services Using Clusterwise Hierarchical Bayes Procedures

Michael Brusch and Daniel Baier

Institute of Business Administration and Economics,
Brandenburg University of Technology Cottbus,
Postbox 101344, 03013 Cottbus, Germany
{m.brusch, daniel.baier}@tu-cottbus.de

Abstract. Conjoint analysis is a widely used method in marketing research. Some problems occur when conjoint analysis is used for complex services where the perception of and the preference for attributes and levels considerably varies among individuals. Clustering and clusterwise estimation procedures as well as Hierarchical Bayes (HB) estimation can help to model this perceptual uncertainty and preference heterogeneity. In this paper we analyze the advantages of clustering and clusterwise HB as well as combined estimation procedures of collected preference data for complex services and therefore extend the analysis of Sentis and Li (2002).

1 Introduction

Conjoint analysis is a "... method that estimates the structure of consumer's preferences ..." (Green and Srinivasan (1978), p. 104). Typically, hypothetical concepts for products or services (attribute-level-combinations) are presented to and rated by a sample of consumers in order to estimate part worths for attribute-levels from a consumer's point of view and to develop acceptable products or services. Since its introduction into marketing in the early 1970s conjoint analysis has become a favored method within marketing research (see, e.g., Green et al. (2001)).

Consequently, conjoint analysis is nowadays a method for which a huge number of applications are known as well as many specialized tools for data collection and analysis have been developed. For part worth estimation, especially clusterwise estimation procedures (see, e.g., Baier and Gaul (1999, 2003)) and Hierarchical Bayes (HB) estimation (see, e.g., Allenby and Ginter (1995), Lenk et al. (1996)) seem to be attractive newer developments.

After a short discussion of specific problems of conjoint analysis when applied to preference measurement for (complex) services (section 2) we propose estimation procedures basing on HB and clustering to reduce these problems (section 3). An empirical investigation (section 4) shows the viability of this proposition.

2 Preference measurement for services

The concepts evaluated by consumers within a conjoint study can be hypothetical products as well as – with an increasing importance during recent years – services. However, services cause special demands on the research design due to their following peculiarities (e.g., Zeithaml et al. (1985), pp. 33): immateriality, integration of an external factor, non-standardization, and perishability.

The main peculiarity is that services cannot be taken into the hands – they are immaterial. This leads to problems during the data collection phase, where hypothetical services have to be presented to the consumer. It has been shown that the "right" description and presentation influences the "right" perception of consumers and consequently the validity of the estimated part worths from the collected data (see, e.g., Ernst and Sattler (2000), Brusch et al. (2002)).

Furthermore, as we all know, the quality of services depends on the producing persons and objects as well as their interaction with persons and objects from the demand side – the so-called external (production) factors. Their synergy, willingness, and quality often cannot be evaluated before consumption. Perceptual uncertainty of the usefulness of different attributes and levels as well as preference heterogeneity is common among potential buyers. Part worth estimates for attribute levels have to take this into account. Intra- and inter-individual variation has to be modelled.

3 Hierarchical Bayes procedures for conjoint analysis

Recently, for modelling this intra- and inter-individual variation, clustering and clusterwise part worth estimation as well as HB estimation have been proposed.

Clustering and clusterwise part worth estimation provide traditional ways to model preference heterogeneity in conjoint analysis (see, e.g., Baier and Gaul (1999, 2003)). The population is assumed to fall into a number of (unknown) clusters or segments whose segment-specific part worths have to be estimated from the collected data.

HB, on the other side, estimates individual part worth distributions by "borrowing" information from other individuals (see, e.g., Baier and Polasek (2003) were for a conjoint analysis setting this aspect of borrowing is described in detail). Preference heterogeneity is not assumed via introducing segments. Instead, the deviation of the individual part worth distributions from a mean part worth distribution is derived from the collected individual data (for methodological details and new developments see, e.g., Allenby et al. (1995), Lenk et al. (1996), Andrews et al. (2002), Liechty et al. (2005)).

The main advantages and therefore the reasons for the attention of HB can be summarized as follows (Orme 2000):

- HB estimation seems (at least) to outperform traditional models with respect to predictive validity.
- HB estimation seems to be robust.

- HB permits – even with little data – individual part worth estimation and therefore allows to model heterogeneity across respondents.
- HB helps differentiating signal from noise.
- HB and its "draws" (replicates) model uncertainty and therefore provide a rich source of information.

If facts and statements about HB are considered, it is not surprising, that the impression results that – especially in case of standard products and services – "HB methods achieve an 'analytical alchemy' by producing information where there is very little data ..." (Sentis and Li (2002), p. 167).

However, question 1: whether this is also true when complex services have to be analyzed, and – in this case – question 2: whether instead of HB or clustering a combination of these procedures should be used are still open.

Our investigation tries to close this gap: Clustering and clusterwise HB as well as combined estimation procedures are applied to collected preference data for complex services. The results are compared with respect to predictive validity. The investigation extends the analysis of Sentis and Li (2002) who observed in a simpler setting that predictive validity (hit rates) were not improved by combining clustering and HB estimation.

4 Empirical investigation

4.1 Research object

For our investigation a complex service is used: an university course of study with new e-learning features, e.g., different possibilities to join the lecture (in a lecture hall or at home using video conferencing) or different types of scripts (printed scripts or multimedia scripts with interactive exercises). Here, complexity is used as term to differentiate from simple aspects of services (e.g., price, opening hours, processing time). Because of this complexity of the attributes and levels we expect perceptual uncertainty (because of difficultly describable attributes and levels) as well as preference heterogeneity among consumers have to be considered.

The research object has four attributes, each with three or four levels (for the structure see Table 1). In total, 15 part worth parameters have to be estimated in our analyzes.

4.2 Research design

A conjoint study is carried out using the nowadays standard tool for conjoint data collection, Sawtooth SoftwareŠs ACA system (Sawtooth Software (2002)), to be precise ACA/Web within SSI/Web (Windows Version 2.0.1b). For our investigation a five-step analysis is used to answer our focused questions.

Step 1 – Analyzing the quality

In our study we had 239 started and 213 finished questionnaires. Standard ACA methodology was used for individual part worth estimation. Standard selection criteria reduced the number of usable respondents to 162 with passably good R^2-measures.

Step 2 – Calculating standardized part worths

The individual part worths were standardized. The attribute level with the lowest (worst) part worth is becoming 0, the best attribute level combination (combination of the best attribute levels of each attribute) 1. In the following, these standardized individual part worths were used for clustering.

Step 3 – Clustering

The sample was divided into two segments (cluster 1 and cluster 2) by means of a cluster analysis and an elbow criterion. The cluster analysis uses Euclidean distances and Ward's method and is based on the standardized individual part worths. From the resulting dendrogram it could be seen that cluster 2 is far more heterogenous than cluster 1.

As shown in Table 1 two clusters with a few differences were found. For example, the different order of the relative importance of the attributes is noticeable. For cluster 1 the most relevant (important) attribute is attribute 3. For cluster 2 – where the relative importance of the attributes is more uniformly distributed – the most relevant attribute is instead attribute 2.

Step 4 – Computing HB utilities

The distribution of individual part worths were computed via aggregated HB as well as via two clusterwise HB part worth estimations. For our analysis, the software ACA/HB from Sawtooth Software, Inc. is used (Sawtooth Software (2006)), the actual most relevant standard tool for conjoint data analysis. Preprocessing in order to segment the available individual data was done via SAS. The following parameters are set:

- 5,000 iterations before using results (burn in),
- 10,000 draws to be used for each respondent,
- no constraints in use,
- fitting pairs & priors, and
- saving random draws.

Thus, 10,000 draws from the individual part worth distribution are available for each respondent from aggregated HB as well as two clusterwise HB estimations resulting in three samples (total sample, cluster 1, cluster 2). These HB utilities will be used to answer our research questions.

Table 1. Conjoint results for the total sample and the clusters

		Total sample (n=162)		Cluster 1 (n=80)		Cluster 2 (n=82)	
		Rel. Imp.	PW	*Rel. Imp.*	PW	*Rel. Imp.*	PW
Attribute 1	Level 1		0.032		0.044		0.020
	Level 2	*18.2 %*	0.117	*14.4 %*	0.080	*21.8 %*	0.154
	Level 3		0.140		0.095		0.184
Attribute 2	Level 1		0.106		0.177		0.036
	Level 2	*27.5 %*	0.157	*28.5 %*	0.183	*26.5 %*	0.132
	Level 3		0.238		0.236		0.240
	Level 4		0.081		0.036		0.124
Attribute 3	Level 1		0.256		0.317		0.196
	Level 2	*29.2 %*	0.145	*33.1 %*	0.131	*25.4 %*	0.159
	Level 3		0.163		0.165		0.161
	Level 4		0.019		0.017		0.021
Attribute 4	Level 1		0.199		0.218		0.181
	Level 2	*25.1 %*	0.122	*24.0 %*	0.068	*26.2 %*	0.174
	Level 3		0.147		0.097		0.195
	Level 4		0.043		0.050		0.035

Rel. Imp. ... relative importance, PW ... part worths

Step 5 – Calculating values for predictive validity

The predictive validity was considered while questioning on the basis of the integration of a specific holdout task. This task included the evaluation of five service concepts, similar to the "calibration concepts" of a usual ACA questionnaire. The respondents were asked for the "likelihood of using". This holdout task was separated from the conjoint task of the ACA questionnaire.

Predictive validity will be measured using two values: the Spearman rank-order correlation coefficient and the first-choice-hit-rate. The Spearman rank-order correlation compares the predicted preference values with the corresponding observed ordinal scale response data from the holdout task. The first-choice-hit-rate is the share of respondents where the stimulus with the highest predicted preference value is also the one with the highest observed preference value.

4.3 Results

The results of our investigation are shown in Tables 2 and 3. Table 2 shows the validity values for the traditional ACA estimation for each partial sample. The validity values are based on the averages of the traditional (standardized) ACA part worths.

As you can see in Table 2 the validity values for cluster 1 are higher than the values for the total sample. Cluster 2 has instead the lowest (worst) validity results.

Table 2. Validity values for the total sample and for the clusters for traditional ACA estimation (using standardized part worths from step 2 at the individual level)

	Total sample (n=161)*	Cluster 1 (n=79)*	Cluster 2 (n=82)
First-choice-hit-rate (using individual data)	62.11 %	73.42 %	51.22 %
Mean Spearman (using individual data)	0.735	0.782	0.689

*. . . *one respondent had missing holdout data and could not be considered*

The validity values shown in Table 3 are based on the HB estimation and are given for the total sample and for the two clusterwise estimations. The clusters are separated after the membership during the estimation (total sample or segment). The description "in total sample" means that the HB utilities of the respondents were computed by "borrowing" information from the total sample (not only from members of the own segment). Thus, the HB estimation happened for all respondents together, but the validity values for the two clusters were computed later separately. On the other hand, the description "in segment" means that the HB utilities of the respondents were computed by "borrowing" information only from members of the own segment (clusterwise HB estimation).

Furthermore, the results in Table 3 are distinguished according to the data basis. The validity values are shown for the computation based on the 10,000 draws (10,000 HB utilities) for each respondent and for the computation based on the mean HB utilities (one HB utility as mean of 10,000 draws (iterations)) for each individual.

From Table 3 it is identifiable that the validity values in cluster 1 are higher and in cluster 2 lower than in the total sample. Further differences between the clusters can be found when looking at the HB estimation basis (joint estimation in the total sample ("in total sample") or clusterwise estimation ("in segment")). Here for cluster 1 the results in the case of a joint estimation are better in most cases than a clusterwise estimation whereas the opposite can be seen in cluster 2.

When comparing the results of Table 3 with those of Table 2 it can be seen that all validity values for the individual averages based HB estimation are higher than for the ACA estimation, regardless which HB estimation basis ("in total sample" or "in segment") is used. In the case of HB estimation using individual draws, a mixed result with respect to validity can be found.

5 Conclusion and outlook

The focused questions of our investigation can be answered. The first question was, whether HB estimation can produce "better results" than traditional part worth estimation when complex services have to be analyzed. This can be affirmed for the usage of individual means, regardless whether the total sample or the segments are

Table 3. Validity values for the total sample and for the clusters for HB estimation ("in total sample": HB estimation at the individual total sample level; "in segment": separate HB estimation at the individual cluster 1 resp. 2 level)

	Total sample (n=161)*	Cluster 1 (n=79)*		Cluster 2 (n=82)	
		In Total Sample	*In Segment*	In Total Sample	*In Segment*
First-choice-hit-rate (using draws, n=10,000)	62.57 %	72.38 %	*72.39 %*	53.12 %	*53.14 %*
Mean Spearman (using draws, n=10,000)	0.727	0.780	*0.778*	0.677	*0.671*
First-choice-hit-rate (using mean draws)	65.22 %	75.95 %	*74.68 %*	54.88 %	*57.32 %*
Mean Spearman (using mean draws)	0.748	0.802	*0.797*	0.696	*0.700*

*... *one respondent had missing holdout data and could not be considered*

considered. Furthermore we were interested whether clusterwise estimation can optimize the "results" of HB estimation. A clear answer is not possible up to now. In our empirical investigation in some cases we had improvements with respect to the validity values (cluster 2) and in some cases not (cluster 1).

This means that our proposition in the paper can help to reduce the problems that occur when service preference measurement via conjoint analysis is the research focus. HB estimation seems to improve validity even in case of complex services with immaterial attributes and levels that cause perceptual uncertainty and preference heterogeneity. However, going further with the more complicated way of performing clusterwise HB estimation doesn't provide automatically better results.

Nevertheless, further comparisons with larger sample sizes and other research objects are necessary. Furthermore, the possibilities of other validity criteria for clearer statements could be used.

References

ALLENBY, G.M. and GINTER, J.L. (1995): Using Extremes to Design Products and Segment Markets. *Journal of Marketing Research, 32, November, 392–403.*

ALLENBY, G.M., ARORA, N. and GINTER, J.L (1995): Incorporating Prior Knowledge into the Analysis of Conjoint Studies. *Journal of Marketing Research, 32, May, 152–162.*

ANDREWS, R.L., ANSARI, A. and CURRIM, I.S. (2002): Hierarchical Bayes Versus Finite Mixture Conjoint Analysis Models: A Comparison of Fit, Prediction, and Partworth Recovery. *Journal of Marketing Research, 39, February, 87–98.*

BAIER, D. and GAUL, W. (1999): Optimal Product Positioning Based on Paired Comparison Data. *Journal of Econometrics, 89, Nos. 1-2, 365–392.*

BAIER, D. and GAUL, W. (2003): Market Simulation Using a Probabilistic Ideal Vector Model for Conjoint Data. In: A. Gustafsson, A. Herrmann, and F. Huber (Eds.): *Conjoint Measurement - Methods and Applications.* Springer, Berlin, 97–120.

BAIER, D. and POLASEK, W. (2003): Market Simulation Using Bayesian Procedures in Conjoint Analysis. In: M. Schwaiger and O. Opitz (Eds.): *Exploratory Data Analysis in Empirical Research.* Springer, Berlin, 413–421.

BRUSCH, M., BAIER, D. and TREPPA, A. (2002): Conjoint Analysis and Stimulus Presentation - a Comparison of Alternative Methods. In: K. Jajuga, A. Sokołowski and H.H. Bock (Eds.): *Classification, Clustering, and Analysis.* Springer, Berlin, 203–210.

ERNST, O. and SATTLER, H. (2000): Multimediale versus traditionelle Conjoint-Analysen. Ein empirischer Vergleich alternativer Produktpräsentationsformen. *Marketing ZFP, 2, 161–172.*

GREEN, P.E. and SRINIVASAN, V. (1978): Conjoint Analysis in Consumer Research: Issues and Outlook. *Journal of Consumer Research, 5, September, 103–123.*

GREEN, P.E., KRIEGER, A.M. and WIND, Y. (2001): Thirty Years of Conjoint Analysis: Reflections and Prospects. *Interfaces 31, 3, part 2, S56–S73.*

LENK, P.J., DESARBO, W.S., GREEN, P.E. and YOUNG, M.R. (1996): Hierarchical Bayes Conjoint Analysis: Recovery of Partworth Heterogeneity from Reduced Experimental Designs. *Marketing Science, 15, 2, 173–191.*

LIECHTY, J.C., FONG, D.K.H. and DESARBO, W.S. (2005): Dynamic models incorporating individual heterogeneity. Utility evolution in conjoint analysis. *Marketing Science, 24, 285–293.*

ORME, B. (2000): Hierarchical Bayes: Why All the Attention? *Quirk's Marketing Research Review, March.*

SAWTOOTH SOFTWARE (2002): ACA System. Adaptive Conjoint Analysis Version 5.0. *Technical Paper Series, Sawtooth Software.*

SAWTOOTH SOFTWARE (2006): The ACA/Hierarchical Bayes v3.0 Technical Paper. *Technical Paper Series, Sawtooth Software.*

SENTIS, K. and LI, L. (2002): One Size Fits All or Custom Tailored: Which HB Fits Better? *Proceedings of the Sawtooth Software Conference September 2001, 167–175.*

ZEITHAML, V.A., PARASURAMAN, A. and BERRY, L.L. (1985): Problems and Strategies in Services Marketing. *Journal of Marketing, 49, 33–46.*

Building an Association Rules Framework for Target Marketing

Nicolas March and Thomas Reutterer

Institute for Retailing and Marketing, Vienna University of Economics and Business Administration, Augasse 2–6, 1090 Vienna, Austria
march@troostwijk.de
thomas.reutterer@wu-wien.ac.at

Abstract. The discovery of association rules is a popular approach to detect cross-category purchase correlations hidden in large amounts of transaction data and extensive retail assortments. Traditionally, such item or category associations are studied on an 'average' view of the market and do not reflect heterogeneity across customers. With the advent of loyalty programs, however, tracking each program member's transactions has become facilitated, enabling retailers to customize their direct marketing efforts more effectively by utilizing cross-category purchase dependencies at a more disaggregate level. In this paper, we present the building blocks of an analytical framework that allows retailers to derive customer segment-specific associations among categories for subsequent target marketing. The proposed procedure starts with a segmentation of customers based on their transaction histories using a constrained version of K-centroids clustering. In a second step, associations are generated separately for each segment. Finally, methods for grouping and sorting the identified associations are provided. The approach is demonstrated with data from a grocery retailing loyalty program.

1 Introduction

One central goal of customer relationship management (CRM) is to target customers with offers that best match their individual consumption needs. Thus, the question of *who* to target with *which* range of products or items emerges. Most previous research in CRM or direct marketing concentrates on the issue *who* to target (for an extensive literature review see, e.g., Prinzie and Van den Poel (2005)). We address both parts of this question and introduce the cornerstones of an analytical framework for customizing direct marketing campaigns at the customer segment level.

In order to identify and to make use of possible cross-selling potentials, the proposed approach builds on techniques for exploratory analysis of market basket data. Retail managers have been interested in better understanding the purchase interdependency structure among categories for quite a while. One obvious reason is that knowledge about correlated demand patterns across several product categories can be exploited to foster cross-buying effects using suitable marketing actions. For example, if customers often buy a particular product A together with article B, it could be useful to promote A in order to boost sales volumes of B, and vice versa. The objective of exploratory market basket analysis is to discover such unknown cross-item correlations from a typically huge collection of purchase transaction data (so-called market baskets) accruing at the retailer's point-of-sale scanning devices (Berry and Linoff (2006)). Among others, algorithms for mining association rules are popular techniques to accomplish this task (cf., e.g. Hahsler et al. (2006)). However, such association rules are typically derived for the entire data set of available retail transactions and thus reflect an 'average' or aggregate view of the market only.

In recent years, many retailers have tried to improve their CRM activities by launching loyalty programs, which provide their members with bar-coded plastic or registered credit cards. If customers use these cards during their payment process, they get a bonus, credits or other rewards. As a side effect, these transactions become personally identifiable by linking them back to the corresponding customers. Thus, retailers are nowadays collecting series of market baskets that represent (more or less) complete buying histories of their primary clientele over time.

2 A segment-specific view of cross-category associations

To exploit the potential benefits offered by such rich information on customers' purchasing behavior within advanced CRM programs, cross-category correlations need to be detected on a more disaggregate (or customer segment) level instead of an aggregate level. Attempts towards this direction are made by Boztug and Reutterer (2007) or Reutterer et al. (2006). The authors employ vector quantization techniques to arrive at a set of 'generic' (i.e., customer-unspecific) market basket classes with internally more distinctive cross-category interdependencies. In a second step they generate a segmentation of households based on a majority voting of each household's basket class assignments throughout the individual purchase history. These segments are proposed as a basis for designing customized target marketing actions.

In contrast to these approaches, the procedure presented below adopts a novel centroids-based clustering algorithm proposed by Leisch and Grün (2006), which bypasses the majority voting step for segment formation. This is achieved by a cross-category effects sensitive partitioning of the set of (non-anonymous) market basket data, which imposes group constraints determined by the household labels associated with each of the market baskets. Hence, during the iterative clustering process the single transactions are "forced" to keep linked with all the other transactions of a specific household's buying history. This results in segments whose members can be characterized by distinctive patterns of cross-category purchase interrelationships.

To get a better feeling of the inter-category purchase correlations within the previously identified segments, association rules derived separately for each segment and evaluated by calculating various measures of significance and interestingness can assist marketing managers for further decision making on targeted marketing actions. Although the within-segment cross-category associations are expected to differ significantly from those generated for the unsegmented data set (because of the data compression step employed prior to the analysis), low minimum thresholds of such measures typically still result in a huge number of potentially interesting associations. To arrive at a clearer and managerially more traceable overview of the various segment-specific cross-category purchase correlations, we arrange them based on a distance concept suggested by Gupta et al. (1999).

The next section characterizes the building blocks of the employed methodology in more detail. Section 4 empirically illustrates the proposed approach using a transaction data set from a grocery retailing loyalty program and presents selected results. Section 5 closes the article with a summary and an outlook on future research.

3 Methodology

The conceptual framework of the proposed approach is depicted in Figure 1 and consists of three basic steps: First, a modified K-centroids cluster algorithm partitions the entire transaction data set and defines K segments of households with an interest in similar category combinations. Secondly, the well-known APRIORI algorithm (Agrawal et al. (1993)) searches within each segment for specific frequent itemsets, which are filtered by a suitable measure of interestingness. Finally, the associations are grouped via hierarchical clustering using a distance measure for associations.

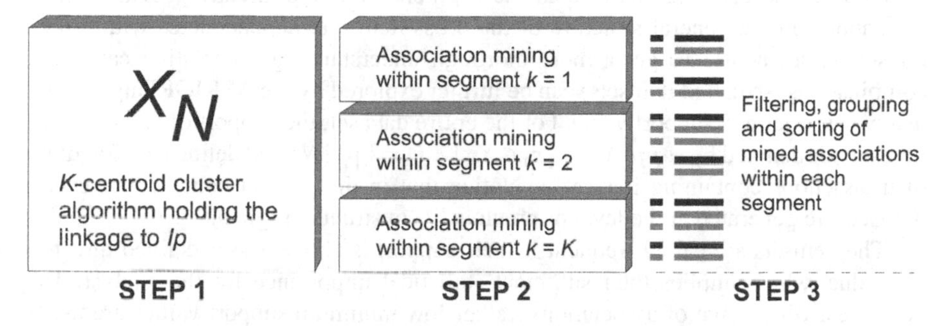

Fig. 1. Conceptual framework of the proposed procedure

Step 1: Each transaction or market basket can be interpreted as a J-dimensional binary vector $x_n = [1,0]^J$ with $j = 1,2\ldots J$ categories. A value of one refers to the presence and a zero to the absence of an item in the market basket. Integrated into a binary matrix X_N, the rows correspond to transactions while each column represents an item. Let the set I_p describe a group constraint indicating the buying history of customer $p = 1,2,\ldots P$ with $\{x_i \in X_N | i \in I_p\}$. The objective function for a modified K-centroids clustering respecting group constraints is (Leisch and Grün (2006)):

$$D(X_N, C_K) = \sum_{p=1}^{P} \sum_{i \in I_p} d(x_i, c(I_p)) \rightarrow \min_{C_K} \quad (1)$$

An iterative algorithm for solving Equation 1 requires calculation of the closest centroid $c(.)$ for each transaction x_i according to the distance measure $d(.)$ at each iteration. To cope with the usually sparse binary transaction data and to make the partition cross-category effects sensitive, the Jaccard coefficient, which gives more weight the co-occurrences of ones rather than common zeros, is used as an appropriate distance measure (cf. Decker (2005)). Notice that in contrast to methods like the K-means algorithm, instead of single transactions groups of market baskets as given by I_p (i.e., customer p's complete buying history) need to be assigned to a minimum distant centroid. This is warranted by a function $f(x_i)$ that determines the centroid closest to the majority of the grouped transactions (cf. Leisch and Grün (2006)).

In order to achieve directly accessible and more intuitively interpretable results, we can calculate cluster-wise means for updating the prototype system instead of optimized canonical binary centroids. This results in an 'expectation-based' clustering solution (cf. Leisch (2006)), whose centroids are equivalent to segment-specific choice probabilities of the corresponding categories. Notice that the segmentation of households is determined such that each customer's complete purchase history points exclusively to one segment. Thus, in the present application context the set of K centroids can be interpreted as prototypical market baskets that summarize the most pronounced item combinations demanded by the respective segment members throughout their purchase history. An illustrative example is provided in Table 1 of the subsequent empirical study.

Step 2: The centroids derived in the segmentation step already provide some indications on the general structure of the cross-item interdependencies within the household segments. To get a more thorough understanding, interesting category combinations (so called itemsets) can be further explored by the APRIORI algorithm using a user defined support value. For the entire data set, the support of an arbitrary itemset A is denoted by $supp(A) = \mid \{x_n \in X_N \mid A \subseteq x_n\} \mid / \mid N \mid$ and defines the fraction of transactions containing itemset A. Notice that in the present context, however, itemsets are generated at the level of previously constructed segments.

The itemsets are called frequent if their support is above a user-defined threshold value, which implies their sufficient statistical importance for the analyst. To generate a wide range of associations, rather low minimum support values are usually preferred. Because not all associations are equally meaningful, an additional measure of interestingness is required to filter the itemsets for evaluation purposes. Since our focus is on itemsets, asymmetric measures like *confidence* or *lift* are less useful (cf. Hahsler (2006)). We advocate here the so-called all-confidence measure introduced by Omiecinski (2003), which is the minimum confidence value for all rules that can be generated from the underlying itemset. Formally it is denoted by $allconf(A) = supp(A)/max_{B \subset A}\{supp(B)\}$ for all frequent subsets B with $B \subset A$.

Step 3: Although the all-confidence measure can assist in reducing the number of itemsets considerably, in practice it can still be difficult to handle several hundreds of remaining associations. For an easier recognition of characteristic inter-item correlations within each segment, the associations can be grouped based on the following Jaccard-like distance measure for itemsets (Gupta et al. (1999)):

$$D(A,B) = 1 - \frac{\mid m(A \cup B) \mid}{\mid m(A) \mid + \mid m(B) \mid - \mid m(A \cup B) \mid} \tag{2}$$

Expression $m(.)$ denotes the set of transactions containing the itemset. From Equation 2 it should be evident that the distance between two itemsets tends to be lower if the involved itemsets occur in many common transactions. This property qualifies the measure to determine specific groups of itemsets that share some common aspects of consumption behavior (cf. Gupta et al. (1999)).

4 Empirical application

The following empirical study illustrates some of the results obtained from the procedure described above. We analyzed two samples of real-world transaction data, each realized by 3,000 members of a retailer's loyalty program. The customers made on average 26 shopping trips over an observational period of one year. Each transaction contains 268 binary variables, which represent the category range of the assortment.

To achieve managerially meaningful results, preliminary screening of the data suggested the following adjustments of the raw data:

1. The purchase frequencies are clearly dominated by a small range of categories, such as fresh milk, vegetables or water (see Figure 2). Since these categories are bought several times by almost every customer during the year under investigation, they provide relatively low information on the differentiated buying habits of the customers. The opposite is supposed to be true for categories with intermediate or lower purchase frequencies. Therefore, we decided to eliminate the upper 52 categories (left side of the vertical line in Figure 2), which occur in more than 10% of all transactions. The resulting empty baskets are excluded from the analysis as well.

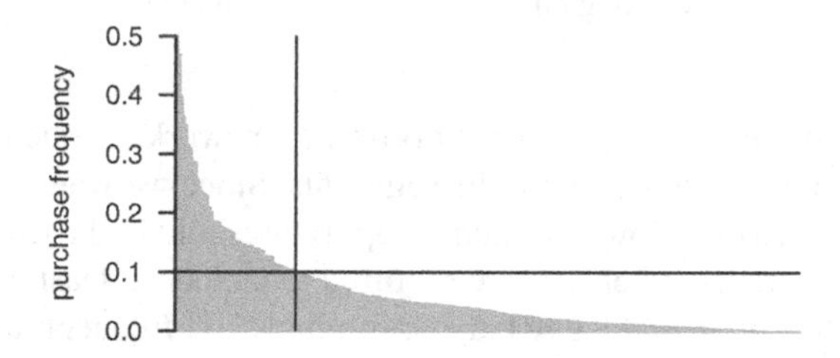

Fig. 2. Distribution of relative category purchase frequencies in decreasing order

2. To include households with sufficiently large buying histories, households with less than six store visits per year were eliminated. In addition, the upper five percentage quantile of households, which use their customer cards extremely often, were deleted.

To find a sufficiently stable cluster solution with a minimum within-sum of distances, the transactions made by the households from the first sample are split into three equal sub samples and clustered up to fifteen times each. In each case, the best solution is kept for the following sub sample to achieve stable results. The converged set of centroids of the third sub sample is used for initialization of the second sample. Commonly used techniques for determination of the number of clusters recommended $K = 11$ clusters as a decent and well-manageable number of household segments. Given these specifications, the partitioning of the second sample using the proposed cluster algorithm detects some segments, which are dominated by category combinations typically bought for specific consumption or usage purposes and other types of categorical similarities. For example, Table 1 shows an extract of a centroid vector including the top six categories in terms of highest conditional purchase probabilities in a segment of households denoted as the "wine segment". A typical market basket arising from this segment is expected to contain red/rosé wines with a probability of 32.3 %, white wines with a probability of 22.5 %, etc. Hence, the labeling "wine segment".

Equally, other segments may be characterized by categories like baby food/care or organic products. On the other hand, there is also a small number of segments with category interrelationships that cannot be easily explained. However, such segments might provide some interesting insights into the interests of households which are so far unknown.

Table 1. Six categories with highest purchase frequencies in the wine segment

No.	Category	Purchase frequency
1.	red / rosé wines	0.3229143
2.	white wines	0.2252356
3.	sparkling wine	0.1225006
4.	condensed milk	0.1206619
5.	appetizers	0.1080211
6.	cooking oil	0.1066422

According to the second step of the proposed framework, frequent itemsets are generated from the transactions within the segments. Since we want to mine a wide range of associations, a quite low minimum support threshold is chosen (e.g., $supp = 1\%$). In addition, all frequent itemsets are required to include at least two categories. Taking this into account, the APRIORI algorithm finds 704 frequent itemsets for the transactions of the wine segment. To reduce the number of associations and to focus on the most interesting frequent itemsets, only the 150 itemsets with highest all-confidence values are considered for grouping according to step 3 of the procedure.

Grouping the frequent itemsets intends to rearrange the order of the generated (segment-specific) associations and to focus the view of the decision maker on characteristic item correlations. The distance matrix derived by Equation 2 is used as input for hierarchical clustering according to the Ward algorithm. Figure 3 shows the dendrogram for the 150 frequent itemsets within the wine cluster. Again, it is not straightforward to determine the correct number of groups g_h. Frequently proposed heuristics based on plotted heterogeneity measures does not help here. Therefore, we pass the distance matrix to the partition around medoid (PAM) algorithm of Kaufman and Rousseeuw (2005) for several g_h values. Using the maximum value of the average silhouette width for a sequence of partitions thirty groups of itemsets are proposed. In Figure 3 the grey rectangles mark two exemplary chosen clusters of associations. The corresponding associations of the right hand group are summarized in Table 2 and clearly indicate an interest of some of the wine households in hard alcoholic beverages.

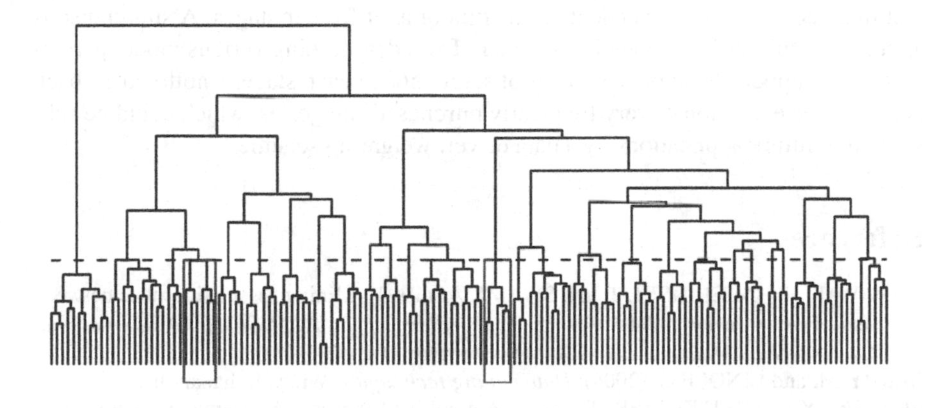

Fig. 3. Dendrogram of 150 frequent itemsets mined from transactions of the wine segment

Table 2. Associations of hard alcoholic beverages within the wine segment

No.	association	support	all-confidence
1.	{brandy, whisky}	0.011	0.23
2.	{brandy, fruit brandy}	0.015	0.18
3.	{fruit brandy, appetizers}	0.018	0.17
4.	{brandy, appetizers}	0.016	0.15
5.	{whisky, fruit brandy}	0.011	0.14

To examine whether the segment-specific associations differ from those generated within the whole data set, we have drawn and analyzed random samples with the same amount of transactions as each of the segments. The comparison of the frequent itemsets mined in the random sample and those from the segment-specific transactions shows that some segment-specific association groups clearly represent a unique characteristic of their underlying household segment. Of course, this is not true in any case. For example, the association group marked by the grey rectangle on the left-hand side in Figure 3 can be found in almost every random sample or segment. It denotes correlations between categories of hygiene products.

5 Conclusion and future work

We presented an approach for identification of household segments with distinctive patterns and subgroups of cross-category associations, which differ from those mined in the entire data set. The proposed framework enables retailers to segment their customers according to their past interest in specific item combinations. The mined segment-specific associations provide a good basis for deriving more responsive recommendations or designing special offers through target marketing activities.

Nevertheless, the stepwise procedure has it's natural limitations imposed by the fact that later steps are dependent on the outcome of former stages. A simultaneous approach would disburden decision makers from determining various model parameters (like support thresholds, number of segments) at each stage. Another drawback is the ad-hoc exclusion of very frequently purchased categories, which could be substituted in future applications by a data driven weighting scheme.

References

AGRAWAL, R., IMIELINSKI, T. and SWAMI, A. (1993): Mining association rules between sets of items in large databases. In: *Proceedings of the ACM SIGMOD International Conference on Management of Data.* Washington D.C., 207–216.

BERRY, M. and LINOFF G. (2004): *Data mining techniques.* Wiley, Indianapolis.

BOZTUG, Y. and REUTTERER, T. (2007): A combined approach for segment-specific analysis of market basket data. In: *European Journal of Operational Research, forthcoming.*

DECKER, R. (2005): Market Basket Analysis by Means of a Growing Neural Network. *The International Review of Retail, Distribution and Consumer Research, 15, 151–169.*

GUPTA, G. K., STREHL, A. and GOSH, J (1999): Distance based clustering of association rules. In: *Intelligent Engineering Systems Through Artificial Neural Networks.* ASME Press, New York, 759–764.

HAHSLER, M., HORNIK, K. and REUTTERER, T. (2006): Implications of Probabilistic Data Modeling for Mining Association Rules. In: M. Spiliopoulou, R. Kruse, C. Borgelt, A. Nürnberger, W. Gaul (Eds.): *From Data and Information Analysis to Knowledge Engineering.* Springer, Heidelberg, 598–605.

KAUFMAN, L. and ROUSSEEUW, P. J. (2005): *Finding Groups in Data: An Introduction to Cluster Analysis.*, Wiley, New York.

LEISCH, F. and GRÜN, B. (2006): Extending standard cluster algorithms to allow for group constraints. In: A. Rizzi, M. Vichi (Eds.): *Compstat 2006, Proceedings in Computational Statistics.* Physica-Verlag, Heidelberg, 885–892.

LEISCH, F. (2006): A toolbox for k-centroids cluster analysis. In: *Computational Statistics and Data Analysis, 51(2), 526–544.*

OMIECINSKI, E. (2003): Alternative Interest Measures for Mining Associations in Databases. In: *IEEE Transactions on Knowledge and Data Engineering, 15(1), 57–69.*

PRINZIE, A. and VAN DEN POEL, D. (2005): Constrained optimization of data-mining problems to improve model performance. A direct-marketing application. In: *Expert Systems with Applications, 29, 630–640.*

REUTTERER, T., MILD, A., NATTER, M. and TAUDES, A. (2006): A dynamic segmentation approach for targeting and customizing direct marketing campaigns. In: *Journal of Interactive Marketing, 20(3), 43–57.*

AHP versus ACA – An Empirical Comparison

Martin Meißner, Sören W. Scholz and Reinhold Decker

Department of Business Administration and Economics, Bielefeld University, 33501 Bielefeld, Germany
{mmeissner, sscholz, rdecker}@wiwi.uni-bielefeld.de

Abstract. The Analytic Hierarchy Process (AHP) has been of substantial impact in business research and particularly in managerial decision making for a long time. Although empirical investigations (e.g. Scholl et al. (2005)) and simulation studies (e.g. Scholz et al. (2006)) have shown its general potential in consumer preference measurement, AHP is still rather unpopular in marketing research.

In this paper, we compare a new online version of AHP with Adaptive Conjoint Analysis (ACA) on the basis of a comprehensive empirical study in tourism which includes 10 attributes and 35 attribute levels. We particularly focus on the convergent and the predictive validity of AHP and ACA. Though both methods clearly differ regarding their basic conception, the resulting preference structures prove to be similar on the aggregate level. On the individual level, however, the AHP approach results in a significantly higher accuracy with respect to choice prediction.

1 Preference measurement for complex products

Conjoint Analysis (CA) is one of the most prominent tools in consumer preference measurement and widely used in marketing practice. However, an often stated problem of full-profile CA is that of dealing with large numbers of attributes. This limitation is of great practical relevance because ideally all attributes and attribute levels that affect individual choice should be included to map a realistic choice process.

Various methods have been suggested to provide more accurate insights into consumer preferences for complex products with many attributes (Green and Srinivasan (1990)). Self-Explicated (SE) approaches, e.g., are used to minimize the information overload by questioning the respondents about each attribute separately. But SE has been criticized for lacking the trade-off perspective underlying CA. For this reason, hybrid methods combining the strengths of SE and full-profile CA have been developed. Sawtooth Software's ACA is a commercially successful computer-based tool facilitating efficient preference measurements for complex products (for details, please see Sawtooth Software (2003)). While several other approaches, such as the hierarchical Bayes extensions of Choice-Based Conjoint Analysis, are available for estimating part-worth utilities on the individual level, ACA is still the standard in preference measurement for products with more than six attributes (Hauser and Toubia (2002), Herrmann et al. (2005)) and widely used in marketing practice (Sawtooth Software (2005)). In this paper, ACA will set a common benchmark for our empirical comparison.

Against this background, we introduce an online version of AHP as an alternative tool for consumer preference measurement in respective settings. Initially, AHP has been developed to analyze complex decision problems by decomposing them hierarchically into better manageable sub-problems. It has been of substantial impact in business research and particularly in managerial decision making for a long time. Empirical investigations (e.g. Scholl et al. (2005)) and simulation studies (e.g. Scholz et al. (2006)) recently demonstrated its general potential in consumer preference measurement. However, to the best of our knowledge, AHP has never been tested in a real-world online consumer survey, even though internet-based surveying gains increasing importance (Fricker et al. (2005)). In this paper, we compare an online version of AHP with ACA by referring to a comprehensive empirical investigation in tourism which includes 10 attributes and 35 attribute levels.

The remainder of the paper is structured as follows: In Section 2, we briefly outline the methodological basis of AHP. Section 3 describes the design of the empirical study. The results are presented in Section 4 and we conclude with some final remarks in Section 5.

2 The Analytic Hierarchy Process – AHP

In AHP, a decision problem, e.g. determining the individually most preferred alternative from a given set of products, has to be arranged in a hierarchy. It is referred to as the "main goal" in the following and represented by the top level of the hierarchy. By decomposing the main goal into several sub-problems, each of them representing the relation of a second level attribute category with the main goal, the complexity of the overall decision problem is reduced. The individual attribute categories, on their part, are broken down into attributes and attribute levels defining "lower" sub-problems. Typically, different alternatives (here: products or concepts) are considered at the bottom level of the hierarchy. But due to the large number of hypothetical products, or rather "stimuli" in the CA terminology, the use of incomplete hierarchies only covering attribute levels, instead of complete stimuli at the bottom level, is advisable.

For the evaluation of summer vacation packages–the objects of investigation in our empirical study–we have structured the decision problem in a 4-level hierarchy. The hierarchical structure displayed in Table 1 reflects the respondents' average perceptions and decomposes the complex product evaluation problem into easy to conceive sub-problems.

First, the respondents have to judge all pairs of attribute levels of each sub-problem on the bottom level of the hierarchy. Then, they proceed with paired comparisons on the next higher level of the hierarchy, an so on. In this way, the respondents are first introduced to the attributes' range and levels.

Table 1. Hierarchical structuring of the vacation package evaluation problem

Attribute category	Attribute	Attribute levels
Vacation spot	Sightseeing offers	1) Many 2) Some 3) Few
	Security concerns	1) Very high 2) High 3) Average
	Climate	1) Subtropical 2) Mediterranean 3) Desert
	Beach	1) Lava sand 2) Sea sand 3) Shingle
Hotel services	Leisure activities	1) Fitness room 2) Lawn sport facilities 3) Aquatic sports facilities 4) Indoor swimming pool 5) Sauna 6) Massage parlor
	Furnishing	1) Air conditioning 2) In-room safe 3) Cable/satellite TV 4) Balcony
	Catering	1) Self-catering 2) Breakfast only 3) Half board 4) Full board 5) All-inclusive
Hotel facilities	Location	1) Near beach 2) Near town
	Type of building	1) Rooming house 2) Hotel complex 3) Bungalow
	Outside facilities	1) Several pools 2) One large pool 3) One small pool

In order to completely evaluate a sub-problem h with n^h elements, $\frac{n^h(n^h-1)}{2}$ pairwise comparisons have to be carried out. Intuitively, the hierarchically decomposition of complex decision problems in many small sub-problems reduces the number of paired comparisons that have to be conducted to evaluate the decision problem.

Each respondent has to provide two responses for each paired comparison. First, the respondent has to state the direction of his or her preference for element i compared to element j with respect to an element h belonging to the next higher level. Second, the strength of his or her preference is measured on a 9-point ratio-scale, where 1 means "element i and j are equal" and 9 means "element i is absolutely preferred to element j" (or vice versa). The respondent's verbal expressions are transformed into priority ratios a_{ij}^h, where a large ratio expresses a distinct preference of i over j in sub-problem h. The reciprocal value $a_{ji}^h = 1/a_{ij}^h$ indicates the preference of element j over i. All pairwise comparisons of one sub-problem measured with respect to a higher level element h are brought together in the matrix $\mathbf{A}^h$ (Saaty (1980)):

$$\mathbf{A}^h = \left(a_{ij}^h\right)_{i,j=1,\ldots,n^h} = \begin{pmatrix} 1 & \ldots & a_{1n^h}^h \\ \vdots & \ddots & \vdots \\ a_{n^h1}^h & \ldots & 1 \end{pmatrix} \qquad \forall\, h \tag{1}$$

Starting from these priority ratios a_{ij}^h, the relative utility values w_i^h are calculated by solving the following eigenvalue problem for each sub-problem h:

$$\mathbf{A}^h \mathbf{w}^h = \lambda_{max}^h \mathbf{w}^h \qquad \forall\, h \tag{2}$$

The normalized principal right eigenvector belonging to the largest eigenvalue λ^h_{max} of matrix $\mathbf{A}^h$ yields the vector $\mathbf{w}^h$, which contains the relative utility values w^h_i for each element of sub-problem h.

An appealing feature of AHP is the computability of a consistency index (CI), which describes the degree of consistency in the pairwise comparisons of a considered sub-problem h. The CI value expresses the relative deviation of the largest eigenvalue λ^h_{max} of matrix $\mathbf{A}^h$ from the number of included elements n^h:

$$CI^h = \frac{\lambda^h_{max} - n^h}{n^h - 1} \qquad \forall\, h \tag{3}$$

To get a notion of the consistency of matrix $\mathbf{A}^h$, CI^h is related to the average consistency index of random matrices RI of the same size. The resulting measure is called the consistency ratio CR^h, with $CR^h = \frac{CI^h}{RI}$. In order to evaluate the degree of consistency for the entire hierarchy, the arithmetic mean of all consistency ratios ACR can be used (Saaty (1980)).

The AHP hierarchy can be represented by an additive model according to multi-attribute value theory. In doing so, the part-worth utilities are determined by multiplying the relative utility values of each sub-problem along the path, from the main goal to the respective attribute level. The attribute importances are calculated by multiplying the relative utility values of the attribute categories with the relative utility values of the attributes with respect to the related category. Then, the overall utility of a product or concept stimulus is derived by summing up the part-worth utilities of all attribute levels characterizing this alternative.

3 Design of the empirical study

The attributes and levels considered in the following empirical study were determined by means of dual questioning technique. Repertory grid and laddering techniques were applied to construct an average hierarchically representation of the product evaluation problem (Scholz and Decker (2007)). Altogether, 200 respondents participated in these pre-studies. The resulting product description design (see Table 1) was used for both the AHP and the ACA survey. The latter was conducted according to the recommendations in Sawtooth Software's recent ACA manual.

Each respondent had to pass either the ACA or the AHP questionnaire to avoid learning effects and to keep the time needed to complete the questionnaire within acceptable limits. Neither ACA nor AHP provide a general measure of predictive validity, which is usually quantified by presenting holdout tasks. If the number of attributes to be considered in a product evaluation problem is high, the use of holdout stimuli is regularly accompanied by the risk of information overload (Herrmann et al. (2005)). The relevant set of attributes was determined for each respondent individually to create a realistic choice setting. Each respondent was shown reduced product stimuli consisting of his or her six most important attributes. Accordingly, the predictive validity was measured by means of a computer administered holdout task similar to the one proposed by Herrmann et al. (2005).

Choice tasks including three holdout stimuli were presented to each respondent after having completed the preference measurement task. One of these alternatives was the best option available for the respective respondent (based on an online estimation of individual part-worth utilities carried out during the interview). The two other stimuli were slight modifications of this best alternative. Each one was generated by randomly changing three attribute levels from the most preferred to the second or third most preferred level.

In the last part of the online questionnaire, each respondent was faced with his or her individual profile of attribute importance estimates. In this regard, the corresponding question "*Does the generated profile reflect your notion of attribute importance?*" had to be answered on a 9-point rating scale ranging from "poor" (= 1) to "excellent" (= 9).

The respondents were invited to participate in the survey via a large public e-mail directory. For practical reasons we sent 50 % more invitations to the ACA than to the AHP survey. We obtained 380 fully completed questionnaires for ACA and 204 for AHP. In both cases, more than 40 % of those who entered the online interview also completed it. Chi-square homogeneity tests show that both samples are structurally identical with respect to socio-demographic variables.

4 Results

The data quality of our samples was assessed by measuring the consistency of the preference evaluation tasks. To evaluate the degree of consistency for the entire hierarchy, *ACR* was used for AHP. In case of ACA the coefficient of determination R^2, measuring the goodness-of-fit of the preference model, was considered. According to both measures, namely $ACR = .17$ and $R^2 = .77$, the internal validity of our study can be rated high. To come up with a fair comparison, we accepted all completed questionnaires and did not eliminate respondents from the samples on the basis of ACA's R^2 or AHP's *ACR*.

As a first step in our empirical investigation, we compared the resulting preference structures on the aggregate level. We transformed the part-worth utilities of both methods such that they sum up to zero for all levels of each attribute to facilitate direct comparisons. The attribute importances were transformed in both cases such that they sum up to one for each respondent. Spearman's rank correlation was used to contrast the convergent validity of AHP with ACA. Table 2 provides the attribute importances and the transformed part-worth utilities of both approaches. The differences regarding the part-worth utilities are rather small. Although both methods are conceptually different, the obvious structural equality points to high convergent validity. The rank correlation between AHP and ACA part-worth utilities equals .90.

In contrast, there are substantial differences between the attribute importances of AHP and ACA on the aggregate level ($r = -.08$). To assess the factual quality of attribute importances, we verified the present results by considering previous empirical studies in the field of tourism. In a recent study by Hamilton and Lau (2004) the *access to the sea or lake* was ranked second among the 10 attributes considered in this study. The importance of the corresponding attribute *location* in our study is higher for AHP than for ACA which favors the values provided by the former. Analogously, the attribute *active sports* (which corresponds to *leisure activities* in our study) was rated as very important by only 6 % of the respondents in a survey by Study Group "Vacation and Travelling" (FUR (2004)). On the other hand, the importance of the attribute *relaxation*, which is similar to *outside facilities* in our study, was highly appreciated. Insofar, the AHP results are in line with the FUR study by awarding high importance to *outside facilities* and lower importance to *leisure activities*.

To find an appropriate external criterion that allows to measure the validity of the resulting individual attribute importances is difficult. We chose the respondents' individual perceptions as an indicator and measured the adequacy of the importance

Table 2. Average attribute importances and part-worth utilities

Category	**ACA**			**AHP**		
Attribute	Importance	Part-worths*		Importance	Part-worths*	
		One	Two		One	Two
		Three	Four		Three	Four
		Five	Six		Five	Six
Vacation spot						
Sightseeing offers	9.51	.24 (1)	.09 (2)	6.19	.21 (1)	.02 (2)
		-.33 (3)			-.23 (3)	
Security concerns	10.87	.36 (1)	.06 (2)	11.86	.53 (1)	-.09 (2)
		-.42 (3)			-.44 (3)	
Climate	11.45	.01 (2)	.36 (1)	9.69	-.13 (2)	.39 (1)
		-.37 (3)			-.26 (3)	
Beach	9.83	-.10 (2)	.35 (1)	5.56	-.09 (2)	.26 (1)
		-.25 (3)			-.17 (3)	
Hotel services						
Leisure activities	11.72	-.20 (6)	-.02 (2)	7.52	-.04 (6)	.02 (2)
		.04 (2)	.20 (1)		-.01 (4)	.01 (3)
		.01 (3)	-.03 (5)		.03 (1)	-.01 (5)
Furnishing	10.15	.10 (1)	-.13 (4)	12.49	.08 (1)	-.09 (4)
		-.03 (3)	.06 (2)		-.01 (3)	.03 (2)
Catering	12.17	-.19 (5)	.03 (3)	13.29	-.07 (5)	-.01 (3)
		.12 (1)	-.07 (4)		.02 (2)	-.04 (4)
		.10 (2)			.10 (1)	
Hotel facilities						
Location	7.78	-.24 (2)	.24 (1)	12.84	-.32 (2)	.32 (1)
Type of building	9.09	.08 (2)	-.22 (3)	8.36	-.03 (2)	-.12 (3)
		.14 (1)			.15 (1)	
Outside facilities	7.40	.25 (1)	.00 (2)	12.11	.28 (1)	-.09 (2)
		-.25 (3)			-.19 (3)	

(* The ranking of attribute levels is depicted in brackets.)

estimates in the last part of the questioning by means of a 9-point rating scale question (see Section 3). Here, AHP was judged significantly better ($p < .01$) with an average value of 7.3 compared to ACA with 6.68. This suggests that AHP yields higher congruence with the individual perceptions than ACA. But since it is not clear to what extent respondents are really aware of their attribute importances, the explanatory power of this indicator has not been fully established.

The predictive accuracy of both methods was checked by comparing the overall utilities of the holdout stimuli with the actual choice in the presented holdout task as explained in Section 3. Both methods were evaluated by two measures: The *first choice hit rate* equals the frequency with which a method correctly predicts the vacation package chosen by the respondents. Here, AHP significantly outperforms ACA with 83.33 % against 60.78 % ($p < .01$). The *overall hit rate* indicates how often a method correctly predicts the rank order of the three holdout stimuli as stated by the respondents. Taking into account that the respondents had to rank alternatives of their evoked sets (i.e. the best and two "near-best" alternatives) the predictive accuracy of both approaches is definitely satisfying. Again, AHP significantly outperforms ACA with an overall hit rate equal to 63.42 % compared to 43.94 % for the latter ($p \leq .01$). For comparison: random prediction would lead to an overall hit rate equal to .16. All in all, AHP shows a significantly higher predictive accuracy for products belonging to the evoked set of the respondents than ACA.

5 Conclusions and outlook

This paper presents an online implementation of AHP for consumer preference measurement in the case of products with larger numbers of attributes. As a first benchmark, we empirically compared AHP with Sawtooth Software's ACA in the domain of summer vacation packages. While both methods yielded high values for internal and convergent validity, AHP significantly outperforms ACA regarding individually tailored holdout tasks generated from the respondents' evoked sets. The results suggest AHP as a promising method for preference-driven new product development.

Further empirical investigations are required to support the results presented here. These should include additional preference measurement approaches, such as SE or Bridging CA (Green and Srinivasan (1990)). Moreover, the implication of different hierarchies have not been fully understood in AHP research (Pöyhönen et al. (2001)). While we conducted extensive pre-studies to come up with an expedient hierarchy, market researchers should be very carefully when structuring their decision problems hierarchically. The application of simple 3-level hierarchies focusing on the main goal, attributes and levels only, and leaving out higher-level attribute categories might be beneficial. These hierarchies would also be reasonable when the product evaluation problem cannot be broken down into 'natural' groups of attribute categories.

References

FRICKER, S., GALESIC, M., TOURANGEAU, R. and YAN, T. (2005): An Experimental Comparison of Web and Telephone Surveys, *Public Opinion Quarterly, 69, 3, 370–392.*

FUR (2004): Travel Analysis 2004 by Study Group "Vacation and Travelling", www.fur.de.

GREEN, P.E. and SRINIVASAN, V. (1990): Conjoint Analysis in Marketing: New Developments with Implications for Research and Practice, *Journal of Marketing, 54, 3–19.*

HAMILTON, J.M. and LAU, M.A. (2004): The Role of Climate Information in Tourist Destination Choice Decision-making, Working Paper FNU-56, Centre for Marine and Climate Research, Hamburg University.

HAUSER, J.R. and TOUBIA, O. (2005): The Impact of Utility Balance and Endogenity in Conjoint Analysis, *Marketing Science, 24, 3, 498–507.*

HERRMANN, A., SCHMIDT-GALLAS, D. and HUBER, F. (2005): Adaptive Conjont Analysis: Understanding the Methodology and Assessing Reliability and Validity. In: A. Gustafsson, A. Herrmann and F. Huber (Eds.): *Conjoint Measurement: Methods and Applications*, Springer, Berlin, 253-278.

PÖYHÖNEN, M., VROLIJK, H. and HÄMÄLÄINEN, R.P. (2001): Behavioral and Procedural Consequences of Structural Variation in Value Trees, *European Journal of Operational Research, 134, 1, 216–227.*

SAATY, T.L. (1980): *The Analytic Hierarchy Process.* McGraw Hill, New York.

SAWTOOTH SOFTWARE (2003): SSI Web v3.5, Sawtooth Software.

SAWTOOTH SOFTWARE (2005): Report on Conjoint Analysis Usage among Sawtooth Software Customers, www.sawtoothsoftware.com, accessed March 2007.

SCHOLL, A., MANTHEY, L., HELM, R. and STEINER, M. (2005): Solving Multiattribute Design Problems with Analytic Hierarchy Process and Conjoint Analysis: An Empirical Comparison. *European Journal of Operational Research, 164, 760–777.*

SCHOLZ, S.W. and DECKER, R. (2007): Measuring the Impact of Wood Species on Consumer Preferences for Wooden Furniture by Means of the Analytic Hierarchy Process. *Forest Products Journal, 57, 3, 23–28.*

SCHOLZ, S.W., MEISSNER, M. and WAGNER, R. (2006): Robust Preference Measurement: A Simulation Study of Erroneous and Ambiguous Judgement's Impact on AHP and Conjoint Analysis. In: H.-O. Haasis, H. Kopfer and J. Schönberger (Eds.): *OR Proceedings 2005*, Springer, Berlin, 613–618.

On the Properties of the Rank Based Multivariate Exponentially Weighted Moving Average Control Charts

Amor Messaoud and Claus Weihs

Fachbereich Statistik, Universität Dortmund, Germany
messaoud@statistik.uni-dortmund.de

Abstract. The rank based multivariate exponentially weighted moving average (rMEWMA) control chart was proposed by Messaoud et al. (2005). It is a generalization, using the data depth notion, of the nonparametric EWMA control chart for individual observations proposed by Hackl and Ledolter (1992). The authors approximated its asymptotic in-control performance using an integral equation and assuming that a sufficiently large reference sample is available. The actual paper studies the effect of the use of reference samples of limited amount of observations on the in-control and out-of-control performances of the proposed control chart. Furthermore, general recommendations for the required reference sample sizes are given so that the in-control and out-of-control performances of the rMEWMA control chart approach their asymptotic counterparts.

1 Introduction

In practice, rMEWMA control charts are used with reference samples of limited amount of observations. In this case, the estimation effect may affect its in-control and out-of-control performances. This issue is discussed in this paper based on the results of Messaoud (2006). In section 2, we review the data depth notion. The rMEWMA control chart is introduced in section 3. The effect of the use of reference samples of limited amount of observations on its in-control and out-of-control performances is studied in section 4.

2 Data depth

Data depth measures how deep (or central) a given point $\mathbf{X} \in \mathbb{R}^d$ is with respect to (w.r.t.) a probability distribution F or w.r.t. a given data cloud $S = \{\mathbf{Y}_1, \ldots, \mathbf{Y}_m\}$. There are several measures for the depth of the observations, such as Mahalanobis depth, simplicial depth, half-space depth, and majority depth of Singh, see Liu et al. (1999). In this work, only the Mahalanobis depth is considered, see section 4.1.

The Mahalanobis depth

The Mahalanobis depth of a given point $\mathbf{X} \in \mathbb{R}^d$ w.r.t. F is defined by

$$MD(F,\mathbf{X}) = \frac{1}{1+(\mathbf{X}-\mu_F)'\Sigma_F^{-1}(\mathbf{X}-\mu_F)},$$

where μ_F and Σ_F are the mean vector and covariance matrix of F, respectively. The sample version of MD is obtained by replacing μ_F and Σ_F with their sample estimates.

3 The proposed *r*MEWMA control chart

Let $\mathbf{X}_t = (x_{1,t},\ldots,x_{d,t})'$ denote the $d \times 1$ vector of quality characteristic measurements taken from a process at the tth time point where $x_{j,t}$, $j = 1, \ldots, d$, is the observation on variate j at time t. Assume that the successive $\mathbf{X}_t$ are independent and identically distributed random vectors. Assume that $m > 1$ independent random observations $\{\mathbf{X}_1, \ldots, \mathbf{X}_m\}$ from an in-control process are available. That is, the *r*MEWMA monitoring procedure starts at time $t = m$.

Let $RS = \{\mathbf{X}_{t-m+1}, \ldots, \mathbf{X}_t\}$ denote a reference sample comprised of the m most recent observations taken from the process at time $t \geq m$. It is used to decide whether or not the process is still in control at time t. The main idea of the proposed *r*MEWMA control chart is to represent each multivariate observation of the reference sample by its corresponding data depth. Thus, the depths $D(RS,\mathbf{X}_i)$, $i = t-m+1, \ldots, t$, are calculated w.r.t. RS.

Now, the same principles proposed by Hackl and Ledolter (1992) are used to construct the *r*MEWMA control chart. Let Q_t^* denote the sequential rank of $D(RS,\mathbf{X}_t)$ among $D(RS,\mathbf{X}_{t-m+1}), \ldots, D(RS,\mathbf{X}_t)$. It is given by

$$Q_t^* = 1 + \sum_{i=t-m+1}^{t} I\Big(D(RS,\mathbf{X}_t) > D(RS,\mathbf{X}_i)\Big), \tag{1}$$

where $I(.)$ is the indicator function. It is assumed that tied data depth measures are not observed. Thus, Q_t^* is uniformly distributed on the m points $\{1,2,\ldots,m\}$. The standardized sequential rank Q_t^m is given by

$$Q_t^m = \frac{2}{m}\left(Q_t^* - \frac{m+1}{2}\right). \tag{2}$$

It is uniformly distributed on the m points $\{1/m-1,\ 3/m-1,\ \ldots,\ 1-1/m\}$ with mean $\mu_{Q_t^m} = 0$ and variance $\sigma_{Q_t^m} = \frac{m^2-1}{3m^2}$, see Hackl and Ledolter (1992).

The control statistic T_t is the EWMA of standardized sequential ranks. It is computed as follows

$$T_t = \min\Big\{B, (1-\lambda)T_{t-1} + \lambda Q_t^m\Big\}, \tag{3}$$

$t = 1, 2, \ldots$, where $0 < \lambda \leq 1$ is a smoothing parameter, $B > 0$ is a reflecting boundary and $T_0 = u$ is a starting value. The process is considered in-control as long as $T_t \geq h$, where $h < 0$ is a lower control limit ($h \leq u \leq B$). Note that the lower-sided *r*MEWMA is considered because the statistic Q_t^m is higher "the better". Indeed, a high value of Q_t^m means that observation $\mathbf{X}_t$ is deep w.r.t. RS which refers to a process improvement. A reflecting boundary is included to prevent the *r*MEWMA control statistic from drifting to one side indefinitely. It is known that EWMA schemes can suffer from an "inertia problem" when there is a process change some time after beginning of monitoring. That is, an EWMA control statistic can have wandered away from a center line in a direction opposite to that of a shift that occurs some time after the start of monitoring. In this unhappy circumstance, an EWMA scheme can take long time to signal. For further details about the design of *r*MEWMA control charts, see Messaoud (2006).

In practice when measurements or other numerical observations are taken, it is often that two or more observations are tied. The most common approach to this problem is to assign to each observation in a tied set the midrank, that is, the average of the ranks reserved for the observations in the tied set.

The statistical design of the *r*MEWMA control chart refers to choices of combinations of λ, h, B and m. It ensures the chart performance meets certain statistical criteria. These criteria are often based on aspects of the run length distribution of the control chart. The run length (RL) of a control chart is a random variable that represents the number of plotted statistics until a signal occurs. The most common measure of control chart performance is the expected value of the run length; i.e. the average run length (ARL). The ARL should be large when the process is statistically in-control (in-control ARL) and small when a shift has occurred (out-of-control ARL). However, conclusions based on in-control and out-of-control ARL alone can be misleading. Knowledge of the in-control and out-of-control RL distributions would provide a comprehensive understanding of the in-control and out-of-control control chart performances. For example, the lower percentiles of the in-control and out-of-control RL distributions give information about the early false alarm rates and the ability to quickly detect an out-of-control condition of a control chart.

The integral equation (4) is used to approximate the asymptotic in-control ARL of *r*MEWMA control charts

$$L(u) = 1 + L(B)\Pr\left(q \geq \frac{B-(1-\lambda)u}{\lambda}\right) + \int_h^B L\Big((1-\lambda)u + \lambda q\Big) f(q)dq, \quad (4)$$

where $f(q)$ is the probability density of the uniform distribution. In this approximation, it is assumed that a sufficiently large reference sample is available and the slight dependence among successive ranks Q_t^m is ignored.

4 Effect of the reference sample size on *r*MEWMA control charts performance

4.1 Simulation Study

Messaoud (2006) conducted a simulation study in order to examine the estimation effect on the desired in-control and out-of-control run length (RL) performances of *r*MEWMA control charts. A desired in-control and out-of-control RL performances mean that the empirical in-control and out-of-control RL distributions approach their asymptotic counterpart. As mentioned, only the Mahalanobis *r*MEWMA control charts are considered.

For the simulation, random independent observations $\{\mathbf{X}_t\}$ are generated from a bivariate normal distribution with mean vector $\mu_0 = (0,0)'$ and variance covariance matrix Σ_X. Note that due to the nonparametric nature of *r*MEWMA control charts, the normality of the observations is not required and any other distribution could be used. The shift scenario in the mean vector from μ_0 to μ_1 is considered to represent the out-of-control process. Its magnitude δ is given by

$$\delta^2 = (\mu_1 - \mu_0)'\Sigma_X^{-1}(\mu_1 - \mu_0). \tag{5}$$

Other out-of-control scenarios are not considered, for example a change in the in-control covariance matrix Σ_X. Note that in the context of multivariate normality, δ is called the noncentrality parameter.

Since the multivariate normal distribution is elliptically symmetrical and the Mahalanobis depth is affine invariant, see Liu et al. (1999), the Mahalanobis *r*MEWMA control charts are directionally invariant. That is, their out-of-control ARL performance depends on a shift in the process mean vector μ only through the value of δ. Thus, without any loss of generality, the shift is fixed in the direction of $\mathbf{e}_1 = (1,0)'$ and the variance covariance matrix Σ_X is taken to be the identity matrix $\mathbf{I}$. For more details about the simulation study, see Messaoud (2006).

4.2 Simulation results

Messaoud (2006) considered the four Mahalanobis *r*MEWMA control charts with $\lambda = 0.05$, 0.1, 0.2 and 0.3. In this paper, only the Mahalanobis *r*MEWMA control chart with $\lambda = 0.3$, $h = -0.551$ and $B = -h$ is studied in detail.

Table 1 shows summary statistics of the in-control ($\delta = 0$) and out-of-control ($\delta \not\equiv 0$) run length (RL) distributions of the Mahalanobis *r*MEWMA control charts based on reference samples of size m = 10, 28, 100, 200, 500, 1000 and 10000 ($m \approx \infty$). Note that the desired in-control ($\delta = 0$) ARL performance is obtained using $m = 28$. This motivates this choice. SDRL is the standard deviation of the run length. $Q(.10)$, $Q(.50)$, and $Q(.90)$ are respectively the 10th, 50th, and 90th percentiles of the in-control and out-of-control RL distributions. In the following, ARL_0 and ARL_1 are used to represent the in-control ($\delta = 0$) and out-of-control (for any $\delta \not\equiv 0$) ARL, respectively. Similarly, $Q_0(q)$ and $Q_1(q)$ refer to the qth percentile of the in-control ($\delta = 0$) and out-of-control (for any $\delta \not\equiv 0$) RL distributions, respectively. Note that $Q_0(.50)$ and $Q_1(.50)$ are respectively the in-control and out-of-control median RL.

Table 1. In-control ($\delta = 0$) and out-of-control ($\delta \not\equiv 0$) run length properties of Mahalanobis *r*MEWMA control charts with $\lambda = 0.3$ and $h = -0.551$ based on reference samples of size m

m		Shift Magnitude δ						
		0.0	0.5	1.0	1.5	2.0	2.5	3.0
	ARL	342.18	341.42	339.42	334.52	326.63	316.80	306.92
	SDRL	338.74	338.62	338.77	338.89	338.54	337.80	337.35
10	$Q(.10)$	38	37	35	30	22	12	5
	$Q(.50)$	238	237	236	230	222	212	201
	$Q(.90)$	786	785	784	779	771	759	749
	ARL	199.77	196.56	183.44	151.96	105.25	59.10	28.04
	SDRL	193.98	193.91	193.13	187.73	169.44	133.97	93.43
28	$Q(.10)$	25	21	9	5	3	3	3
	$Q(.50)$	140	137	124	86	12	5	4
	$Q(.90)$	456	452	438	399	325	205	44
	ARL	185.15	170.90	118.21	43.15	9.17	4.32	3.47
	SDRL	176.11	175.05	162.56	104.09	31.47	4.67	0.91
100	$Q(.10)$	24	15	6	4	3	3	3
	$Q(.50)$	133	118	40	10	5	4	3
	$Q(.90)$	414	398	329	124	12	6	5
	ARL	188.05	160.85	76.68	15.85	5.88	4.02	3.36
	SDRL	177.44	173.60	131.29	40.39	4.87	1.49	0.75
200	$Q(.10)$	23	14	6	4	3	3	3
	$Q(.50)$	138	99	25	8	5	3	3
	$Q(.90)$	420	389	234	26	10	6	4
	ARL	196.22	138.36	38.11	10.40	5.47	3.92	3.32
	SDRL	185.11	163.28	63.83	8.37	2.85	1.34	0.69
500	$Q(.10)$	24	14	6	4	3	3	3
	$Q(.50)$	141	79	21	8	5	3	3
	$Q(.90)$	445	350	78	20	9	6	4
	ARL	199.35	119.63	29.83	9.91	5.38	3.88	3.31
	SDRL	192.86	140.93	32.33	7.28	2.73	1.31	0.67
1000	$Q(.10)$	24	13	6	4	3	3	3
	$Q(.50)$	141	73	20	8	5	3	3
	$Q(.90)$	455	280	65	19	9	6	4
	ARL	201.00	99.02	26.16	9.58	5.29	3.85	3.29
	SDRL	197.71	98.23	23.12	6.66	2.61	1.26	0.65
∞	$Q(.10)$	24	13	6	4	3	3	3
	$Q(.50)$	141	68	19	8	5	3	3
	$Q(.90)$	459	223	56	18	9	5	4

NOTE: ARL = average run length
SDRL = standard deviation of run length distribution
$Q(q)$ = qth percentile of run length distribution

Performance of *r*MEWMA control charts based on small reference samples

Table 1 shows that the ARL_0 performance of the *r*MEWMA control chart is approximately equal to the desired ARL_0 of 200 using $m = 28$. Moreover, $Q_0(.10)$, $Q_0(.50)$ and $Q_0(.90)$ are approximately equal to their asymptotic counterparts. However, Table 1 shows that the ARL_1, $Q_1(.50)$, and $Q_1(.90)$ values of this control chart are much larger than the ARL_1, $Q_1(.50)$ and $Q_1(.90)$ values of *r*MEWMA control charts with larger values of m. Therefore, even though that using *relatively* small reference samples achieves the desired in-control RL performance, this choice reduces considerably the *r*MEWMA control charts ability to quickly detect an out-of-control condition.

Performance of *r*MEWMA control charts based on moderate and large reference samples

In the following, the *r*MEWMA control charts based on moderate and large reference samples are considered, i.e., $m = 100, 200, 500$, and 1000.

In-Control case ($\delta = 0$)

Table 1 shows that the ARL_0 values of the *r*MEWMA control charts based on reference samples of size $m = 100, 200, 500$ and 1000 are shorter than the desired ARL_0 of 200. That is, these control charts produce more false alarms than expected. However, interpretation based on the ARL_0 values alone can be misleading. The $Q_0(.90)$ values given in Table 1 indicate that the larger percentiles of the in-control RL distributions affect the ARL_0 values.

For example, consider the *r*MEWMA control chart with $m = 200$. Its ARL_0 value is 6.44% shorter than its asymptotic value, see Table 1. Table 1 shows that the $Q_0(.10)$ value is approximately equal to its asymptotic value of 24. The $Q_0(.50)$ value is equal to 138. It is slightly shorter than its asymptotic value of 141. That is, the control chart produce in average a false alarm within 138 observations with a probability of 0.5 and within 141 observations with the same probability when $m \approx \infty$. Thus, the control chart does not suffer from the problem of early false alarms. However, the $Q_0(.90)$ value is equal to 420. It is much shorter than its asymptotic value of 459. This implies that the larger percentiles affect the ARL_0 value.

Now we will focus on the probabilities of the occurrence of early false alarms. As mentioned, these probabilities are reflected in the lower percentiles of the in-control RL distributions. The 5th, 10th, 20th, 30th, 40th and 50th percentiles of the in-control RL distributions of the *r*MEWMA control charts with reference samples of size $m = 100, 200, 500$ and 1000 are nearly the same as their asymptotic values, see Messaoud (2006). Only the $Q_0(.40)$ and $Q_0(.50)$ values of the *r*MEWMA control charts with $100 \leq m \leq 200$ are slightly shorter than their asymptotic values.

Therefore, we can conclude that the observed decreases in the ARL_0 values in Table 1 are caused by the shorter values of the larger percentiles. Practitioners should not fear for the problem of early false alarms when reference samples of size $m \geq 100$ observations are used.

Out-of-control case ($\delta \neq 0$)

Table 1 shows that the ARL_1 values of the *r*MEWMA control charts are larger than their asymptotic counterparts. However, interpretation based on the ARL_1 values alone may lead to inaccurate conclusions. Thus, the lower percentiles and the median of the out-of-control RL distributions are investigated. They provide useful information about the ability of *r*MEWMA control charts to quickly detect an out-of-control condition.

First, we investigate the out-of-control RL performance of the *r*MEWMA control charts for shifts of magnitude $\delta \geq 1.5$. Table 1 shows that the $Q_1(.10)$ and $Q_1(.50)$ values are nearly the same as their asymptotic values. However, the $Q_1(.90)$ values are larger than their asymptotic values. That is, the ARL_1 values are affected by some long runs. For example, consider the *r*MEWMA control chart with reference sample of size $m = 100$. Its ARL_1 value for detecting a shift of magnitude $\delta = 1.5$ is 350.42% larger than its asymptotic value of 9.58. Table 1 shows that the $Q_1(.10)$ and $Q_1(.50)$ values are nearly the same as their asymptotic counterparts. However, the ARL_1 value is affected by some long runs. The $Q_1(.90)$ value is equal to 124. It is much larger than its asymptotic value of 18. Therefore, we can conclude that the estimation effect does not affect the ability of the *r*MEWMA control chart with $\lambda = 0.3$ to quickly detect shifts of magnitude $\delta \geq 1.5$ when reference samples of size $m \geq 100$ are used.

Now we investigate the out-of-control RL performance of the *r*MEWMA control charts for shifts of magnitude $\delta = 0.5$ and 1.0. The lower percentiles of the out-of-control RL distributions of *r*MEWMA control charts with $100 \leq m \leq 200$ are larger than their asymptotic values, see Messaoud (2006). That is, the estimation effect affects the sensitivity of these control charts to react to shifts of magnitude $\delta \leq 1.0$. For *r*MEWMA control charts with $500 \leq m \leq 1000$, the lower percentiles of the out-of-control RL distribution are nearly the same or slightly larger than the asymptotic values. Therefore, we can conclude that using reference samples of size $m \geq 500$ ensures that the *r*MEWMA control chart with $\lambda = 0.3$ perform like one with sufficiently large reference samples, i.e., $m \approx \infty$. Its ability to quickly detect an out-of-control condition is not affected.

Sample size requirements

Note that similar results are observed for *r*MEWMA control charts with $\lambda = 0.05$, 0.1 and 0.2, see Messaoud (2006). Therefore, we can conclude that using large reference samples of size $m \geq 500$ will reduce the estimation effect on the in-control and out-of-control RL performances of *r*MEWMA control charts. The early false alarms produced by the *r*MEWMA control charts and the early detection of out-of-control conditions are mainly used to evaluate their in-control and out-of control performances. The reader should be aware that the sample size recommendation may differ for other out-of-control scenarios. For example, a shift in the in-control covariance matrix.

5 Conclusion

In this work, the estimation effect on the performance of the *r*MEWMA control chart is studied. General recommendations for the required reference sample sizes are given so that the in-control and out-of-control RL performances of *r*MEWMA control chart approach their asymptotic counterparts. As noted, only the shift scenario in the mean vector is considered to represent the out-of-control process. The required large reference samples of size $m \geq 500$ observations should not be a problem for the applications of *r*MEWMA monitoring procedures. Nowadays, advances in data collection activities as well as the computational power of digital computers have increased the available data sets in many industrial processes. However, practitioners should not neglect the estimation effect on the in-control and out-of-control performances of the *r*MEWMA control charts if for some industrial applications forming large reference samples might be problematic.

Acknowledgements

This work has been supported by the Collaborative Research Centre "Reduction of Complexity in Multivariate Data Structures" (SFB 475) of the German Research Foundation (DFG).

References

HACKL, P. and LEDOLTER, J. (1992): A New Nonparametric Quality Control Technique. *Communications in Statistics-Simulation and Computation 21, 423–443.*

LIU, R. Y., PARELIUS, J. M., and SINGH, K. (1999): Multivariate Analysis by Data Depth: Descriptive Statistics, Graphics and Inference (with discussion). *The Annals of Statistics, 27, 783–858.*

MESSAOUD, A. (2006): Monitoring Strategies for Chatter Detection in a Drilling Process. PhD Dissertation, Department of Statistics, University of Dortmund.

MESSAOUD, A., THEIS, W., WEIHS, C. and HERING, F. (2005): Application and Use of Multivariate Control Charts in a BTA Deep Hole Drilling Process. In: C. Weihs, and W. Gaul (Eds.): *Classification- The Ubiquitous Challenge.* Springer, Berlin-Heidelberg, 648-655.

Are Critical Incidents Really Critical for a Customer Relationship? A MIMIC Approach

Marcel Paulssen[1] and Angela Sommerfeld[2]

Institut für Marketing, Humboldt-Universität zu Berlin
10178 Berlin, Germany
{paulssen, sommerfeld}@wiwi.hu-berlin.de

Abstract. With increasing duration of a relationship the probability that customers experience specifically negative interaction episodes but also very positive interaction episodes increases. A key question that has not been investigated in the literature concerns the impact of these extreme interaction experiences, referred to as Critical Incidents (CIs) on the quality and strength of consumer-firm relationships. In a sample of customers in a service setting we first demonstrate that indeed the number of negative (positive) CIs possess a negative (positive) and asymmetric impact on measures of relationship quality (satisfaction, trust) and measure of relationship strength (loyalty). Second using a MIMIC approach we further shed light on the question which particular incidents are really critical for a customer firm relationship and which have to be prevented with priority.

1 Introduction

Customers' interaction experiences with a service provider vary widely, ranging from remarkable positive to remarkable negative experiences. Especially extreme interaction experiences in the service process might significantly influence the customer-firm relationship. These extreme interaction experiences are therefore referred to as Critical Incidents (CIs). Several methods to record and measure CIs exist. The most commonly used is the Critical Incident Technique (CIT), which records CIs through interviews. The method's major advantage in comparison to e.g. traditional attribute based measures of satisfaction lies in the collection of experiences from the respondents' perspective. Thus service quality perceptions are not measured with ratings on predefined attributes but captured in the customer's own words (Edvardsson, 1992). An extension of the CIT is the Sequential Incident Technique (SIT). Following the process of service delivery and consumption, CIs as well as usual events are collected to inform about crucial points within this process. Besides its costliness it remains unclear if this method is suitable for understanding customer satisfaction and its' application is limited to standardized processes (Stauss & Weinlich, 1997). Both methods assume that the collected CIs are indeed critical for the customer-company relationship but do not assess their criticality. This shortcoming is alleviated with the Switching Path Analysis Technique (SPAT), which retrospectively asks customers for experiences accountable for their provider switch (Roos, 2002). Although this method clearly collects incidents that are truly critical, it is only applicable to respondents who have just switched - a strong limitation for recruiting respondents and a reason for the rare application of SPAT.

In contrast more than 140 studies appeared in the marketing literature applying the conventional CIT following its introduction by Bitner, Booms, and Tetreault (1990) to the marketing community. Even though the CIT is the most applicable method and also the most widely applied technique, current CIT studies suffer from severe methodological weaknesses. A current review by Gremler (2004) on the usage of the CIT in marketing highlights frequent shortcomings of its existing applications. Specifically multiple incidents occurring in the same context and multiple occurrence of the same CI are generally not collected. Besides, many studies restricted their collection to negative CIs and one CI per respondent. An alarming 38% of these studies do not report any type of reliability statistic. Furthermore the relevance of CIs for a customer relationship is assumed but not assessed, since CIs are not linked to key relationship constructs such as satisfaction, trust and loyalty. Following Gremler's (2004) call we conducted a study without these mentioned shortcomings. In contrast to existing studies rather than merely assuming we will explicitly model the impact of the number of experienced CIs on relationship outcomes notably trust and loyalty. Further we will apply a Multiple Indicators and Multiple Causes (MIMIC) approach to model the impact of the category of the experienced CIs on relationship outcomes in order to understand which interaction episodes were particularly damaging or supporting for a marketing relationship.

2 Hypotheses

During their relationship with a service provider some customers might constantly experience encounters characterized by expected employee behaviors and reactions. In addition to these neutral encounters, other customers experience remarkably delighting or upsetting interactions episodes. These unexpected CIs are a source of dis/satisfaction (Bitner et al., 1990, p. 83), and are assumed to impact the overall evaluation of the service. The negative effect of experiencing one negative CI versus none on satisfaction has already been confirmed (Odekerken-Schröder et al., 2000). However, explicit tests regarding the influence of the number of experienced CIs and their valence (positive / negative) on overall satisfaction are lacking. The overall evaluation of the service is based on the quality of regular service performances as well as extreme positive and negative experiences. Therefore, we argue that the number of prior positive (negative) CIs has a positive (negative) influence on service satisfaction. We therefore propose:

H_1: *The number of positive critical incidents impacts positively on satisfaction.*
H_2: *The number of negative critical incidents impacts negatively on satisfaction.*

CIs possess a high diagnosticity for drawing inferences regarding the partner's intentionality and disposition and therefore the status of the relationship itself (Ybarra & Stephan, 1999; Fiske, 1980). Even though highly counterintuitive, the only present study nullifies an effect of CIs on trust (Odekerken-Schröder et al., 2000). Still CIs are exactly those "moments of truth" that relationship partner's use to make inferences about the intentionality of the relationship other and can therefore be either trust building or trust destroying. We therefore propose that:

H_3: The number of positive critical incidents impacts positively on trust.

H_4: The number of negative critical incidents impacts negatively on trust.

Building on Kahneman and Tversky's prospect theory (1979) a broad body of research has demonstrated the consistent asymmetrical impact of negative information, attributes, and events (Baumeister et al., 2001). CIs are by definition clearly above or below a neutral reference point, thus are either perceived to be very positive or very negative. In accordance with findings on the asymmetric impact of negative events in the psychological literature (e.g. Taylor, 1991), we propose:

H_5: The number of negative critical incidents impacts more strongly on satisfaction than the number of positive critical incidents.

H_6: The number of negative critical incidents impacts more strongly on trust than the number of positive critical incidents.

Numerous studies have demonstrated that both trust and satisfaction are determinants of repurchase intentions (Fornell et al., 1996; Geyskens et al., 1999; Morgan & Hunt, 1994; Szymanski & Henard, 2001). Because findings concerning these relationships are almost unanimous we do not further elaborate on them and propose:

H_7: Trust in the service provider increases loyalty to the service.

H_8: Satisfaction with the service increases loyalty to the service.

Furthermore Singh and Sirdeshmukh (2000) argue that satisfactory experiences with a product or service are likely to reinforce expectations of competent performance in the future, whereas below-expectation performance are expected to reduce trust, thus we propose:

H_9: Satisfaction with a focal product increases trust in the manufacturer.

In the process of forming loyalty intentions previous studies have documented the mediating role of satisfaction (Szymanski & Henard, 2001) and trust (e.g. Morgan & Hunt, 1994). Building on these findings from the marketing literature, we propose a similar mediating mechanism for the experienced number of CIs on loyalty intentions.

H_{10}: The impact of critical incidents on loyalty is fully mediated via trust and satisfaction.

We need to note an important control variable neglected in the previous literature on CIs: mood. Cognitive processes are affected by mood (e.g. Forgas, 1995). In the context of the present study two mechanism merit attention. First, mood congruent recall (e.g. Eich et al., 1994) implies that mood influences the frequency of recalled CIs, thus respondents in a positive (negative) mood will be more (less) likely to recall positive CIs. Second, affect infusion (e.g. Forgas, 1995) implies that mood affects judgments as satisfaction and trust, thus a positive mood will additionally enhance (reduce) these ratings. The interplay of these two effects might result in inflated correlations between number of recalled incidents and measures of relationship quality. To exclude this possibility we control for respondents' mood states.

3 Method

The empirical investigation was conducted with customers of repair departments of a major German car manufacturer. In 5 different outlets, in a metropolitan area, customers entering the store were asked to take part in a satisfaction survey. 207 customers agreed to participate and 191 of them had prior experiences with the service department. Accordingly 191 face-to-face interviews were conducted, consisting of a fully-structured and a semi-structured part. In the opening fully-structured part, participants were first asked to indicate their current mood, followed by questions capturing their satisfaction, trust and intention to stay loyal with the service department. Afterwards, in the semi-structured part the interview respondents were asked to talk about any CI concerning the repair department. This part of the interview followed the widely used procedure of Bitner et al. (1990). Thus, the central question was: *"Think of your experiences with the repair department. Can you remember particularly good or bad experiences during your contacts with the repair department?"*. It has to be mentioned, that in contrast to most CIT studies, neither the number of CIs was restricted nor the valence of the critical incident as positive or negative. All customer reports were recorded.

4 Results

Customers that answered all relevant questions ($N = 146$) were eligible for hypotheses testing. These respondents were predominantly male (86%), on average 52 years old (standard deviation=12.2), and reported in total 185 CIs, of which 78 were positive. The hypothesized models were estimated with LISREL 8.52 (Jöreskog & Sörbom, 2001). In the first step we tested our hypothesis concerning the impact of the number of CIs on relationship outcomes (satisfaction, trust, and loyalty). The basic model exhibits an excellent fit with: $\chi^2(29) = 20.45$, $p = .88$, Root Mean Square Error of Approximation ($RMSEA$) $= 0.0$, and Comparative Fit Index (CFI) $= 1.00$. Overall, the model explains 31% of variance in satisfaction, 53% in trust, and 82% in loyalty. Except for the influence of positive CIs on trust (H_3) ($\gamma_{21} = .02$, $p > .05$), all hypotheses were supported. Results show that the number of experienced positive CIs per respondent possesses a positive impact on satisfaction with the repair department (H_1) ($\gamma_{31} = .26$, $p < .01$). The number of experienced negative CIs impacts negatively on satisfaction with the repair department (H_2) ($\gamma_{32} = -.47$, $p < .01$) and trust in the service provider (H_4) ($\gamma_{22} = -.36$, $p < .01$). The expected influences of satisfaction on trust (H_9) ($\beta_{23} = .47$, $p < .01$) and loyalty (H_8) ($\beta_{13} = .51$, $p < .01$), as well as trust on loyalty (H_7) ($\beta_{12} = .49$, $p < .01$), were also confirmed.

Next, we tested our hypotheses concerning the asymmetric impact of the number of positive and negative experienced CIs on satisfaction and trust. A model with gamma coefficients constrained to be equal results in a non-significant decrease in model fit for satisfaction ($\chi^2_d(1) = .39$) and a significant decrease in model fit for trust ($\chi^2_d(1) = 7.4$, $p < 0.01$). Thus the results confirm the asymmetric impact of CIs on trust (H_6), but not on satisfaction (H_5). Then, we tested whether the influence of CIs is fully mediated via satisfaction and trust following the approach of Baron and Kenny (1986). Both positive and negative CIs have a significant impact on loyalty, which drops to zero when the indirect effects via satisfaction and trust are included in the model. Thus the full mediation hypothesis is supported (H_{10}).

In order to show that previous findings hold, when controlling for mood, mood was added to the model, allowing it to impact all constructs. The resulting model fits the data well ($\chi^2(64) = 82.87$, $p = .06$, $RMSEA = 0.04$, and $CFI = .99$) and confirms the need to control findings for respondents' current mood. Mood significantly affects the number of negative CIs experienced (or better recalled) with $\gamma_{1\,5} = .36$, $p < .01$, judgments of satisfaction ($\gamma_{1\,3} = -.27$, $p < .01$), and trust ($\gamma_{1\,2} = -.29$, $p < .01$). Only positive CIs ($\gamma_{1\,4} = -.17$, $p > .05$) were not influenced by respondents' current mood. As expected, mood did not influence loyalty intentions ($\gamma_{1\,1} = -.04$, $p > .05$). Although mood showed the proposed influences, all previously reported findings hold while controlling for mood.

After we have confirmed that the number of positive and negative CIs impacts customer-firm relationship constructs, we address the question which categories of CIs are especially critical for the relationship. Therefore the 185 reported events were coded by three independent judges. The first judge developed an initial classification scheme and subsequently assigned the CIs into this initial scheme. The second and third judges had been advised to question this classification scheme, while assigning all obtained 185 CIs into the scheme. The resulting intercoder reliability was acceptable as assessed by various indices: *percentage of agreement*: .80, *Cohen's Kappa*: .76, and *Perreault und Leigh*: .77. Disagreement regarding categories and assignment of individual CIs to categories was resolved through a discussion of the unclear incidents with an expert from the automotive industry. The classification process revealed 7 negative CI categories (e.g. *low speed of service*) and seven positive CI categories. Since three categories had been experienced by less than 3% of the customers, they were excluded from the analysis, thus 11 categories remained eligible for the analyses (6 negative, 5 positive).

Instead of testing a model with the aggregate number of experienced CIs, the different CI categories experienced by the respondents were included in a MIMIC model. Basically each respondent can be described with a vector of zeros and ones indicating which particular incident type he has experienced in his relationship with the service department of his dealerships. These binary incident category variables are then related to relationship outcomes with a MIMIC approach, an approach tailored to deal with dichotomous independent variables (Bollen, 1989). The basis for model estimation here are, unlike ordinary Structural Equation Models, biserial correlations (Jöreskog & Sörbom, 2001, p. 240). The proposed MIMIC model (see figure 1) exhibits an excellent fit with: $\chi^2(138) = 117.26$, $p = .90$, $RMSEA = 0.0$, and $CFI = 1.00$. The model confirmed that not all CIs are indeed critical for the customer-firm relationship, and those CI categories that are critical varied in the strength of their impact. *Breaking a promise* and *experiencing poor quality of repair work* influence solely satisfaction ratings ($\gamma_{36} = -.23$, $p < .01$ and $\gamma_{39} = -.32$, $p < .01$), whereas CIs classified as *showing no goodwill* and *restriction to basic service* lowered customers trust in the service provider ($\gamma_{28} = -.14$, $p < .01$ and $\gamma_{2\,10} = -.14$, $p < .01$). The incident category which should be primarily avoided is *negative behaviors toward the customer*, since it clearly has the most damaging impact on the customer-firm relationship, due to its dual influence on trust ($\gamma_{2\,11} = -.27$, $p < .01$) and satisfaction ($\gamma_{3\,11} = -.26$, $p < .01$). Interestingly, only one of the positive CI categories (offering additional service) impacts on satisfaction with the repair department ($\gamma_{33} = .23$, $p < .01$) and none impacts on trust.

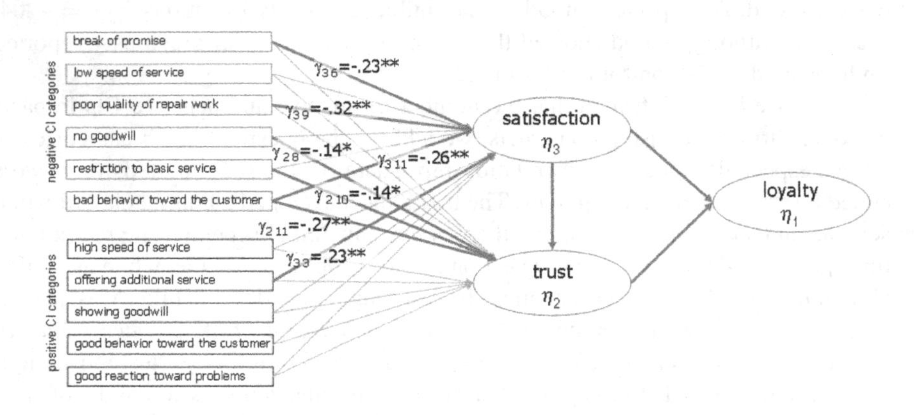

Fig. 1. MIMIC model: CI categories and their impact on relationship measures, significant path coefficients are depicted.

5 Discussion

Even though several papers in the marketing literature have raised the question whether and which incidents are really critical for a customer-firm relationship (Edvardsson & Strandvik, 2000) ours is the first study to explicitly address this question. In the present study, we conducted CI interviews without restricting valence and number of incidents reported, and assessed their impact on measures of relationship quality. Our results confirm that positive and negative incidents possess a partially asymmetric impact on satisfaction and trust. Negative incidents have particularly damaging effects on a relationship through their strong impact on trust (total causal effect: 0.58). These results are in stark contrast to Odekerken-Schröder et al.'s (2000) conclusion, that CIs do not play a significant role for developing trust. Further the damage inflicted by negative incidents can hardly be "healed" with very positive experiences, since the total causal effect of the number of positive incidents on trust is substantially smaller (0.12). Thus, management should clearly put emphasis on avoiding negative interaction experiences. The employed MIMIC approach followed Gremler's call (2004, p. 79) to "determine which events are truly critical to the long-term health of the customer-firm relationship" and revealed which specific incident categories have a particular strong impact on relationship health and should be avoided with priority, such as negative behavior toward the customer. The collected vivid verbatim stories from the customer's perspective provide very concrete information for managers and can be easily communicated to train customer-contact personnel (Zeithaml & Bitner, 2003; Stauss & Hentschel, 1992). For further studies, as pointed out by one of the reviewers, an alternative evaluation possibility would be to measure the experienced severity of the experienced CI-categories instead of their mere occurrence.

References

BARON, R. and KENNY, D. (1986): The Moderator-Mediator Variable Distinction in Social Psychological Research: Conceptual, Strategic and Statistical Considerations. *Journal of Personality and Social Psychology*, 51 (6), 1173-1182.

BAUMEISTER, R. F., BRATSLAVSKY, E., FUNKENAUER, C., and VOHS, K. D. (2001): Bad is Stronger than good. *Review of General Psychology*, 5 (4), 323-370.

BITNER, M. J., BOOMS, B. H., and TETREAULT, M. S. (1990): The Service Encounter - Diagnosing Favorable and Unfavorable Incidents. *Journal of Marketing*, 54(1), 71-84.

BOLLEN, K. A. (1989): Structural Equations with Latent Variables. New York: Wiley.

EDVARDSSON, B. (1992). Service Breakdowns: A Study of Critical Incidents in an Airline. *International Journal of Service Industry Management, 3(4)*, 17-29.

EDVARDSSON, B., and STRANDVIK, T. (2000): Is a Critical Incident Critical for a Customer Relationship? *Managing Service Quality*, 10(2), 82-91.

EICH E, MACAULAY D., and RYAN L. (1994): Mood Dependent Memory for Events of the Personal Past. *Journal of Experimental Psychology - General*, 123 (2), 201-215.

FISKE, S. (1980): Attention and Weight in Person Perception - the Impact of Negative and Extreme Behaviour. *Journal of Personality and Social Psychology*, 38 (6), 889-906.

FORGAS, J. P. (1995): Mood and Judgment: The Affect Infusion Model (AIM). *Psychological Bulletin*, 117 (1), 39-66.

FORNELL, C., JOHNSON, M. D., ANDERSON, E. W., CHA, J., and BRYANT, B. E. (1996): The American Customer Satisfaction Index: Nature, Purposes, and Findings. *Journal of Marketing*, 60 (October), 7-18.

GEYSKENS, I., STEENKAMP, J-B. E. M., and KUMAR, N. (1999): A Meta-Analysis of Satisfaction in Marketing Channel Relationships. *Journal of Marketing Research*, 36 (May), 223-238.

GREMLER, D. (2004): The Critical Incident Technique in Service Research. *Journal of Service Research*, 7(1), 65-89.

JÖRESKOG, K. and SÖRBOM, D. (2001): LISREL 8: User's Reference Guide. Chicago: Scientific Software International.

KAHNEMAN, D. and TVERSKY, A. (1979): Prospect Theory - Analysis of Decision under Risk. *Econometrica*, 47(2), 263-291.

MORGAN, R. M. and HUNT, S. D. (1994): The commitment-trust theory of relationship marketing. *Journal of Marketing*, 58(3), 20-38.

ODEKERKEN-SCHRÖDER, G., van BIRGELEN, M., LEMMINK, J., de RUYTER, K., and WETZELS, M. (2000): Moments of Sorrow and Joy: An Empirical Assessment of the Complementary Value of Critical Incidents in Understanding Customer Service Evaluations. *European Journal of Marketing, 34*(1/2), 107-125.

ROOS, I. (2002): Methods of Investigating Critical Incidents - A Comparative Review. *Journal of Service Research*, 4 (3), 193-204.

SINGH, J. and SIRDESHMUKH, D. (2000): Agency and Trust Mechanisms in Consumer satisfaction and loyalty judgments. *Journal of the Academy of the Marketing Science*, 28 (1), 150-167.

STAUSS, B. and HENTSCHEL, B. (1992): Attribute-based versus Incident-based Measurement of Service Quality: Results of an Empirical Study in the German Car Service Industry. In: P. Kunst and J. Lemmink, (Eds.), *Quality Management in Services* (59-78).

STAUSS, B. and WEINLICH, B. (1997): Process-oriented Measurement of Service Quality - Applying the Sequential Incident Technique. *European Journal of Marketing*, 31 (1), 33-55.

SZYMANSKI, D. M. and HENARD, D. H. (2001): Customer Satisfaction: A Meta-Analysis of the Empirical Evidence. *Academy of Marketing Science*, 29(1), 16-35.

TAYLOR, S. (1991): Asymmetrical Effects of Positive and Negative Events: The Mobilization-Minimization Hypothesis. *Psychological Bulletin*, 110 (1), 67-85.

YBARRA, O. and STEFAN, W. G, (1999): Attributional Orientations and the Prediction of Behavior: The Attribution-Prediction Bias. *Journal of Personality and Social Psychology*, 76 (5), 718-728.

ZEITHAML, V. A. and BITNER, M. J. (2003): *Services Marketing: Integrating Customer Focus across the Firm* (3rd ed.). New York: McGraw-Hill.

Heterogeneity in the Satisfaction-Retention Relationship – A Finite-mixture Approach

Dorian Quint and Marcel Paulssen

Humboldt-Universität zu Berlin, Institut für Industrielles Marketing-Management,
Spandauer Str. 1, 10178 Berlin, Germany
dorian.quint@inbox.com,
paulssen@wiwi.hu-berlin.de

Abstract. Despite the claim that satisfaction ratings are linked to actual repurchase behavior, the number of studies that actually relate satisfaction ratings to actual repurchase behavior is limited (Mittal and Kamakura 2001). Furthermore, in those studies that investigate the satisfaction-retention link customers have repeatedly been shown to defect even though they state to be highly satisfied. In a dramatic illustration of the problem Reichheld (1996) reports that while around 90% of industry customers report to be satisfied or even very satisfied, only between 30% to 40% actually repurchase. In this contribution, the relationship between satisfaction and retention was examined using a sample of 1493 business clients in the market of light transporters of a major European market. To examine heterogeneity in the satisfaction-relationship, a finite-mixture approach was chosen to model a mixed logistic regression. The subgroups found by the algorithm do differ with respect to the relationship between satisfaction and loyalty, as well as with respect to the exogenous variables. The resulting model allows us to shed more light on the role of the numerous moderating and interacting variables on the satisfaction-loyalty link in a business-to-business context.

1 Introduction

It has been one of the fundamental assumptions of relationship marketing theory that customer satisfaction has a positive impact on retention[1]. Satisfaction was supposed to be the only necessary and sufficient condition for attitudinal loyalty (stated repurchase behavior) and the more manifest retention (actual repurchase behavior) and has been used as an indicator for future profits (Reichheld 1996, Bolton 1998). However, this seemingly undisputed relationship could not be fully confirmed by empirical studies (Gremler and Brown 1996). Further research points out that there can be a large gap between one-time satisfaction and repurchase behavior. Not always leads an intention to repurchase (i.e. the statement in a questionnaire) to an actual repurchase and continuous repurchasing might exist without satisfaction because of mere price settings (see Söderlund and Vilgon 1999, Morwitz 1997). What is more, only

[1] I.e. Anderson et al. (2004), Bolton (1998), Söderlund and Vilgon (1999).

a small number of studies has actually examined repurchase behavior instead of the easier to get repurchase intentions (Bolton 1998, Mittal and Kamakura 2001, Rust and Zahorik 1993). The tenor of these studies is that the link between satisfaction and retention is clearly weaker than the link between satisfaction and loyalty.

Many other factors were discovered to have an influence on retention. Also more technical issues like common method variance, mere measurement effects or simply unclear definitions added to raise doubt on the importance and the exact magnitude of the contribution of satisfaction (Reichheld 1996, Söderlund/Vilgon 1999, Giese/Cote 2000). Another reason for the weak relationship between satisfaction and retention is that it may not be a simple linear one, but one moderated by several different variables. Several studies have already studied the effect of moderating variables on the satisfaction-loyalty link (e.g. Homburg and Giering 2001). However, the great majority of empirical studies in this field measured repurchase intentions instead of objective repurchase behavior (Seiders et al. 2005). Thus, the conclusion from prior work is that considerable heterogeneity is present that might explain the often surprisingly weak overall relationship.

An important contribution has been put forth by Mittal and Kamakura (2001). They combined the concepts of response biases and different thresholds[2] into their model to capture individual differences between respondents. Based on their results they created a customer group where repurchase behavior was completely unrelated to levels of stated satisfaction. However, their approach fails to identify real existing groups that have a distinctive relationship between satisfaction and retention. For example, if model results show that older people have a lower threshold and thus repurchase with a higher probability given a certain level of satisfaction, this is not the full story. Other factors, measured or unmeasured, might set off the age effect. In order to find groups with distinctive relationships between satisfaction and retention, we have explicitly chosen a finite-mixture[3] approach, which results in a mixed-logistic regression setup. This model type basically consists of G logistic regressions – one for each latent group. This way, each case i is assigned to a group with a unique relation between the two constructs of interest. However, in a Bernoulli case like this (see McLachlan and Peel 2000, p.163ff), identifiability is not given. The necessary and sufficient condition for identifiability is $G^{max} \leq \frac{1}{2}(m+1)$, where m is the number of Bernoulli trials. For $m = 1$ no ML-regression can be estimated. But Follmann and Lambert (1991) prove theoretical identifiability of a special case of binary ML-regressions. Only the thresholds λ are allowed to vary over the groups, while all remaining regression parameters are equal for all groups. According to Theorem 2 of Follmann and Lambert (1991) theoretical identifiability then depends only on the maximal number of different values of one covariate N^{max} given the values of all other covariates are held constant. The maximal number of components is then given by $G^{max} = \sqrt{N^{max}+2} - 1$. Thus, the theorem restricts the choice of the variables,

[2] In our model thresholds are tolerance levels and can be conceived as the probability of repurchase given all other covariates are zero.

[3] For an overview on finite-mixture models, see McLachlan and Peel (2002) and the references therein.

but ultimately helps building a suitable model for the relationship under investigation. In our final model we also included so-called concomitant covariate variables, which help to understand latent class membership and enhance interpretability of each group or class. This is achieved by using a multinomial regression of the latent class variable c on these variables x:

$$P(c_{gi} = 1|x_i) = \frac{e^{\alpha_g + \gamma'_g x_i}}{\sum_{l=1}^{G} e^{\alpha_l + \gamma'_l x_i}} = \frac{e^{\alpha_g + \gamma'_g x_i}}{1 + \sum_{l=1}^{G-1} e^{\alpha_l + \gamma'_l x_i}}. \quad (1)$$

Here α is a $(G-1)$-dimensional vector of logit constants and Γ a $(G-1) \times Q$ matrix of logit coefficients. The last group G serves as a standardizing reference group with $\alpha_G = 0$ and $\gamma_G = 0$. This results in a model of a mixed logistic regression with concomitant variables:

$$P(y_i = 1|x_i) = \sum_{g=1}^{G} P(c_{gi} = 1|x_i) P(y_i = 1|c_{gi} = 1, x_i),$$

with

$$P(y_i = 1|c_{gi} = 1, x_i) = \frac{e^{-\lambda_g + \beta_g x_i}}{1 + e^{-\lambda_g + \beta_g x_i}}. \quad (2)$$

2 The Model

To analyze the relationship between satisfaction and retention with a ML-regression, data is being used from a major European light truck market in a B2B environment. This data entails all major brands, which makes it possible to identify brand switchers and loyal customers. All respondents bought at least one light truck between two and four months before filling in the questionnaire. Out of all respondents who replied to all relevant questions only those were retained who bought the new truck as a replacement for their old one – resulting in 1493 observations. The satisfaction-retention link is now being operationalized in Mplus 4.0 using the response-bias-effect introduced by Mittal and Kamakura (2001), which enables us to use Theorem 2 of Follmann and Lambert (1991). Following Paulssen and Birk (2006) only demographic and by brand moderated demographic response-bias-effects are estimated in our model. The resulting equation for the latent satisfaction in logit is then:

$$sat_i^* = \beta_1 sat_i + \beta_2 sat_i * cons_i + \beta_3 sat_i * age_i + \beta_4 sat_i * brand_i + \beta_5 sat_i * cons_i * brand_i + \beta_6 sat_i * age_i * brand_i + \varepsilon_i.$$

The satisfaction-retention link for a latent class g can then be written as[4]:

[4] Here age stands for the standardized stated age, cons for consideration set and brand indicates a specific brand.

$$P(\text{Retention} = y_i | c_{gi} = 1, sat, cons, age, brand) = P(sat_i^* > \lambda_g)$$
$$= \frac{e^{-\lambda_g + sat^*}}{1 + e^{-\lambda_g + sat^*}}.$$

The latent class variable c is being regressed on the concomitant variables using a multinomial regression. As concomitant variables we used: Length of ownership of the replaced van (standardized), Ownership (self-employed 0, company 1), Brand of replaced van (other brands 0, specific "brand 1" 1), Consideration Set of other brands than the owned one (empty 0, at least one other brand 1) and Dealer (not involved in talks 0, involved 1). The model was estimated for several numbers of latent classes, with the theoretical maximum of classes being five. The fit indices for this model series can be found in table 1. All four ML-models possess a better fit than a simple logistic regression, but show a mixed picture. The AIC allows for a model with four classes and BIC allows for only one. To decide on the number of classes, the adjusted BIC was used, which allows for three classes[5]. This model was estimated using 500 random starting values and 500 iterations as recommended by Muthen and Muthen (2006, p.327). The Log-Likelihood of the chosen model is not reproduced in only nine out of 100 sequences, which, according to Muthen and Muthen (2006, p.325), points clearly toward a global maximum.

Table 1. Model Fit

criterion	Simple LR	$G=1$	$G=2$	$G=3$	$G=4$
Log-Likelihood	-971.280	-928.104	-902.727	-888.346	-873.472
AIC	1946.559	1870.209	1833.454	1814.692	1802.945
BIC	1957.176	1907.369	1907.774	1915.554	1951.584
Adjusted BIC	1950.823	1885.132	1863.300	1855.196	1862.636
Entropy	–	–	0.531	0.563	0.885

Entropy for the chosen model is 0.563, which indicates modest separation of the classes. As can be clearly seen in table 2, the discriminatory power is mixed with class 2 being well separated (0.821), while classes 1 and 3 are not perfectly separable.

Table 2. Miss-classification matrix

	1	2	3
1	0.762	0.063	0.175
2	0.179	0.821	0.000
3	0.328	0.000	0.672

[5] See Nylund et al. 2006.

The results of this model are shown in table 3. The thresholds of latent classes 2 and 3 were fixed after the first models we used showed extreme values for them, resulting in a probability of repurchase of 0% respectively 100%. This means that for both groups repurchase probability is independent of the values of the covariates. In this way the algorithm eventually works as a filter and puts those respondents who repurchase or do not repurchase independent of their satisfaction into separate groups. Thus, the only unfixed threshold is 3.174 for latent class 1. This class has a weight of 49.4%, while class 2 has 27.7% and class 3 represents 29.9% of the respondents. The estimated value for β_1 is 0.944 and is, like all other coefficients, significant on the 5% level. The value for β_1 represents the main effect of satisfaction with the previous van in case all other covariates are zero. In this case the odds ratio for repurchasing the same brand is increased by $e^{0.944} = 2.57$, which means satisfaction has a positive effect on the odds of staying with the same brand versus buying another brand. The estimates β_2 and β_3 correspond to response-bias-effects in case, the brand is not the specific brand 1. Both estimates are significant, meaning that response bias is present. The interpretation of the beta-coefficients is similar as before in that all other covariates are assumed to be zero. When considering only respondents who had previously a van of brand 1, that is *brand* = 1, things change. The effect for age, given the consideration set is empty, becomes $0.147 - 0.131 = 0.016$ almost completely wiping out the influence of response bias. For the covariate consideration set results are analogous: Given a sample-average age the response--bias-effect for respondents who replaced a van by brand 1 collapses to $-0.244 + 0.254 = 0.01$. As to the multinomial logistic regression of the latent class variable c on the concomitant variables, class 3 has been chosen to be the reference class. The constants α_g can be used to compute the probabilities of class membership for each respondent, who has an average length of ownership, who are self-employed, had not replaced a van of brand 1, did not consider another brand and who were not involved in talks with the dealer. For this group class membership for class g is $e^{\alpha_g}/(1+\Sigma_{l=1}^{2} e^{\alpha_l})$[6]. The probability of class membership in class 1 increases with increasing length of ownership. For low lengths of about one year, probability of membership is highest for class 3. However, probability of membership in class 2 is hardly influenced by the length of ownership. Self-employed respondents have a probability of belonging to class 1 of more than 80% despite the non-significance of the owner variable. The influences on class membership for the other concomitant variables can be explained analogously.

This model with three latent classes fits the data better than a simple linear regression of retention on satisfaction. The latter results in a marginal Nagelkerke-R^2 of a bad 0.063. Now, if we look again at table 2, we might make a hard allocation of respondents to class 1, despite the fact that separation of the classes is not perfect.

[6] The probability of belonging to class 1 is 67.94%, for class 2 17.94% and for class 3 14.12%. If the values of all concomitant variables are 1, the corresponding probabilities become 65.83%, 27% and 7.17%. If all other values of the concomitant variables are 0, a change from 0 to 1 in the brand variable, means that the odds to belong to class 1 compared to class 3 are just $e^{1.455} = 4.28$.

Table 3. ML-regression results

Variable	Value	Std.error	Z-Statistic
Response Bias for all classes			
Satisfaction	0.944	0.164	5.749*
Age*Satisfaction	0.147	0.046	3.230*
Consideration*Satisfaction	-0.244	0.113	-2.157*
Brand 1*Satisfaction	-0.367	0.100	-3.673*
Age*Brand 1*Satisfaction	-0.131	0.056	-2.349*
Consideration*Brand 1*Satisfaction	0.254	0.123	2.075*
Thresholds			
λ_1 Threshold	3.174	0.963	3.297*
λ_2 Threshold	15.000	–	–
λ_3 Threshold	-15.000	–	–
Class 1: Concomitant Variables	**Value**	**Std.error**	**Z-Statistic**
α_1 Constant	1.573	1.272	1.237
Length	1.131	0.432	2.620*
Owner	-1.995	1.130	-1.765
Brand 1	1.455	0.635	2.292*
Consideration	0.141	0.445	0.316
Dealer	0.912	0.746	1.223
Class 2: Concomitant Variables			
α_2 Constant	0.243	1.085	0.224
Length	1.049	0.443	2.369*
Owner	-0.440	1.009	-0.436
Brand 1	0.275	0.500	0.549
Consideration	1.199	0.299	4.004*
Dealer	-0.213	0.370	-0.577

* significant on the 5% level

For class 1 we then arrive at a very good Nagelkerke-R^2 value of 0.509. This means that the estimated model basically works as a filter leaving one group of respondents with a very strong relation between satisfaction and retention and two smaller groups with no relation at all. At this point the classes of the final model shall be interpreted. While average satisfaction ratings are essentially the same (6.77, 6.82 and 6.60 for classes 1 to 3), the relation between satisfaction and retention is very different. As indicated above, class 1 describes a filtered link between satisfaction with the replaced van and retention. This class contains predominantly respondents who are self-employed, who were involved in talks with the dealer, who had a long length of ownership of their previous van and who drove a van of brand 1. In this class increasing satisfaction corresponds to a higher retention rate. This means in turn that marketing measures to increase retention via satisfaction campaigns are feasible for this group. Respondents of class 2 considered brands other than the brand of their replaced van prior to their purchase decision, which increased the number of choices they had for making the purchase decision. However, this class can also be considered as being influenced by other factors than were observed in our study. These factors might further explain why the retention rate is zero, although some members were in fact satisfied with their replaced van. It is easy to imagine that a large number of reasons, including pure coincidence, can lead to such a behavior. The third class, where respondents repurchase independent of their satisfaction, has at least one distinctive feature. This class is dominated by very short lengths of ownership, which might be explained by the presence of leasing contracts.

3 Discussion

Previous studies have examined customer characteristics as moderating effects of the satisfaction-retention link. In order to further investigate this, we built on a model developed by Mittal and Kamakura (2001) that we expanded by including manufacturer and company characteristics as additional moderating variables. Previous research did not fully investigate the moderating role of manufacturer/brand and company characteristics on the satisfaction retention link. Furthermore, by applying a concomitant logit mixture approach we applied a new research method to this problem. Our results imply that similar to findings of Mittal and Kamakura (2001) customer groups exist where repurchase behavior is completely invariant to rated satisfaction. In the largest customer group a strong relationship between satisfaction and repurchase was present. Respondents in this group were self-employed, participated in dealer talks and kept their commercial vehicles longer than members of the other classes. It is notable that for respondents who stated they were self-employed and participated in dealer talks the satisfaction-retention relationship is strong, indicating that those respondents had substantial leverage on decision making. That is, these respondents immediately punished bad performance of the incumbent brand and switched to other brands. For respondents that worked for companies other factors (purchasing policies of the company, satisfaction from other members of the buying center) than their stated satisfaction may play a role. It also seems to be necessary that the respondent had a significant involvement in the buying process as indicated by his participation in dealer talks. This result also points to limitation of the often applied key informant approach – key informants have to be carefully screened. It does not suffice to ask whether they participate in certain business decisions.

References

ANDERSON, E. W., FORNELL, C., MAZVANCHERYL, S. K. (2004): Customer Satisfaction and Shareholder Value. *Journal of Marketing, 68, 172–185.*

BOLTON, R. N. (1998): A Dynamic Model of the Duration of the Customer's Relationship with a Continuous Service Provider: The Role of Satisfaction. *Marketing Science, 17, 45–65.*

FOLLMANN, D. A., LAMBERT, D. (1991): Identifiability of finite mixtures of logistic regression models. *Journal of Statistical Planning and Inference, 27, 375–381.*

GIESE, J. L., COTE, J. A. (2000): Defining Consumer Satisfaction. *Academy of Marketing Science Review, 2000, 1–24.*

GREMLER, D. D., BROWN, S. W. (1996): Service Loyalty: Its Nature, Importance, and Implications. *Advancing Service Quality: A Global Perspective. International Service Quality Association, 171–180.*

HOMBURG, C., GIERING, A. (2001): Personal Characteristics as Moderators of the Relationship Between Customer Satisfaction and Loyalty: An Empirical Analysis. *Psychology & Marketing, 18, 43-Ű66.*

MCLACHLAN, G., PEEL, D. (2000): *Finite Mixture Models.* Wiley, New York.

MITTAL, V., KAMAKURA, W. A. (2001): Satisfaction, Repurchase Intent, and Repurchase Behavior: Investigating the moderating Effect of Customer Characteristics. *Journal of Marketing Research, 38, 131–142.*

MORWITZ, V. G. (1997): Why Consumers Don't Always Accurately Predict Their Own Future Behavior. *Marketing Letters, 8, 57–70.*

MUTHEN, L. K., MUTHEN, B. O. (2006): *Mplus User's Guide.* Fourth issue, Los Angeles.

NYLUND, K. L., ASPAROUHOV, T., MUTHEN, B. (2006): Deciding on the number of classes in latent class analysis and growth mixture modeling. A Monte Carlo simulation study. *Accepted by Structural Equation Modeling.*

PAULSSEN, M., BIRK, M. (2006): It's not demographics alone! How demographic, company characteristics and manufacturer moderate the satisfaction retention link. *Humboldt-Universität zu Berlin, Wirtschaftswissenschaftliche Fakultät.* Working Paper.

REICHHELD, F. F. (1996): Learning from Customer Defections. *Harvard Business Review, 74, 56–69.*

RUST, R. T., ZAHORIK, A. J. (1993): Customer Satisfaction, Customer Retention, and Market Share. *Journal of Retailing, 69, 193–215.*

SEIDERS, K., VOSS, G. B., GREWAL, D., GODFREY A. L. (2005): Do Satisfied Customers Buy More? Examining Moderating Influences in a Retailing Context. *Journal of Marketing, 68, 26–43.*

SÖDERLUND, M., VILGON, M. (1999): Customer Satisfaction and Links to Customer Profitability: An Empirical Examination of the Association Between Attitudes and Behavior. *Stockholm School of Economics, Working Paper Series in Business Administration, Nr. 1999:1.*

An Early-Warning System to Support Activities in the Management of Customer Equity and How to Obtain the Most from Spatial Customer Equity Potentials

Klaus Thiel and Daniel Probst

Department of Analytical Customer Relationship Management
CRM, T-Online, T-Com, DTAG
T-Online Allee 1, 64295 Darmstadt, Germany

Abstract. This paper will show, how the usage of an early-warning system, which has been developed and implemented for a big internet service provider, can detect customer equity potentials respectively risks and how to use this information to launch special customer treatment depending on strategic customer control dimensions in order to increase customer equity. The strategic customer control dimensions are: customer lifetime value, customer lifecycle and customer behaviour types. The development of the customer control dimensions depends on the availability of relevant customer data. Thus, from the huge amount of available customer data, relevant attributes have been selected. In order to reduce complexity and use standardised processes the raw-data is aggregated, for example into clusters. We will demonstrate by means of a real-life example the detection of spatial customer equity risks and the launch of customer equity increasing treatment using the early-warning system in interaction with the mentioned strategic customer control dimensions.

1 Introduction[1]

In the b2c-sector a continuing increase in competition, growing customer expectations, variety seeking and an erosion of margin can be observed. The solution of successful enterprises is a paradigm change from product-centred to customer-centred organisations combined with the long-run objective of customer equity (CE) maximisation.

In this paper we will show both theoretically and by means of a real-life example, the detection of CE risks at a very early stage as well as the launch of CE increasing treatment using the early-warning system (EWS) in interaction with the strategic customer control dimensions (SCCD). The SCCD are customer lifetime value (CLV), customer lifecycle (CLC) and customer behaviour types (CBT).

[1] The content of this article originates from various projects. All projects have been executed under the leadership or substantial cooperation of the aCRM department, with the department head Klaus Thiel.

In section 2 we will briefly describe the SCCD CLV, CLC and CBT. Furthermore, we will show that space is another important dimension which has to be considered. Section 3 will deal with the methodology and the features of the EWS. Finally, we will show the operation of the EWS and its interaction with the SCCD by means of a real-life example.

2 Strategic customer control dimensions

In order to regain competitiveness many enterprises in the b2c-sector conduct a strategic change from product-centred to customer-centred organisations (Peppers and Rogers (2004)). Due to the huge amount of customer data available in the b2c-sector, one big challenge for managers is to find the balance between standardised processes and individual customer value management. In order to follow the strategic guideline of our enterprise (change from a product-centred to a customer-centred organisation) and in order to reduce complexity in multivariate data-structures we derive relevant customer insight from the huge amount of data and aggregate the data to the relevant SCCD CLV, CLC and CBT. The mentioned SCCD are also discussed in relevant literature (e.g. Peppers and Rogers (2004), Gupta and Lehmann (2005), Stauss (2000)). We have implemented the SCCD in the customer data-warehouse (DWH) in order to use it for operative campaign management. Thanks to a reduced complexity in data structures we can mainly use standardised processes for customer interaction. All processes and analyses are aligned and approved with the data-protection department and are conform with national and supranational data-protection guidelines (Steckler and Pepels (2006)).

2.1 Customer lifetime value

On the one hand, the CLV is a target and controlling variable. The total sum of CLV based on all customers, the so-called CE has to be maximised and is permanently controlled in the course of time. On the other hand, the CLV is also applied as a treatment variable. This means, that certain types of treatment in the customer value management framework are launched depending on the CLV.

In order to enable customer value management, we have developed and implemented the following CLV formula according to Gupta and Lehmann (2005):

$$CLV_c = inv_c + \sum_{t=1}^{T_k} \frac{m_{c,t} \cdot r_{k,t}^t}{(1+i)^t} \quad c = 1,2,\ldots,12; k = 1,2,\ldots,K; t = 1,2,\ldots,T_k. \quad (1)$$

The index t denotes years, c denotes single customers and k denotes cohorts. CLV_c is the CLV of customer c, inv_c is the investment from the enterprise in the customer at the beginning of the relationship and $m_{c,t}$ is the annual margin of customer c. The calculatory interest rate i is stipulated by financial controlling. In order to avoid large variations of the retention-density we calculate it on cohort level instead of single customer level. All customer who have the same set-up date (quarter and year) as well as the same set-up channel of their relationship with the enterprise belong to the same cohort k and get the cohort specific retention-density $r_{k,t}^{t}$ at time t. The retention-density of cohort k at time t is calculated by: $r_{k,t}^{t} = \frac{1}{(1-p_{k,t})^{t}}$, where $p_{k,t}$ is the churn-density of cohort k. Finally, $p_{k,t}$ is calculated by: $p_{k,t} = \frac{h_{k,t}}{n_k}$, where $h_{k,t}$ denotes all customers of cohort k who have churned during period t and n_k is the total cohort size at the beginning of the observation period t. Every month the values of $m_{c,t}$ and $r_{k,t}^{t}$ are dynamically re-calculated and assumed as constant until the end of the cohort specific lifetime T_k. In order to maximise the CE, the total sum of CLV based on all customers has to be maximised:

$$CE = \sum_{c=1}^{C} CLV_c \rightarrow \max. \tag{2}$$

As mentioned, the CLV is also applied as a treatment variable. In the framework of customer value management all customers are assigned to six different value classes: A, B, C, D, E, F. Approximately 2% of the top-customers are in class A. All customers with a negative CLV belong to class F. The remaining customers are uniformly distributed into the classes B to E. Depending on the customer value class, incentives such as hardware discount, birthday emails with a gift, invitation to special events, service levels on customer touch-points etc. are offered to the customer.

2.2 Customer lifecycle

In addition to the CLV, we have also developed and implemented a CLC in order to support customer value management. The whole CLC consists of the following six customer life phases: set-up-phase, finding-phase, growth-phase, risk-phase, churn-phase and after-care-phase. The individual phases are initialised by various triggers. Depending on the trigger-status, each customer is assigned to his or her individual phase. For example, the finding-phase starts, respectively ends, after three, respectively twenty internet sessions. Each phase contains various phase-specific customer types of treatment. During the set-up-phase the customer receives special technical support, in the finding-phase we show the customer the product variety and during the growth-phase we focus on cross- and up-selling offers. We launch churn-prevention offers during the risk-phase and win-back offers in case of churn. After a successful win-back we try to bind the customer more closely during the after-care phase.

2.3 Customer behaviour types

In order to receive a better understanding of the customer needs we have developed and implemented CBT in an iterative procedure. In the first step we apply cluster analysis procedures (hierarchical and partition-based procedures) in order to identify groups of customers with homogeneous behaviour (Hand et al. (2001)). The second step consists of the validation of the clusters with the use of discriminant analysis procedures (Hand et al. (2001)). As the CBT are used for strategic purposes, stability in the medium-term of time is necessary. For this purpose we conduct stability tests with the use of previous campaign data in the third step. In the final step, we execute a market research in order to profile the clusters. This result in the following eight CBT: occasional users, stay-at-homes, after eights, conservatives, professionals, fun seekers, super surfers and 24-7 uppers and downers. Concerning customer value management, the communication design depends on the CBT, for instance. We observe significantly better response rates when we use a communication design depending on CBT rather than a general one. In addition, we trigger the development of new products as well as further developments of existing products depending on the CBT.

2.4 Interaction between the different strategic customer controlling dimensions

The three SCCD are hard-coded in the DWH. That means that each customer is assigned to a CLV class, a CLC phase and a CBT. The combination of the mentioned six CLV classes, six CLC phases and eight CBT results in 288 permutations, which could be assigned with different business rules. Due to business restrictions, at the moment we are working with approximately 150 permutations which are assigned with specific business recommendations. In the case of customer interaction (inbound and outbound), the stored information in DWH is linked with the corresponding business recommendation. This recommendation is available for the staff at the customer touch-point. The objective, in medium-term, is the integration of specific business rules for all 288 permutations.

2.5 Spatial differences

Apart from the SCCD already described, space is another important dimension concerning customer value management. To merely consider national distinctions in the customer base would cause an oversimplification of regional differences (Cliquet (2006)). If you compare measured values like population density per square kilometre with a range from 40 to 3973 or spending capacity per person with a range from €10,050 to €26,440 (infas GEOdaten (2006)) it becomes obvious, that successful customer equity management relies on spatial differentiation. This approach requires a coherent system of different regional levels, which is determined by logical relations between the different levels. The most suitable choice is the official AGS-Structure (Amtlicher Gemeindeschlüssel), which underlies a permanent update process by the Statistical Federal Office. This structure describes levels from the Federal State (Bundesland, AGS02) over the District (Kreis, AGS05) to the City (Stadt, AGS08), providing for each level an explicit identifier. Private data providers improve this structure by means of their own, more detailed levels beneath the City level (e.g. Kreis-Gemeinde-Schlüssel (KGS)). The resulting micro-geographical levels allow appropriate ways of analysing and steering.

2.6 Interaction between strategic customer controlling dimensions and space

In order to build up a framework which allows the maximisation of CE the combination between the SCCD CLV, CLC and CBT and the different regional levels from AGS02 up to AGS08 is necessary. The further multiplication of the 288 permutations regarding the SCCD, with 439 Districts (AGS05) in Germany, for example, would produce 126,432 permutations. This results in too much complexity and cannot automatically be assigned with different business rules in a real-life application. How have we solved this problem? First, if we detect significantly different regional developments then we temporarily overrule the implemented business rule only for the concerned customers by a new one. The new business rule can consist of an adjustment of the existing rule (e.g.: the price for the offered product-bundle is now €9.95 instead of €19.95), or can consist of a completely new rule (e.g.:instead of a launch of a new product-bundle we offer 12-month free-of-charge usage of the existing product). For all unconcerned customers the existing business rule will still be valid. After the various regional developments have disappeared we will switch off the overruling. Second, if we discover that certain regional developments are different in the long run, then we will analyse the developments in more depth. Depending on the results, two decisions are feasible: a) we adjust or enlarge one or more of the SCCD to cover the regional developments; b) the overruling remains valid until the different regional developments have disappeared.

3 Early-warning system

The necessity for an EWS has particularly arise from increasing competition. Many new competitors in the telecommunications market first focused on regional offers (e.g. Telecom Italia (e.g. Alice), Netcologne, Telebel) which resulted in massive regional churn und regional loss of customer equity. Regarding this, spatial analyses and usable rankings for managing customer equity were overdue. Apart from the wide range of accessibility, the EWS is described by its comprehensive spatial structure and an easy-to-understand way of presenting information. Finally, it can be applied as a digital, interactive map of Germany, which shows regions in different colours, depending on their risk conditions. Especially, by using cartographical imagery, it is possible to present complex figures in a compact way, offering an easy-to-use steering wheel to top management. The next two subsections will introduce the methodology and architecture of the EWS.

3.1 Methodology

The EWS visualizes risky areas regarding churn-density, new-customer-density as well as market-density on each regional level, AGS02 up to AGS08. The churn-density on regional level a in region b at time t is calculated by:

$$l_{a(b),t} = \frac{u_{a(b),t}}{w_{a(b),t}} \quad a = 1,2,\ldots,A; b = 1,2,\ldots,B; t = 1,2,\ldots,T; \tag{3}$$

where $u_{a(b),t}$ denotes all customers who have churned during period t on regional level a in region b and $w_{a(b),t}$ denotes all existing customers at the beginning of the observation period t on regional level a in region b. In analogy, the regional new-customer-density is defined by the ratio of all customers who enter the enterprise during period t on regional level a in region b to all existing customers in period t on regional level a in region b. Finally, the regional market-density is defined by the ratio of all existing customers of the enterprise to all customers with a contract with an internet service provider in Germany in period t on regional level a in region b. In the final step, on each regional level the indicator variables churn-indicator, market-share-indicator and new-customer-indicator are calculated. 20% of the regions with the highest churn-density get the churn-indicator one, whereas 80% of the regions with the lowest churn-density get the churn-indicator zero. 20% of the regions with the lowest market-density get the market-share-indicator one, whereas 80% of the regions with the highest market-density get the market-share-indicator zero. Finally, 20% of the regions with the lowest new-customer-density get the new-customer-indicator one, whereas 80% of the regions with the lowest new-customer-density get the new-customer-indicator zero. Every week the three mentioned indicators are dynamically re-calculated. Depending on the permutation and under consideration of the strategic business framework we have developed and implemented the following risk matrix with the corresponding risk level in the EWS (Figure 1). The risk level vary from one (green), over four (yellow) up to eight (red).

Churn-Indicator Regional Churn-Density	Market-Share-Indicator Regional Market-Density	New-Customer-Indicator Regional New-Customer-Density
0	0	0
0	0	1
0	1	0
0	1	1
1	0	0
1	0	1
1	1	0
1	1	1

Risk-Level
1
2
3
4
5
6
7
8

Fig. 1. Risk Matrix and Risk Level

3.2 EWS-Architecture

The big challenge in the context of spatial marketing is the easy-to-use presentation of multidimensional customer data in spatial units as a reliable support for management decisions (Schüssler (2006)). In order to process the necessary customer data, acquired information must be implemented at a very early stage in all relevant business processes. For that reason the EWS is completely integrated in the DWH which allows the access to all sources of relevant data. That means, in order to receive the described risk levels on each regional level in each region, every week 40 million data fields have to be provided and processed. According to the guidelines of data protection, the whole process uses anonymous data and is optimised by the aggregation per regional level to the principles of data-avoidance and data-economy (Steckler and Pepels (2006)).

As already mentioned, the intent is to observe and understand regional developments at an early stage in order to react quickly. When individual permissions have been obtained, then customer contact can be made by the use of campaign management. With the help of an implemented DWH process designed to add the AGS- and KGS-Identifiers, the introduced statistics und risk levels are computed by SAS software (SAS Institute Inc.). The results are automatically written in a special database (Oracle Spatial) that is able to store not only alphanumerical objects, but also spatial data (Brinkhoff (2005)). This Oracle Spatial combines the results and the spatial data (e.g. the boundary of the City of Darmstadt and its risk rank) in one place, so that following processes can interact with the database based on industrial standards (Brinkhoff (2005)). By using a web-map-service (WMS) to present and transport the results to the user, the complexity is significantly reduced. Depending on an authorisation system, everyone in the company can access the EWS by browser via the intranet. So, this architecture concentrates rules and data in one point and thus guarantees the single source of truth for related questions. Without any technical obstacles and a low potential for misinterpretation, the EWS offers regional customer insight for further analysis, regional campaigns as well as decision support for the top-management.

4 Empirical example

With the use of the EWS (cf. Section 3) we analysed the market situation as it was in March 2006 in ten Districts (AGS05) in and around Berlin. We detected the following situation regarding the risk ranking (cf. Figure 1): one with risk eight, four with risk six, two with risk five and finally two with risk four. In the final step, we adjusted the existing business rules derived from the SCCD (cf. Section 2) and launched special types of pilot treatment (e.g. special product bundles with a reduced price, special customer service, at home product installation free of charge etc.) in the ten mentioned Districts from April until September 2006. Table 1 shows the change in the key figures during the observation period from March 2006 until October 2006 for the ten Districts in and around Berlin.

The results show improvements regarding churn-density (reduced by 27%) and new-customer-density (increased by 12%) in comparison to the initial situation as in March 2006. As the total churn amount was still increasing, the market-density dropped by 2.2%. In spite of this development CE increased by a mere 0.2%. More detailed analysis reveals an increasing customer stock (sum of all customers of the enterprise) by 0.6% during the observation period. This means, in comparison with the initial situation in March 2006 the average CLV dropped but the CE increased.

Table 1. Change after the launch of special treatment

Variable	Change
Churn-Density	-27%
New-Customer-Density	+12%
Market-Density	-2.2%
Customer-Equity	+0.2%

Additional, we have to draw attention to the control group. After observing the mentioned situation in the Districts in and around Berlin, we randomly selected three further Districts in the surrounding area. The chosen Districts were still being treated in the conventional way. In comparison to the focused Districts undergoing special pilot treatment, we observed a reduced market-density by 5%, a reduced CE by 0.8% and a reduced customer stock by 1.5%. The comparison reveals that the special types of pilot treatment work successfully.

5 Conclusion

We have shown that the SCCD allow a better understanding of customer behaviour and customer needs. In addition, we have pointed out, by means of a real-life example, that the EWS enables the detection of CE potentials respectively risks. Finally, we have illustrated that the interaction between EWS and SCCD allows the deduction of CE inreasing types of treatment.

On the one hand, the SCCD substantially reduce complexity in multivariate data-structures and allows the use of standardised processes in the case of customer interaction. On the other hand, the interaction between SCCD and different regional levels increases complexity considerably. It remains to be examined whether the operation of a completely automatic customer value management depending on all possible permutations of three or more SCCD in interaction with all regions on all regional levels will be possible. The implementation of such an application in real-life business will be very important, because complexity in data structures is continuously rising due to an increasing amount of customer data and customer micro-segments.

References

BRINKHOFF, T. (2005): *Geodatenbanksysteme in Theorie und Praxis: Einführung in objektrelationale Geodatenbanken unter besonderer Berücksichtigung von Oracle Spatial.* Herbert Wichmann Verlag, Heidelberg.

CLIQUET, G. (Ed.) (2006): *Geomarketing. Methods and Strategies in Spatial Marketing.* Geographical Information System Series, London.

GUPTA, S. and LEHMANN, D.R. (2005): *Managing Customers as Investments: The Strategic Value of Customers in the Long Run.* Wharton, NY.

HAND, D., MANNILA, H. and SMYTH, P (2001): *Principles of Data Mining.* The MIT Press Cambridge, Massachusetts.

infas GEOdaten GmbH, Bonn: http://www.infas-geodaten.de

PEPPERS, D and ROGERS, M. (2004): *Managing Customer Relationships: A Strategic Framework.* Wiley, Hoboken.

SAS INSTITUTE Inc., SAS Campus Drive Cary, USA, http://www.sas.com

SCHÜSSLER, F. (2006): *Geomarketing. Anwendungen Geographischer Informationssysteme im Einzelhandel.* Tectum, Marburg.

STAUSS, B. (2000): Perspektivenwandel: Vom Produkt-Lebenszyklus zum Kundenbeziehungs-Lebenszyklus. *Thexis, 17, Nr. 2, S. 15-18.*

STECKLER, B und PEPELS, W. (Hrsg.) (2006): *Das Recht im Direktmarketing.* Erich Schmidt, Berlin.

Classifying Contemporary Marketing Practices

Ralf Wagner

SVI Chair for International Direct Marketing
DMCC - Dialog Marketing Competence Center,
University of Kassel, Germany
rwagner@wirtschaft.uni-kassel.de

Abstract. This paper introduces a finite-mixture version of the adjacent-category logit model for the classification of companies with respect to their marketing practices. The classification results are compared to conventional K-means clustering, as established for clustering marketing practices in current publications. Both, the results of this comparison as well as a canonical discriminant analysis, emphasize the opportunity to offer fresh insights and to enrich empirical research in this domain.

1 Introduction

Although emerging markets and transition economies are attracting increasing attention in marketing, Pels and Brodie (2004) argue that conventional marketing knowledge is not valid for these markets per se. Moreover, Burgess and Steenkamp (2006) claim that emerging markets offer unexploited research opportunities due to their significant departures from the assumptions of theories developed in the Western world, but call for more rigorous research in this domain. But, the majority of studies concerned with marketing in transition economies are either qualitative descriptive or restricted to simple cluster analysis. This paper seizes the challenge by:

- introducing a finite-mixture approach facilitating the fitting of a response model and the clustering of observations simultaneously,
- investigating whether or not the Western-type distinction between marketing mix management and relationship management holds for groups of companies from Russia and Lithuania, and
- exploring the consistency of the marketing activities.

The remainder of this paper is structured as follows. The next section provides a description of the research approach, which is embedded in the *Contemporary Marketing Practices* (CMP) Project. In the third section, a finite-mixture approach is introduced and criteria for determining the number of clusters in the data are discussed. The data and the results of this study are outlined in section 4, and section 5 concludes with a discussion of these results.

2 Knowledge on interactive marketing

2.1 The research approach

In their review of reasons for the current evolution of marketing Vargo and Lusch (2004) propose that all economies will change to service economies and that this change will foster the switch from transactional to relational marketing. Another argument for emphasizing relational, particularly interactive marketing elements, is supported by the distinction of B2B from B2C markets. Unfortunately, these claims are based on deductive argumentation, literature analysis, and case studies, but are hardly supported by empirical analyses. Subsequently we aim to investigate the manner as to how marketing elements are combined in transition economies.

For this purpose we distinguish the conventional transaction marketing approach (by means of managing the marketing-mix of product, place, price, and promotion) from four types of customer relationship and customer dialogue-oriented marketing (Coviello et al. (2002)):

Transaction Marketing (TM): Companies attract and satisfy potential buyers by managing the marketing mix. They actively manage communication 'to' buyers in a mass-market. Moreover, the buyer-seller transactions are discrete and arm's-length.

Database Marketing (DM): Database technology enables the creation of individual relationships with customer. Companies aim to retain identified customers, although marketing is still 'to' the customer. Relationships as such are not close or interpersonal, but are facilitated and personalized through the use of database technology.

E-Marketing (EM): Vendors use the Internet and other interactive technologies to sell products and services. The focus is on creating and mediating the dialogue between the organization and identified customers.

Interaction Marketing (IM): A Face-to-face interaction between individuals maintains a communication process truly 'with' the customer. Companies invest resources to develop a mutually beneficial and interpersonal relationship.

Network Marketing (NM): All Marketing activities are embedded in the activities of a network of companies. All partners commit resources to develop their company's position in the network of company level relationships.

For the empirical investigation of the relevance of these marketing types, a survey approach has been developed. The next subsection describes the results concerning marketing practices in transition economies already gained through this survey approach.

2.2 Benchmark: Already known clusters in transition economies

Two studies of marketing practices in emerging markets have been published. The first study, by Pels & Brodie (2004), points out five distinct clusters of marketing practices in the emerging Argentinean economy. This clustering is made up by applying the K-means algorithm to the respondent's ratings describing their organization's marketing activities. Working with the same questionnaires and applying the K-means algorithm, Wagner (2005) revealed three clusters of marketing practices in the Russian transition economy. In the first study, the number of clusters was chosen with respect to the interpretability of the results, whereas in the second study, the number of clusters was estimated using the GAP-criterion. Both studies are restricted to the identification of groups of organizations with similar marketing practices, but do not address the relationships between particular elements of the marketing and relationship mix. To tackle these issues, a mixture regression model is introduced in the next section. In order to provide an assessment of the improvement due to the employment of more sophisticated methods, we will outline the results of applying K-means clustering to the data at hand as well.

3 A Finite Mixture approach for classifying marketing practices

3.1 Response Model

Mixture modeling enables the identification of the structure underlying the patterns of variables and the partition of the observations n $(n = 1,\ldots,N)$ into groups or segments s $(s = 1,\ldots,S)$ with a similar response structure simultaneously. Assuming that each group is made up by a different generating process, π_{ns} refers to the probability for the nth observation to originate from the generating process of group s. Let $\sum_{s=1}^{S} \pi_{ns} = 1$ with $\pi_{ns} \geq 0 \, \forall n,s$ then the density of the observed response data is given by:

$$f(y_n|\tilde{\mathbf{x}}_n, \mathbf{x}_n, \tilde{\theta}, \Theta) = \sum_{s=1}^{S} \pi_{ns}(k = s|\tilde{\mathbf{x}}_n, \tilde{\theta}) f_s(y_n|s, \mathbf{x}_n, \theta_s) \quad (1)$$

with
k ... nominal latent variable $(k = 1,\ldots,S)$
y_n ... scalar response variable
$\tilde{\mathbf{x}}_n$ ... vector of variables influencing the latent variable k (covariates)
$\tilde{\theta}$ ... vector of parameters quantifying the impact of $\tilde{\mathbf{x}}_n$ on k
$\mathbf{x}_n$ ... vector of variables influencing y_n (predictors)
θ_s ... vector of parameters quantifying the impact of $\mathbf{x}_n$ on y_n in segment s
Θ ... matrix of parameter vectors θ_s

From equation 1, it is obvious, that this model differs from conventional latent class regression models because of the covariates $\tilde{\mathbf{x}}_n$ and the corresponding parameter vector $\tilde{\theta}$, which enable an argumentation of the segment membership. Thus, the covariates $\tilde{\mathbf{x}}_n$ differ from the $\mathbf{x}_n$, because the elements $\tilde{\mathbf{x}}_n$ are assumed to have impact on y_n by means of causality in the response structure, but only by means of segment membership. For the application of clustering marketing practices, this feature seems to be relevant to capture the differences between organizations offering goods or services and serving B2C or B2B markets.

The distribution of conditional densities $f_s(y_n|s, \mathbf{x}_n, \theta_s)$ might be chosen from the exponential family, e.g., normal, Poisson, or binomial distribution. For the subsequent application of analyzing an ordinal response using a logit approach, the canonical links are binominal (cf. McCullagh and Nelder (1989)). The link function of the adjacent-category logit model, with $r = 1, \ldots, R$ response categories, is given by:

$$\log\left(\frac{P_s(y = r+1|\mathbf{x})}{P_s(y = r|\mathbf{x})}\right) = \theta^*_{0rs} + \mathbf{x}'_n \phi_{rs} \theta_s \quad . \quad (r = 2, \ldots, R; s = 1, \ldots, S) \tag{2}$$

with $\theta^*_{0rs} = \theta_{01s} - \theta_{0rs} \; \forall s$. Thus, the comparison of adjacent categories equals the estimation of binary logits. In order to utilize the information of the ordinal alignment of the categories, a score ϕ_{rs} for each category r is introduced, so that $\theta_{qrs} = -\phi_{rs}\theta_q \; \forall q, s$ with $\phi_{1s} > \phi_{2s} > \cdots > \phi_{Rs}$ (Anderson (1984)). Consequently, the probability of choosing the category k is:

$$P_s(Y_n = r|s, \mathbf{x}_n) = \frac{\exp(\theta^*_{0rs} - \mathbf{x}'_n \phi_{rs} \theta_s)}{\sum_{l=1}^{R} \exp(\theta^*_{0ls} - \mathbf{x}'_n \phi_{ls} \theta_s)}. \tag{3}$$

Noticeably, the number of parameters to be estimated is highly affected by the number of segments S, but increases just by one for each category.

3.2 Criteria for deciding on the number of clusters

To determine the optimal number of clusters from the structure of a given data set, distortion-based methods, such as the GAP-criterion or the Jump-criterion, have been found to be efficient for revealing the correct number (see Wagner et al. (2005) for a detailed discussion). In contrast to partitioning cluster algorithms, the fitting of response models usually involves the maximization of the likelihood function.

$$\ln L = \sum_{n=1}^{N} \sum_{s=1}^{S} \left(z_{ns} \ln f_s(y_n|s, \mathbf{x}_n, \theta_s) + z_{ns} \ln \pi_{ns}(k = s|\tilde{\mathbf{x}}_n, \tilde{\theta}) \right) \tag{4}$$

with

$$z_{ns} = \begin{cases} 1, & \text{if observation } n \text{ in segment } s, \\ 0, & \text{otherwise} \end{cases} \qquad (n = 1, ..., N, \quad s = 1, ..., S)$$

Using this likelihood, the optimal number of clusters can be determined by minimizing the Akaike Information Criterion:

$$AIC = -2 \ln L + 2Q \qquad \text{with } Q \ldots \text{ number of parameters} \tag{5}$$

Systematic evaluations of competing criteria revealed that the modified Akaike Information Criterion,

$$MAIC = -2 \ln L + 3Q \tag{6}$$

outperforms AIC as well as other criteria such as, e.g., BIC or CAIC (Andrews and Currim (2003)).

4 Empirical application

4.1 Data description and preprocessing

The data are gathered in using the standardized questionnaires developed within the CMP project. The first sample of $n_1 = 32$ observations was generated in the course of postgraduate management training in Moscow. A second sample of $n_2 = 40$ observations was gathered in cooperation with the European Bank for Reconstruction and Development. This sample differs from the first one because it includes organizations based in St. Petersburg and a smaller town Yaroslavl located 250 km north-east of Moscow. A third sample of $n_3 = 28$ observations was gathered on the lines of the first sample in the course of postgradual management training, but in Lithuania covering organizations based in the capital, Vilnius, and the city of Kaunas. Because of this particular structure of the data under consideration, we expect the data set to comprise observations from–statistically spoken–different generating processes. Thus, the mixture approach outlined in section 3.1 should fit the data better than simple approaches.

Each of the five marketing concepts (as depicted in subsection 2.1) is measured using nine Likert-type scaled items describing the nature of buyer-seller relations, the managerial intention, the spending of marketing budgets and the type of staff engaged in marketing activities (cf. Wagner (2005) for details). According to the proceeding of Coviello et al. (2002), a factor score is computed from each of the nine sets of indicators. These make up the vectors of predictors $\mathbf{x}_n$. As outlined in section 2, two major reasons for introducing the new relational marketing concepts, DM, EM, IM, and NM, are the increasing importance of service marketing and the differences between industrial und consumer markets. Therefore, two binary variables indicating whether the organization n offers services and serves industrial markets are included as covariates $\tilde{\mathbf{x}}_n$ according to equation 1. Additionally, the questionnaires comprise a variable capturing an ordinal rating of the organization's overall commitment to transactional marketing. This is the endogenous variable, y_n, of the response model. The approach allows the quantification of combinations of marketing concepts as well as revealing substitutive relations.

4.2 Results

Table 1 depicts the results of fitting the model for 1 to 5 segments.

Table 1. Model's fit with different numbers of segments

S	$\log L$	AIC	$MAIC$	Class. Err.	pseudo R^2
1	-131.49	280.97	289.97	.00	.23
2	-117.04	276.09	297.09	.09	.81
3	-100.96	267.92	300.92	.08	.92
4	-89.76	269.52	314.52	.11	.92
5	-82.62	279.24	336.24	.11	.95

It is obvious from the table that the optimal number of segments according to the *AIC* und pseudo R^2 is 3. But, the *MAIC* does not confirm this advice, which holds for other criteria (e.g., BIC) as well. This result is surprising with respect to the discussion in subsection 3.2 (see Andrews and Currim (2003) for an explanation of data scenario's impact). Table 2 provides the parameter estimates for the predictors and the covariates of the response model.

Table 2. Parameter estimates for predictors and covariates

	Segment 1	Segment 2	Segment 3	Wald-Statistic
inner segment R^2	.87	.81	.89	–
θ_{01s}	-17.01	-2.63	-7.68	19.25
θ_{02s}	-7.96	6.51	1.97	biased
θ_{03s}	5.45	2.76	7.82	biased
θ_{04s}	8.93	2.10	2.59	biased
θ_{05s}	10.59	-8.75	-4.70	biased
TM score	3.67	4.05	.88	10.29
DM score	2.71	1.03	-7.87	8.94
EM score	-2.44	-.64	1.37	6.62
IM score	1.15	.23	6.52	5.17
NM score	-.44	.29	2.99	3.03
Intercept Covariates	.83	-.82	-.01	6.97
B2B markets	-.32	-1.36	1.68	6.86
services	.20	1.90	-2.10	8.81

Obviously, the model fits with all three segments. The organizations assigned to segment 1 are offering goods and services to Russian consumers rather than to business customers. As expected, the score for TM is positively related to the dependent variable (overall commitment towards transactional marketing), but the parameters for DM and IM are positive as well. Thus, the organizations combine TM with these relational marketing elements, but substitute for EM and NM. The organizations in segment 2 are offering services to consumer markets. In contrast to Western-type marketing folklore, the estimated parameter for TM is the highest positive parameter for this segment. The organizations in segment 3 are selling industrial goods. In line with conventional theory, the TM score is positive, but not substantial. Moreover, IM appears to be most important for these organizations and they have the highest parameter of all three segments for NM. So far, the results match the theory, but interestingly, these organizations refrain from engaging in DM.

Table 3 provides a comparison of *K*-means clustering with the classification of the response model. The number of three clusters in *K*-means clustering has been chosen to achieve a grouping comparable to the finite-mixture approach, but is also confirmed by the GAP-criterion.

Table 3. Comparison with K-means clustering

	Segment 1	Segment 2	Segment 3	Total
Cluster I	27	13	10	50
Cluster II	9	4	1	14
Cluster III	19	8	9	36
Total	55	25	20	100

Surprisingly, the grouping with respect to the response structure differs completely from the grouping conventional K-means clustering. Particularly, cluster I, which covers half of all observations, spreads over all segments and, inversely, the organizations of segment 1 are matched to all three clusters. This result is confirmed by the projection of the clustering solutions in a plane of two canonical discriminant axes depicted in figure 1.

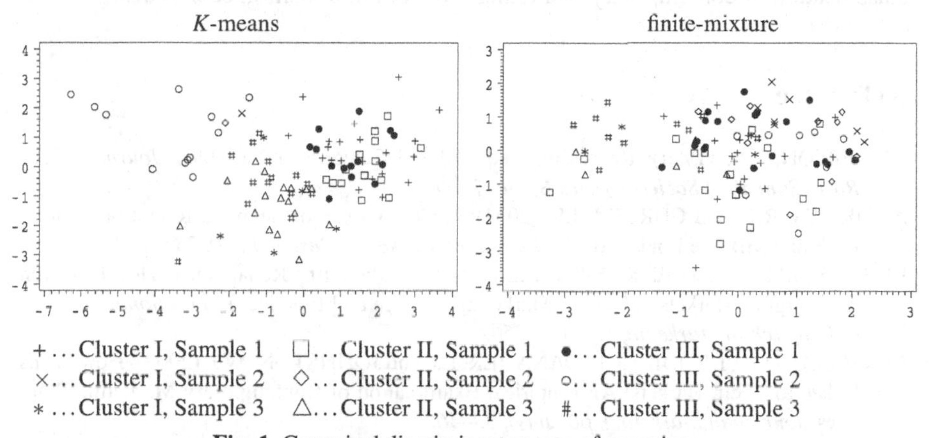

Fig. 1. Canonical discriminant spaces of groupings

The horizontal axis in the left-hand figure accounts for 82.03 % of the data variance, the vertical axis for the remaining 17.97 % of this solution (Wilks' $\Lambda = .15$, F-statistic $= 21.17$). Here, the clusters are well separated and non-overlapping. Cluster I and cluster III comprise observations from all three samples, while cluster II is dominated by observations from the Lithuanian sample. In the right-hand figure, the horizontal axis accounts for 89.63 % of the data variance, the vertical axis for the remaining 17.97 % of this solution (Wilks' $\Lambda = .56$, F-statistic$= 4.40$). The structure of segments is not reproduced by the canonical discriminant analysis, although the same predictors and covariates were used. All segments are highly overlapping. Thus, it is argued that the grouping, with respect to the response structure, offers new insights into the structure underlying the data, which can not be revealed by the clustering methods prevailing in current marketing research publications.

5 Conclusions

The finite-mixture approach introduced in this paper facilitates the fitting of a response model and the clustering of observations simultaneously. It reveals a structure underlying the data, which has been shown not to be feasible by conventional clustering algorithms. Analyzing the relevance of the new relationship paradigm for marketing in transition economies (Russia and Lithuania), this study clarifies that the borderline is not simply described by distinguishing services vs. goods markets or industrial vs. consumer markets.

The empirical results give reasons for rethinking the relevance of Western marketing folklore for transition economies, because only the marketing practices of one group of organizations targeting industrial customers fit the Western-type guidelines. Thus, this study confirms conjectures drawn from previous studies, but takes advantage of a more rigorous approach for an interpretation rather than a description of classification of contemporary marketing practices in transition economies.

References

ANDERSON, J.A. (1984): Regression and Ordered Categorical Variables. *Journal of the Royal Statistical Society (Series B), 46, 1–30.*

ANDREWS, R.L. and CURRIM, I.S. (2003): A Comparison of Segment Retention Criteria for Finite Mixture Logit Models. *Journal of Marketing Research, 40, 235–243.*

BURGES, S.M. and STEENKAMP, J.-B.E.M. (2006): Marketing Renaissance: How Research in Emerging Markets Advances Marketing Science and Practice. *Internationals Journal of Research in Marketing, 23, 337–356.*

COVIELLO, N.E., BRODIE, R.J., DANAHER, P.J. and JOHNSTON, W.J. (2002): How Firms Relate to Their Markets: An Empirical Examination of Contemporary Marketing Practices. *Journal of Marketing, 66 (July), 33–46.*

McCULLAGH, P. and NELDER, J.A. (1989): *Generalized Linear Models.* $2^n d$ ed., Chapmann & Hall, Boca Raton.

PELS, J. and BRODIE, R.J. (2004): Profiling Marketing Practice in a Transition Economy: The Argentine Case. *Journal of Global Marketing, 17, 67–91.*

VARGO, S.L. and LUSCH, R.F. (2004): Evolving to a New Dominant Logic for Marketing. *Journal of Marketing, 68, 1–17.*

WAGNER, R. (2005): Contemporary Marketing Practices in Russia. *European Journal of Marketing, 39, 199–215.*

WAGNER, R., SCHOLZ, S.W. and DECKER, R. (2005): The Number of Clusters in Market Segmentation. In: D. Baier, R. Decker, and L. Schmidt-Thieme (Eds.): *Data Analysis and Decision Support.* Springer, Berlin, 157–176.

Part VII

Banking and Finance

Predicting Stock Returns with Bayesian Vector Autoregressive Models

Wolfgang Bessler and Peter Lückoff

Center for Finance and Banking, Licher Strasse 74, 35394 Giessen, Germany
{Wolfgang.Bessler, Peter.Lueckoff}@wirtschaft.uni-giessen.de

Abstract. We derive a vector autoregressive (VAR) representation from the dynamic dividend discount model to predict stock returns. This valuation approach with time-varying expected returns is augmented with macroeconomic variables that should explain time variation in expected returns and cash flows. The VAR is estimated by a Bayesian approach to reduce some of the statistical problems of earlier studies. This model is applied to forecasting the returns of a portfolio of large German firms. While the absolute forecasting performance of the Bayesian vector-autoregressive model (BVAR) is not significantly different from a naive no-change forecast, the predictions of the BVAR are better than alternative time-series models. When including past stock returns instead of macroeconomic variables, the forecasting performance becomes superior relative to the naive no-change forecast especially over longer horizons.

1 Introduction

The prediction of asset returns has been a pivotal area of research in financial economics since the beginning of the last century. For many decades the common academic belief was that asset prices followed a random walk in the short and in the long run. In contrast, linear regression studies in the late 1980s and 1990s and more recent extensions of these studies show a certain degree of predictability (Fama and French (1988) and Hodrick (1992), Cremers (2002), Avramov and Chordia (2006), respectively). Predictability, however, does not necessarily imply market inefficiency (Kaul (1996)). Rather, time-varying expected returns can lead to return predictability in a risk-averse world that is consistent with rational behavior and efficient markets. In addition, time-varying expected returns may explain the excess volatility puzzle which states that asset prices fluctuate too widely to be rationally explained by variation in fundamentals. Investors are believed to be relatively less risk avers during boom periods, thus, demanding only a low risk premium whereas they are relatively more risk avers during recessions requiring a higher risk premium. The objective of this study is to add to our understanding of stock price fluctuations by deriving a vector autoregressive (VAR) form of the dynamic dividend discount model for predicting stock returns. This model is augmented with macroeconomic variables in order to explain the time variation in expected returns and cash flows.

In the next section we review the literature and in section 3 we derive a BVAR model from the dynamic dividend discount model in order to explain the conditional distribution of returns over time. This approach combines various extensions of the early linear regression studies. The empirical results are presented in section 4 and the last section concludes the paper.

2 Literature review

Many studies have tried to explain expected stock returns with fundamental variables. An overview of the literature is provided in Table 1. Most of these studies

Table 1. Comparison of variables used in linear regressions.

Linear Regressions														
Author(s)	Sample	1	2	3	4	5	6	7	8	9	10	11	12	13
Chen et al. (1986)	1953–1983	X					X	X	X	X	X		X	X
Campbell (1987)	1959–1983	X						X	X		X			
Harvey (1989)	1941–1987	X	X				X	X		X	X	X		
Ferson (1990)	1947–1985	X						X	X		X			
Ferson and Harvey (1991)	1959–1986	X	X				X	X		X	X			X
Ferson and Harvey (1993)	1970–1989	X						X					X	X
Whitelaw (1994)	1953–1989		X				X	X			X			
Pesaran and Timmermann (1995)	1954–1992		X	X				X	X		X		X	X
Pontiff and Schall (1998)	1926–1994		X				X	X		X				
Ferson and Harvey (1999)	1963–1994		X				X	X		X	X			
Bossaerts and Hillion (1999)	1956–1995	X	X	X	X			X		X		X		X
Cremers (2002)	1994–1998	X	X	X		X	X	X	X	X	X	X	X	X

Reproduced from Cremers (2002, p. 1226). For references see Cremers (2002).

Variables: 1 – lagged return; 2 – dividend yield; 3 – P/E-ratio; 4 – payout ratio; 5 – trading volume; 6 – default spread; 7 – yield on T-bill; 8 – change in yield on T-bill; 9 – term spread; 10 – yield spread between overnight fixed income security and T-bill; 11 – january dummy; 12 – growth rate of industrial production; 13 – change in inflation or unexpected inflation.

were able to detect some degree of predictability. In these cases predictability is usually defined as at least one non-zero coefficient in a regression model using one or a combination of lagged fundamental variables (Table 1) to predict future returns (Kaul (1996)). Some of these studies used overlapping observations which may have resulted in autocorrelated residuals. In addition, a small sample bias could have been caused by the endogeneity of lagged explanatory variables (Hodrick (1992)). If the dependent and at least one of the independent variables are non-stationary, the relationship between expected returns and fundamental variables might be spurious

(Ferson and Sarkissian (2003)). Unfortunately, the small sample bias and a spurious regression reinforce each other.

To mitigate the problems due to these biases, several modifications have been suggested in the literature (Hodrick (1992)). In order to correct the inferences we introduce lags of explanatory variables which then results in a VAR representation. Moreover, determining the best predictors is difficult as various studies find different variables to be good predictors. This observation indicates that data mining might be at work. More recent studies use Bayesian approaches in order to integrate more variables and condition inferences on the whole set of potential predictors (Avramov (2002), Cremers (2002), Avramov and Chordia (2006)). These studies still find predictability based on both forecasting errors and profitability of investment strategies.

In recent years, Bayesian methods are becoming more prominent in finance such as asset pricing, portfolio choice, and performance evaluation of mutual funds. An important advantage of the Bayesian methods is that they yield the complete distribution of the model parameters. Thus, estimation risk can be incorporated into asset allocation decisions. Furthermore, applying Bayesian techniques in asset allocation decisions leads to much more stable portfolio weights than with classical portfolio optimization (Barberis (2000)).

3 Model

3.1 Dynamic dividend discount model

The forecasting equation is derived from the dividend discount model with time-varying expected returns:

$$P_t = E_t\left[\sum_{t=1}^{\infty} \frac{D_t}{(1+R_t)^t}\right] \tag{1}$$

As this equation is non-linear, we use the log-linear approximation of Campbell and Shiller (1989) and take expectations:

$$p_t = \frac{k}{1-\rho} + E_t\left\{\sum_{i=0}^{\infty} \rho^i \left[(1-\rho)d_{t+1+i} - r_{t+1+i}\right]\right\} \tag{2}$$

where k and ρ are constants and lower case letters denote logarithms. Subtracting equation (2) from the logarithmic dividend results in:

$$d_t - p_t = -\frac{k}{1-\rho} + E_t\left[\sum_{i=0}^{\infty} \rho^i \left(r_{t+1+i} - \Delta d_{t+1+i}\right)\right] \tag{3}$$

From equation (3) it can be seen that the current (logarithmic) dividend yield is a good predictor for future expected returns. The intuition behind that equation is that the dividend yield itself is a stationary variable. Thus, if the variable is above its long-run mean either the price has to increase, the dividend has to fall or both effects have to occur at the same time. Because prices usually fluctuate more widely than dividends, a subsequent price change rather than a dividend change is more likely.

3.2 Bayesian vector autoregressive model

The dynamic dividend discount model can be augmented by macroeconomic variables that are able to explain expected returns and cash flows over time. The resulting model can be written in a general VAR form:

$$y_t = \Phi_1 y_{t-1} + z_t \tag{4}$$

The vector y contains returns, dividend yields, and macroeconomic variables. Note that in general any VAR(p) model can be stacked and represented as VAR(1) model as in equation (4). We introduce prior information into the model by using a Bayesian approach. This imposes structure on the model in a flexible way and downsizes the impact of shocks on our forecasts as shocks do not tend to repeat themselves in the same manner in the future. Furthermore, a larger number of variables and lags can be included than in the classical case without the threat of overfitting. BVAR models were used to predict business cycles which tend to be the main driver of our valuation equation (Litterman (1986)). In addition, BVAR models have proved to be valuable in other applications such as the prediction of foreign exchange rates (Sarantis (2006)).

In order to keep the definition of the prior capable, we impose prior means that suggest a random walk structure for the stock price and three hyperparameters for the prior precision as suggested by Litterman (1986). This assumes more explanatory power of own lags in contrast to other variables and decreasing explanatory power with increasing lag length for each equation in the VAR. In the estimation we follow Litterman (1986) and use his extension of the mixed-estimation approach.

4 Empirical study

To evaluate our model we compare its forecasting quality with five benchmark model types. We form an equally-weighted portfolio of ten arbitrarily chosen DAX-stocks for the period from 01:1992 to 01:2005 based on monthly returns. In order to simulate a real-time forecasting strategy we use 59 rolling windows of 90 months in order to estimate (and in some cases optimize) the model and predict the portfolio returns for a horizon of one to 15 months out-of-sample. As benchmark models we use (1) a random walk as naive forecast, (2) an AR(1) model as simple time-series model, (3) dynamic ARIMA models that are optimized for each rolling window using the Schwartz-Bayes criterion (differing in the maximum lag length), (4) linear regressions (static and dynamic versions using a stepwise regression approach), and (5) classical VAR models.

We employ the dividend yield of the portfolio and a total of 30 different macroeconomic variables such as interest rates, sentiment indicators, implied volatilities and foreign exchange rates in order to construct 28 different model specifications of our model types. As the results are comparable to our general findings, we focus on the models using the dividend yield, the change in GDP and the change in the unemployment rate as predictors. To judge the quality of our forecasts, we apply a direct approach in that we use squared forecasting errors as well as mean squared errors (MSE).

4.1 1-step forecasts

First, we replicate the results from previous studies that found predictability based on the significance of the predictor variables. As presented in Figure 1 the dividend yield is significant for almost 60 % of the sample and the change in GDP is significant for the entire sample. The change in the unemployment rate becomes insignificant after the first rolling window. The time-varying pattern of the t-values and the co-

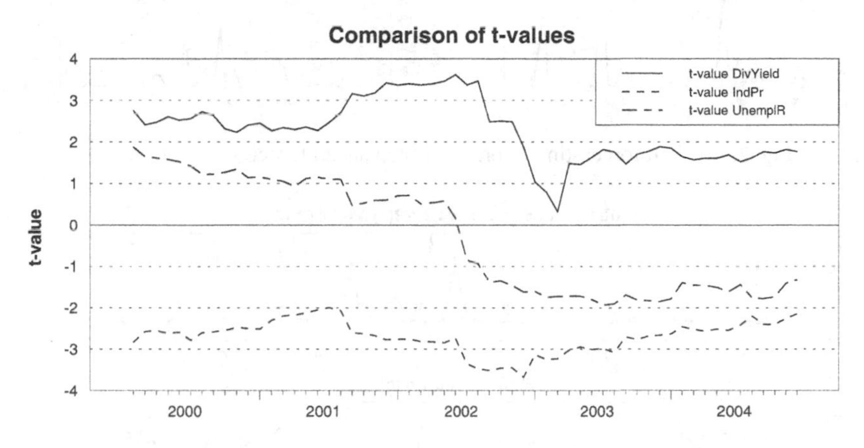

Fig. 1. Comparison of t-values of linear regression over time

efficients of the predictors (not reported) are in line with Bessler and Opfer (2004). Thus, model uncertainty is not a static problem but rather needs to be taken into account over time as well. A comparison of squared errors over time for the six models reveals that the AR(1) and the BVAR yield the best results (Figure 2). All models show a similar pattern with peaks occurring in those months when returns are both very high in magnitude and negative most of the time. Thus, such extreme returns cannot be predicted with these models. By taking a closer look at the two models with the best forecasting performance, i. e. the AR(1) and the BVAR, two issues should be noted. The difference in squared errors between the AR(1) and the BVAR in the upper panel of Figure 3 reveals that the BVAR is superior in normal markets. Nevertheless, the AR(1) performs better in down markets. This can be explained with the increasing (positive) autocorrelation in returns during market turmoil. The change in autocorrelation is not reflected in the BVAR as its forecasts do not respond to shocks as quickly as the AR(1) forecasts. However, a comparison of the MSE for the six models indicates that none of the models produces significantly better forecasts than a naive forecast.

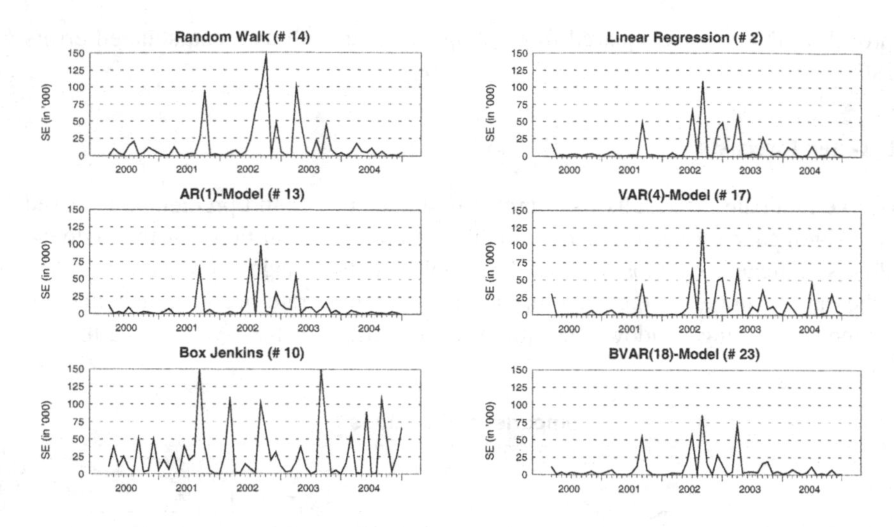

Fig. 2. Squared forecasting errors for 1-step ahead forecasts over time

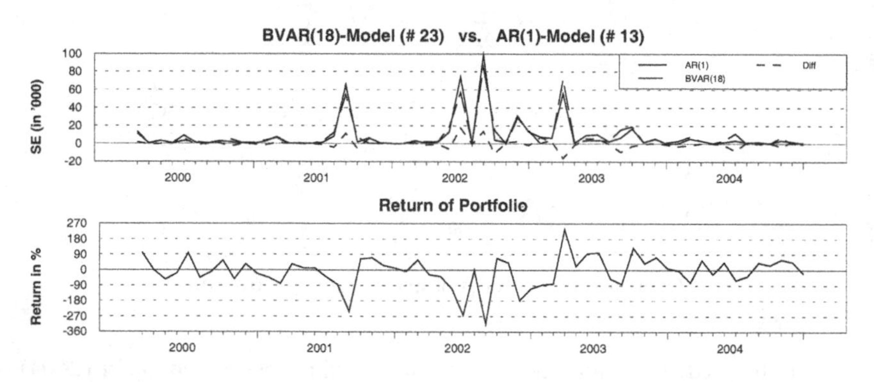

Fig. 3. Comparison of squared forecasting errors for BVAR and random walk

4.2 1- to 15-step forecasts

By looking at longer forecasting horizons of up to 15 months, the dominance of the BAVR becomes even more pronounced (Figure 4). However, the simple AR(1) still provides comparable results.

4.3 Single stocks as variables

An interesting result emerges when we substitute the macroeconomic variables in the BVAR with the return series of the ten stocks of the portfolio. The forecasting performance of the BVAR based on single stock returns improves significantly. For example, over a forecasting horizon of 12 months the MSE of the BVAR is about 3 percentage points smaller than the MSE of a naive forecast. The superior results

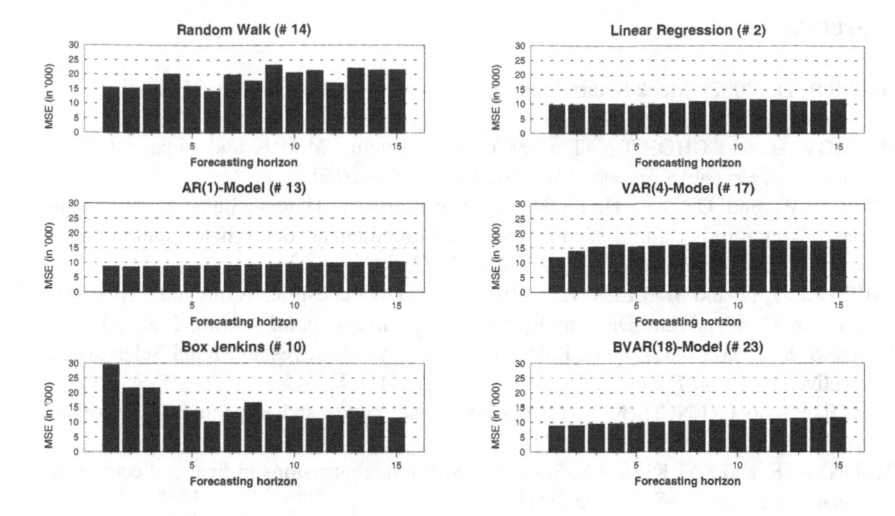

Fig. 4. Squared forecasting errors for 1- to 15-step ahead forecasts

based on single stocks rather than macroeconomic variables can not be explained by a decoupling of returns from macroeconomic factors during the rise and fall of the new economy era. The MSE for the subsample including the downturn up to 03:2003 is only about 1 percentage point smaller than the naive forecast's MSE over all forecasting horizons. In contrast, for the subsample from 04:2003 onwards, the MSE is at least 2 percentage points smaller than that of a naive forecast and again the lowest for a 12 months horizon (6.5 percentage points smaller) which implies a high degree of predictability.

5 Conclusion and outlook

The objective of this study was to evaluate the forecasting performance of BVAR models for stock returns relative to five benchmark models. Our results suggest that even if we can reproduce the predictability results of earlier studies based on the significance of parameters none of the models based on macroeconomic variables is capable of predicting stock returns as measured by forecasting errors. However, there is a certain degree of predictability of the BVAR when we use the returns of single stocks instead of macroeconomic variables. Thus, it seems worthwhile to take a closer look at the cross-correlation structure of stock returns over monthly horizons. For future studies we suggest to use asymmetric weighting matrices for the prior that take into account the differences between industries (cyclical vs. non-cyclical) and sizes of the companies. Alternatively, our methodology could be extended to an application on bond markets as it is derived from a simple present-value relation.

References

AVRAMOV, D. (2002): Stock Return Predictability and Model Uncertainty. *Journal of Financial Economics, 64, 423–458.*

AVRAMOV, D. and CHORDIA, T. (2006): Asset Pricing Models and Financial Market Anomalies. *Review of Financial Studies, 19, 3, 1001–1040.*

BESSLER, W. and OPFER, H. (2004): Eine Empirische Untersuchung zur Bedeutung makroökonomischer Einflussfaktoren auf Aktienrenditen am deutschen Kapitalmarkt. *Finanzmarkt und Portfoliomanagement, 4, 412–436.*

CAMPBELL, J. D. and SHILLER, R. J. (1989): The Dividend-price Ratio and Expectations of Future Dividends and Discount Factors. *Review of Financial Studies, 1, 3, 195–228.*

CREMERS, K. J. M. (2002): Stock Return Predictability: A Bayesian Model Selection Perspective. *Review of Financial Studies, 15, 4, 1223–1249.*

FAMA, E. F. and FRENCH, K. R. (1988): Dividend Yields and Expected Stock Returns. *Journal of Financial Economics, 22, 3–25.*

FERSON, W. E. and SARKISSIAN, S. (2003): Spurious regressions in financial economics? *Journal of Finance, 58, 4, 1393-Ű1412.*

HODRICK, R. J. (1992): Dividend Yields and Expected Stock Returns: Alternative Procedures for Inference and Measurement. *Review of Financial Studies, 5, 3, 357–386.*

KAUL, G. (1996): Predictable Components in Stock Returns. In: G. S. Maddala, C. R. Rao (Eds.): *Statistical Methods in Finance.* Elsevier Science, Amsterdam, 269-296.

LITTERMANN, R. B. (1986): Forecasting with Bayesian Vector Autoregressions Ű Five Years of Experience. *Journal of Business and Economic Statistics, 4, 1, 25-Ű38.*

SARANTIS, N. (2006): On the Short-term Predictability of Exchange Rates - A BVAR Time-varying Parameters Approach. *Journal of Banking and Finance, 30, 2257-Ű2279.*

The Evaluation of Venture-Backed IPOs – Certification Model versus Adverse Selection Model, Which Does Fit Better?

Francesco Gangi and Rosaria Lombardo

Faculty of Economics, Department of Strategy and Quantitative Methods
Second University of Naples, Italy
{francesco.gangi, rosaria.lombardo}@unina2.it

Abstract. In this paper we aim to investigate the consistency of the certification model against the adverse selection model with respect to the operational performances of venture-backed (VB) IPOs. We analyse a set of economic-financial variables an italian IPOs sample between 1995 and 2004. After non-parametric tests, to take into account the non-normal, multivariate nature of the problem, we propose a non-parametric regression model, i.e. Partial Least Squares, as appropriate investigative tool.

1 Introduction

In financial literature the performance evaluation of venture backed IPOs has stimulated an important debate. Two are the main theoretical approaches. The first one has pointed out the certification role and the value added services of venture capitalists. The second one has emphasized the negative effects of adverse selection and opportunistic behaviours on IPOs under-performance, especially with respect to the timing of the IPOs.
In different studies (Wang *et al.*, 2003; Brau *et al.*, 2004; Del Colle *et al.*, 2006) parametric tests and Ordinary Least Squares regression have been proposed as investigative tools. In this work we investigate complicated effects of adverse selection and conflict of interests by non-parametric statistical approaches. Underlining the non-normal data distribution, we propose as appropriate instruments non-parametric tests and Partial Least Squares regression model (PLS; Tenenhaus, 1998; Durand, 2001). At first we test if the differences of operational performances are significant between the pre-IPOs sample and post-IPOs sample. Next, given the complicated multivariate nature of the problem, we study the dependence relationships of firm performance (measured by ROE) from quantitative and qualitative variables of context (like market conditions).

2 The theoretical financial background: the certification model and the adverse selection model

The common denominator of theoretical approaches on venture capitalist role is represented by the asymmetric information management. On one hand, the certification model considers an efficient solution of this question, due to scouting process and activism of private equity agents. More specifically, the certification model takes into account the selection capacity and the monitoring function of venture capitalists that allow to make better resources allocation and better control systems than other financial solutions (Barry *ed al.*, 1990; Sahlman, 1990; Magginson e Weiss, 1991; Jain e Kini, 1995; Rajan e Zingales, 2004). Consequently, this model predicts good performances of venture backed firms, even better than non backed ones. The causes of this effect ought to be: more stable corporate relations; strict covenants; frequent operational control activities; board participation; stage financing options. These aspects should compensate the incomplete descriptive contractual structure that follows every transaction, allowing a more efficient management of the asymmetric information problem. So, venture backed IPOs should generate good performances in terms of growth, profitability and financial robustness, even better if they are compared with non backed ones.

On the other hand, IPOs under-performance could be related to adverse selection processes, even if these companies are participated by a venture capitalist. In this case two related aspects should be considered. The first one is that not necessarily the best firms are selected by venture capital agents. The second one is that the timing of IPO cannot coincide with a new cycle of growth or with an increase in profitability. Relatively to the first matter, some factors could determine a disincentive to accept the venture capital way in, such as latent costs, loose of control rights and income sharing. At the same time, the quotation option could not match an efficient signal towards the market. According to the packing order theory, the IPO choice can be neglected or rejected at all by the firms that are capable to create value by themselves, without the financial support of a fund or the stock exchange. At first, low quality company, could receive more incentives to the quotation if the value assigned by the market exceeds inside expectations, especially during bubble periods (Benninga, 2005; Coakley et al. 2004). In this situation, venture capitalist could assume an insider approach too, for example stimulating an anticipated IPO, as described by the grandstanding model (Gompers, 1996; Lee and Wahal, 2004). At second, venture capitalists could be in conflict of interests towards the market when they have to accelerate the capital turnover. This is a big question if the venture capitalist operate like an intermediary of resources obtained during the fund raising process. In this case, the venture capitalist assumes a double role: he is a principal with respect to the target company; but he is an agent with respect to the fund, configuring a more complex, onerous, therefore less efficient agency nexus model. So the hypothesis is that a not efficient management of asymmetric information can also explain the VB IPOs under-performance, confuting the assumption of superior IPOs results compared to non- VB IPOs (Wang et. Al, 2003; Brau et al., 2004). The opportunistic behaviours of previous shareholders could not be moderated by venture capitalist's

Table 1. Wilcoxon Signed Rank Test in VB IPOs: Test1=H0: $Me_{T1} = Me_{T2}$; Test2= H0:$Me_{T1} = Me_{T3}$; Test3= H0:$Me_{T2} = Me_{T3}$

Ratios	Me_{T1}	Me_{T2}	Me_{T3}	Test1	Test2	Test3
ROS	9.97	7.34	5.39	-0.87	-1.78*	-1.66*
ROE	9.75	6.84	-1.51	-1.16	-1.91*	-1.66**
ROI	7.33	6.69	3.30	-1.35	-1.79*	-1.73*
Leverage	292.67	79.75	226.96	-3.29***	-0.09	-2.54***

Table 2. Mann-Whitney Test comparison in VB IPOs: Test1=H0: $Me_{VB_{T2}} = Me_{NonVB_{T2}}$; Test2=H0:$Me_{VB_{T3}} = Me_{NonVB_{T3}}$

Ratios	VB_{T2}	$NonVB_{T2}$	VB_{T3}	$NonVB_{T3}$	Test1	Test2
ROS	9.52	3.93	5.39	2.16	103	116
ROE	6.7	3.83	3.3	2.01	111	110
ROI	6.85	3.83	-1.51	1.5	113	105
Leverage	79.75	72.52	226.96	88.28	120	58**

governance solutions. Furthermore, venture capitalists could even incentive a speculative approach to maximize and anticipate the way out from low quality companies, dimming their hypothetical certification function.

3 Data set and non-parametric hypothesis tests

The study of the Italian venture backed IPOs is based on a sample of 17 companies listed from 1995 to 2004. The universe consists of 28 manufacturing companies that have gone public after the way in of a formal venture capitalist with a minority participation. In addition to the principal group, we have composed a control sample represented by non-venture backed IPOs comparable by industries and size. The performance analysis is based on balance sheets ratios. In particular, the study assumes the profitability and the financial robustness as the main parameters to evaluate operational competitiveness before and after the quotation. Ratios are referred to three critical moments, or terms of the factor, called events, consisting in deal-year (**T1**), IPO-year (**T2**) and first year post-IPO (**T3**). At first we test the performance differences of balance sheet ratios within the venture backed IPOs with respect to the three events (T1, T2, T3). Successively we test significant difference between the two independent samples of VB IPOs and non-VB IPOs. For the particular sample characteristics (non-normal distribution and eteroschedasticity) we consider non-parametric tests like Wilcoxon signed rank test (Wilcoxon and Wilcox, 1964) for paired dependent observations and Mann-Whitney test (Mann and Whitney, 1947) for comparisons of independent samples. Coherently with the adverse selection model, we test

if the venture backed companies show an operational underperformance between the pre-IPO and post-IPO phases.
Subsequently, coherently to the certification model, we test if the venture backed companies have the best performance if compared with non venture backed IPOs. The statistics of VB IPOs show an underperformance trend of venture backed companies during the three defined terms. In particular, all the profitability ratios decline constantly. Moreover, we find an high level of leverage (Debt/Equity) at the deal moment, and in the first year post-IPO the financial robustness goes down again very rapidly. So the prediction of a re-balancing effect on financial structure has been considered only with respect to the IPOs events (see table 1). The results of Wilcoxon Signed Rank Test have been reported in table 1. The null hypothesis is confirmed for profitability parameters comparing ratio medians of T1 and T2 moments, whereas the differences between ratio medians of T1 and T3 and T2 and T3 are significant (the significant differences are marked by the symbols: *=10%, **=5%, ***=1%). So the profitability breakdown is mainly a post-IPO problem, with a negative effect of leverage. These results suggests that venture capitalists do not add value in the post-IPO period, otherwise, the adverse selection moderates the certification function and the best practice effects expected from venture capital solutions.
Furthermore we test the hypothesis that VB IPOs generate superior operating performance compared with non-venture IPOs. Using the Mann-Whitney test, we compare IPO-ratios of the two independent samples. The findings show no significant difference between the samples at the IPO-time and at the first year post-IPO; only the leverage level shows an higher growth in the venture group than in non-venture one, confirming the contraction of financial robustness and the loss of the re-balancing effect on financial structure produced by the IPOs (see table 2). In conclusion the test results are more consistent with the adverse selection theory.
Underlining the multivariate, non-normal nature of the problem, after hypothesis tests, we propose to investigate VB performance by a suitable non-parametric regression model.

4 Multivariate investigation tools: Partial Least squares regression model

In presence of a low-ratio of observations to variables and in case of multicollinearity in the predictors, a natural extension of the multiple linear regression is PLS regression model. It has been promoted in the chemiometrics literature as an alternative to ordinary least squares (OLS) in the poorly or ill-conditioned problems (Tenenhaus, 1998). Let $\mathbf{Y}$ be the categorical n,q response matrix and $\mathbf{X}$ the n,p matrix of the predictors observed on the same n statistical units. The resulting transformed predictors are called latent structures or latent variables. In particular, PLS chooses the latent variables as a series of orthogonal linear combinations (under a suitable constraint) that have maximal covariance with linear combinations of $\mathbf{Y}$. PLS constructs a sequence of centered and uncorrelated exploratory variables, i.e. the PLS (latent)

components $(\mathbf{t}^1,...,\mathbf{t}^A)$. Let $\mathbf{E}_0 = \mathbf{X}$ and $\mathbf{F}_0 = \mathbf{Y}$ be the design and response data matrices, respectively. Define $\mathbf{t}^k = \mathbf{E}_{k-1}\mathbf{w}^k$ and $\mathbf{u}^k = \mathbf{F}_{k-1}\mathbf{c}^k$, where the weighting unit vectors $\mathbf{w}^k$ and $\mathbf{c}^k$ are computed by maximizing the covariance between linear compromises of the updated predictor and response variables, $max[cov(\mathbf{t}_k, \mathbf{u}_k)]$.
Update the new variables $\mathbf{E}_k$ and $\mathbf{F}_k$ as the residuals of the least-squares regression on the component previously computed.
The number A of the retained latent variables, also called the model dimension, is usually estimated by cross-validation (CV).

Two particular properties make PLS attractive and establish a link between the geometrical data analysis and the usual regression. First, when $A = rank\,\mathbf{X}$,

$$PLS(\mathbf{X},\mathbf{Y}) \equiv \{OLS(\mathbf{X},\mathbf{Y}^j)\}_{j=1,...,q},$$

if the OLS regression exists.
Second, the principal component analysis, PCA, of $\mathbf{X}$ can be viewed as the "self-PLS" regression of $\mathbf{X}$ onto itself,

$$PLS(\mathbf{X},\mathbf{Y}=\mathbf{X}) \equiv PCA(\mathbf{X}).$$

PLS regression model has the following properties: efficient in spite of low ratio of observations on column dimension of **X**; efficient in the multi-collinear context for predictors (concurvity); robust against extreme values of predictors (local polynomials). The PLS regression model examines the predictors of ROE at IPO-year (T2) as variables of performance of VB IPOs companies. The predictor variables are: one quantitative (the leverage measured at the year of the venture capital way in, **LEVERAGET1**) and four qualitative: 1) the short time interval between the deal and the IPO time (1 year by-deal, **1Yby deal**; 2 year by-deal, **2Yby deal**); 2) the size of companies listed, (**SME; Large**); 3) the trend of Milan Stock Exchange, (Hot Market **Hotmkt**, Normal Market, **NORMmkt**); 4) the origin of fund, (**Bank Fund**; non-Bank Fund, **N-Bank Fund**). The building-model stage consists of finding a balance between goodness of fit and prediction and thriftness. The goodness of fit is valued by $R^2(A)$, in our study is equal to 60%, and the thriftness by PRESS criterion, the dimension space suggested by PRESS is $A = 1$. By PLS regression we want to verify the effects of some variables which could subtend opportunistic approaches. Moreover, the analysis is concentrated on the effect of independent variables that could allow the recognition of a conflict of interests between venture agents and the new stockholders. The importance of each predictors on the response is evaluated by looking at regression coefficients (β) whose graphical representation is given in figure 1. For example the regression coefficient value of leverage at the deal-time is a predictor of under-performance in the IPO year ($\beta_{\mathbf{LEVERAGET1}} = -0.36$). This finding is consistent with the assumption that adverse selection at the deal reflect its effects when the target firm is listed, especially when the gap between these two moments is very short. We could also say that pre-IPO poorly performing firms continue to produce bad performance afterward too.
Concerning the qualitative predictors, the interval time ($\beta_{\mathbf{1Ybydeal}} = -0.17$) and the

firm size ($\beta_{\mathbf{SME}} = -0.17$) are useful variables to capture the influence of a too early quotation, similarly to the grandstanding approach. The market trend ($\beta_{\mathbf{HOTmkt}} = -0.13$) is useful to verify the impact of a speculative bubble on IPOs performance. Furthermore, the origin of fund ($\beta_{\mathbf{FundBank}} = -0.17$) it's necessary to evaluate the potential conflict of interest of an agent that covers a double role: banking and venture financing. All these variables summarize the risk of an adverse selection pro-

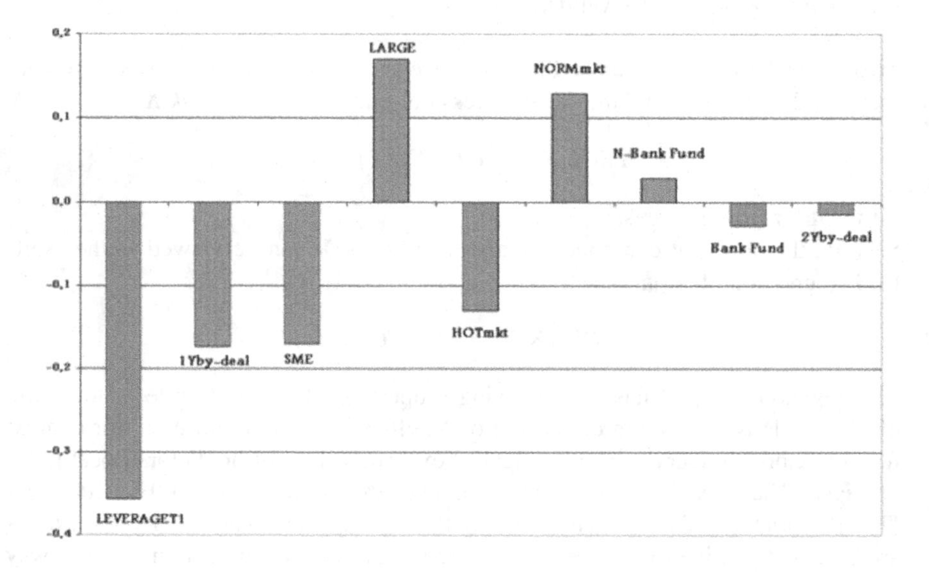

Fig. 1. Decreasing Influence of Qualitative and Quantitative Predictors on ROE-T2.

cess and speculative approach that can contrast the certification function of venture capitalist investments. So, in the first place the leverage, reached after the venture capitalist way in, is the most negative predictor of ROE at IPO time. In the second place, the shorter are the time intervals between the deal and IPO time, the worst is the influence on ROE. In the third place, the firm size **SME** is a relevant predictor too. In fact, smaller and less structured enterprises have a negative incidence on IPOs operating performance. In the fourth place, even the market trend seems to assume a significant role to explain the VB IPO under-performance. More specifically, hot issues **HOTmkt** determine a negative effect on ROE. Finally, in a less relevant position there is the fund origin **Fund Bank**, for this variable the theoretical assumption is confirmed too, because of the negative influence of bank based agents. In synthesis we can say that ROE under-performance depends from the following predictors: **LEVERAGE1, 1Yby deal, HOTmkt, SME**. So, coherently with inferential tests, the PLS findings related to the IPO segment of the Italian Private Equity Market move away the venture finance solution from the theoretical certification function.

5 Conclusion

The results of the non-parametric tests as well as the more complete multivariate dependence model show that operational performances of VB IPOs are significantly consistent with the adverse selection and opportunistic model. Specifically, a large part of IPOs under-performance is due to the leverage "abuse" at the Deal-Time, and the PLS regression shows that too early quotation by-deal, hot issues and small firm size are all predictors of profitability falls. Probably we should rethink a "romantic" vision about the venture capitalist role: sometimes he is simply an agent in conflict of interest, or he has not always the skill to select the best firms for the financial market. Obviously there are a lot of implications for further research and developments of this work. An international comparison with other financial systems and a further supply and demand analysis ought to be carried out.

Acknowledgments

This work was supported by SUN-University funds 2006, responsible Rosaria Lombardo and Francesco Gangi. The paper was written by both authors in particular sections 1,2,3,5 are mainly attributed at Francesco Gangi and section 4 at Rosaria Lombardo.

References

BARRY, C., MUSCARELLA, C., PEAVY, J. and VETSUYPENS, M. (1990): The role of venture capital in the creation of public companies. Evidence from the going public process. *Journal of Financial Economics, 27, pp. 447-471.*

BENNINGA, S., HELMANTEL, M. and SARIG, O. (2005): The timing of initial public offering. *Journal of Financial Economics, 75, pp. 115-132.*

BRAU, J., BROWN, R. and OSTERYOUNG, J. (2004): Do venture capitalists add value to small manufacturing firms? An empirical analysis of venture and non-venture capital-backed initial public offerings. *Journal of Small Business Management, 42, pp. 78-92.*

COAKLEY, J., HADASS, L. and WOOD, A. (2004): Post-IPO operating performance, venture capitalists and market timing. *Department of Accounting, Finance and Management, University of Essex, pp. 1-32.*

DEL COLLE, D.M., FINALI RUSSO, P. and GENERALE, A. (2006): The causes and consequences of venture capital financing. An analysis based on a sample of italian firms. *Temi di discussione Banca d'Italia, 6-45.*

DURAND, J.F. (2001): Local Polynomial additive regression through PLS and Splines: PLSS. *Chemometrics & Intelligent Laboratory Systems, 58, pp. 235-246.*

GOMPERS, P. (1996): Grandstanding in the venture capital industry. *Journal of Financial Economics, 42, pp. 1461-1489.*

JAIN, B. and KINI, O. (1995): Venture capitalist participation and the post-issue operating performance of IPO firms.*Managerial and Decision Economics, 16, pp. 593-606.*

LEE, P. and WAHAL, S. (2004): Grandstanding, certification and the underpricing of venture capital backed IPOs. *Journal of Financial Economics, 73, pp. 375-407.*

MANN, H.B. and WHITNEY, D.R. (1947): On a test of whether one of 2 random variables is stochastically larger than the other. *Annals of mathematical statistics, 18, pp. 50-60.*

MEGGINSON, W., WEISS, K. (1991): Venture capital certification in initial public offerings. *Journal of Finance, 46, pp. 879-903.*

RAJAN, R. G. and ZINGALES, L. (2003): *Saving capitalism from the capitalists.* Einaudi, Torino.

SAHLMAN, W.A. (1990): The structure and governance of venture-capital organizations.*Journal of Financial Economics, 27.*

TENENHAUS, M. (1998): *La Regression PLS, Theorie et Pratique.* Editions Technip, Paris.

WANG, C., WANG, K. and LU, Q. (2003): Effects of venture capitalists' participation in listed companies.*Journal of Banking & Finance, 27, pp. 2015-2034.*

WILCOXON, F. and WILCOX, A.R. (1964): *Some rapid approximate statistical procedures.* Lederle Lab., Pearl River N.Y.

Using Multiple SVM Models for Unbalanced Credit Scoring Data Sets

Klaus B. Schebesch[1] and Ralf Stecking[2]

[1] Faculty of Economics, University "Vasile Goldiş", Arad, Romania
kbsbase@gmx.de

[2] Faculty of Economics, University of Oldenburg, D-26111 Oldenburg, Germany
ralf.w.stecking@uni-oldenburg.de

Abstract. Owing to the huge size of the credit markets, even small improvements in classification accuracy might considerably reduce effective misclassification costs experienced by banks. Support vector machines (SVM) are useful classification methods for credit client scoring. However, the urgent need to further boost classification performance as well as the stability of results in applications leads the machine learning community into developing SVM with multiple kernels and many other combined approaches. Using a data set from a German bank, we first examine the effects of combining a large number of base SVM on classification performance and robustness. The base models are trained on different sets of reduced client characteristics and may also use different kernels. Furthermore, using censored outputs of multiple SVM models leads to more reliable predictions in most cases. But there also remains a credit client subset that seems to be unpredictable. We show that in unbalanced data sets, most common in credit scoring, some minor adjustments may overcome this weakness. We then compare our results to the results obtained earlier with more traditional, single SVM credit scoring models.

1 Introduction

Classifier combinations are used in the hope of improving the out-of-sample classification performance of single *base* classifiers. It is well known (Duin and Tax (2000), Kuncheva (2004), Koltchinskii et al. (2004)), that the results of such combiners can be both better or worse than expensively trained single models and also that combiners can be superior when used on relatively sparse empirical data. In general, as the base models are less powerful (and inexpensive to produce), their combiners tend to yield much better results. However, this advantage is decreasing with the quality of the base models (e.g. Duin and Tax (2000)). Our past credit scoring single-SVM classifiers concentrate on misclassification performance obtainable by different SVM kernels, different input variable subsets and financial operating characteristics (Schebesch and Stecking (2005a,b), Stecking and Schebesch (2006), Schebesch and Stecking (2007)). In credit scoring, classifier combination using such base models may be very useful indeed, as small improvements in classification accuracy matter especially in the case of unbalanced (e.g. with more good than bad credit clients) data sets and as *fusing* models on different inputs may be required by practice. Hence, the paper presents in sections 2 and 3 model combinations with base models on all available inputs using single classifiers with six different kernels for unbalanced data sets, and finally in section 4 SVM model combinations of base models on randomly selected input subsets using the same kernel classifier placing some emphasis on correcting overtraining which may also result from model combinations.

2 SVM models for unbalanced data sets

The data set used is a **sample** of 658 clients for a building and loan credit with a total number of 40 input variables. This sample is drawn from a much larger population of 17158 credit clients in total. Sample and population do not have the same share of good and bad credit clients: the majority class is **undersampled** (drawn less frequently from the poulation than the opposite category) to get a more balanced data set. In our case the good credit clients share 93.3% of the population, but only 50.9% of the the sample. In the past, a variety of SVM models were constructed in order to forecast the defaulting behavior of new credit clients, but without taking into account the *sampling bias* systematically. For balanced data sets SVM with six different kernel functions were already evaluated. Detailed information about kernels, hyperparameters and tuning can be found in (Stecking and Schebesch (2006)).

In case of unbalanced data sets the SVM approach can be described as follows: Let $f_k(x) = \langle \Phi_k(x), w_k \rangle + b_k$ be the output of the kth SVM model for unknown pattern x, with b_k a constant, Φ_k the (usually unknown) feature map which lifts points from the input space $\mathbf{X}$ into feature space $\mathcal{F}$, hence $\Phi : \mathbf{X} \rightarrow \mathcal{F}$. The weight vector w_k is defined by $w_k = \sum_i \alpha_i y_i \Phi_k(x_i)$ with α_i the dual variables $(0 \leq \alpha_i \leq C(y_i))$, and y_i be the binary output of input pattern x. For unbalanced data sets the usually unique upper bound C for α_i is replaced by two output class dependent cost factors $C(-1)$ and $C(+1)$. Different cost factors penalize for example false classified bad credit clients stronger than false classified good credit clients. Note also, that $\langle \Phi_k(x), \Phi_k(x_i) \rangle = K(x, x_i)$, where K is a kernel function, for example $K(x, x_i) = \exp\left(-s\|x - x_i\|^2\right)$, i.e. the well known RBF kernel with user specified kernel parameter s.

Multiple SVM models and combination

In previous work (Schebesch and Stecking (2005b)) SVM **output regions** were defined in the following way: (1) if $|f_k(x)| \geq 1$, then x is called a *typical* pattern with low classification error, (2) if $|f_k(x)| < 1$, then x is a *critical* pattern with high classification error. Combining SVM models for classification we calculate sign $\left(\sum_k y_k^*\right)$ with $y_k^* = +1$ while $f_k(x) \geq 1$ and $y_k^* = -1$ while $f_k(x) \leq -1$, which means: SVM model k has zero contribution for its critical patterns. For illustrative purpose we combine two SVM models (RBF and second degree polynomial) and mark nine regions (see figure 1): typical/typical regions are `I, III, VII, IX`, critical/critical

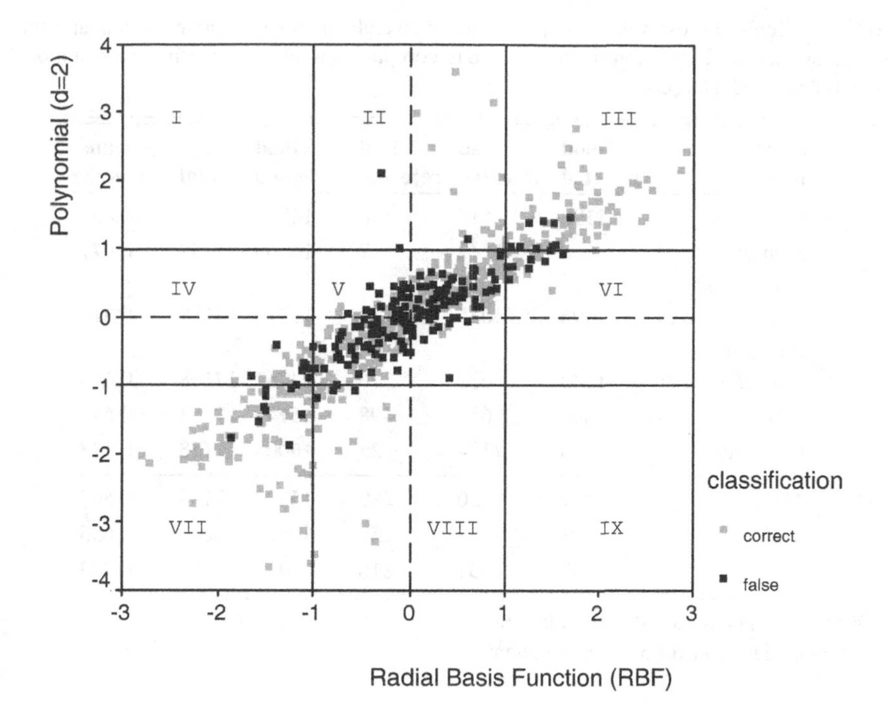

Fig. 1. Combined predictions of SVM with *(i)* polynomial kernel $K(x,x_i) = (\langle x,x_i \rangle + 5)^2$ and *(ii)* RBF kernel $K(x,x_i) = \exp\left(-0.05\|x-x_i\|^2\right)$. Black (grey) boxes represent false (correct) classified credit clients. Nine regions (`I`-`IX`) are defined s.t. the SVM output of both models.

region is `V` and typical/critical regions are `II, IV, VI, VIII`. Censored classification uses only typical/typical regions (with a classification error of 10.5 %) and typical/critical regions (where critical predictions are set to zero) with a classification error of 18.8 %. For the critical/critical region `V` no classification is given, as the expected error within this region would be 39.7 %. For this combination strategy the number of unpredictable patterns is quite high (360 out of 658). However, by enhancing the diversity and by increasing the number of SVM models used in combinations, the number of predictable patterns will also increase. 0.1cm2.4mm

3 Multiple SVM for unbalanced data sets in practice

Table 1 shows the classification results of six single SVM and three multiple SVM models using tenfold cross validation. Single models are built with the credit scoring data sample of 658 clients using SVM kernel parameters from (Stecking and Schebesch (2006)) and varying cost factors $C(+1) = 0.3 \times C(-1)$ from (Schebesch and Stecking (2005a)), allowing for higher classification accuracy towards *good* credit clients. The classification results obtained are weighted by $w = \frac{16009}{335}$ for *good*

Table 1. Tenfold cross validation performance (weighted) for six single models and three combination models. *g-means metric* is used to compare classification performance of models with unbalanced data sets.

SVM model	Good rejected	Bad accepted	Bad rejected	Good accepted	Total	g-means metric
Linear	1195	619	530	14814	17158	0.653
Sigmoid	0	1142	7	16009	17158	0.077
Polynomial (2nd degree)	96	1035	114	15913	17158	0.314
Polynomial (3rd degree)	1242	633	516	14767	17158	0.643
RBF	860	651	498	15149	17158	0.640
Coulomb	0	1124	25	16009	17158	0.148
M1	287	850	299	15722	17158	0.505
M2 *	191	331	256	12425	13203	0.655
M3 **	3548	331	818	12425	17158	0.743

* 3955 credit clients could not be classified.
** Default class *bad* for 3955 credit clients.

and by $w = \frac{1149}{323}$ for *bad* credit clients to get an estimation of the "true" population performance for all models. For **unbalanced** data sets, where one class dominates the other, the total error (= sum of *Good rejected* and *Bad accepted* divided through the *total* number of credit clients) is not an appropriate measure. The **g-means metric** has been favored by several researchers (Akbani et al. (2004), Kubat and Matwin (1997)) instead. If the accuracy for good and bad cases is measured separately as $a^+ =$ good accepted / (good accepted + good rejected) and $a^- =$ bad rejected / (bad rejected + bad accepted), respectively, then the geometric mean for both accuracy measures is $g = \sqrt{a^+ \cdot a^-}$. The g-means metric tends to favor a *balanced accuracy* in both classes. It can be seen from table 1, that in terms of g-means metric SVM with *linear* ($g = 0.653$), *polynomial third degree* ($g = 0.643$) and *RBF* kernel ($g = 0.640$) dominate the single models. Additionally, three multiple models are suggested: *M1* ($g = 0.505$) combines the real output values of the six single models and takes sign $\left(\sum_{k=1}^{6} f_k(x)\right)$ for class prediction. *M2* ($g = 0.655$) only uses output values with $|f_k(x)| \geq 1$ for combination, leaving 3955 credit clients with $\forall k : |f_k(x)| < 1$ unclassified (see also critical/critical region V from figure 1). It is strongly suggested to use refined classification functions (e.g. with more detailed credit client information) for these cases. Alternatively, as a very simple strategy, one may introduce *M3* (default class combination) instead. Default class *bad* (rejecting all 3955 unclassified credit clients) leads to $g = 0.743$ which is highest for all models.

4 Combination of SVM on random input subsets

Previous work on input subset selection from our credit client data (Schebesch and Stecking (2007)) suggests that SVM models using around half or even less of the full input set can lead to good performance in terms of credit client misclassification error. This is especially the case when the inputs chosen for the final model are determined by the rank their contribution to above average base models from a population of several hundred models using random input subsets. Now we proceed to combine the outputs of such reduced input base SVM models. Input subsets are sampled randomly from $m = 40$ inputs, leading to subpopulations with base models using $r = 5, 6, ..., 35$ input variables respectively. For each r we draw 60 independent input subsets from the original m inputs, resulting in a population of $31 \times 60 = 1860$ base models. These differently informed (or weak) base models are trained and validated by highly automated model building with a minimum of SVM parameter variation and with timeout, hence we expect them to be sub-optimal in general. The real valued SVM base model outputs (see also previous sections) $f_{(rj)}(x) \in \mathbb{R}^N$ are now indexed such that (rj) is the jth sample with r inputs. These outputs are used twofold:

- fusing them into simple (additive) combination rules, and
- using them as inputs to *supervised* combiners.

A supervised combiner can itself be a SVM or some other optimization method (Koltchinskii et al. (2004)), using *some* $\{f_{(rj)}\}$ as inputs and the original data labels $y \in \{-1, +1\}^N$ as reference outputs. Potential advantages of random input subset base model combination are expected to occur for *relatively sparse* data and to diminish for large N. As our $N = 658$ cases are very few in order to reliably detect any specific nonlinearity in $m = 40$ input dimensions, our data are clearly sparse. Using combiners on outputs of weak base models may easily outperform all the base models but it may also conceal some overtraining. In order to better understand this issue we evaluate combiners on two sets S_A and S_B of base model outputs. Set S_A contains the usual output vectors of *trained and validated* base models when passed through a prediction seep over x, i.e. $\{f_{(rj)}\}$ for all (rj). At first set S_B is initialized to S_A. Then it is *corrected by the validation sweep* which includes all misclassifications of a model occurring when passed through the leave-one-out error validation. If $f_{(rj)i}^{-i}$ is the output of model (rj) at case $i \in \{1, ..., N\}$ and if this base model is *effectively retrained* on data (x, y) but without case i, then S_A and S_B may differ at entry $(rj)i$. This especially includes the cases when $f_{(rj)i}^{-i} f_{(rj)i} < 0$ and therefore may also contain *new* misclassifications $f_{(rj)i}^{-i} y_i < 0$. Hence combiners on subsets of S_B should lead to a more conservative (stringent) prediction of expected out-of-sample performance. We first inquire into the effects of additive output combination on misclassification error using both sets S_A and S_B. The following simple rules (which have no tunable parameters) are used: (1) LINCUM, which adds the base model outputs as they occur during the sweep over input numbers r, i.e. finally combining all the 1860 base models, (2) I60, which adds the outputs of base models with the same number of inputs r respectively, producing 31 combiners, and (3) E3, adding the outputs of the three best (elite) base models from each sorted list of the l-1-o errors. This again leads to 31 combiners. The experiments shown in fig.2 (lhs plot) indicate that, when used on

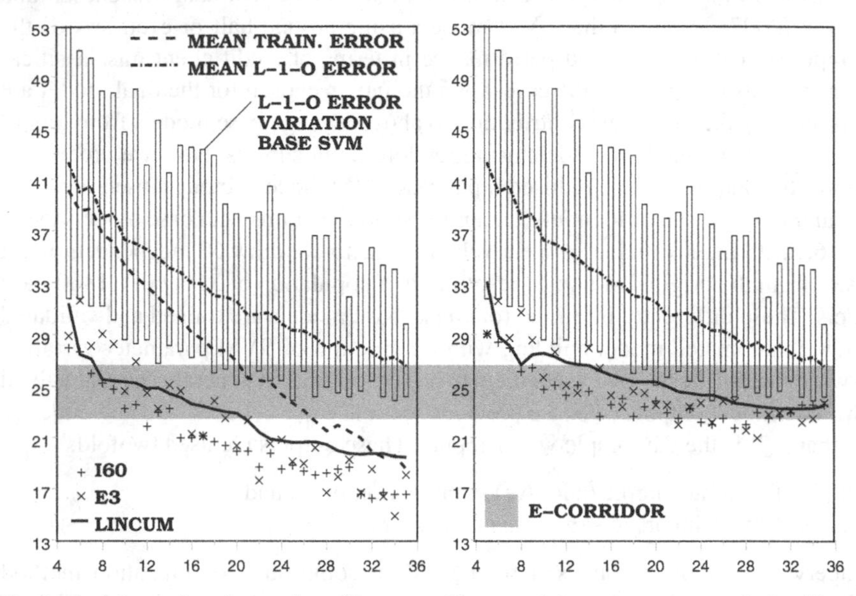

Fig. 2. Each inset plots the number of inputs r of base models against percent misclassification error of base models and of combiners. Lhs plot: training and validation errors of base models and errors of their simple combiners on set S_A. Rhs plot: validation errors of base models (from lhs plot, for comparison) and errors of simple combiners on Set S_B. For further explanation see the main text.

outputs from set S_A, these simple combiners consistently outperform the minimum l-1-o errors from the respective base model subpopulation, with I60 having the best results on outputs of base models which use small input subsets. However the level of l-1-o misclassification errors seems to be very low, especially when compared to the errors obtainable by extensively trained full input set models (i.e. around 23-24%) from previous work. Hence, in fig.2 (rhs plot) we report the l-1-o errors for the same simple combiners when using outputs from set S_B. Now the errors clearly shift upwards, with the bulk of combiners within the error corridor of extensively trained full input models (E-corridor, the shaded area within the plots). A benefit still exists as the errors of the combiners remain in general well below (only in some cases very near to) the minimum l-1-o-errors of the respective base models populations. With increasing r rule LINCUM relies on information increasingly similar to what a full input set model would see. Hence, it should (and for the set S_B it actually is) tending towards an error level within the E-corridor. Next, the outputs of the base models used by the combiners E3 and I60 are now used by 31 *supervised* combiners respectively (also SVM models with RBF-kernels, for convenience). These models denoted by SVM(E3) and SVM(I60) are then trained on subsets from S_A and subjected to l-1-o error validation (fig.3, lhs plot). Compared to the simple combiners they display still further partial improvement, but with (validated) errors very low for bigger input subsets ($r > 16$) that this now appears quite an improbable out-of-sample error range prediction for our credit scoring data. Training and validating SVM(E3) and SVM(I60) on subsets from S_B instead (fig.3, rhs plot) remedies the problem. Most of the l-1-o errors are shifted back into the E-corridor. In this case there is no advantage for SVM(E3) on S_B over simple E3 on S_B and also no improvement for SVM(I60) on S_B for $r > 16$. However, for small r SVM(I60) on S_B seems to predict the E-corridor. For somewhat bigger r this is in fact also the case for simple rule I60 on S_A and for SVM(I60) on S_A. Note that combination procedures with such characteristics can be

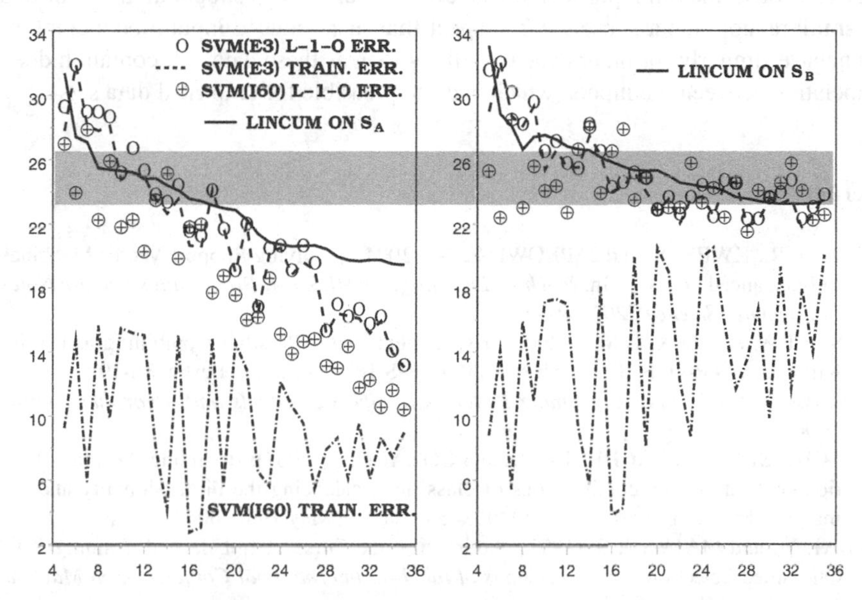

Fig. 3. Axes description same as in fig 2. Lhs plot: training and validation errors of supervized combiners on set S_A. Simple rule LINCUM from 2 (lhs) for comparison. Rhs plot: training and validation errors of supervized combiners on set S_B. Simple rule LINCUM from 2 (rhs) for comparison. of the validated base models. For further explanation see the main text.

very useful for problems with large feature dimensions which contains deeply hidden redundancy in the data, i.e. which cannot be uncovered by explicit variable selection (sometimes this is addressed by *manifold learning*). Censored outputs as described in the previous sections can be easily included. Also note that there is no tuning and no need for other complementary procedures like, for instance, combining based on input or output dissimilarities.

5 Conclusions and outlook

In this paper we examined several combination strategies for SVM. A simple addition of output values leads to medium performance. By including region information the classification accuracy (as measured by g-means metric) rises considerably, while for a number of credit clients no classification is possible. For unbalanced data sets we propose to introduce *default classes* to overcome this problem. Finally, simple combinations of outputs of *weak* base SVM classifiers on random input subsets yield misclassification errors comparable to extensively trained full input set models and they also seem to improve on them. Training and validating supervised combiners on the outputs of the base models seems to confirm this result. However combiners also tend to overtrain! The more appropriate way of using combiners is to use base model outputs corrected by the validation sweep. Many of these combiners are at least as good as the full input set models, even on base model subpopulations formed by **small** random input subsets. We suspect that such reduced input model combiners hehave similarly for other data as well, as long as these data still contain hidden association between the inputs (which is quite plausible for empirical data sets).

References

AKBANI, R., KWEK, S. and JAPKOWICZ, N. (2004): Applying Support Vector Machines to Imbalanced Datasets. In: *Machine Learning: ECML 2004, Proceedings Lecture Notes in Computer Science 3201*. 39-50.

DUIN, R.P.W. and TAX, D.M.J. (2000): Experiments with Classifier Combining Rules. In: Kittler, J. and Roli, F. (Eds.): MCS 2000, LNCS 1857. Springer, Berlin, 16-19.

KUNCHEVA, L.I. (2004): *Combining Pattern Classifiers: Methods and Algorithms*. Wiley 2004.

KOLTCHINSKII, V., PANCHENKO, D. and LOZANO, F. (2004): Bounding the generalization error of convex combinations of classifiers: balancing the dimensionality and the margins. From: `arXiv:math PR/0405345` posted on May 19th 2004.

KUBAT, M. and MATWIN, S. (1997): Addressing the Curse of Imbalanced Training Sets: One-Sided Selection. In: *Proceedings of the 14th International Conference on Machine Learning*. 179-186.

SCHEBESCH, K.B. and STECKING, R. (2005a): Support Vector Machines for Credit Scoring: Extension to Non Standard Cases. In: Baier, D. and Wernecke, K.-D. (Eds.): *Innovations in Classification, Data Science and Information Systems*. Springer, Berlin, 498-505.

SCHEBESCH, K.B. and STECKING, R. (2005b): Support vector machines for credit applicants: detecting typical and critical regions. *Journal of the Operational Research Society, 56(9), 1082-1088.*

SCHEBESCH, K.B. and STECKING, R. (2007): Selecting SVM Kernels and Input Variable Subsets in Credit Scoring Models. In: Decker, R. and Lenz, H.-J. (Eds.): *Advances in Data Analysis*. Springer, Berlin, 179-186.

STECKING, R. and SCHEBESCH, K.B. (2006): Comparing and Selecting SVM-Kernels for Credit Scoring. In: Spiliopoulou, M., Kruse, R., Borgelt, C., Nürnberger, A., Gaul, W. (Eds.): *From Data and Information Analysis to Knowledge Engineering*. Springer, Berlin, 542-549.

Part VIII

Business Intelligence

Comparison of Recommender System Algorithms Focusing on the New-item and User-bias Problem

Stefan Hauger[1], Karen H. L. Tso[2] and Lars Schmidt-Thieme[2]

[1] Department of Computer Science, University of Freiburg
Georges-Koehler-Allee 51, 79110 Freiburg, Germany
hauger@informatik.uni-freiburg.de

[2] Information Systems and Machine Learning Lab, University of Hildesheim
Samelsonplatz 1, 31141 Hildesheim, Germany
{tso,schmidt-thieme}@ismll.uni-hildesheim.de

Abstract. Recommender systems are used by an increasing number of e-commerce websites to help the customers to find suitable products from a large database. One of the most popular techniques for recommender systems is collaborative filtering. Several collaborative filtering algorithms claim to be able to solve i) the new-item problem, when a new item is introduced to the system and only a few or no ratings have been provided; and ii) the user-bias problem, when it is not possible to distinguish two items, which possess the same historical ratings from users, but different contents. However, for most algorithms, evaluations are not satisfying due to the lack of suitable evaluation metrics and protocols, thus, a fair comparison of the algorithms is not possible.

In this paper, we introduce new methods and metrics for evaluating the user-bias and new-item problem for collaborative filtering algorithms which consider attributes. In addition, we conduct empirical analysis and compare the results of existing collaborative filtering algorithms for these two problems by using several public movie datasets on a common setting.

1 Introduction

A Recommender system is a type of customization tool in e-commerce that generates personalized recommendations, which match with the taste of the users. Collaborative filtering (CF) (Sarwar et al. (2000, 2001)) is a popular technique used in recommender systems. It is used to predict the user interest for a given item based on user profiles. The concept of this technique is that the user, who received a recommendation for some sorts of items, would prefer the same items as other individuals with a similar mind set.

However, besides its simplicity, one of the shortcomings of CF are the new-item or cold-start problem. If no ratings are given for new items, it is difficult for standard CF algorithms to determine their own clusters by using rating similarity and thus they fail to give accurate predictions. Another problem is the user-bias from historical ratings (Kim and Li (2004)), which occurs when two items, based on historical ratings

Item-User-Matrix

	User 1	User 2	User 3	User 4
Item 1	5	3	4	3
Item 2	1	3	4	3
Item 3	4	3	4	3
Item 4		2	1	3
Item 5		2	1	3
Item 6	5	3	4	3

Item-Attribute-Matrix

	Att. 1	Att. 2
Item 1	1	0
Item 2	0	1
Item 3	1	0
Item 4	0	1
Item 5	1	0
Item 6	1	0

Fig. 1. User-Bias Example

have the same opportunity to be recommended to a user, but additional information shows that one item belongs to a group which is preferred by the user and the other not. For example, as shown in Figure 1, by applying CF, the probabilities that item 4 and 5 to be recommended for user 1 are equal. When the attributes are also taken into consideration, it can be observed that items 1, 3 and 6 which belong to attribute 1 are rated higher than user 1 than item 2 which belongs to attribute 2. Thus, user 1 has a preference for items related to attribute 1 over items related to attribute 2. Subsequently, by the CF algorithm, a higher probability should be assigned to item 5, which is more attached to attribute 1, than to item 4, which is related to attribute 2.

Recommender system algorithms that incorporate attributes claim to solve the user-bias and the new-item problem, however, no good evaluation techniques exist. For that reason, in this paper, we make the following contributions: (i) we introduce new methods and metrics for evaluating these problems and (ii) through a common experimental setting, we present evaluation results for three existing CF algorithms, which do not take attributes into account, namely user-based CF (Sarwar et al. (2000)), item-based CF (Sarwar et al. (2001)) and Gaussian aspect model by Hofmann (2004) as well as an approach, which takes attributes into account, by Kim & Li (2004). In the next section, we present the related work. In section 3, a brief description of the aspect model by Hofmann and the approach by Kim & Li will be presented. An introduction of the evaluation techniques for the new-item and the user-bias problem will follow in section 4. Section 5 consists of results on the empirical evaluations we have conducted and in section 6 we present the conclusions of the results and discuss possible future work.

2 Related works

Evaluating CF algorithms is not anything novel as there have already been relatively standard measures for evaluating the CF algorithms. Most of the evaluations done on CF focus on the overall performance of the CF algorithms (Breese et al. (1998), Sarwar et al. (2000), Herlocker et al. (2004)). However, as mentioned in the previous section, CF suffers from several shortcomings which are the new-item problem, also known as the cold-start problem, as well as the user-bias problem. It has been claimed that incorporating attributes could help to alleviate these drawbacks (Kim and Li (2004)). In fact, there exist many approaches for combining content information with CF (Burke (2002), Melville et al. (2002), Kim and Li (2004), Tso and Schmidt-Thieme (2005)). However, there has been lack of suitable evaluations which compute comparative analysis of attribute-aware and non attribute-aware CF algorithms, focusing on these two problems.

Schein et al. (2002) have already discussed methods and metrics for the new-item problem, in which they have introduced a performance metric called CROC curve. However, this metric is only suitable for the new-item problem. In this paper, we use standard performance metric, but introduce new protocols for evaluating the new-item and the user-bias problems. Hence, this evaluation setting allows users to compare the results with standard CF evaluation metrics, which does not restrict to evaluate only the new-item problem, but also on the user-bias problem. In addition, we compare the predicting accuracy of various collaborative filtering algorithms in this evaluation setting.

3 Observed approaches

In this section, we present a brief description of the two state-of-the-art CF models: the aspect model by Hofmann (2004) and the approach by Kim & Li (2004).

Aspect model by Hofmann

Hofmann (2004) specified different versions of the aspect model regarding the collaborative filtering domain. In this paper, we focus on the Gaussian model, because it shows the best prediction accuracy for non-specific problems. He uses the aspect model to identify the hidden semantic relationship among item y and users u, by using a latent class variable z, which represents the user clusters associated with each observation pair of a user and an item. In the aspect model, the users and items are considered as independent from each other and every observation can be described by a quartet $< u,y,v,z >$, where v denotes the rating user u has given to item y. For every observation quartet, the probability is then computed as follows:

$$P(u,y,v,z) = P(v|y,z)\, P(z|u)\, P(u)$$

The focus of our evaluation in this paper is on the Gaussian pLSA model, in which $P(v|y,z)$ is represented by the Gaussian density function. In the gaussian pLSA model, every combination of z and an item y has a location parameter $\mu_{y,z}$ and a scale parameter $\sigma_{y,z}$. The probability of the rating, v is then:

$$P(v|y,z) = P(v;\mu_{y,z},\sigma_{y,z}) = \frac{1}{\sqrt{2\pi}\sigma_{y,z}} exp\left[-\frac{(v-\mu_{y,z})^2}{2\sigma_{y,z}^2}\right]$$

As z is unobserved, Hoffmann used the Expectation Maximization (EM) algorithm to learn the two model parameters: $P(v|y,z)$ and $P(z|u)$. The EM algorithm has two main steps. The first step is computation of the Expectation (E-Step), which is done by computing the variation distribution Q over the latent variable z. The second step is Maximization (M-Step), in which the model parameters are updated by using the Q distribution computed in the previous E-Step. These two steps are executed until it converges to a local optimal limit. The EM steps for the Gaussian pLSA model are:

E-Step:

$$Q(z;u,y,v,\theta) = \frac{P(z|u)P(v;\mu_{y,z},\sigma_{y,z})}{\sum_{z'} P(z'|u)P(v;\mu_{y,z},\sigma_{y,z})}$$

M-Step:

$$P(z|u) = \frac{\sum_{<u',y>:u'=u} Q(z;u',y,v,\theta)}{\sum_{z'} \sum_{<u',y>:u'=u} Q(z';u',y,v,\theta)}$$

The location and scale parameters would also have to be updated.

Analogously, the same model can be applied by representing the latent class variable z, not as the user communities but as item cluster.

Approach by Kim and Li

The approach by Kim & Li (2004) seeks to solve the problem of user-bias and the new-item with the help of item attributes. They have incorporated attributes of movies such as genre, actors, years, etc. to collaborative filtering. It is expected that when attributes are considered, it is possible to recommend a new item based on just the user's fondness of the attributes, even though no user has voted for the item.

Kim & Li have a rather similar model as the aspect model by Hoffmann, yet there are several differences. First, class z associates only with the item, but not with the users in contrast to the pLSA model by Hofmann. Note that, the latent class z in this approach is regarded as an item clusters, instead of the user communities. Furthermore, they have applied some heuristic techniques to compute the corresponding model parameters, which can be done in two steps. First, using attributes, they clustered the items in different cliques with a simple K-means clustering algorithm. After clustering the items, they computed the probability of every item, i.e. the value indicating how much the item belongs to every clique. Then, an item-clique matrix with all the probabilities is derived. In the second step, the original item-user matrix is extended with the item-clique matrix, thus the attribute-cliques are just used as normal users.

Class z is built with the help of the extended item-user matrix. Every class z consists of a number of items of high similarity. The quality of class z is responsible for the accuracy of the later prediction of the use vote. A K-Medoids clustering algorithm using the Pearson's Correlation is used to compute the classes. After clustering the items into class z, a new item for each class z is created using the arithmetic mean. This new item is then the representative vector of the class z.

With the help of these representative items and a group matrix, which stores the membership of every item of the item-user matrix, it is possible to compute the expected vote for a user. In calculating the prediction, it is assumed that class z satisfies the Gaussian distribution. Let V_y be the rating vector of item y, V_z the representative vector of item cluster z, $ED(\cdot)$ the Euclidean distance, $v_{u,y}$ the user u's vote on item

y and U_z the set of items, which are in the same item cluster z, then the membership degree $p(z|y)$ and the mean rating, $\mu_{u,z}$, of user u on class z can be calculated as follows:

$$p(z|y) = \frac{1/ED(V_y, V_z)}{\sum_{z'=1}^{k} 1/ED(V_y, V_{z'})} \qquad \mu_{u,z} = \frac{\sum_{y \in U_z} v_{u,y} p(z|y)}{\sum_{y \in U_z} p(z|y)}$$

4 Evaluation protocols

New-item problem

To evaluate the prediction accuracy, we use a protocol which deletes one vote randomly from every user in the dataset, the so-called, AllBut1 protocol (Breese et al. 1998). The new-item problem is evaluated by a protocol similar to the AllBut1 protocol. Likewise, this protocol also deletes existing votes and builds up the model, which is to be evaluated with the reduced dataset. The new items are created by deleting all votes for a randomly selected item. After this is done for the required number of items, one vote is deleted from each user as in the AllBut1 protocol. This protocol has the advantage that the results of the new items can be compared with the results for past-rated items. Mean Absolute Error (MAE) is used as metrics in our experiments.

User-bias problem

The user-bias problem occurs, when two items have the same rating, but one item belongs to a group of items, which have not been given a good vote by the user, whereas the other item belongs to a group, which was in contrast given a good vote by the user; then the item, which belongs to the good-rated group, should be recommended.

To find a pair of items for an user, all the items, which are rated by the user, are taken into consideration and grouped two times. Once in item groups with equal rating and the second time in items groups with equal attributes. The historical vote vectors of these pairs of items of the users are then compared, excluding the vote of the observed user. In the next step, we select all pairs of items, which are in the same group of equally rated items and different group of attributes. One pair, which is to be predicted, is randomly chosen and deleted from the dataset. This is then done for all users in the database.

For each of these 'user-biased' pairs, the vote prediction for these pairs are computed and compared with the four collaborative filtering algorithms we use in our experiments. MAE metric is used to evaluate the predicting accuracy.

5 Evaluation and experimental results

Two datasets are used for our experiments - the EachMovie, containing 2,558,871 votes from 61,132 users on 1,623 movies, and the MovieLens100k dataset, containing 100,000 ratings from 943 users on 1,682 movies. The datasets also contain genre information for every movie in binary presentation, which we used as attributes. The EachMovie dataset contains 10 different genres, MovieLens contains 18. We conduct for both datasets 10 samples, in which 10 trials were run. For each sample 1500 movies are selected, whereas a 1000 users in EachMovie and 600 users in MovieLens are selected. and 20 neighbours for MovieLens and EachMovie for both user- and the item-based CF. No normalization is used in the aspect model and z is set to 40 for both datasets. In the Kim & Li approach, we used 20 attribute-groups and 40 item clusters for both datasets. We have selected the above parameter settings, because they were reported as the parameters which have given the best results in former experiments by the corresponding authors.

At first, we compared four observed approaches, namely the user-based CF, item-based CF, aspect model and Kim & Li approach, using the AllBut1 protocol. In Figures 2 and 3, the aspect model performs the best, the approach by Kim & Li is only slightly worse, while the user- and item-based CF algorithms perform the worst.

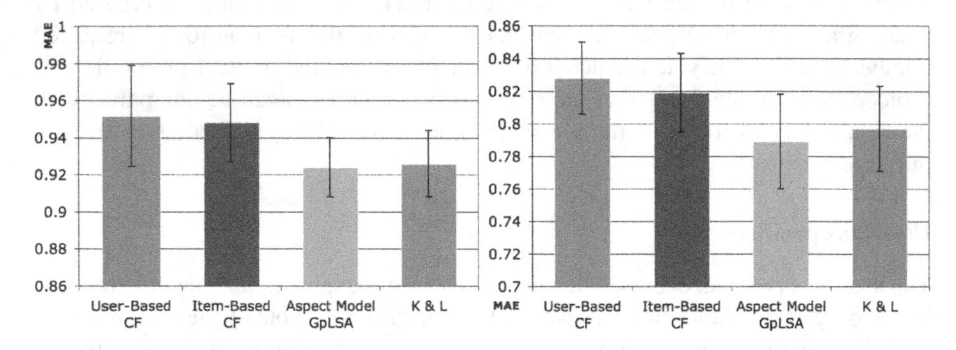

Fig. 2. AllBut1 using EachMovie.

Fig. 3. AllBut1 using MovieLens.

New-item problem

The results of the new-item problem are presented in Figure 4 and 5. Comparing the performance achieved by the algorithms, which use no attributes and the Kim & Li approach, we can see that the performance of the Kim & Li approach is only negligibly affected when more new items are added, while the predicting accuracy of the other approaches becomes much worse. This phenomenon is in line with our expectations, because it is not possible for algorithms, which do not take the attributes into account, to find any relations between new items and already rated items. As for the Kim & Li approach, there is no difficulty to assign an unrated item to an item cluster, because it includes the attributes. The average standard deviation is about 0.03.

User-bias problem

In the experiments of the user-bias problem, the number of items for prediction is between 60 to 70% of the total number of items, which is a representative amount.

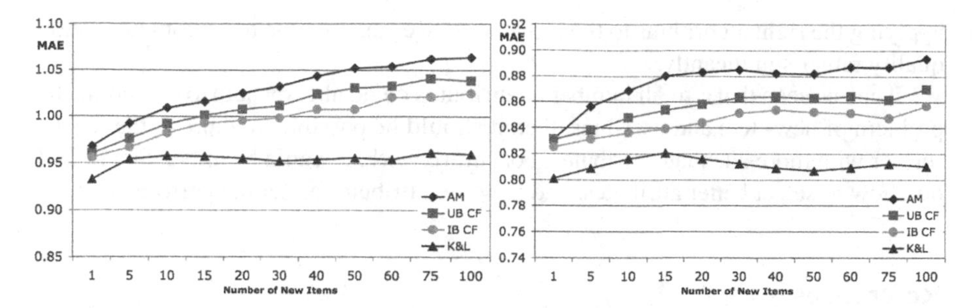

Fig. 4. New-Item using EachMovie.

Fig. 5. New-Item using MovieLens.

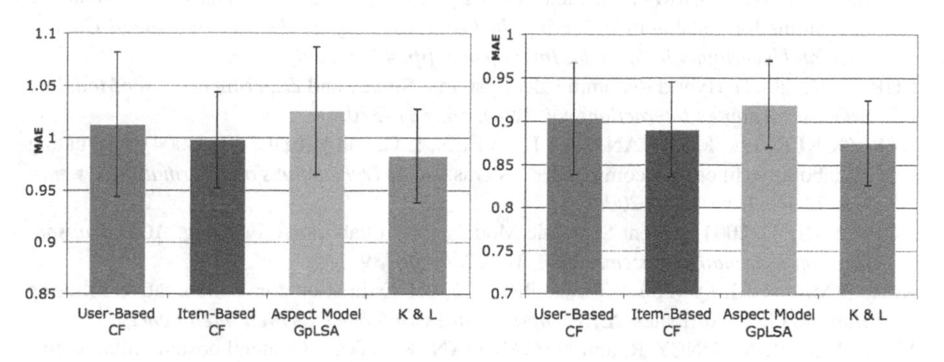

Fig. 6. User-Bias using EachMovie.

Fig. 7. User-Bias using MovieLens.

Besides, as shown in Figures 6 and 7, our expectations are confirmed. Only the approach by Kim & Li can mine the difference between two items with the same historical rating, but belong to different attributes; while the other approaches do not have any possibility to find the type of items the user likes because they do not take attributes into consideration. It is interesting to see that the aspect model, which performs best in general, performs worst to the user- and item-based CF when special problems such as the user-bias and new-item problem are considered.

6 Conclusion

The aim of this paper is to show that the new-item problem and user-bias problem can be solved with the help of attributes. We have used three CF algorithms, which do not use any attributes, and one approach, which takes the attribute information into account to compute the recommendations in our evaluation. Our evaluations have shown that it is possible to solve the new-item problem and user-bias problem with the help of attributes. In general, the approach by Kim & Li can not surpass the aspect model, but it can solve specific problems of new-item and user-bias more effectively. Especially for the new-item problem, where in the reality it is not uncommon to have 30-50 new items being injected to the database. Hence, we can conclude that by applying the right algorithms to the right cases, we can improve the recommendation quality rather significantly.

It can be seen that a small number of attributes could already help to overcome the problem of new-item and user-bias, then it should be possible to improve the results further with more adequate attributes. For future work, it would be interesting to find out, how to select better attributes, and how the attributes affect the performance.

References

BREESE, J.S., HECKERMAN, D., and KADIE, C. (1998): Empirical analysis of predictive algorithms for collaborative filtering. *In Proceedings of the Fourteenth Annual Conference on Uncertainty in Artificial Intelligence, pp. 43–52, July 1998.*

BURKE, R. (2002): Hybrid Recommender Systems: Survey and Experiments. *User Modeling and User-Adapted Interaction. vol. 12(4), pp. 331–370.*

HERLOCKER, J.L., KONSTAN, J.A., TERVEEN, L.G. and RIEDL, J.T. (2004): Evaluating collaborative filtering recommender systems. *ACM Transactions on Information Systems, vol. 22, no. 1, pp. 5–53, 2004.*

HOFMANN, T. (2004): Latent Semantic Models for Collaborative Filtering. *ACM Transactions on Information Systems, 2004, Vol 22(1), pp. 89–115.*

KIM, B.M. and LI, Q. (2004): Probabilistic Model Estimation for Collaborative Filtering Based on Item Attributes. *IEEE International Conference on Web Intelligence.*

MELVILLE, P., MOONEY, R. and NAGARAJAN, R. (2002): Contentboosted collaborative filtering. *In Proceedings of Eighteenth National Conference on Artificial Intelligence (AAAI-2002), pp. 187–192.*

SARWAR, B.M., KARYPIS, G., KONSTAN, J.A. and RIEDL, J. (2000): Analysis of recommendation algorithms for e-commerce. *In Proceedings of the Second ACM Conference on Electronic Commerce (ECÕ00), 2000, pp. 285–295.*

SARWAR, B.M., KARYPIS, G., KONSTAN, J.A. and RIEDL, J. (2001): Itembased collaborative filtering recommendation algorithms. *In Proceedings of the 10th international conference on World Wide Web. New York, NY, USA: ACM Press, 2001, pp. 285–295.*

SCHEIN, A.I., POPESCUL, A., UNGAR, L.H. and PENNOCK, D.M. (2002):Methods and metrics for cold-start recommendations. *In Proceedings of the 25th annual international ACM SIGIR conference on Research and development in information retrieval. New York, NY, USA: ACM Press, 2002, pp. 253–260.*

TSO, K. and SCHMIDT-THIEME L. (2005): Attribute-aware Collaborative Filtering. *In Proceedings of 29th Annual Conference of the Gesellschaft für Klassifikation (GfKl) 2005, Magdeburg, Springer.*

Collaborative Tag Recommendations

Leandro Balby Marinho and Lars Schmidt-Thieme

Information Systems and Machine Learning Lab (ISMLL)
Samelsonplatz 1, University of Hildesheim, D-31141 Hildesheim, Germany
{marinho,schmidt-thieme}@ismll.uni-hildesheim.de

Abstract. With the increasing popularity of collaborative tagging systems, services that assist the user in the task of tagging, such as tag recommenders, are more and more required. Being the scenario similar to traditional recommender systems where nearest neighbor algorithms, better known as collaborative filtering, were extensively and successfully applied, the application of the same methods to the problem of tag recommendation seems to be a natural way to follow. However, it is necessary to take into consideration some particularities of these systems, such as the absence of ratings and the fact that two entity types in a rating scale correspond to three top level entity types, i.e., user, resources and tags. In this paper we cast the tag recommendation problem into a collaborative filtering perspective and starting from a view on the plain recommendation task without attributes, we make a ground evaluation comparing different tag recommender algorithms on real data.

1 Introduction

The process of building the Semantic Web (Berners-Lee et al. 2001) is currently an area of high activity. Both the theory and technology to support it have been already defined and now one must fill this structure with life. In spite of the sounding simplicity, this task actually represents the biggest challenge towards its realization, i.e., adding semantic annotation to Web documents and resources in order to provide knowledge access instead of unstructured material. Annotation represents an extra effort which certainly will not be voluntarily done without good reasons. In this sense, it is necessary to incentive and educate the user into this practice, e.g., showing the benefits that can be achieved through it and alleviating the extra burden with the recommendation of relevant annotations. With the recent appearing and increasing popularity of the so called collaborative tagging systems this is finally possible (Golber et al. (2005)).

Recommending tags can serve various purposes, such as: increasing the chances of getting a resource annotated (or tagged) and reminding a user what a resource is about. Furthermore, lazy annotating users would not need to come up with a tag themselves but just select the ones readily available in the recommendation list according to what they think is more suitable for the given resource.

Tag recommender systems recommend relevant tags for an untagged user resource. Relevant here can assume different perspectives, for example, a tag can be judged relevant to a given resource according to the society point of view, through the opinion of experts in the domain or even based on the personal profile of an individual user. The question would be, which concept of relevance would the user prefer the most when using tag recommender services. This paper attempts to address this question through the following contributions: (i) formulation of the tag recommendation problem and the introduction of a collaborative filtering-based tag recommender algorithm, (ii) presentation of a simple protocol for tag recommender evaluation (iii) and (iv) a ground and quantitative evaluation on real-life data comparing different tag recommender algorithms.

2 Related work

The literature regarding the specific problem of collaborative tag recommendation is still sparse. The majority of the recent research work about collaborative tagging systems and folksonomies is concerned in devising approaches to better structure the data for browsing and searching where the recommendation problem is sometimes only highlighted as a potential property to be further explored in future work (Mika (2005), Hotho et al. (2006), Brooks and Montanez (2006), Heymann and Garcia-Molinay (2006)). We briefly describe below the works specifically investigating the problem of collaborative tag recommendation.

Autotag (Mishne (2006)) is a tool that suggests tags for weblog posts using collaborative filtering methods. Given a new weblog post, posts which are similar to it are identified through traditional information retrieval similarity measures. Next, the tags assigned to these posts are aggregated creating a ranked list of likely tags. Despite the collaborative filtering scenario, there is no real personalization because the user is not taken directly into account. Furthermore, the evaluation is done in a semi-automatically fashion where the assumption of tag relevance for a given resource is defined to some extent by human experts.

Xu et al. (2006) introduce a collaborative tag suggestion algorithm based on a set of general criteria to identify high quality tags. Some of the considered criteria are: high coverage of multiple facets to ensure good recall, least effort to reduce the cost involved in browsing, and high popularity to ensure tag quality. A goodness measure for tags, derived from collective user authorities, is iteratively adjusted by a reward-penalty algorithm, which also incorporates other sources of tags, e.g., content-based auto-generated tags. There is no quantitative evaluation.

Benz et al. (Benz et al. (2006)) introduce a collaborative approach for bookmark classification based on a combination of nearest-neighbor-classifiers. Two separate kinds of recommendations are generated: Keyword recommendations on the one hand, i.e. which keywords to use for annotating a new bookmark, and a recommendation of a classification on the other hand. The keyword recommender can be regarded as a collaborative tag recommender but its just a component of the overall algorithm, and therefore there is no information about its effectiveness as a stand-alone tool.

The state-of-the-art tag recommenders in practice are services that provide the most-popular tags used by the society for a particular resource (Fig. 2). This is usually done by means of tag clouds where the most frequently used tags are depicted in a larger font or otherwise emphasized.

The approaches described above address important aspects of the problem, but there is still a lack regarding quantitative evaluation on basic tag recommender algorithms. Furthermore, there is no common or agreed protocol where the different algorithms should be compared.

3 Recommender Systems

Recommender systems (RS) recommend products to customers based on ratings or past customer behavior. In general, RS predict ratings of items or suggest a list of unknown items to the user. They usually take the users, items and the ratings of items into account. A recommender system can be briefly formulated as:

- A set of users U
- A set of items I
- A set $S \subseteq \mathbb{R}$ of possible ratings where $r : U \times I \rightarrow S$ is a partial function that associates ratings to user/item pairs. In datasets r typically is represented as a list of tuples $(u, i, r(u, i))$ with $u \in U, i \in I$ and r defined for the domain $dom_r \subseteq U \times I$
- Task: In recommender systems the recommendations are for a given user $u \in U$ a set $\tilde{I}(u) \subseteq I$ of items. Usually $\tilde{I}(u)$ is computed by first generating a ranking on the set of items according to some quality or relevance criterion, from which then the top n elements are selected (see Eq. 2 below).

In CF, for m users and n items, the user profiles are represented in a user-item matrix $\mathbf{X} \in \mathbb{R}^{m \times n}$. The matrix can be decomposed into row vectors:

$$\mathbf{X} := [\mathbf{x}_1, ..., \mathbf{x}_m]^\top \text{ with } \mathbf{x}_u := [x_{u,1}, ..., x_{u,n}]^\top, \text{ for } u := 1, \ldots, m,$$

where $x_{u,i}$ indicates that user u rated item i by $x_{u,i} \in \mathbb{R}$. Each row vector $\mathbf{x}_u$ corresponds thus to a user profile representing the item ratings of a particular user. This decomposition leads to user-based CF.

The matrix can alternatively be represented by its column vectors:

$$\mathbf{X} := [\mathbf{x}_1, ..., \mathbf{x}_m] \text{ with } \mathbf{x}_i := [x_{i,1}, ..., x_{i,m}]^\top, \text{ for } i := 1, \ldots, n,$$

where each column vector $\mathbf{x}_i$ corresponds to a specific item's ratings by all m users. This representation leads to item-based recommendation algorithms.

The pairwise similarities between users is usually computed by means of vector similarity:

$$sim(\text{prof}_u, \text{prof}_v) := \frac{\langle \text{prof}_u, \text{prof}_v \rangle}{\| \text{prof}_u \| \| \text{prof}_v \|} \quad (1)$$

where $u, v \in U$ are two users and $prof_u$ and $prof_v$ are their profile vectors.

Let $B \subseteq I$ be the basket of items of the active user $u \subseteq U$ and N_u his/her best-neighbors. The topN recommendations usually consists of a list of items ranked by decreasing frequency of occurrence in the ratings of the neighbors:

$$\tilde{I}(u) := \underset{i \in I}{\operatorname{arg\,max}^n} |\{v \in N_u \mid i \in r_{v,i}\}| \tag{2}$$

where $B \cap \tilde{I}(u) := \emptyset$ and n is the size of the recommendation list.

The brief discussion above refers only to the user-based CF case, since it is the focus of our work. Moreover, we consider only the recommendation task since in collaborative tagging systems there are no ratings and therefore no prediction. For a detailed description about the item-based CF algorithm see Deshpande et al. (2004).

4 Tag Recommender Systems

Tag recommender systems recommend relevant tags for a given resource. As already discussed in section 1, the notion of relevance here can assume different perspectives and is usually hard to judge what concept of relevance would be preferable to a particular user. Collaborative tagging systems usually allow the users to see the most popular tags used for a given resource. This can be thought of a social-based tag recommender service since it represents the society opinion as a whole. Through CF we can measure the extent to which personalized notions of tag relevance are preferable in comparison with the socialized ones.

Collaborative tagging systems are usually composed of users, resources and tags and allow users to assign tags to resources. What is considered a resource depends on the type of the system, e.g. URLs (del.icio.us[1]), pictures (Flickr[2]), music(Last.fm[3]), etc. A tag recommender system can be formulated as follows:

- A set of users U
- A set of resources R
- A set of tags T
- A function $s : U \times R \rightarrow \tilde{T}$ associating tags to user/resources pairs, where $\tilde{T} \subseteq T$ and s is defined for the domain $dom_s \subseteq U \times R$
- Task: In tag recommender systems the recommendations are for a given user $u \in U$ and a resource $r \in R$ a set $\tilde{T}(u,r) \subseteq T$ of tags. As well as in the traditional formulation (section 3), $\tilde{T}(u,r)$ can also be computed by first generating a ranking on the set of tags according to some quality or relevance criterion, from which then the top n elements are selected (see Algo.1 below).

When comparing the formulation above with the one in section 3, we observe that CF cannot be applied directly. This is due to the additional dimension represented by

---

[1] http://del.icio.us
[2] http://www.flickr.com
[3] http://www.last.fm

T. Either we use more complex methods do deal directly with it or reduce it to a lower dimensional space where we could apply CF. We follow the latter one.

To this end we take all the two dimensional projections of the original matrix preserving the user information. Letting $K := |U|$, $M := |I|$ and $L := |T|$, the projections result in two user profile matrices: a user-resource $K \times M$ matrix $\mathbf{X}$ and a user-tag $K \times L$ matrix $\mathbf{Y}$. In collaborative tagging systems there is usually no rating information. The only information available is whether or not a resource and/or a tag occurred with the user. This can be encoded in the binary matrices $\mathbf{X} \in \{0,1\}^{k \times m}$ and $\mathbf{Y} \in \{0,1\}^{k \times l}$ indicating occurrence, e.g. $x_{k,m} = 1$ and $y_{k,l} = 1$, or non-occurrence of resources and tags with the users. Now we have the required setup to apply collaborative filtering.

The algorithm starts selecting the users who have tagged the resource in question. Next, the pairwise similarity computation is performed (Eq.1). Notice that now we have two possible setups in which the neighborhood can be formed, either based on the profile matrix $\mathbf{X}$ or $\mathbf{Y}$. The neighborhood's tags for the resource in question are aggregated and weighted based on the neighbors' similarities with the active user. Next the weights of each particular tag are summed up and the recommendation list is ranked by decreasing value of the summed weights. Ties are broken by smaller index. The overall CF procedure for tag recommendations is summarized in Algo.1.

Algorithm 1 CF for tag recommendations

- Given a new and/or untagged resource $r \in R$ for the active user $u \in U$
- Let A $:= \{v \subseteq U \mid s_{v,r} \not\equiv \emptyset\}$ denote the set of users who have tagged r where s is a function associating tags to user/resources pairs
  - Find k best neighbors:

$$N_u := \operatorname*{arg\,max}^{k}_{v \in A} sim(\text{prof}_u, \text{prof}_v)$$

  - Output the top n tags:

$$\tilde{T}(u,r) := \operatorname*{arg\,max}^{n}_{t \in T} \sum_{v \in N_u} sim(\text{prof}_u, \text{prof}_v)\delta(v,r,t)$$

where $\delta(v,r,t) := 1$ if $(v,r,t) \in U \times R \times T$ and 0 else.

5 Experimental setup and results

For our experiments we used the data made available by the Audioscrobbler[4] system, a music engine based on a collection of music profiles. These profiles are built through the use of the company's flagship product, Last.fm, a system that provides personalized radio stations for its users and updates their profiles using the music they listen to and also makes personalized artist recommendations. In addition, Audioscrobbler exposes large portions of data through their web services API.

[4] http://www.audioscrobbler.net

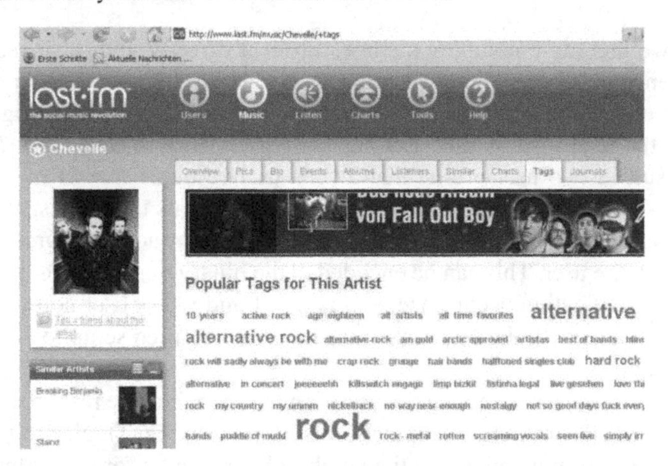

Fig. 1. Most popular tags for a given artist

Here we considered only the resources with 10 or more tag assignments. This gave us 2.917 users, 1.853 artists (playing the role of resources), 2.045 tags and 219.702 instances ((user, resource, tag) triples).

We evaluated four tag recommenders: (i) a *most global frequent tags*, which recommend the most used tags in the sample dataset, (ii) a *most popular tag by resource*, which recommends the most used tags for a particular resource (in our case an artist), (iii) a *user-resource-based CF*, which computes the neighborhood based on the user-resource matrix and (iv) a *user-tag-based CF*, which computes the neighborhood based on the user-tag matrix. Notice that (ii) represents the state-of-the-art recommender used in practice (Fig.1).

To evaluate the recommenders we used a variant of the leave-one-out holdout estimation that we named leave-tags-out. The idea is to choose a resource at random for each user in the test set and hide the tags attached to it. The algorithm must try to predict the hidden tags. To count the hits made by the algorithms we used the usual recall measure,

$$recall^{macro}(\mathcal{D}) := \frac{1}{\mid \mathcal{D} \mid} \sum_{1=1}^{|\mathcal{D}|} \frac{\mid Y_i \cap Z_i \mid}{\mid Y_i \mid} \tag{3}$$

where $\mathcal{D}$ is the test set, Y_i the true tags and Z_i the predicted ones. Since the precision is forced by taking into account only a restricted number n of recommendations there is no need to evaluate precision or F1 measures, i.e., for this kind of scenario precision is just the same as recall up to a multiplicative constant. Each algorithm was evaluated 10 times for n=10 (size of recommendation list) and the results averaged (Fig. 2).

Looking at the Figure 2 we see that the *most popular by resource* recommender reached a surprisingly high recall and that the *user-resource-based CF* did not perform significantly better than that. The good results of the *most popular by resource* algorithm can in part be explained by the fact that this service is already available by

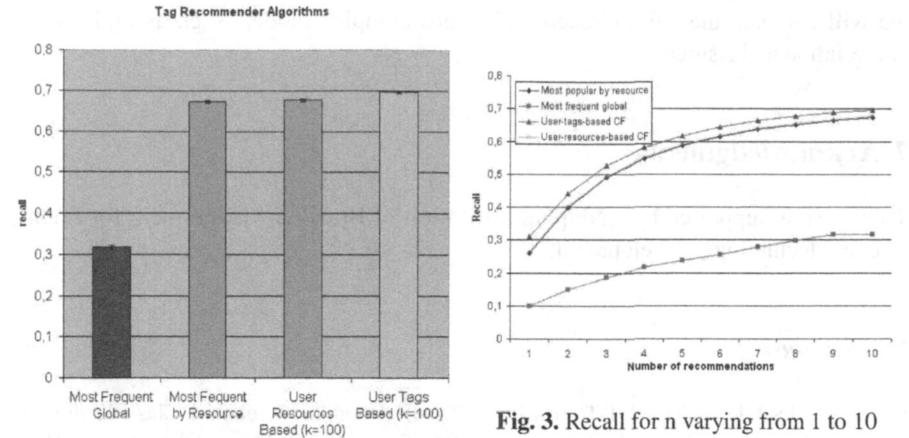

Fig. 2. Recall of tag recommenders for n=10

Fig. 3. Recall for n varying from 1 to 10

the system. Besides that, it shows the strong influence of the society's vocabulary on the user's personal opinion. In the other hand, the *user-tag-based CF* recommender performed at least 2% better[5] than both the *most-popular tag by resource* and *user-resource-based CF*. Also notice that the improvement is consistent for different values of n (Fig. 3). The best k-neighbors values were estimated through successive runnings where k was incremented until a point where no more improvements in the results were observed.

6 Conclusions

In this paper we applied CF to the tag recommendation problem and made a quantitative evaluation of its performance in comparison with other simpler tag recommenders. Furthermore, we used a simple and suitable protocol with which further approaches can be compared.

Despite the already good results of the baseline algorithms, the straightforward CF based on the user-tag profile matrix showed a significant improvement. This shows that users with similar tag vocabulary tend to tag alike, which indicates a preference for personalized tag recommendation services.

It is also notorious the reasonable good results achieved by the *most global frequent tags* recommender, which indicates its adequacy for cold-start related problems, where just a few tags are available in the system.

In future work we plan to reproduce the same experiments with different datasets from different domains to confirm the results here presented. We also want to refine the CF algorithms exploring different combinations between the user similarities obtained from the two profile matrices, i.e., user-resources and user-tags. Moreover,

[5] T-test for a significance level of 0.05.

we will compare the CF approach with more complex models such as multi-label and relational classifiers.

7 Acknowledgments

This work is supported by CNPq, an institution of Brazilian Government for scientific and technologic development.

References

BENZ, D., TSO, K., SCHMIDT-THIEME, L. (2006): Automatic Bookmark Classification: A Collaborative Approach. In: Proceedings of the Second Workshop on Innovations in Web Infrastructure (IWI 2006), Edinburgh, Scotland.

BERNERS-LEE, T., HENDLER, J. and LASSILA, O. (2001): "Semantic Web", Scientific American, May 2001.

BROOKS, C. H., MONTANEZ, N. (2006): Improved annotation of the blogosphere via autotagging and hierarchical clustering. New York, NY, USA : ACM Press, WWW '06: Proceedings of the 15th international conference on World Wide Web : 625Ů632.

DESHPANDE, M. and KARYPIS, G. (2004): Item-based top-n recommendation algorithms. ACM Transactions on Information Systems, 22(1):1-34.

GOLBER, S., HUBERMAN, B.A. (2005): "The Structure of Collaborative Tagging System", Information Dynamics Lab: HP Labs, Palo Alto, USA, available at: http://arxiv.org/abs/cs.DL/0508082

HEYMANN, P. and GARCIA-MOLINAY, H. (2006): Collaborative Creation of Communal Hierarchical Taxonomies in Social Tagging Systems. Technical Report InfoLab 2006-10, Department of Computer Science, Stanford University, Stanford, CA, USA, April 2006.

HOTHO, A., JAESCHKE, R., SCHMITZ, C., STUMME, G. (2006): Information Retrieval in Folksonomies: Search and Ranking. Heidelberg : Springer , The Semantic Web: Research and Applications 4011 : 411-426.

MIKA, P. (2005): Ontologies Are Us: A Unified Model of Social Networks and Semantics. In: Y. Gil, E. Motta, V. R. Benjamins and M. A. Musen (Eds.), ISWC 2005, vol. 3729 of LNCS, pp. 522Ű536. Springer-Verlag, Berlin Heidelberg.

MISHNE, G. (2006): AutoTag: a collaborative approach to automated tag assignment for weblog posts. New York, NY, USA : ACM Press , WWW '06: Proceedings of the 15th international conference on World Wide Web : 953Ů954.

SARWAR, B., KARYPIS, G., KONSTAN, J. and REIDL, J. (2001): Item-based collaborative filtering recommendation algorithms. In Proceedings of the 10th international conference on World Wide Web. New York, NY, USA: ACM Press, pp. 285-295.

XU, Z., FU, Y., MAO J., SU, D. (2006): Towards the Semantic Web: Collaborative Tag Suggestions. Edinburgh, Scotland: Proceedings of the Collaborative Web Tagging Workshop at the WWW 2006.

Applying Small Sample Test Statistics for Behavior-based Recommendations

Andreas W. Neumann and Andreas Geyer-Schulz

Institute of Information Systems and Management, Universität Karlsruhe (TH),
76128 Karlsruhe, Germany
{a.neumann, geyer-schulz}@iism.uni-karlsruhe.de

Abstract. This contribution reports on the development of small sample test statistics for identifiying recommendations in market baskets. The main application is to lessen the cold start problem of behavior-based recommender systems by faster generating quality recommendations out of the first small samples of user behavior. The derived methods are applied in the area of library networks but are generally applicable in any consumer store setting. Analysis of market basket size at different organisational levels of German research library networks reveals that at the highest network level market basket size is considerably smaller than at the university level. The overall data volume is considerably higher. These facts motivate the development of small sample tests for the identification of non-random sample patterns. As in repeat-purchase theory the independent stochastic processes are modelled. The small sample tests are based on modelling the choice-acts of a decision maker completely without preferences by a multinomial model and combinatorial enumeration over a series of increasing event spaces. A closed form of the counting process is derived.

1 Introduction

Recommender systems are lately becoming standard features at online stores. As shown by the revealed preference theory of Paul A. Samuelson (1948) (1938a) (1938b) customer purchase data reveals the preference structure of decision makers. It is the best indicator of interest in a specific product and outperforms surveys with respect to reliability significantly. A behavior-based recommender system reads observed user behavior (e. g. purchases) as input, then aggregates and directs the resulting recommendations to appropriate recipients. One of the main mechanism design problems of behavior-based recommender systems is the cold start problem. A certain amount of usage data has to be observed before the first recommendations can be computed. Starting with recommendations drawn from almost similar applications in general is a bad idea since it can not be guaranteed that the usage patterns of customers in these applications are identical. Behavior-based recommendations are best suited to the user group whose usage data is used to generate the very same recommendations. Thus, to lessen the cold start problem small sample test statistics are needed to faster generate quality recommendations out of the first small samples of user behavior. The main problem is to determine which co-purchases occur randomly and which show a relationship between two products. In this contribution we apply the derived methods to usage data from scientific libraries. The methods and algorithms are generally applicable in any consumer store setting. For an overview on recommender systems e. g. see Adomavicius and Tuzhilin (2005).

2 The ideal decision maker: The decision maker without preferences

Modelling the preference structure of decision makers in a classical way leads to causal models which explain the choice of the decision maker, allow prediction of future behavior and to infer actions of the seller to influence/change the choice of the decision maker (e. g. see Kotler (1980)). In the library setting causal modelling of the preference structure of decision-makers would require the identification (and estimation) of such a model which explains the choice of a decision maker or of a homogeneous group of decision makers (a customer segment) for each of the more than 10 000 000 books (objects) in a library meta catalog. Solving the model identification problem requires selecting the subset of relevant variables out of $2^{10000000}$ subsets in the worst case in an optimal way. While a lot of research has investigated automatic model selection, e. g. by Theil's R^2 or Akaike's information criterion (AIC) (for further references see Maddala (2001) pp. 479–488), the problem is still unsolved.

The idea to ignore interdependencies between system elements for large systems has been successfully applied in the derivation of several laws in physics. The first example is the derivation of Boltzmann's famous H-theorem where the quantity H which he defined in terms of the molecular velocity distribution function behaves exactly like the thermodynamic entropy (see Prigogine (1962)). In the following, we ignore the interdependencies between model variables completely. For this purpose, we construct an ideal decision maker without preferences. Such an ideal decision maker can be regarded as a prototype of a group of homogeneous decision makers without preferences against which groups of decision makers with preferences can be tested. For a group of ideal decision makers, this is obvious, for a group of decision makers with preferences the principle of self-selection (Spence (1974), Rothschild and Stiglitz (1976)) grants homogeneity. The ideal decision maker draws k objects (each object represents a co-purchase (a pair of books)) out of an urn with n objects with replacement at random and – for simplicity – with equal probability. The number of possible co-purchases – and thus the event space – is unknown.

In marketing several conceptual models which describe a sequence of sets (e. g. total set $\supseteq$ awareness set $\supseteq$ consideration set $\supseteq$ choice set, Kotler (1980) p. 153) have been developed to describe this situation (Narayana and Markin (1975), Spiggle and Sewall (1987)). Narayana and Markin have investigated the size of the awareness set for several branded products empirically. E. g., they report a range from 3–11 products with an average of 6.5 in the awareness set for toothpaste and similar results for other product categories. This allows the conjecture that the event space size is larger than k and in the worst case bounded by k-times the maximal size of the awareness set.

A survey of the statistical problems (e. g. violation of the independence of irrelevant alternatives assumption, biases in estimating choice models etc.) related to this situation can be found in Andrews and Srinivasan (1995) or Andrews and Manrai (1998). Recent advances in neuroimaging even allowed experimental proof of the influence of branding on brain activity in a choice situation which leads to models which postulate interactions between reasoning and emotional chains (e. g. Deppe et al. (2005), Bechara et al. (1997)). As result of the sampling process of an ideal decision maker we observe a histogram with at most k objects with the drawing frequencies summing to k. For each event space in k to n, the distribution of the drawing frequencies is a partition of k, the set of all possible distributions is given by enumerating all possible partitions of k for this event space. The probability of observing a specific partition in a specific event space is the sum of the probabilities of all sample paths of length k leading to this partition. The probability distribution of partitions drawn by an ideal decision maker in a specific event space $n > k$ serves as the base of the small sample test statistic in section 6. For the theory of partitions see Andrews (1976).

3 Library meta catalogs: An exemplary application area

For evaluation purposes we apply our techniques in the area of meta catalogs of scientific libraries. Due to transaction costs the detailed inspection of documents in the online public access catalog (OPAC) of a library can be put on a par with a purchase incidence in a consumer store setting. A market basket consists of all documents that have been co-inspected by one user within one session. To answer the question, which co-inspections occur non-randomly, for larger samples we apply an algorithm based on calculating inspection frequency distribution functions following a logarithmic series distribution (LSD) (Geyer-Schulz et al. (2003a)). Such a recommender system is operational at the OPAC of the university library of Karlsruhe (UBKA) since June 2002 (Geyer-Schulz et al. (2003b)) and within Karlsruhe's Virtual Catalog (KVK), a meta catalog searching 52 international catalogs, since March 2006. These systems are fully operational services accessible by the general public, for further information on how to use these see Participate! at http://reckvk.em.uni-karlsruhe.de/.

Table 1. Statistical Properties of the Data (Status of 2007-02-19)

	UBKA	KVK
Number of total documents in catalog	1,000,000	$>$ 10,000,000
Number of total co-inspected documents	527,363	255,248
Average market basket size	4.9	2.9
Av. aggregated co-inspections per document	117.4	5.4

Table 1 shows some characteristics of the UBKA and KVK usage data. Because of the smaller market basket size, the shorter observation period, and the much higher (unknown) number of total documents in the meta catalog KVK, the average aggregated co-inspections per document in the KVK is very small. Due to sample size constraints methods using statistical tests on distributions (like LSD) are only reliably applicable with many co-inspections. Special small sample statistics are needed to compute recommendations out of samples of few co-inspections. Our methods are based on the assumption that all documents in the catalog have the same probability of being co-inspected. In real systems generally this assumption does not hold, but especially when starting to observe new catalogs no information about the underlying distribution of the inspection processes of documents is known. Finally, recommendations are co-inspections that occur significantly more often then predicted in the case of the assumption being true.

4 Mathematical notation

For the mathematical formulation we use the following notation. The number of total documents $n+1$ in the catalog is finite but unknown (this leaves n documents as possible co-inspections for each document D in the catalog). Recommendations are computed separately for each document D. Each user session (market basket) contains all documents that the user inspected within that session, multiple inspections of the same document are counted as one. All user sessions are aggregated. The aggregated set $C(D)$ contains all documents, that at least one user has inspected together with D. The number of co-inspections with D of all elements of $C(D)$ is known, this histogram is called $H(D)$, it is the outcome of the multinomial experiment. When removing all documents with no inspections from $H(D)$ and then re-writing the number of co-inspections as a sum, it can be intepreted as an integer partition of k with the number of co-inspections of each co-inspected document as the summands. k is the number of non-aggregated co-inspections (multiple inspections in different sessions are counted separately). E. g. $4+1+1$ is an integer partition of $k=6$ and shows that the corresponding document D has been co-inspected in at least 4 (the highest number) different sessions with 3 (the number of summands) other documents, with the first document 4-times and with the second and third one time each.

5 POSICI: Probability Of Single Item Co-Inspections

The first method we introduce is based on the following question. What is the probability $p_j(n)$ that at least one other document has been co-inspected exactly j-times with document D? To answer the question we use the setup of the multinomial distribution directly. Let $(N_1,\ldots,N_n)$ be the vector of the number of times document i $(1 \leq i \leq n)$ was co-inspected with D. Then $(N_1,\ldots,N_n) \sim \mathcal{M}(k;q_1,\ldots,q_n)$, $q_i = \frac{1}{n}$, $1 \leq i \leq n$. Now define $A_i = \{N_i = j\}$. By applying the inclusion-exclusion principle we can now compute:

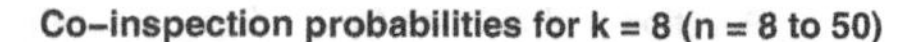

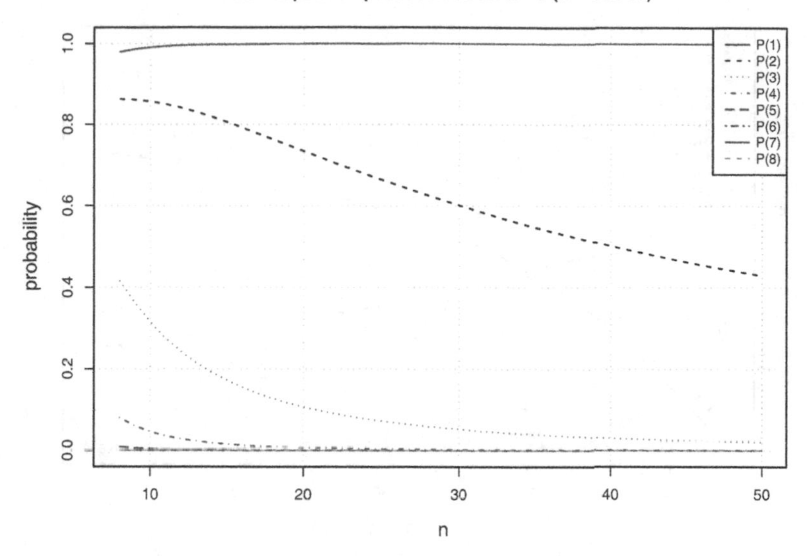

Fig. 1. Inspection probabilities $p_j(n)$ for $k = 8$ and growing n in POSICI.

$$p_j(n) = P\left(\bigcup_{i=1}^{n} A_i\right) = \sum_{\nu=1}^{n} (-1)^{\nu-1} \sum_{1 \le i_1 < \ldots < i_\nu \le n} P(A_{i_1} \cap \ldots \cap A_{i_\nu}) \tag{1}$$

Since many of the summands on the right hand side are known to be equal to zero, this equation can be implemented quite efficiently. Figure 1 shows $p_j(n)$ for $k = 8$ and growing n. In general $\lim_{n\to\infty} p_1(n) = 1$ and $\lim_{n\to\infty} p_j(n) = 0$ for $j = 2, 3, \ldots$ holds. Further on, $p_j(n)$ is decreasing in j for all n. Based on these probablities we define the POSICI Recommendation Generating Algorithm:

1. Let D be the document for which recommendations are calculated.
2. Let $n = k$ and t be a fixed chosen acceptance threshold ($0 < t < 1$).
3. Determine $j_0 = \min_{j=2,\ldots,k} \{ j \mid p_j(n) < t\, p_1(n) \}$.
4. Recommend all documents that have been co-inpected with D at least j_0-times.

Thus, e. g. in the setting of figure 1 and $t = 0.2$ all documents that have been co-inspected at least 4-times are being recommended. POSICI is built on the theory, that co-inspections other than j-times add more noise than information about the incentive to co-inspect the current document j-times.

6 POMICI: Probability Of Multiple Items Co-Inspections

The second method is derived from the question: What is the probability $p_{part}(n)$ that the partition corresponding to the complete histogram $H(D)$ of all co-inspections

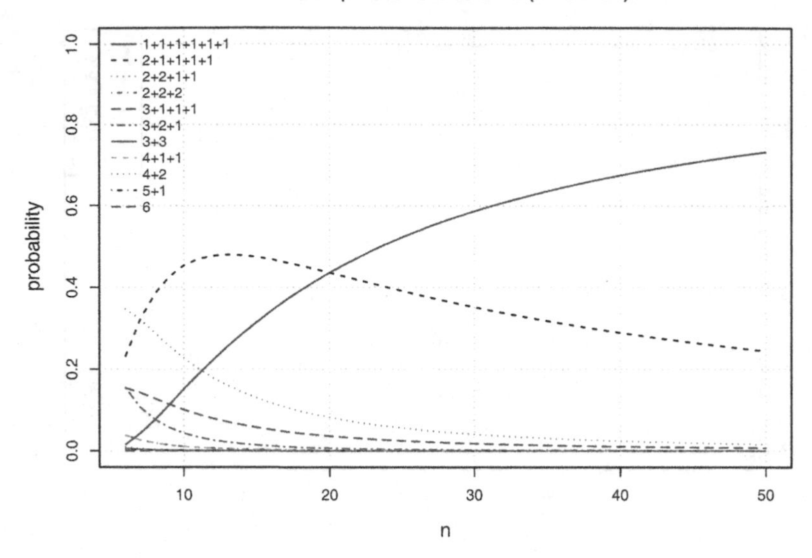

Fig. 2. Inspection probabilities $p_{part}(n)$ for $k = 6$ and growing n in POMICI.

with D occurs? To answer this question we re-formulate the problem in an algebraic setting. Let X be the set of words of length k from an alphabet of n letters, and l_i the number of letters (i. e. documents), that occur exactly i-times in $x \in X$ (i. e. in $H(D)$). First we examine the actions of the group $G = S_n \times S_k$ on the set X, and then the actions of the stabilizer subgroup G_x on the set S_n for the identitiy $id \in S_n$. By applying two times the orbit-stabilizer theorem together with Lagrange's theorem from group theory ($|G| = |Gx|\,|G_x| = |Gx|\,|G_x id|\,|G_{xid}|$) and then some counting arguments we come to the solution:

$$p_{part}(n) = \frac{|Gx|}{|X|} = \frac{|G|}{|X|\,|G_x id|\,|G_{xid}|} = \frac{n!\,k!}{n^k \left(n - \sum_{i=1}^{k} l_i\right)!\, \prod_{j=1}^{k} l_j!\,(j!)^{l_j}} \tag{2}$$

In general $\lim_{n\to\infty} p_{1+\dots+1}(n) = 1$ and $= 0$ for all other partitions holds. As can be seen exemplary in figure 2, only above a certain n the order by probability of the partitions is stable. We use the smallest of these n to construct the POMICI Recommendation Generating Algorithm:

1. Let D be the document for which recommendations are calculated.
2. Let t be a fixed chosen acceptance threshold ($0 < t < 1$).
3. Let n_D be the smallest integer, after which the order by probability of the partitions for $n \geq n_D$ is stable.
4. Let s be the largest integer that occurs in the partition with the highest probability below $t\, p_{1+\dots+1}(n_D)$.
5. For all partitions *part* with $p_{part}(n_D) < t\, p_{1+\dots+1}(n_D)$ do

a) Recommend all documents from $H(D)$ that have been co-inspected at least s-times.

Thus, e. g. in the setting of figure 2 and $t = 0.05$ all documents that have been observed within the partitions $3+2+1$, $3+3$, $4+1+1$, $4+2$, $5+1$ or 6 and have been co-inspected at least 3 times are being recommended ($n_D = 21$, $p_{1+\cdots+1}(21) = 0.4555$). Note, that this choice of n_D indicates a risk-averse decision maker. POMICI is built on the theory, that the distribution of co-inspections other than j-times reveals more information than noise about the incentive to co-inspect the current document j-times.

7 POSICI vs. POMICI

Since both methods are based on a homogeneous group of decision makers modeled by the underlying uniform multivariate distribution, a direct connection between them exists. The sum of the probabilities of all partitions from POMICI with at least one product that was co-inspected exactly j-times is equal to the probability in POSICI, that there exists at least one product, that was co-inspected exactly j-times. In other words, we get from POMICI to POSICI by aggregating all partitions that only differ in the noise area defined in the POSICI underlying preference theory. Thus, equation 2 can also be used instead of the inclusion-exclusion principle to calculate the probability in equation 1.

By setting the threshold t for the POSICI and POMICI algorithms respectively, the number of generated recommendations can be adjusted for both methods. As can be seen in figure 3, when the total number of recommendations is equal, POMICI generally generates longer recommendation lists for fewer documents than POSICI.

8 Conclusions and further research

POSICI and POMICI are based on different assumptions in the underlying preference theory. To determine which method leads to qualitatively better recommendations in a specific setting the following question has to be answered. When does the partition tail of smaller integers resembles noise and when incentive behavior? One way to answer the question lies in the human evaluation of larger data sets. This is planned for the library application.

Two ways to enhance the algorithms appear to be promising. First, if the overall inspection probability of documents is known (through large behavior data sets), both methods can be extended to be based on an underlying non-uniform multinomial distribution. This can not be applied in the case of a cold start but can be useful in the scenario of very small market baskets covering a large part of the total documents. Second, portraying the additions of further co-purchases ($k \rightarrow k+1$) as a Markov-process enables us to calculate the probability of a product with currently low co-inspections to develop into high co-inspections, thus a reliable recommendation.

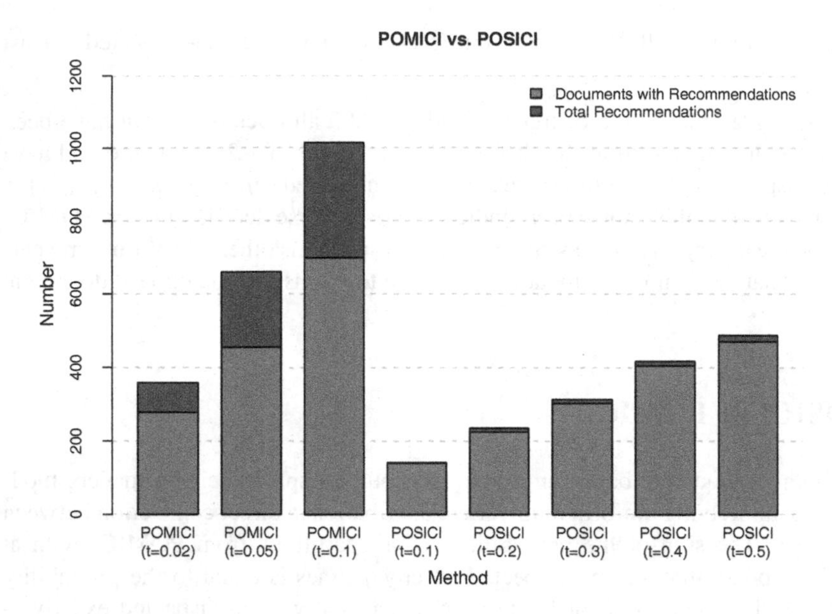

Fig. 3. Number of generated recommendations for all documents with $k \leq 15$ on the KVK data for various t.

Acknowledgement We gratefully acknowledge the funding of the project "Recommender Systems for Meta Library Catalogs" by the Deutsche Forschungsgemeinschaft.

References

ADOMAVICIUS, G. and TUZHILIN, A. (2005): Toward the Next Generation of Recommender Systems: A Survey of the State-of-the-Art and Possible Extensions. *IEEE Transactions on Knowledge and Data Engineering, 17(6), 734-749.*

ANDREWS, G.E. (1976): *The Theory of Partitions.* Addison-Wesley, Reading.

ANDREWS, R.L. and MANRAI, A.K. (1998): Simulation experiments in choice simplifikation: The effects of task and context on forecasting performance. *Journal of Marketing Research, 35(2), 198–209.*

ANDREWS, R.L. and SRINIVASAN, T.C. (1995): Studying consideration effects in empirical choice models using scanner panel data. *Journal of Marketing Research, 32(1), 30–41.*

BECHARA, A., DAMASIO, H., TRANEL, D., and DAMASIO, A.R. (1997): Deciding Advantageously Before Knowing the Advantageous Strategy. *Science, 257(28), 1293–1295.*

DEPPE, M., SCHWINDT, W., KUGEL, H., PLASSMANN, H., and KENNING, P. (2005): Nonlinear Response Within the Medial Prefrontal Cortex Reveal When Specific Implicit Information Influences Economic Decision Making. *Journal of Neuroimaging, 15(2), 171–182.*

GEYER-SCHULZ, A. and HAHSLER, M. and NEUMANN, A. and THEDE, A. (2003a): Behavior-Based Recommender Systems as Value-Added Services for Scientific Libraries. In: H. Bozdogan: *Statistical Data Mining & Knowledge Discovery*. Chapman & Hall / CRC, Boca Raton, 433–454.

GEYER-SCHULZ, A. and NEUMANN, A. and THEDE, A. (2003b): An Architecture for Behavior-Based Library Recommender Systems. *Journal of Information Technology and Libraries, 22(4).*

KOTLER, P. (1980): *Marketing management: analysis, planning, and control.* Prentice-Hall, Englewood Cliffs.

MADDALA, G.S. (2001): *Introduction to Econometrics.* John Wiley, Chichester.

NARAYANA, C.L. and MARKIN, R.J. (1975): Consumer Behavior and Product Performance: An Alternative Conceptualization. *Journal of Marketing, 39(4), 1–6.*

PRIGOGINE, I. (1962): *Non-equilibrium statistical mechanics.* John Wiley & Sons, New York, London.

ROTHSCHILD, M. and STIGLITZ, J. (1976): Equilibrium in Competitive Insurance Markets: An Essay on the Economics of Imperfect Information. *Quarterly Journal of Economics, 90(4), 629–649.*

SAMUELSON, P.A. (1938a): A Note on the Pure Theory of Consumer's Behaviour. *Economica, 5(17), 61–71.*

SAMUELSON, P.A. (1938b): A Note on the Pure Theory of Consumer's Behaviour: An Addendum. *Economica, 5(19), 353–354.*

SAMUELSON, P.A. (1948): Consumption Theory in Terms of Revealed Preference. *Economica, 15(60), 243–253.*

SPENCE, M.A. (1974): *Market Signaling: Information Transfer in Hiring and Related Screening Processes.* Harvard University Press, Cambridge, Massachusetts.

SPIGGLE, S. and SEWALL, M.A. (1987): A Choice Sets Model of Retail Selection. *Journal of Marketing, 51(2), 97–111.*

Part IX

Text Mining, Web Mining, and the Semantic Web

Classifying Number Expressions in German Corpora

Irene Cramer[1], Stefan Schacht[2], Andreas Merkel[2]

[1] Dortmund University, Germany
irene.cramer@uni-dortmund.de
[2] Saarland University, Germany
{stefan.schacht, andreas.merkel}@lsv.uni-saarland.de

Abstract. Number and date expressions are essential information items in corpora and therefore play a major role in various text mining applications. However, so far number expressions were investigated in a rather superficial manner. In this paper we introduce a comprehensive number classification and present promising, initial results of a classification experiment using various Machine Learning algorithms (amongst others AdaBoost and Maximum Entropy) to extract and classify number expressions in a German newspaper corpus.

1 Introduction

In many natural language processing (NLP) applications such as Information Extraction and Question Answering number expressions play a major role, e.g. questions about the altitude of a mountain, the final score of a football match, or the opening hours of a museum make up a significant amount of the users' information need. However, common Named Entity task definitions do not consider number and date/time expressions in detail (or as in the Conference on Computational Natural Language Learning (CoNLL) 2003 (Tjong Kim Sang (2003) do not incorporate them at all). We therefore present a novel, extended classification scheme for number expressions, which covers all Message Understanding Conference (MUC) (Chinchor (1998a)) types but additionally includes various structures not considered in common Named Entity definitions. In our approach, numbers are classified according to two aspects: their function in the sentence and their internal structure. We argue that our classification covers most of the number expressions occurring in text corpora. Based on this classification scheme we have annotated the German CoNLL 2003 data and trained various machine learning algorithms to automatically extract and classify number expressions. We also plan to incorporate the number extraction and classification system described in this paper into an open domain Web-based Question Answering system for German. As mentioned above, the recognition of certain date, time, and number expressions is especially important in the context of Information Extraction and Question Answering. E. g. the MUC Named Entity definitions (Chinchor (1998b)) include the following basic types: date, time (<TIMEX>)

as well as monetary amount and percentage (<NUMEX>), and thus fostered the development of extraction systems able to handle number and date/time expressions. Famous Information Extraction systems developed in conjunction with MUC are e.g. FASTUS (Appelt et al. (1993)) or LaSIE (Humphreys et al. (1998)). At that time, many researchers used finite-state approaches to extract Named Entities. More recent Named Entity definitions, such as CoNLL 2003 (Tjong Kim Sang (2003)), aiming at the development of Machine Learning based systems, however, again excluded number and date expressions. Nevertheless, due to the increasing interest in Question Answering and the TREC QA tracks (Voorhees et al. (2000)), recently, a number of research groups investigate various techniques to fast and accurately extract information items of different types form text corpora and the Web, respectively. Many answer typologies naturally include number and date expressions, e.g. the ISI Question Answer Typology (Hovy et al. (2002)). Unfortunately, in the corresponding papers only the whole Question Answering System's performance is specified, we therefore could not detect any performance values, which would be directly comparable to our results. A very interesting and partially comparable (they only consider a small fraction of our classification) work (Ahn et al. (2005)) investigates the extraction and interpretation of time expressions. Their reported accuracy values range between about 40% and 75%.

Paper Plan: This paper is structured as follows. Section 2 presents our classification scheme and the annotation. Section 3 deals with the features and the experimental setting. Section 4 analyzes the results and comments on the future perspectives.

2 Classification of number expressions

Many researchers use regular expressions to find numbers in corpora, however, most numbers are part of a larger construct such as '2,000 miles' or 'Paragraph 249 Bürgerliches Gesetzbuch'. Consequently, the number without its context has no meaning or is highly ambiguous (2,000 miles vs. 2,000 cars). In applications such as Question Answering it is therefore necessary to detect this additional information. Table 1 shows example questions that obviously ask for number expressions as answers. The examples clearly indicate that we are not looking for mere digits but multi-word units or even phrases consisting of a number and its specifying context. Thus, a number is not a stand-alone information and, as the examples show, might not even look like a number at all. This paper therefore proposes a novel, extended classification that handles number expressions similar to Named Entities and thus provides a flexible and scalable method to incorporate these various entity types into one generic framework. We classify numbers according to their internal structure (which corresponds to their text extension) and their function (which corresponds to their class).

We also included all MUC types to guarantee that our classification conforms with previous work.

Table 1. Example Questions and Corresponding Types

Q: How far is the Earth from Mars?	miles? light-years?
Q: How high is building X?	meters? floors?
Q: What are the opening hours of museum X?	daily from 9 am to 5 pm
Q: How did Dortmund score against Cottbus last weekend?	2:3

2.1 Classification scheme

Based on Web data and a small fraction of online available German newspaper corpora (Frankfurter Rundschau[1] and die tageszeitung[2]) we deduced 5 basic types: `date` (including date and time expressions), `number` (covering count and measure expressions), `itemization` (rank and score), `formula`, and `isPartofNE` (such as street number or zipcode). As further analyses of the corpora showed most of the basic types naturally split into sub-types, which also conforms to the requirements imposed on the classification by our applications. The final classification thus comprises the 30 classes shown in table 2. The table additionally gives various examples and a short explanation of the class' sense and extension.

2.2 Corpora and annotation

According to our findings in Web data and newspaper corpora we developed guidelines which we used to annotate the German CoNLL 2003 data. To ensure a consistent and accurate annotation of the corpus, we worked every part over in several passes and performed a special reviewing process for critical cases. Table 3 shows an exemplary extract of the data. It is structured as follows: the first column represents the token, the second column its corresponding lemma and the third column its part-of-speech, the fourth column specifies the information produced by a chunker. We did not change any of these columns. In column five, typically representing the Named Entity tag, we added our own annotation. We replaced the given tag if we found the tag `O` (`=other`) and appended our classification in all other cases.[3] While annotating the corpora we met a number of challenges:

- **Preprocessing:** The CoNLL 2003 corpus exhibits a couple of erroneous sentence and token boundaries. In fact, this is much more problematic for the extraction of number expressions than for Named Entity Recognition, which is not surprising, since it inherently occurs more frequently in the context of numbers.
- **Very complex expressions:** We found many `date.relative` and `date.regular` expressions, which are extremely complex types in terms of length, internal structure, as well as possible phrasing and therefore difficult to extract and classify. In addition, we also observed very complex `number.amount` contexts and a couple of broken sports score tables, which we found very difficult to annotate.

[1] http://www.fr-online.de/

[2] http://www.taz.de/

[3] Our annotation is freely available for download. However, we cannot provide the original CoNLL 2003 data, which you need to reconstruct our annotation.

Table 2. Overview of Number Classes

Name of Sub-Type	Examples	Explanation
date.period	for 3 hours, two decades	time/date period, start and end point not specified
date.regular	weekdays 10 am to 6 pm	expressions like opening hours etc.
date.time	at around 11 o'clock	common time expressions
date.time.period	6-10 am	duration, start and end specified
date.time.relative	in two hours	relative specification tie: e.g. now
date.time.complete	17:40:34	time stamp
date.date	October 5	common date expressions
date.date.period	November 22-29, Wednesday to Friday, 1998/1990	duration, start and end specified
date.date.relative	next month, in three days	relative specification tie: e.g. today
date.date.complete	July 21, 1991	complete date
date.date.day	on Monday	all weekdays
date.date.month	last November	all months
date.date.year	1993	year specification
number.amount	4 books, several thousand spectators	count, number of items
number.amount.age	aged twenty, Peter (27)	age
number.amount.money	1 Mio Euros, 1,40	monetary amount
number.amount.complex	40 children per year	complex counts
number.measure	18 degrees Celsius	measurements not covered otherwise
number.measure.area	30.000 acres	specification of area
number.measure.speed	30 mph	specification of speed
number.measure.length	100 km bee-line, 10 meters	specification of length, altitude, ...
number.measure.volume	43,7 l of rainfall, 230.000 cubic meters of water	specification of capacity
number.measure.weight	52 kg sterling silver, 3600 barrel	specification of weight
number.measure.complex	67 l per square mile, 30x90x45 cm	complex measurement
number.percent	32 %, 50 to 60 percent	percentage
number.phone	069-848436	phone number
itemization.rank	third rank	ranking e.g. in competition
itemization.score	9 points, 23:26 goals	score e.g. in tournament
formula.variables	$\prod \cos(x)$	generic equations
formula.parameters	$y = 4.132 * x^3$	specific equations

- **Ambiguities:** In some cases we needed a very large context window to disambiguate the expressions they annotated. Additionally, we even found examples which we could not disambiguate at all. E.g. *über 3 Jahre* with the possible translations *more than 3 years* or *for 3 year*. In German such structures are typically disambiguated by prosody.
- **Particular text type:** A comparison between CoNLL and the corpora we used to develop our guidelines showed that there might be a very particular style. We also had the impression that the CoNLL training and test data differ with respect to type distribution and style. We therefore based our experiments on the complete data and performed cross-validation.

We think that the thus annotated corpora represent a valuable resource, especially, given the well-known data sparseness for German.

Table 3. Extract of the Annotated CoNLL 2003 Data

Am	am	APPRART	I-PC	date.date.complete
14.	14.	ADJA	I-NC	date.date.complete
August	August	NN	I-NC	date.date.complete
1922	@card@	CARD	I-NC	date.date.complete
rief	rufen	VVFIN	I-VC	O
er	er	PPER	I-NC	O
den	d	ART	B-NC	O
katholischen	katholisch	ADJA	I-NC	O
Gesellenverein	Gesellenverein	NN	I-NC	O
ins	ins	APPRART	I-PC	O
Leben	Leben	NN	I-NC	O
.	.	$.	O	O

Furthermore, our findings during the annotation process again emphasized the need of an integrated concept of number expressions and Named Entities: we found 467 `isPartofNE` items, which are extremely difficult to classify without any hint about proper names in the context window.

3 Experimental evaluation

3.1 Features

Our features (see table 4 for details) are adapted from those reported in previous work on Named Entity Recognition (e.g. Bikel et al. (1997), Carreras et al. (2003)). We based the extraction on a very simple and fast analysis of the tokens combined with shallow grammatical clues. To additionally capture information about the context we used a sliding window of five tokens (the word itself, the previous two, the following two).

3.2 Classifiers

To get a feeling for the expectable performance, we conducted a preliminary test by experimenting with Weka (Witten et al. (2005)). For this purpose we ran the Weka implementations of a Decision Tree, k-Nearest Neighbor, and Naive Bayes algorithm with the standard settings and no preprocessing or tuning. Because of previous, promising experiences with AdaBoost (Carreras et al. (2003)) and Maximum Entropy in similar tasks, we decided to also apply these two classifiers. We used the `maxent` implementation of the Maximum Entropy algorithm[4]. For the experiments with AdaBoost we used our own `C++` implementation, which we tuned for large sparse feature vectors with binary entries.

[4] http://www2.nict.go.jp/x/x161/members/mutiyama/software.html

Table 4. Overview of Features Used

feature group	features
only digit strings	2-digit integer [30-99], other 2-digit integer, 4-digit integer [1000-2100], other 4-digit integer, other integer
digit and non-digit strings	1-digit or 2-digit followed by point, 4-digit with central point or colon, any digit sequence with point, colon, comma, comma and point, hyphen, slash, or other non-digit character
non-digit strings	any character sequence max length 3, any character sequence, followed by point, any character sequence with slash, any character sequence
grammar	part-of-speech tag, lemma
window	all features mentioned above for window +/-2

3.3 Results

The performance of the Decision Tree, k-Nearest Neighbor, Naive Bayes, and Maximum Entropy algorithm is on average mediocre, as Table 5 reveals. On the contrary, our AdaBoost implementation shows satisfactory or even good f-measure values for almost all cases and thus significantly outperforms the rest of the classifiers.

Table 5. Overview of the F-Measure Values (AB: AdaBoost, DT: Decision Tree, KNN: k-Nearest Neighbor, ME: Maximum Entropy, NB: Naive Bayes)

class	AB	DT	KNN	ME	NB	class	AB	DT	KNN	ME	NB
other	0.99	0.99	0.98	0.99	0.97	itemization.score	**0.83**	0.43	0.40	0.78	0.04
date	**0.37**	0.13	0.21	0.24	0.19	number	**0.64**	0.00	0.08	0.00	0.00
date.date	0.67	0.73	0.67	**0.74**	0.09	number.amount	0.33	0.53	0.25	**0.67**	0.26
date.date.complete	0.72	0.61	**0.74**	0,49	0.20	number.amount.age	**0.62**	0.28	0.14	0.45	0.02
date.date.day	**0.53**	0.15	0.14	0.20	0.06	number.amount.complex	**0.09**	0.00	0.00	0.00	0.00
date.date.month	**0.37**	0.05	0.08	0.24	0.00	number.amount.money	**0.82**	0.45	0.28	0.79	0.30
date.date.period	**0.43**	0.38	0.36	0.45	0.09	number.measure	**0.22**	0.16	0.00	0.17	0.00
date.date.relative	0.54	0.36	0.16	**0.59**	0.00	number.measure.area	**0.88**	0.10	0.00	0.40	0.00
date.date.year	**0.82**	0.73	0.58	0.76	0.60	number.measure.complex	**0.34**	0.21	0.19	0.22	0.09
date.regular	0.49	0.43	0.37	**0.54**	0.14	number.measure.length	**0.69**	0.17	0.11	0.39	0.01
date.time	**0.87**	0.76	0.67	0.83	0.45	number.measure.speed	**0.91**	0.17	0.18	0.00	0.00
date.time.period	0.41	0.40	**0.46**	0.38	0.31	number.measure.volume	**0.66**	0.06	0.00	0.00	0.00
date.time.relative	**0.38**	0.02	0.07	0.00	0.00	number.measure.weight	**0.49**	0.00	0.00	0.00	0.00
itemization	0.21	**0.28**	0.23	0.17	0.12	number.percent	**0.83**	0.32	0.10	0.56	0.06
itemization.rank	**0.84**	0.31	0.23	0.70	0.00	number.phone	**0.96**	0.85	0.89	0.95	0.65

Table 5 also shows that there are classes with a consistently poor performance, such as `number.amount.complex`, `number. measure`, or `itemization`, and a consistently good performance, such as `number.phone` or `date.date.year`. We think that this correlates with the amount of data as well as the heterogeneity of the classes. For instance, `number.measure` and `itemization` items occur indeed frequently in the corpus but these two classes are–according to our definition–'garbage collectors' and therefore much less homogenous. In contrast, there are classes, such as `date.time.period` or `date.regular`, with rather low f-measure values but a very precise definition; we admittedly suspect that the annotation of these types in our corpora might be inconsistent or inaccurate. We also suppose that there are number expressions which exhibit an exceedingly large variety of phrasing. As a matter of fact, these are inherently difficult to learn if the data do not feature sufficient coverage.

Table 6. Overview of the Precision Values (AB: AdaBoost)

class	AB	class	AB	class	AB
other	0.98	date.time	0.88	number.amount.complex	0.39
date	0.61	date.time.period	0.54	number.percent	0.87
date.date	0.75	date.time.relative	0.50	number.phone	0.96
date.date.complete	0.79	itemization	0.34	number.measure	0.70
date.date.day	0.83	itemization.rank	0.88	number.measure.area	0.93
date.date.month	0.79	itemization.score	0.91	number.measure.length	0.85
date.date.year	0.85	number	0.81	number.measure.speed	0.94
date.date.relative	0.73	number.amount	0.48	number.measure.volume	0.76
date.date.period	0.65	number.amount.age	0.79	number.measure.weight	0.56
date.regular	0.68	number.amount.money	0.89	number.measure.complex	0.65

Fortunately, there are is a number of classes with a pretty high f-measure value–that is more than 0.8–for at least one of the five classifiers, e.g. `date.date.year`, `itemization.rank`, and `number.phone`. More importantly there are, as Table 6 shows, only six classes with a precision value of less than 0.6. We are therefore very confident to be able to successfully integrate the AdaBoost implementation of our number extraction component into a Web-based open domain Question Answering System, since in a Web-based framework the focus tends to be on precision rather than coverage or recall.

4 Conclusions and future work

We presented a novel, extended number classification and developed guidelines to annotate a German newspaper corpus accordingly. On the basis of our annotated data we have trained and tested five classification algorithms to automatically extract and classify them with promising evaluation results. However, the accuracy is still low for some classes, especially for the small or heterogenous ones. But we feel confident to improve our system by incorporating selected training data, especially, in the case of small classes. To find the weak points in our system, we plan to perform a detailed analysis of all number types and their precision, recall, and f-measure values. We also consider a revision of our annotation, because there still might be inconsistently and inaccurately annotated sections in the corpus. As mentioned above, the CoNLL 2003 data exhibit a typical newspaper style, which might limit the applicability of our system to particular corpus types (although, initial experiments with Web data do not support this skepticism). We therefore intend to augment our training data with Web texts annotated according to our guidelines. In addition, we plan to experiment with an expanded feature set and several pre-processing methods such as feature selection and normalization. Research in the area of Named Entity extraction shows that multiple classifier systems or the concept of multi-view learning might be especially effective in our application. We therefore plan to investigate several classifier combinations and also take a hybrid approach–combining grammar rules and statistical methods–into account. We plan to integrate our number extraction system into a Web-based open domain Question Answering system for German and hope to improve the coverage and performance of the answer types processed. While there is still room for improvement, we think–considering the complexity of our task–the achieved performance is surprisingly good.

References

AHN, D., FISSAHA ADAFRE, S. and DE RIJKE, M. (2005): Recognizing and Interpreting Temporal Expressions in Open Domain Texts. *S. Artemov et al. (eds): We Will Show Them: Essays in Honour of Dov Gabbay, Vol 1., College Publications.*

APPELT, D., BEAR, J., HOBBS, J., ISRAEL, D., KAMEYAMA, M., STICKEL, M. and TYSON, M. (1993): FASTUS: A Cascaded Finite-State Tranducer for Extracting Information from Natural-Language Text. *SRI International.*

BIKEL, D., MILLER, S., SCHWARTZ, R. and WEISCHEDEL, R. (1997): Nymble: a high-performance learning name-finder. *Proceedings of 5th ANLP.*

CARRERAS, X., MÀRQUEZ, L. and PADRÓ, L. (2003): A Simple Named Entity Extractor using AdaBoost. *Proceedings of CoNLL-2003*

CHINCHOR, N. A. (1998a): Overview of MUC-7/MET-2. *Proceedings of the Message Understanding Conference 7.*

CHINCHOR, N. A. (1998b): MUC-7 Named Entity Task Definition (version 3.5) *Proceedings of the Message Understanding Conference 7.*

HOVY, E. H., HERMJAKOB, U. and RAVICHANDRAN, D. (2002): A Question/Answer Typology with Surface Text Patterns. *Proceedings of the DARPA Human Language Technology conference (HLT).*

HUMPHREYS, K., GAIZAUSKAS, R., AZZAM, S., HUYCK, C., MITCHELL, B. CUNNINGHAM, H. and WILKS, Y. (1998): University of Sheffield: Description of the LaSIE-II System as Used for MUC-7. *Proceedings of the 7th Message Understanding Conference (MUC-7).*

TJONG KIM SANG, E. F. and DE MEULDER, F. (2003): Introduction to the CoNLL Shared Task: Language-Independent Named Entity Recognition. *Proceedings of the Conference on Computational Natural Language Learning.*

VOORHEES, E. and TICE, D. (2000): Building a Question Answering Test Collection. *Proceedings of SIGIR-2000.*

WITTEN, I. H. and FRANK, E. (2005): *Data Mining: Practical machine learning tools and techniques.* 2nd Edition, Morgan Kaufmann, San Francisco.

Non-Profit Web Portals - Usage Based Benchmarking for Success Evaluation

Daniel Delić and Hans-J. Lenz

Institut für Produktion, Wirtschaftsinformatik und Operations Research,
Freie Universität Berlin, Germany
delic@zedat.fu-berlin.de
hjlenz@wiwiss.fu-berlin.de

Abstract. We propose benchmarking users' navigation patterns for the evaluation of non-profit Web portal success and apply multiple-criteria decision analysis (MCDA) for this task. Benchmarking provides a potential for success level estimation, identification of best practices, and improvement. MCDA enables consistent preference decision making on a set of alternatives (i. e. portals) with regard to the multiple decision criteria and the specific preferences of the decision maker (i. e. portal provider). We apply our method to non-profit portals and discuss the results.

1 Introduction

Portals within an integrated environment provide users with information, links to information sources, services, and productivity and community supporting features (e. g., email, calendar, groupware, and forum). Portals can be classified according to their main purpose into, e. g., community portals, business or market portals, or information portals. In this paper we focus on non-profit information portals.

Usage of non-profit portals is for free in general. Nevertheless, they cause costs. This makes success evaluation an important task in order to optimize the service quality given usually limited resources. The interesting questions are: (1) what methods and criteria should be applied for success measurement, and (2) what kind of evaluation referent should be employed for the interpretation of results. Simple usage statistics, usage metrics (indicators) as well as navigation pattern analysis have been proposed for such a task, usually within the framework of a goal-centered evaluation or an evaluation of improvement relative to past performance. Goal-centered evaluation, however, requires knowledge of desired performance levels. Defining such levels in the context of non-profit portal usage may due to lack of knowledge or experience be a difficult task. For instance, how often has a page to be requested in order to be considered as successful? On the other hand, evaluation of improvement is incomplete because it does not provide information about the success level at all. Benchmarking, on the contrary, does not require definition of performance levels in advance. Furthermore, it has proved suitable for success level estimation, identification of best practices, and improvement (Elmuti and Kathawala (1997)).

We present our approach of non-profit information portal success evaluation, based on benchmarking usage patterns from several similar portals by applying MCDA. The applied measurement criteria are not based on common e-commerce customer lifecycle measures such as acquisition or conversion rates (Cutler and Sterne (2000)). Thus the criteria are especially suitable for (but not limited to) the analysis of portals that offer their contents for free and without the need for users to register. At such portals due to anonymity or privacy directives it is often difficult to track customer relationships over several sessions. This is a common case with non-profit portals.

The paper is organized as follows: in Section 2 we give a brief overview over related work. In Section 3 the method is described. In Section 4 we present a case study and discuss the results. Section 5 contains some conclusions.

2 Related work

Existing usage analysis approaches can be divided into three groups: analysis of (1) simple traffic- and time-based statistics, (2) session based metrics and patterns, and (3) sequential usage patterns.

Simple statistics (Hightower et al. (1998)) are, for instance, the number of *hits* for a certain period or for a certain page. However, those figures are of limited use because they do not contain information about dependencies between a user's requests during one visit (i. e. *session*).

Session based metrics are applied in particular for commercial site usage, e. g., customer *acquisition* and *conversion* rates (Berthon et al. (1996), Cutler and Sterne (2000)) or *micro conversion rates* such as *click-to-basket* and *basket-to-buy* (Lee et al. (1999)). Data mining methods can deliver interesting information about patterns and dependencies between page requests. For example, association rule mining may uncover pages which are requested most commonly together in the users' sessions (Srivastava et al. (2000)). Session based analysis with metrics and data mining gives a quite well insight into dependencies between page requests. What is missing, is the explicit analysis of the users' sequences of page requests.

With sequential analysis the traversal paths of users can be analyzed in detail and insights gained about usage patterns, such as "Over which paths users get from page A to B?". Thus "problematic" paths and pages can be identified (Berendt and Spiliopoulou (2000)).

The quality of the interpretation of results depends considerably on the employed evaluation referent. For commercial sites existing market figures can be used. For non-profit portals such "market" figures in general do not exist. Alternative approaches are proposed: Berthon et al. (1996) suggest to interpret measurement results w. r. t. the goals of the respective provider. However, this implies that the provider himself is able to specify realistic goals. Berendt and Spiliopoulou (2000) measure success by comparing usage patterns of different user groups and by analyzing performance outcomes relative to past performance. While this is suitable for the identification of a site's weak points and for its improvement, neither the overall success level of the site nor the necessity for improvement can be estimated in this way. Hightower et al. (1998) propose a comparative analysis of usage among similar Web sites based on a simple statistical analysis. As already mentioned above, simple statistics alone are of limited information value.

3 Method

Our goal is to measure a portal's success of providing information content pages. Moreover, we want to identify weak points and find possibilities for improvement. The applied benchmarking criteria are based on sequential usage pattern analysis. Our approach consist of three steps: (1) preprocessing the page requests, (2) defining the measurement criteria, and (3) developing the MCDA model.

3.1 Preprocessing page requests

Big portals, especially those with highly dynamic content, can contain many thousands of pages. In general, for such portals usage patterns at the individual page level do not occur frequently enough for the identification of interesting patterns. Therefore the single page requests as defined by their URI in the log are mapped to a predefined concept hierarchy, and the dependencies between concepts are analyzed. Various types of concept hierarchies can be defined, e.g., based on content, service, or page type (Berendt and Spiliopoulou (2000), Spiliopoulou and Pohle (2001)). We define a concept hierarchy based on page types (Fig. 1).[1] The page requests then are mapped (i. e. classified) according to their URI if possible. If the URI does not contain sufficient information, the text to link ratios of the corresponding portal pages are analyzed and the requests are mapped accordingly.[2] Homepage requests are mapped to concept H, all other requests are mapped according to the descriptions in Table 1.

3.2 Measurement criteria

We concentrate on the part of the navigation paths between the first request for page type H (homepage) and the first consecutive request for a target page type from the set $TP = \{\text{M}, \text{MNI}, \text{MNINE}\}$. Of interest is whether or not users navigating from the homepage reach those target pages, and how their traversal paths look like. Sequential usage pattern analysis (Berendt and Spiliopoulou (2000), Spiliopoulou and Pohle (2001)) is applied.

[1] The page type definitions are partly adapted from Cooley et al. (1999).

[2] Therefore a training set is created manually by the expert and then analyzed by a classification learning algorithm.

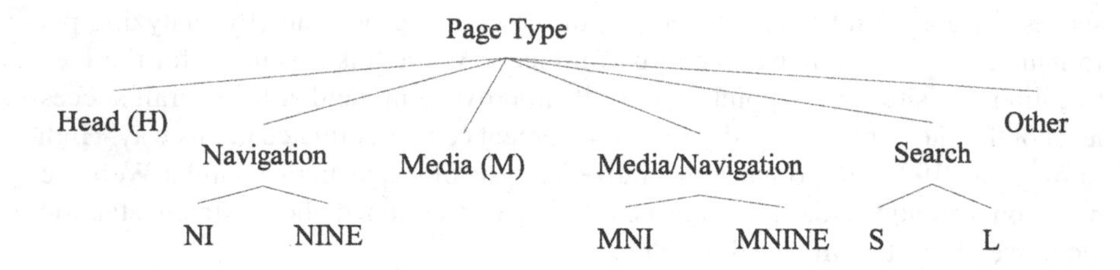

Fig. 1. Page type based concept hierarchy.

A log portion is a set $S = \{s_1, s_2, ..., s_N\}$ of sessions. A session is a set $s = \{r_1, r_2, ..., r_L\}$ of page requests. All sessions $s \in S$ containing at least one request which is related to concept H, denoted as $con(r_i) = \mathrm{H}$ for $i = 1, ..., L$, are of interest. These sessions are termed *active* sessions: $S^{\mathrm{ACT}} = \{s \in S \mid con(r_i) = \mathrm{H} \wedge r_i \in s\}$.

Let $seq(s) = \langle a_1 a_2 ... a_n \rangle$ denote the sequence, i. e. an ordered list, of all page requests in session s. Then $sseq = \langle b_1 b_2 ... b_m \rangle$ is a *subsequence* of $seq(s)$, denoted as $sseq \Vdash seq(s)$, iff there exist an i and $a_i, ..., a_{(i+m-1)} \in s$ and $a_{(i+j-1)} = b_j, \forall j = 1, ..., m$. The *subsequence* of a user's clickpath in a session which is of interest starts with $con(b_1) = \mathrm{H}'$ and ends with $con(b_m) = p$, with $m = \min_{i=1,2,...,n}\{i \mid con(r_i) = p \wedge p \in TP\}$. H$'$ denotes the first occurrence of H in $seq(s)$, and p is the first subsequent occurrence of a request for a target page type from the set TP. We denote this subsequence $\langle b_1 ... b_m \rangle$ as $\mathrm{H}' \star p$ where $\star$ is a wildcard for all in between requests. We want to analyze navigation based usage patterns only. Thus all sessions containing $\mathrm{H}' \star p$ with requests for page types L and S *not* part of the sequence are of interest. These sessions are called *positive* sessions w. r. t. the considered target page type p: $S_p^{\mathrm{POS}} = \{s \in S \mid \mathrm{H}' \star p \Vdash seq(s) \wedge con(r_i) \neq \{\mathrm{L}, \mathrm{S}\}\}, \forall r_i \Vdash \mathrm{H}' \star p$.

Definition 1. *The effectiveness of requests for a page of type p over all active sessions is defined by*

$$eff(p) = \frac{| S_p^{POS} |}{| S^{ACT} |} \tag{1}$$

The *effectiveness* ratio shows in how many active sessions requests for a page of type p occur. A low value may indicate a problem with those pages.

Definition 2. *Let $length(H', p)_s$ denote the length of a sequence $H' \star p$ in $s \in S$, given by the number of its non-H' elements. Then the efficiency of requests for a page of type p over all respective positive sessions is defined by*

$$efc(p) = \frac{| S_p^{POS} |}{\sum_{s \in S_p^{POS}} length(H', p)_s}. \tag{2}$$

The *efficiency* ratio shows how many pages on average are requested in the positive sessions before the first request for a target page of type p occurs. A low value stands for long click paths on average which in turn may indicate a problem for the users with reaching those pages.

3.3 Development of the MCDA model

The MCDA method applied is *Simple Additive Weighting* (SAW). SAW is suitable for decision problems with multiple alternatives based on multiple (usually conflicting) criteria. It allows consistent preference decision making on a set $A = \{a_1, a_2, ..., a_s\}$ of alternatives, a set $C = \{c_1, c_2, ..., c_l\}$ of criteria and their corresponding weights $W = \{w_1, w_2, ..., w_l\}$ ($\sum_l w_l = 1$). The latter reflect the decision maker's preference for each criterion. SAW aggregates the criteria c_j based outcome values x_{ij} for an alternative a_i into an overall utility score $U^{\mathrm{SAW}}(a_i)$. The goal is to obtain a ranking of the alternatives according to their utility scores. Firstly, the outcome values x_{ij} are normalized to the interval $[0,1]$ by applying a value function $u_j(x_{ij})$. Following, the utility score for each alternative is derived by $U^{\mathrm{SAW}}(a_i) = \sum_{j=1}^{l} w_j \cdot u_j(x_{ij}), \forall a_i \in A$. For SAW the criteria based outcome values must be at least of an ordinal scale and the decision maker's preference order relation on them must be complete and transitive. For a more detailed introduction we refer to Figueira et al. (2005), Lenz and Ablovatski (2006).

Table 1. Page types

Concept	Page Type	Purpose	Characteristics
H	Head	Entry page for the considered portal area	Topmost page of the focused site hierarchy or sub-hierarchy
M	Media	Provides information content represented by some form of media such as text or graphics	High text to link ratio
NI, NINE	Navigation	Provides links to site internal (NI) or to site internal and external (NINE) targets	Small text to link ratio
MNI, MNINE	Media/Navigation	Provides some (introductory) information and links to further information sources	Medium text to link ratio
S	Search Service	Provides search service	Contains a search form
L	Search Result	Provides search results	Contains a result list

We use $eff(p)$ and $efc(p)$ as measurement criteria (see Fig. 2) for the portal success evaluation. Within this context we give the following definition of portal success:

Definition 3. *The success level of a non-profit portal in providing information content pages w. r. t. the chosen criteria and weights is determined by its utility score relative to the utility scores of all other considered portals $a \in A$.*

According to Definition 3 the portal with the highest utility score, denoted as a^*, is the most successful: $U^{SAW}(a^*) \geq U^{SAW}(a_i)$ with $a^* \neq a_i$ and $a^*, a_i \in A$.

4 Case study

The proposed approach is applied to a case study of four German eGovernment portals. Each portal belongs to a different German state. Their contents and services are mainly related to state specific topics about schools, education, educational policy etc. One of the main target user groups are teachers.

Preprocessed[3] log data from November 15th and 19th of 2006 from each server are analyzed. The numbers of active sessions in the respective log portions are 746 for portal 1, 2168 for portal 2, and 4692 for portal 3. The obtained decision matrix is shown in Fig. 2. The main decision criteria are the p requests with the subcriteria $eff(p)$ and $efc(p)$. The corresponding utility score function for this two level structure of criteria is $U^{\text{SAW}}(a_i) = \sum_{j=1}^{3} w_j \cdot \left[\sum_{k=1}^{2} w_{jk} \cdot u_{jk}(x_{ijk})\right], \forall a_i \in A$.

	M (0.33)		MNI (0.33)		MNINE (0.33)		U^{SAW}
	eff (0.83)	efc (0.17)	eff (0.83)	efc (0.17)	eff (0.83)	efc (0.17)	
Portal 1	0.1126	0.2054	0.2815	0.3518	0.6408	0.7685	0.36
Portal 2	0.1425	0.2050	0.1836	0.2079	0.1965	0.2338	0.18
Portal 3	0.0058	0.2455	0.0254	0.2459	0.3382	0.4175	0.15

Fig. 2. Decision matrix (with weights in brackets)

The interpretation of results is carried out from the perspective of portal provider 2 (denoted as p2). Thus, the weights are set according to the preferences of p2. As can be seen from the decision matrix (Fig. 2), M, MNI, and MNINE requests are equally important to p2. However, effectiveness of requests is considerably more important to p2 than efficiency, i. e. it is more important that users find (request) the pages at all, than that they do that within the shortest possible paths.

The results show that portal 1 exhibits a superior overall performance over the two others. According to Definition 3 portal 1 is clearly the most successful w. r. t. the considered criteria and weights. Several problems for portal 2 can be identified. Efficiency values for MNI and MNINE requests are lower (i. e. the users' clickpaths are longer) than for the two other portals. The effectiveness value of MNINE requests is the lowest. This indicates a problem with those pages. As a first step towards identifying possible causes we apply the sequence mining tool WUM (Spiliopoulou and Faulstich (1998)) for visualizing the usage patterns containing MNINE requests. The results show that those patterns contain many NI and NINE requests in between. A statistical analysis of consecutive NINE requests confirms these findings. As it can be seen from Fig. 3 the percentage frequency $n(X = x)/N \cdot 100$ (for $x = 1, 2, ..., 5$) of sessions with one or several consecutive NINE requests is significantly higher for portal 2. Finally, a manual inspection of the portal's pages uncovers many navigation pages (NI, NINE) containing only very few links and nothing else. Such pages are the cause for a deep and somewhat "too complicated" hierarchical structure of the portal site which might cause users to abandon it before reaching any MNINE page.

[3] For a detailed description on preprocessing log data refer to Cooley et al. (1999).

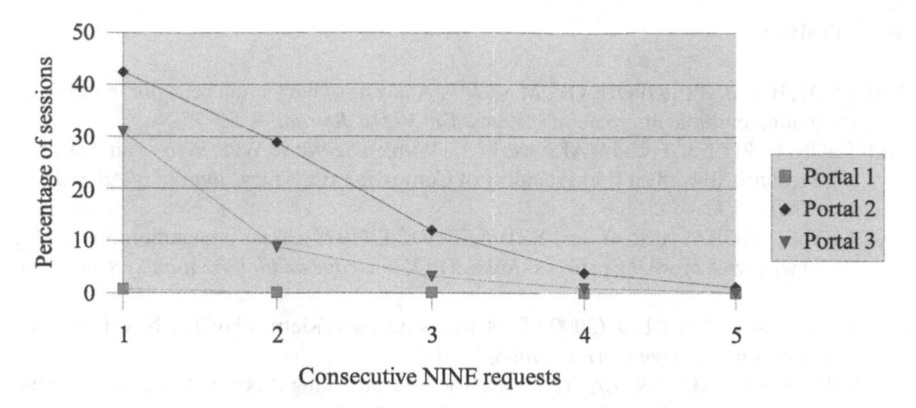

Fig. 3. NINE request distribution

We recommend to p2 to flatten the hierarchical structure by reducing the number of NI, NINE pages by, e. g., merging several consecutive "few-link" NI, NINE pages into one page where possible. Another solution could be to use more MNINE pages for navigation purposes instead of NINE pages (as it is the quite successful strategy of portal 1).

5 Conclusions

A multi-criteria decision model for success evaluation of information providing portals based on the users' navigation patterns is proposed. The objective is to estimate a portal's performance, identify weak points, and derive possible approaches for improvement. The model allows a systematic comparative analysis of the considered portal alternatives on basis of the decision maker's preferences. Furthermore, the model is very flexible. Criteria can be added or excluded according to the evaluation task at hand. In practice, this approach can be a useful tool that helps a portal provider to evaluate and improve its success, especially in areas where no common "market figures" or other success benchmarks exist.

However, a prerequisite for this approach is the existence of other similar portals which can serve as benchmarks. This is a limiting factor, since (1) there simply may not exist similar portals or (2) other providers are not willing (e. g., due to competition) or able (e. g., due to capacity) to cooperate.

Future research will include the analysis of patterns with more than one target page type request in one session. We also plan to analyze and compare the users' search behavior to get hints on the quality of the portals' search engines. Finally, the usage based MCDA model will be extended by a survey to incorporate user opinions.

References

BERENDT, B. and SPILIOPOULOU, M. (2000): Analysis of navigation behavior in web sites integrating multiple information systems. *The VLDB Journal, 9, 56–75.*

BERTHON, P., PITT, L. F. and WATSON, R. T. (1996): The World Wide Web as an Advertising Medium: Toward an Understanding of Conversion Efficiency. *Journal of Advertising Research, 36(1), 43–55.*

COOLEY, R., MOBASHER, B. and SRIVASTAVA, J. (1999): Data preparation for mining world wide web browsing patterns. *Journal of Knowledge and Information Systems, 1(1), 5–32.*

CUTLER, M. and STERNE, J. (2000): E-Metrics - Business Metrics For The New Economy. *http://www.netgen.com (current Mai 8, 2004).*

ELMUTI, D. and KATHAWALA, Y. (1997): The Benchmarking Process: Assessing its Value and Limitations. *Industrial Management, 39(4), 12–20.*

FIGUEIRA, J., GRECO, S. and EHRGOTT, M. (2005): *Multiple Criteria Decision Analysis: State of the Art Surveys.* Springer Science + Business Media, Boston.

HIGHTOWER, C., SIH, J. and TILGHMAN, A. (1998): Recommendations for Benchmarking Web Site Usage among Academic Libraries. *College & Research Libraries, 59(1), 61–79.*

LEE, J., HOCH, R., PODLASECK, M., SCHONBERG, E. and GOMORY, S. (2000): Analysis and Visualization of Metrics for Online Merchandising. In: *Lecture Notes in Computere Science, 1836/2000, 126–141.* Springer, Berlin.

LENZ, H.-J. and ABLOVATSKI, A. (2006): MCDA — Multi-Criteria Decision Making in e-Commerce. In: G. D. Riccia, D. Dubois, R. Kruse, and H.-J. Lenz (eds.): *Decision Theory and Multi-Agent Planning.* Springer, Vienna.

SPILIOPOULOU, M. and FAULSTICH L. C. (1998): WUM: A Web Utilization Miner. In: *EDBT Workshop WebDB 98.* Valencia, Spain.

SPILIOPOULOU, M. and POHLE, C. (2001): Data Mining for Measuring and Improving the Success of Web Sites. *Journal of Data Mining and Knowledge Discovery, 5(1), 85–114.*

SRIVASTAVA, J., COOLEY, R., DESHPANDE, M. and TAN, P.-N. (2000): Web Usage Mining: Discovery and Applications of Usage Patterns from Web Data. *SIGKDD Explorations, 1(2), 12–23.*

Text Mining of Supreme Administrative Court Jurisdictions

Ingo Feinerer and Kurt Hornik

Department of Statistics and Mathematics,
Wirtschaftsuniversität Wien, A-1090 Wien, Austria
{h0125130, Kurt.Hornik}@wu-wien.ac.at

Abstract. Within the last decade text mining, i.e., extracting sensitive information from text corpora, has become a major factor in business intelligence. The automated textual analysis of law corpora is highly valuable because of its impact on a company's legal options and the raw amount of available jurisdiction. The study of supreme court jurisdiction and international law corpora is equally important due to its effects on business sectors.

In this paper we use text mining methods to investigate Austrian supreme administrative court jurisdictions concerning dues and taxes. We analyze the law corpora using R with the new text mining package **tm**. Applications include clustering the jurisdiction documents into groups modeling tax classes (like income or value-added tax) and identifying jurisdiction properties. The findings are compared to results obtained by law experts.

1 Introduction

A thorough discussion and investigation of existing jurisdictions is a fundamental activity of law experts since convictions provide insight into the interpretation of legal statutes by supreme courts. On the other hand, text mining has become an effective tool for analyzing text documents in automated ways. Conceptually, clustering and classification of jurisdictions as well as identifying patterns in law corpora are of key interest since they aid law experts in their analyses. E.g., clustering of primary and secondary law documents as well as actual law firm data has been investigated by Conrad et al. (2005). Schweighofer (1999) has conducted research on automatic text analysis of international law.

In this paper we use text mining methods to investigate Austrian supreme administrative court jurisdictions concerning dues and taxes. The data is described in Section 2 and analyzed in Section 3. Results of applying clustering and classification techniques are compared to those found by tax law experts. We also propose a method for automatic feature extraction (e.g., of the senate size) from Austrian supreme court jurisdictions. Section 4 concludes.

2 Administrative Supreme Court jurisdictions

2.1 Data

The data set for our text mining investigations consists of 994 text documents. Each document contains a jurisdiction of the Austrian supreme administrative court (Verwaltungsgerichtshof, VwGH) in German language. Documents were obtained through the legal information system (Rechtsinformationssystem, RIS; `http://ris.bka.gv.at/`) coordinated by the Austrian Federal Chancellery. Unfortunately, documents delivered through the RIS interface are HTML documents oriented for browser viewing and possess no explicit metadata describing additional jurisdiction details (e.g., the senate with its judges or the date of decision). The data set corresponds to a subset of about 1000 documents of material used for the research project "Analyse der abgabenrechtlichen Rechtsprechung des Verwaltungsgerichtshofes" supported by a grant from the Jubiläumsfonds of the Austrian National Bank (Oesterreichische Nationalbank, OeNB), see Nagel and Mamut (2006). Based on the work of Achatz et al. (1987) who analyzed tax law jurisdictions in the 1980s this project investigates whether and how results and trends found by Achatz et al. compare to jurisdictions between 2000 and 2004, giving insight into legal norm changes and their effects and unveiling information on the quality of executive and juristic authorities. In the course of the project, jurisdictions especially related to dues (e.g., on a federal or communal level) and taxes (e.g., income, value-added or corporate taxes) were classified by human tax law experts. These classifications will be employed for validating the results of our text mining analyses.

2.2 Data preparation

We use the open source software environment R for statistical computing and graphics, in combination with the R text mining package **tm** to conduct our text mining experiments. R provides premier methods for clustering and classification whereas **tm** provides a sophisticated framework for text mining applications, offering functionality for managing text documents, abstracting the process of document manipulation and easing the usage of heterogeneous text formats.

Technically, the jurisdiction documents in HTML format were downloaded through the RIS interface. To work with this inhomogeneous set of malformed HTML documents, HTML tags and unnecessary white space were removed resulting in plain text documents. We wrote a custom parsing function to handle the automatic import into **tm**'s infrastructure and extract basic document metadata (like the file number).

3 Investigations

3.1 Grouping the jurisdiction documents into tax classes

When working with larger collections of documents it is useful to group these into clusters in order to provide homogeneous document sets for further investigation by experts specialized on relevant topics. Thus, we investigate different methods known in the text mining literature and compare their results with the results found by law experts.

k-means Clustering

We start with the well known k-means clustering method on term-document matrices. Let $\mathrm{tf}_{t,d}$ be the frequency of term t in document d, m the number of documents, and df_t is the number of documents containing the term t. Term-document matrices M with respective entries $\omega_{t,d}$ are obtained by suitably weighting the term-document frequencies. The most popular weighting schemes are *Term Frequency* (*tf*), where $\omega_{t,d} = \mathrm{tf}_{t,d}$, and *Term Frequency Inverse Document Frequency* (*tf-idf*), with $\omega_{t,d} = \mathrm{tf}_{t,d} \log_2(m/\mathrm{df}_t)$, which reduces the impact of irrelevant terms and highlights discriminative ones by normalizing each matrix element under consideration of the number of all documents. We use both weightings in our tests. In addition, text corpora were stemmed before computing term-document matrices via the **Rstem** (Temple Lang, 2006) and **Snowball** (Hornik, 2007) R packages which provide the Snowball stemming (Porter, 1980) algorithm.

Domain experts typically suggest a basic partition of the documents into three classes (income tax, value-added tax, and other dues). Thus, we investigated the extent to which this partition is obtained by automatic classification. We used our data set of about 1000 documents and performed k-means clustering, for $k \in \{2,\ldots,10\}$. The best results were in the range between $k = 3$ and $k = 6$ when considering the improvement of the within-cluster sum of squares. These results are shown in Table 1. For each k, we compute the agreement between the k-means results based on the term-document matrices with either *tf* or *tf-idf* weighting and the expert rating into the basic classes, using both the Rand index (Rand) and the Rand index corrected for agreement by chance (cRand). Row "Average" shows the average agreement over the four ks. Results are almost identical for the two weightings employed. Agreements are rather low, indicating that the "basic structure" can not easily be captured by straightforward term-document frequency classification.

Table 1. Rand index and Rand index corrected for agreement by chance of the contingency tables between k-means results, for $k \in \{3,4,5,6\}$, and expert ratings for *tf* and *tf-idf* weightings.

	Rand		cRand	
k	*tf*	*tf-idf*	*tf*	*tf-idf*
3	0.48	0.49	0.03	0.03
4	0.51	0.52	0.03	0.03
5	0.54	0.53	0.02	0.02
6	0.55	0.56	0.02	0.03
Average	0.52	0.52	0.02	0.03

We note that clustering of collections of large documents like law corpora presents formidable computational challenges due to the dimensionality of the term-document matrices involved: even after stopword removal and stemming, our about 1000 documents contained about 36000 different terms, resulting in (very sparse) matrices with about 36 million entries. Computations took only a few minutes in our cases. Larger datasets as found in law firms will require specialised procedures for clustering high-dimensional data.

Keyword based Clustering

Based on the special content of our jurisdiction dataset and the results from k-means clustering we developed a clustering method which we call *keyword based clustering*. It is inspired by simulating the behaviour of tax law students preprocessing the documents for law experts. Typically the preprocessors skim over the text looking for discriminative terms (i.e., keywords). Basically, our method works in the same way: we have set up specific keywords describing each cluster (e.g., "income" or "income tax" for the income tax cluster) and analyse each document on the similarity with the set of keywords.

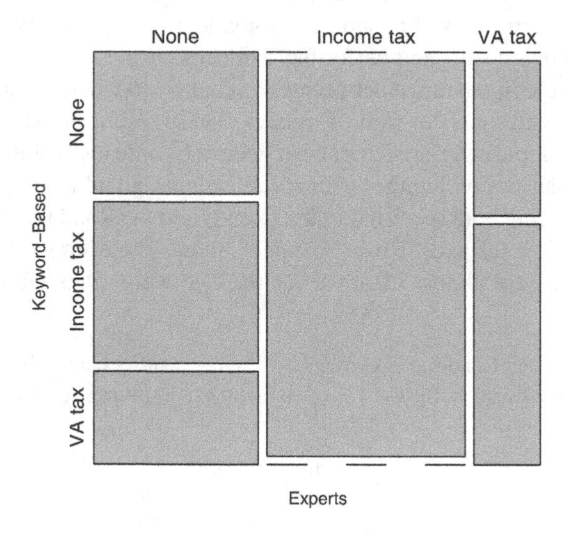

Fig. 1. Plot of the contingency table between the keyword based clustering results and the expert rating.

Figure 1 shows a mosaic plot for the contingency table of cross-classifications of keyword based clustering and expert ratings. The size of the diagonal cells (visualizing the proportion of concordant classifications) indicates that the keyword based clustering methods works considerably better than the k-means approaches, with a

Rand index of 0.66 and a corrected Rand index of 0.32. In particular, the expert "income tax" class is recovered perfectly.

3.2 Classification of jurisdictions according to federal fiscal code regulations

A further rewarding task for automated processing is the classification of jurisdictions into documents dealing and into documents not dealing with Austrian federal fiscal code regulations (Bundesabgabenordnung, BAO).

Due to the promising results obtained with string kernels in text classification and text clustering (Lodhi et al., 2002; Karatzoglou and Feinerer, 2007) we performed a "C-svc" classification with support vector machines using a full string kernel, i.e., using

$$k(x,y) = \sum_{s \in \Sigma^*} \lambda_s \cdot \nu_s(x) \cdot \nu_s(y)$$

as the kernel function $k(x,y)$ for two character sequences x and y. We set the decay factor $\lambda_s = 0$ for all strings $|s| > n$, where n denotes the document lengths, to instantiate a so-called full string kernel (full string kernels are computationally much better natured). The symbol Σ^* is the set of all strings (under the Kleene closure), and $\nu_s(x)$ denotes the number of occurrences of s in x.

For this task we used the **kernlab** (Karatzoglou et al., 2006; Karatzoglou et al., 2004) R package which supports string kernels and SVM enabled classification methods. We used the first 200 documents of our data set as training set and the next 50 documents as test set. We compared the 50 received classifications with the expert ratings which indicate whether a document deals with the BAO by constructing a contingency table (confusion matrix). We received a Rand index of 0.49. After correcting for agreement by chance the Rand index floats around at 0. We measured a very long running time (almost one day for the training of the SVM, and about 15 minutes prediction time per document on a 2.6 GHz machine with 2 GByte RAM).

Therefore we decided to use the classical term-document matrix approach in addition to string kernels. We performed the same set of tests with *tf* and *tf-idf* weighting, where we used the first 200 rows (i.e, entries in the matrix representing documents) as training set, the next 50 rows as test set.

Table 2. Rand index and Rand index corrected for agreement by chance of the contingency tables between SVM classification results and expert ratings for documents under federal fiscal code regulations.

	tf	*tf-idf*
Rand	0.59	0.61
cRand	0.18	0.21

Table 2 presents the results for classifications obtained with both *tf* and *tf-idf* weightings. We see that the results are far better than the results obtained by employing string kernels.

These results are very promising, and indicate the great potential of employing support vector machines for the classification of text documents obtained from jurisdictions in case term-document matrices are employed for representing the text documents.

3.3 Deriving the senate size

Table 3. Number of jurisdictions ordered by senate size obtained by fully automated text mining heuristics. The percentage is compared to the percentage identified by humans.

Senate size	0	3	5	9
Documents	0	255	739	0
Percentage	0.000	25.654	74.346	0.000
Human Percentage	2.116	27.306	70.551	0.027

Jurisdictions of the Austrian supreme administrative court are obtained in so-called senates which can have 3, 5, or 9 members, with size indicative of the "difficulty" of the legal case to be decided. (It is also possible that no senate is formed.) An automated derivation of the senate size from jurisdiction documents would be highly useful, as it would allow to identify structural patterns both over time and across areas. Although the formulations describing the senate members are quite standardized it is rather hard and time-consuming for a human to extract the senate size from hundreds of documents because a human must read the text thoroughly to differ between senate members and auxiliary personnel (e.g., a recording clerk). Thus, a fully automated extraction would be very useful.

Since most documents contain standardized phrases regarding senate members (e.g., "The administrative court represented by president Dr. X and the judges Dr. Y and Dr. Z ... decided ...") we developed an extraction heuristic based on widely used phrases in the documents to extract the senate members. In detail, we investigate punctuation marks and copula phrases to derive the senate size. Table 3 summarizes the results for our data set by giving the total number of documents for senate sizes of zero (i.e., documents where no senate was formed, e.g., due to dismissal for want of form), three, five, or nine members. The table also shows the percentages and compares these to the aggregated percentages of the full data set, i.e., $n > 1000$, found by humans. Figure 2 visualizes the results from the contingency table between machine and human results in form of an agreement plot, where the observed and expected diagonal elements are represented by superposed black and white rectangles, respectively. The plot indicates that the extraction heuristic works very well. This is supported by the very high Rand index of 0.94 and by the corrected Rand index of 0.86.

Further improvements could be achieved by saving identified names of judges in order to identify them again in other documents. Of course, ideally information such as senate size would be provided as metadata by the legal information system, perhaps even determined automatically by text mining methods for "most" documents (with a per-document measure of the need for verification by humans).

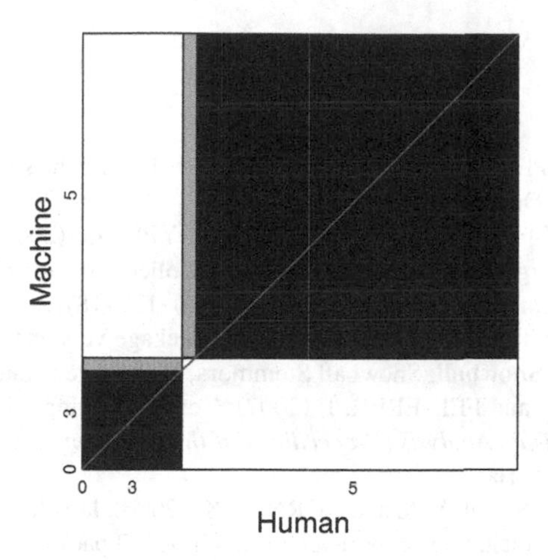

Fig. 2. Agreement plot of the contingency table between the senate size reported by text mining heuristics and the senate size reported by humans.

4 Conclusion

In this paper we have presented approaches to use text mining methods on (supreme administrative) court jurisdictions. We performed k-means clustering and introduced keyword based clustering which works well for text corpora with well defined formulations as found in tax law related jurisdictions. We saw that the clustering works well enough to be used as a reasonable grouping for further investigation by law experts. Second, we investigated the classification of documents according to their relation to federal fiscal code regulations. We used both string kernels and term-document matrices with *tf* and *tf-idf* weighting as input for support vector machine based classification techniques. The experiments unveiled that employing term-document matrices yields both superior performance as well as fast running time. Finally, we considered a situation typical in working with specialized text corpora, i.e., we were looking for a specific property in each text corpus. In detail we derived the senate size of each jurisdiction by analyzing relevant text phrases considering punctuation marks, copulas and regular expressions. Our results show that text mining methods can clearly aid legal experts to process and analyze their law document corpora, offering both considerable savings in time and cost as well as the possibility to conduct investigations barely possible without the availability of these methods.

Acknowledgments

We would like to thank Herbert Nagel for providing us with information and giving us feedback.

References

ACHATZ, M., KAMPER, K., and RUPPE H. (1987): Die Rechtssprechung des VwGH in Abgabensachen. Orac Verlag, Wien.

CONRAD, J., AL-KOFAHI, K., ZHAO, Y. and KARYPIS, G. (2005): Effective Document Clustering for Large Heterogeneous Law Firm Collections. In: *10th International Conference on Artificial Intelligence and Law (ICAIL)*. 177–187.

FEINERER, I. (2007): **tm**: Text Mining Package, R package version 0.1-2.

HORNIK, K. (2007): **Snowball**: Snowball Stemmers, R package version 0.0-1.

KARATZOGLOU, A. and FEINERER, I. (2007): Text Clustering with String Kernels in R. In: *Advances in Data Analysis (Proceedings of the 30th Annual Conference of the GfKl)*. 91–98. Springer-Verlag.

KARATZOGLOU, A., SMOLA, A. and HORNIK, K. (2006): **kernlab**: Kernel-based machine learning methods including support vector machines, R package version 0.9-1.

KARATZOGLOU, A., SMOLA, A., HORNIK, K. and ZEILEIS, A. (2004): **kernlab** — An S4 Package for Kernel Methods in R. *Journal of Statistical Software, 11(9), 1–20.*

LODHI, H., SAUNDERS, C., SHAWE-TAYLOR, J., WATKINS, C., and CRISTIANINI, N. (2002): Text classification using string kernels. *Journal of Machine Learning Research, 2, 419–444.*

NAGEL, H. and MAMUT, M. (2006): Rechtsprechung des VwGH in Abgabensachen 2000–2004.

PORTER, M. (1980): An algorithm for suffix stripping. *Program, 14(3), 130–137.*

R DEVELOPMENT CORE TEAM (2006): R*: A Language and Environment for Statistical Computing.* R Foundation for Statistical Computing, Vienna, Austria. ISBN 3-900051-07-0, URL `http://www.R-project.org/`.

SCHWEIGHOFER, E. (1999): *Legal Knowledge Representation, Automatic Text Analysis in Public International and European Law*. Kluwer Law International, Law and Electronic Commerce, Volume 7, The Hague. ISBN 9041111484.

TEMPLE LANG, D. (2006): **Rstem**: Interface to Snowball implementation of Porter's word stemming algorithm, R package version 0.3-1.

Supporting Web-based Address Extraction with Unsupervised Tagging

Berenike Loos[1] and Chris Biemann[2]

[1] European Media Laboratory GmbH,
Schloss-Wolfsbrunnenweg 33, 69118 Heidelberg, Germany
berenike.loos@eml-d.villa-bosch.de

[2] University of Leipzig, NLP Department,
Johannisgasse 26, 04103 Leipzig, Germany
biem@informatik.uni-leipzig.de

Abstract. The manual acquisition and modeling of tourist information as e.g. addresses of points of interest is time and, therefore, cost intensive. Furthermore, the encoded information is static and has to be refined for newly emerging sight seeing objects, restaurants or hotels. Automatic acquisition can support and enhance the manual acquisition and can be implemented as a run-time approach to obtain information not encoded in the data or knowledge base of a tourist information system. In our work we apply unsupervised learning to the challenge of web-based address extraction from plain text data extracted from web pages dealing with locations and containing the addresses of those. The data is processed by an unsupervised part-of-speech tagger (Biemann, 2006a), which constructs domain-specific categories via distributional similarity of stop word contexts and neighboring content words. In the address domain, separate tags for street names, locations and other address parts can be observed. To extract the addresses, we apply a Conditional Random Field (CRF) on a labeled training set of addresses, using the unsupervised tags as features. Evaluation on a gold standard of correctly annotated data shows that unsupervised learning combined with state of the art machine learning is a viable approach to support web-based information extraction, as it results in improved extraction quality as compared to omitting the unsupervised tagger.

1 Introduction

When setting up a Natural Language Processing (NLP) system for a specific domain or a new task, one has to face the acquisition bottleneck: creating resources such as word lists, extraction rules or annotated texts is expensive due to high manual effort. Even in times where rich resource repositories exist, these often do not contain material for very specialized tasks or for non-English languages and, therefore, have to be created ad-hoc whenever a new task has to be solved as a component of an application system. All methods that alleviate this bottleneck mean a reduction in time and cost. Here, we demonstrate that unsupervised tagging substantially increases performance in a setting where only limited training resources are available.

As an application, we operate on automatic address extraction from web pages for the tourist domain.

1.1 Motivation: Address extraction from the web

In an open-domain spoken dialog system, the automatic learning of ontological concepts and corresponding relations between them is essential as a complete manual modeling of them is neither practicable nor feasible due to the continuously changing denotation of real world objects. Therefore, the emergence of new entities in the world entails the necessity of a method to deal with those entities in a spoken dialog system as described in Loos (2006).

As a use case to this challenging problem we imagine a user asking the dialog system for a newly established restaurant in a city, e.g. ("How do I get to the *Auerstein*"). So far, the system does not have information about the object and needs the help of an incremental learning component to be able to give the demanded answer to the user. A classification as well as any other information for the word "Auerstein" are hitherto not modeled in the knowledge base and can be obtained by text mining methods as described in Faulhaber et al. (2006). As soon as the object is classified and located in the system's domain ontology, it can be concluded that it is a building and that all buildings have addresses. At this stage the herein described work comes into play, which deals with the extraction of addresses in unstructured text. With a web service (as part of the dialog system's infrastructure) the newly found address for the demanded object can be used for a route instruction.

Even though structured and semi-structured texts such as online directories can be harvested as well, they often do not contain addresses of new places and do, therefore, not cover all addresses needed. However, a search in such directories can be used in combination with the method described herein, which can be used as a fallback solution.

1.2 Unsupervised learning supporting supervised methods

Current research in supervised approaches to NLP often tries to reduce the amount of human effort required for collecting labeled examples by defining methodologies and algorithms that make a better use of the training set provided. Another promising direction to tackle this problem is to empower standard learning algorithms by the addition of unlabeled data together with labeled texts. In the machine learning literature, this learning scheme has been called semi-supervised learning (Sarkar and Haffari, 2006). The underlying idea behind our approach is that syntactic and semantic similarity of words is an inherent property of corpora, and that it can be exploited to help a supervised classifier to build a better categorization hypothesis, even if the amount of labeled training data provided for learning is very low. We emphasize that every contribution to widening the acquisition bottleneck is useful, as long as its application does not cause more extra work than the contribution is worth. Here, we provide a methodology to plug an unsupervised tagger into an address extraction system and measure its contribution.

2 Data preparation

In our semi-supervised setting, we require two different data sets: a small, manually annotated dataset used for training our supervised component, and a large, unannotated dataset for training the unsupervised part of the system. This section describes how both datasets were obtained. For both datasets we used the results of Google queries for places as restaurants, cinemas, shops etc. To obtain the annotated data set, 400 of the resulting Google pages for the addresses of the corresponding named entities were annotated manually with the labels: `street`, `house`, `zip` and `city`, all other tokens received the label `O`.

As the unsupervised learning method is in need of large amounts of data, we used a list with about 20,000 Google queries each returning about 10 pages to obtain an appropriate amount of plain text. After filtering the resulting 700 MB raw data for German language and applying cleaning procedures as described in (Quasthoff et al., 2006) we ended up with about 160 MB totaling 22.7 million tokens. This corpus was used for training the unsupervised tagger.

3 Unsupervised tagging

3.1 Approach

Unlike in standard (supervised) tagging, the unsupervised variant relies neither on a set of predefined categories nor on any labeled text. As a tagger is not an application of its own right, but serves as a pre-processing step for systems building upon it, the names and the number of syntactic categories is very often not important.

The system presented in Biemann (2006a) uses Chinese Whispers clustering (Biemann, 2006b) on graphs constructed by distributional similarity to induce a lexicon of supposedly non-ambiguous words with respect to part of speech (PoS) by selecting only safe bets and excluding questionable cases from the category building process. In this implementation two clusterings are combined, one for high and medium frequency words, the other collecting medium and low frequency words. High and medium frequency words are clustered by similarity of their stop word context feature vectors: a graph is built, including only words that are endpoints of high similar pairs. Clustering this graph of typically 5,000 vertices results in several hundred clusters, which are subsequently used as PoS categories. To extend the lexicon, words of medium and low frequency are clustered using a graph that encodes similarity of significant neighbor co-occurrences (as defined in Dunning, 1993). Both clusterings are mapped by overlapping elements into a lexicon that provides PoS information for some 50,000 words.

For obtaining a clustering on datasets of this size, an effective algorithm like Chinese Whispers is crucial. Increased lexicon size is the main difference between this and other approaches (e.g. (Schütze, 1995), (Freitag , 2004)), that typically operate with 5,000 words. Using the lexicon, a trigram tagger with a morphological extension is trained, which can be used to assign tags to all tokens in a text. The tag sets obtained with this method are usually more fine-grained than standard tag sets and reflect syntactic as well as semantic similarity. In Biemann (2006a), the tagger output was directly evaluated against supervised taggers for English, German and Finnish via information-theoretic measures. While it is possible to relatively compare the performance of different components of a system or different systems along this scale, it does only give a poor impression on the utility of the unsupervised tagger's output. Therefore, an application-based evaluation is undertaken here.

3.2 Resulting tagset

As described in Section 2, we had a relatively small corpus in comparison to previous work with the same tagger, that typically operates on about 50 million tokens. Nonetheless, the domain specifity of the corpus leads to an appropriate tagging, which can be seen in the following examples from the resulting tag set (numbers in brackets give the words in the lexicon per tag):

1. Nouns: *Verhandlungen, Schritt, Organisation, Lesungen, Sicherung,...* (800)
2. Verbs: *habe, lernt, wohnte, schien, hat, reicht, suchte...* (191)
3. Adjectives: *französischen, künstlerischen, religiösen...* (142)
4. locations: *Potsdam, Passau, Innsbruck, Ludwigsburg, Jena...* (320)
5. street names: *Bismarckstr, Leonrodstr, Schillerstr, Ungererstr...* (150)

On the one hand, big clusters are formed that contain syntactic tags as shown for the example tags 1 to 3. Items 4 and 5 show that not only syntactic tags are created by the clustering process, but also domain specific tags, which are useful for an address extraction. Note that the actual tagger is capable of tagging all words, not only words in the lexicon – the number of words in the lexicon are merely the number of types used for training. We emphasize that the comparatively small training corpus (usually, 50M–500M tokens are employed) leaves room for improvements, as more training text showed to have a positive impact on tagging quality in previous studies.

4 Experiments and evaluation

This section describes the supervised system, the evaluation methodology and the results we obtained in a comparative evaluation of either providing or not providing the unsupervised tags.

4.1 Conditional random field tagger

We perceived address extraction as a tagging task: labels indicating `city`, `street`, `house` number, `zip` code or other (`O`) from the training set are learned and applied to unseen examples. Note that this is not comparable to a standard task like Named Entity Recognition (cf. Roth and van den Bosch, 2002), since we are only interested in labeling the address of the target location, and not other addresses that might be contained in the same document. Rather, this is an instance of Information Extraction (see Grishman, 1997). For performing the task, we train the MALLET tagger (McCallum, 2002), which is based on Conditional Random Fields (CRFs, see Lafferty et al. 2001). CRFs define a conditional probability distribution over label sequences given a particular observation sequence. CRFs have been proven to have equal or superior performance at tagging tasks as compared to other systems like Hidden Markov Models or the Maximum Entropy Framework. The flexibility of CRFs to include arbitrary, non-independent features allows us to supply unsupervised tags or no tags to the system without changing the overall architecture. The tagger can operate on a different set of features ranging over different distances. The following features per instance are made available to the CRF:

- word itself
- relative position to target name
- unsupervised tag

We experimented with different orders as well as with different time shifts.

CRF order

The order of the CRF defines how many preceding labels are used for the determination of the current label. An order of 1 means that only the previous label is used, order 2 allows for the usage of two previous labels etc. As higher orders mean more information, which is in turn supported by fewer training examples, an optimum at some small order can be expected.

Time shifting

Time shifting is an operation that allows the CRF to use not only the features for the current position, but also features from surrounding positions. This is reached by copying the features from surrounding positions, indicating what relative position they were copied from. As with orders, an optimum can be expected for some small range of time shifting, exhibiting the same information/sparseness trade-off. For illustration, the following listing shows an original training instance with time shift 0, as well as the same instance with time shifts -2, -1, 0, 1, 2, for the scenario with unsupervised tags. Note that relative positions are not copied in time-shifting because of redundancy. The following items show these shifts:

- **shift 0:**
  - Extrablatt 0 T115 `o`
  - 53 1 T215 `house`
  - Hauptstr 2 T64 `street`
  - Heidelberg 3 T15 `city`
  - 69117 4 T215 `zip`
- **shift 1:**
  - 1 -1:Extrablatt -1:T115 0:53 0:T215 1:Hauptstr 1:T64 `house`

  - 2 -1:53 -1:T215 0:Hauptstr 0:T64 1:Heidelberg 1:T15 `street`
- **shift 2:**
  - 1 -2:Cafe -2:T10 -1:Extrablatt -1:T115 0:53 0:T215 1:Hauptstr 1:T64 2:Heidelberg 2:T15 `house`
  - 2 -2:Extrablatt -2:T115 -1:53 -1:T215 0:Hauptstr 0:T64 1:Heidelberg 1:T15 2:69117 2:T215 `street`

In the example for shift 0 a full address with all features is shown: word, relative position to target "Extrablatt", unsupervised tag and classification label. For exemplifying shifts 1 and 2, only two lines are given, with -2:, -1:, 0:, 1: and 2: being the relative position of copied features. In the scenario without unsupervised tags all features "T<number>" are omitted.

4.2 Evaluation methodology

For evaluation, we split the training set into 5 equisized parts and performed 5 sub-experiments per parameter setting and scenario, using 4 parts for training and the remaining part for evaluation in a 5-fold-cross-validation fashion. The split was performed per target location: locations in the test set were never contained in the training. To determine our system's performance, we measured the amount of correctly classified, incorrectly classified (false positives) and missed (false negatives) instances per class and report the standard measures Precision, Recall and F1-measure as described in Rijsbergen (1979). The 5 sub-experiments were combined and checked against the full training set.

4.3 Results

Our objective is to examine to what extent the unsupervised tagger influences classification results. Conducting the experiments with different CRF parameters as outlined in Section 4.1, we found different behaviors for our four target classes: whereas for `street` and `house` number, results were slightly better in the second order CRF experiments, the first order CRF scored clearly higher for `city` and `zip` code. Restricting experiments to first order CRFs and regarding different shifts, a shift of 2 in both directions scored best for all classes except `city`, where both shift 0 and 1 resulted in slightly higher scores. The best overall setting, therefore, was determined to be the first order CRF with a shift of 2. For this setting, Figure 1 presents the results in terms of precision, recall and F1.

What can be observed not only from Figure 1 but also for all parameter settings is the following: Using unsupervised tags as features as compared to no tagging leads to a slightly decreased precision but a substantial increase in recall, and always affects the F1 measure positively. The reason can be sought in the generalization power of the tagger: having at hand syntactic-semantic tags instead of merely plain words, the system is able to classify more instances correctly, as the tag (but not the word) has occurred with the correct classification in the training set before. Due to overgeneralization or tagging errors, however, precision is decreased. The effect is

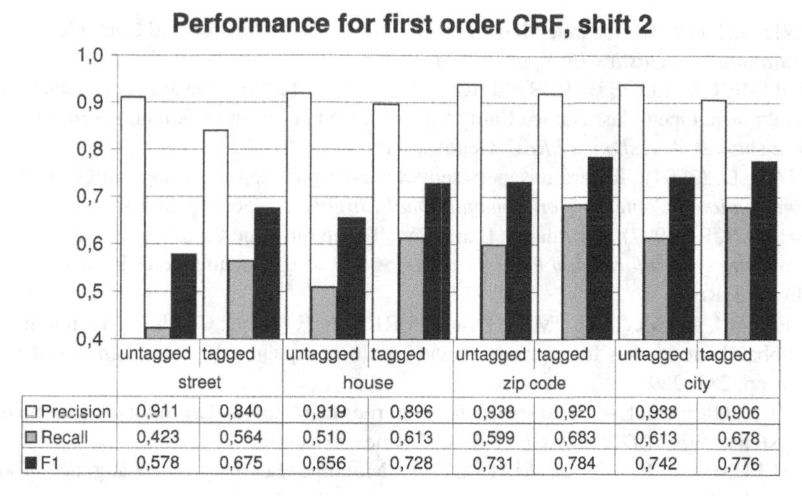

	street		house		zip code		city	
	untagged	tagged	untagged	tagged	untagged	tagged	untagged	tagged
□ Precision	0,911	0,840	0,919	0,896	0,938	0,920	0,938	0,906
▨ Recall	0,423	0,564	0,510	0,613	0,599	0,683	0,613	0,678
■ F1	0,578	0,675	0,656	0,728	0,731	0,784	0,742	0,776

Fig. 1. Results in precision, recall and F1 for all classes, obtained with first order CRF and a shift of 2.

strongest for `street` with a loss of 7% in precision with a recall boost of 14%. In general, unsupervised tagging clearly helps at this task, as a little loss in precision is more than compensated with a boost in recall.

5 Conclusion and further work

In this research we have shown that the use of large, unannotated text can improve classification results on small, manually annotated training sets via building a tagger model with unsupervised tagging and using the unsupervised tags as features in the learning algorithm. The benefit of unsupervised tagging is especially significant in domain-specific settings, where standard pre-processing steps such as supervised tagging do not capture the abstraction granularity necessary for the task, or simply no tagger for the target language is available. For further work, we aim at combining the possibly several addresses per target location. Given the evaluation values obtained with our method, the task of dynamically extracting addresses from web-pages to support address search for the tourist domain is feasible and a valuable, dynamic add-on to directory-based address search.

References

BIEMANN, C. (2006a): Unsupervised Part-of-Speech Tagging Employing Efficient Graph Clustering. *Proc. COLING/ACL-06 SRW*, Sydney, Australia.

BIEMANN, C. (2006b): Chinese Whispers - an Efficient Graph Clustering Algorithm and its Application to Natural Language Processing Problems. *Proceedings of the HLT-NAACL-06 Workshop on Textgraphs*, New York, USA.

DUNNING, T. (1993): Accurate Methods for the Statistics of Surprise and Coincidence. *Computational Linguistics 19(1), pp. 61–74.*

FAULHABER A., LOOS B., PORZEL R., MALAKA, R. (2006): Towards Understanding the Unknown: Open-class Named Entity Classification in Multiple Domains. *Proceedings of the Ontolex Workshop at LREC*, Genova, Italy

FREITAG, D. (2004): Toward unsupervised whole-corpus tagging. *Proceedings of the 20th International Conference on Computational Linguistics*, Geneva, Switzerland

GRISHMAN, R. (1997): Information Extraction: Techniques and Challenges. In Maria Teresa Pazienza (ed.) *Information Extraction.* Springer-Verlag, Lecture Notes in Artificial Intelligence, Rome

LAFFERTY, J. and McCALLUM, A. K. and PEREIRA, F. (2001): Conditional random fields: Probabilistic models for segmenting and labeling sequence data. *Proceedings of ICML-01*, pp. 282–289.

LOOS, B. (2006): On2L – A Framework for Incremental Ontology Learning in Spoken Dialog Systems. *Proc. COLING/ACL-06 SRW*, Sydney, Australia

MCCALLUM, A. K. (2002): MALLET: A Machine Learning for Language Toolkit. http://mallet.cs.umass.edu.

QUASTHOFF, U., RICHTER, M. and BIEMANN, C. (2006): Corpus Portal for Search in Monolingual Corpora. *Proceedings of LREC-06*, Genoa, Italy

ROTH, D. and VAN DEN BOSCH, A. (Eds.) (2002): Proceedings of the Sixth Workshop on Computational Language Learning (CoNNL-02), Taipei, Taiwan.

SARKAR, A. and HAFFARI, G. (2006): Inductive Semi-supervised Learning Methods for Natural Language Processing. *Tutorial at HLT-NAACL-06*, NYC, USA.

SCHÜTZE, H. (1995): Distributional part-of-speech tagging. *Proceedings of the 7th Conference on European chapter of the Association for Computational Linguistics*, Dublin, Ireland

VAN RIJSBERGEN, C. J. (1979): *Information Retrieval, 2nd edition.* Dept. of Computer Science, University of Glasgow.

A Two-Stage Approach for Context-Dependent Hypernym Extraction

Berenike Loos[1] and Mario DiMarzo[2]

[1] European Media Laboratory GmbH,
Schloss-Wolfsbrunnenweg 33, 69118 Heidelberg, Germany
berenike.loos@eml-d.villa-bosch.de
[2] Universität Heidelberg, Institute of Computer Science, Germany
mario.di_marzo@urz.uni-heidelberg.de

Abstract. In this paper we present an unsupervised method to deal with the classification of out-of-vocabulary words in open-domain spoken dialog systems. This classification is vital to ameliorate the human-computer interaction and to be able to extract additional information, which can be presented to the user. We propose a two-stage approach for interpreting named entities in a document corpus: to cluster documents dealing with a particular named entity and to classify it with the help of structural and contextual information in these documents. The idea is to take the resulting websites from a search engine queried for a named entity as documents and to cluster those which are semantically similar. Named entities can then be classified with the information contained in the clusters. Our evaluation showed that the precision of the classification task was as high as 64.47%.

1 Introduction

Open-domain spoken dialog systems need to deal with the classification of out-of-vocabulary (OOV) words to be able to give the user the requested information and to ameliorate the human-computer interaction. Therefore, an approach is needed which semantically classifies those OOV words. For our approach we worked with named entities, as these are the class of words which are most likely to be new to the dialog system.

The presented approach combines a clustering of a document corpus with a method to find hypernyms of named entities in document clusters. For a list of named entities denominating locations in German cities the resulting web pages of the Google search engine are cached (e.g. Lotus, Merlin etc.).

With the help of a document clustering the websites are divided into clusters of similar contents. These clusters are then used for an approach of hypernym extraction to classify the named entities. An example could be the named entity "Lotus", which is not only a restaurant in Heidelberg but also the trademark of a car and a software. Our approach would split the resulting website texts into three clusters and classify the named entities depending on the textual context.

In the next section the steps for the document clustering task are presented and in Section 3 the consecutive hypernym extraction is described.

2 Document clustering

For a separation of the hypernym candidates of the named entity it is necessary to have the documents of the corpus in different groups according to the context.

We apply the cluster algorithms *Clique* and the non-hierarchical *Single-Link*. The Clique Algorithm takes documents into the same cluster which have pairwise similarity to each other. The result are many small clusters which share some documents. In the Single-Link algorithm a document only needs some kind of similarity to one of the documents of the cluster. The result are comparatively few big clusters. (See Subsection 2.2 for a detailed description of the single steps processed by the algorithms.)

The evaluation will therefore show into which direction to go with respect to clustering approaches in the future.

2.1 Data preparation

In the preprocessing step standard term vectors where established using the Porter Stemmer. The similarity is calculated with the cosine coefficient as shown in Formula (1).

$$\cos(\vec{x},\vec{y}) = \frac{\vec{x}\cdot\vec{y}}{|\vec{x}|\cdot|\vec{y}|} = \frac{\sum_{i=1}^{n} x_i y_i}{\sqrt{\sum_{i=1}^{n} x_i^2}\cdot\sqrt{\sum_{i=1}^{n} y_i^2}} \tag{1}$$

The higher the calculated value the more similar are two documents to each other. The similarity between all possible combinations results in the Document Document Relation Matrix (DDRM).

The Document-Document Similarity Matrix (DDSM) can be prepared with the help of the DDRM and a threshold value. The result is a boolean matrix, which has entries of one for similarity values which are higher than or equal to the threshold and zero for cases in which it is lower.

2.2 The clustering algorithms

The clustering algorithms applied with the DDSM in this work are *Single-Link* and *Clique*.

The Single-Link algorithm as described by Kowalski (1997) works in four steps: First, it chooses a document d of the remaining documents and adds it to a new document cluster. Second, it adds all documents which are similar to d according to the DDSM to the recent cluster. Third, it performs the second step for each document which was added to the recent cluster. And last, if there are no more documents which will be added to the recent cluster, it performs the first step, otherwise it terminates.

The Clique Clustering algorithm described by Koch (2001) finds document clusters by creating a seed-list with similar documents starting with an initial document. As soon as the seed-list consists of all similar documents it is declared to be a cluster. This procedure is done for all documents and, therefore, all finally belong to any of the created clusters.

These two algorithms were chosen as the resulting clusters are quite different to each other. Clique differs significantly from Single Link, since Clique produces smaller and more clusters than Single Link. A cluster established by Clique contains always pairwise similar documents. Hence, all documents within the cluster are similar to each other. In order to add a document *d* into a Single-Link cluster, it is sufficient that *d* is similar to only one of the documents belonging to the recent cluster.

3 Hypernym extraction

According to Lyons (1977) hyponymy is the relation which holds between a more specific lexeme (i.e. a hyponym) and a more general one (i.e. a hypernym). E.g. animal is a hypernym of cat. Hypernym Extraction (HE) is applied in cases where the hypernym of a given noun or named entity has to be found for example as part of an ontology learning framework.

After the documents of the corpus are divided into different clusters the HE can take place separately for all of the clusters. For this approach a Part-of-Speech Tagger provides the part-of-speech tags for all terms. The hypernyms of named entities are generally nouns and therefore only nouns are considered in the extraction. Three approaches were therefore considered resulting in three vectors, which are lateron consolidated: the frequency of a term in the neighborhood of the named entity; the distance of a term to the named entity; and the existence of a lexico-syntactic pattern indicating the hypernym/hyponym relationship as proposed by Hearst (1992).

Hearst used the notion of the hypernym/ hyponym relationship pragmatically when referring to named entities and similar to Miller et al. (1990) who stated that "a concept represented by a lexical item L_0 is said to be a hypernym of the concept represented by a lexical item L_1 if native speakers of English accept sentences constructed from the frame *An* L_0 *is a kind of* L_1. Here L_1 is the *hypernym* of L_0 and the relationship is reflexive and transitive, but not symmetric."

No distinction is made between the relationship of nouns and named entities to more general terms. This stands in contrast to the terminology of ontologies. Here this relationship is *is-a* between classes (corresponding to nouns on language level) and *instance-of* for the relation between classes and instances(corresponding to named entities).

3.1 Term frequency

For each of the clusters a unique list of nouns occurring in the documents belonging to a cluster is extracted. This list contains all possible nouns (hypernym candidates) and, therefore, serves as a basis to establish the Named-Entity-Term-Vector (NETV). The NETV is a vector, which contains a value for each noun (hypernym candidate) in the unique list. The value is calculated by the cosine coefficient (as shown in Formula 1) and signifies the co-occurrence of a hypernym candidate and the named entity based on term frequency.

3.2 Term distance

The term distance approach takes the notion into account that smaller distances between hypernym candidate and named entity signify a more probable hypernym relation. Hence, smaller distances are considered to be more valuable and are, therefore, preferred.

An example is the following German sentence:

- Das *Hotel Auerstein* befindet sich verkehrstechnisch günstig im nördlichen Heidelberger Stadtteil Handschuhsheim. (In English: The *Hotel Auerstein* is located in direct access from the city center of Heidelberg in the northern neighborhood Handschuhsheim.)

Therefore, a NETV of dimension p can be built, where p is the number of terms in the unique-list. The entries for the vector are computed by calculating the distance weights as described in the following: First, a parameter value for the highest possible distance of a hypernym candidate and the named entity is identified as shown in Figure 3 in the Evaluation Section. It appeared that the results are most promising for the distance of $p = 8$.

The average distance weight v_n of the pairwise occurrence of a hypernym n and the named entity i is calculated according to Formula 2, where w_i is the weight of the named entity.

$$v_n = \frac{\sum^{\#i} w_{i,n}}{\#i} \tag{2}$$

As the NE occurs more than once in the documents, all occurrences and their neighborhood have to be taken into account for the calculation. Therefore, an average value of all distances $d_{i,n}$ between any i and the occurrence of a hypernym candidate n in the neighborhood of i are calculated which are defined by parameter p. The single distance weights are calculated with Formula 3.

$$w_{i,n} = \begin{cases} 1 - \frac{|d_{i,n}|-1}{p} & , d_{i,n} < p \\ 0 & , else \end{cases} \tag{3}$$

3.3 Lexico-syntactic patterns

To take not only statistical methods into account, we tested the results for lexico-syntactic patterns according to Hearst (1992). Therefore, we developed a boolean named-entity-term-vector. Even though the detection of lexico-syntactic patterns is not frequent, the probability that once found patterns are correct is high.

3.4 Weighting and consolidation

From the three described methods for hypernym extraction result three NETVs with the same dimension, which are consolidated to one vector. As the probability of correctness for once found lexico-syntactic Patterns is high, the weighting of them is also high. Nonetheless, the weighting of the others is taken into account even if a lexico-syntactic pattern is found.

The following formula serves for the calculation of the consolidated NETV, where h is the NETV for the lexico-syntactic patterns, for the term frequency f, for the term distance b and w_1, w_2, w_3 the weights, which are used as parameters in the evaluation:

$$k = \frac{w_1 \cdot h + w_2 \cdot f + w_3 \cdot b}{3} \tag{4}$$

According to the entries of the consolidated NETV the most probable hypernym candidate can be chosen.

4 Evaluation

For the evaluation setup we extracted websites from Google for 90 named entities, which resulted in 90 corpora with each including 10 to 20 documents. For a Gold-standard all of these were annotated manually for hypernyms by two annotators. Furthermore, they marked the corpora which include ambiguous named entities, which can be used for the evaluation of the document clustering task. These documents were clustered manually for similar documents.

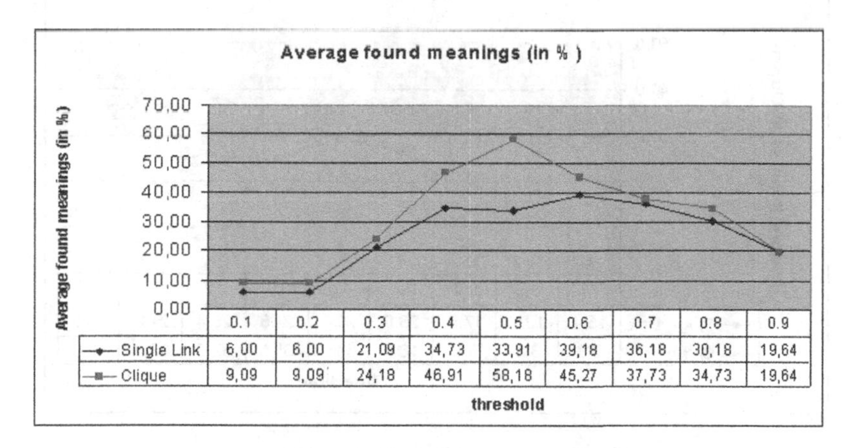

	0.1	0.2	0.3	0.4	0.5	0.6	0.7	0.8	0.9
Single Link	6,00	6,00	21,09	34,73	33,91	39,18	36,18	30,18	19,64
Clique	9,09	9,09	24,18	46,91	58,18	45,27	37,73	34,73	19,64

Fig. 1. Percentage of found meanings for Single-Link and Clique

4.1 Evaluation of the clustering task

For the clustering task it appeared that the choice of the clustering algorithm was important for the results and was, therefore, chosen as a parameter. Furthermore, the choice of a good threshold value is important for the establishment of the DDSM. This parameter is referred to as *threshold*. For testing it was evaluated for the range between 0 and 1 with increments of 0.1.

Two metrics were responsible for the evaluation of the clustering task: The probability that all different meanings for a named entity were found with the application of a clustering approach with a specific threshold value and the recall of automatically correctly clustered documents. The first one is referred to as *average found meanings* in the following.

Figure 1 shows the results for *average found meanings* for the two cluster algorithms depending on the threshold value. We averaged over all named entities we had. *Found meanings* refers to the clusters in which the meaning was contained and could therefore be found in a later hypernym extraction. It appeared that for a threshold value of 0.5 the results of Clique outperformed Single-Link considerably as well as for the recall (as shown in Figure 2). For the recall we calculated how many of the documents which are assigned to one cluster should actually be there.

For the analysis of an optimal threshold value it is necessary that only clusterings are analyzed which consist of clusters indicated by manual annotation to be clusters. The precision of the clustering task has to be 100% as only this can yield reliable results for the hypernym extraction.

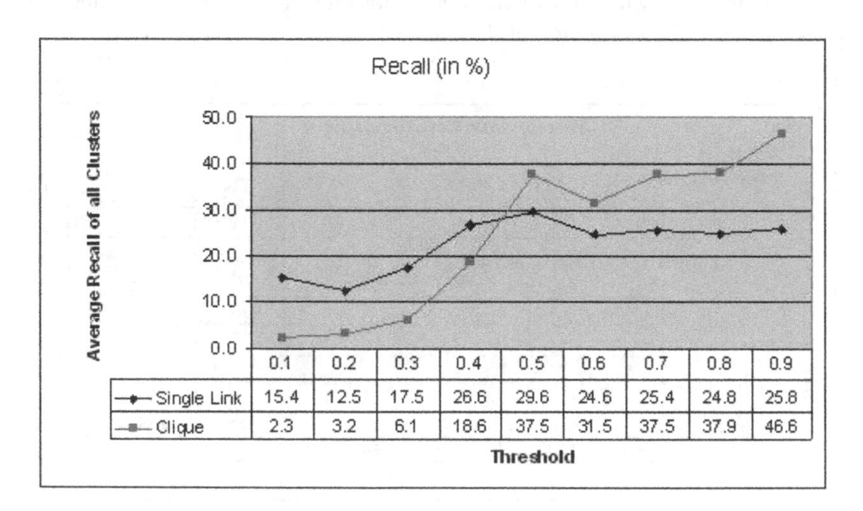

	0.1	0.2	0.3	0.4	0.5	0.6	0.7	0.8	0.9
Single Link	15.4	12.5	17.5	26.6	29.6	24.6	25.4	24.8	25.8
Clique	2.3	3.2	6.1	18.6	37.5	31.5	37.5	37.9	46.6

Fig. 2. Recall for Single-Link and Clique

4.2 Evaluation of the Hypernym Extraction task

For the Hypernym Extraction (HE) task the formula for weighting the nouns in the neighborhood of the NE yields the best results. This parameter is referred to as *neighborRelevance*.

The evaluation of the *neighborRelevance* parameter showed that a window of eight words surrounding the NE yielded the best results as shown in Figure 3. This means, that if a window of eight words surrounding the named entity is chosen, the best results are attained. Nonetheless, it should be taken into account that the analysis of shorter snippets is cheaper and therefore also the comparatively good results for a value of 4 should be kept in mind for performance reasons. The formula for the calculations is described in Subsection 3.2.

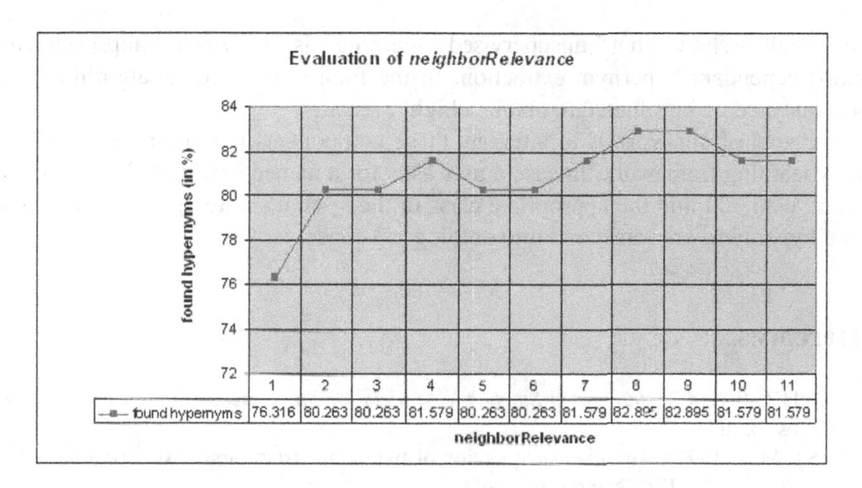

Fig. 3. Evaluation for neighbor relevance

The precision for the HE task depending on the value of the parameter *amountOfExtractedHypernyms*, which refers to the number of hypernyms given by the HE module, were 64.47% for value 1, 77.63% for value 2 and 84.21% for value 3. The results vary from the ones of the evaluation for neighbor relevance due to slightly changed parameter values. Overall we had results which outperformed earlier developed methods as described in Faulhaber et al. (2006) for hypernym extraction by about 4% (absolute).

Table 1 shows the results for the best parameter choice according to our evaluation for a combination of the modules for clustering and HE which we obtained by empirical evaluation. These results of parameter values are not only of interest for the described approaches but also generally for the tasks of document clustering and hypernym extraction. The parameter *maxWeight* is the sum of the three parameters *hearstWeight*, *termDistanceWeight* and *termFrequenceWeight*.

Table 1. Parameter Value Selection

Parameter	Value
Algorithm	Clique
Threshold	0.5
maxWeight	30
termFrequenceWeight	16
termDistanceWeight	11
hearstWeight	2
neighborRelevance	8

5 Conclusion and future work

The results show that unsupervised learning is a viable approach for context-dependent hypernym extraction. In the future more cluster algorithms are to be analyzed and evaluated to obtain a higher recall.

The goal of our work is to integrate these components into an incremental ontology learning framework. In case a user asks for a named entity not known to the system, it should find the appropriate class in the system's ontology. Therefore, the found hypernyms are transfered into ontological concepts.

References

GALLWITZ, F. (2002): *Integrated Stochastic Models for Spontaneous Speech Recognition.* Logos, Berlin.

HEARST, M.A.(1992): Automatic acquisition of hyponyms from large text corpora. In *Proceedings of COLING*, Nantes, France.

FAULHABER, A. LOOS B., PORZEL R., MALAKA, R. (2006): Towards Understanding the Unknown: Open-class Named Entity Classification in Multiple Domains. In *Proceedings of the Ontolex Workshop at LREC*, Genova, Italy.

KOCH, I. (2001): Enumerating all connected maximal common subgraphs in two graphs. *Theoretical Computer Science*, 250(1-2):1–30.

KOWALSKI, G. (1997): *Information Retrieval Systems: Theory and Implementation.* Kluwer Academic Publishers, USA.

LYONS, J. (1977): *Semantics.* University Press, Cambridge, MA.

MILLER, G., BECKWITH, R., FELLBAUM, C., GROSS, D. and MILLER, K. (1990): Introduction to wordnet: An on-line lexical database. *Journal of Lexicography*, 3(4):235–244, January.

Analysis of Dwell Times in Web Usage Mining

Patrick Mair[1] and Marcus Hudec[2]

[1] Department of Statistics and Mathematics and ec3
Wirtschaftuniversität Wien
Augasse 2-6, 1090 Vienna, Austria
patrick.mair@wu-wien.ac.at

[2] Department of Scientific Computing and ec3
University of Vienna
Universitätsstr. 5, 1010 Vienna, Austria
marcus.hudec@univie.ac.at

Abstract. In this contribution we focus on dwell times a user spends on various areas of a web site within a session. We assume that dwell times may be adequately modeled by a Weibull distribution which is a flexible and common approach in survival analysis. Furthermore we introduce heterogeneity by various parameterizations of dwell time densities by means of proportional hazards models. According to these assumptions the observed data stem from a mixture of Weibull densities. Estimation is based on EM-algorithm and model selection may be guided by BIC. Identification of mixture components corresponds to a segmentation of users/sessions. A real life data set stemming from the analysis of a world wide operating eCommerce application is provided. The corresponding computations are performed with the `mixPHM` package in R.

1 Introduction

Web Usage Mining focuses on the analysis of visiting behavior of users on a web site. Common starting point are the so called click-stream data which are derived from web-server logs and may be viewed as the electronic trace a user leaves on a web site. Adequate modeling of the dynamics of browsing behavior is of particular relevance for the optimization of eCommerce applications. Recently Montgomery et al. (2004) proposed a dynamic multinomial probit model of navigation patterns which lead to an remarkable increase of conversion rates. Park and Fader (2004) developed multivariate exponential-gamma models which enhance cross-site customer acquisition. These papers indicate the potential that such approaches offer for web-shop providers.

In this paper we will focus on modeling dwell times, i.e., the time a user spends for viewing a particular page impression. They are defined by the time span between two subsequent page requests and can be calculated by taking the difference between the two logged time points when the page request have been issued. For the analysis of complex web sites which consist of a large number of pages it is often reasonable to reduce the number of different pages by aggregating individual page-impressions to semantically related page categories reflecting meaningful regions of the web site.

Analysis of dwell times is an important source of information with regard to the relevance of the content for different users and the effectiveness of the page in attracting visitors. In this paper we are particularly interested in segmentation of users into various groups which exhibit a similar behavior with regard to the dwell times they spend on various areas of the site. Such a segmentation analysis is an important step towards a better understanding of the way a user interacts on a web site. It is therefore of relevance with regard to the prediction of user behavior as well as for a user-specific customization or even personalization of web sites.

2 Model specification and estimation

2.1 Weibull mixture model

Since survival analysis focuses on duration times until some event occurs (e.g. the death of a patient in medical applications) it seems straightforward to apply these concepts to the analysis of dwell times in web usage mining applications.

With regard to dwell time distributions we assume that they follow a Weibull distribution with density function $f(t) = \lambda\gamma t^{\gamma-1}\exp(-\lambda t^{\gamma})$, where λ is a scale parameter and γ the shape parameter. For modeling the heterogeneity of the observed population, we assume K latent segments of sessions. While the Weibull assumption holds within all segments, different segments exhibit different parameter values. This leads to the underlying idea of a *Weibull mixture model*. For each page category p $(p = 1, \ldots, P)$ under consideration the resulting mixture has the following form

$$f(t_p) = \sum_{k=1}^{K} \pi_k f(t_p; \lambda_{pk}, \gamma_{pk}) = \sum_{k=1}^{K} \pi_k \lambda_{pk} \gamma_{pk} t_p^{\gamma_{pk}-1} \exp(-\lambda_{pk} t_p^{\gamma_{pk}}) \qquad (1)$$

where t_p represents the dwell time on page category p with mixing proportions π_k which correspond to the relative size of each segment.

In order to reduce the number of parameters involved we impose restrictions on the hazard rates of different components of the mixture respectively pages. An elegant way of doing this is offered by the concept of *Weibull proportional hazards models* (WPHM). The general formulation of a WPHM (see e.g., Kalbfleisch and Prentice (1980)) is

$$h(t; Z) = \lambda\gamma t^{\gamma-1}\exp(Z\beta). \qquad (2)$$

where Z is a matrix of covariates, and β are the regression parameters. The term $\lambda\gamma t^{\gamma-1}$ is the baseline hazard rate $h_0(t)$ due to the Weibull assumption and $h(t;Z)$ hazard proportional to $h_0(t)$ resulting from the regression part in the model.

2.2 Parsimonious modeling strategies

We propose five different models with respect to different proportionality restrictions in the hazard rates as to reduce the number of parameters. In the `mixPHM` package by Mair and Hudec (2007) the most general model is called `separate`: The WPHM is computed for each component and page separately. Hence, the hazard of session i belonging to component k ($k = 1,\dots,K$) on page category p ($p = 1,\dots,P$) is

$$h(t_{i,p};1) = \lambda_{k,p}\gamma_{k,p}t_{i,p}^{\gamma_{k,p}-1}\exp(\beta 1). \tag{3}$$

The parameter matrices can be represented jointly as

$$\Lambda = \begin{pmatrix} \lambda_{1,1} & \dots & \lambda_{1,P} \\ \vdots & \ddots & \vdots \\ \lambda_{K,1} & \dots & \lambda_{K,P} \end{pmatrix} \tag{4}$$

for the scale parameters and

$$\Gamma = \begin{pmatrix} \gamma_{1,1} & \dots & \gamma_{1,P} \\ \vdots & \ddots & \vdots \\ \gamma_{K,1} & \dots & \gamma_{K,P} \end{pmatrix} \tag{5}$$

for the shape parameters. Both the scale and the shape parameters can vary freely and there is no assumption of hazard proportionality in the `separate` model. In fact, the parameters ($2 \times K \times P$ in total) are the same as they were estimated directly by using a Weibull mixture model.

Next, we impose a proportionality assumption across the latent components. In the classification version of the EM-algorithm (see next section) in each iteration step we have a "crisp" assignment of each session to a component. Thus, if we consider this component vector g as main effect in the WPHM, i.e., $h(t;g)$, we impose proportional hazards for the components across the pages (`main.g` in `mixPHM`). Again, the elements of the matrix Λ of scale parameters can vary freely, whereas the shape parameter matrix reduces to the vector $\Gamma = (\gamma_{1,1},\dots,\gamma_{1,P})$. Thus, the shape parameters are constant over the components and the number of parameters is reduced to $K \times P + P$.

If we impose page main effects in the WPHM, i.e., $h(t;p)$ or `main.p`, respectively, as before, the elements of Λ are not restricted at all but this time the shape parameters are constant over the pages, i.e., $\Gamma = (\gamma_{1,1},\dots,\gamma_{1,K})$. The total number of parameters is now $K \times P + K$.

For the main-effects model $h(t;g+p)$ we impose proportionality restrictions on both Λ and Γ such that the total number of parameters is reduced to $K+P$. For the scale parameter matrix proportionality restrictions of this `main.gp` model hold row-wise as well as column-wise:

$$\Lambda = \begin{pmatrix} \lambda_1 & c_2\lambda_1 & \dots & c_P\lambda_1 \\ \vdots & \vdots & \ddots & \vdots \\ \lambda_K & c_2\lambda_K & \dots & c_P\lambda_K \end{pmatrix} = \begin{pmatrix} \lambda_1 & \dots & \lambda_P \\ d_2\lambda_1 & \dots & d_2\lambda_P \\ \vdots & \ddots & \vdots \\ d_K\lambda_1 & \dots & d_K\lambda_1 \end{pmatrix}. \tag{6}$$

The c- and d-scalars are proportionality constants over the pages and components, respectively. The shape parameters are constant over the components and pages. Thus, Γ reduces to one shape parameter γ which implies that the hazard rates are proportional over components and pages.

To relax the rather restrictive assumption with respect to Λ we can extend the main effects model by the corresponding component-page interaction term, i.e., $h(t; g * p)$. In mixPHM notation this model is called int.gp. The elements of Λ can vary freely whereas Γ is again reduced to one parameter only, leaving us with a total number of parameters of $K \times P + 1$. With respect to the hazard rate this relaxation implies again proportional hazards over components and pages.

2.3 EM-estimation of parameters

In order to estimate such mixtures of WPHM, we use the EM-Algorithm (Dempster et al. (1977), McLachlan and Krishnan (1997)). In the E-Step we establish the expected likelihood values for each session with respect to the K components. At this point it is important to take into account the probability that a session i of component k visits page p, denoted by $Pr_{k,p}$, which is estimated by the corresponding relative frequency. The elements of the resulting $K \times P$ matrix are model parameters and have to be taken into account when determining the total number of parameters. The resulting likelihood $\tau_{k,p}(s_i)$ for session i being in component k for each page p individually, is

$$\tau_{k,p}(s_i) = \begin{cases} f(y_p; \hat{\lambda}_{k,p}, \hat{\gamma}_{k,p}) Pr_{k,p}(s_i) & \text{if } p \text{ was visited by } s_i \\ 1 - Pr_{k,p}(s_i) & \text{if } p \text{ was not visited by } s_i \end{cases} \tag{7}$$

To establish the joint likelihood, a crucial assumption is made: independence of the dwell times over page-categories. To make this assumption feasible, a well-advised page categorization must be established. For instance, if some page-categories would be hierarchical, the independence assumption would not hold. Without this independence assumption, a multivariate mixture Weibull model would have to be fitted which takes into account the covariance structure of the observations. This would require that each session must have a full observation vector of length p, i.e, each page category is visited within each session which seems not to be realistic within the context of dwell times in web usage mining.

However, for a reasonable independence assumption the likelihood over all pages that session i belongs to component k is given by

$$L_k(s_i) = \prod_{p=1}^{P} \tau_{k,p}(s_i). \tag{8}$$

Thus, by looking at each session i separately, a vector of likelihood values $\Psi_i = (L_1(s_i), L_2(s_i), \ldots, L_k(s_i))$ results.

At this point, the M-step is carried out. The mixPHM package provides three different methods. The classical version of the EM-algorithm (*maximization EM*;

`EMoption = "maximization"` in `mixPHM`) computes the posterior probabilities that session i belongs to group k and does not make a group assignment within each iteration step but rather updates the matrix of posterior probabilities Q. A faster EM-version is proposed by Celeux and Govaert (1992) which they call *classification EM* (`EMoption = "classification"` in `mixPHM`): Within each iteration step a group assignment is performed due to $\sup_k(\Psi_i)$. Hence, the computation of the posterior matrix is not needed. A randomized version of the M-step considers a combination of the approaches above: After the computation of the posterior matrix Q, a randomized group assignment is performed due to the corresponding probability values (`EMoption = "randomization"`).

As usual, the joint likelihood L is updated at each EM-iteration l until a certain convergence criterion ε is reached, i.e., $\left|L^{(l)} - L^{(l-1)}\right| < \varepsilon$. Theoretical issues about the EM-convergence in Weibull mixture models can be found in Ishwaran (1996) and Jewell (1982).

3 Real life example

In this section we use a real dataset of a large Austrian company which runs a web-shop to demonstrate our modeling approach. We restrict empirical analysis to a subset of 333 buying-sessions and 7 page-categories we perform a dwell time based clustering with corresponding proportionality hazard assumptions by using the `mixPHM` package in R (R Development Core Team, 2007).

```
   bestview checkout service figurines jewellery landing search
6        16      592      30        12       183       0     13
15      136      157       0       139       430      11      0
23      428     2681      17      2058      2593      56    186
37      184      710      52        12       450      34      0
61        0      874     307       570         6      25     53
```

The above extract of the data matrix shows the dwell times of 5 sessions, while we coded non-visited page categories as 0's.

We start with a rather exploratory approach to determine an appropriate proportionality model with an adequate number of clusters K. By using the `msBIC` statement we can accomplish such a heuristic model search.

```
> res.bic <- msBIC(x,K=2:5,method="all")
> res.bic

Bayes Information Criteria
Survival distribution: Weibull

            K = 2    K = 3    K = 4    K = 5
separate 23339.27 23202.23 23040.01 22943.11
main.g   23355.66 23058.25 22971.86 22863.43
main.p   23503.73 23368.77 23165.60 23068.47
int.gp   23572.21 23422.51 23305.63 23075.76
main.gp  23642.74 23396.51 23271.72 23087.64
```

It is obvious that the main.g model with $K = 5$ components fits quite well compared to the other models (if we fit models for $K > 5$ the *BIC*'s do not decrease perspicuously anymore). For the sake of demonstration of the imposed hazard proportionalities, we compare this model to the more flexible separate model. First, we fit the two models again by using the phmclust statement which is the core routine of the mixPHM package. The matrices of shape parameters Γ_{sep} and Γ_g, respectively, for the first 5 pages (due to limited space) are:

```
> res.sep <- phmclust(x,5,method="separate")
> res.sep$shape[, 1:5]

           bestview checkout   service figurines jewellery
Component1 3.686052 2.692687 0.8553160 0.9057708 1.2503048
Component2 1.327496 3.393152 1.6260679 0.9716507 0.9941698
Component3 1.678135 2.829635 1.0417360 1.0706117 0.6902553
Component4 1.067241 1.847353 0.9860697 0.9339892 0.6321027
Component5 1.369876 2.030376 1.4565000 0.6434554 1.2414859

> res.g <- phmclust(x,5,method="main.g")
> res.g$shape[, 1:5]

           bestview checkout  service figurines jewellery
Component1 1.362342 2.981528 1.116042 0.7935599 0.9145463
Component2 1.362342 2.981528 1.116042 0.7935599 0.9145463
Component3 1.362342 2.981528 1.116042 0.7935599 0.9145463
Component4 1.362342 2.981528 1.116042 0.7935599 0.9145463
Component5 1.362342 2.981528 1.116042 0.7935599 0.9145463
```

The shape parameters in the latter model are constant across components. As a consequence, page-wise within group hazard rates can vary freely for both models, while the group-wise within page hazard rates can cross only for the separate model (see Figure 1).

From Figure 2 it is obvious that the hazards are proportional across components for each page. Note that due to space limitations, in both plots we only used three selected pages to demonstrate the hazard characteristics. The hazard plots allow to asses the relevance of different page categories with respect to cluster formation. Similar plots for dwell time distributions are available.

4 Conclusion

In this work we presented a flexible framework to analyze dwell times on web pages by adopting concepts from survival analysis to probability based clustering. Unobserved heterogeneity is modeled by mixtures of Weibull distributed dwell times. Application of the EM-algorithm leads to a segmentation of sessions.

Since the Weibull distribution is rather highly parameterized it offers a sizeable amount of flexibility for the hazard rates. A more parsimonious modeling may either be achieved by posing proportionality restrictions on the hazards or making use of simpler distributional assumptions (e.g., for constant hazard rates). The

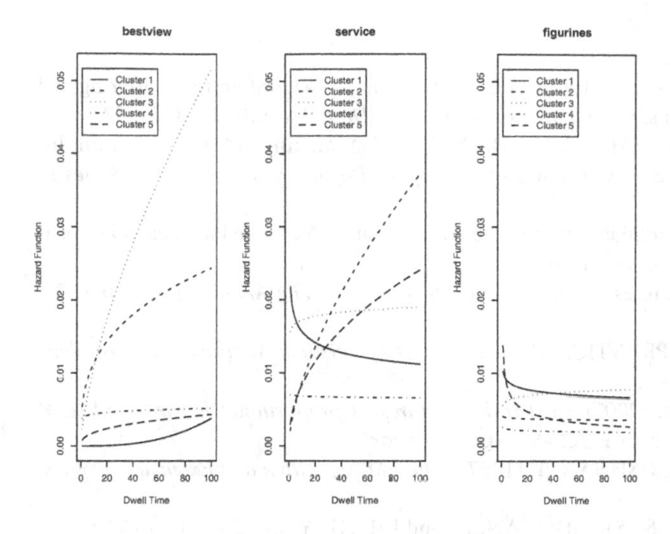

Fig. 1. Hazard Plot for Model `separate`

`mixPHM` package covers therefore additional survival distributions such as Exponential, Rayleigh, Gaussian, and Log-logistic.

A segmentation of sessions as it is achieved by our method may serve as a starting point for optimization of a website. Identification of typical user behavior allows an efficient dynamic modification of content as well as an optimization of adverts for different groups of users.

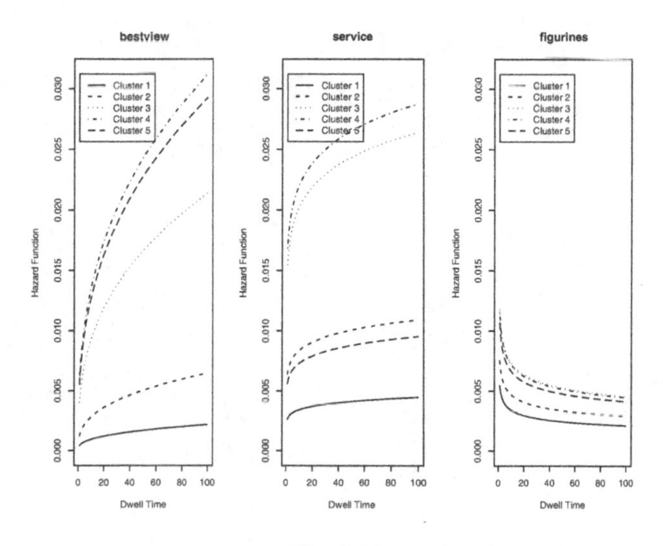

Fig. 2. Hazard Plot for Model `main.g`

References

CELEUX, G., and GOVAERT, G. (1992). A Classification EM Algorithm for Clustering and Two Stochastic Versions. *Computational Statistics & Data Analysis, 14, 315–332.*

DEMPSTER, A.P., LAIRD, N.M. and RUBIN, D.B. (1977). Maximum Likelihood from Incomplete Data via the EM-Algorithm. *Journal of the Royal Statistical Society, Series B, 39, 1–38.*

ISHWARAN, H. (1996). Identifiability and Rates of Estimation for Scale Parameters in Location Mixture Models. *The Annals of Statistics, 24, 1560-1571.*

JEWELL, N.P. (1982). Mixtures of Exponential Distributions. *The Annals of Statistics, 24, 479–484.*

KALBFLEISCH, J.D. and PRENTICE, R.L. (1980): *The Statistical Analysis of Failure Time Data*. Wiley, New York.

MAIR, P. and HUDEC, M. (2007). *mixPHM: Mixtures of proportional hazard models.* R package version 0.5.0: `http://CRAN.R-project.org/`

MCLACHLAN, G.J. and KRISHNAN, T. (1997). *The EM Algorithm and Extensions.* Wiley, New York.

MONTGOMERY, A.L., LI, S., SRINIVASAN, K. and LIECHTY, J.C. (2004). Modeling online browsing and path analysis using clickstream data. *Marketing Science, 23, 579–595.*

PARK, Y. and FADER, P.S. (2004). Modeling browsing behavior at multiple websites. *Marketing Science, 23, 280–303*

R Development Core Team. (2007). *R: A Language and Environment for Statistical Computing*. Vienna, Austria. (ISBN 3-900051-07-0)

New Issues in Near-duplicate Detection

Martin Potthast and Benno Stein

Bauhaus University Weimar
99421 Weimar, Germany
{martin.potthast, benno.stein}@medien.uni-weimar.de

Abstract. Near-duplicate detection is the task of identifying documents with almost identical content. The respective algorithms are based on fingerprinting; they have attracted considerable attention due to their practical significance for Web retrieval systems, plagiarism analysis, corporate storage maintenance, or social collaboration and interaction in the World Wide Web.

Our paper presents both an integrative view as well as new aspects from the field of near-duplicate detection: (*i*) *Principles and Taxonomy.* Identification and discussion of the principles behind the known algorithms for near-duplicate detection. (*ii*) *Corpus Linguistics.* Presentation of a corpus that is specifically suited for the analysis and evaluation of near-duplicate detection algorithms. The corpus is public and may serve as a starting point for a standardized collection in this field. (*iii*) *Analysis and Evaluation.* Comparison of state-of-the-art algorithms for near-duplicate detection with respect to their retrieval properties. This analysis goes beyond existing surveys and includes recent developments from the field of hash-based search.

1 Introduction

In this paper two documents are considered as near-duplicates if they share a very large part of their vocabulary. Near-Duplicates occur in many document collections, from which the most prominent one is the World Wide Web. Recent studies of Fetterly *et al.* (2003) and Broder *et al.* (2006) show that about 30% of all Web documents are duplicates of others. Zobel and Bernstein (2006) give examples which include mirror sites, revisions and versioned documents, or standard text building blocks such as disclaimers. The negative impact of near-duplicates on Web search engines is threefold: indexes waste storage space, search result listings can be cluttered with almost identical entries, and crawlers have a high probability of exploring pages whose content is already acquired.

Content duplication also happens through text plagiarism, which is the attempt to present other people's text as own work. Note that in the plagiarism situation document content is duplicated at the level of short passages; plagiarized passages can also be modified to a smaller or larger extent in order to obscure the offense.

Aside from deliberate content duplication, copying happens also accidentally: in companies, universities, or public administrations documents are stored multiple times, simply because employees are not aware of already existing previous work (Forman *et al.* (2005)). A similar situation is given for social software such as customer review boards or comment boards, where many users publish their opinion about some topic of interest: users with the same opinion write essentially the same in diverse ways since they read not all existing contributions.

A solution to the outlined problems requires a reliable recognition of near-duplicates – preferably at a high runtime performance. These objectives compete with each other, a compromise in recognition quality entails deficiencies with respect to retrieval precision and retrieval recall. A reliable approach to identify two documents d and d_q as near-duplicates is to represent them under the vector space model, referred to as $\mathbf{d}$ and $\mathbf{d}_q$, and to measure their similarity under the l_2-norm or the enclosed angle. d and d_q are considered as near-duplicates if the following condition holds:

$$\varphi(\mathbf{d}, \mathbf{d}_q) \geq 1 - \varepsilon \quad \text{with } 0 < \varepsilon \langle 1,$$

where φ denotes a similarity function that maps onto the interval $[0,1]$. To achieve a recall of 1 with this approach, each pair of documents must be analyzed. Likewise, given d_q and a document collection D, the computation of the set D_q, $D_q \subset D$, with all near-duplicates of d_q in D, requires $O(|D|)$, say, linear time in the collection size. The reason lies in the high dimensionality of the document representation $\mathbf{d}$, where "high" means "more than 10": objects represented as high-dimensional vectors cannot be searched efficiently by means of space partitioning methods such as kd-trees, quad-trees, or R-trees but are outperformed by a sequential scan (Weber *et al.* (1998)). By relaxing the retrieval requirements in terms of precision and recall the runtime performance can be significantly improved. Basic idea is to estimate the similarity between d and d_q by means of fingerprinting. A fingerprint, F_d, is a set of k numbers computed from d. If two fingerprints, F_d and F_{d_q}, share at least κ numbers, $\kappa \leq k$, it is assumed that d and d_q are near-duplicates. I. e., their similarity is estimated using the Jaccard coefficient:

$$\frac{|F_d \cap F_{d_q}|}{|F_d \cup F_{d_q}|} \geq \frac{\kappa}{k} \quad \Rightarrow \quad P\big(\varphi(\mathbf{d}, \mathbf{d}_q) \geq 1 - \varepsilon\big) \text{ is close to 1}$$

Let $F_D = \bigcup_{d \in D} F_d$ denote the union of the fingerprints of all documents in D, let $\mathcal{D}$ be the power set of D, and let $\mu : F_D \to \mathcal{D},\ x \mapsto \mu(x)$, be an inverted file index that maps a number $x \in F_D$ on the set of documents whose fingerprints contain x; $\mu(x)$ is also called the postlist of x. For document d_q with fingerprint F_{d_q} consider now the set $\hat{D}_q \subset D$ of documents that occur in at least κ of the postlists $\mu(x)$, $x \in F_{d_q}$. Put another way, $\hat{D}_q$ consists of documents whose fingerprints share a least κ numbers with F_{d_q}. We use $\hat{D}_q$ as a heuristic approximation of D_q, whereas the retrieval performance, which depends on the finesse of the fingerprint construction, computes as follows:

$$prec = \frac{\hat{D}_q \cap D_q}{\hat{D}_q}, \qquad rec = \frac{\hat{D}_q \cap D_q}{D_q}$$

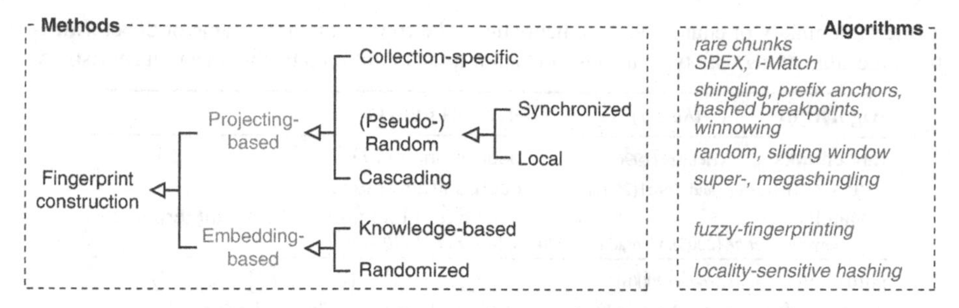

Fig. 1. Taxonomy of fingerprint construction methods (left) and algorithms (right).

The remainder of the paper is organized as follows. Section 2 gives an overview of fingerprint construction methods and classifies them in a taxonomy, including so far unconsidered hashing technologies. In particular, different aspects of fingerprint construction are contrasted and a comprehensive view on their retrieval properties is presented. Section 3 deals with evaluation methodologies for near-duplicate detection and proposes a new benchmark corpus of realistic size. The state-of-the-art fingerprint construction methods are subject to an experimental analysis using this corpus, providing new insights into precision and recall performance.

2 Fingerprint construction

A chunk or an n-gram of a document d is a sequence of n consecutive words found in d.[1] Let C_d be the set of all different chunks of d. Note that C_d is at most of size $|d| - n$ and can be assessed with $O(|d|)$. Let $\mathbf{d}$ be a vector space representation of d where each $c \in C_d$ is used as descriptor of a dimension with a non-zero weight.

According to Stein (2007) the construction of a fingerprint from $\mathbf{d}$ can be understood as a three-step-procedure, consisting of dimensionality reduction, quantization, and encoding:

1. Dimensionality reduction is realized by projecting or by embedding. Algorithms of the former type select dimensions in $\mathbf{d}$ whose values occur unmodified in the reduced vector $\mathbf{d}'$. Algorithms of the latter type reformulate $\mathbf{d}$ as a whole, maintaining as much information as possible.
2. Quantization is the mapping of the elements in $\mathbf{d}'$ onto small integer numbers, obtaining $\mathbf{d}''$.
3. Encoding is the computing of one or several codes from $\mathbf{d}''$, which together form the fingerprint of d.

Fingerprint algorithms differ primarily in the employed dimensionality reduction method. Figure 1 organizes the methods along with the known construction algorithms; the next two subsections provide a short characterization of both.

[1] If the hashed breakpoint chunking strategy of Brin *et al.* (1995) is applied, n can be understood as expected value of the chunk length.

Table 1. Summary of chunk selection heuristics. The rows contain the name of the construction algorithm along with typical constraints that must be fulfilled by the selection heuristic σ.

Algorithm (Author)	**Selection heuristic** $\sigma(c)$
rare chunks (Heintze (1996))	c occurs once in D
SPEX (Bernstein and Zobel (2004))	c occurs at least twice in D
I-Match (Chowdhury *et al.* (2002), Conrad *et al.* (2003), Kołcz *et al.* (2004))	$c = d$; excluding non-discriminant terms of d
shingling (Broder (2000))	$c \in \{c_1, \ldots, c_k\}, \; \{c_1, \ldots, c_k\} \subset_{rand} C_d$
prefix anchor (Manber (1994))	c starts with a particular prefix, or
(Heintze (1996))	c starts with a prefix which is infrequent in d
hashed breakpoints (Manber (1994))	$h(c)$'s last byte is 0, or
(Brin *et al.* (1995))	c's last word's hash value is 0
winnowing (Schleimer *et al.* (2003))	c minimizes $h(c)$ in a window sliding over d
random (misc.)	c is part of a local random choice from C_d
one of a sliding window (misc.)	c starts at word i mod m in d; $1 \leq m \leq \lvert d \rvert$
super- / megashingling (Broder (2000) / Fetterly *et al.* (2003))	c is a combination of hashed chunks which have been selected with shingling

2.1 Dimensionality reduction by projecting

If dimensionality reduction is done by projecting, a fingerprint F_d for document d can be formally defined as follows:

$$F_d = \{h(c) \mid c \in C_d \text{ and } \sigma(c) = true\},$$

where σ denotes a selection heuristic for dimensionality reduction that becomes true if a chunk fulfills a certain property. h denotes a hash function, such as MD5 or Rabin's hash function, which maps chunks to natural numbers and serves as a means for quantization. Usually the identity mapping is applied as encoding rule. Broder (2000) describes a more intricated encoding rule called supershingling.

The objective of σ is to select chunks to be part of a fingerprint which are best-suited for a reliable near-duplicate identification. Table 1 presents in a consistent way algorithms and the implemented selection heuristics found in the literature, whereas a heuristic is of one of the types denoted in Figure 1.

2.2 Dimensionality reduction by embedding

An embedding-based fingerprint F_d for a document d is typically constructed with a technique called "similarity hashing" (Indyk and Motwani (1998)). Unlike standard hash functions, which aim to a minimization of the number of hash collisions, a similarity hash function $h_\varphi : \mathbf{D} \to U$, $U \subset \mathbf{N}$, shall produce a collision with a high probability for two objects $\mathbf{d}, \mathbf{d}_q \in \mathbf{D}$, iff $\varphi(\mathbf{d}, \mathbf{d}_q) \geq 1 - \varepsilon$. In this way h_φ downgrades a fine-grained similarity relation quantified within φ to the concept "similar or not similar", reflected by the fact whether or not the hashcodes $h_\varphi(\mathbf{d})$ and $h_\varphi(\mathbf{d}_q)$ are

Table 2. Summary of complexities for the construction of a fingerprint, the retrieval, and the size of a tailored chunk index.

Algorithm	Runtime Construction	Runtime Retrieval	Chunk length	Finger-print size	Chunk index size
rare chunks	$O(\|d\|)$	$O(\|d\|)$	n	$O(\|d\|)$	$O(\|d\| \cdot \|D\|)$
SPEX $(0 < r \langle 1)$	$O(\|d\|)$	$O(r \cdot \|d\|)$	n	$O(r \cdot \|d\|)$	$O(r \cdot \|d\| \cdot \|D\|)$
I-Match	$O(\|d\|)$	$O(k)$	$\|d\|$	$O(k)$	$O(k \cdot \|D\|)$
shingling	$O(\|d\|)$	$O(k)$	n	$O(k)$	$O(k \cdot \|D\|)$
prefix anchor	$O(\|d\|)$	$O(\|d\|)$	n	$O(\|d\|)$	$O(\|d\| \cdot \|D\|)$
hashed breakpoints	$O(\|d\|)$	$O(\|d\|)$	$E(\|c\|) = n$	$O(\|d\|)$	$O(\|d\| \cdot \|D\|)$
winnowing	$O(\|d\|)$	$O(\|d\|)$	n	$O(\|d\|)$	$O(\|d\| \cdot \|D\|)$
random	$O(\|d\|)$	$O(k)$	n	$O(k)$	$O(\|d\| \cdot \|D\|)$
one of sliding window	$O(\|d\|)$	$O(\|d\|)$	n	$O(\|d\|)$	$O(\|d\| \cdot \|D\|)$
super- / megashingling	$O(\|d\|)$	$O(k)$	n	$O(k)$	$O(k \cdot \|D\|)$
fuzzy-fingerprinting	$O(\|d\|)$	$O(k)$	$\|d\|$	$O(k)$	$O(k \cdot \|D\|)$
locality-sensitive hashing	$O(\|d\|)$	$O(k)$	$\|d\|$	$O(k)$	$O(k \cdot \|D\|)$

identical. To construct a fingerprint F_d for document d a small number of k variants of h_φ are used:

$$F_d = \{h_\varphi^{(i)}(\mathbf{d}) \mid i \in \{1, \ldots, k\}\}$$

Two kinds of similarity hash functions have been proposed, which either compute hashcodes based on knowledge about the domain or which ground on domain-independent randomization techniques (see again Figure 1). Both similarity hash functions compute hashcodes along the three steps outlined above: An example for the former is fuzzy-fingerprinting developed by Stein (2005), where the embedding step relies on a tailored, low-dimensional document model and where fuzzification is applied as a means for quantization. An example for the latter is locality-sensitive hashing and the variants thereof by Charikar (2002) and Datar *et al.* (2004). Here the embedding relies on the computation of scalar products of $\mathbf{d}$ with random vectors, and the scalar products are mapped on predefined intervals on the real number line as a means for quantization. In both approaches the encoding happens according to a summation rule.

2.3 Discussion

We have analyzed the aforementioned fingerprint construction methods with respect to construction time, retrieval time, and the resulting size of a complete chunk index. Table 2 compiles the results.

The construction of a fingerprint for a document d depends on its length since d has to be parsed at least once, which explains that all methods have the same complexity in this respect. The retrieval of near-duplicates requires a chunk index μ as described at the outset: μ is queried with each number of a query document's fingerprint F_{d_q}, for which the obtained postlists are merged. We assume that both the lookup time and the average length of a postlist can be assessed with a constant for either method.[2] Thus the retrieval runtime depends only on the size k of a fingerprint. Observe that the construction methods fall into two groups: methods whose fingerprint's size increases with the length of a document, and methods where k is independent of $|d|$. Similarly, the size of μ is affected. We further differentiate methods with fixed length fingerprints into these which construct small fingerprints where $k \leq 10$ and those where $10 \langle k < 500$. Small fingerprints are constructed by fuzzy-fingerprinting, locality-sensitive hashing, supershingling, and I-Match; these methods outperform the others by orders of magnitude in their chunk index size.

3 Wikipedia as evaluation corpus

When evaluating near-duplicate detection methods one faces the problem of choosing a corpus which is representative for the retrieval situation and which provides a realistic basis to measure both retrieval precision and retrieval recall. Today's standard corpora such as the TREC or Reuters collection have deficiencies in this connection: In standard corpora the distribution of similarities decreases exponentially from a very high percentage at low similarity intervals to a very low percentage at high similarity intervals. Figure 2 (right) illustrates this characteristic at the Reuters corpus. This characteristic allows only precision evaluations since the recall performance depends on very few pairs of documents. The corpora employed in recent evaluations of Hoad and Zobel (2003), Henzinger (2006), and Ye *et al.* (2006) lack in this respect; moreover, they are custom-built and not publicly available. Conrad and Schriber (2004) attempt to overcome this issue by the artificial construction of a suitable corpus.

Wikipedia corpus:

Property	**Value**
documents	6 Million
revisions	80 Million
size (uncompressed)	1 terabyte

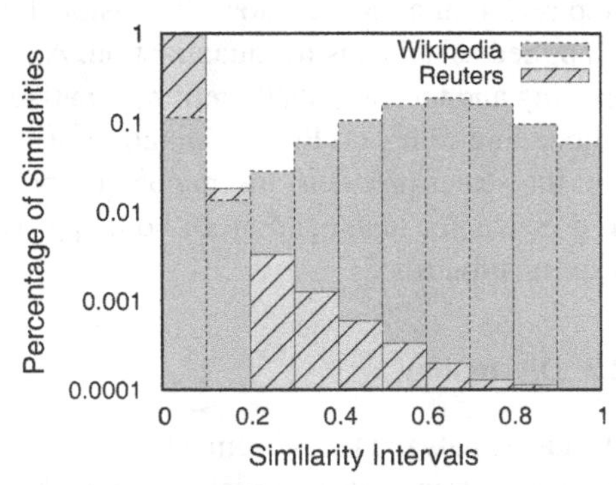

Fig. 2. The table (left) shows order of magnitudes of the Wikipedia corpus. The plot contrasts the similarity distribution within the Reuters Corpus Volume 1 and the Wikipedia corpus.

[2] We indexed all English Wikipedia articles and found that an increase from 3 to 4 in the chunk length implies a decrease from 2.42 to 1.42 in the average postlist length.

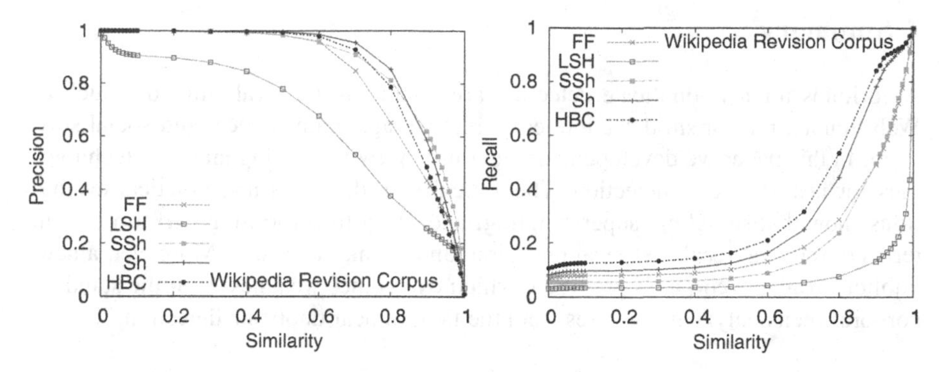

Fig. 3. Precision and recall over similarity for fuzzy-fingerprinting (FF), locality-sensitive hashing (LSH), supershingling (SSh), shingling (Sh), and hashed breakpoint chunking (HBC).

We propose to use the Wikipedia Revision Corpus for near-duplicate detection including all revisions of every Wikipedia article.[3] The table in Figure 2 shows selected order of magnitudes of the corpus. A preliminary analysis shows that an article's revisions are often very similar to each other with an expected similarity of about 0.5 to the first revision. Since the articles of Wikipedia undergo a regular rephrasing, the corpus addresses the particularities of the use cases mentioned at the outset. We analyzed the fingerprinting algorithms with 7 Million pairs of documents, using the following strategy: each article's first revision serves as query document d_q and is compared to all other revisions as well as to the first revision of its immediate successor article. The former ensures a large number of near-duplicates and hence improves the reliability of the recall values; rationale of the latter is to gather sufficient data to evaluate the precision (cf. Figure 2, right-hand side).

Figure 3 presents the results of our experiments in the form of precision-over-similarity curves (left) and recall-over-similarity curves (right). The curves are computed as follows: For a number of similarity thresholds from the interval $[0; 1]$ the set of document pairs whose similarity is above a certain threshold is determined. Each such set is compared to the set of near-duplicates identified by a particular fingerprinting method. From the intersection of these sets then the threshold-specific precision and recall values are computed in the standard way.

As can be seen in the plots, the chunking-based methods perform better than similarity hashing, while hashed breakpoint chunking performs best. Of those with fixed size fingerprints shingling performs best, and of those with small fixed size fingerprints fuzzy-fingerprinting and supershingling perform similar. Note that the latter had both 50 times smaller fingerprints than shingling which shows the possible impact of theses methods on the size of a chunk index.

[3] http://en.wikipedia.org/wiki/Wikipedia:Download, last visit on February 27, 2008

4 Summary

Algorithms for near-duplicate detection are applied in retrieval situations such as Web mining, plagiarism detection, corporate storage maintenance, and social software. In this paper we developed an integrative view to existing and new technologies for near-duplicate detection. Theoretical considerations and practical evaluations show that shingling, supershingling, and fuzzy-fingerprinting perform best in terms of retrieval recall, retrieval precision, and chunk index size. Moreover, a new, publicly available corpus is proposed, which overcomes weaknesses of the standard corpora when analyzing use cases from the field of near duplicate detection.

References

BERNSTEIN, Y. and ZOBEL, J. (2004): A scalable system for identifying co-derivative documents, *Proc. of SPIRE '04*.

BRIN, S., DAVIS, J. and GARCIA-MOLINA, H. (1995): Copy detection mechanisms for digital documents, *Proc. of SIGMOD '95*.

BRODER, A. (2000): Identifying and filtering near-duplicate documents, *Proc. of COM '00*.

BRODER, A., EIRON, N., FONTOURA, M., HERSCOVICI, M., LEMPEL, R., MCPHERSON, J., QI, R. and SHEKITA, E. (2006): Indexing Shared Content in Information Retrieval Systems, *Proc. of EDBT '06*.

CHARIKAR, M. (2002): Similarity Estimation Techniques from Rounding Algorithms, *Proc. of STOC '02*.

CHOWDHURY, A., FRIEDER, O., GROSSMAN, D. and MCCABE, M. (2002): Collection statistics for fast duplicate document detection, *ACM Trans. Inf. Syst.,20*.

CONRAD, J., GUO, X. and SCHRIBER, C. (2003): Online duplicate document detection: signature reliability in a dynamic retrieval environment, *Proc. of CIKM '03*.

CONRAD, J. and SCHRIBER, C. (2004): Constructing a text corpus for inexact duplicate detection, *Proc. of SIGIR '04*.

DATAR, M., IMMORLICA, N., INDYK, P. and MIRROKNI, V. (2004): Locality-Sensitive Hashing Scheme Based on p-Stable Distributions, *Proc. of SCG '04*.

FETTERLY, D., MANASSE, M. and NAJORK, M. (2003): On the Evolution of Clusters of Near-Duplicate Web Pages, *Proc. of LA-WEB '03*.

FORMAN, G., ESHGHI, K. and CHIOCCHETTI, S. (2005): Finding similar files in large document repositories, *Proc. of KDD '05*.

HEINTZE, N. (1996): Scalable document fingerprinting, *Proc. of USENIX-EC '96*.

HENZINGER, M. (2006): Finding Near-Duplicate Web Pages: a Large-Scale Evaluation of Algorithms, *Proc. of SIGIR '06*.

HOAD, T. and ZOBEL, J. (2003): Methods for Identifying Versioned and Plagiarised Documents, *Jour. of ASIST, 54*.

INDYK, P. and MOTWANI, R. (1998): Approximate Nearest Neighbor—Towards Removing the Curse of Dimensionality, *Proc. of STOC '98*.

KOŁCZ, A., CHOWDHURY, A. and ALSPECTOR, J. (2004): Improved robustness of signature-based near-replica detection via lexicon randomization, *Proc. of KDD '04*.

MANBER, U. (1994): Finding similar files in a large file system, *Proc. of USENIX-TC '94*.

SCHLEIMER, S., WILKERSON, D. and AIKEN, A. (2003): Winnowing: local algorithms for document fingerprinting, *Proc. of SIGMOD '03*.

STEIN, B. (2005): Fuzzy-Fingerprints for Text-based Information Retrieval, *Proc. of I-KNOW '05*.

STEIN, B. (2007): Principles of Hash-based Text Retrieval, *Proc. of SIGIR '07*.

WEBER, R., SCHEK, H. and BLOTT, S. (1998): A Quantitative Analysis and Performance Study for Similarity-Search Methods in High-Dimensional Spaces, *Proc. of VLDB '98*.

YE, S., WEN, J. and MA, W. (2006): A Systematic Study of Parameter Correlations in Large Scale Duplicate Document Detection, *Proc. of PAKDD '06*.

ZOBEL, J. and BERNSTEIN, Y. (2006): The case of the duplicate documents: Measurement, search, and science, *Proc. of APWeb '06*.

Comparing the University of South Florida Homograph Norms with Empirical Corpus Data

Reinhard Rapp

Universitat Rovira i Virgili, GRLMC, Tarragona, Spain
reinhard.rapp@urv.cat

Abstract. The basis for most classification algorithms dealing with word sense induction and word sense disambiguation is the assumption that certain context words are typical of a particular sense of an ambiguous word. However, as such algorithms have been only moderately successful in the past, the question that we raise here is if this assumption really holds. Starting with an inventory of predefined senses and sense descriptors taken from the University of South Florida Homograph Norms, we present a quantitative study of the distribution of these descriptors in a large corpus. Hereby, our focus is on the comparison of co-occurrence frequencies between descriptors belonging to the same versus to different senses, and to the effects of considering groups of descriptors rather than single descriptors. Our findings are that descriptors belonging to the same sense co-occur significantly more often than descriptors belonging to different senses, and that considering groups of descriptors effectively reduces the otherwise serious problem of data sparseness.

1 Introduction

Resolving semantic ambiguities of words is among the core problems in natural language processing. Many applications, such as text understanding, question answering, machine translation, and speech recognition suffer from the fact that – despite numerous attempts (e.g. Kilgarriff and Palmer, 2000; Pantel and Lin, 2002; Rapp, 2004) – there is still no satisfactory solution to this problem. Although it seems reasonable that the statistical approach is the method of choice, it is not obvious what statistical clues should be looked at, and how to deal with the omnipresent problem of data sparseness.

In this situation, rather than developing another algorithm and adding it to the many that already exist, we found it more appropriate to systematically look at the empirical foundations of statistical word sense induction and disambiguation (Rapp, 2006). The basic assumption underlying most if not all corpus-based algorithms is the observation that each sense of an ambiguous word seems to be associated with certain context words. These context words can be considered to be indicators of this particular sense. For example, context words such as *grow* and *soil* are typical of the *flora* meaning of *plant*, whereas *power* and *manufacture* are typical of its industrial meaning. Being associated implies that the indicators should co-occur significantly more often than expected by chance in the local contexts of the respective ambiguous word. Looking only at local contexts can be justified by the observation that for humans in almost all cases the local context suffices to achieve an almost perfect disambiguation performance, which implies that the local context carries all essential information.

If there exist several indicators of the same sense, then it can not be ruled out, but it is probably unlikely that they are mutually exclusive. As a consequence, in the local contexts of an ambiguous word indicators of the same sense should have co-occurrence frequencies that are significantly higher than chance, whereas for indicators relating to different senses this should not be the case or, if so, only to a lesser extend.

Our aim in this study is to quantify this effect by generating statistics on the co-occurrence frequencies of sense indicators in a large corpus. Hereby, our inventory of ambiguous words, their senses, and their sense indicators is taken from the *University of South Florida Homograph Norms* (USFHN), and the co-occurrence counts are taken from the *British National Corpus* (BNC). As previous work (Rapp, 2006) showed that the problem of data sparseness is severe, we also propose a methodology for effectively dealing with it.

2 Resources

For the purpose of our study a list of ambiguous words is required together with their senses and some typical indicators of each sense. As described in Rapp (2006), such data was extracted from the USFHN. These norms were compiled by collecting the associative responses given by test persons to a list of 320 homographs, and by manually assigning each response to one of the homograph's meanings. Further details are given in Nelson et al. (1980).

For the current study, from this data we extracted a list of all 134 homographs where each comes together with five associated words that are typical of its first sense, and another five words that are typical of its second sense. The first ten entries in this list are shown in Table 1. Note that for reasons to be discussed later we abandoned all homographs where either the first or the second sense did not receive at least five different responses. This was the case for 186 homographs, which is the reason that our list comprises only 134 of the 320 items. As in the norms the homographs were written in uppercase letters only, we converted them to that spelling of uppercase and lowercase letters that was found to have the highest occurrence frequency in the BNC. In the few cases where subjects had responded with multi word units, these were disregarded unless one of the words carried almost all of the meaning.

Another resource that we use is the BNC, which is a balanced sample of written and spoken English that comprises about 100 million words. As described in Rapp (2006), this corpus was used without special pre-processing, and for each of the

134 homographs concordances were extracted comprising text windows of particular widths (e.g. ± 10 words around the given word).

Table 1. Some homographs and the top five associations for their two main senses.

HOMOG.	SENSE 1	SENSE 2
bar	drink beer tavern stool booze	crow bell handle gold press
beam	wood ceiling house wooden building	light laser sun joy radiate
bill	pay money payment paid me	John Uncle guy name person
block	stop tackle road buster shot	wood ice head cement substance
bluff	fool fake lie call game	cliff mountain lovely high ocean
board	wood plank wooden ship nails	chalk black bill game blackboard
bolt	nut door lock screw close	jump run leap upright colt
bound	tied tie gagged rope chained	jump leap bounce up over
bowl	cereal dish soup salad spoon	ball pins game dollars sport
break	ruin broken fix tear repair	out even away jail fast

3 Approach

For each homograph we were interested in three types of information. One is the average intra-sense association strength for sense 1, i.e. the average association strength between all possible pairs of words belonging to sense 1. Another is the average intra-sense association strength for sense 2, which is calculated analogously. And a third is the average inter-sense association strength between senses 1 and 2, i.e. the average association strength between all possible pairs of words under the condition that the two words in each pair must belong to different senses. Using the homograph *bar*, Figure 1 illustrates this by making explicit all pairs of associations that are involved in the computation of the average strengths. Hereby the association strength a_{ij} between two words i and j is computed as the number of lines in the concordance where both words co-occur (f_{ij}) divided by the product of the concordance frequencies f_i and f_j of the two words:

$$a_{ij} = \frac{f_{ij}}{f_i \cdot f_j}$$

This formula normalizes for word frequencies and thereby avoids undesired effects resulting from their tremendous variation. In cases where the denominator was zero we assigned a score of zero to the whole expression. Note that the counts in the denominator are observed word frequencies within the concordance, not within the entire corpus.

Whereas the above formula computes association strengths for single word pairs, what we are actually interested in are the three types of average association strengths a_{ij} as depicted in Figure 1. For ease of reference, in the remainder of the paper we use the following notation:

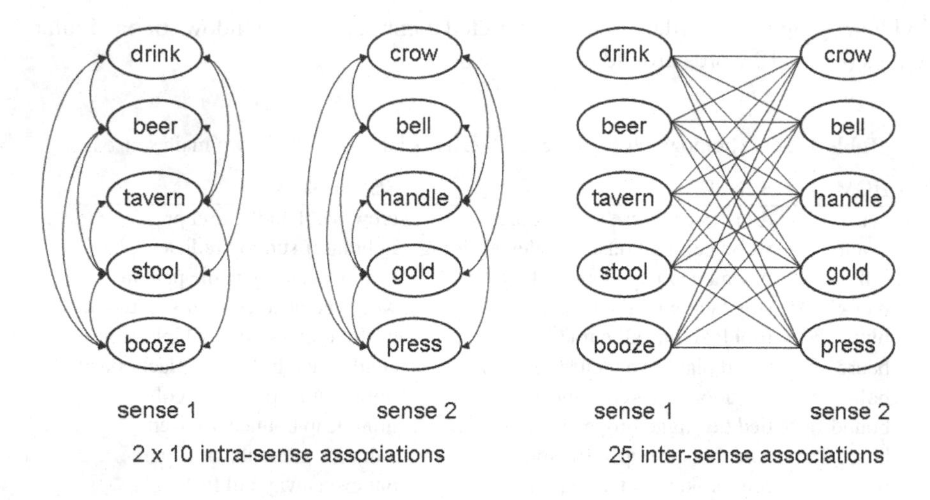

Fig. 1. Computation of average intra-sense and inter-sense association strengths exemplified using the associations to the homograph *bar*.

$S1 =$ average a_{ij} over the 10 word pairs relating to sense 1
$S2 =$ average a_{ij} over the 10 word pairs relating to sense 2
$SS =$ average a_{ij} over the 25 word pairs relating to senses 1 and 2

The reasoning behind computing average scores is to minimize the problem of data sparseness by taking many observations into account. An important feature of our setting is that if we – as described in the next section – increase the number of words that we consider (5 per sense in the example of Figure 1), the number of possible pairs increases quadratically, which means that this should be an effective measure for solving the sparse-data problem. Note that we could not sensibly go beyond five words in this study, as the number of associations provided in the USFHN is rather limited for each sense, so that there is only a small number of homographs where more than five associations are provided for the two main senses.

When comparing the scores $S1$, $S2$, and SS, what should be our expectations? Most importantly, as discussed in the introduction, same-sense co-occurrences should be more frequent than different-sense co-occurrences, thus both $S1$ and $S2$ should be larger than SS. But should $S1$ and $S2$ be at the same level or not? To answer this question, recall that $S1$ relates to the main sense of the respective homograph, and $S2$ to its secondary sense. If both senses are similarly frequent, then both scores are based on equally good data and can be expected to be at similar levels. However, if the vast majority of cases relates to the main sense, and if the secondary sense occurs only a few times in the corpus (an example being the word *can* with its frequent verb and infrequent noun sense), then the co-occurrence counts – which are always based on the entire concordance – would mainly reflect the behavior of the main sense, and might be only marginally influenced by the secondary sense. As will be shown in the next section, for our data $S2$ turns out to be at about the same level as $S1$. Note, however, that this could be an artefact of our choice of homographs, as we had to pick those where the subjects provided at least five different associative responses to each sense.

4 Results and discussion

Following the procedure as described in the previous section, Table 2 shows the first 10 out of 134 results for the homograph-based concordances of width ±100 with five associations being considered for each sense (for a list of these associations see Table 1). In all but two cases the values for $S1$ and $S2$ are – as expected – both larger than those for SS. Only the homographs *bill* and *break* behave unexpectedly. In the first case this may be explained by our way of dealing with capitalization (see section 2), in the second it is probably due to continuation associations such as *break – out* and *break – away*, which are not specifically dealt with in our system.

Note that the above qualitative considerations are only meant to give an impression of some underlying sophistications which make it unrealistic to expect an overall accuracy of 100%. Nevertheless, exact quantitative results are given in Table 3. For several concordance widths, this table shows the number of homographs where the results turn out to be as expected, i.e. $S1 > SS$ and $S2 > SS$ (columns 3 and 4). Perfect results would be indicated by values of 134 (i.e. the total number of homographs) in each of these two columns. The table also shows the number of cases where $S1 > S2$, and – as an additional information – the number of cases where $S1$ and $S2$, $S2$ and SS, and $S1$ and $S2$ are equal. As $S1$, $S2$, and SS are averages over several floating point values, their equality is very unlikely except for the case when all underlying co-occurrence scores are zero, which is only true if data is very sparse. Thus the equality scores can be seen as a measure of data sparseness. As data sparseness affects all three equality scores likewise, they can be expected to be at similar levels. Nevertheless, to confirm this expectation empirically, scores are shown for all three.

Table 2. Results for the first ten homographs (numbers to be multiplied by 10^{-6}).

Homograph	$S1$	$S2$	SS	Homograph	$S1$	$S2$	SS
bar	223	199	37	board	205	799	53
beam	1305	1424	123	bolt	1794	3747	962
bill	166	95	202	bound	675	692	139
block	194	945	112	bowl	327	644	25
bluff	934	2778	226	break	156	63	95

In Table 3, for each concordance width we also distinguish four cases where each relates to a different number of associations (or sense indicators) considered. Whereas so far we always assumed that for each homograph we take five associations into account that relate to its first, and another five associations that relate to its second sense (as depicted in Figure 1), it is of course also possible to reduce the number of associations considered to four, three, or two. (A reduction to one is not possible as in this case the intra-sense association strengths $S1$ and $S2$ are not defined.) This is what we did to obtain comparative values that enable us to judge whether an increase in the number of associations considered actually leads to the significant gains in accuracy that can be expected if our analysis from the previous section is correct.

Having described the meaning of the columns in Table 3, let us now look at the actual results. As mentioned above, the last three columns of the table give us information on data sparsity. For the concordance width of ± 1 word their values are fairly close to 134, which means that most co-occurrence frequencies are zero with the consequence that the results of columns 3 to 5 are not very informative. Of course, when looking at language use, this result comes not so unexpected, as the direct neighbors of content words are often function words, so that adjacent co-occurrences involving other content words are rare.

If we continue to look at the last three columns of Table 3, but now consider larger concordance widths, we see that the problem of data sparseness steadily decreases with larger widths, and that it also steadily decreases when we consider more associations. At a concordance width of ± 100 and when looking at a minimum of four associations, the problem of data sparsity seems to be rather small.

Next, let us look at column 5 ($S1 > S2$) which must be interpreted in conjunction with the last column ($S1 = S2$). In all cases its values are fairly close to its complement $S1 < S2$, which is not in the table but can be computed from the other two columns. For example, for the concordance width of ± 100 for the column $S1 > S2$ we get the readings 60, 67, 65, and 68 from Table 3, and can compute the corresponding values of 58, 61, 66, and 64 for $S1 < S2$. Both sequences appear very similar. Interpreted linguistically, this means that intra-sense association strengths tend to be similar for the primary and the secondary sense, at least for our selection of homographs.

Let us finally look at columns 3 and 4 of Table 3, which should give us an indication whether our co-occurrence based methodology has the potential to work if used in a system for word sense induction or disambiguation. Both columns indicate that we get improvements with larger context widths (up to 100) and when considering more associations. At a context width of ± 100 words and when considering all five associations the value for $S1 > SS$ reaches its optimum of 114. With two undecided cases, this means that the count for $S1 < SS$ is 18, i.e. the ratio of correct to incorrect cases is 6.33. This corresponds to a 85% accuracy, which appears to be a good result. However, the corresponding ratio for $S2$ is only 2.77, which is considerably worse and indicates that some of our previous discussion concerning the weaknesses of secondary senses (cf. section 3) – although not confirmed when comparing $S1$ to $S2$ – seems not unfounded. In future work, it would be of interest to explore if there is a relation between the relative occurrence-frequency of a secondary sense and its intra-sense association strength.

What we like best about the results is the gain in accuracy when the number of associations considered is increased. At the concordance width of ± 100 words we get 77 correct predictions ($S1 > SS$) when we take two associations into account, 97 with three, 108 with four, and 114 with five. The corresponding sequence of ratios ($S1 > SS$ / $S1 < SS$) looks even better: 1.88, 3.13, 4.70, and 6.33. This means that

Table 3. Overall quantitative results for several concordance widths and various numbers of associations considered.

Width	Assoc.	$S1 > SS$	$S2 > SS$	$S1 > S2$	$S1 = SS$	$S2 = SS$	$S1 = S2$
±1	2	1	0	1	131	132	133
	3	2	1	2	126	127	131
	4	3	3	3	123	123	128
	5	4	4	4	120	120	126
±3	2	15	10	13	105	108	107
	3	23	22	22	94	97	93
	4	36	30	29	74	79	79
	5	41	42	33	67	65	64
±10	2	34	32	31	78	70	75
	3	59	53	54	45	51	42
	4	85	64	67	24	35	19
	5	90	67	68	18	26	14
±30	2	59	54	47	43	39	45
	3	84	66	65	22	25	21
	4	95	86	65	16	14	11
	5	101	86	70	10	10	6
±100	2	77	74	60	18	22	16
	3	97	86	67	9	11	6
	4	108	94	65	4	4	3
	5	114	97	68	2	4	2
±300	2	85	77	68	8	11	10
	3	96	85	66	2	5	3
	4	105	94	63	1	1	2
	5	102	103	68	1	1	1
±1000	2	75	76	65	2	6	5
	3	82	89	59	1	3	1
	4	87	88	68	0	1	0
	5	90	92	68	0	0	1

with increasing number of associations the quadratic increase of possible word pairs leads to considerable improvements

5 Conclusions and future work

Our experiments showed that associations belonging to the same sense of a homograph have significantly higher co-occurrence counts than associations belonging to different senses. However, the big challenge is the omnipresent problem of data sparsity, which in many cases will not allow us to reliably observe this in a corpus. Our results suggest two strategies to minimize this problem: One is to look at the optimal window-size which in our setting was somewhat larger than average sentence length but is likely to depend on corpus size. The other is to increase the number of associations considered, and to look at the co-occurrences of all possible pairs of associations. Since the number of possible pairs increases quadratically with the number of words that are considered, this should have a strong positive effect on the sparse-data problem, which could be confirmed empirically. Both strategies to deal with the sparse-data problem can be applied in combination, seemingly without undesired interaction.

With the best settings of the two parameters, we obtained an accuracy of about 85%. This indicates that the statistical clues considered have the potential to work. In addition, we see numerous possibilities for further improvement: These include increasing the number of associations looked at, using a larger corpus, optimizing window size in a more fine-grained manner than presented in Table 3, trying out other association measures such as the log-likelihood ratio, and to use automatically generated associations instead of those produced by human subjects. Automatically generated associations have the advantage that they are based on the corpus used, so with regard to the sparse-data problem a better behavior can be expected.

Having shown how in our particular framework looking at groups of related words rather than looking at single words can significantly reduce the problem of data sparseness due to the quadratic increase in the number of possible relations, let us mention some more speculative implications of such methodologies: Our guess is that an analogous procedure should also be possible for other core problems in statistical language processing that are affected by data sparsity. On the theoretical side, the elementary mechanism of quadratic expansion would also be an explanation for the often unrivalled performance of humans, and it may eventually be the key to the solution of the poverty-of-the-stimulus problem.

Acknowledgments

This research was supported by a Marie Curie Intra-European Fellowship within the 6th Framework Programme of the European Community.

References

KILGARRIFF, A.; PALMER, M. (eds.) (2000). *International Journal of Computers and the Humanities. Special Issue on SENSEVAL*, 34(1–2), 2000.

NELSON, D.L.; MCEVOY, C.L.; WALLING, J.R.; WHEELER, J.W. (1980). The University of South Florida homograph norms. *Behavior Research Methods & Instrumentation* 12(1), 16–37.

PANTEL, P.; LIN, D. (2002). Discovering word senses from text. In: *Proceedings of ACM SIGKDD*, Edmonton, 613–619.

RAPP, R. (2004). A practical solution to the problem of automatic word sense induction. In: *Proceedings of the 42nd Annual Meeting of the Association for Computational Linguistics*, Comp. Vol., 195–198.

RAPP, R. (2006). Are word senses reflected in the distribution of words in text. In: Grzybek, P.; Köhler, R. (eds.): *Exact Methods in the Study of Language and Text. Dedicated to Professor Gabriel Altmann on the Occasion of his 75th Birthday*. Berlin: Mouton de Gruyter, 571–582.

Content-based Dimensionality Reduction for Recommender Systems

Panagiotis Symeonidis

Aristotle University, Department of Informatics,
Thessaloniki 54124, Greece
symeon@csd.auth.gr

Abstract. Recommender Systems are gaining widespread acceptance in e-commerce applications to confront the information overload problem. Collaborative Filtering (CF) is a successful recommendation technique, which is based on past ratings of users with similar preferences. In contrast, Content-based Filtering (CB) exploits information solely derived from document or item features (e.g. terms or attributes). CF has been combined with CB to improve the accuracy of recommendations. A major drawback in most of these hybrid approaches was that these two techniques were executed independently. In this paper, we construct a feature profile of a user based on both collaborative and content features. We apply Latent Semantic Indexing (LSI) to reveal the dominant features of a user. We provide recommendations according to this dimensionally-reduced feature profile. We perform experimental comparison of the proposed method against well-known CF, CB and hybrid algorithms. Our results show significant improvements in terms of providing accurate recommendations.

1 Introduction

Collaborative Filtering (CF) is a successful recommendation technique. It is based on past ratings of users with similar preferences, to provide recommendations. However, this technique introduces certain shortcomings. For instance, if a new item appears in the database, there is no way to be recommended before it is rated.

In contrast, Content-Based filtering (CB) exploits only information derived from document or item features (e.g., terms or attributes). Latent Semantic Indexing (LSI) has been extensively used in the CB field, in detecting the latent semantic relationships between terms and documents. LSI constructs a low-rank approximation to the term-document matrix. As a result, it produces a less noisy matrix which is better than the original one. Thus, higher level concepts are generated from plain terms.

Recently, CB and CF have been combined to improve the recommendation procedure. Most of these hybrid systems are process-oriented: they run CF on the results of CB and vice versa. CF exploits information from the users and their ratings. CB exploits information from items and their features. However being hybrid systems, they miss the interaction between user ratings and item features.

In this paper, we construct a feature profile of a user to reveal the duality between users and features. For instance, in a movie recommender system, a user prefers a movie for various reasons, such as the actors, the director or the genre of the movie. All these features affect differently the choice of each user. Then, we apply Latent Semantic Indexing Model (LSI) to reveal the dominant features of a user. Finally, we provide recommendations according to this dimensionally-reduced feature profile. Our experiments with a real-life data set show the superiority of our approach over existing CF, CB and hybrid approaches.

The rest of this paper is organized as follows: Section 2 summarizes the related work. The proposed approach is described in Section 3. Experimental results are given in Section 4. Finally, Section 5 concludes this paper.

2 Related work

In 1994, the GroupLens system implemented a CF algorithm based on common users preferences. Nowadays, this algorithm is known as user-based CF. In 2001, another CF algorithm was proposed. It is based on the items' similarities for a neighborhood generation. This algorithm is denoted as item-based CF.

The Content-Based filtering approach has been studied extensively in the Information Retrieval (IR) community. Recently, Schult and Spiliopoulou (2006) proposed the Theme-Monitor algorithm for finding emerging and persistent ŞthemesŤ in document collections. Moreover, in IR area, Furnas et al. (1988) proposed LSI to detect the latent semantic relationship between terms and documents. Sarwar et al. (2000) applied dimensionality reduction for the user-based CF approach.

There have been several attempts to combine CB with CF. The Fab System (Balabanovic et al. 1997), measures similarity between users after first computing a content profile for each user. This process reverses the CinemaScreen System (Salter et al. 2006) which runs CB on the results of CF. Melville et al. (2002) used a content-based predictor to enhance existing user data, and then to provide personalized suggestions though collaborative filtering. Finally, Tso and Schmidt-Thieme (2005) proposed three attribute-aware CF methods applying CB and CF paradigms in two separate processes before combining them at the point of prediction.

All the aforementioned approaches are hybrid: they either run CF on the results of CB or vice versa. Our model, discloses the duality between user ratings and item features, to reveal the actual reasons of their rating behavior. Moreover, we apply LSI on the feature profile of users to reveal the principal features. Then, we use a similarity measure which is based on features, revealing the real preferences of the user's rating behavior.

3 The proposed approach

Our approach constructs a feature profile of a user, based on both collaborative and content features. Then, we apply LSI to reveal the dominant features trends. Finally, we provide recommendations according to this dimensionally-reduced feature profile of the users.

3.1 Defining rating, item and feature profiles

CF algorithms process the rating data of the users to provide accurate recommendations. An example of rating data is given in Figures 1a and 1b. As shown, the example data set (Matrix R) is divided into a training and test set, where I_{1-12} are items and U_{1-4} are users. The null cells (no rating) are presented with dash and the rating scale is between [1-5] where 1 means strong dislike, while 5 means strong like.

Definition 1 *The rating profile $R(U_k)$ of user U_k is the k-th row of matrix R.*

For instance, $R(U_1)$ is the rating profile of user U_1, and consists of the rated items I_1,I_2,I_3,I_4,I_8 and I_{10}. The rating of a user u over an item i is given from the element $R(u,i)$ of matrix R.

	I_1	I_2	I_3	I_4	I_5	I_6	I_7	I_8	I_9	I_{10}	I_{11}	I_{12}
U_1	5	3	5	4	-	1	-	3	-	5	-	-
U_2	3	-	-	-	4	5	1	-	5	-	-	1
U_3	1	-	5	4	5	-	5	-	-	3	5	-

(a)

	I_1	I_2	I_3	I_4	I_5	I_6	I_7	I_8	I_9	I_{10}	I_{11}	I_{12}
U_4	5	-	1	-	-	4	-	-	3	-	-	5

(b)

	f_1	f_2	f_3	f_4
I_1	1	1	0	0
I_2	1	0	0	0
I_3	1	0	1	1
I_4	1	0	0	1
I_5	0	1	1	0
I_6	0	1	0	0
I_7	0	0	1	1
I_8	0	0	0	1
I_9	0	1	1	0
I_{10}	0	0	0	1
I_{11}	0	0	1	1
I_{12}	0	1	0	0

(c)

Fig. 1. (a) Training Set ($n \times m$) of Matrix R, (b) Test Set of Matrix R, (c) Item-Feature Matrix F

As described, content data are provided in the form of features. In our running example illustrated in Figure 1c for each item we have four features that describe its characteristics. We use matrix F, where element $F(i, f)$ is one, if item i contains feature f and zero otherwise.

Definition 2 *The item profile $F(I_k)$ of item I_k is the k-th row of matrix F.*

For instance, $F(I_1)$ is the profile of item I_1, and consists of features F_1 and F_2. Notice that this matrix is not always boolean. Thus, if we process documents, matrix F would count frequencies of terms.

To capture the interaction between users and their favorite features, we construct a feature profile composed of the rating profile and the item profile.

For the construction of the feature profile of a user, we use a positive rating threshold, P_τ, to select items from his rating profile, whose rating is not less than this value. The reason is that the rating profile of a user consists of ratings that take values from a scale(in our running example, 1-5 scale). It is evident that ratings should be "positive", as the user does not favor an item that is rated with 1 in a 1-5 scale.

Definition 3 *The feature profile $P(U_k)$ of user U_k is the k-th row of matrix P whose elements $P(u,f)$ are given by Equation 1.*

$$P(u,f) = \sum_{\forall R(u,i) > P_\tau} F(i,f) \tag{1}$$

In Figure 2, element $P(U_k,f)$ denotes an association measure between user U_k and feature f. In our running example (with $P_\tau = 2$), $P(U_2)$ is the feature profile of user U_2, and consists of features f_1, f_2 and f_3. The correlation of a user U_k over a feature f is given from the element $P(U_k, f)$ of matrix P. As shown, feature f_2 describe him better, than feature f_1 does.

	f_1	f_2	f_3	f_4
U_1	4	1	1	4
U_2	1	4	2	0
U_3	2	1	4	5

(a)

	f_1	f_2	f_3	f_4
U_4	1	4	1	0

(b)

Fig. 2. User-Feature matrix P divided in (a) Training Set ($n \times m$), (b) Test Set

3.2 Applying SVD on training data

Initially, we apply Singular Value Decomposition (SVD) on the training data of matrix P that produces three matrices based on Equation 2, as shown in Figure 3:

$$P_{n \times m} = U_{n \times n} \cdot S_{n \times m} \cdot V'_{m \times m} \tag{2}$$

4	1	1	4
1	4	2	0
2	1	4	5

$P_{n \times m}$

-0.61	0.28	-0.74
-0.29	-0.95	-0.12
-0.74	0.14	0.66

$U_{n \times n}$

8.87	0	0	0
0	4.01	0	0
0	0	2.51	0

$S_{n \times m}$

-0.47	-0.28	-0.47	-0.69
0.11	-0.85	-0.27	0.45
-0.71	-0.23	0.66	0.13
-0.52	0.39	-0.53	0.55

$V'_{m \times m}$

Fig. 3. Example of: $P_{n \times m}$ (initial matrix P), $U_{n \times m}$ (left singular vectors of P), $S_{n \times m}$ (singular values of P), $V'_{m \times m}$ (right singular vectors of P).

3.3 Preserving the principal components

It is possible to reduce the $n \times m$ matrix S to have only c largest singular values. Then, the reconstructed matrix is the closest rank-c approximation of the initial matrix P as it is shown in Equation 3 and Figure 4:

$$P^*_{n \times m} = U_{n \times c} \cdot S_{c \times c} \cdot V'_{c \times m} \tag{3}$$

$P^*_{n \times i}$

2.69	0.57	2.22	4.25
0.78	3.93	2.21	0.04
3.17	1.38	2.92	4.78

$U_{n \times c}$

-0.61	0.28
-0.29	-0.95
-0.74	0.14

$S_{c \times c}$

8.87	0
0	4.01

$V'_{c \times m}$

-0.47	-0.28	-0.47	-0.69
0.11	-0.85	-0.27	0.45

Fig. 4. Example of: $P^*_{n \times m}$ (approximation matrix of P), $U_{n \times c}$ (left singular vectors of P^*), $S_{c \times c}$ (singular values of P^*), $V'_{c \times m}$ (right singular vectors of P^*).

We tune the number, c, of principal components (i.e., dimensions) with the objective to reveal the major feature trends. The tuning of c is determined by the information percentage that is preserved compared to the original matrix.

3.4 Inserting a test user in the c-dimensional space

Given the current feature profile of the test user u as illustrated in Figure 2b, we enter pseudo-user vector in the c-dimensional space using Equation 4. In our example, we insert U_4 into the 2-dimensional space, as shown in Figure 5:

$$\mathbf{u_{new}} = \mathbf{u} \cdot V_{m \times c} \cdot S^{-1}_{c \times c} \tag{4}$$

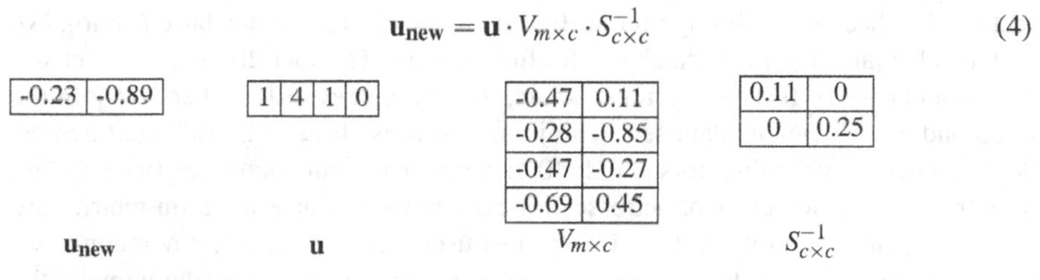

$\mathbf{u_{new}}$

-0.23	-0.89

$\mathbf{u}$

1	4	1	0

$V_{m \times c}$

-0.47	0.11
-0.28	-0.85
-0.47	-0.27
-0.69	0.45

$S^{-1}_{c \times c}$

0.11	0
0	0.25

Fig. 5. Example of: $\mathbf{u_{new}}$ (inserted new user vector), $\mathbf{u}$ (user vector), $V_{m \times c}$ (two left singular vectors of V), $S^{-1}_{c \times c}$ (two singular values of inverse S).

In Equation 4, $\mathbf{u_{new}}$ denotes the mapped ratings of the test user $\mathbf{u}$, whereas $V_{m \times c}$ and $S^{-1}_{c \times c}$ are matrices derived from SVD. This $\mathbf{u_{new}}$ vector should be added in the end of the $U_{n \times c}$ matrix which is shown in Figure 4.

3.5 Generating the Neighborhood of users/items

In our model, we find the k nearest neighbors of pseudo user vector in the c-dimensional space. The similarities between train and test users can be based on Cosine Similarity. First, we compute the matrix $U_{n \times c} \cdot S_{c \times c}$ and then we perform vector similarity. This $n \times c$ matrix is the c-dimensional representation for the n users.

3.6 Generating the top-N recommendation list

The most often used technique for the generation of the top-N list, is the one that counts the frequency of each positively rated item inside the found neighborhood, and recommends the N most frequent ones. Our approach differentiates from this technique by exploiting the item features. In particular, for each feature f inside the found neighborhood, we add its frequency. Then, based on the features that an item consists of, we count its weight in the neighborhood. Our method, takes into account the fact that, each user has his own reasons for rating an item.

4 Performance study

In this section, we study the performance of our Feature-Weighted User Model (FRUM) against the well-known CF, CB and a hybrid algorithm. For the experiments, the collaborative filtering algorithm is denoted as CF and the content-based algorithm as CB. As representative of the hybrid algorithms, we used the Cinemascreen Recommender Agent (SALTER et al. 2006), denoted as CFCB. Factors that are treated as parameters, are the following: the neighborhood size (k, default value 10), the size of the recommendation list (N, default value 20) and the size of train set (default value 75%). P_τ threshold is set to 3. Moreover, we consider the division between training and test data. Thus, for each transaction of a test user we keep the 75% as hidden data (the data we want to predict) and use the rest 25% as not hidden data (the data for modeling new users). The extraction of the content features has been done through the well-known internet movie database (imdb). We downloaded the plain imdb database (ftp.fu-berlin.de - October 2006) and selected 4 different classes of features (genres, actors, directors, keywords). Then, we join the imdb and the Movielens data sets. The joining process lead to 23 different genres, 9847 keywords, 1050 directors and 2640 different actors and actresses (we selected only the 3 best paid actors or actresses for each movie). Our evaluation metrics are from the information retrieval field. For a test user that receives a top-N recommendation list, let R denote the number of *relevant recommended items* (the items of the top-N list that are rated higher than P_τ by the test user). We define the following: *Precision* is the ratio of R to N.*Recall* is the ratio of R to the total number of relevant items for the test user (all items rated higher than P_τ by him). In the following, we also use $\mathrm{F}_1 = 2 \cdot \mathrm{recall} \cdot \mathrm{precision}/(\mathrm{recall} + \mathrm{precision})$. F_1 is used because it combines both precision and recall.

4.1 Comparative results for CF, CB, CFCB and FRUM algorithms

For the CF algorithms, we compare the two main cases, denoted as user-based (UB) and item-based (IB) algorithms. The former constructs a user-user similarity matrix while the latter, builds an item-item similarity matrix. Both of them, exploit the user ratings information(user-item matrix R). Figure 6a demonstrates that IB compares favorably against UB for small values of k. For large values of k, both algorithms converge, but never exceed the limit of 40% in terms of precision. The reason is that as the k values increase, both algorithms tend to recommend the most popular items. In the sequel, we will use the IB algorithm as a representative of CF algorithms.

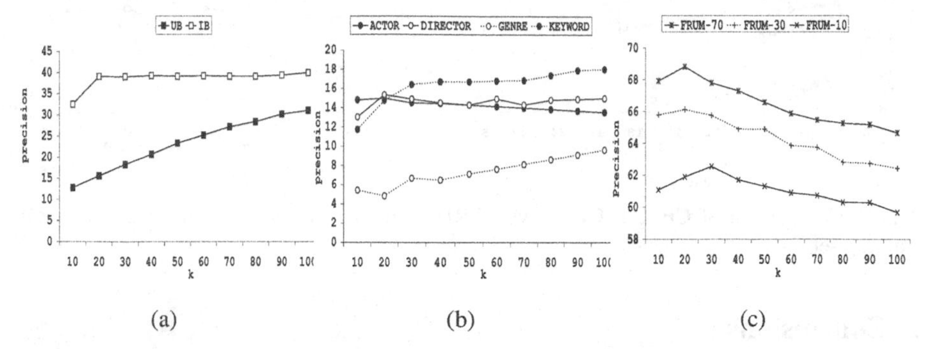

(a) (b) (c)

Fig. 6. Precision vs. k of: (a) UB and IB algorithms, (b) 4 different feature classes, (c) 3 different information percentages of our FRUM model

For the CB algorithms, we have extracted 4 different classes of features from the imdb database. We test them using the pure content-based CB algorithm to reveal the most effective in terms of accuracy. We create an item-item similarity matrix based on cosine similarity applied solely on features of items (item-feature matrix F). In Figure 6b, we see results in terms of precision for the four different classes of extracted features. As it is shown, the best performance is attained for the "keyword" class of content features, which will be the default feature class in the sequel.

Regarding the performance of our FRUM, we preserve, each time, a different fraction of principal components of our model. More specifically, we preserve 70%, 30% and 10% of the total information of initial user-feature matrix P. The results for precision vs. k are displayed in Figure 6c. As shown, the best performance is attained with 70% of the information preserved. This percentage will be the default value for FRUM in the sequel.

In the following, we test FRUM algorithm against CF, CB and CFCB algorithms in terms of precision and recall based on their best options. In Figure 7a, we plot a precision versus recall curve for all four algorithms. As shown, all algorithms' precision falls as N increases. In contrast, as N increases, recall for all four algorithms increases too. FRUM attains almost 70% precision and 30% recall, when we recommend a top-20 list of items. In contrast, CFCB attains 42% precision and 20% recall. FRUM is more robust in finding relevant items to a user. The reason is two-fold:(i) the sparsity has been downsized through the features and (ii) the LSI application reveals the dominant feature trends.

Now we test the impact of the size of the training set. The results for the F_1 metric are given in Figure 7b. As expected, when the training set is small, performance downgrades for all algorithms. FRUM algorithm is better than the CF, CB and CFCB in all cases. Moreover, low training set sizes do not have a negative impact on measure F_1 of the FRUM algorithm.

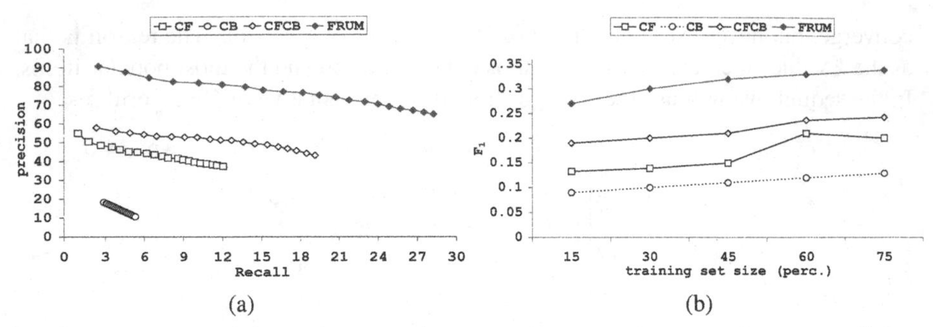

(a) (b)

Fig. 7. Comparison of CF, CB, CFCB with FRUM in terms of (a) precision vs. recall (b) training set size.

5 Conclusions

We propose a feature-reduced user model for recommender systems. Our approach builds a feature profile for the users, that reveals the real reasons of their rating behavior. Based on LSI, we include the pseudo-feature user concept in order to reveal his real preferences. Our approach outperforms significantly existing CF, CB and hybrid algorithms. In our future work, we will consider the incremental update of our model.

References

BALABANOVIC, M. and SHOHAM, Y. (1997): Fab: Content-based, collaborative recommendation, *ACM Communications*,volume 40,number 3,66-72

FURNAS, G. and DEERWESTER, et al. (1988): Information retrieval using a singular value decomposition model of latent semantic structure, *SIGIR* , 465-480

MELVILLE, P. and MOONEY R. J. and NAGARAJAN R. (2002): Content-Boosted Collaborative Filtering for Improved Recommendations, *AAAI*, 187-192

SALTER, J. and ANTONOPOULOS, N. (2006): CinemaScreen Recommender Agent: Combining Collaborative and Content-Based Filtering *Intelligent Systems Magazine*, volume 21, number 1, 35-41

SARWAR, B. and KARYPIS, G. and KONSTAN, J. and RIEDL, J. (2000) Application of dimensionality reduction in recommender system-A case study", *ACM WebKDD Workshop*

SCHULT, R and SPILIOPOULOU, M. (2006) : Discovering Emerging Topics in Unlabelled Text Collections *ADBIS 2006*, 353-366

TSO, K. and SCHMIDT-THIEME, L. (2005) : Attribute-aware Collaborative Filtering, *German Classification Society GfKl 2005*

Part X

Linguistics

The Distribution of Data in Word Lists and its Impact on the Subgrouping of Languages

Hans J. Holm

Hannover, Germany
HJJHolm@web.de

Abstract. This work reveals the reason for the bias in the separation levels computed for natural languages with only a small amount of residues; as opposed to stochastically normal distributed test cases like those presented in Holm (2007a). It is shown how these biased data can be correctly projected to true separation levels. The result is a partly new chain of separation for the main Indo-European branches that fits well to the grammatical facts, as well as to their geographical distribution. In particular it strongly demonstrates that the Anatolian languages did not part as first ones and thereby refutes the Indo-Hittite hypothesis.

1 General situation

Traditional historical linguists use a priori to look upon quantitative methods with suspicion, because they argue that only those agreements can decide the question, which are supposed to stem exclusively from their direct ancestor, the so-called 'common innovations', or synapomorphies, in biological terminology (Hennig 1966). However, this seemingly perfect concept has in over a hundred years of research brought about anything but agreement on even a minimum of groupings (e.g. those in Hamp 2005). There is no grouping, which is not debated in one or more ways. 'Lumpers' and 'splitters' are at work, as with nearly all proposed language families.

Quantitative attempts (cf. Holm (2005), to be updated (2007b), for an overview) have not proved to be superior: First, all regard such trivial results like distinguishing e.g. Greek from Germanic as a proof; secondly, many of them are fixated on a mechanistic rate (or "clock") assumption for linguistic changes (what is not our focus here); and worst, mathematicians, biologists, and even some linguists retreat to the too loose view that the amount of agreements is a direct measure of relatedness. Elsewhere I have demonstrated that this assumption is erroneous because these researchers overlook the dependence of this surface phenomenon from at least three stochastic parameters, the "proportionality trap" (cf. Holm (2003); Swofford et al. (1996:487)).

2 Special situation

2.1 Recapitulation: What is the proportionality trap?

Definition: If a "mother" language splits into (two) daughter languages, these are "genealogically related". At the point of split, or era of separation, both start with the full amount of inherited features. However, they will soon begin to differentiate by loss or replacement of these original features. These individual replacements occur

- independently (!) from each other,
- by new irregular socio-psychological impacts in history. They are therefore non-deterministic in that the next state of the environment is partially but not fully determined by the previous state. Least, they are
- irreversible, because, when a feature is changed, it will normally never reappear.

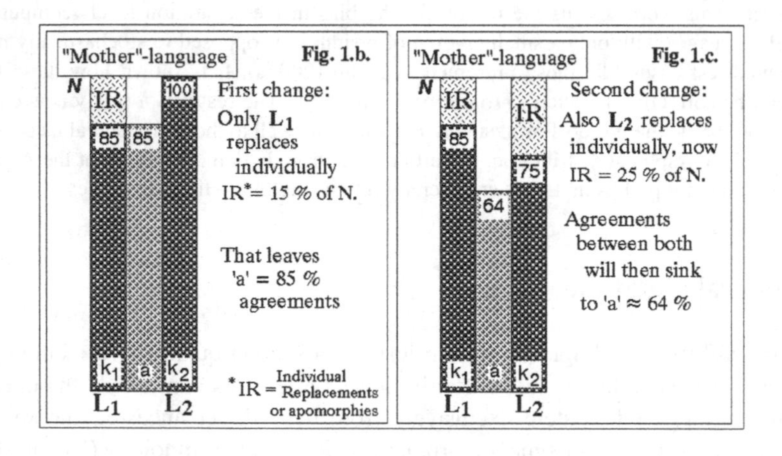

Fig. 1. Different agreements, same relationship(!)

Because of these properties we have to regard linguistic change mathematically as a stochastic process by draws without replacement. Any statistician will immediately recognize this as the hypergeometric distribution. In word lists, we in fact have all four parameters (cf. Fig. 1) needed as follows:

- The amount of inherited features k_i and k_j (or residues, cognates, symplesiomorphies) regarded as preserved from the common ancestor of any two languages L_i and L_j;
- the amount of shared agreements '$a_{i,j}$' between them;
- the amount N of their common features at the time of separation (the universe), not visible in the surface structure of the data. Exactly this invisible universe N are we seeking for, because it represents the underlying structure, the amount of features, which must have been present in both languages at the era of their

separation in the past. And again any mathematician has the solution: This "separation level" N for each pair of branches can be estimated by the 2nd momentum of the hypergeometric, the maximum likelihood estimator, transposed to

$$\hat{N} = k_i \times k_j / a_{i,j}$$

Since changes can only lower the number of common features, a higher separation level must lie earlier in time, and thus we obtain a ranked chain of separation in a family of languages.

2.2 Applications up to now

The first one to propose and apply this method was the British mathematician D.G. Kendall (1950) with the Indo-European data of Walde/Pokorny (1926-32). It has then independently been extensively applied to the data of thc improved dictionary of Pokorny (1959) by Holm (2000, passim). The results seemed to be convincing, in particular for the North-Western group, and also for the relation of Greek and the Indo-Iranian group. The late separations of Albanian, Armenian, and Hittite could well have been founded in their central position and therefore did not appear suspicious.

Only when in a further application to Mixe-Zoquean data by Cysouw et al. (2006) a resembling observation occurred that only languages with few known residues appeared to separate late, a systematic bias could be suspected. Cysouw et al. discarded the SLR-method, because their results partly contradicted the subgrouping of Mixe-Zoquean as inferred by traditional methods of two historical linguists (which in fact did not completely agree with each other). In a presentation Cysouw (2004) suspected that the "unbalanced amount of available data distorts the estimates", and "Error 1: they are grouped together, because of many shared retentions." However, this only demonstrates that the basics explained above are not correctly understood, since the hypergeometric does just not rest on one parameter alone.

In this study we will use the most modern and acknowledged Indo-European data base, the "Lexikon der indogermanischen Verben" (Rix et al. (2002), henceforth LIV-2. I am very obliged to the authors for sending me the digitalized version, which in fact only enabled me to quantify the contents in acceptable time. The reasons for this tremendous undertaking were:

- The commonplace (though seldom mentioned) in linguistics that verbs are much lesser borrowed than nouns, what is not taken into account by any quantitative work up to now.
- The more trustworthy combined work of a team at an established department of Indo-European under the supervision of a professional historical linguist should guarantee a very high standard, moreover in this second edition.
- Compared with the in many parts outdated Pokorny, we have now much better knowledge of the Anatolian and Tocharian languages.

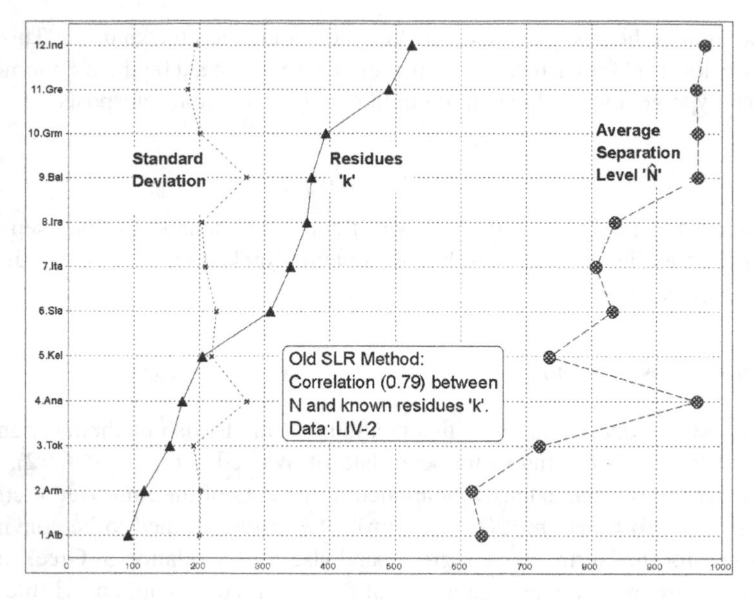

Fig. 2. Unwanted dependence N from k in LIV-2 list

3 The bias

3.1 Unwanted dependence

Nevertheless, these much better data, not suspicious of poor knowledge, displayed the same bias as the other ones, as demonstrated in Fig. 2, which presents the correlation between the residues 'k' and the corresponding '$\hat{N}$'s in falling order. Thus we have a problem. The reason for this bias, opposite to Cysouw et al., could not lie in a poor knowledge of the data, nor could it lie in the algorithm, as I have additionally tested in hundreds of random cases, some of which published in Holm (2007a). Consequently, the reason had to be found in the word lists alone, the properties of which we will have to inspect now with closer scrutiny:

3.2 Revisiting the properties of word lists

The effects of scatter on the subgrouping problem as well as its handling has been intensively investigated by Holm (2007a). This study as well as textbooks suggest that the sum of residues 'k' should exceed at least 90 % of the universe 'N', and a single 'k' must not fall below 20 %. However, since the LIV-2 database is big enough to guarantee a low scatter, there must be something else, overlooked up to now. A first hint has already been given by D.G. Kendall (1950:41), who noticed that "One must, however, assume that along a given segment of a given line of descent the chance of survival is the same for every root exposed to risk, and one must also assume that the several roots are exposed to risk independently". The latter condition is the easier part, since linguists would agree that changes in the lexicon or grammar occur independently of each other. (The so-called push-and-pull chains are mainly a phonetic symptom and of lesser interest here). The real problem is the first condition, since the chance of survival is not at all the same for any feature, and every word has its own history. For our purpose, we must not necessarily deal with the reasons for these changes in detail. Could the reason for the observed bias perhaps be found in a distribution that contradicts the conditions of the hypergeometric, and perhaps other quantitative approaches, too?

3.3 The distribution in word lists

For that purpose, the 1195 reconstructed verbal roots of the LIV-2 are entered as "characters" into a spreadsheet, while the columns contain the 12 branches of Indo-European. We let then sum up the cross totals into a new column, containing now the frequency for every row. After sorting to these, we get twelve different blocks or slices, one for each frequency. By counting out every language per slice, we get 12 matrices, of which we enter the arithmetic means into the final table. Let us now have a closer look at the plot of this as Fig. 3.

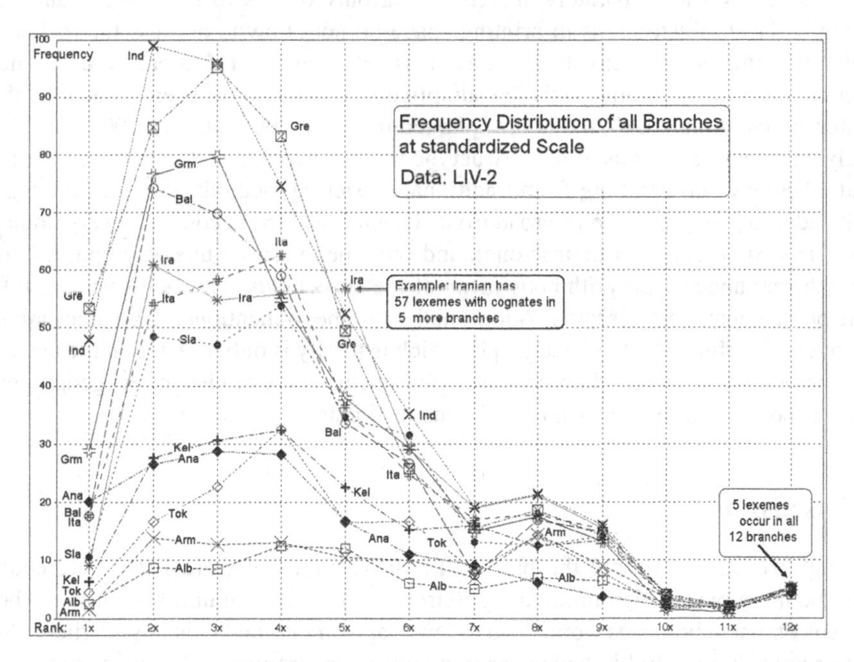

Fig. 3. All frequencies of the LIV-2 data

3.4 Detecting the reason

Immediately we observe to the right hand the few verbs which occur in many languages, growing up to the left with the many verbs occurring in fewer languages, breaking down to the special case of verbs occurring in one language only. To find out the reason of the false correlation between these curves and the bias with the smaller represented languages, we must first identify the connections with our formula: $\hat{N}$ depends on the product of the residues 'k' of any language, as represented here as the area below their curve. This product is then divided by their agreements 'a', which are naturally represented by the frequencies (bottom line). And here - not easily to detect - lies the key: The more to the right hand, the higher the agreements per residue. Further: The smaller the *sum* of residues of a branch (area below its curve), the higher is the proportion of agreements, ending in a false lower separation level. So far we have located the problem. But we are still far from a solution.

4 Solution and operationalization

Since the bias will turn up every time one employs the *total* of the data, we must compute every slice separately, thereby using only data with the same chance of being replaced. Restrictions in printing space do not allow to address the different options of implementations. In any case, it is extremely useful to pre-order the numerical outcome according to the presumptive next neighbor of every branch by the historical-bundling ("Bx") method, explained in Holm (2000:84-5, or 2005:640). Finally, the matrix allows us to reconstruct the tree. To avoid the attractions, deletions, and other new bias resulting from traditional clustering methods, it is methodologically advisable to proceed on a broad front. That means, first to combine every branch with its next neighbor (if their is one), and only then proceed this way uphill, finding the next node by the arithmetic mean of the cross fields. This helps very well to rule out the unavoidable scatter. Additionally, the above-mentioned Bx-values are in particular helpful to "flatten" the graph, which naturally is only ordered in the one direction of descent, but might represent different clusters or circles in real geography, what is not displayable in such two-dimensional graph.

5 Discussion

Though we have ruled out the bias in the distributions of word lists, there could well be more bias hidden in the data: Extremely different cultural backgrounds between compared branches would lead to fewer agreements and thus false earlier split. An example are the Baltic languages in a cultural environment of hunter- and gatherer communities vs. the Anatolian languages in an environment of advanced civilizations. Secondly, there may be left differences in the reliability of the research itself. Third, it is well-known that the relative position of the branches give raise to more innovations in the center vs. more conservative behavior in peripheral positions ("Saumlage").

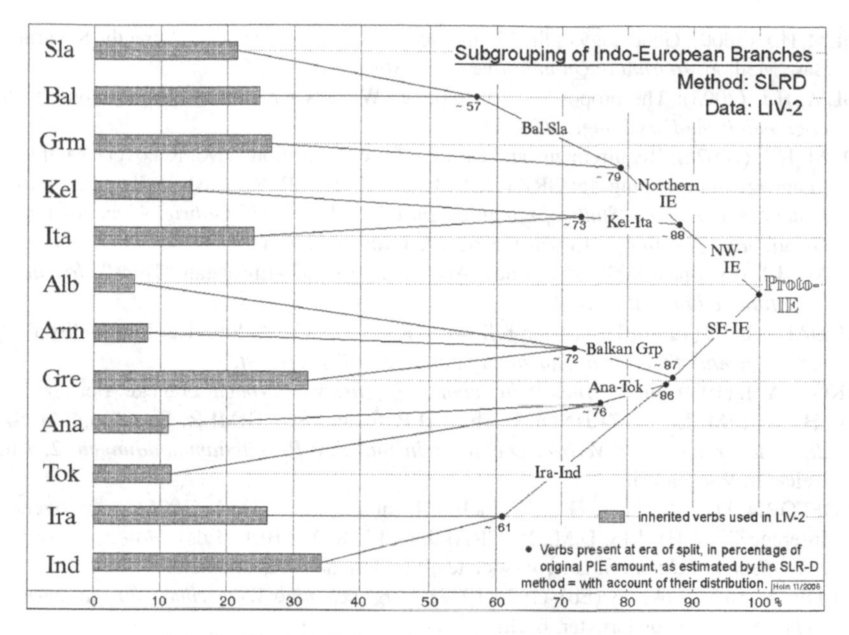

Fig. 4. New SLRD-based tree of the main IE branches

6 Conclusions

Linguistically, this study clearly refutes an early split of/from Anatolian and thereby the "Indo-Hittite" Hypothesis. Methodologically, the former "Separation-Level Recovery method" is updated to one accounting for the Distribution (SLRD). The insights gained should prevent everybody from trusting methods not regarding the hypergeometric behavior of language change, as well as the distribution in word lists. People must not be dazzled by apparently good results, which regularly appear, due alone to very strong signals in the data, or simply by chance.

References

CYSOUW, M. (2004): *email.eva.mpg.de/ cysouw/pdf/cysouwWIP.pdf*

CYSOUW, M., WICHMANN, S. and KAMHOLZ, D. (2006): A critique of the separation base method for genealogical subgrouping, with data from Mixe-Zoquean. *Journal of Quantitative Linguistics, 13(2-3), 225–264.*

EMBLETON, S.M. (1986): *Statistics in historical linguistics* [Quantitative Linguistics 30]. Brockmeyer, Bochum.

GRZYBEK, P., and R. KÖHLER (Eds). (2007): *Exact Methods in the Study of Language and Text* [Quantitative Linguistics 62]. De Gruyter Berlin.

HAMP, E.P. (1998): "Whose were the Tocharians? Linguistic subgrouping and Diagnostic Idiosyncrasy" *The Bronze Age and Early Iron Age Peoples of Eastern Central Asia. Vol. 1:307-46.* Edited by Victor H. Mair. Washington DC: Institute for the Study of Man.

HOLM, H.J. (2000): Genealogy of the Main Indo-European Branches Applying the Separation Base Method. *Journal of Quantitative Linguistics, 7-2, 73–95.*

HOLM, H.J. (2003): The proportionality trap; or: What is wrong with lexicostatistics? *Indogermanische Forschungen 108, 38–46.*

HOLM, H.J. (2007a): Requirements and Limits of the Separation Level Recovery Method in Language Subgrouping. In: GRZYBEK, P. and KÖHLER, R. (Eds), *Viribus Quantitatis. Exact Methods in the Study of Language and Text. Festschrift Gabriel Altmann zum 75. Geburtstag*. Quantitative Linguistics 62. De Gruyter, Berlin.

HOLM, H.J. (to appear 2007b): The new Arboretum of Indo-European "Trees". *Journal of Quantitative Linguistics* 14-2.

KENDALL, D.G. (1950): Discussion following Ross, A.S.C., Philological Probability Problems. *Journal of the Royal Statistical Society, Ser. B 12, p. 49f.*

POKORNY, J. (1959): *Indogermanisches etymologisches Wörterbuch.* Francke, Bern.

RIX, H., KÜMMEL, M., ZEHNDER, Th., LIPP, R. and SCHIRMER, B. (2001): *Lexikon der indogermanischen Verben. Die Wurzeln und ihre Primärstammbildungen.* 2. Aufl. Reichert, Wiesbaden.

SWOFFORD, D.L., OLSEN, G.J., Waddell, P.J., and HILLIS, D.M. (1996): "Phylogenetic Inference". In: HILLIS, D.M., M. CRAIG, and B.K. MABLE (Eds). *Molecular Systematics, Second Edition*. Sinauer Associates, Sunderland MA, Chapter 11.

WALDE, A., and J. Pokorny (Ed). (1926-1932): *Vergleichendes Wörterbuch der indogermanischen Sprachen.* de Gruyter, Berlin.

Quantitative Text Analysis Using L-, F- and T-Segments

Reinhard Köhler and Sven Naumann

Linguistische Datenverarbeitung, University of Trier, Germany
{koehler, naumsven}@uni-trier.de

Abstract. It is shown that word length and other properties of linguistic units display a lawful behavior not only in form of distributions but also with respect to their syntagmatic arrangement in a text. Based on L-segments (units of constant or increasing lengths), F-segments, and T-segments (units of constant or increasing frequency or polytextuality respectively), the dynamic behavior of segment patterns is investigated. Theoretical models are derived on the basis of plausible assumptions on influences of the properties of individual units on the properties of their constituents in the text. The corresponding hypotheses are tested on data from 66 German texts of four authors and two different genres. Experiments with various characteristics show promising properties which could be useful for author and/or genre discrimination.

1 Introduction

Most quantitative studies in linguistics are almost exclusively based on a "bag of words" model, i.e. they disregard the syntagmatic dimension, the arrangement of the units in the course of the given text. Gustav Herdan underlined this fact as early as 1966; he called the two different types of study "language in the mass vs. language in the line" (Herdan (1966), p. 423). Only very few investigations have been carried out so far with respect to sequences of properties of linguistic units (cf. Hřebíček (2000), Andersen (2005), Köhler (2006) and Uhlířová (2007)). A special approach is time series studies (cf. Pawlowski (2001)), but the application of such methods to studies of natural language is not easy to justify and is connected with methodological problems of assigning numerical values to categorical observations. The present paper approaches the problem of the dynamic behavior of sequences of properties without limitation to a specific grouping such as word pairs or N-grams. It starts from the general hypothesis that sequences in a text are organized in lawful patterns rather than chaotically or according to a uniform distribution. There are several possibilities to define units such as phrases or clauses which could be used to find patterns or regularities. They suffer, however, from several disadvantages: (1) They do not provide an appropriate granularity. While words are too small, sentences seem to be too large units to unveil syntagmatic patterns with quantitative methods. (2) Linguistic units are inherently connected to specific grammar models. (3) Their application leads again to a "bag of units" model and thus to a loss of syntagmatic information. Therefore, we establish a different unit for our present project: A unit based on the rhythm of the property under study itself. We will demonstrate the approach using word length as an illustrative example. We define an **L-segment** as a maximal sequence of monotonically increasing numbers, where these numbers represent the lengths of adjacent words of the text. Using this definition, we segment a given text in a left to right fashion starting with the first word. In this way, a text can be represented as an uninterrupted sequence of L-segments. Thus, the text fragment (1) is segmented as shown by the L-segment sequence (2)if word length is measured in terms of syllable number:

(1) *Word length studies are almost exclusively devoted to the problem of distributions.*
(2) (1-1-2) (1-2-4) (3) (1-1-2) (1-4)

This kind of segmentation is similar to Boroda's F-motiv for musical "texts". Boroda (1982) defined his F-motiv in an analogous way but with respect to the duration of the notes of a musical piece. The advantage of such a definition is obvious: Any text can be segmented in an objective, unambiguous, and exhaustive way, i.e. it guaranties that each element will be assigned to exactly one unit. Furthermore, it provides units of an appropriate granularity and it can be applied iteratively, i.e. sequences of L-segments and their lengths can be studied etc., thus the unit can be scaled over a principally unlimited range of sizes and granularities, limited only by text length. Analogically, F- and T-segments are formed by monotonically increasing sequences of frequency and polytextuality values. Other units than words can be used as basic units, such as morphs, syllables, phrases, clauses, sentences, and other properties such as polysemy, synonymy, age etc. can be used for analogous definitions. Our study concentrated on L-, F-, and T-segments, and we will report here mainly on the findings using L-segments. While the focus of our interest lies on basic scientific questions such as "why do these units show their specific behavior?" we also had a look at the possible use of our findings for purposes such as text genre classification and authorship determination.

2 Data

For the present study we compiled a small corpus by selecting 66 documents from the *Projekt Gutenberg-DE* (http://gutenberg.spiegel.de/). The corpus consists of 30 poems and 36 short stories, written by 4 different German authors between the late 18th and the early 20th century:

Text length varies between 90 and 8500 *running word forms* (RWF). As to be expected, the poems tend to be considerably shorter than the narrative texts: The average length of the poems is 446 RWF and 3063 RWF of the short stories.

Table 1. Text numbers in the corpus with respect to genre and author

	Brentano	Goethe	Rilke	Schnitzler	$\sum$
poetry	10	10	10	-	30
prose	2	9	10	15	36

3 Distribution of segment types

Starting from the hypothesis that L-, F- and T-segments are not only units which are easily defined and easy to determine, but also posses a certain psychological reality i.e. that they play a role in the process of text generation, it seems plausible to assume that these units display a lawful distributional behaviour similar to the well-known linguistic units such as words or syntactic constructions (c.f. Köhler (1999)). A first confirmation - however on data from only a single Russian text - was found in (Köhler (2007)). A corresponding test on the data of the present study corroborates the hypothesis. Each of the 66 texts shows a rank-frequency distribution of the 3 kinds of segment patterns according to the Zipf-Mandelbrot distribution, which was fitted to the data in the following form:

$$P_x = \frac{(b+x)^{-a}}{F(n)}, \qquad x = 1,2,3,\ldots,n$$
$$a \in \mathbb{R}$$
$$b > -1$$
$$n \in \mathbb{N}$$
$$F(n) = \sum_{i=1}^{n}(b+i)^{-a} \tag{1}$$

Figure 1 shows the fit of this distribution to the data of one of the texts on the basis of

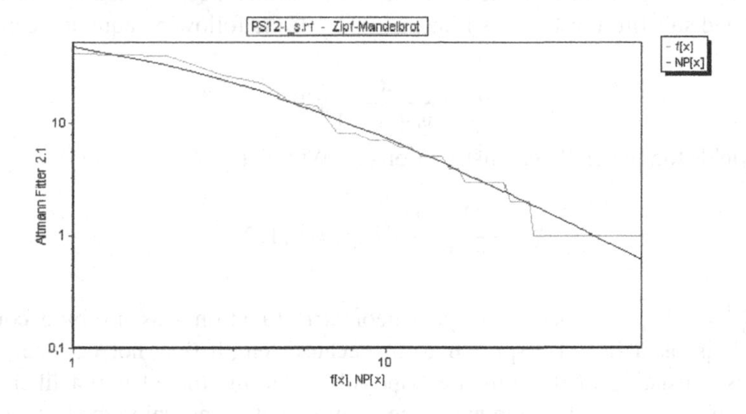

Fig. 1. Rank-Frequency Distribution of L-Segments

L-segments on a log-log scale. In this case, the goodness-of-fit test yielded $P(\chi^2) \approx$

1.0 with 92 degrees of freedom. $N = 941$ L-segments were found in the text forming $x_{max} = 112$ different patterns. Similar results were obtained for all three kinds of segments and all texts. Various experiments with the frequency distributions show promising differences between authors and genres. However, these differences alone do not yet allow for a crisp discrimination.

4 Length distribution of L-segments

As a consequence of our general hypothesis, not only the segment types but also the length of the segments should follow lawful patterns. Here, we study the distribution of L-segment length. First, a theoretical model is set up on the basis of three plausible assumptions:

1. There is a tendency in natural language to form compact expressions. This can be achieved at the cost of more complex constituents on the next level. An example is the following: The phrase "*as a consequence*" consists of 3 words, where the word "*consequence*" has 3 syllables. The same idea can be expressed using the shorter expression "consequently", which consists of only 1 word of 4 syllables. Hence, more compact expressions on one level go along with more complex expressions on the next level. Here, the consequence of the formation of longer words is relevant. The variable K will represent this tendency.
2. There is an opposed tendency, viz. word length minimization. It is a consequence of the same tendency of effort minimization which is responsible for the first tendency but now considered on the word level. We will denote this requirement by M.
3. The mean word length in a language can be considered as constant, at least for a certain period of time. This constant will be represented by q.

According to a general approach proposed by Altmann (cf. Altmann and Köhler (1996)) and substituting $k = K - 1$ and $m = M - 1$, the following equation can be set up:

$$P_x = \frac{k+x-1}{m+x-1} q P_{x-1} \quad (2)$$

which yields the hyper-Pascal distribution (cf. Wimmer and Altmann (1999)):

$$P_x = \frac{\binom{k+x-1}{x}}{\binom{m+x-1}{x}} q^x P_0, x = 0, 1, 2, \ldots \quad (3)$$

with $P_0^{-1} = {}_2F_1(k, 1; m; q)$ - the hyper-geometric function - as norming constant. Here, (3) is used in a 1-displaced form because length 0 is not defined, i.e. L-segments consisting of 0 words are impossible. As this model is not likely to be adequate also for F- and T-segments - the requirements concerning the basic properties frequency and polytextuality do not imply interactions between adjacent levels - a simpler one can be set up. Due to length limitations to our contribution in this volume we will not describe the appropriate model for these segment types but it can be said here that their length distributions can be modeled and explained by the hyper-Poisson distribution.

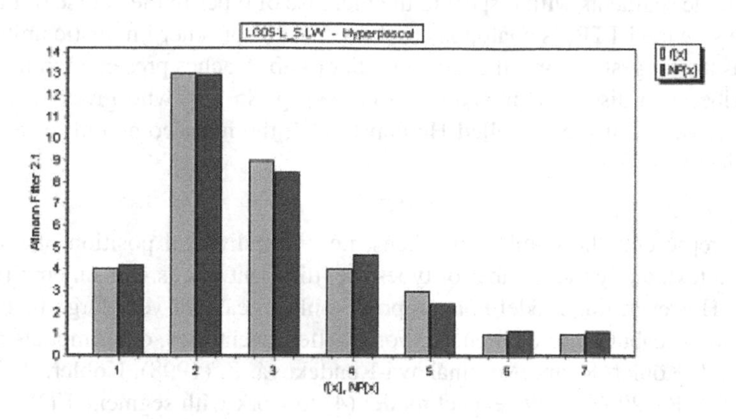

Fig. 2. Theoretical and empirical distribution of L-segments in a poem

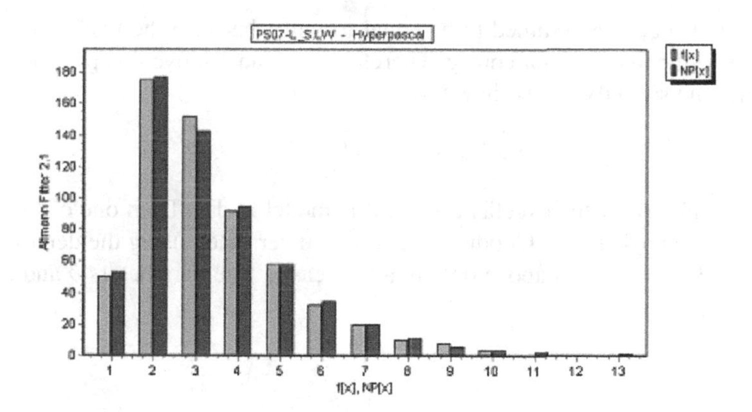

Fig. 3. Theoretical and empirical distribution of L-segments in a short story

The empirical tests on the data from the 66 texts support our hypothesis with good and very good 2 values. Figures 2 and 3 show typical graphs of the theoretical and empirical distributions as modeled using the hyper-Pascal distribution. Figure 2 is an example of poetry; Figure 3 shows a narrative text. Good indicators of text genre or authors could not yet be found on the basis of these distributions. However, only a few of the available characteristics have been considered so far. The same is true of the corresponding experiments with F- and T-segments.

5 TTR studies

Another hypothesis investigated in our study is the assumption that the dynamic behavior of the segments with respect to the increase of types in the course of the given text, the so-called TTR, is analogous to that of words or other linguistic units. Word TTR has the longest history; the large number of approaches presented in linguistics is described and discussed in (Altmann (1988), p. 85-90), who gives also a theoretical derivation of the so-called Herdan model, the most commonly used one in linguistics:

$$y = x^a, \quad (4)$$

where x represents the number of tokens, i.e. the individual position of a running word in a text, and y the number of types, i.e. different words. a is an empirical parameter. However, this model is appropriate only in case of very large inventories, such as the vocabulary of a language. For smaller inventories, other models must be derived (cf. Köhler, R. and Martináková-Rendeková, Z. (1998), Köhler, R. (2003a) and Köhler, R. (2003b)). We expect model (4) to work with segment TTR, an equation, which was derived by Altmann (1980) for the Menzerath-Altmann Law and later in the framework of synergetic linguistics:

$$y = ax^b e^{cx}, c < 0. \quad (5)$$

The value of a can be assumed to be equal to unity, because the first segment of a text must be the first type, of course. Therefore, we can remove this parameter from the model and simplify (4) as shown in (5):

$$y = e^{-c} x^b e^{cx} = x^b e^{c(x-1)}, c < 0. \quad (6)$$

Figures 4 and 5 show the excellent fits of this model to data from one of the poems and one of the prose texts. Goodness-of-fit was determined using the determination coefficient R^2, which was above 0.99 in all 66 cases. The parameters b and c of the

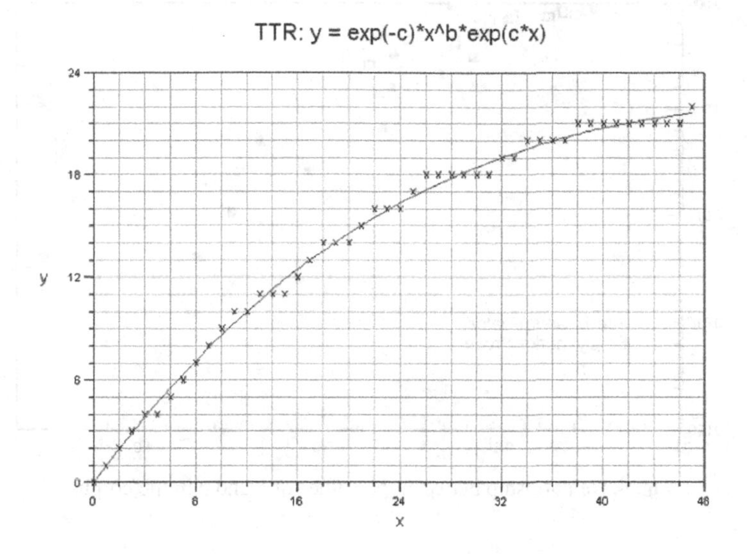

Fig. 4. L-segment TTR of a poem

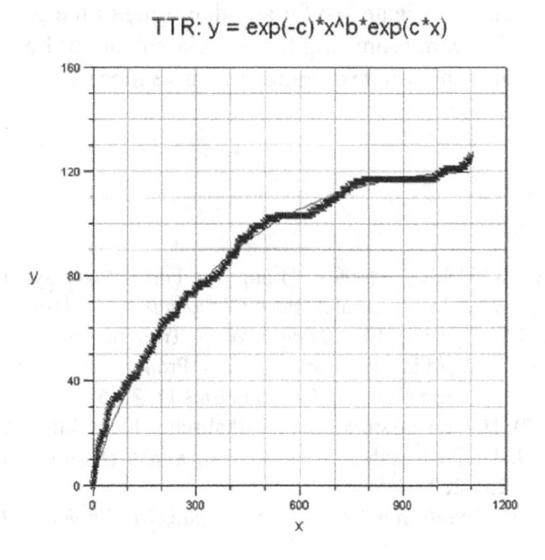

Fig. 5. L-segment TTR of a short story

TTR model turned out to be quite promising characteristics of text genre and author. They are not likely to discriminate these factors sufficiently when taken alone but seem to carry a remarkable amount of information. Figure 6 shows the relationship between the parameters b and c.

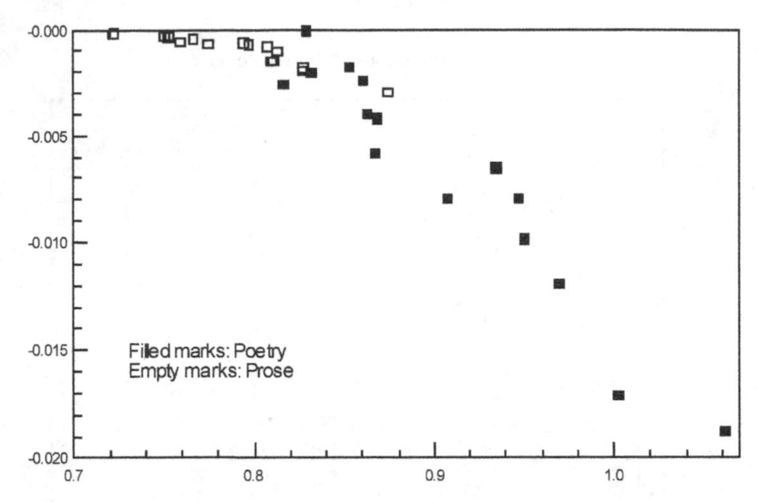

Fig. 6. Relationship between the values of b and c in the corpus

6 Conclusion

Our study has shown that L-, F- and T-Segments on the word level display a lawful behavior in all aspects investigated so far and that some of the parameters, in particular those of the TTR, seem promising for text classification. Further investigations on more languages and on more text genres will give more reliable answers to these questions.

References

ALTMANN, G. and KÖHLER, R. (1996): "Language Forces? and Synergetic Modelling of Language Phenomena. In: P. Schmidt [Ed.]: *Glottometrika 15. Issues in General Linguistic Theory and The Theory of Word Length.* WVT, Trier, 62-76.

ANDERSEN, S. (2005): Word length balance in texts: Proportion constancy and word-chain-lengths in Proust's longest sentence. Glottometrics 11, 32-50.

BORODA, M. (1982): Häufigkeitsstrukturen musikalischer Texte. In: J. Orlov, M. Boroda, G. Moisei and I. Nadarejŝvili [Eds.]: *Sprache, Text, Kunst. Quantitative Analysen.* Brockmeyer, Bochum, 231-262.

HERDAN, G. (1966): *The advanced Theory of Language as Choice and Chance.* Springer, Berlin et al., 423.

KÖHLER, R. (1999): Syntactic Structures. Properties and Interrelations. *Journal of Quantitative Linguistics 6, 46-57.*

KÖHLER, R. (2000): A study on the informational content of sequences of syntactic units. In: L.A. Kuz'min [Ed.]: Jazyk, glagol, predlo?enie. K 70-letiju G. G. Sil'nitskogo. Smolensk, S. 51-61.

KÖHLER, R. and G. ALTMANN (2000): Probability Distributions of Syntactic Units and Properties. *Journal of Quantitative Linguistics 7/3, S.189-200.*

KÖHLER, R. (2006b): Word length in text. A study in the syntagmatic dimension. To appear.

KÖHLER, R. (2006a): The frequency distribution of the lengths of length sequences. In: J. Genzor and M. Bucková [Eds.]: *Favete linguis. Studies in honour of Victor Krupa.* Slovak Academic Press, Bratislava, 145-152.
UHLÍHOVÁ, L. (2007): Word frequency and position in sentence. To appear.
WIMMER, G. and ALTMANN, G. (1999): Thesaurus of Univariate Discrete Probability Distributions. Stamm, Essen.

Projecting Dialect Distances to Geography: Bootstrap Clustering vs. Noisy Clustering

John Nerbonne[1], Peter Kleiweg[1], Wilbert Heeringa[1] and Franz Manni[2]

[1] Alfa-informatica, University of Groningen, Netherlands
{j.nerbonne, p.c.j.kleiweg, w.j.heeringa}@rug.nl
[2] Musée de l'Homme, Paris, France
manni@mnhn.fr

Abstract. Dialectometry produces aggregate DISTANCE MATRICES in which a distance is specified for each pair of sites. By projecting groups obtained by clustering onto geography one compares results with traditional dialectology, which produced maps partitioned into implicitly non-overlapping DIALECT AREAS. The importance of dialect areas has been challenged by proponents of CONTINUA, but they too need to compare their findings to older literature, expressed in terms of areas.

Simple clustering is unstable, meaning that small differences in the input matrix can lead to large differences in results (Jain et al. 1999). This is illustrated with a 500-site data set from Bulgaria, where input matrices which correlate very highly ($r = 0.97$) still yield very different clusterings. Kleiweg et al. (2004) introduce COMPOSITE CLUSTERING, in which random noise is added to matrices during repeated clustering. The resulting borders are then projected onto the map.

The present contribution compares Kleiweg et al.'s procedure to resampled bootstrapping, and also shows how the same procedure used to project borders from composite clustering may be used to project borders from bootstrapping.

1 Introduction

We focus on dialectal data, examined at a high level of aggregation, i.e. the average linguistic distance between all pairs of sites in large dialect surveys. It is important to seek groups in this data, both to examine the importance of groups as organizing elements in the dialect landscape, but also in order to compare current, computational work to traditional accounts. Clustering is thus important as a means of seeking groups in data, but it suffers from instability: small input differences can lead to large differences in results, i.e., in the groups identified.

We investigate two techniques for overcoming the instability in clustering techniques, bootstrapping, well known from the biological literature, and "noisy" clustering, which we introduce here. In addition we examine a novel means of projecting the results of (either technique involving) such repeated clusterings to the geographic map, arguing that it is better suited to revealing the detailed structure in dialectological distance matrices.

2 Background and motivation

We assume the view of dialectometry (Goebl, 1984 *inter alia*) that we characterize dialects in a given area in terms of an aggregate distance matrix, i.e. an assignment of a linguistic distance d to each pair of sites s_1, s_2 in the area $D_l(s_1, s_2) = d$. Linguistic distances may be derived from vocabulary differences, differences in structural properties such as syntax (Spruit, 2006), differences in pronunciation, or otherwise. We ignore the derivation of the distances here, except to note two aspects. First, we derive distances via individual linguistic items (in fact, words), so that we are able to examine the effect of sampling on these items. Second, we focus on true distances, satisfying the usual distance axioms, i.e. having a minimum at zero: $\forall s_1 D(s_1, s_1) = 0$; symmetry: $\forall s_1, s_2 D(s_1, s_2) = D(s_2, s_1)$; and the triangle inequality: $\forall s_1 s_2 s_3 D(s_1, s_2) \leq D(s_1, s_3) + D(s_3, s_2)$ (see (Kruskal 1999:22). We return to the issue of whether the distances are ULTRAMETRIC in the sense of the phylogenetic literature below.

We focus here on how to analyze such distance matrices, and in particular how to detect areas of relative similarity. While multi-dimensional scaling has undoubtedly proven its value in dialectometric studies (Embleton (1987), Nerbonne et al. (1999)), we still wish to detect DIALECT AREAS, both in order to examine how well areas function as organizing entities in dialectology, and also in order to compare dialectometric work to traditional dialectology in which dialect areas were seen as the dominant organizing principle.

CLUSTERING is a standard way in which to seek groups in such data, and it is applied frequently and intelligently to the results of dialectometric analyses. The research community is convinced that the linguistic varieties are hierarchically organized; thus, e.g., the urban dialect of Freiburg is a sort of Low Alemannic, which is in turn Alemannic, which is in turn Southern German, etc. This means that the techniques of choice have been different varieties of hierarchical clustering (Schiltz (1996), Mucha and Haimerl (2005)).

Hierarchical clustering is most easily understood procedurally: given a square distance matrix of size $n \times n$, we seek the smallest distance in it. Assume that this is the distance between i and j. We then fuse the two elements i and j, obtaining an $n-1$ square matrix. One needs to determine the distance from the newly added $i+j$ element to all remaining k, and there are several alternatives for doing this, including nearest neighbor, average distance, weighted average distance, and minimal variance (Ward's method). See Jain et al. (1999) for discussion. We return in the discussion section to the differences between the clustering algorithms, but in order to focus on the effects of bootstrapping and "noisy" clustering, we use only weighted average (WPGMA) in the experiments below.

The result of clustering is a DENDROGRAM, a tree in which the history of the clustering may be seen. For any two leaf nodes in the dendrogram we may determine

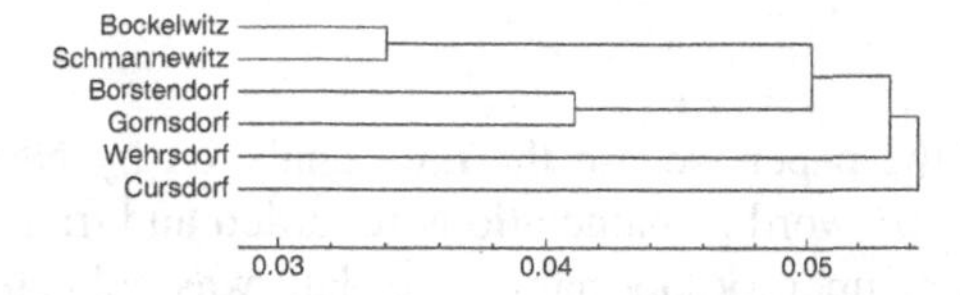

Fig. 1. An Example Dendrogram. Note the cophenetic distance is reflected in the horizontal distance from the leaves to the encompassing node. Thus the cophenetic distance between Borstendorf and Gornsdorf is a bit more than 0.04.

the point at which they fuse, i.e. the smallest internal node which contains them both. In addition, we record the COPHENETIC DISTANCE: this is the distance from one subnode to another at the point in the algorithm at which the subnodes fused.

Note that the algorithms depend on identifying *minimal* elements, which leads to instability: small changes in the input data can lead to very different groups' being identified (Jain et al., 1999). Nor is this problem merely "theoretical". Figure 2 shows two very different cluster results which from genuine, extremely similar data (the distance matrices correlated at $r = 0.97$).

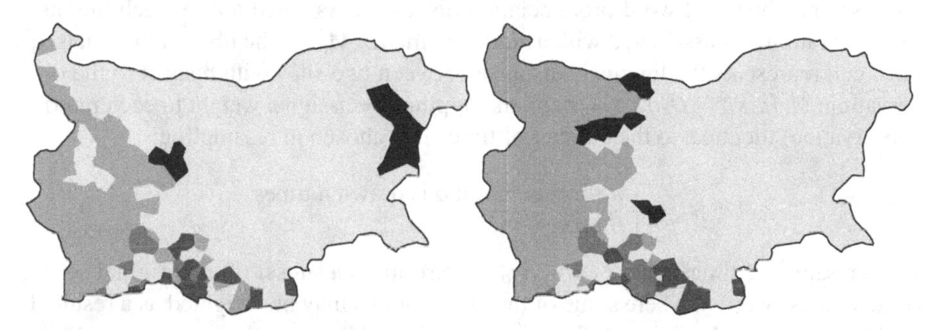

Fig. 2. Two Bulgarian Datasets from Osenova et al. (to appear). Although the distance matrices correlated nearly perfectly ($r = 0.97$), the results of WPGMA clustering differ substantially. Bootstrapping and noisy clustering resolve this instability.

Finally, we note that the distances we shall cluster do *not* satisfy the ultrametric axiom: $\forall s_1 s_2 s_3 D(s_1, s_2) \leq max\{D(s_2, s_3), D(s_1, s_3)\}$ (Page and Holmes (2006:26)). Phylogeneticists interpret data satisfying this axiom temporally, i.e., they interpret data points clustered together as later branches in an evolutionary tree. The dialectal data undoubtedly reflects historical developments to some extent, but we proceed from the premise that the social function of dialect variation is to signal geographic provenance, and that similar linguistic variants signal similar provenance. If the signal is subject to change due to contact or migration, as it undoubtedly is, then similarity could also result from recent events. This muddies the history, but does not change the socio-geographic interpretation.

2.1 Data

In the remainder of the paper we use the data analyzed by Nerbonne and Siedle (2005) consisting of 201 word pronunciations recorded and transcribed at 186 sites throughout all of contemporary Germany. The data was collected and transcribed by researchers at Marburg between 1976 and 1991. It was digitized and analyzed in 2003–2004. The distance between word pronunciations was measured using a modified version of edit distance, and full details (including the data) are available. See Nerbonne and Siedle (2005).

3 Bootstrapping clustering

The biological literature recommends the use of bootstrapping in order to obtain stable clustering results (Felsenstein, 2004: Chap. 20). Mucha and Haimerl (2005) and Manni et al. (2006) likewise recommend bootstrapping for the interpretation of clustering applied to dialectometric data.

In bootstrapped clustering we resample the data, using replacement. In our case we resample the set of word-pronunciation distances. As noted above, each linguistic observation o is associated with a $site \times site$ matrix M_o. In the observation matrix, each cell represents the linguistic distance between two sites with respect to the observation: $M_o(s,s') = D(o_s, o_{s'})$. In bootstrapping, we assign a weight to each matrix (observation) identical to the number of times it is chosen in resampling:

$$w_o = \begin{cases} n \text{ if observation } o \text{ is drawn n times} \\ 0 \text{ otherwise} \end{cases}$$

If we resample I times, then $I = \sum_o w_o$. The result is a subset of the original set of observations (words), where some of the observations may be weighted as a resulted of the resampling. Each resampled set of words yields a new distance matrix $M_{i \in I}$, namely the average distances of the sites using the weighted set of words obtained via bootstrapping.

We apply clustering to each M_i obtained via bootstrapping, recording for each group of sites encountered in the dendrogram (each set of leaves below some node) both that the group was encountered, and the cophenetic distance of the group (at the point of fusion). This sounds as if it could lead to a combinatorial problem, but fortunately most of the 2^{180} possible groups are never encountered.

In a final step we extract a COMPOSITE DENDROGRAM from this collection, consisting of all of the groups that appear in a majority of the clustering iterations, together with their cophenetic distance. See Fig. 3 for an example.

4 Clustering with noise

Clustering with noise is also motivated by the wish to prevent the sort of instability illustrated in Fig. 2. To cluster with noise we assume a single distance matrix, from which it turns out to be convenient to calculate variance (among all the distances). We then specify a small noise ceiling c, e.g. $c = \sigma/2$, i.e. one-half standard deviation of distances in the matrix. We then repeat 100 times or more: add random amounts of noise r to the matrix (i.e., different amounts to each cell), allowing r to vary uniformly, $0 \leq r \leq c$.

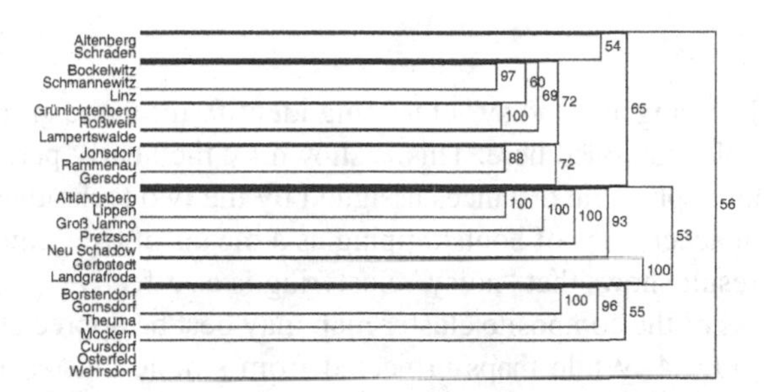

Fig. 3. A Composite Dendrogram where labels indicate how often a groups of sites was clustered and the (horizontal) length of the brackets reflects mean cophenetic distance.

If we let M_i stand in this case for the matrix obtained by adding noise (in the i-th iteration), then the rest of the procedure is identical to bootstrapping. We apply clustering to M_i and record the groups clustered together with their cophenetic distances, just as in Fig. 3.

5 Projecting to geography

Since dialectology studies the geographic variation of language, it is particularly important to be able to examine the results of analyses as these correspond to geography.

In order to project the results of either bootstrapping or noisy clustering to the geographic map, we use the customary Voronoi tessellation (Goebl (1984)), in which each site is embedded in a polygon which separates it from other sites optimally. In this sort of tiling there is exactly one border running between each pair of adjacent sites, and bisecting the imaginary line linking the two. To project mean cophenetic distance matrices onto the map we simply draw the Voronoi tessellation in such a way that the darkness of each line corresponds to the distance between the two sites it separates. See Fig. 4 for examples of maps obtained by bootstrapping two different clustering algorithms. These largely corroborate scholarship on German dialectology (König 1991:230–231).

Unlike dialect area maps these COMPOSITE CLUSTER MAPS reflect the variable strength of borders, represented by the border's darkness, reflecting the consensus cophenetic distance between the adjacent sites.

Haag (1898) (discussed by Schiltz (1996)) proposed a quantitative technique in which the darkness of a border was reflected by the number of differences counted in a given sample, and similar maps have been in use since. Such maps look similar to the maps we present here, but note that the borders we sketch need not be reflected in *local* differences between the two sites. The clustering can detect borders even where differences are gradual, when borders emerge only when many sites are compared.[1]

6 Results

Bootstrapping clustering and "noisy" clustering identify the same groups in the 186-site German sample examined here. This is shown by the nearly perfect correlation between the mean cophenetic distances assigned by the two techniques ($r = 0.997$). Given the general acceptance of bootstrapping as a means of examining the stability of clusters, this result shows that "noisy" clustering is as effective.

The usefulness of the composite cluster map may best be appreciated by inspecting the maps in Fig. 4. While maps projected from simple clustering (see Fig. 2) merely partition an area into non-overlapping subareas, these composite maps reflect a great deal more of the detailed structure in the data. The map on the left was obtained by bootstrapping using WPGMA.

Although both bootstrapping and adding noise identifies stable groups, neither removes the bias of the particular clustering algorithm. Fig. 4 compares the bootstrapped results of WPGMA clustering with unweighted clustering (UPGMA, see Jain (1999)). In both cases bootstrapping and noisy clustering correlate nearly perfectly, but it is clear that the WPGMA is sensitive to more structure in the data. For example, it distinguishes Bavaria (in southeastern Germany) from the Southwest (Swabia and Alemania). So the question of the optimal clustering method for dialectal data remains. For further discussion see `http://www.let.rug.nl/kleiweg/kaarten/MDS-clusters.html`.

7 Discussion

The "noisy"clustering examined here requires that one specify a parameter, the noise ceiling, and, naturally, one prefers to avoid techniques involving extra parameters. On the other hand it is applicable to single matrices, unlike bootstrapping, which requires that one be able to identify components to be selected in resampling. Both techniques require that one specify a number of iterations, but this is a parameter of convenience. Small numbers of iterations are convenient, and large values result in very stable groupings.

---

[1] Fischer (1980) discusses adding a contiguity constraint to clustering, which structures the hypothesis space in a way that favors clusterings of contiguous regions. Since we use the projection to geography to spot linguistic anomalies—dialect islands, but also field worker and transcriber errors—we do not wish to push the clustering in a direction that would hide these anomalies.

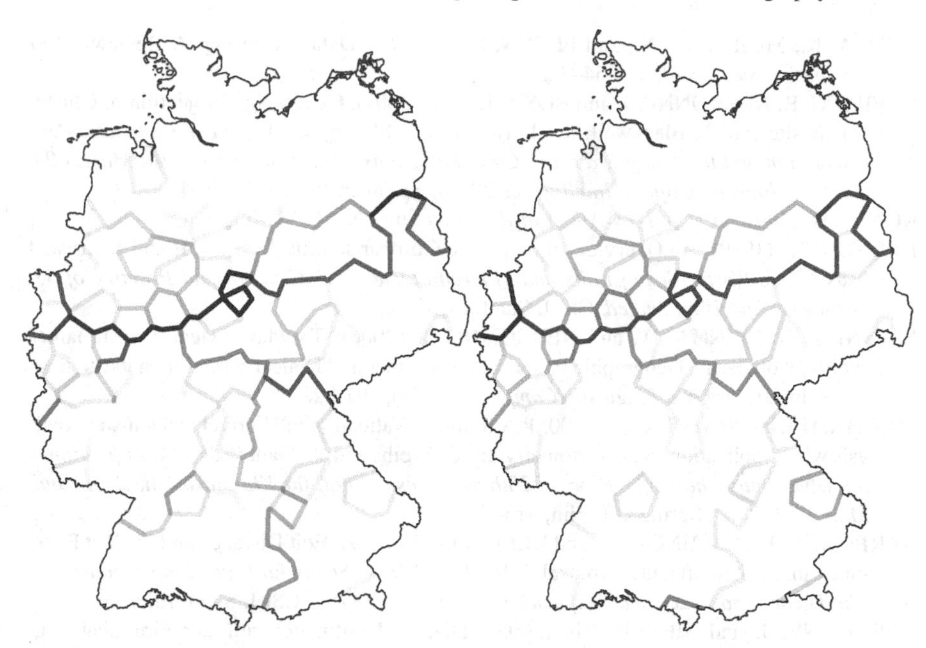

Fig. 4. Two Composite Cluster Maps, on the left one obtained by bootstrapping using weighted group average clustering, and on the right one obtained by unweighted group average. We do not show the maps obtained using "noisy" clustering, as these are indistinguishable from the maps obtained via bootstrapping. The composite distance matrices correlate nearly perfectly ($r = 0.997$) when comparing bootstrapping and "noisy" clustering.

Acknowledgments

We are grateful to Rolf Backofen, Hans Holm and Wilbert Heeringa for discussion, to Hans-Joachim Mucha and Edgar Haimerl for making their 2007 paper available before GfKl 2007, and to two anonymous GfKl referees.

References

EMBLETON, S. (1987): Multidimensional Scaling as a Dialectometrical Technique. In: R. M. Babitch (Ed.) *Papers from the Eleventh Annual Meeting of the Atlantic Provinces Linguistic Association*, Centre Universitaire de Shippagan, New Brunswick, 33-49.

FELSENSTEIN, J. (2004): *Inferring Phylogenies*. Sinauer, Sunderland, MA.

FISCHER, M. (1980): Regional Taxonomy: A Comparison of Some Hierarchic and Non-Hierarchic Strategies. *Regional Science and Urban Economics* 10, 503–537.

GOEBL, H. (1984): *Dialektometrische Studien: Anhand italoromanischer, rätoromanischer und galloromanischer Sprachmaterialien aus AIS und ALF* 3 Vol. Max Niemeyer, Tübingen.

HAAG, K. (1898): *Die Mundarten des oberen Neckar- und Donaulandes*. Buchdruckerei Egon Hutzler, Reutlingen.

JAIN, A. K., MURTY, M. N., and FLYNN, P. J. (1999): Data Clustering: A Review. *ACM Computing Surveys* 31(3), 264–323.

KLEIWEG, P., NERBONNE, J. and BOSVELD, L. (2004): Geographic Projection of Cluster Composites. In: A. Blackwell, K. Marriott and A. Shimojima (Eds.) *Diagrammatic Representation and Inference. 3rd Intn'l Conf, Diagrams 2004. Cambridge, UK, Mar. 2004. (Lecture Notes in Artificial Intelligence* 2980). Springer, Berlin, 392-394.

KÖNIG, W. (1991, [1]1978): *DTV-Atlas zur detschen Sprache*. DTV, München.

KRUKSAL, J. (1999): An Overview of Sequence Comparison. In: D. Sankoff and J. Kruskal (Eds.) *Time Warps, String Edits and Macromolecules: The Theory and Practice of Sequence Comparison, 2nd ed.* CSLI, Stanford, 1–44.

MANNI, F. HEERINGA, W. and NERBONNE, J. (2006): To what Extent are Surnames Words? Comparing Geographic Patterns of Surnames and Dialect Variation in the Netherlands. In *Literary and Linguistic Computing* 21(4), 507-528.

MUCHA, H.J. and HAIMERL, E. (2005): Automatic Validation of Hierarchical Cluster Analysis with Application in Dialectometry. In: C. Weihs and W. Gaul (Eds.) *Classification—the Ubiquitous Challenge. Proc. of 28th Mtg Gesellschaft für Klassifikation, Dortmund, Mar. 9–11, 2004*. Springer, Berlin, 513–520.

NERBONNE, J., HEERINGA, W. and KLEIWEG, P. (1999): Edit Distance and Dialect Proximity. In: D. Sankoff and J. Kruskal (Eds.) *Time Warps, String Edits and Macromolecules: The Theory and Practice of Sequence Comparison, 2nd ed.* CSLI, Stanford, v-xv.

NERBONNE, J. and SIEDLE, Ch. (2005): Dialektklassifikation auf der Grundlage aggregierter Ausspracheunterschiede. *Zeitschrift für Dialektologie und Linguistik* 72(2), 129-147.

PAGE, R.D.M., and HOLMES, E.C. (2006): *Molecular Evolution: A Phylogenetic Approach.* ([1]1998) Blackwell, Oxford.

SCHILTZ, G. (1996): German Dialectometry. In: H.-H. Bock and W. Polasek (Eds.) *Data Analysis and Information Systems: Statistical and Conceptual Approaches. Proc. of 19th Mtg of Gesellschaft für Klassifikation, Basel, Mar. 8–10, 1995*. Springer, Berlin, 526–539.

SPRUIT, M. (2006): Measuring Syntactic Variation in Dutch Dialects. In J. Nerbonne and W. Kretzschmar, Jr. (Eds.) *Progress in Dialectometry: Toward Explanation*. Special issue of *Literary and Linguistic Computing* 21(4), 493–506.

Structural Differentiae of Text Types – A Quantitative Model

Olga Pustylnikov and Alexander Mehler

Faculty of Linguistics and Literature Study,
University of Bielefeld, Germany
{Olga.Pustylnikov, Alexander.Mehler}@uni-bielefeld.de

Abstract. The categorization of natural language texts is a well established research field in computational and quantitative linguistics (Joachims 2002). In the majority of cases, the vector space model is used in terms of a *bag of words* approach. That is, lexical features are extracted from input texts in order to train some categorization model and, thus, to attribute, for example, authorship or topic categories. Parallel to these approaches there has been some effort in performing text categorization not in terms of lexical, but of structural features of document structure. More specifically, quantitative text characteristics have been computed in order to derive a sort of structural text signature which nevertheless allows reliable text categorizations (Kelih & Grzybek 2005; Pieper 1975). This *"bag of features"* approach regains attention when it comes to categorizing websites and other document types whose structure is far away from the simplicity of tree-like structures. Here we present a novel approach to structural classifiers which systematically computes structural signatures of documents. In summary, we present a text categorization algorithm which in the absence of any lexical features nevertheless performs a remarkably good classification even if the classes are thematically defined.

1 Introduction

An alternative way to categorize documents apart from the well established " *bag of words*" approach is to categorize by means of structural features. This approach functions in absence of any lexical information utilizing quantitative characteristics of documents computed from the logical document structure.[1] That means that markers like content words are completely disregarded. Features like distributions of sections, paragraphs, sentence length etc. are considered instead.

Capturing structural properties to build a classifier assumes that given category separations are reflected by structural differences. According to Biber (1995) we can expect that functional differences correlate with structural and formal representations of text types. This may explain good overall results in terms of *F-Measure*[2].

[1] See also Mehler et al. (2006).

[2] The harmonic mean of *precision* and *recall* is used here to measure the overall success of the classification

However, the *F-Measure* gives no information about the quality of the investigated categories. That is, no a prior knowledge about the suitability of the categories for representing homogenous classes and for applying them in machine learning tasks is provided. Since natural language categories e.g. in form of web documents or other textual units arise not necessarily with a well defined structural representation available it is important to know how the classifier behaves dealing with such categories.

Here, we investigate a large number of existing categories, thematic classes or *rubrics* taken from a 10 years newspaper corpus of *Süddeutsche Zeitung* (SZ 2004) whereas a rubric represents a recurrent part of the newspaper like `sportst' or `tv-newst'. We test systematically their goodness in a structural classifier framework asking more specifically for a maximal subset of all rubrics which gives an *F-Measure* above a predefined *cut-off* $c \in [0,1]$ (e.g. $c = 0.9$). We evaluate the classifier in the way allowing to exclude possible drawbacks with respect to:

- the categorization model used (here SVM[3] and Cluster Analysis),[4]
- the text representation model used (here the *bag of features* approach) and
- the structural homogeneity of categories used.

The first point relates to distinguishing supervised and unsupervised learning. That is, we perform these sorts of learning although we do not systematically evaluate them comparatively with respect to all possible parameters. Rather, we investigate the potential of our features evaluating them with respect to both scenarios. The representation format (vector representation) is restricted by the model used (e.g. SVM). Thus, we concentrate on the third point and apply an *iterative categorization procedure* (ICP)[5] to explore the structural suitability of categories. In summary, our experiments have twofold goals:

1. to study given categories using the ICP in order to filter out structurally inconsistent types and
2. to make judgements about the structural classifier's behavior dealing with categories of different size and quality levels.

2 Category selection

The 10 years corpus of the SZ used in the present study contains 95 different rubrics. The frequency distribution of these rubrics shows an enormous inequality for the whole set (See Figure 1). In order to minimize the calculation effort we reduce the initial set of 95 rubrics to a smaller subset according to the following criteria.

1. First, we compute the mean μ and the standard deviation σ for the whole set.

---

[3] Support Vector Machines.

[4] Supervised vs. unsupervised respectively.

[5] See sec. 4.

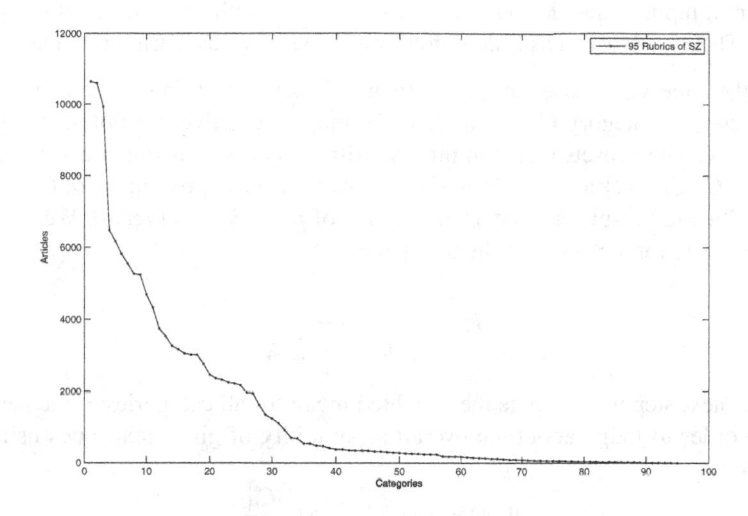

Fig. 1. Categories/Articles-Distribution of 95 Rubrics of SZ.

2. Second, we pick out all rubrics R with the cardinality $|R|$ (the number of examples within the corpus) ranging between the interval:

$$\mu - \sigma/2 < |R| < \mu + \sigma/2$$

This selection method allows to specify a window around the mean value of all documents leaving out the unusual cases.[6] Thus, the resulting subset of 68 categories is selected.

3 The evaluation procedure

The data representation format for the subset of rubrics uses a vector representation (*bag of features* approach) where each document is represented by a feature vector.[7] The vectors are calculated as structural signatures of the underlying documents. To avoid drawbacks (See Sec. 1) caused by the evaluation method in use, we compare three different categorization scenarios:

1. Supervised scenario by means of SVM-light[8],
2. Unsupervised scenario in terms of Cluster Analysis and
3. Finally, a baseline experiment based on random clustering.

[6] The method is taken from Bock (1974). Rieger (1989) uses it to identify above-average agglomeration steps in the clustering framework. Gleim et al. (2007) successfully applied the method to develop quality filters for wiki articles.

[7] See Mehler et al. (2007) for a formalization of this approach.

[8] Joachims (2002).

Consider an input corpus K and a set of categories $\mathbb{C}$ with the number of categories $|\mathbb{C}| = n$. Then we proceed as follows to evaluate our various learning scenarios:

- For the supervised case we train a binary classifier by treating the negative examples of a category $C_i \in \mathbb{C}$ as $K \setminus [C_i]$ and the positive examples as a subset $[C_i] \subseteq K$. The subsets C_i are in this experiment pairwise disjunct and we define $\mathbb{L} = \{[C_i] | C_i \in \mathbb{C}\}$ as a partition of positive and negative examples of C_i. Classification results are obtained in terms of *precision* and *recall*. We calculate the *F-score* for a class C_i in the following way:

$$F_i = \frac{2}{\frac{1}{\text{recall}_i} + \frac{1}{\text{precision}_i}}$$

  In the next step we compute the weighted mean for all categories of the partition $\mathbb{L}$ in order to judge about the overall separability of given text types using the *F-Measure*:

$$\text{F-Measure}(\mathbb{L}) = \sum_i^n \frac{|C_i|}{|K|} F_i$$

- In the case of unsupervised experiments we approach as follows: The unsupervised procedure evaluates different variants of Cluster Analysis (hierarchical, k-means) trying out several linkage possibilities (complete, single, average, weighted) in order to achieve the best performance. Similar to the supervised case best clustering results are presented in terms of *F-Measure* values.
- Finally, the random baseline is calculated by preserving the original category sizes and by mapping articles randomly to them. Results of random clustering help to check the success of both learning scenarios. Thus, clusterings close to the random baseline indicate either a failure of the cluster algorithm or that the separability of the text types can't be well separated by structure.

In summary, we check the performance of structural signatures within two learning scenarios – supervised and unsupervised – and compare the results with the random clustering baseline. Next Section describes the *incremental categorization procedure* (ICP) to investigate the structural homogeneity of categories.

4 Exploring the structural homogeneity of text types by means of the Iterative Categorisation Procedure (ICP)

In this Section we return to the question mentioned at the beginning. Given a *cut-off* $c \in [0,1]$ (e.g. $c = 0.9$) we ask for the maximal subset of rubrics allowing to achieve an *F-Measure* value $F > c$. Decreasing the *cut-off* c successively we get a rank ordering of rubrics ranging from the best contributors to the worst ones. The ICP allows to determine a result set of maximal size n with the maximal internal homogeneity compared to all candidate sets in question. Starting with a given set of input categories to be learned we proceed as follows:

1. **Start:** Select a seed category $C \in A$ and set $A_1 = \{C\}$. The rank r of C equals $r(C) = 1$. Now repeat:
2. **Iteration** ($i > 1$)**:** Let $B = A \setminus A_{i-1}$. Select the category $C \in B$ which when added to A_{i-1} maximizes the F-Measure value among all candidate extensions of A_{i-1} by means of single categories of B. Set $A_i = A_{i-1} \cup \{C\}$ and $r(C) = i$.
3. **Break off:** The iteration algorithm terminates if either

   i) $A \setminus A_i = \emptyset$ or

   ii) the F-Measure value of A_i is smaller than a predefined cut-off or

   iii) the F-Measure value of A_i is smaller than the one of the operative baseline.

   If none of these stop conditions holds repeat step (2).

The kind of ranking described here is more informative than the *F-Measure* value alone. That is, the *F-Measure* gives global information about the overall separability of categories. The ICP in contrast, provides additional local information about the weights of single categories with respect to the overall performance. This information allows to check the suitability of single categories to serve as structural prototypes. Knowledge about the homogeneity of each category provides a deeper insight into the possibilities of our approach.

In the next Section the rankings of the ICP applied to supervised and unsupervised learning and compared with the random clustering baseline are presented. In order to exclude a dependence of the structural approach on one of the learning methods, we also apply the *best-of-unsupervised-ranking* to the supervised scenario and compare the outcomes. That means, we use exactly the same range having performed best in the unsupervised experiment for SVM learning.

5 Results

Table 1 gives an overview about the categories used. From the total number of 95 rubrics 68 were selected using the selection method described in Section 2, 55 were considered in unsupervised, 16 in supervised experiments. The common subset used in both cases consists of 14 categories.

The Y-axis of Figure 2 represents the *F-Measure* values and the X-axis the rank order of categories iteratively added to the seed set. The supervised scenario (upper curve) performs best ranging around the value of 1.0. The values of the unsupervised case decrease more rapidly (the third curve from above). The unsupervised best-of-ranking categorized with the supervised method (second curve from above) lies between the best results of the two methods. The lower curve represents the results of random clustering.

6 Discussion

According to Figure 2 we can see, that all *F-Measure* results lie high above the baseline of random clustering. All the subsets are well separated by their document

Table 1. Corpus Formation (by Categories).

Category Set	Number
Total	95
Selected Initial Set	68
Unsupervised	55
Supervised	16
Unsupervised ∩ Supervised	14

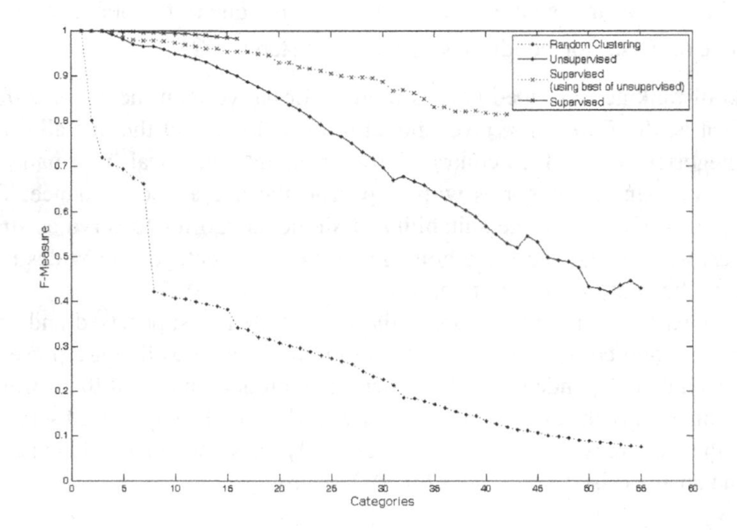

Fig. 2. The F-Measure Results of All Experiments

structure which indicates a potential of structure-based categorizations. The point here was to observe the decrease of the *F-Measure* value while adding new categories.

The supervised method shows the best results remaining stable with a growing number of additional categories. The unsupervised method shows a more rapid decrease but is less time consuming. Cluster Analysis succeeds to rank 55 rubrics whereas SVM-light ranks only 16 within the same time span.

In order to compare the performance of both methods (supervised vs. unsupervised) more precisely we ran the supervised categorization based on the *best-off-ranking* of the unsupervised case. The resulting curve remains longer stable than the unsupervised one. Since the order and the features of categories are equal, the resulting difference indicates an overall better accuracy of SVM compared to Cluster Analysis.

One assumption for the success of the structural classifier was that the performance may depend on the article size, that is, on the representativeness of a category. To account for this, we compared the category size of the *best-off-rankings* of both

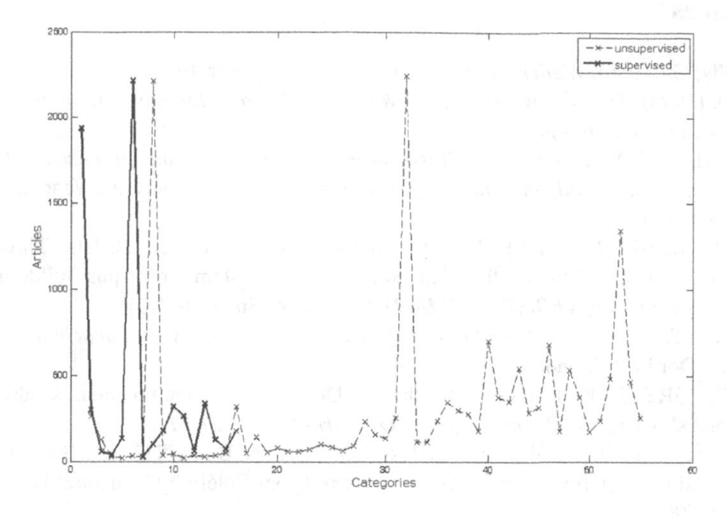

Fig. 3. Categories/Articles-Distribution of Sets used in Supervised/Unsupervised Experiments.

experiments. Figure 3 shows a high variability in size, which indicates that the size factor does not influence the classifier.

7 Conclusion

In this paper we presented experiments which shed light on the possibilities of a classifier operating with structural signatures of text types. More specifically, we investigated the ability of the classifier to deal with a large number of natural language categories of different size and quality. The *best-off-rankings* showed that different evaluation methods (supervised/unsupervised) prefer different combinations of categories to achieve the best separation. Furthermore, we could see that the overall difference in performance of two methods depends rather on the method used than on the combination of categories.

Another interesting finding is that the structural classifier seems not to depend on category size allowing a good categorization of small, less representative categories. That fact motivates to use logical document (or any other kind of) structure for machine learning tasks and to extend the framework to more demanding tasks, when it comes to deal with, e.g., web documents.

References

ALTMANN, G. (1988): *Wiederholungen in Texten*. Brockmeyer, Bochum.

BIBER, D. (1995): *Dimensions of Register Variation: A Cross-Linguistic Comparison*. University Press, Cambridge.

BOCK, H.H. (1974): *Automatische Klassifikation. Theoretische und praktische Methoden zur Gruppierung und Strukturierung von Daten (Cluster-Analyse)*. Vandenhoeck & Ruprecht, Göttingen.

GLEIM, R.; MEHLER, A.; DEHMER, M.; PUSTYLNIKOV, O. (2007): Isles Through the Category Forest — Utilising the Wikipedia Category System for Corpus Building in Machine Learning. In: *WEBIST '07, WIA(2)*. Barcelona, Spain, 142-149.

JOACHIMS, T. (2002): *Learning to classify text using support vector machines*. Kluwer, Boston/Dordrecht/London.

KELIH, E.; GRZYBEK, P. (2005): Satzlänge: Definitionen, Häufigkeiten, Modelle (Am Beispiel slowenischer Prosatexte). In: *LDV-Forum 20(2), 31-51.*

MEHLER, A.; GEIBEL, P.; GLEIM, R.; HEROLD, S.; JAIN, B.; PUSTYLNIKOV, O. (2006): Much Ado About Text Content. Learning Text Types Solely by Structural Differentiae. In: *OTT'06.*

MEHLER, A.; GEIBEL, P.; PUSTYLNIKOV, O.; HEROLD, S. (2007): Structural Classifiers of Text Types. To appear in: LDV Forum.

PIEPER, U. (1975): Differenzierung von Texten nach Numerischen Kriterien. In: *Folia Linguistica VII, 61-113.*

RIEGER, B. (1989): *Unscharfe Semantik: Die empirische Analyse, quantitative Beschreibung, formale Repräsentation und prozedurale Modellierung vager Wortbedeutungen in Texten.* Peter Lang, Frankfurt a. M.

SÜDDEUTSCHER VERLAG (2004). *Süddeutsche Zeitung 1994-2003. 10 Jahre auf DVD.* München.

Part XI

Data Analysis in Humanities

Scenario Evaluation Using Two-mode Clustering Approaches in Higher Education

Matthias J. Kaiser, Daniel Baier

Institute of Business Administration and Economics,
Brandenburg University of Technology Cottbus,
Postbox 101344, 03013 Cottbus, Germany
{mjkaiser, daniel.baier}@tu-cottbus.de

Abstract. Scenario techniques have become popular tools for dealing with possible futures. Driving forces of the development (the so-called key factors) and their possible projections into the future are determined. After a reduction of the possible combinations of projections to a set of consistent and probable candidates for possible futures, traditionally one-mode cluster analysis is used for grouping them. In this paper, two-mode clustering approaches are proposed for this purpose and tested in an application for the future of eLearning in higher education. In this application area, scenario techniques are a very young and promising methodology.

1 Introduction: Scenario analysis

Since its first applications for business prognostication (e.g., Kahn, Wiener (1967), Meadows et al. (1972), Schwartz (1991)), scenario techniques have become popular tools for governmental and corporate planners in order to deal with possible futures ("scenarios") and to support decisions in the face of uncertainty. Nowadays, in many research areas scenario analysis is an attractive tool with a huge variety of applications (e.g., Götze (1993), Mißler-Behr (2002), Welfens et al. (2004), van der Heijden (2005), Pasternack (2006), Ringland (2006)). However, for higher education, the application of scenario analysis is new (e.g., Sprey (2003)). Different methodological approaches have been proposed, most of them using (roughly) four stages (e.g., Coates (2000), Phelps et al. (2001)):

- In a first stage, the scope of the scenario analysis has to be defined including the focal issues (e.g. influence areas) and the driving forces for them (social, economic, political, environmental, technological factors). After a reduction of these driving forces with respect to relevance, importance, and inter-connection, a list of so-called key factors results (e.g., A, B, C).
- Then, in the second stage, alternative projections (possible levels) for these key factors (e.g., A1, A2, A3, B1, B2) have to be determined. By combining these projections, a database of candidates for possible futures (e.g., (A1,B1,C1,...), (A1,B2,C1,...)) is available. Additionally, the consistency for pairs of projections

(e.g., (A1,B1), (A1,B2)) and the probability/realism of single projections within the time span under research has to be rated.

- Then, in a third stage, the candidates in the database have to be evaluated on basis of their projections' pairwise consistency and probability. Using rankings and/or cut-off values or similar approaches, the database is reduced to a set of consistent and probable candidates. Finally, the reduced set of candidates (the so-called first mode), described by their projections w.r.t. the key factors (the so-called second mode), is grouped via cluster analysis into a small number of candidate groups, the so-called "scenarios". In an unrelated second step these candidate groups have to be analyzed to find out which projections best characterize them. Recently, new fuzzy clustering approaches have been proposed for dealing with this identification problem (see e.g. Mißler-Behr (1993), (2002)).
- Finally, in a fourth stage, strategic options how to deal with the selected possible futures ("scenarios") have to be developed.

In this paper we develop new two-mode clustering approaches for simultaneously grouping candidates and projections in the third stage. The new approach bases on Baier et al. (1997)'s two-mode additive clustering procedure for simultaneous market segmentation and structuring with overlapping and non-overlapping cases.

2 Two-Mode clustering (for scenario evaluation)

2.1 The model

As in Baier et al. (1997), the following notation is used (see Krolak-Schwerdt, Wiedenbeck (2006) for a recent comparison of similar additive clustering approaches): i=1,...,I is an index for first mode objects (e.g., preselected consistent and probable candidates (A1,B1,C1,...) or (A1,B2,C1,...) from stage two). j=1,...,J is an index for second mode objects (e.g., projections A1, A2, A3, ...). k=1,...,K is an index for first mode clusters (cluster of candidates) and l=1,...,L an index for second mode clusters (clusters of projections). $\mathbf{S} = (s_{ij})_{I\times J}$ is a matrix of (observed) associations between first and second mode objects ($s_{ij} \in \boldsymbol{R}\ \forall i,j$). With association values of 1 – if the projection is part of the candidate – or 0 – if the projection is not part of the candidate –, $\mathbf{S}$ is a binary data matrix (see, e.g., Li (2005) for an analysis of binary data using two-mode clustering).

Model parameters are the following: $\mathbf{P}=(p_{ik})_{I\times K}$ is a binary matrix describing first mode cluster membership with p_{ik}=1 if first mode object i belongs to first mode cluster k and =0 otherwise. $\mathbf{Q}=(q_{jl})_{J\times L}$ is a binary matrix describing second mode cluster membership with q_{jl}=1 if second mode object j belongs to second mode cluster l and =0 otherwise. $\mathbf{W}=(w_{kl})_{K\times L}$ is a matrix of weights ($w_{kl} \in \boldsymbol{R}\ \forall k,l$).

In order to provide results where candidates are members of one and only one scenario whereas projections are allowed to be member of none, one, or more than one scenario, additional assumptions are necessary: The first mode membership matrix $\mathbf{P}$ is restricted to be non-overlapping (i.e. $\sum_{k=1}^{K} p_{ik} = 1\ \forall i$) whereas for the second mode membership matrix **Q** no such restrictions hold. **Q** is allowed to be overlapping.

2.2 Parameter estimation

The parameters are determined in order to minimize the objective function

$$Z = \sum_{i=1}^{I}\sum_{j=1}^{J}(s_{ij} - \hat{s}_{ij})^2 \text{ with } \hat{s}_{ij} = \sum_{k=1}^{K}\sum_{l=1}^{L} p_{ik} w_{kl} q_{jl} \quad \forall i,j, \tag{1}$$

or, equivalently, to maximize the variance accounted for

$$\text{VAF} = 1 - Z/\sum_{i=1}^{I}\sum_{j=1}^{J}(s_{ij} - \bar{s})^2 \text{ with } \bar{s} = \sum_{i=1}^{I}\sum_{j=1}^{J} s_{ij}/(IJ) \tag{2}$$

on the basis of the underlying model **S** = **PWQ'** + error.

In our approach, an alternating least squares procedure is applied. The different sets of model parameters (**P**, **W**, and **Q**) are initialized and alternatingly improved w.r.t. Z. Alternatively, a Bayesian model formulation could be used (see DeSarbo et al. (2005) in a market structuring setting). However, for our approach, we first discuss the iterative steps for obtaining improved estimates for selected model parameters when estimates for the remaining sets of model parameters are given. Finally, the complete procedure is presented.

a) Estimation of **P** *for given* **W** *and* **Q**: Set

$$p_{ik} = \begin{cases} 1 \text{ if } \sum_{j=1}^{J}(s_{ij} - \sum_{l=1}^{L} w_{kl} q_{jl})^2 = \min_{1 \le \kappa \le K} \{\sum_{j=1}^{J}(s_{ij} - \sum_{l=1}^{L} w_{\kappa l} q_{jl})^2\} \\ 0 \text{ otherwise} \end{cases} \quad \forall i,k. \tag{3}$$

b) Estimation of **Q** *and* **W** *for given* **P**: Using (for l=1,...,L selected)

$$Z = \sum_{i=1}^{I}\sum_{j=1}^{J}(\underbrace{s_{ij} - \sum_{k=1}^{K} \sum_{l'=1 \wedge l' \neq l}^{L} p_{ik} w_{kl'} q_{jl'}}_{=: \, s_{ijl}} - \sum_{k=1}^{K} p_{ik} w_{kl} q_{jl})^2 \tag{4}$$

(s_{ijl} is constant w.r.t. $q_{1l},\ldots,q_{Jl},w_{1l},\ldots,w_{Kl}$), estimates of **Q** and **W** can be obtained by starting from initial values and alternatingly improving the parameter estimates for second mode cluster $l = 1,\ldots,L$ via

$$q_{jl} = \begin{cases} 1 \text{ if } \sum_{i=1}^{I}(s_{ijl} - \sum_{k=1}^{K} p_{ik} w_{kl})^2 < \sum_{i=1}^{I}(s_{ijl})^2 \\ 0 \text{ otherwise} \end{cases} \quad \forall j \tag{5}$$

and minimizing

$$\sum_{i=1}^{I}\sum_{j=1}^{J}(s_{ijl}-\sum_{k=1}^{K}p_{ik}w_{kl}q_{jl})^2 \text{ via OLS w.r.t. } \{w_{1l},\ldots,w_{Kl}\} \tag{6}$$

(OLS=ordinary least squares regression).

Thus, our estimation procedure can be described as follows:

1. Determine initial estimates of **P**, **W**, and **Q**. Compute Z.
2. Repeat
   Improve the estimates of **P** using *a)*.
   Improve the estimates of **Q** and **W** using *b)*.
   Until Z cannot be improved any more.

For applying the above model and algorithms for scenario evaluation, additionally, the first and second mode clusters can be linked by setting K=L and restricting **W** to an identity matrix. This can be achieved by initialization and by omitting the corresponding algorithmic steps where **W** is updated. In the following section, this approach (with K=L and **W** restricted to an identity matrix) is applied in stage three of a scenario analysis in higher education.

3 Example: Scenario evaluation in higher education

3.1 Stage One: Defining the scope of the analysis

Currently, at many universities, the concrete future of higher education and how to deal with this uncertainty is unclear. Whereas some developments like the demographics (older and fewer Germans), the ongoing of the Bologna-process (more standardization and Europe-wide exchange in higher education), the importance of better and life-long education, or the higher competition between universities for funds and talented students seem to be predictable, other developments are highly uncertain (see, e.g., Michel (2006), Opaschowski (2006), Schulmeister (2006)).

Especially for universities that plan to invest in technical teaching and learning environments and/or plan to attract more students for distance learning - this is unbearable. Therefore, our main research question deals with the future of higher education. As a focal time point we use the year 2020. Also, this analysis is used as an application example for our new two-mode clustering approach.

In the first stage of our scenario analysis, basing on a Delphi-study on the future of eLearning, acceptance and preferences surveys, and other research projects at our institute (e.g. Göcks (2006)) as well as from other research institutes (e.g. Cuhls et al. (2002), Opaschowski (2006)) (university) internal as well as (university) external influencing factors on higher education were identified and possible projections for the near future were described.

Moreover, using expert workshops with teachers, students, people from university administration and government, these lists and descriptions were extended and modified, resulting in six areas of influence and thirty influencing factors (see figure 1) with a total of 73 detailed described projections w.r.t. these influencing factors.

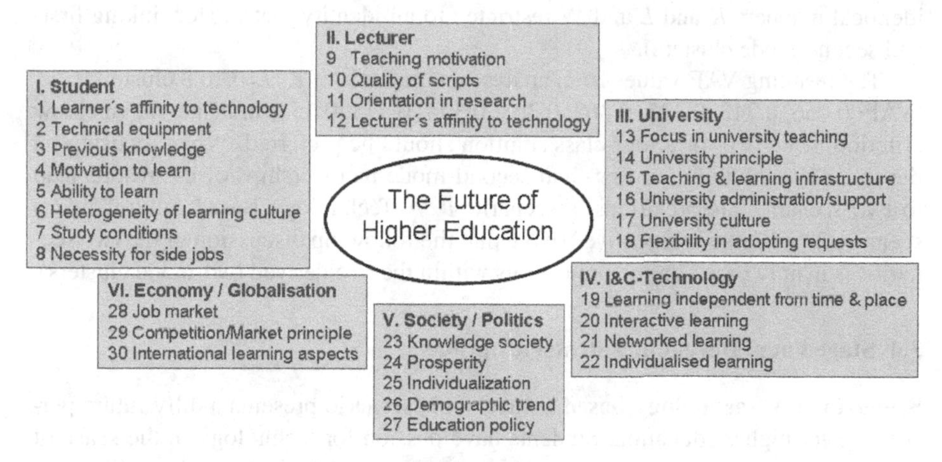

Fig. 1. Influencing factors overview

3.2 Stage Two: Creating a database of candidates

In the second stage of scenario analysis, these thirty influencing factors were reduced to 12 key factors for the ongoing analysis. We did this by filtering redundant aspects and indirect dependencies. Additionally, we used scoring methods and evaluation aspects from a group of scientific experts and analyzed relevant scientific sources (see, e.g., Kröhnert et al. (2004), Michel (2006)). Furthermore, the alternative projections for each key factor were reduced and specified in detail (resulting in one page text for each projection). As a result, a database of $2^{11}3^{1}$=6,144 candidates (all possible combinations of the 2-3 projections for each of the 12 key factors) for possible futures was available.

Additionally, the pairwise consistency of these projections was evaluated using values ranging from 1="totally inconsistent" to 9="totally consistent". Consequently, as discussed in the theoretical introduction, a consistency value was calculated for each candidate (e.g. (A1,B2,C3,...)) as the mean pairwise consistency of its pairs of projections (e.g. (A1,B2), (A1,C3), (B2,C3),...).

3.3 Stage Three: Evaluating, selecting, and clustering candidates

In a third stage the database was first reduced and then clustered. For reduction, the so-called "'complete combination scanning"' was used, what means that for each pair of projections that candidate with the highest mean pairwise consistency was kept for further analysis. The reduction resulted into 286 candidates.

The binary descriptions of these candidates resulted into a binary database **S** with 286 rows and 25 columns. This database was – in the follow-up analysis – subjected to the two-mode clustering approaches for scenario evaluation from section 2.2 with identical numbers K and L and **W** restricted to an identity matrix (for linking first- and second-mode clusters).

The resulting VAF-values from analyses with totals of K=L=1 to 8 clusters (VAF=0.056, 0.243, 0.325, 0.362, 0.363, 0.394, 0.448, 0.452) indicate via an elbow criterion that a two- or a four-class solution should be preferred. When focusing on the two-class solution, the first- and second-mode memberships of the results lead to two scenario interpretations, a scenario 1 "A Technology Based Future" and a scenario 2 "A Worse Perspective" (Note that the follow-up discussion of the two scenarios is mainly based on the projections within the two derived two-mode clusters).

3.4 Stage Four: Developing strategic options

Scenario 1: A Technology based future: This scenario presents a dilly future perspective for higher education. Students have passion for technology in the sense of education technologies and learning software. They are motivated to learn like conscientious learners. The university lecturers see a greater importance in giving lectures than in doing research.

The traditional lecture forms will be enhanced by eLearning components like online teaching and blended learning scenarios. There will be a unity of traditional and new lesson forms. The future will contain state universities as well as private ones in the education market.

The learning infrastructure and administration environment (technology, buildings, networks, etc.) will be excellent. Because of hard competition in the education market, the universities are very flexible and try to be better than their competitors. They are able to assimilate new aspects and trends in learning innovations (like eLearning) very quickly. The usage of information and communication technologies is established very well and in higher education eLearning aspects are used very often.

eLearning aspects help to enforce individualised learning for better results in the studies of each student. These facts will be supported by a high level of education awareness in the whole society in addition. The importance of job market issues forces the students to acquire an additional expertise in languages, soft skills, and other competences.

Scenario 2: A worse perspective: The second extreme scenario presents us the complete opposite to scenario 1. The future in higher education is not very attractive. No interested and committed students in the study courses, lecturers with little interest in teaching, no changes in traditional ways of teaching and no private education suppliers in the market. Universities have resources to offer an optimal learning environment and infrastructure (library, internal working places, etc.). No flexibility will prevailed at the universities and no eLearning technologies will be used. The consequence is that no individualized learning will be offered. Education is no longer an emphasis from the society point of view.

When analyzing the four-class solution, the above results are supported: Again the two extreme scenarios could be found, but now two additional in-between scenarios are available. These two scenarios mainly differ from the above two w.r.t. the university principle (state, private, or mixed) and the importance of job market issues on the teaching contents and environment (high or low influence).

4 Conclusions

In this paper, we have introduced new two-mode clustering approaches for scenario evaluation. It fits naturally in the traditional four-stage-approach to scenario analysis by alternatively analyzing the database of consistent candidates for possible futures. In contrast to the traditional one-mode clustering approaches for this purpose, the two-mode approach quite naturally develops clusters of candidates and describing projections. No follow-up decisions concerning fuzzy memberships of candidates or memberships of projections have to be made.

References

BAIER, D., GAUL, W., and SCHADER, M. (1997): Two-Mode Overlapping Clustering With Applications to Simultaneous Benefit Segmentation and Market Structuring. In: Klar, R. and Opitz, O. (Eds.), *Classification and Knowledge Organization*. Springer, Heidelberg, 557–566.

COATES, J. F. (2000): Scenario Planning. *Technological Forecasting and Social Change, 65, 115-123.*

CUHLS, C., BLIND, K., and GRUPP, H. (2002): *Innovations for our Future. Delphi '98: New Foresight on Science and Technology*. Physica-Verlag, Heidelberg.

DESARBO, W.S., FONG, D., and LIECHTY, J. (2005): Two-Mode Cluster Analysis via Hierarchical Bayes. In: Baier, D. and Wernecke, W. (Eds.), *Innovations in Classification, Data Science, and Information Systems*. Springer, Heidelberg, 19–29.

GÖCKS, M. S. (2006): *Betriebswirtschaftliche eLearning-Anwendungen in der universitären Ausbildung*. Shaker, Aachen.

GÖTZE, U. (1993): *Szenario-Technik in der strategischen Unternehmensplanung*. 2nd Edition, DUV, Wiesbaden.

KAHN, H. and WIENER, A. J. (1967): *The Year 2000: A Framework for Speculation on the Next Thirty-Three Years*. Macmillan, New York.

KRÖHNERT, S., VAN OLST, N., and KLINGHOLZ, R. (2004): *Deutschland 2020: Die demographische Zukunft der Nation*. Berlin-Institut für Bevölkerung und Entwicklung, Berlin.

KROLAK-SCHWERDT, S., WIEDENBECK, M. (2006): The Recovery Performance of Two-Mode Clustering Methods: Monte Carlo Experiment. In: Spiliopoulou, M. et al. (Eds.), *From Data and Information Analysis to Knowledge Engineering*. Springer, Heidelberg, 190–197.

LI, T. (2005): A General Model for Clustering Binary Data. In: *Conference on Knowledge Discovery and Data Mining (KDD) 2005*. Chicago, 188-197.

MEADOWS, D., RANDERS, J. and BEHRENS, W. (1972): *The Limits to Growth*. Universe, New York.

MICHEL, L. P. (2006): *Digitales Lernen: Forschung - Praxis - Märkte*. Books on Demand, Norderstedt.

MISSLER-BEHR, M. (1993): *Methoden der Szenarioanalyse*. DUV, Wiesbaden.

MISSLER-BEHR, M. (2002): Fuzzy Scenario Evaluation. In: Gaul, W. and Ritter, G. (Eds.): *Classification, Automation, a. New Media*. Springer, Berlin, 351-358.

OPASCHOWSKI, H. W. (2006): *Deutschland 2020: Wie wir morgen leben - Prognosen der Wissenschaft*. 2nd Edition, Verlag für Sozialwissenschaften, Wiesbaden.

PASTERNACK, G. (2006): *Die wirtschaftlichen Aussichten der ostdt. Braunkohlenwirtschaft bis zum Jahr 2020: Eine Szenario-Analyse*. Kovac, Hamburg.

PHELPS, R., CHAN, C., and KAPSALIS, S.C. (2001): Does Scenario Planning Affect Performance? Two Exploratory Studies. *Journal of Business Research, 51, 223–232*.

RINGLAND, G. (2006): *Scenario Planning*. John Wiley, Chichester.

SCHULMEISTER, R. (2006): *eLearning: Einsichten und Aussichten*. Oldenbourg, München.

SCHWARTZ, P. (1991): The Art of the Long View. Doubleday, Philadelphia.

SPREY, M. (2003): *Zukunftsorientiertes Lernen mit der Szenario-Methode*. Klinkhardt, Bad Heilbrunn.

VAN DER HEIJDEN, K. (2005): *Scenarios: The Art of Strategic Conversation*. 2nd Edition, John Wiley, Chichester.

WELFENS, P. J. J. (2004): *Internetwirtschaft 2010: Perspektiven und Auswirkungen*. Physica, Heidelberg.

Visualization and Clustering of Tagged Music Data

Pascal Lehwark, Sebastian Risi and Alfred Ultsch

Databionics Research Group, Philipps University Marburg, Germany
pascal@indiji.com, sebastian.risi@gmail.com,
ultsch@informatik.uni-marburg.de

Abstract. The process of assigning keywords to a special group of objects is often called tagging and becomes an important character of community based networks like Flickr, YouTube or Last.fm. This kind of user generated content can be used to define a similarity measure for those objects. The usage of Emergent-Self-Organizing-Maps (ESOM) and U-Map techniques to visualize and cluster this sort of tagged data to discover emergent structures in collections of music is reported. An item is described by the feature vector of the most frequently used tags. A meaningful similarity measure for the resulting vectors needs to be defined by removing redundancies and adjusting the variances. In this work we present the principles and first examples of the resulting U-Maps.

1 Introduction

The increased interest in folksonomies like Flickr, Last.fm, YouTube, del.icio.us and other community based networks shows, that tagging is already used by many users to discover new material and becomes a collaborative way of classifying items, being controlled by the creator and consumer of the content. One popular way to visualize tag relations is the use of *tag clouds*. They are used to visualize the most used tags on a website. More frequently used tags have a larger font and they are normally ordered alphabetically. For our study we chose to analyse the data provided by the music community *Last.fm*, an internet radio featuring a music recommendation system. The users can assign tags to artists and browse the content via tags allowing them to only listen to songs tagged in a certain way.

Tags make it possible to organize the media (artists and songs) in a semantic way and states a useful base for discovering new music. Because of the huge amount of artists and songs, an intuitive user interface is required to avoid losing the overview. We propose the *Emergent-Self-Organizing-Map (ESOM)* (Ultsch (2003)) to cluster tagged data because it has some advantages over other clustering algorithms. It is topology preserving and combined with the *U-Map* it provides a visually appealing user interface and an intuitive way of exploring new content. The remainder of this paper is organized as follows. First some related work on tagged data, clustering music and documents with the ESOM is presented. Then we describe the main learning algorithm of the ESOM in section 3 together with the U-Map visualization. Next, the dataset is presented together with the used methods of data preparation. In section 5 we present our experimental results. We round off the paper by giving the conclusion in section 6 together with future research directions.

2 Related work

There has been some work on enhancing the user interface based on tags and we will briefly mention some here. Flickr uses *Flickr clusters* which can provide related tags to a popular tag, grouped into clusters. Begelman (2006) uses clustering algorithms to find strongly related tags visualizing them as a graph. Hassan-Montero et al. (2006) propose a method for an improved tag cloud and a technique to display these tags with clustering based layout.

The ESOM has already been used successfully to visualize collections of music, photos and on clustering documents. Most of these works have in common that they cluster the data based on features extracted directly from the media. An example is MusicMiner (Mörchen (2005)) which uses the *timbre distance*, a measure based on frequency analysis of audio data. The *websom project* (Kaski (1998)) is an ESOM based approach in free text mining. Here each document is encoded as a histogram of word categories which are formed by the ESOM algorithm based on the similarities in the contexts of the words.

Although our approach is different because we are not using information that can be extracted from the objects' raw data itself but instead user generated content, the works mentioned previously show that the ESOM is a powerful tool in visualizing high dimensional data.

3 Emergent Self Organizing Maps

The ESOM is an artificial neural network that performs a mapping from a high dimensional data space R^n onto a two-dimensional grid of neurons. The unsupervised training process is partly motivated by how visual information is handled in the cerebral cortex of the mammalian brain and equals a regression of an ordered set of model vectors $m_i \in R^n$ into the space of observation vectors $x \in R^n$ by performing the following process:

$$m_i(t+1) = m_i(t) + h_{c(x),i}(x(t) - m_i(t))$$

where t is the sample index of the regression step, whereby the regression is performed recursively for each presentation of a sample of x. Index c, the *bestmatching unit* (BMU) or *winner*, is defined by the condition

$$||x(t) - m_c(t)|| \leq ||x(t) - m_i(t)|| \forall i$$

The so called *neighbourhood function h* is often taken to be the Gaussian

$$h_{c(x),i} = \alpha(t)exp(-\frac{||r_i - r_c||^2}{2\sigma^2(t)})$$

where $0 < \alpha(t) < 1$ is the learning-rate factor, which decreases monotonically with the regression steps, r_i and r_c are the vectorial locations in the display grid and $\sigma(t)$ corresponds to the width of the neighbourhood function, which is also decreasing monotonically with the regression steps. For a more detailed discussion of the SOM see Kaski (1997).

U-Map visualization

The U-Map (Ultsch (2003)) is constructed on top of the map of ESOM. The *U-Height* for each neuron n_i equals the accumulated distances of n_i to its immediate neighbors $N(i)$. It is calculated as follows:

$$\text{U-Height}(n_i) = \sum_{j \in N(i)} d(m_i, m_j)$$

where $d(x,y)$ is the distance function used in the SOM algorithm to construct the map and $N(i)$ denotes the indices of the immediate neighbours of neuron i.

A single U-Height shows the local distance structure of the corresponding neuron. The overall structure of densities emerges, if a global view of a U-Map is regarded. A U-Map is usually displayed as a three dimensional landscape and has become a standard tool to display the distance structures of the ESOM. Therefore the U-Map delivers a 'landscape' of the distance relationships of the input data in the data space. It has the property that weight vectors of neurons with large U-Heights are very distant from other vectors in the data space and that weight vectors of neurons with small U-Heights are surrounded by other vectors in the data space. Outliers and other possible cluster structures can easily be recognized. U-Maps have been used in a number of applications to detect new and meaningful knowledge in data sets.

4 Data

We extracted 1200 artists from the Last.fm website together with the 250 most frequently used tags like *rock*, *pop*, *metal*, etc.

4.1 Peparation of the datasets

Before the ESOM can be trained, special demands have to be fulfilled. Tags from the Last.fm dataset which do not stand for a certain kind of music genre, like *seen-live*, *favourite albums*, etc. were excluded. Highly correlated tags were condensed to a single feature. For the preparation of the tagged data we used a modification of the *Inverse Document Frequency (IDF)*.

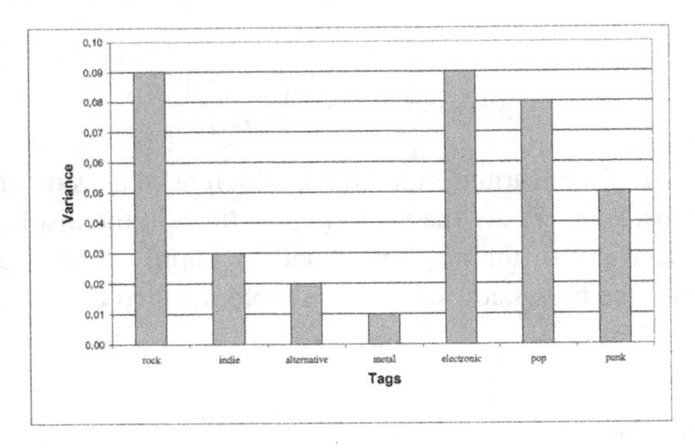

Fig. 1. The variances of the seven most popular tags

Last.fm provides the number of people ($t_{ij} = tagcount_{ij}$) that have used a specific tag i for an artist j. We scaled t_{ij} to the range of $[0, 1]$. Then we slightly modified the term frequency to be more appropriate for tagged data:

$$\text{tf}_{ij} = \frac{t_{ij}}{\sum_k t_{kj}}$$

with the denominator being the accumulated frequencies over all tags used for artist j. The *IDF* of tag i is defined as

$$\text{idf}_i = log \frac{|D|}{\sum_k t_{ik}}$$

with $|D|$ being the total number of artists in the collection and $\sum_k t_{ik}$ being the accumulated frequencies of tag i over all artists. The resulting importance of tag i for artist j is given by

$$\text{tfidf}_{ij} = \text{tf}_{ij}\text{idf}_i$$

As can be seen in figure 1 all the tags of the Last.fm dataset differ a lot in variance but for a meaningful comparison of the variables these variances have to be adjusted. For this purpose we used the *empirical cumulative distribution function (ECDF)*. The idea behind the ECDF is to assign a probability of $\frac{1}{n}$ to each of the n observations in the sample. The final tag frequencies are then given by

$$\text{tfidf}_{ij}^{ECDF} = \frac{|\text{tfidf}_{ik} \leq \text{tfidf}_{ij}|}{n}, k = 1..n$$

The adjusted variances after applying ECDF can be seen in figure 2. The accumulated tag frequencies of the Last.fm dataset can be seen in figure 3.

Finally we optain the feature vector w_j for artist j as

$$w_j = (\text{tfidf}_{1j}^{ECDF}, .., \text{tfidf}_{nj}^{ECDF})$$

In the context of self organizing maps, two different measures have been proposed to compute the similarity between two feature vectors w_i and w_j. The first

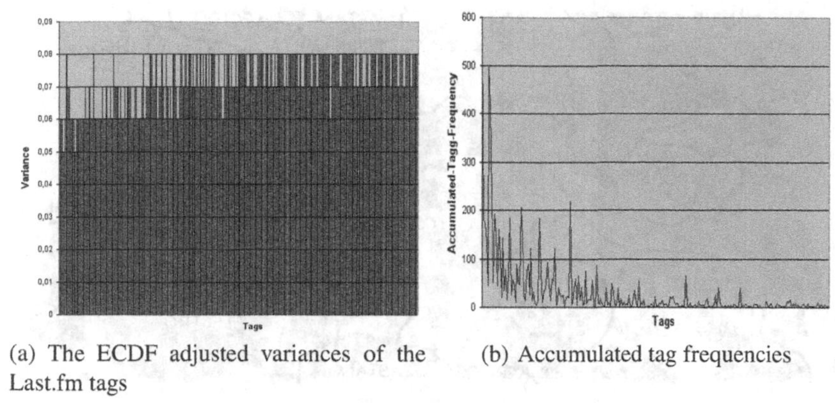

(a) The ECDF adjusted variances of the Last.fm tags

(b) Accumulated tag frequencies

Fig. 2. Tag variances

method uses the familiar *euclidean distance*, while the second approach is based on the *cosine similarity*

$$cos(w_i, w_j) = \frac{w_i^t w_j}{\|w_i\| \, \|w_j\|}$$

This method emphasizes the relative values that each dimension has within each vector and not their overall length. Two vectors can have a value of zero even if their euclidean distance is arbitrarily large. A SOM model which uses the cosine similarity instead of euclidean distance has also been proposed by Kohonen (1982), introduced as *Dot-Product-SOM*, and has been succesfully used for document clustering problems. For the close analogy to tag spaces we decided to use this model rather than the standard model based on the euclidean distance.

Note, that the update function changes to

$$m_i(t+1) = \frac{m_i(t) + h_{c(x),i}(t)x(t)}{\|m_i(t) + h_{c(x),i}(t)x(t)\|}$$

Although the training process slows down due to the normalization at each step, the search for the bestmatch is very fast and simple.

5 Experimental results

We trained a 80x50 emergent self organizing map using 50 epochs with the preprocessed data using the *Databionics ESOM tool* (Ultsch and Mörchen, 2005). A *toroid topology* was used to avoid border effects.

The U-Map in figure 4 (visualized using *Spin3D*) can be interpreted as height values on top of the usually two dimensional grid of the ESOM, leading to an intuitive paradigm of a landscape. Clearly defined borders between clusters, where

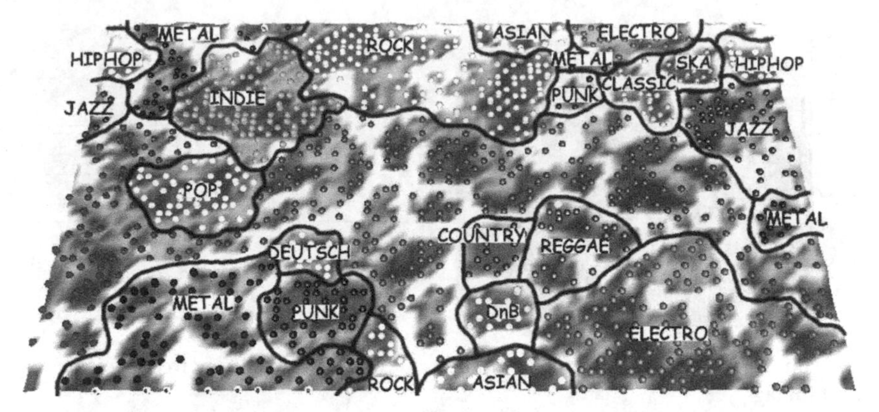

Fig. 3. Resulting U-Map. Note that the map is toroid, for example, the metal cluster is not split but spreaded over boundaries.

Fig. 4. Zoom of cluster *rock* to illustrate the good innercluster quality.

large distances in data space are present, are visualized in form of high mountains. Smaller intra cluster distances or borders of overlapping clusters form smaller hills. Homogeneous regions of data space are placed in valleys.

Detailed inspection of the map shows a very good conservation of the intercluster relations between the different music genres. One can observe smooth transitions between clusters like *metal*, *rock*, *indie* and *pop*.
In figure 5 we show a detailed view of the cluster *rock*.

The innercluster relations, e.g. the relations between genres like *hard rock*, *classic rock*, *rock and roll* and *modern rock* are very well preserved. This property also holds for the other clusters.

An interesting area is the little cluster *metal* next to the cluster *classic*. A precise examination revealed the reason for this cluster not being part of the big cluster *metal*. The cluster *classic* contains the former classic artists like Ludwig van Beethoven on the lower right edge with a transition to newer artists of the classical genre when moving to the upper left. The neighbouring artists of the minicluster *metal* are bands like Apocalyptica and Therion which use a lot of classical elements in their songs.

6 Conclusion and future work

Our goal was to find a visualization method that fits the need and constraints of browsing collections of tagged data. A high dimensional feature vector of 250 dimensions is hard to grasp and clustering can reveal groups of similar objects based on their tags. The global organization of the tagged artists worked really well and in contrast to other clustering algorithms, soft transitions between the groups of similar tagged artist can be seen. The modified *Inverse Document Frequency* turned out to be a good preparation method when working with tagged data. It is however essential for the ESOM that the feature vectors are not to sparse and that the overlap between them is not to low. These problems occurred in experiments with the photo community *flickr* where information about tags is only binary (a tag occurs or not) without information about the tag frequencies.

We showed that the ESOM enables the user to navigate through the high dimensional space in an intuitive way. Future work could include combining the clustering of artists and their songs and an automatic playlist generation system from regions and paths on the map. The maps presented here can be seen in color and high resolution at www.indiji.com/musicsom.

References

BEGELMAN, G., KELLER, P. and SMADJA, F. Automated Tag Clustering: Improving search and exploration in the tag space http://www.rawsugar.com/lab.

HASSAN-MONTERO, Y., HERRERO-SOLANA, V. Improving Tag-Clouds as Visual Information Retrieval Interfaces To appear: International Conference on Multidisciplinary Information Sciences and Technologies, InSciT2006, Merida, Spain, 2006.

KASKI, S., HONKELA, T. LAGUS and K., KOHONEN, T. WEBSOM–self-organizing maps of document collections. Neurocomputing, volume 21, pages 101-117, 1998

KASKI, S, KANGASZ, J. and KOHONEN, T. Bibliography of self-organizing map (SOM) papers: 1981-1997

KOHONEN, T. Self-Organized Formation of Topological Correct Feature Maps Biological Kybernetics Vol. 43, pp59-69, 1982.

MATHES, A. Folksonomies Ű Cooperative Classification and Communication Through Shared Metadata. http://www.adammathes.com/academic/ computermediatedcommunication/folksonomies.html.

MILLEN, D., FEINBERG, J. Using Social tagging to Improve Social Navigation Workshop on the Social Navigation and Community based Adaptation Technologies, 2006

MILLEN, D., FEINBERG, J. and KERR, B. Social Bookmarking in the Enterprise. Social Computing, Vol. 3, No. 9. Nov. 2005.

MÖRCHEN, F., ULTSCH, A., NÖCKER, M. and STAMM, C. Visual mining in music collections In Proceedings 29th Annual Conference of the German Classification Society (GfKl 2005), Magdeburg, Germany, Springer, Heidelberg, 2005

MÖRCHEN, F., ULTSCH, A., THIES, M., LÖHKEN, I., NÖCKER, M., STAMM, C., EFTHYMIOU, N. and KÜMMERER, M. MusicMiner: Visualizing timbre distances of music as topographical maps Technical Report No. 47, Dept. of Mathematics and Computer Science, University of Marburg, Germany, 2005

ROBERTSON, S., Understanding Inverse Document Frequency: On theoretical arguments for IDF Journal of Documentation 60 no. 5, pp 503Ű520

ULTSCH, A. Self-organizing neural networks for visualization and classification In Proc. GfKl, Dortmund, Germany, 1992.

ULTSCH, A. Maps for the Visualization of high dimensional Data Spaces In: Yamakawa T (eds) Proceedings of the 4th Workshop on Self-Organizing Maps, 225-230, 2003.

ULTSCH, A. U*-matrix: a tool to visualize clusters in high dimensional data. Technical report, Departement of Mathematics and Computer Science, Philipps-University Marburg, 2003.

ULTSCH, A. ,HERRMANN, L. The architecture of Emergent Self-Organizing Maps to reduce projection errors Proc ESANN, Brugges, pp. 1-6. 2005

ULTSCH, A., MÖRCHEN, F. ESOM-Maps. tools for clustering, visualization, and classification with Emergent SOM. Technical Report 46, CS Department, University Marburg, Germany, 2005.

ZHAO, Y., KARYPIS, G. Criterion Functions for Document Clustering Experiments and analysis. Machine Learning, in press, 2003.

Effects of Data Transformation on Cluster Analysis of Archaeometric Data

Hans-Joachim Mucha[1], Hans-Georg Bartel[2] and Jens Dolata[3]

[1] Weierstraß-Institut für Angewandte Analysis und Stochastik (WIAS), Mohrenstraße 39, 10117 Berlin, Germany
mucha@wias-berlin.de

[2] Institut für Chemie, Humboldt-Universität zu Berlin, Brook-Taylor-Straße 2, 12489 Berlin, Germany
hg.bartel@yahoo.de

[3] Landesamt für Denkmalpflege Rheinland-Pfalz, Abt. Archäologie, Amt Mainz, Große Langgasse 29, 55116 Mainz, Germany
dolata@ziegelforschung.de

Abstract. In archaeometry the focus is mainly on chemical analysis of archaeological artifacts such as glass objects or pottery. Usually the artefacts are characterized by their chemical composition. Here the focus is on cluster analysis of compositional data. Using Euclidean distances cluster analysis is closely related to principal component analysis (PCA) that is a frequently used multivariate projection technique in archaeometry. Since PCA and cluster analysis based on Euclidean distances are scale dependent, some kind of "appropriate" data transformation is necessary. Some different techniques of data preparation will be presented. We consider the log-ratio transformation of Aitchison and the transformation into ranks in more detail. From the statistical point of view the latter is a robust method.

1 Introduction

Often the archaeometric data we analyze are measured with respect to the chemical compositions of many variables that usually have quite different scales. For example, Mucha et al. (2001) investigated a data set of ancient coarse ceramics by cluster analysis, where the set of 19 variables consists of nine oxides and ten trace elements (see below Section 6). The former are given in percent and the latter are measured in parts per million (ppm). Hence some kind of treatment of the data is necessary since PCA and cluster analysis based on Euclidean distances are scale dependent. Without some standardization, the Euclidean distances can be fully dominated by the variable in the more sensitive units. However, as we will see below, an inappropriate data transformation can result in covering the differences between well-separated groups (clusters). Moreover it can produce outliers.

Besides different scales of the variables, often problems with outliers and with long-tailed (skew) distributions of the variables were addressed in the archaeometric data, see recently Baxter (2006). Figure 1 shows an example taken from Baxter and Freestone (2006) (see also below Section 5). This is discrete data rather than metric data: the measurements are given as 0.01, 0.02 and so on. The usual way of dealing with outliers seems to be omitting them, see for instance Baxter (2006) and Baxter and Freestone (2006). Another more objective way is using transformation into ranks, as it will be shown below.

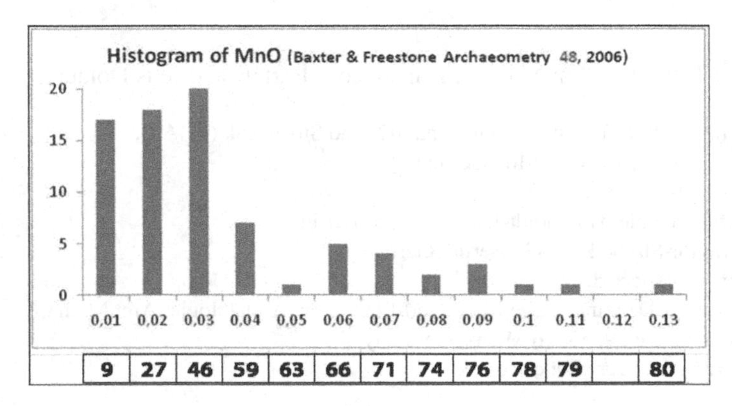

Fig. 1. The frequency plot of MnO of 80 objects shows a skew density. Additionally, at the bottom the corresponding rank values are shown.

Indeed, the performance of multivariate statistical methods like cluster analysis and PCA is often seriously affected by these two main problems: scale dependence and outliers. Concerning PCA see Baxter (1995) and Baxter and Freestone (2006). Therefore data transformations and outlier treatment are highly recommended by these authors.

Here different data transformations will be presented and compared. Our investigation shows that especially nonparametric transformations like the transformation of the data into a matrix of ranks for subsequent multivariate statistical analysis give good and for archaeologists reasonable results. We consider two data sets: the compositional data of colourless Romano-British vessel glass where the variables measured sum to 100%, and the sub-compositional data of Roman bricks and tiles from the Rhine area where the variables measured sum to approximately 100%.

2 Data transformation in archaeometry

Let I objects $\mathbf{x}_i$ be on hand for J variables. That is, a data matrix $\mathbf{X} = (x_{ij})$ with elements $x_{ij} \geq 0$ is under investigation. For compositional data, Aitchison (1986) recommended the log-ratio transformation

$$y_{ij} = \log(x_{ij}/g(\mathbf{x}_i)) \;, \tag{1}$$

where $g(\mathbf{x}_i) = (x_{i1}x_{i2}\dots x_{iJ})^{1/J}$ is the geometric mean of the ith object. This transformation is restricted to values $x_{ij} > 0$. Baxter and Freestone (2006) criticized that Aitchison argued that all others transformations are "meaningless" and "inappropriate" for compositional data. The authors presented the failure of PCA for different data sets based on the log-ratio transformation. In Section 5 below the failure of cluster analysis methods based on the log-ratio transformation will be presented.

The transformation of the variables by

$$y_{ij} = (x_{ij} - \overline{x}_j)/s_j \tag{2}$$

is known as standardization. Herein $\overline{x}_j$ and s_j are the mean and standard deviation of variable j, respectively. The new variables $\mathbf{y}_j$ has mean equals 0 and variance equals 1. The logarithmic transformations

$$y_{ij} = \log(x_{ij}) \tag{3}$$

or

$$y_{ij} = \log(x_{ij} + 1) \tag{4}$$

can handle skew densities, where (3) is restricted to values $x_{ij} > 0$, as the log-ratio transformation (1). Here the meaning of differences is changed.

3 Transformation into ranks

The multivariate statistical analysis based on ranks rather than based on the original data solves the problems of different scales and skewness. The influence of outliers is removed in the univariate case. In the multivariate case, the influence of outliers is highly reduced usually but theoretically the problem of outliers remains to some degree (Rohatch et al. (2006)).

Table 1. Measurements and the corresponding ranks of MnO

Value	0.01	0.02	0.03	0.04	0.05	0.06	0.07	0.08	0.09	0.10	0.11	0.13
Frequency	17	18	20	7	1	5	4	2	3	1	1	1
Rank	9	26.5	45.5	59	63	66	70.5	73.5	76	78	79	80

Transformation into ranks is quite simple: one replaces the measurements by their ranks $1, 2, \dots, I$ where I is the number of observations. The mean of each of the new rank order variables become the same: $(I+1)/2$. Moreover, the variance of each of the new variables become the same: $(I^2 - 1)/12$. In case of multiple values we recommend to average the corresponding ranks (Figure 1). Table 1 contains both the original values and the ranks of MnO of the 80 objects (see also Figure 1, data source: Baxter and Freestone (2006)).

Mucha (1992) presented a successful application of partitioning cluster analysis based on rank data. Also, Mucha (2007) investigated the stability of hierarchical clustering based on rank data. The aim of this paper here is to show that cluster analysis based on rank data gives good results and that it can outperform log-ratio cluster analysis.

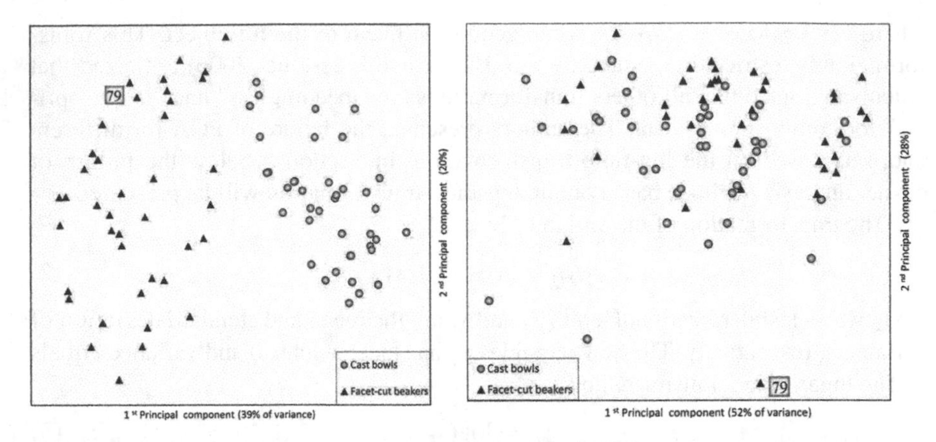

Fig. 2. PCA plot of groups of Romano-British vessel glass based on ranks (left hand side), and PCA plot of group membership based on log-ratio transformed data (right).

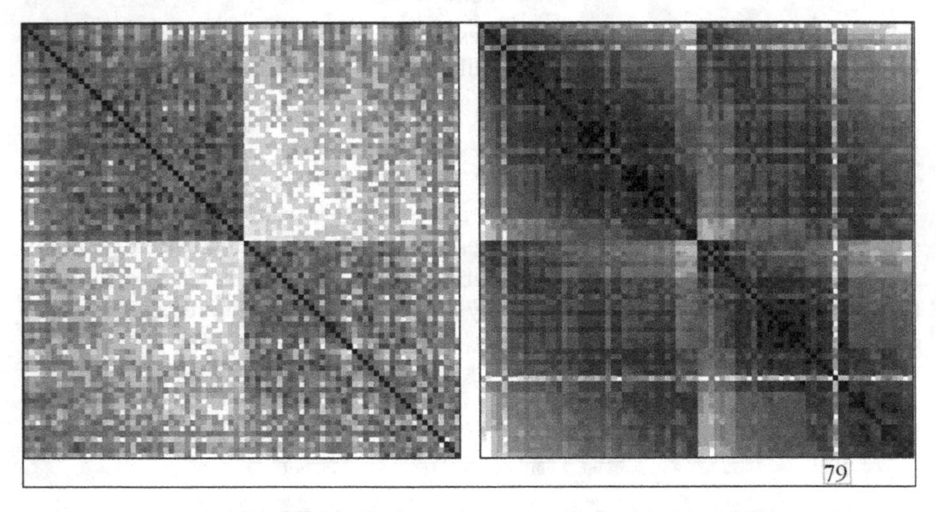

Fig. 3. Fingerprint of the true Euclidean distances of rank data (left) and of log-ratio transformed data (right). (Small distances are marked by dark gray, great distances by light gray.)

4 Distances and cluster analysis

Henceforth let us focus on the squared Euclidean distances in cluster analysis because PCA is based on the same distance measure and the PCA plots are very popular in archaeometry (Baxter (1995), Baxter and Freestone (2006)). Cluster analysis and PCA are multivariate statistical methods that are based on distance measures. Further let us restrict to the well-known hierarchical *Ward's* method (Späth (1985)). It is the simplest of the model-based Gaussian clustering methods that are applied by Papageorgiou et al. (2002) for finding groups of artefacts.

In case of the log-ratio transformation (1) the squared Euclidean distance between two objects i and h is

$$d(\mathbf{x}_i, \mathbf{x}_h) = \sum_{j=1}^{J} (y_{ij} - y_{ih})^2 = \sum_{j=1}^{J} (\log \frac{x_{ij}}{g(\mathbf{x}_i)} - \log \frac{x_{hj}}{g(\mathbf{x}_h)})^2. \quad (5)$$

Often it is called Aitchison distance. Appropriate clustering techniques for squared Euclidean distances are the the partitioning K-means method (Mucha (1992)) and the hierarchical *Ward's* method, as mentioned already above.

5 Romano-British vessel glass classified

This is simulated data based on real data of colourless Romano-British vessel glass (Baxter et al. (2005)). Details and the complete source can be taken from Baxter and Freestone (2006). This example is based on two groups that are well-known different.

Group 1 consists of 40 cast bowls with high amounts of Fe_2O_3. Group 2 also consists of 40 objects: this is a collection of facet-cut beakers with low Al_2O_3. In Figure 2 at the left hand side, the two groups are shown in the first plane of the PCA based on rank data. This projection gives a good approximation of the distances between objects. Axis 1 (39%) and axis 2 (20%) are highly significant (see Lebart et al. (1984) for tables of significance of eigenvalues of PCA). The *Ward's* method finds the true groups without any error. The same optimum clustering result is obtained when using the transformation (4).

In Figure 2 at the right hand side, the two groups are presented by the PCA plot after the data transformation by (1). This transformation produces outliers such as the object 79 that is drawn additionally. The PCA is based on the Aitchison distance measure (5). In the two-dimensional projection the distances are approximative ones. The *Ward's* method never finds the true two groups. Table 2 at the left hand side shows the very low correspondence between the given groups and the clusters found. The same bad cluster analysis result is obtained when using the transformation (3). The transformation (2) performs here much better: the *Ward's* method results in 5 errors only (see Table 2 at the right hand side). The corresponding PCA-plot of the standardized data using (2) is published as Figure 8 by Baxter and Freestone (2006). There is no outlier in this plot as well as in the plot of Figure 2 at the left hand side.

Table 2. True groups versus clusters

True Groups	*Ward's* method Cluster 1	with (1) Cluster 2	*Ward's* method Cluster 1	with (2) Cluster 2
Cast bowls	27	**9**	**37**	3
Facet-cut beakers	**33**	8	2	**38**

Figure 3 compares two fingerprints of the Euclidean distances of rank data (left hand side) and of log-ratio transformed data (right), respectively. Here the objects are sorted first by group and then within the group by the first principal component based on rank analysis and by the first principal component based on log-ratio scaling, respectively. The fingerprint at the right hand side shows no clear class structure. Additionally, the outlier 79 is marked at the bottom. The corresponding high distance values to all the remaining objects build the eye-catching column and row in light gray, respectively.

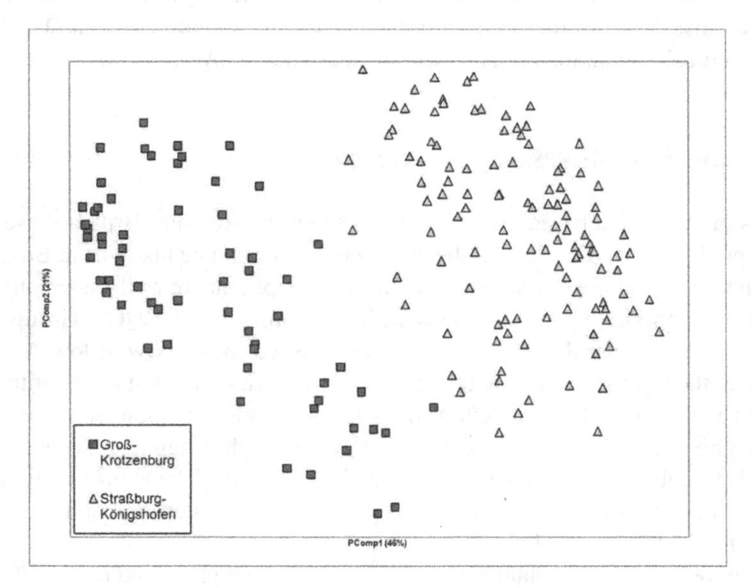

Fig. 4. PCA plot of group membership based on rank data.

6 Roman bricks and tiles classified

Roman bricks and tiles from the Rhine area are described by 19 chemical elements that were measured using X-Ray Fluorescence Analysis (XRF). All the chemical measurements were performed by G. Schneider of the Freie Universität Berlin. Two well-known locations of production are Groß-Krotzenburg and Straßburg-Königshofen (Dolata (2000)). In this reference the author published the complete data source. It is possible to confirm the two well-known groups by cluster analysis based on rank data?

Figure 4 shows the PCA plot of the two groups based on rank data. The hierarchical *Ward's* method method finds the true groups without any error.

In Figure 5 the two groups are shown by the PCA projection based on the data transformation (1). Here *Ward's* method finds the groups but one error occurs: the outlier at the bottom at the left hand side coming from Straßburg-Königshofen is misclassified.

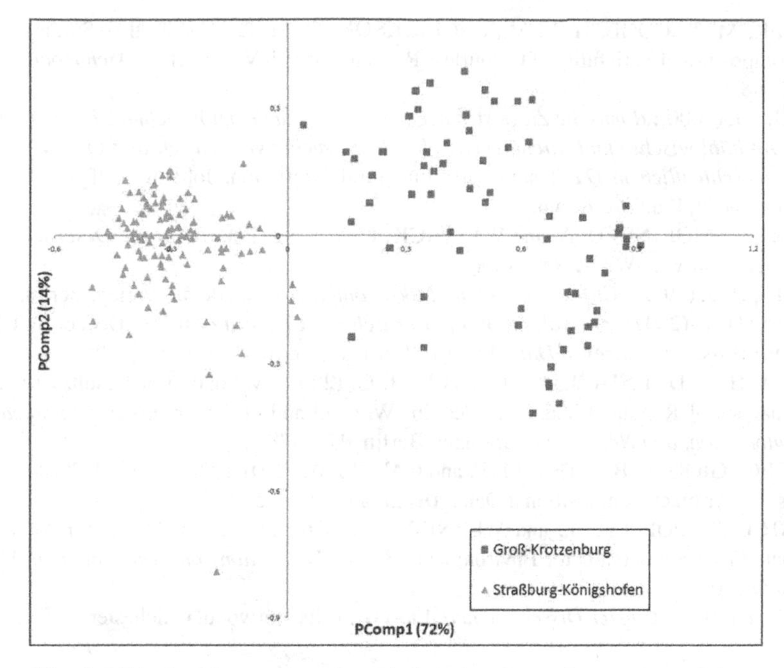

Fig. 5. PCA plot of group membership based on log-ratio transformed data.

7 Summary

There are different data transformations in use in archaeometry with advantages and disadvantages. Comparison of different data transformations based on simulated and real data shows that transformation into ranks is useful in the case of outliers and skew densities. However, most of the quantitative information is lost by going to ranks. From archaeological point of view rank analysis gives reasonable results. Other transformations (like AitchisonŚs log-ratio or (3)) are highly affected by outliers, skew densities and values near 0. Therefore finding the true groups by cluster analysis fails in the case of glass data. Moreover, new artificial outliers can be produced by transformations such as (1) and (3) in case of measurements near zero.

References

AITCHISON, J. (1986): *The Statistical Analysis of Compositional Data.* Chapman and Hall, London.

BAXTER, M. J. (1995): Standardization and Transformation in Principal Component Analysis, with Applications to Archaeometry. *Applied Statistics, 44, 513–527.*

BAXTER, M. J. (2006): A Review of Supervised and Unsupervised Pattern Recognition in Archaeometry. *Archaeometry, 48, 671–694.*

BAXTER, M. J. and FREESTONE, I. C. (2006): Log-ratio Compositional Data Analysis in Archaeometry. *Archaeometry, 48, 511–531.*

BAXTER, M. J., COOL, H. E. M., and JACKSON, C. M. (2005): Further Studies in the Compositional Variability of Colourless Romano-British Vessel Glass. *Archaeometry, 47, 47–68.*

DOLATA, J. (2000): *Römische Ziegelstempel aus Mainz und dem nördlichen Ober-germanien - Archäologische und archäometrische Untersuchungen zu chrono-logischem und baugeschichtlichem Quellenmaterial.* Inauguraldissertation, Johann Wolfgang Goethe-Universität, Frankfurt/Main.

LEBART, L., MORINEAU, A. and WARWICK, K. M. (1984): *Multivariate Descriptive Statistical Analysis.* Wiley, New York.

MUCHA, H.-J. (1992): *Clusteranalyse mit Mikrocomputern.* Akademie Verlag, Berlin.

MUCHA, H.-J. (2007): *On Validation of Hierarchical Clustering.* In: R. Decker and H.-J. Lenz (Eds.): *Advances in Data Analysis*, Springer, Berlin, 115–122.

MUCHA, H.-J., DOLATA, J., and BARTEL, H.-G. (2001): Validation of Results of Cluster Analysis of Roman Bricks and Tiles. In: W. Gaul and G. Ritter (Eds.): *Classification, Automation, and New Media.* Springer, Berlin, 471–478.

PAPAGEORGIOU, I., BAXTER, M. J., and CAU, M. A. (2001): Model-based Cluster Analysis of Artefact Compositional Data. *Archaeometry, 43, 571–588.*

ROHATCH, T., PÖPPEL, G., and WERNER, H. (2006): Projection Pursuit for Analyzing Data From Semiconductor Environments. *IEEE Transactions on Semiconductor Manufacturing, 19, 87–94.*

SPÄTH, H. (1985): *Cluster Dissection and Analysis.* Ellis Horwood, Chichester.

Fuzzy PLS Path Modeling: A New Tool For Handling Sensory Data

Francesco Palumbo[1], Rosaria Romano[2] and Vincenzo Esposito Vinzi[3]

[1] University of Macerata, Italy
francesco.palumbo@unimc.it
[2] University of Copenhagen, Denmark
rro@life.ku.dk
[3] ESSEC Business School of Paris, France
vinzi@essec.fr

Abstract. In sensory analysis a panel of assessors gives scores to blocks of sensory attributes for profiling products, thus yielding a three-way table crossing assessors, attributes and products. In this context, it is important to evaluate the panel performance as well as to synthesize the scores into a global assessment to investigate differences between products. Recently, a combined approach of fuzzy regression and PLS path modeling has been proposed. Fuzzy regression considers crisp/fuzzy variables and identifies a set of fuzzy parameters using *optimization* techniques. In this framework, the present work aims to show the advantages of fuzzy PLS path modeling in the context of sensory analysis.

1 Introduction

In sensory analysis a panel of assessors gives scores to blocks of sensory attributes for profiling products, thus yielding a three-way table crossing assessors, attributes and products. This type of data are characterized by three different sources of complexity: complex structure of relations among the variables (different blocks), three directions of information (samples, assessors, attributes) and influential human beings' involvement (assessors' evaluations).

Structural Equation Models (SEM) (Bollen, 1989) consist of a network of causal relationships among Latent Variables (LV) defined by blocks of Manifest Variables (MV). The main idea behind SEM is that the features on which the analysis would focus cannot be properly measured and are determined through the measured variables. In a recent contribution (Tenenhaus and Esposito-Vinzi, 2005), SEM have been successfully used to analyze sensory data. When SEM are based on the scores of a set of assessors, they are generally based on the mean scores. However, it is important to analyze if there exist individual differences between assessors. Even if assessors are carefully trained to adopt the same yardstick, this cannot completely protect us against their single sensibility.

When human estimation is influential and the observations cannot be described accurately but we can give only an approximate description of them, fuzzy approach is more useful and convenient than the classical one (Zadeh, 1965). Fuzzy sets allow us coding and treating many different kinds of *imprecise* data. Recently, a fuzzy approach to SEM has been proposed (Romano, 2006) and successively used for comparing different SEM (Romano and Palumbo, 2006*b*).

The present paper proposes to use the new fuzzy structural equation models for handling the different sources of information and uncertainty arising from sensory data. First a brief introduction to the methodology of reference (Romano, 2006) will be given. Then an application to data from sensory profiling will be presented.

2 Fuzzy PLS path modeling

Fuzzy PLS Path Modeling is a new methodology to dealing with system complexity. It allows us taking into account both complexity in information codification and in structures of relations among the variables. *Fuzzy codification* and *structural equations* are combined to handling these different sources of complexity, respectively.

The strategy allowing imprecision in codification for reducing complexity is appropriately expressed by Zadeh's *principle of incompatibility* (Zadeh, 1973). The main idea is that the traditional techniques for analyzing systems are not well suited to dealing with human systems. In human thinking, the key elements are not numbers but classes of objects or concepts in which the membership of each element to the class is gradual (fuzzy) rather than sharp. For instance, the concept of *sweet coffee* does not correspond to an exact amount of sugar in the coffee. But it is possible to define the classes *sweet coffee, normal coffee, bitter coffee*.
On the other hand, the descriptive complexity of a system can also be reduced by breaking the system into its appropriate subsystems. This is the general principle behind Structural Equation Models (SEM) (Bollen, 1989). The basic idea is that different subsets of variables are the expression of different concepts, belonging to the same phenomenon. These concepts are named *latent variables* (LV) as they are not directly observable but measurable by means of a set of *manifest variables* (MV). The aim of SEM is to study the system of relations between each LV and its MV, and among the different LV inside the system. Considering one by one each part forming the whole system, and analyzing the relations among the different parts, the system complexity is reduced allowing a better description of the main system characteristics.

F-PLSPM consists in introducing fuzzy models inside SEM, by means of a two-stage procedure. This allows dealing with system complexity using both an approach which is tolerant to imprecision and a well suited methodology to link the different parts into which the system may be decomposed.

2.1 Interval data, fuzzy data and fuzzy models

It is very common to measure statistical variables in terms of single-values. However, for many reasons, and in many situations *exact measures* are very hard (or even

impossible) to achieve.
A rigorous study of *interval data* is given by *Interval Analysis* (Alefeld and Herzenberger, 1987). In this framework, an *interval value* is a bounded subset of real numbers $[x] = [\underline{x}, \overline{x}]$, formally:

$$[x] = \{x \in \mathbb{R} |\ \underline{x} \leq x \leq \overline{x}\} \tag{1}$$

where $\underline{x}$ and $\overline{x}$ are called *lower* and *upper* bound, respectively. Alternatively, an *interval value* may by expressed in terms of *width* (or *radius*), x_w, and *center* (or *midpoint*), x_c: $x_w = \frac{1}{2}|\overline{x} - \underline{x}|$ and $x_c = \frac{1}{2}|\overline{x} + \underline{x}|$.

A fuzzy set is a codification of the information allowing us to represent vague concepts expressed in natural language. Formally, given the *universe of objects* Ω, ω as the generic element, a *fuzzy set* $\tilde{A}$ in Ω is defined as a set of ordered pairs:

$$\tilde{A} = \{(\omega, \mu_{\tilde{A}}(\omega)) | \omega \in \Omega\} \tag{2}$$

where the value $\mu_{\tilde{A}}(\omega_0)$ expresses the *membership degree* for a generic element $\omega_0 \in \Omega$. The larger the value of $\mu_{\tilde{A}}(\omega)$, the higher the degree of membership of ω in $\tilde{A}$. If the *membership function* is permitted to have only the values 0 and 1 then the *fuzzy* set is reduced to a classical *crisp* set. The universal set Ω may consist of discrete (ordered and non ordered) objects or it can be a continuous space.

A fuzzy set in the real line that satisfies both the conditions of *normality* and *convexity* is a *fuzzy number*.
It must be normal so that the statement "real number close to r" is fully satisfied by r itself, i.e. $\mu_A(r) = 1$. In addition, all its $\alpha-$cuts for $\alpha \not\equiv 0$ must be closed intervals so that the arithmetic operations on *fuzzy sets* can be defined in terms of operations on closed intervals. On the other hand, if all its $\alpha-$cuts are closed intervals, it follows that the *fuzzy number* is a convex *fuzzy set*.
In *possibility theory* (Zadeh, 1978), a branch of fuzzy set theory, fuzzy numbers are described by *possibility distributions*.
A *possibility distribution* $\pi_{\tilde{A}}(\omega)$ is a function which satisfies the following conditions (Tanaka and Guo, 1999): *i)* there exists an ω such that $\pi_{\tilde{A}}(\omega) = 1$ (normality); *ii)* $\alpha-$cuts of fuzzy numbers are convex; *iii)* $\pi_{\tilde{A}}(\omega)$ is piecewise continuous.
Particular fuzzy numbers are the *symmetrical fuzzy numbers* whose *possibility distribution* may be denoted as:

$$\pi_{\tilde{A}_i}(\omega) = max\left(0, 1 - \left|\frac{\omega - c_i}{r_i}\right|^q\right) \tag{3}$$

Specifically, (3) corresponds to *triangular* fuzzy numbers when $q = 1$, to *square root* fuzzy numbers when $q = 1/2$ and *parabolic* fuzzy numbers when $q = 2$. It is easy to show that (3) corresponds to *intervals* when $q = +\infty$.
It is worth noticing that fuzzy variables are associated with possibility distributions in the similar way that random variables are associated with probability distributions. Furthermore, *possibility distributions* are numerically equal to membership functions (Zadeh, 1978).

In the early 80's, Tanaka proposed the first fuzzy linear regression model, moving on from *fuzzy sets theory* and *possibility theory* (Tanaka *et al.*, 1980). The functional relation between dependent and independent variables is represented as a fuzzy linear function whose parameters are given by *fuzzy numbers*. Tanaka proposed the first *Fuzzy Possibilistic Regression* (FPR) using the following fuzzy linear model with crisp input and fuzzy parameters:

$$\tilde{y}_n = \tilde{\beta}_0 + \tilde{\beta}_1 x_{n1} + \ldots + \tilde{\beta}_p x_{np}, + \ldots + \tilde{\beta}_P x_{nP} \tag{4}$$

where the parameters are symmetric triangular fuzzy numbers denoted by $\tilde{\beta}_p = (c_p; w_p)_L$ with c_p and w_p as center and the spread, respectively.
Differently from statistical regression, the deviations between data and linear models are assumed to depend on the vagueness of the parameters and not on measurement errors. The basic idea of Tanaka's approach was to minimize the uncertainty of the estimates, by minimizing the total spread of the fuzzy coefficients. Spread minimization must be pursued under the constraint of the inclusion of the whole given data set, which satisfies a degree of belief α $(0 < \alpha < 1)$ defined by the decision maker.
The estimation problem is solved via a mathematical programming approach, where the objective function aims at minimizing the spread parameters, and the constraints guarantee that observed data fall inside the fuzzy interval:

$$\text{minimize} \sum_{n=1}^{N} \sum_{p=0}^{P} w_p |x_{np}| \tag{5}$$

subject to the following constraints:

$$\begin{aligned}
&\left(c_0 + \textstyle\sum_{p=1}^{P} c_p x_{np}\right) + (1-\alpha)\left(w_0 + \textstyle\sum_{p=1}^{P} w_p |x_{np}|\right) \geq y_n \\
&\left(c_0 + \textstyle\sum_{p=1}^{P} c_p x_{np}\right) - (1-\alpha)\left(w_0 + \textstyle\sum_{p=1}^{P} w_p |x_{np}|\right) \leq y_n \\
&w_p \geq 0, c_p \in R, x_{n0} = 1, n = (1,\ldots,N), p = (1,\ldots,P)
\end{aligned}$$

where $x_{n0} = 1$ $(n = 1,\ldots,N)$, $w_p \geq 0$ and $c_p \in R$ $(p = 1,\ldots,P)$.

2.2 The F-PLSPM algorithm

The F-PLSPM follows the component based approach SEM-PLS, alternatively defined PLS Path Modeling (PLS-PM) (Tenenhaus *et al.*, 2005). The reason is that fuzzy regression and PLS path modeling share several characteristics. They are both *soft modeling* and *data oriented* approaches.

Specifically, fuzzy regression joins PLS-PM in its final step, allowing for a *fuzzy structural model* (see, Figure 1) but a still *crisp measurement model*. This connection implies a two stage estimation procedure:

- *stage* 1: latent variables are estimated according to the PLS-PM estimation procedure (Wold, 1982);

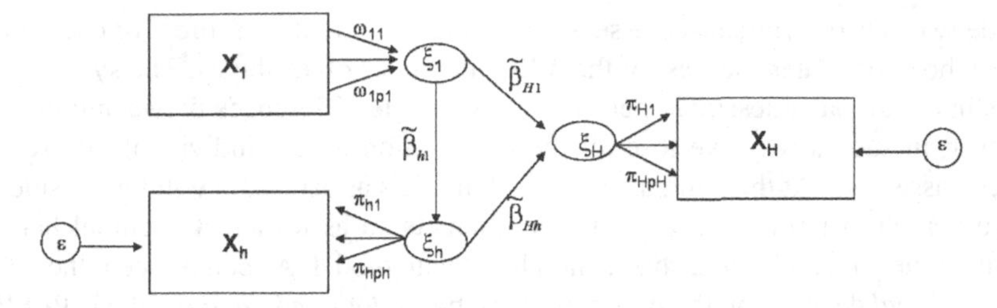

Fig. 1. Fuzzy path model representation

- *stage* 2: FPR on the estimated latent variables is performed so that the following *fuzzy structural model* is obtained:

$$\xi_h = \tilde{\beta}_{h0} + \sum_{h'} \tilde{\beta}_{hh'} \xi_{h'} \tag{6}$$

  where $\tilde{\beta}_{hh'}$ refers to the generic *fuzzy path coefficient*, ξ_h and $\xi_{h'}$ are adjacent latent variables and $h, h' \in [1, \ldots, H]$ vary according to the model complexity.

It is worth noticing that the *structural model* from this procedure is different with respect to the traditional *structural model*. Here the path coefficients are fuzzy numbers and there is no error term, as a natural consequence of a FPR. In the analysis of a statistical model one should always, in one way or another, take into account the goodness of fit, above all in comparing different models. The proposal is then to use the FPR. The estimation of fuzzy parameters, instead of single-valued (crisp) parameters, permits us to gather both the structural and the residual information. The characteristic to embed the residual in the model via fuzzy parameters (Tanaka and Guo, 1999) permits to evaluate the differences between assessors (panel performance) as well as the reproducibility of each assessor (assessor performance) (Romano and Palumbo, 2006b).

3 Application

The data set comes from sensory profiling of 14 cheese samples by a panel of 12 assessors on the basis of twelve attributes in two replicates.
The final data matrix consists of 336 rows (12 assessors × 14 samples × 2 replicates) and 12 columns (attributes: intensity odour, acidic odour, sun odour, rancid odour, intensity flavour, acidic flavour, sweet flavour, salty flavour, bitter flavour, sun flavour, metallic flavour, rancid flavour). Two blocks of variables describe the latent variables *odour* and *flavour*. First the hierarchical PLS model proposed by Tenenhaus and Vinzi (2005) will be used to estimate a global model after averaging over the assessors and the replicates (see, Figure 2). Thus, collapsing the data structure into a two-way table (samples × attributes). Then fuzzy PLS path modeling will

provide two sets of synthesized assessments: the overall latent scores for each product and the partial latent scores for the different blocks of attributes. The synthesis of scores into a global assessment permits to investigate differences between products. However, in such a way, we lose all the information on the individual differences between assessors. At this aim, as many path models as assessors will be considered and compared in terms of fuzzy path coefficients so as to detect eventual heterogeneity in the panel. Figure 2 shows the global path model. As can be seen, the latent variable *global* depends on the two latent variables *odour* and *flavour*. The F-PLSPM

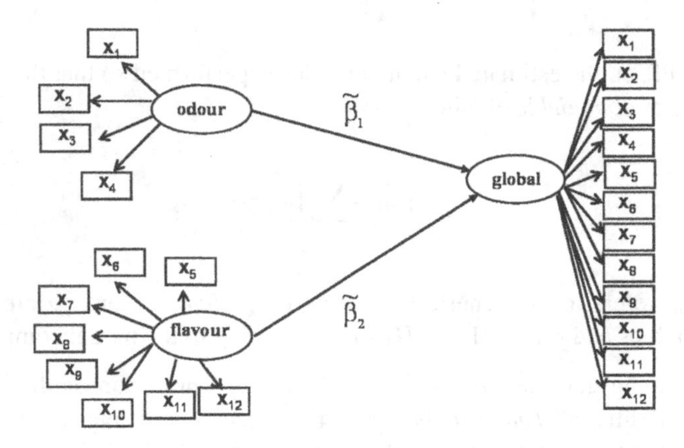

Fig. 2. Global model

algorithm is used to estimate the *fuzzy path coefficients* ($\tilde{\beta}_1$ and $\tilde{\beta}_2$). Crisp path coefficients in Table 1 show that the global quality of the products mostly depends on the *flavour* rather than on the *odour*. Furthermore, fuzzy path coefficients describe a worse panel performance for the *flavour* emphasized by a more imprecise estimate (wider fuzzy interval). Therefore, the F-PLSPM algorithm enriches the results of the classical PLSPM crisp approach by providing information on the imprecision of path coefficients. At the same time, the coherence of results is granted as the crisp estimates are comprised within the fuzzy intervals.

Table 1. Global Model Path Coefficients

Latent Variable	**crisp path coefficients**	**fuzzy path coefficients**
Odour	0.4215	[0.3952;0.4517]
Flavour	0.6283	[0.6043;0.7817]

The most interesting result coming from the proposed approach is in Figure 3, which compares the interval valued estimates on the different assessors.

Figure 3 reports the fuzzy path coefficients for the 12 local models referred to each assessor. By looking within each plot (*flavour* and *odour*) separately, the assessor performance and the coherence between assessors can be evaluated: a) the wider

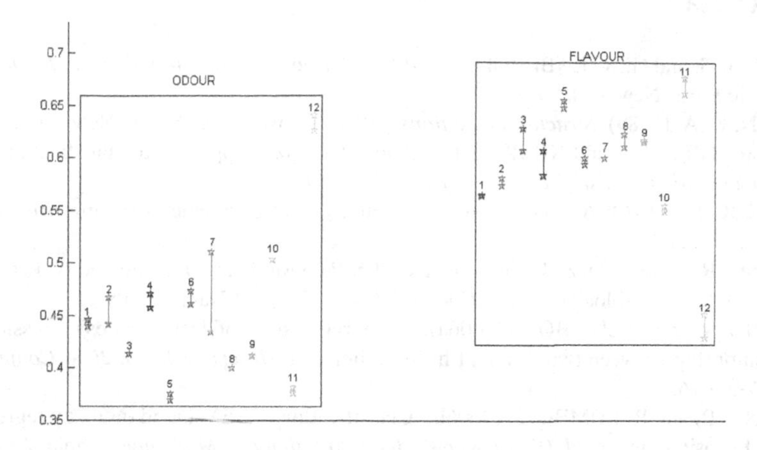

Fig. 3. Local fuzzy path coefficients

the interval, the less consistent is the assessor; b) the closer the intervals between them, the more coherent are the assessors. In the example, for the odour, assessor 7 is the least consistent assessor while assessor 12, being positioned far away from the rest of the assessors, is the least coherent as compared to the panel. Finally, by comparing the two plots, differences in the way each assessor perceives flavour and odour may be detected: for instance, assessor 7 is the most imprecise for the odour while it is extremely consistent for the flavour; assessor 12 is similarly consistent for both flavour and odour but, in both cases, it is in clear disagreement with the panel (a much higher influence of the odour as opposed to a much lower influence of the flavour).

4 Conclusion

The joint use of PLS component-based approach to structural equation modeling and fuzzy possibilistic regression has yielded promising results in the framework of sensory data analysis. Namely, while taking into account the multi-block feature of sensory data, the proposed Fuzzy-PLSPM leads to a fuzzy estimation of the path coefficients. Such an estimation provides information on the precision of the classical estimates and allows a thorough comparison of the sensory evaluations *between* assessors and *within* assessors for different products. Future directions of research aim to extend the fuzzy approach also to the measurement model by introducing an appropriate fuzzy possibilistic regression in the external estimation phase of the PLSPM algorithm. This further development has a twofold interest: allowing for fuzzy input data; yielding fuzzy estimates of the loadings, of the outer weights and, as a consequence, of the latent variable scores, thus embedding the measurement error that naturally affects sensory assessments.

References

ALEFELD, G. and HERZENBERGER, J. (1983): *Introduction to Interval computation.* Academic Press, New York.

BOLLEN, K. A. (1989): *Structural equations with latent variables.* Wiley, New York.

COPPI, R., GIL, M.A. and KIERS, H.L. (2006): The fuzzy approach to statistical analysis. *Computational statistics & data analysis, 51 (1), 1–14.*

JÖRESKOG K. (1970): A general method for analysis of covariance structure. *Biometrika, 57, 239–251.*

ROMANO, R. (2006): Fuzzy Regression and PLS Path Modeling: a combined two-stage approach for multi-block analysis. *Doctoral Thesis*, Univ. of Naples, Italy.

ROMANO, R. and PALUMBO, F. (2006a): Fuzzy regression and least squares regression: the relationship between two different fitting criteria. *Abstracts of the SIS2006 Conference, 2, 693–696.*

ROMANO, R. and PALUMBO, F. (2006b): Classification of SEM based on fuzzy regression. In: Esposito-Vinzi et *al.* (Eds.): *Knowledge Extraction and Modeling.* Tilapia, Anacapri, 67-68.

TANAKA, H., UEIJIMA, S. and ASAI, K. (1980): Fuzzy linear regression model. *IEEE Transactions Systems Man Cybernet, 10, 2933–2938.*

TANAKA, H. and GUO, P. (1999) *Possibilistic Data Analysis for Operations Research.* Physica-Verlag, Wurzburg.

TENENHAUS, M. and ESPOSITO VINZI, V. (2005): PLS regression, PLS path modeling and generalized Procrustean analysis: a combined approach for multiblock analysis. *Journal of Chemometrics, 19 (3), 145–153.*

TENENAHUS, M., ESPOSITO VINZI, V., CHATELIN, Y.-M. and LAURO, C. (2005): PLS path modeling *Comp. Stat. and Data Anal. 48, 159–205.*

WOLD, H. (1982) Soft modeling: the basic design and some extensions. In: K.G. Joreskog and H. Wold (Eds.): *Systems under Indirect Observation, Vol. Part II.* North-Holland, Amsterdam, 1-54.

ZADEH, L. (1965): Fuzzy Sets. *Information and Control, 8, 338–353.*

ZADEH, L. (1973): Outline of a new approach to the analysis of complex systems and decision processes. *IEEE Trans. Systems Man and Cybernet, 1, 28–44.*

Automatic Analysis of Dewey Decimal Classification Notations

Ulrike Reiner

Verbundzentrale des Gemeinsamen Bibliotheksverbundes (VZG)
37077 Göttingen, Germany
ulrike.reiner@gbv.de

Abstract. The Dewey Decimal Classification (DDC) was conceived by Melvil Dewey in 1873 and published in 1876. Nowadays, the DDC serves as a library classification system in about 138 countries worldwide. Recently, the German translation of the DDC was launched, and since then the interest in DDC has rapidly increased in German-speaking countries. The complex DDC system (Ed. 22) allows to synthesize (to build) a huge amount of DDC notations (numbers) with the aid of instructions. Since the meaning of built DDC numbers is not obvious – especially to non-DDC experts – a computer program has been written that automatically analyzes DDC numbers. Based on Songqiao Liu's dissertation (Liu (1993)), our program decomposes DDC notations from the main class 700 (as one of the ten main classes). In addition, our program analyzes notations from all ten classes and determines the meaning of every semantic atom contained in a built DDC notation. The extracted DDC atoms can be used for information retrieval, automatic classification, or other purposes.

1 Introduction

While searching for books, journals, or web resources, you will often come across numbers such as "025.1740973", "016.02092", or "720.7073". What do they mean? Librarian professionals will identify these strings as numbers (notations) of the Dewey Decimal Classification (DDC), which is named after its creator, Melvil Dewey. Originally, Dewey designed the classification for libraries, but in the meantime DDC has also been discovered for classifying the web or other resources. The DDC is used, among others, because it has a long-standing tradition and is still up to date: in order to cope with scientific progress, it is currently under development by a ten-member international board (the Editorial Policy Committee, EPC). While the first edition, which was published in 1876, only comprised a few pages, the current 22nd edition of the DDC spans a four-volume work with almost 4,000 pages. Today, the DDC contains approx. 48,000 DDC notations and about 8,000 instructions. The DDC notations are enumerated in the schedules and tables of the DDC. With the aid of the instructions mentioned above, human classifiers can build new synthesized notations (numbers) if these are not specifically listed in the DDC schedules. This way, an enormous amount of synthesized DDC notations has been built intellectually over the last 130 years. These mostly unused notations are contained in library catalogues – like a hidden treasure. They can be considered as belonging to the "Deep Lib", one of the subsets of the "Deep Web" (Bergman (2001)). Can these notations be made accessible for information retrieval purposes with reasonable effort?

Our answer to this question consists in the automatic analysis of notations of the DDC. The analysis program we have developed determines all DDC notations (together with their corresponding captions) contained in a synthesized (built) DDC notation. Before we go into details of the automatic analysis of DDC notations in section 3, section 2 provides the basis for the analysis. In section 4, the results are presented, and section 5 draws a conclusion.

2 DDC notations

Notations play an important role in the DDC:

"Notation is the system of symbols used to represent the classes in a classification system. ... The notation provides a universal language to identify the class and related classes, regardless of the fact that different words or languages may be used to describe the class." (http://www.oclc.org/dewey/versions/ddc22/intro.pdf)

The following picture serves as an example for the aforesaid. Class C is represented by the notation 025.43 or, respectively, by the captions of three different languages:

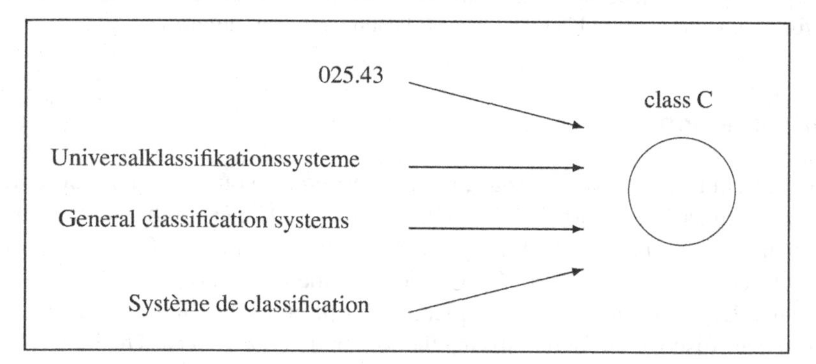

Fig. 1. Class C represented by notation 025.43 or by several captions

In compliance with the DDC system, the automatic analysis of notations of the DDC is carried out in the VZG (*V*erbundZentrale des *G*emeinsamen Bibliotheksverbundes) project Colibri (*CO*ntext generation and *LI*nguistic tools for *B*ibliographic *R*etrieval *I*nterfaces). The goal of this project is to enrich title records on the basis of the DDC to improve retrieval. The analysis of DDC notations is conducted under the following research questions (which are also posed in a similar way in Liu (1993), p. 18): Q1. Is it possible to automatically decompose molecular DDC notations into atomic DDC notations? Q2. Is it possible to improve automatic classification and retrieval by means of atomic DDC notations? An atomic DDC notation is a semantically indecomposable string (of symbols) that represents a DDC class. A molecular DDC notation is a string that is syntactically decomposable into atomic DDC notations.

DDC notations can be found at several places in the DDC. In DDC summaries, the notations for the main classes (or tens), the divisions (or hundreds), and the sections (or thousands) are enumerated. Other notations are listed in the schedules ("DDC schedule notations") or tables ("DDC table notations") or internal tables. DDC schedules are "the series of DDC numbers 000-999, their headings (captions), and notes." (Mitchell (1996), p. lxv). A DDC table is "a table of numbers that may be added to other numbers to make a class number appropriately specific to the work being classified" (Mitchell (1996), p. lxv). Further notations are contained in the "Relative Index" of the DDC. The frequency distributions of schedule (table) notations are shown in Fig. 2 (Fig. 3), while schedno0 is short hand for DDC schedule notations beginning with 0, schedno1 for DDC schedule notations beginning with 1, etc. The captions for the main classes are: 000: Computer science, information & general works; 100: Philosophy & psychology; 200: Religion; 300: Social sciences; 400: Language; 500: Science; 600: Technology; 700: Arts & recreation; 800: Literature; 900: History & geography. As illustrated by Fig. 2, DDC notations are not distributed uniformly: the most schedule notations can be found in the class "Technology", followed by the notations in the class "Social sciences". The fewest notations belong to the class "Philosophy & psychology". With regard to the table notations (Fig. 3), the 7,816 Table 2 notations ("Geographic Areas, Historical Periods, Persons") stand out, whereas, in contrast, the quantities of all other table notations are comparatively small (Table 1: Standard Subdivisions; Table 3: Subdivisions for the Arts, for Individual Literatures, for Specific Literary Forms; Table 4: Subdivisions of Individual Languages and Language Families; Table 5: Ethnic and National Groups; Table 6: Languages).

As mentioned before, DDC notations that are not explicitly listed in the schedules can be built by using DDC instructions. This process is called "notational synthesis" or "number building". Its results are synthesized DDC notations (molecular DDC notations) that usually only DDC experts are able to interpret. But with the aid of our computer program "DDC analyzer", the meaning of molecular DDC notations is revealed and the determined atomic DDC notations can be used, among others, to answer question Q2.

3 Automatic analysis of DDC notations

The GBV Union Catalog *GVK* (*G*emeinsamer *V*erbund*K*atalog, http://gso.gbv.de/) contains 3,073,423 intellectually DDC-classified title records (status: July, 2004). After the automatic elimination of segmentation marks, obviously incorrect DDC notations (3.8 per cent of all DDC notations), and duplicate DDC notations, a total of 466,134 different DDC notations is available for the automatic analysis of

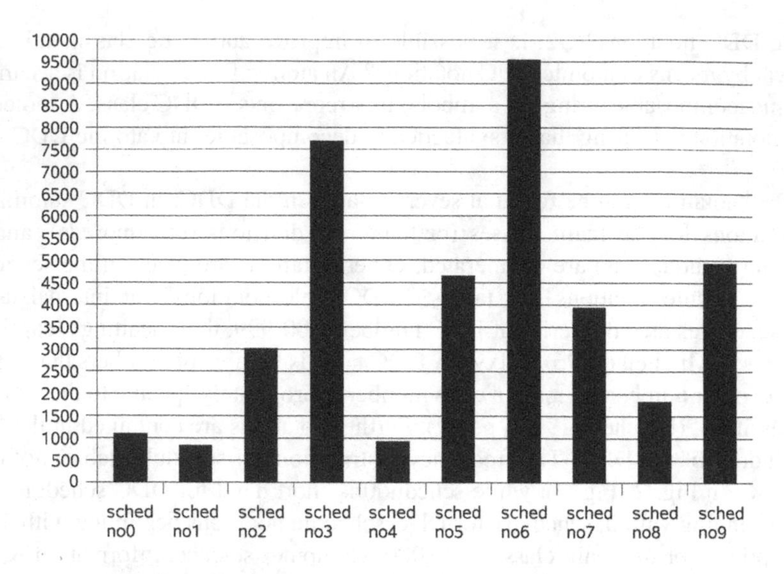

Fig. 2. Frequency distribution of DDC schedule notations

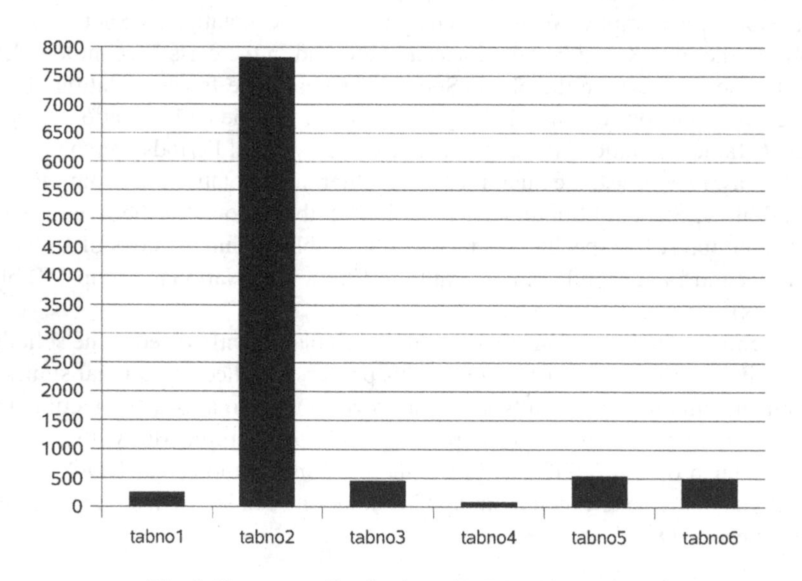

Fig. 3. Frequency distribution of DDC table notations

DDC notations. This set of all GVK DDC notations serves as input data for the DDC analyzer. The frequency of DDC schedule notations is as follows (in descending order): those beginning with 3 (189,246), with 9 (62,115), with 7 (52,632), with 6 (51,704), with 5 (33,649), with 0 (23,946), with 2 (20,888), with 8 (20,678), with 4 (6,680), and with 1 (4,596). The arity of DDC notations of all GVK DDC notations is Gaussian distributed with a maximum at 10, i.e. most DDC notations have approx. arity 10, the shortest DDC notation has arity 1, the longest DDC notation has arity 29. Other important input data for the DDC analyzer we used were the 600 DDC numbers given in Liu's dissertation. These 600 DDC numbers that we call "Liu's sample" were randomly selected from class 700 from the OCLC database by Liu.

As a member of the Consortium DDC German, we have access to the machine-readable data of the 22nd edition of the DDC system. These data are stored in an xml file. The English electronic web version is available as WebDewey (http://connexion.oclc.org/), the German pendant as MelvilClass (http://services.ddc-deutsch.de/melvilclass-login). For our purpose, only the relevant data of the xml file, which contains the expert knowledge of the DDC system, are extracted and stored in a "knowledge base". Here, DDC notations, descriptors, and descriptor values are stored in consecutive fields, while facts and rules – as we call them – are represented in a very similar way:

T1–093-T1–099+021#<ba4r2>#Statistics
025.17#<na1r1>##025.17#025.341-025.349#025.34#####
025.344#<hat>#Electronic resources

The three example lines of the knowledge base should be read as follows: *Fact*: T1–093-T1–099+021 has the caption "Statistics". *Rule*: Add to base number 025.17 the numbers following 025.34 in 025.341-025.349. *Fact*: 025.344 has the caption "Electronic resources". '#' serves as field separator. The xml tags that are given in angle brackets stand for: "ba4" ("beginning of add table (all of table number)"), "na1" ("add note (part of schedule number)") and "hat" ("hierarchy at class"). "r1" and "r2", which follow "na1" or, respectively, "ba4", stand for the first two macro rules. The knowledge base contains 48,067 facts and 8,033 rules. The 8,033 rules can be generalized to macro rules. While Liu (1993) defined 17 (macro) rules for the decomposition for class 700, we defined 25 macro rules for all DDC classes.

Our program, the DDC analyzer, works as follows: after initializing variables, it reads the knowledge base and, triggered by one or more DDC notations to be analyzed, executes the analysis algorithm. The number of correct and incorrect DDC notations is counted. For a DDC notation, there are two phases to the analyzing process including: determining the facts from left to right (phase 1) and determining the facts via rules from left to right (phase 2). After checking which output format has to be printed, the result is printed as a DDC analysis diagram or as a DDC analysis result set. After all DDC notations have been analyzed, the number of totally/partially analyzed DDC notations is printed. There are different reasons for a partially analyzed DDC notation: either the implementation of the DDC analyzer is incorrect/incomplete or the DDC notation is incorrectly synthesized or a part of the DDC system itself is incorrect.

4 Results

To demonstrate our progress in comparison with Liu's work, we compare his decomposition result with our DDC analysis diagram for the 37th molecular DDC notation of his sample:

Liu (1993), pp. 99–100
720.7073 has been decomposed as follows:
720: Architecture
0707: Geographical treatment
73: United States
The title of this book is:
#aVoices in architectural education: #bcultural politics and
The subject headings for this book are:
#aArchitecture #xStudy and teaching #zUnited States.
#aArchitecture and state #zUnited States.

Reiner (2007a), p. 49
`720.7073` <liu_37_to_analyze; length: 8>
`7-----` Arts & recreation <hatzen>
`72----` Architecture <hatzen>
`720---` Architecture <hat>
`-0.7--` Education, research, related topics <T1–07>
`-0.707-` Geographic treatment <T1–0707>
`--.-7-` North America <na4r7span:T1–0701-T1–0709:T2–7>
`--.-73` United States <na4r7span:T1–0701-T1–0709:T2–73>

The information given in angle brackets should be read as follows: "hatzen" is the concatenation of "hat" ("hierarchy at class") and "zen" ("zen built entry (main tag)"). "T1–" stands for "table 1", "T2–" for "table 2", "na4" for "add note (add of table number)", "r7" for "macro rule 7", "span" for "span of numbers", and ":" for "delimiter". As you can see, while Liu decomposes the synthesized DDC notation into three chunks, our DDC analysis diagram shows the finest possible analysis of the molecular DDC notation. The fine analysis provides the advantage of uncovering additional captions: "Arts & recreation", "Architecture", "North America", and "Education, research, related topics".

A DDC analysis diagram contains analysis and synthesis information: 1. the molecular DDC notation to be analyzed; 2. an identifier (name) and the length of the molecular DDC notation; 3. the sequence and position of the digits within the molecular DDC notation; 4. the Dewey dot at position 4; 5. the relevant parts of the molecular DDC notation for each analysis step; 6. the corresponding caption for every atomic DDC notation; 7. the parts irrelevant for the respective analysis step marked with "-"; 8. the type of the applied facts and rules that appear in angle brackets. In case it has been explained how to read the given information mentioned in 8., every synthesis step can be reproduced. While DDC analysis diagrams are intended for human experts, the DDC analysis result set can be used for data transfer. Currently, we distinguish three kinds of analysis result sets. The first one is a set of DDC <notation;caption> tuples:

7;Arts & recreation
72;Architecture
720;Architecture
T1–07;Education, research, related topics
T1–0707;Geographic treatment
T2–7;North America
T2–73;United States

The second one delivers all DDC notations contained in a synthesized number:

liu_37:720.7073;7;72;720;T1-07;T1-0707;T2-7;T2-73

The third analysis result set is in MAB2 format:

705a△a720.7073△p72△cT1–070△f0707△g73

All 600 analyzed DDC notations of Liu's sample have been compared accordingly with the results of Liu (1993). It turns out that Liu's decompositions can be reproduced. Minor differences result from printing errors in his dissertation and the usage of different (20th/22nd) DDC editions. After 14 years, 36 DDC notations of Liu's sample are out of date because of relocations and discontinuations. As far as the analysis of the 466,134 GVK DDC notations of all DDC classes is concerned, currently 297,782 (168,352) DDC notations can be totally (partially) analyzed, i.e. 63.9 per cent (36.1 per cent) are totally (partially) analyzed. In some DDC classes, the analyzing degree is even higher, which means that, e.g., 87 per cent of the 51,704 DDC notations of the class "Technology" (600) can be totally analyzed.

5 Conclusion

In 1993, Liu showed that DDC synthesized class numbers of main class 700 can be decomposed automatically. Our program analyzes notations from all ten main classes. Compared to Liu's approach, our analysis procedure delivers more information, which is furthermore presented in a new way. Since Liu's expert-evaluated results are reproduced, we can (statistically) infer that our DDC analyzer works correctly with high probability. Increasing the quantity of DDC notations totally analyzed will be the next step. The results can be used to improve (multilingual) DDC information retrieval or DDC automatic classification systems. On the basis of analysis diagrams, DDC tutorials or expert systems could be developed to support teaching of DDC number building or to control the quality of built DDC numbers.

References

BERGMAN, M., K.: The Deep Web: Surfacing Hidden Value. *The Journal of Electronic Publishing, Volume 7, Issue 1, August 2001.* Online: http://www.press.umich.edu/jep/07-01/bergman.html

MITCHELL, J.,S. (ed.) (1996): *Dewey Decimal Classification and Relative Index. Ed. 21, Volumes 1-4.* Forest Press, OCLC, Inc., Albany, New York, 1996. (http://connexion.oclc.org/).

LIU, S. (1993): *The Automatic Decomposition of DDC Synthesized Numbers.* Ph.D. diss., University of California, Graduate School of Library and Information Science, Los Angeles, 1993.

REINER, U. (2005): *VZG-Projekt Colibri – DDC-Notationsanalyse und -synthese.* September 2004 - Februar 2005. VZG-Colibri-Bericht 2/2004. Verbundzentrale des Gemeinsamen Bibliotheksverbundes (VZG), Göttingen, 2005.

REINER, U. (2007a): *Automatische Analyse von Notationen der Dewey-Dezimalklassifikation.* 31st Annual Conference of the German Classification Society on Data Analysis, Machine Learning, and Applications – Librarian Workshop: Subject Indexing and Library Science. March 7-9, 2007, Freiburg i. Br., Germany. (http://www.gbv.de/vgm/info/biblio/01VZG/06Publikationen/2007/pdf/pdf_2835.pdf).

REINER, U. (2007b): *Automatische Analyse von DDC-Notationen und DDC-Klassifizierung von GVK-Plus-Titeldatensätzen.* Workshop zur Dewey-Dezimalklassifikation "DDC-Einsichten und -Aussichten 2007". March 1, 2007, SUB Göttingen, Germany. (http://www.gbv.de/vgm/info/biblio/01VZG/06Publikationen/2007/pdf/pdf_2836.pdf).

A New Interval Data Distance Based on the Wasserstein Metric

Rosanna Verde and Antonio Irpino

Dipartimento di Studi Europei e Mediterranei
Seconda Universitá degli Studi di Napoli - Caserta (CE), Italy
{rosanna.verde, antonio.irpino}@unina2.it

Abstract. Interval data allow statistical units to be described by means of interval values, whereas their representation by single values appears to be too reductive or inconsistent, that is, unable to keep the uncertainty usually inherent to the observed data. In the present paper, we present a novel distance for interval data based on the Wasserstein distance between distributions. We show its interesting properties within the context of clustering techniques. We compare the obtained results using the dynamic clustering algorithm, taking into consideration different distance measures in order to justify the novelty of our proposal.[1]

1 Introduction

The representation of data by means of intervals of values is becoming more and more frequent in different fields of application. In general, an interval description depends on the uncertainty that affects the observed values of a phenomenon. The uncertainty can be considered as the inability to obtain true values depending on ignorance of the model that regulates the phenomenon. This uncertainty can be of three types: *randomness*, *vagueness* or *imprecision* (Coppi et al., 2006). Randomness is present when it is possible to hypothesize a probability distribution of the outcomes of an experiment, or when the observation is affected by an error component that is modeled as a random variable (i.e., white noise in a Gaussian distribution). Vagueness is related to an unclear fact or whether the concept even applies. Imprecision is related to the difficulty of accurately measuring a phenomenon. While randomness is strictly related to a probabilistic approach, vagueness and imprecision have been widely treated using fuzzy set theory, as well as the interval algebra approach. Probabilistic, fuzzy and interval algebra sometimes overlap in treating interval data. In the literature, interval algebra and fuzzy theory are treated very closely, especially in defining dissimilarity measures for the comparison of values affected by uncertainty expressed by intervals.

Interval data have even been studied in Symbolic Data Analysis (SDA) (Bock and Diday (2000)), a new domain related to multivariate analysis, pattern recognition and

[1] The present paper has been supported by the LC^3 Italian research project.

artificial intelligence. In this framework, in order to take into account the variability and/or the uncertainty inherent to the data, the description of a single unit can assume multiple values (bounded sets of real values, multi-categories, weight distributions), where intervals are a particular case. In SDA, several dissimilarity measures have been proposed. Chavent and Lechevallier (2002) and Chavent et al. (2006) proposed Hausdorff L_1 distances, while De Carvalho et al. (2006) proposed L_q distances and De Souza et al. (2004) an adaptive L_2 version. It is worth noting that these measures are based essentially on the boundary values of the compared intervals. These distances have been mainly proposed as criterion functions in clustering algorithms to partition a set of interval data where a cluster structure can be assumed.

In the present paper, we present some dissimilarity (or distance) functions, proposed in fuzzy, symbolic data analysis and probabilistic contexts to compare intervals of real values. Finally, we introduce a new metric based on the Wasserstein distance that respects all the classical properties of a distance and, being based on the quantile functions associated with the interval distributions, seems particularly able to keep the whole information contained in the intervals and not only on the bounds.

The structure of the paper is as follows: in Section 2, some families of distances for interval data, arising by different contexts, are shown. In Section 3, the new distance, based on a probabilistic metric, as the L_2 version of the Monge-Kantorovich-Wasserstein-Gini metric between quantile functions (usually known as Wasserstein metric), is introduced. In Section 4, the performances of the proposed distance are compared in a clustering process of a set of interval data. The obtained partition is compared to an expert one. Therefore, performing the clustering by Adaptive L_2 as well as Hausdorff L_1 distance, the proposed distance provides again the best results in terms of the Correct Rand Index. Section 5 closes the paper with some remarks and perspectives.

2 A brief survey of the existing distances

According to Symbolic Data Analysis, an interval variable X is a correspondence between a set E of units and a set of closed intervals $[a,b]$, where $a \leq b$ and $a,b \in \mathbb{R}$. Without losing in generality, the notation is quite the same for the interval algebra approach.
Let A and B be two intervals described, respectively, by $[a,b]$ and $[u,v]$. $d(A,B)$ can be considered as a distance if the main properties that define a distance are achieved: $d(A,A) = 0$ (reflexivity), $d(A,B) = d(B,A)$ (symmetry) and $d(A,B) \leq d(A,C) + d(C,B)$ (triangular inequality).

Hereinafter, we present some of the most used distances for interval data. The main properties of such measures are even underlined.

Tran and Duckstein distance between intervals. Some of these distances have been developed within the framework of the fuzzy approach. One of these is the distance defined by Tran and Duckstein (2002) that has the following formulation:

$$d_{TD}(A,B) = \int_{-1/2}^{1/2}\int_{-1/2}^{1/2} \left\{\left[\left(\tfrac{a+b}{2}\right) + x(b-a)\right] - \left[\left(\tfrac{u+v}{2}\right) + y(v-u)\right]\right\}^2 dxdy = \\ = \left[\left(\tfrac{a+b}{2}\right) - \left(\tfrac{u+v}{2}\right)\right]^2 + \tfrac{1}{3}\left[\left(\tfrac{b-a}{2}\right)^2 + \left(\tfrac{v-u}{2}\right)^2\right] \quad (1)$$

In practice, they consider the expected value of the distance between all the points belonging to interval A and all those points belonging to interval B. In their paper, they ensure that it is a distance, but it is easy to observe that the distance does not satisfy the first properties mentioned above. Indeed, the distance of an interval by itself is equal to zero only if the interval is thin:

$$d_{TD}(A,A) = \left[\left(\tfrac{a+b}{2}\right) - \left(\tfrac{a+b}{2}\right)\right]^2 + \tfrac{1}{3}\left[\left(\tfrac{b-a}{2}\right)^2 + \left(\tfrac{b-a}{2}\right)^2\right] = \tfrac{2}{3}\left(\tfrac{b-a}{2}\right)^2 \geq 0 \quad (2)$$

Hausdorff-based distances. The most common distance used for the comparison of two sets is the Hausdorff distance [2]. Considering two sets A and B of points of $\mathbb{R}^n$, and a distance $d(x,y)$ where $x \in A$ and $y \in B$, the Hausdorff distance is defined as follows:

$$d_H(A,B) = \max\left(\sup_{x\in A}\inf_{y\in B} d(x,y), \sup_{y\in B}\inf_{x\in A} d(x,y)\right) \quad (3)$$

If $d(x,y)$ is the L_1 *City block* distance, then Chavent et al. (2002) proved that

$$d_H(A,B) = \max(|a-u|, |b-v|) = \left|\tfrac{a+b}{2} - \tfrac{u+v}{2}\right| + \left|\tfrac{b-a}{2} - \tfrac{v-u}{2}\right| \quad (4)$$

An analytical formulation of this metric using the Euclidean distance has been devised (Book, 2005).

L_q distances between the bounds of intervals. A family of distances between intervals has been proposed by De Carvalho et al. (2006). Considering a set of interval data described into a space $\mathbb{R}^p$, the metric of norm q is defined as:

$$d_{L_q}(A,B) = \left(\sum_{j=1}^{p} |a-u|^q + |b-v|^q\right)^{1/q}. \quad (5)$$

They also showed that if the norm is L_∞ then $d_{L_\infty} = d_H$ (in L_1 norm).

The same measure was extended (De Carvalho (2007)) to an adaptive one in order to take into account the variability of the different clusters in a dynamical clustering process.

3 Our proposal: Wasserstein distance

If we suppose a uniform distribution of points, an interval of reals $A(t) = [a, b]$ can be expressed as the following type of function:

[2] The name is related to Felix Hausdorff, who is well-known for the separability theorem on topological spaces at the end of the 19th century.

$$A(t) = [a,\, b] = a + t\,(b - a) \qquad 0 \leq t \leq 1. \tag{6}$$

If we consider a description of the interval by means of its midpoint m and radius r, the same function can be rewritten as follows:

$$A(t) = m + r(2t - 1) \qquad 0 \leq t \leq 1. \tag{7}$$

Then, the squared Euclidean distance between homologous points of two intervals $A = [a,b]$ and $B = [u,v]$, or described by the *midpoint-radius* notation $A = (m_A, r_A)$ and $B = (m_B, r_B)$, is defined as follows:

$$\begin{aligned} d_W^2(A,B) &= \int_0^1 [A(t) - B(t)]^2\,dt = \int_0^1 [(m_A - m_B) + (r_A - r_B)(2t_j - 1)]^2\,dt = \\ &= (m_A - m_B)^2 + \tfrac{1}{3}(r_A - r_B)^2 \end{aligned} \tag{8}$$

In this case, we assume that the points are uniformly distributed between the two bounds. From a probabilistic point of view, this is similar to comparing two uniform density functions $U(a,b)$ and $U(u,v)$. In this way, we may use the Monge-Kantorivich-Wasserstein-Gini metric (Gibbs and Su, (2002)). Let Ψ be a distribution function; Ψ^{-1} is the corresponding quantile function. Given two univariate random variables ψ_A and ψ_B, the Wasserstein-Kantorovich distance is defined as:

$$d(\psi_A, \psi_B) = \int_0^1 \left|\Psi_A^{-1} - \Psi_B^{-1}\right| dt \tag{9}$$

In Barrio et al. (1999), the L_2 version (defined as Wasserstein distance) of this distance was proposed to study the weak convergence of distributions.

$$d_W(\psi_B, \psi_B) = \left[\int_0^1 \left(\Psi_A^{-1}(t) - \Psi_B^{-1}(t)\right)^2 dt\right]^{\frac{1}{2}} \tag{10}$$

In our context, it is possible to prove that:

$$d_W(U(a,b), U(u,v)) = \sqrt{(\mu_A - \mu_B)^2 + (\sigma_A - \sigma_B)^2} \tag{11}$$

where $\mu_A = \frac{a+b}{2}$ (resp. $\mu_B = \frac{u+v}{2}$) and $\sigma_A = \sqrt{\frac{(b-a)^2}{12}}$ (resp.$\sigma_A = \sqrt{\frac{(v-u)^2}{12}}$). In general, given two densities ψ_A and ψ_B with the first two finite moments: $\mu_A = E(A)$ (resp. $\mu_B = E(B)$), $\sigma_A = \sqrt{VAR(A)}$ (resp. $\sigma_B = \sqrt{VAR(B)}$) and $Corr_{QQ}$ as the correlation of the quantiles of Ψ_A and Ψ_B, Irpino and Romano (2007) proved that the (10) can be decomposed as:

$$d_W^2(\psi_A, \psi_B) = (\mu_A - \mu_B)^2 + (\sigma_A - \sigma_B)^2 + 2\sigma_A\sigma_B\left[1 - Corr_{QQ}(\Psi_A, \Psi_B)\right] \tag{12}$$

The proposed decomposition allows the effect of the two densities on the distance generated by different location, different size and different shape to be considered.

In order to calculate the distance between two elements described by p interval variables, we propose the following extension of the distance to the multivariate case in the sense of Minkowski:

$$d_W(\mathbf{A},\mathbf{B}) = \sqrt{\sum_{j=1}^{p}\left(\left|\frac{a_j+b_j}{2}-\frac{u_j+v_j}{2}\right|^2+\frac{1}{3}\left|\frac{b_j-a_j}{2}-\frac{v_j-u_j}{2}\right|^2\right)} \tag{13}$$

4 Dynamic clustering algorithm using different criterion functions

In this section, we present the effect of using different distances as the allocation function for the dynamic clustering of a temperature dataset. The Dynamic Clustering Algorithm (DCA) (Diday (1971)) represents a general reference for unsupervised, not hierarchical and iterative, clustering algorithms. In particular, DCA simultaneously looks for the partition of the set of data and the representation of the clusters. The main contributions to the clustering of interval data have been presented in the framework of symbolic data analysis, especially for defining a way to represent the clusters by means of *prototypes* (Chavent et al. (2006)). In the literature, several authors indicate how to compute *prototypes*. In particular, Verde and Lauro (2000) proposed that the *prototype* of a cluster must be considered as an element having the same properties of the clustered elements. In such a way, a cluster of intervals is described by a single *prototypal* interval, in the same way as a cluster of points is represented by its barycenter.
Let E be a set of n data described by p interval variables X_j $(j=1,\ldots,p)$. The general DCA looks for the partition $P \in P_k$ of E in k classes, among all the possible partitions P_k, and the vector $L \in L_k$ of k prototypes representing the classes in P, such that, the following Δ fitting criterion between L and P is minimized:

$$\Delta(P^*,L^*) = Min\{\Delta(P,L) \mid P \in P_k, L \in L_k\}. \tag{14}$$

Such a criterion is defined as the sum of dissimilarity or distance measures $\delta(x_i,G_h)$ of fitting between each object x_i belonging to a class $C_h \in P$ and the class representation $G_h \in L$:

$$\Delta(P,L) = \sum_{h=1}^{k}\sum_{x_i \in C_h}\delta(x_i,G_h).$$

A prototype G_h associated to a class C_h is an element of the space of the description of E, and it can be represented as a vector of intervals. The algorithm is initialized by generating k random clusters or, alternatively, k random prototypes. Generally, the criterion $\Delta(P,L)$ is based on an additive distance on the p descriptors.
In the present paper, we present an application based on a dynamic clustering of a real-world data set. The data set used in our experiments is the interval temperature dataset shown in Table 1, which was previously used as a benchmark interval data for cluster analysis in De Carvalho (2007), Guru and Kiranagi (2005) and Guru et

Table 1. The temperature dataset

City	Jan	Feb	Mar	...	Oct	Nov	Dec
Amsterdam	[-4,4]	[-5,3]	[2,12]	...	[5,15]	[-1,4]	[-1,4]
Athens	[6,12]	[6,12]	[8,16]	...	[16,23]	[11,18]	[8,14]
Bahrain	[13,19]	[14,19]	[17,30]	...	[24,31]	[20,26]	[15,21]
Bombay	[19,28]	[19,28]	[22,30]	...	[24,32]	[24,30]	[25,30]
...	...	...	...	...	...	...	...
Tokyo	[0,9]	[0,10]	[3,13]	...	[13,21]	[8,16]	[2,12]
Toronto	[-8,-1]	[-8,-1]	[-4,4]	...	[6,14]	[-1,17]	[-5,1]
Vienna	[-2,1]	[-1,3]	[1,8]	...	[7,13]	[2,7]	[1,3]
Zurich	[-11,9]	[-8,15]	[-7,18]	...	[5,23]	[0,19]	[-11,8]

al. (2004). We performed a dynamic clustering using as the allocation function the Hausdorff L_1 distance, the L_2 of De Carvalho et al. (2006), the De Carvalho adaptive distance (De Souza et al. (2004)) and the L_2 Wasserstein one alternatively. We chose to obtain a partition into four clusters, and we compared the resulting partition to that a priori one given by experts using the Corrected Rand Index. The expert classification were the following (Guru et al. (2004)): **Class 1** (Bahrain, Bombay, Cairo, Calcutta, Colombo, Dubai, Hong Kong, Kula Lampur, Madras, Manila, Mexico, Nairobi, New Delhi, Sidney); **Class 2** (Amsterdam, Athens, Copenhagen, Frankfurt, Geneva, Lisbon, London, Madrid, Moscow, Munich, New York, Paris, Rome, San Francisco, Seoul, Stockholm, Tokyo, Toronto, Vienna, Zurich); **Class 3** (Mauritius); **Class 4** (Tehran).

Using the three different allocation functions, we obtained 3 optimal partitions into 4 clusters (Tab.). 2). On the basis of the dynamic clustering, we evaluated the obtained partitions with respect to the a priori ones using the Corrected Rand Indices (Hubert and Arabie, (1985)).

5 Conclusion and perspectives

Interval descriptions can be derived from measurements subject to error ($\mu \pm e$). If they are assumed to be (probabilistic) models for the error term, Hausdorff distances are not influenced by the distribution of values and the L_q implicitly considers that all the information is equally concentrated on the bounds of intervals. The Wasserstein distance permits the different position, variability and shape of the compared distributions to be evaluated and taken separately into account, clearing way for interpreting data results. With a few modifications, it can also be used for the comparison of two fuzzy numbers measured by LR fuzzy variables. Further, being an Euclidean distance, it is easy to show that the Wasserstein distance satisfies the König-Huygens theorem for the decomposition of inertia. This allows us to apply the usual indices based on the comparison between the *inter* and the *intra* groups' inertia for the evaluation and the interpretation of the results of a clustering or of a classification procedure.

Table 2. Clusters obtained using different allocation functions. Last row: Corrected Rand Index (CRI) of the obtained partition compared with the expert partition

c	L_2 Wasserstein	Adaptive L_2	Hausdorff L_1 distance
1	Bahrain Bombay Cairo Calcutta Colombo Dubai HongKong KulaLumpur Madras Manila NewDelhi	Bahrain Bombay Calcutta Colombo Dubai HongKong KulaLumpur Madras Manila NewDelhi	Bahrain Dubai HongKong NewDelhi Cairo MexicoCity Nairobi
2	Amsterdam Copenhagen Frankfurt Geneva London Moscow Munich Paris Stockholm Toronto Vienna Zurich	Amsterdam Copenhagen Frankfurt Geneva London Moscow Munich Paris Stockholm Toronto Vienna	Amsterdam Copenhagen Frankfurt Geneva London Moscow Munich Paris Stockholm Toronto Vienna Zurich
3	Mauritius MexicoCity Nairobi Sydney	Cairo Mauritius MexicoCity Nairobi Sydney	Bombay Calcutta Colombo KulaLumpur Madras Manila Mauritius Sydney
4	Athens Lisbon Madrid New York Rome SanFrancisco Seoul Tehran Tokyo	Athens Lisbon Madrid New York Rome SanFrancisco Seoul Tehran Tokyo Zurich	Athens Lisbon Madrid NewYork Rome SanFrancisco Seoul Tehran Tokyo
CRI	0.53	0.49	0.46

On the other hand, a lot of effort is required for the extension of the distance to the multivariate case. Indeed, here we just proposed an extension (in the sense of Minkowski) of the distance under the hypothesis of independence between the descriptors of a multidimensional interval datum.

References

BARRIO, E., MATRAN, C., RODRIGUEZ-RODRIGUEZ, J. and CUESTA-ALBERTOS, J.A. (1999): Tests of goodness of fit based on the L2-Wasserstein distance. *Annals of Statistics , 27, 1230-1239.*

COPPI, R., GIL, M.A., and KIERS, H.A.L. (2006): The fuzzy approach to statistical analysis. *Computational statistics and data analysis, 51, 1-14.*

BOCK, H.H. and DIDAY, E., (2000): *Analysis of Symbolic Data, Exploratory Methods for Extracting Statistical Information from Complex Data.* Springer-Verlag, Heidelberg.

CHAVENT, M., and LECHEVALLIER, Y. (2002): Dynamical clustering algorithm of interval data: optimization of an adequacy criterion based on Hausdorff distance. In: Sokokowsky, A., Bock H. H. (Eds.): *Classification, Clustering and Data Analysis*, Springer, Heidelberg, 53–59.

CHAVENT, M., DE CARVALHO, F.A.T., LECHEVALLIER, Y., and VERDE, R. (2006): New clustering methods for interval data, *Computational statistics, 21, 211–229.*

DE CARVALHO, F.A.T. (2007): Fuzzy c-means clustering methods for symbolic interval data.*Pattern Recognition Letters, 28, 423–437*

DE CARVALHO, F.A.T., BRITO, P., and BOCK, H. (2006): Dynamic clustering for interval data based on L2 distance. *Computational Statistics*, 21, 2, 231-250

DE SOUZA, R. M. C. R. and DE CARVALHO, F. DE A. T. (2004): Clustering of Interval-Valued Data Using Adaptive Squared Euclidean Distances. In *Proc. of ICONIP 2004, 775-780.*

DIDAY, E. (1971): La meéthode des Nueées dynamiques. *Rev. Statist. Appl. 19 (2), 19–34.*

GIBBS, A.L. and SU, F.E. (2002): On choosing and bounding probability metrics, *International Statistical Review, 70, 419.*

GURU, D. S. and KIRANAGI, B. B. (2005): Multivalued type dissimilarity measure and concept of mutual dissimilarity value for clustering symbolic patterns. *Pattern Recognition, 38, 1, 151-156.*

GURU, D. S., KIRANAGI, B. B. and NAGABHUSHAN, P. (2004): Multivalued type proximity measure and concept of mutual similarity value useful for clustering symbolic patterns. *Pattern Recognition Letters, 25, 10, 1203-1213.*

HUBERT, L. and ARABIE, P. (1985): Comparing partitions. *Journal of Classification, 2, 193–218.*

IRPINO, A. and ROMANO, E. (2007): Optimal histogram representation of large data sets: Fisher vs piecewise linear approximations *Revue des Nouvelles Technologies de l'Information, RNTI-E-9, 99–110.*

TRAN, L. and DUCKSTEIN, L. (2002): Comparison of fuzzy numbers using a fuzzy distance measure, *Fuzzy Sets and Systems, 130, 331–341.*

VERDE, R. and LAURO, N. (2000): Basic choices and algorithms for symbolic objects dynamical clustering, in: *XXXIIe Journées de Statistique,Fés, Maroc, Societé Française de Statistique, 38–42.*

Keywords

Author Index